Lecture Notes in Computer Science　　10169

Commenced Publication in 1973
Founding and Former Series Editors:
Gerhard Goos, Juris Hartmanis, and Jan van Leeuwen

More information about this series at http://www.springer.com/series/7407

Petr Tichavský · Massoud Babaie-Zadeh
Olivier J.J. Michel · Nadège Thirion-Moreau (Eds.)

Latent Variable Analysis and Signal Separation

13th International Conference, LVA/ICA 2017
Grenoble, France, February 21–23, 2017
Proceedings

 Springer

Editors
Petr Tichavský
Institute of Information Theory
   and Automation
Prague
Czech Republic

Massoud Babaie-Zadeh
Sharif University of Technology
Tehran
Iran

Olivier J.J. Michel
Grenoble-Alpes University
Grenoble
France

Nadège Thirion-Moreau
Toulon University
Toulon
France

ISSN 0302-9743   ISSN 1611-3349   (electronic)
Lecture Notes in Computer Science
ISBN 978-3-319-53546-3   ISBN 978-3-319-53547-0   (eBook)
DOI 10.1007/978-3-319-53547-0

Library of Congress Control Number: 2017931540

LNCS Sublibrary: SL1 – Theoretical Computer Science and General Issues

Printed on acid-free paper

This Springer imprint is published by Springer Nature
The registered company is Springer International Publishing AG
The registered company address is: Gewerbestrasse 11, 6330 Cham, Switzerland

Preface

This volume gathers the full articles presented at the 13th International Conference on Latent Variable Analysis and Signal Separation, LVA/ICA 2017, which was held during February 21–23, 2017, and was hosted by the Grenoble-Alpes University, in Grenoble, France, at the GreEn-Er School of Engineering.

Since its inception in 1999, under the name "Independent Component Analysis and Blind Source Separation (ICA)," the series of LVA conferences (held approximately every 18 months) have attracted hundreds of researchers and practitioners. The conference has continuously broadened its horizons and scope of applications. The LVA/ICA research topics encompass a wide range of general mixtures of latent variable models but also theories and tools drawn from a great variety of disciplines such as signal and image processing, applied statistics, machine learning, linear and multilinear algebra, numerical analysis, optimization, etc. These research areas are of interest to numerous application fields ranging from audio, telecommunications, food industry, biochemistry, to biomedical engineering or observation sciences to cite a few. Thus it offers very exciting interdisciplinary interactions. It also constitutes a multi-disciplinary discussion forum for scientists and engineers where they can gain access to a broad understanding of the state of the research in the field, keep up to date with active research areas, discover or address the main theoretical challenges, but also face real-world problems and share experiences.

This edition of the conference also marks a return to its roots, since the first edition was held in Aussois, which, like Grenoble, is located in France in the Rhône Alpes region. This volume of Springer's *Lecture Notes in Computer Science* (LNCS) continues the tradition, which began in ICA 2004 (held in Granada, Spain), of publishing the conference proceedings in this form.

For this 13th issue of the LVA-ICA international conference, 58 full papers were submitted to regular sessions and to special sessions. Each submission of a regular full paper was peer reviewed by at least two members of our Technical Program Committee (TPC) or by competent additional reviewers assigned by the TPC members. Most papers received three reviews. From these 58 submitted papers, 53 were accepted as oral (31 papers) and poster (22 papers) presentations. The conference program included two special sessions: "From Source Positions to Room Properties: Learning Methods for Audio Scene Geometry Estimation" and "Latent Variable Analysis in Observation Sciences," proposed and chaired by R. Gribonval (Inria Rennes, France), and Y. Deville (Toulouse University, France). Regular topics included: theoretical developments (dictionary and manifold learning, optimization algorithms, performance analysis, etc.); audio signal processing applications; tensor-based methods for blind signal separation; signal processing for physics, biology, and biomedical applications; and sparsity aware signal processing.

The Organizing Committee was pleased to invite three leading experts in these fields for keynote lectures:

- Sharon Gannot (Bar-Ilan University, Israel),
- Olivier François (Grenoble-Alpes University, France),
- José Bioucas-Dias (University of Lisbon, Portugal).

Aware of the growing interest in emerging, as well as in classic LVA-related topics among novice and veteran researchers alike, the Organizing Committee decided to precede the conference by a one-day advanced Winter School on LVA and Advanced Data, with the support of LabEx PERSYVAL (Grenoble-Alpes University), including plenary lectures given by:

- Nikos Sidiropoulos (University of Minnesota, USA),
- Jean-François Cardoso (Paris-Saclay University, France),
- Pierre Comon (Grenoble-Alpes University, France),
- Christian Jutten (Grenoble-Alpes University, France).

The LVA-ICA conference was followed by a special one-day workshop organized with the support of European Research Council projects DECODA and CHESS.

The conference also provided a forum for the sixth community-based Signal Separation Evaluation Campaign (SiSEC 2017). SiSEC 2017 successfully continued the series of evaluation campaigns initiated at ICA 2007, in London. This year's SiSEC campaign featured five round robin tests: four audio tasks consisting of underdetermined speech and music mixtures (UND), two-channel mixtures of speech and real-world background noise (BGN), professionally produced music recordings (MUS), asynchronous recordings of speech mixtures (ASY), and one biomedical task on recording of the sounds generated by the heart (BIO).

The success of the LVA/ICA 2017 conference was the result of the hard work of many people and the support of many sponsors (Grenoble-Alpes University, Toulon University, CNRS, Région Auvergne-Rhône Alpes, Agglomération de Grenoble), whom we wish to warmly thank here. First, we wish to thank all the authors and all the members of the TPC, without whom this high-quality volume would not exist.

We also want to express our gratitude to the members of the International LVA Steering Committee for their continued support to the conference, as well as to the SiSEC 2017 organizers and finally to the local Organizing Committee.

February 2017

Petr Tichavský
Massoud Babaie-Zadeh
Olivier Michel
Nadège Thirion-Moreau

Organization

General Chairs

Olivier J.J. Michel	Grenoble-Alpes University, Grenoble-INP, France
Nadège Thirion-Moreau	University of Toulon and Aix Marseille, France

Program Chairs

Petr Tichavský	Institute of Information Theory and Automation, Czech Republic
Massoud Babaie-Zadeh	Sharif University of Technology, Tehran, Iran

Organization

GIPSA-Lab, Grenoble-Alpes University, France
LSIS, Toulon and Aix Marseille Universities, France

Special Sessions Organizers

Rémi Gribonval	Inria, Rennes, France
Yannick Deville	Toulouse University, France

SiSEC Chair

Antoine Liutkus	Inria, Nancy, France

Sponsors

CNRS, France
Grenoble-Alpes University, France
Grenoble-INP
Toulon and Aix Marseille Universities, France
European Research Council projects CHESS (C. Jutten) and DECODA (P. Comon)
Région Auvergne-Rhône Alpes, France
La Métro, Grenoble

International Steering Committee

Tülay Adali	University of Maryland, Baltimore County, USA
Andrzej Cichocki	Riken Brain Science Institute, Japan

Lieven De Lathauwer   K.U. Leuwen, Belgium
Rémi Gribonval   Irisa, Rennes, France
Christian Jutten   Grenoble-Alpes University, France
Shoji Makino   University of Tsukuba, Japan
Nobutaka Ono   National Institute of Informatics/SOKENDAI, Japan
Mark Plumbley   University of Surrey, UK
Paris Smaragdis   University of Illinois at Urbana-Champaign, USA
Petr Tichavský   The Czech Academy of Sciences, Czech Republic
Emmanuel Vincent   Inria Nancy, France
Arie Yeredor   Tel Aviv University, Israel

Program Committee

Tülay Adali   University of Maryland, Baltimore County, USA
Sophie Achard   CNRS, Grenoble, France
Shoko Araki   NTT Communication Science Laboratories, Japan
Pierre Comon   CNRS, Grenoble, France
Massoud Babaie-Zadeh   Sharif University of Technology, Tehran, Iran
David Brie   University of Lorraine, France
Marc Castella   Telecom SudParis, France
Charles Casimiro Cavalcante Federal University of Ceará, Brazil
A. Taylan Cemgil   Boğaziçi University, Turkey
Jeremy Cohen   University of Mons, Belgium
Sergio Cruces   University of Seville, Spain
Guanghui Cheng   University of Chengdu, China
Otto Debals   K.U. Leuven, Belgium
Yannick Deville   University of Toulouse, France
Nicolas Dobigeon   University of Toulouse, France
Sharon Gannot   Bar-Ilan University, Israel
Nicolas Gillis   University of Mons, Belgium
Martin Haardt   Ilmenau University of Technology, Germany
Mariya Ishteva   University of Brussel, Belgium
Zbyněk Koldovský   Technical University of Liberec, Czech Republic
Ivica Kopriva   Rudjer Boskovich Institute, Croatia
Dana Lahat   Grenoble-Alpes University, France
Elmar Lang   University of Regensburg, Germany
Olivier Michel   Grenoble-Alpes University, France
Sebastian Miron   University of Lorraine, France
Ali Mansour   ENSTA Bretagne, France
Ali Mohammad Djafari   CNRS, Paris, France
Eric Moreau   University of Toulon, France
Francesco Nesta   Conexant Systems, USA
Nobutaka Ono   National Institute of Informatics/SOKENDAI, Japan
Anh-Huy Phan   RIKEN, Japan
Ronald Phlypo   Grenoble-Alpes University, France
Mark Plumbley   University of Surrey, UK

Contents

Sparsity-Aware Signal Processing

Tensor Approaches

Higher-Order Block Term Decomposition
for Spatially Folded fMRI Data

Christos Chatzichristos[1,2(✉)], Eleftherios Kofidis[1,3], Yiannis Kopsinis[1,4],
Manuel Morante Moreno[1,2], and Sergios Theodoridis[1,2,5]

[1] Computer Technology Institute & "Press Diophantus" (CTI), Patras, Greece
{chatzichris,morante}@cti.gr
[2] Department of Informatics and Telecommunications,
National and Kapodistrian University of Athens, Athens, Greece
stheodor@di.uoa.gr
[3] Department of Statistics and Insurance Science,
University of Piraeus, Piraeus, Greece
kofidis@unipi.gr
[4] LIBRA MLI Ltd., Edinburgh, UK
kopsinis@ieee.org
[5] IAASARS, National Observatory of Athens, 15236 Penteli, Greece

Abstract. The growing use of neuroimaging technologies generates a
massive amount of biomedical data that exhibit high dimensionality.
Tensor-based analysis of brain imaging data has been proved quite effective in exploiting their multiway nature. The advantages of tensorial
methods over matrix-based approaches have also been demonstrated
in the context of functional magnetic resonance imaging (fMRI) data
analysis. However, such methods can become ineffective in demanding
scenarios, involving, e.g., strong noise and/or significant overlapping of
activated regions. This paper aims at investigating the possible gains
that can be obtained from a better exploitation of the spatial dimension,
through a higher (than 3)-order tensor modeling of the fMRI signals. In
this context, a higher-order Block Term Decomposition (BTD) is applied,
for the first time in fMRI analysis. Its effectiveness in handling strong
instances of noise is demonstrated via extensive simulation results.

Keywords: fMRI · Tensors · BTD · CPD · TPICA

1 Introduction

Functional Magnetic Resonance Imaging (fMRI) is a noninvasive technique for
studying brain activity. During an fMRI experiment, a series of brain images
is acquired, while the subject performs a set of tasks responding to external
stimuli. Changes in the blood-oxygen-level dependent (BOLD) signal are used
to examine activation in the brain [1]. The localization of the activated brain
areas is a challenging "cocktail party" problem and our goal is to distinguish

© Springer International Publishing AG 2017
P. Tichavský et al. (Eds.): LVA/ICA 2017, LNCS 10169, pp. 3–15, 2017.
DOI: 10.1007/978-3-319-53547-0_1

those areas as well as activation patterns (time courses) through some Blind Source Separation (BSS) (decomposition) method [2].

In fMRI studies, the structure of the data involves multiple modes, such as trial and subject, in addition to the intrinsic modes of time and space [3]. The use of multivariate bi-linear (matrix-based) methods through concatenating different modes, has been, up to recently, the state of the art for extracting information concerning spatial and temporal features. Such methods have also been extended to multi-subject experiments [4,5]. After acquiring an fMRI image at a time instance n (Fig. 1), the three-dimensional data (referred to here as *folded* data) are reshaped (unfolded) into a sequence of vectors $t_{n=1,2,...,N}$. These N vectors (images at different time instants) are stacked together to form a matrix, $\mathbf{W}_1$. Similarly, the data coming from a second subject (multi-subject case) are stacked together to form another matrix, $\mathbf{W}_2$, and the two matrices are joined together to form $\mathbf{W}_{12}$; the latter will then be decomposed by some BSS method. Thus, the $5th$-order problem of a multi-subject fMRI experiment has been transformed into a second-order problem. This unfolding of higher-order data into two-way arrays leads to decompositions that are inherently non-unique, and, most importantly, can result in a loss of the multi-way linkages.

The multi-way nature of the data can be retained in multilinear models, which, in general (a) produce representations that are unique [6], (b) can improve the ability of extracting spatiotemporal modes of interest, and (c) facilitate subsequent interpretations, that are neurophysiologically meaningful [3]. In the example of Fig. 1, instead of being concatenated into a matrix $\mathbf{W}_{12}$, the matrices $\mathbf{W}_1$ and $\mathbf{W}_2$ can be used to form a third-order tensor; hence, a tensor decomposition method can be mobilized for the BSS task [7]. The results from such methods seem to be very promising and provide, in most cases, better spatial and temporal localization of the activity, compared to the matrix-based approaches. However, still, these methods inherit the initial step of the spatial unfolding of the data into a *vector* t_n from their matrix-based counterparts. That is, somehow, they do not fully exploit the multiway nature of the obtained data, which seems not to be the natural thing to do for the task at hand.

It is, therefore, of interest to explore the application of higher-order models and investigate possible benefits from fully exploiting the underlying spatial information. In order to use all the spatial information in the available data, our kick off point will be to bypass the initial step of unfolding the data into

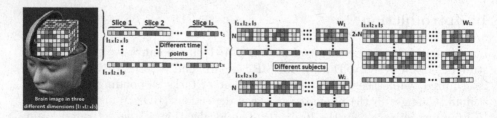

Fig. 1. Brain images unfolded in vectors and stacked in matrices.

$\mathbf{t}_n$ vectors. Furthermore, the Block-Term Decomposition (BTD) model [8–10] will be adopted, for a first time in fMRI analysis, in view of its higher modeling potential and its increased robustness to rank estimation errors. Through extensive simulation results, it will be demonstrated that the proposed method can overcome drawbacks of the state-of-the art tensorial techniques, improving the accuracy of the decomposition results even in cases of challenging scenarios.

1.1   Notation

Vectors, matrices and higher-order tensors are denoted by bold lower-case, upper-case and calligraphic upper-case letters, respectively. The transpose of a given matrix, $\mathbf{A}$, is written as $\mathbf{A}^T$. An entry of a vector $\mathbf{a}$, a matrix $\mathbf{A}$, or a tensor $\mathcal{A}$ is denoted by a_i, $a_{i,j}$, $a_{i,j,k}$, etc. The column-wise Khatri-Rao product of two matrices, $\mathbf{A} \in \mathbb{R}^{I \times R}$ and $\mathbf{B} \in \mathbb{R}^{J \times R}$, is denoted by $\mathbf{A} \odot \mathbf{B} = [\mathbf{a}_1 \otimes \mathbf{b}_1, \mathbf{a}_2 \otimes \mathbf{b}_2, \ldots, \mathbf{a}_R \otimes \mathbf{b}_R]$, with $\mathbf{a}_i, \mathbf{b}_i$ being the ith columns of $\mathbf{A}, \mathbf{B}$, respectively. The outer product is denoted by $\circ$. For an Nth-order tensor, $\mathcal{A} \in \mathbb{R}^{I_1 \times I_2 \times \cdots \times I_N}$, $\mathbf{A}_{\times n}$ denotes its mode-n unfolded (matricized) version, that results from mapping the tensor element with indices $(i_1, i_2, \ldots, i_N)$ to a matrix element (i_n, j), with $j = 1 + \sum_{k=1, k \neq n}^{N}[(i_k - 1) \prod_{m=1, m \neq n}^{k-1} I_m]$ for $N > 2$.

2   Tensorial fMRI Analysis

2.1   Canonical Polyadic Decomposition (CPD)

The Canonical Polyadic Decomposition (CPD) (or PARAFAC) [11] approximates a tensor of fMRI data, $\mathcal{T} \in \mathbb{R}^{I \times J \times K}$, by a sum of R vector outer products,

$$\mathcal{T} = \sum_{r=1}^{R} \mathbf{a}_r \circ \mathbf{b}_r \circ \mathbf{c}_r + \mathcal{E}, \tag{1}$$

where $\mathcal{E}$ stands for the modeling error tensor. Equivalently,

$$\mathbf{T}_{\times 1} = \mathbf{A}(\mathbf{C} \odot \mathbf{B})^T + \mathbf{E}_{\times 1}, \tag{2}$$

where $\mathbf{A} = [\mathbf{a}_1, \mathbf{a}_2, \ldots, \mathbf{a}_R]$ contains the factors of the spatial (voxel) activity of the R sources (components) and the similarly defined matrices, $\mathbf{B}$ and $\mathbf{C}$, correspond to the associated time courses and subjects, respectively. The sum-of-squares of the residual $\mathcal{E}$ is minimized to determine the latent factors in the R terms. The main advantage of the CPD, besides its simplicity, is the fact that it is unique up to permutation and scaling under mild conditions [12]. Uniqueness of CPD is crucial to its application in fMRI. In fact, it was demonstrated [13, 14] that CPD with fMRI data is robust to overlaps (spatial and/or temporal) as well as noise. On the other hand, the result of CPD is largely dependent on the correct estimation of the tensor rank R [15, 16], which is known to be a difficult task in tensor modeling.

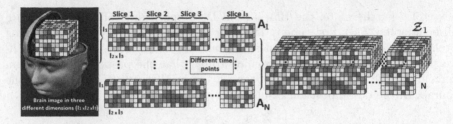

Fig. 2. Brain images unfolded in matrices, and stacked in tensors.

2.2   Tensor Probabilistic Independent Component Analysis (TPICA)

Independent Component Analysis (ICA) has demonstrated promising results in the characterization of single-subject fMRI data [4]. TPICA, as proposed in [17], is essentially a hybrid of the Probabilistic ICA (PICA) [18] method and the CPD method for multi-subject cases. Given a tensor of fMRI data, $\mathcal{T}$, TPICA factorizes it as:

$$\mathbf{T}_{\times 1} \approx \mathbf{AM}^T, \tag{3}$$

where the rows of $\mathbf{A}$ are assumed to be a sample of independent, non-Gaussian [18] random variables and $\mathbf{M} = \mathbf{C} \odot \mathbf{B}$ is a Khatri-Rao structured mixing matrix. TPICA computes the decomposition of the tensor in two steps: an ICA step, which estimates $\mathbf{M}$ and $\mathbf{A}$, and, at a second step, a Khatri-Rao factorization of $\mathbf{M}$ (using Singular Value Decomposition (SVD)) to compute the factors $\mathbf{B}$ and $\mathbf{C}$. TPICA is more robust than CPD to rank estimation inaccuracies but it exhibits inferior performance in the presence of overlap in the sources and/or strong noise [13,14].

3   Block Term Decomposition (BTD) for fMRI

Phan et al. [19,20] proved that, unfolding noisy data to low order tensors generally results in loss of accuracy in the respective decomposition. The extent of this loss in accuracy depends on the degree of collinearity of the columns in the unfolded mode. Furthermore, as pointed out in [21], the ability of multiway fits to make more robust predictions, compared to their two-way counterparts, seems to grow with the noise level. The use of higher-order tensors could, therefore, improve the result of the decomposition, both in terms of accuracy and robustness in cases of strong noise.

A way to benefit from the findings mentioned before is to adopt an alternative type of data unfolding, instead of reshaping the whole brain volume into a vector $\mathbf{t}_n$ (Fig. 1). For the proposed unfolding, we adopt the mode-1 (frontal) matricization of the respective data tensor, $\mathbf{A}_n$ (Fig. 2) (other modes of matricization can also be used). By stacking the N matrices together, a 3rd-order tensor, $\mathcal{Z}_1$, is formed (Fig. 2). Finally, for n different subjects, a 4-th order tensor is created by stacking together all three-order $\boldsymbol{\mathcal{Z}}_n$ tensors. CPD with such type of unfolding will succeed only in cases where the assumption of rank-1 is correct (S_1 source in Fig. 3). In cases of spatial activation of higher rank (S_2 source), CPD results in phantoms (from the multiplication of x_1 with y_2, x_2 with y_1 and y_3, etc.). The different language centers obtained from language generation tasks are examples of such sources (Fig. 4). The constraints of CPD can be proved to be restrictive in such cases.

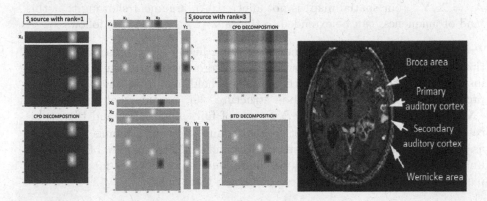

Fig. 3. Decomposition of sources S_1 and S_2.

Fig. 4. Language centers [22].

On the other hand, it seems less restrictive to decompose the tensor in terms of low rank factors, which, however, are not necessarily of rank one; this enhances the potential for modeling more general phenomena [10]. As an alternative to CPD, the use of Block Term Decomposition (BTD) [8–10] in the folded higher-order tensor is investigated in this work. The rank-$(L_r, L_r, 1)$ BTD of a tensor, $\boldsymbol{\mathcal{T}} \in \mathbb{R}^{I_1 \times I_2 \times I_3}$, into a sum of rank-$(L_r, L_r, 1)$ terms is given by

$$\boldsymbol{\mathcal{T}} = \sum_{r=1}^{R} \mathbf{A}_r \circ \mathbf{b}_r = \sum_{r=1}^{R} (\mathbf{X}_r \mathbf{Y}_r^T) \circ \mathbf{b}_r, \tag{4}$$

where the matrix $\mathbf{A}_r = \mathbf{X}_r \mathbf{Y}_r^T \in \mathbb{R}^{I_1 \times I_2}$ has rank L_r. BTD (3rd-order) has been successfully applied in modeling epileptic seizures in electro-encephalograms [23] and proved capable of modeling nonstationary (in frequency or in space) seizures, better than other techniques. A generalization of BTD has been also used in EEG Motor Imagery Data (MID) [24]. BTD has not been previously applied in fMRI analysis, to the best of the authors' knowledge. In this work, it is proposed to decompose the *four*-order data using the rank-$(L_r, L_r, 1, 1)$ BTD:

$$\boldsymbol{\mathcal{T}} = \sum_{r=1}^{R} \mathbf{A}_r \circ \mathbf{b}_r \circ \mathbf{c}_r = \sum_{r=1}^{R} (\mathbf{X}_r \mathbf{Y}_r^T) \circ \mathbf{b}_r \circ \mathbf{c}_r. \tag{5}$$

3.1 Uniqueness

It was proven in [8] that BTD in (4) is (essentially) unique provided that the matrices $\begin{bmatrix} \mathbf{X}_1 \, \mathbf{X}_2 \cdots \mathbf{X}_R \end{bmatrix}$ and $\begin{bmatrix} \mathbf{Y}_1 \, \mathbf{Y}_2 \cdots \mathbf{Y}_R \end{bmatrix}$ are full column rank and the matrix $\mathbf{B} = \begin{bmatrix} \mathbf{b}_1 \, \mathbf{b}_2 \cdots \mathbf{b}_R \end{bmatrix}$ does not contain collinear columns, up to the following indeterminacies. Scaling and permutation, as in CPD, and the simultaneous post-multiplication of $\mathbf{X}_r$ by a nonsingular matrix $\mathbf{F}$ with the pre-multiplication of $\mathbf{Y}_r$ by $\mathbf{F}^{-1}$; an indeterminacy which does not affect our result since the matrix $\mathbf{A}_r = \mathbf{X}_r \mathbf{Y}_r^T$ (our spatial map) is not affected. An argument showing that this kind of uniqueness can be extended to the rank-$(L_r, L_r, 1, 1)$ case follows.

Proof. As suggested in [6], the uniqueness of a higher-order tensor decomposition can be shown through a reduction to a third-order tensor, which is "the first instance of multilinearity for which uniqueness holds and from which uniqueness propagates by virtue of Khatri-Rao structure" [6]. Assume that the matrices $\begin{bmatrix} \mathbf{X}_1 \, \mathbf{X}_2 \cdots \mathbf{X}_R \end{bmatrix}$ and $\begin{bmatrix} \mathbf{Y}_1 \, \mathbf{Y}_2 \cdots \mathbf{Y}_R \end{bmatrix}$ are of full column rank and the matrices $\mathbf{B}, \mathbf{C} = \begin{bmatrix} \mathbf{c}_1 \, \mathbf{c}_2 \cdots \mathbf{c}_R \end{bmatrix}$ do not contain collinear or null columns (a realistic assumption for matrices that represent time and subjects). In view of the above, uniqueness for (5) can be proved via the uniqueness of its three-mode counterpart, where the fourth mode is nested into the third one:

$$\mathcal{T} = \sum_{r=1}^{R} \mathbf{A}_r \circ \mathbf{g}_r, \tag{6}$$

where $\mathbf{G} = \begin{bmatrix} \mathbf{g}_1 \, \mathbf{g}_2 \cdots \mathbf{g}_R \end{bmatrix} = \mathbf{B} \odot \mathbf{C}$ is the Khatri-Rao product of two matrices of no null nor collinear columns, and hence does not have null or collinear columns either [25, Proposition 1]. Following Theorem 4.1 of [9], since the matrices $\begin{bmatrix} \mathbf{X}_1 \, \mathbf{X}_2 \cdots \mathbf{X}_R \end{bmatrix}$ and $\begin{bmatrix} \mathbf{Y}_1 \, \mathbf{Y}_2 \cdots \mathbf{Y}_R \end{bmatrix}$ have full column rank and the matrix $\mathbf{G}$ has no collinear columns, the decomposition is essentially unique.

4 Simulation Results

Two different simulation studies are presented, with scenarios and data reproduced from [13,14]. Rank determination in the experiments is performed with the aid of the Core Consistency Diagnostic (CorConDia) method [15] and the triangle method [16]. The estimated rank of the decomposition increases as the signal to noise ratio (SNR) levels decrease significantly, because some peaks of the noise get higher amplitude than the useful signal and are recognized as a source [13]. For BTD, the Non Linear Least Squares method of the Structured Data Fusion (SDF) toolbox [26] is employed, as implemented in Tensorlab 3.0 [27].

4.1 Simulation of a Perception Study

In this study, the data of three subjects were simulated under the assumptions of a simplified version of a realistic perception study [14]. The simulated data

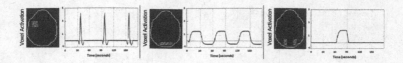

Fig. 5. The three sources used in the first experiment.

are a 60×50 axial slice of voxel activity from somewhere near the level of Brocas area. The second and third components of Fig. 5 represent the visual and motion perception components respectively, which have 50% of shared active voxels. The data from each subject contained all the three sources with different activation levels. The activity within each active voxel was randomly sampled from a Uniform [0.8, 1.2] distribution for each replication of each simulation condition. White Gaussian noise was added. Following [13], the SNR is defined as the Frobenius norm of the signal divided by the Frobenius norm of the noise.

With all the methods, the estimated rank of the decomposition was $R = 3$ for high SNR, while for SNR = 0.05 for CPD and BTD, $R = 4$. The result of the decomposition is insensitive to overestimation of L_r. Moreover, in cases of underestimation (L_r smaller than the true value) the result gets better when we increase L_r. Note that higher L_r values result in increased complexity. A compromise between accuracy and complexity shall be made. For BTD, $L_r = 3$, has been selected for all r. In Figs. 5, 6 and 7, the performance gain of BTD as the noise power increases can be observed (mean of 30 runs). TPICA starts failing at values of SNR lower than 0.12 while CPD, following the findings of [14], maintain the almost perfect separation of the sources at values of SNR ≈ 0.1. The fact that BTD identifies the sources correctly at the overlapping areas even with SNR = 0.05 must be emphasized.

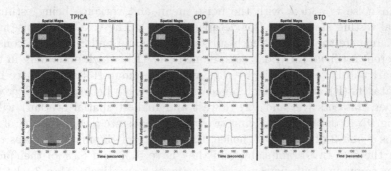

Fig. 6. Decomposition of the data (SNR = 0.12).

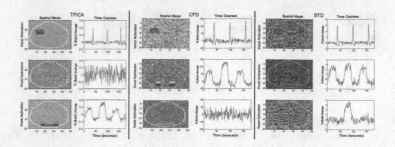

Fig. 7. Decomposition of the data (SNR = 0.08).

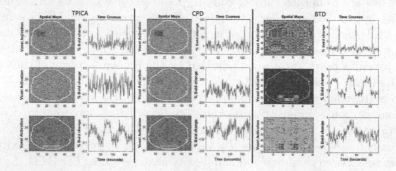

Fig. 8. Decomposition of the data (SNR = 0.05).

4.2   Multi-slice Simulation

The signal consists of artificial voxel activation maps (of three different slices), time patterns and activation strengths for three subjects. White random Gaussian noise is added, with the noise mean and variance being estimated from real resting state fMRI (for details, see [17]). The voxel-wise noise mean and variance are the same for each subject. Beckmann and Smith [17] consider five different artificial fMRI datasets, named (A)–(E), which differ only in their signal part and have no spatial overlap, while Stegeman [13] added three more fMRI datasets (F)–(H) with high percentage of overlap between the sources. In this paper, datasets (A) and (G) (lowest and highest spatial overlap) will be used (Fig. 9). In dataset G, the first two spatial maps are a combination of spatial maps 1 and 2 of dataset A, hence, the activity of the first two maps takes place in the first two brain slabs, and they have 51% and 63% of their active voxels in common (high percentage of overlap), respectively. Map 3, the time courses **B** and the noise instance are the same as in dataset A. Having been convolved with a canonical haemodynamic response function, the time courses and the spatial maps consist of three different spatiotemporal processes, which are present in every subject with different power. The performance evaluation is based on the Pearson correlation. We computed the correlation between the time courses obtained from the different methods and the actual ones, and the

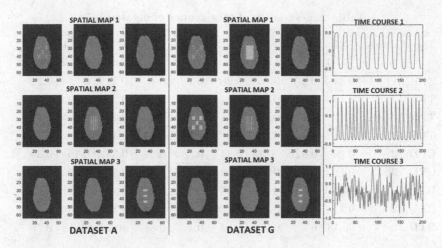

Fig. 9. Spatial maps of dataset A (left), G (center) and the common time courses (right).

correlation between the spatial maps acquired and the real "averaged" spatial maps computed with Ordinary Least Squares (OLS) regression (similarly to [13,17]).

Dataset A. In CPD and BTD, four components were extracted whereas in TPICA three (estimated rank, as also exhibited in [13]). Tables 1 and 2 present the mean correlation (over 30 runs) of spatial maps and time courses (respectively) of the different decompositions at three different SNR values. Note the stability in the performance of BTD compared to the other two methods. Furthermore, the different effect of noise in TPICA and CPD can be readily observed. In TPICA, the correlation between the estimated spatial map and the "true"

Table 1. Correlation of spatial maps of dataset A.

Maps	SNR = 0.08			SNR = 0.06			SNR = 0.04		
	Map 1	Map 2	Map 3	Map 1	Map 2	Map 3	Map 1	Map 2	Map 3
TPICA M1	0.68	0.1	0.1	0.54	−0.12	0.15	0.34	0.19	0.18
TPICA M2	0.1	0.99	0.1	0.12	0.89	0.1	0.22	0.76	0.18
TPICA M3	0.1	0.1	0.96	0.11	0.11	0.82	0.19	0.2	0.72
CPD M1	0.74	−0.2	−0.12	0.69	−0.34	−0.28	0.48	−0.41[a]	0.30
CPD M2	0.18	0.97	−0.12	−0.28	0.9	−0.15	−0.34	0.68	0.19
CPD M3	−0.21	−0.12	0.99	−0.29	0.15	0.88	0.4	−0.2	0.7
BTD M1	0.90	0.12	−0.14	0.80	0.14	−0.15	0.68	−0.29	−0.18
BTD M2	0.11	0.96	0.12	0.18	0.92	−0.12	−0.29	0.87	−0.19
BTD M3	−0.12	0.1	0.99	−0.2	0.12	0.94	−0.23	−0.19	0.84

[a] At the fourth component the crosstalk is also relatively high

Table 2. Correlation of time courses of dataset A.

Time Courses	SNR = 0.08			SNR = 0.06			SNR = 0.04		
	Tcs 1	Tcs 2	Tcs 3	Tcs 1	Tcs 2	Tcs 3	Tcs 1	Tcs 2	Tcs 3
TPICA T1	0.34	0	0	0.28	0	0.1	0.2	0	0.22
TPICA T2	0	0.92	0	0	0.89	0.12	0	0.70	0.16
TPICA T3	0.1	0.1	0.96	0.11	0.11	0.82	0.19	0.2	0.72
CPD T1	0.58	0	0.18	0.5	0.15	−0.2	0.4	0.2	0.2
CPD T2	0.20	0.92	−0.24	0.34	0.88	0.28	0.35	0.75	0.39
CPD T3	0.18	0	0.95	−0.2	0.12	0.85	0.22	−0.23	0.68
BTD T1	0.68	0	0	0.65	0.1	0.12	0.58	−0.21	0.22
BTD T2	0.11	0.92	0.18	−0.24	0.89	0.22	0.23	0.85	0.28
BTD T3	−0.18	0	0.95	−0.2	0.1	0.88	−0.22	−0.18	0.78

one decreases dramatically as the level of the noise gets higher, while in CPD the decrease is slower but with a significant increase of the cross-talk (correlation among "wrong" spatial maps). The correlations compared to those in [13] are slightly higher, since in addition to the intra-cranial voxels all the voxels inside the minimum rectangle around the brain mask were used.

Dataset G. Tables 3 and 4 exhibit the drawback of TPICA compared to CPD and BTD in cases of high overlap, and the gains obtained by BTD against the other two methods, in the presence of high levels of noise. Even with relatively high SNR (=0.12), TPICA fails to correctly separate the sources, as the second slice of the spatial maps of the first two sources is almost the same. At SNR = 0.08, the maximum correlation of map 2 occurs with spatial map 1, instead of the correct spatial map 2. This is due to the fact that TPICA fails in distinguishing

Table 3. Correlation of spatial maps of dataset G.

Maps	SNR1 = 0.15			SNR2 = 0.12			SNR3 = 0.08[a]		
	Map 1	Map 2	Map 3	Map 1	Map 2	Map 3	Map 1	Map 2	Map 3
TPICA M1	0.85	0.75	0.11	0.65	0.47	0.12	0.55	0.29	0.14
TPICA M2	0.69	0.89	0.09	0.59	0.48	0.12	0.54	0.32	0.11
TPICA M3	0.11	0.12	0.98	0.12	0.12	0.94	0.20	0.19	0.88
CPD M1	0.96	−0.20	−0.12	0.87	0.44	0.21	0.75	−0.47	0.22
CPD M2	−0.48	0.94	0.12	0.51	0.90	−0.12	−0.54	0.72	0.29
CPD M3	−0.11	0.11	0.98	0.24	0.11	0.94	0.25	0.28	0.90
BTD M1	0.96	−0.15	0.1	0.94	0.23	0.12	0.88	−0.32	0.13
BTD M2	−0.25	0.94	0.18	0.28	0.92	0.18	−0.27	0.88	0.19
BTD M3	−0.12	0.17	0.95	0.13	0.18	0.95	0.14	0.2	0.92

[a]TPICA in SNR = 0.08 often fails to separate 3 components

Table 4. Correlation of time courses of dataset G.

Time Courses	SNR = 0.08			SNR = 0.06			SNR = 0.04		
	Tcs 1	Tcs 2	Tcs 3	Tcs 1	Tcs 2	Tcs 3	Tcs 1	Tcs 2	Tcs 3
TPICA T1	0.96	−0.54	0.13	0.54	0.43	0.2	0.40	0.14	0.2
TPICA T2	−0.31	0.80	0.11	0.45	0.50	−0.13	0.38	0.23	−0.16
TPICA T3	0.14	−0.12	0.94	0.20	0.24	0.93	0.22	−0.50	0.60
CPD T1	0.96	−0.24	−0.11	0.84	0.26	0.2	0.62	−0.50	−0.2
CPD T2	−0.30	0.92	0.21	0.32	0.840	−0.23	−0.38	0.53	−0.26
CPD T3	0.11	0.11	0.94	0.14	0.14	0.93	−0.22	0.3	0.65
BTD T1	0.96	0.21	0.11	0.94	0.22	0.12	0.88	0.32	0.18
BTD T2	0.20	0.92	−0.11	−0.22	0.890	−0.13	0.23	0.83	−0.16
BTD T3	0.1	0.11	0.96	−0.14	0.15	0.91	0.21	−0.29	0.85

the two different sources, and source 2 is identified as noise (mixture of all sources). On the other hand, CPD and BTD recognize correctly all the sources at high SNR. At lower SNR, CPD cannot distinguish the second slice of the two first sources, while BTD still gives an almost perfect result. As the noise level increases, the difference in the performance of the decompositions increases in favor of BTD.

5   Conclusions

In this paper, a novel approach to fMRI signal separation was presented. It is based on a higher-order unfolding of the data, combined with the use of BTD for tensor decomposition, in an effort to better exploit the original 3D spatial structure of the data. Extensive simulation results demonstrated the enhanced robustness of the proposed method in the presence of noise, as compared with CPD and TPICA-based decompositions. In cases of spatial and temporal overlap, both CPD and BTD give better results than TPICA, albeit at the cost of higher sensitivity to the rank estimate. In terms of computational cost, the complexity of BTD is higher than CPD. It was also observed that when the rank, L_r, of the BTD increases, the result of the decomposition is equally good or even better, at the expense of higher complexity.

Acknowledgments. The authors would like to thank Profs. A. Stegeman and N. Helwig for providing the datasets used in [13] and [14], respectively, and Prof. S. Van Huffel for her critical comments on earlier version of this paper. Constructive comments from the reviewers are also gratefully acknowledged. This research has been funded by the European Union's Seventh Framework Programme (H2020-MSCA-ITN-2014) under grant agreement No. 642685 MacSeNet.

References

1. Lindquist, M.A.: The statistical analysis of fMRI data. Stat. Sci. **23**, 439–464 (2008)
2. Theodoridis, S.: Machine Learning: A Bayesian and Optimization Perspective. Academic Press, Boston (2015)
3. Andersen, A.H., Rayens, W.S.: Structure-seeking multilinear methods for the analysis of fMRI data. NeuroImage **22**, 728–739 (2004)
4. Calhoun, V.D., Adalı, T.: Unmixing fMRI with independent component analysis. IEEE Eng. Med. Biol. Mag. **25**, 79–90 (2006)
5. Andersen, K.W., Mørup, M., Siebner, H., Madsen, K.H., Hansen, L.K.: Identifying modular relations in complex brain networks. In: IEEE International Workshop on Machine Learning for Signal Processing (MLSP) (2012)
6. Sidiropoulos, N., Bro, R.: On the uniqueness of multilinear decomposition of N-way arrays. J. Chemom. **14**, 229–239 (2000)
7. Cichocki, A., Mandic, D., De Lathauwer, L., Zhou, G., Zhao, Q., Caiafa, C., Phan, H.A.: Tensor decompositions for signal processing applications: from two-way to multiway component analysis. IEEE Signal Process. Mag. **32**, 145–163 (2015)
8. De Lathauwer, L.: Decompositions of a higher-order tensor in block terms-Part I: lemmas for partitioned matrices. SIAM J. Matrix Anal. Appl. **30**, 1022–1032 (2008)
9. De Lathauwer, L.: Decompositions of a higher-order tensor in block terms-Part II: definitions and uniqueness. SIAM J. Matrix Anal. Appl. **30**, 1033–1066 (2008)
10. Lathauwer, L.: Block component analysis, a new concept for blind source separation. In: Theis, F., Cichocki, A., Yeredor, A., Zibulevsky, M. (eds.) LVA/ICA 2012. LNCS, vol. 7191, pp. 1–8. Springer, Heidelberg (2012). doi:10.1007/978-3-642-28551-6_1
11. Harshman, R.A.: Foundations of the PARAFAC procedure: models and conditions for an "explanatory" multi-modal factor analysis. UCLA Work. Papers in Phonetics, pp. 1–84(1970)
12. Kruskal, J.B.: Three-way arrays: rank and uniqueness of trilinear decompositions, with application to arithmetic complexity and statistics. Linear Algebra Appl. **18**(2), 95–138 (1977)
13. Stegeman, A.: Comparing Independent Component Analysis and the PARAFAC model for artificial multi-subject fMRI data. Unpublished Technical report, University of Groningen (2007)
14. Helwig, N.E., Hong, S.: A critique of tensor probabilistic independent component analysis: implications and recommendations for multi-subject fmri data analysis. J. Neurosci. Methods **2**, 263–273 (2013)
15. Bro, R., Kiers, H.: A new efficient method for determining the number of components in PARAFAC models. J. Chemom. **17**(5), 274–286 (2003)
16. Castellanos, J.L., Gmez, S., Guerra, V.: The triangle method for finding the corner of the L-curve. Appl. Numer. Math. **43**(4), 359–373 (2002)
17. Beckmann, C., Smith, S.: Tensorial extensions of independent component analysis for multisubject fMRI analysis. NeuroImage **25**, 294–311 (2005)
18. Beckmann, C., Smith, S.: Probabilistic independent component analysis for functional magnetic resonance imaging. IEEE Trans. Med. Imaging **23**(2), 137–152 (2004)
19. Phan, A.H., Tichavsky, P., Cichocki, A.: CANDECOMP/PARAFAC decomposition of high-order tensors through tensor reshaping. IEEE Trans. Signal Process. **61**(19), 4847–4860 (2013)

20. Tichavsky, P., Phan, A.H., Koldovsky, Z.: Cramér-Rao-induced bounds for CAN-DECOMP/PARAFAC tensor decomposition. IEEE Trans. Signal Process. **61**(8), 1986–1997 (2013)
21. Norgaard, L.: Classification and prediction of quality and process parameters of thick juice and beet sugar by fluorescence spectroscopy and chemometrics. Zuckerindustrie **120**(11), 970–981 (1995)
22. Phillips, N.C.: Gasthuisberg University Hospital raises fMRI to new level with Intera 3.0 T. http://netforum.healthcare.philips.com/
23. Hunyadi, B., Camps, D., Sorber, L., Van Paesschen, W., De Vos, M., Van Huffel, S., De Lathauwer, L.: Block term decomposition for modelling epileptic seizures. EURASIP J. Adv. Signal Process. (2014). doi:10.1186/1687-6180-2014-139
24. Phan, A.H., Cichocki, A., Zdunek, R., Lehky, S.: From basis components to complex structural patterns. In: IEEE International Conference on Acoustics, Speech and Signal Processing (ICASSP), Vancouver (2013)
25. Brie, D., Miron, S., Caland, F., Mustin, C.: An uniqueness condition for the 4-way CANDECOMP/PARAFAC model with collinear loadings in three modes. In: IEEE International Conference on Acoustics, Speech and Signal Processing (ICASSP), Prague (2011)
26. Sorber, L., Barel, M.V., De Lathauwer, L.: Structured data fusion. IEEE J. Sel. Topics Signal Process. **9**, 586–600 (2015)
27. Vervliet, N., Debals, O., Sorber, L., Van Barel, M., De Lathauwer, L.: Tensorlab user guide (2016). http://www.tensorlab.net

Modeling Parallel Wiener-Hammerstein Systems Using Tensor Decomposition of Volterra Kernels

Philippe Dreesen[1](✉), David T. Westwick[2], Johan Schoukens[1],
and Mariya Ishteva[1]

[1] Department VUB-ELEC, Vrije Universiteit Brussel (VUB), Brussels, Belgium
philippe.dreesen@gmail.com
[2] Department of Electrical and Computer Engineering,
University of Calgary, Calgary, Canada

Abstract. Providing flexibility and user-interpretability in nonlinear system identification can be achieved by means of block-oriented methods. One of such block-oriented system structures is the parallel Wiener-Hammerstein system, which is a sum of Wiener-Hammerstein branches, consisting of static nonlinearities sandwiched between linear dynamical blocks. Parallel Wiener-Hammerstein models have more descriptive power than their single-branch counterparts, but their identification is a non-trivial task that requires tailored system identification methods. In this work, we will tackle the identification problem by performing a tensor decomposition of the Volterra kernels obtained from the nonlinear system. We illustrate how the parallel Wiener-Hammerstein block-structure gives rise to a joint tensor decomposition of the Volterra kernels with block-circulant structured factors. The combination of Volterra kernels and tensor methods is a fruitful way to tackle the parallel Wiener-Hammerstein system identification task. In simulation experiments, we were able to reconstruct very accurately the underlying blocks under noisy conditions.

Keywords: Tensor decomposition · System identification · Volterra model · Parallel Wiener-Hammerstein system · Block-oriented system identification · Canonical polyadic decomposition · Structured data fusion

1 Introduction

System identification is the art of building dynamical models from noisy measurements of input and output data. Linear system identification is a well-established discipline [11,13,20] and has yielded successful applications in a wide variety of fields. In the last decades, the use of nonlinear models has become more important in order to capture the nonlinear effects of the real world. Many different nonlinear identification methods have been proposed [6,17], but very often these solutions are either tailored to a specific application, or are too complex to understand or study.

© Springer International Publishing AG 2017
P. Tichavský et al. (Eds.): LVA/ICA 2017, LNCS 10169, pp. 16–25, 2017.
DOI: 10.1007/978-3-319-53547-0_2

In this paper, we will draw ideas from two nonlinear system identification approaches and try to combine the benefits of both. The first approach tackles the disadvantage of increased complexity of nonlinear models by considering block-oriented models [7], which combine flexibility with user-interpretation by interconnecting linear dynamical blocks and static nonlinear functions. Unfortunately, even simple block-oriented models, such as Wiener (cascade of linear-nonlinear), Hammerstein (cascade nonlinear-linear) or Wiener-Hammerstein (cascade linear-nonlinear-linear) require an iterative optimization on a non-convex objective function, and identification procedures that are tailored towards a specific block-structure [16]. The second approach that we will use is the Volterra model [1,14], which is an extension of the well-known impulse response model for linear dynamical systems. Volterra models take into account higher-order polynomial nonlinearities and can thus be seen as a generalization of the Taylor series expansion for nonlinear dynamical systems. The advantages of Volterra models are that any fading-memory system can be approximated to an arbitrary degree of accuracy [1,15] and the parameter estimation task is a linear problem. The disadvantages are that the resulting models contain a very high number of parameters and thus cannot be given physical interpretation.

We generalize the earlier results of [4,5] on Wiener-Hammerstein system identification using tensor decompositions in two ways: First, we show that the case of parallel branches for a fixed degree d gives rise to a canonical polyadic decomposition with block-structured factors. The study of parallel Wiener-Hammerstein systems is useful, as they are universal approximators, whereas single-branch Wiener-Hammerstein systems are not [12,14]. Second, we jointly consider Volterra kernels of several degrees by means of the structured data fusion framework of [18] and solve the problem as a joint structured tensor decomposition. By simultaneously decomposing several Volterra kernels, the available information is used maximally. The presented method is implemented by means of structured data fusion [18] using Tensorlab 3.0 [21], and is validated on simulation experiments.

The paper is organized as follows. In Sect. 2 we illustrate the link between the Volterra kernels and tensors and introduce the canonical polyadic decomposition for tensors. Section 3 illustrates how the Volterra kernels of a parallel Wiener-Hammerstein system have a natural connection to the canonical polyadic decomposition with block-circulant factors. This ultimately leads to a joint structured canonical polyadic decomposition of the Volterra kernels of several orders that solve the parallel Wiener-Hammerstein identification task. We validate the method on numerical simulation examples in Sect. 4. In Sect. 5 we draw the conclusions.

2   Volterra Kernels, Tensors and Tensor Decomposition

2.1   The Volterra Model for Nonlinear Systems

We consider single-input-single-output systems that map an input u at time instance k onto an output y at time instance k. The Volterra series expansion generalizes the well-known convolution operator for linear dynamical systems to the nonlinear case. Essentially the Volterra model of a nonlinear system expresses the output $y(k)$ as a polynomial function of time-shifted input variables $u(k), u(k-1), \ldots, u(k-m)$, with m denoting the memory length of the model. Formally we can write the Volterra model as

$$y(k) = \sum_{d=1}^{D} \left(\sum_{s_1,\ldots,s_d=0}^{m} H_d(s_1,\ldots,s_d)u(k-s_1)\cdots u(k-s_d) \right), \qquad (1)$$

where $H_d(\cdot,\ldots,\cdot)$ denotes Volterra kernel of degree d. The Volterra series expansion allows for representing a large class of nonlinear systems up to an arbitrary degree of accuracy [1].

2.2   From Polynomials to Tensors

It is well-known that multivariate homogeneous polynomials can be identified to symmetric higher-order tensors [3]. For instance, we may represent the quadratic polynomial $p(x_1, x_2) = 5x_1^2 - 8x_1x_2 + x_2^2$ as a matrix multiplied from both sides by a vector containing x_1 and x_2 as

$$p(x_1, x_2) = 5x_1^2 - 8x_1x_2 + x_2^2$$
$$= \begin{bmatrix} x_1 & x_2 \end{bmatrix} \begin{bmatrix} 5 & -4 \\ -4 & 1 \end{bmatrix} \begin{bmatrix} x_1 \\ x_2 \end{bmatrix}.$$

In general, we may thus write a (nonhomogeneous) polynomial as

$$p(x_1,\ldots,x_n) = p_0 + \mathbf{x}^T \mathbf{p}_1 + \mathbf{x}^T \mathbf{P}_2 \mathbf{x} + \mathcal{P}_3 \times_1 \mathbf{x}^T \times_2 \mathbf{x}^T \times_3 \mathbf{x}^T + \ldots, \qquad (2)$$

where $\mathbf{x} = \begin{bmatrix} x_1 \ldots x_n \end{bmatrix}^T$ and $\times_n$ is the n-mode product defined as follows. Let $\mathcal{X}$ be a $I_1 \times I_2 \times \cdots \times I_N$ tensor, and let $\mathbf{u}^T$ be an $1 \times I_n$ row vector, then we have

$$(\mathcal{X} \times_n \mathbf{u}^T)_{i_1 \cdots i_{n-1} i_{n+1} \cdots i_N} = \sum_{i_n=1}^{I_n} x_{i_1 i_2 \cdots i_N} u_{i_n}.$$

Notice that the result is a tensor of order $N-1$, as mode n is summed out.

2.3   Canonical Polyadic Decomposition

It is often useful to decompose a tensor into simpler components, and for the proposed method we will use the canonical polyadic decomposition. The canonical polyadic decomposition [2,8,10] (also called CanDecomp or PARAFAC) expresses the tensor $\mathcal{T}$ as a sum of rank-one terms as

$$\mathcal{T} = \sum_{i=1}^{R} \mathbf{a}_r \circ \mathbf{b}_r \circ \mathbf{c}_r,$$

with ∘ denoting the outer product and the number of components R denoting the CP rank of the tensor $\mathcal{T}$. Often a short-hand notation is used as

$$\mathcal{T} = [\![\mathbf{A}, \mathbf{B}, \mathbf{C}]\!],$$

where $\mathbf{A} = \begin{bmatrix} \mathbf{a}_1 \cdots \mathbf{a}_R \end{bmatrix}$ and $\mathbf{B}$ and $\mathbf{C}$ are defined likewise.

3 Parallel Wiener-Hammerstein as Tensor Decomposition

For self-containment we will first review and rephrase a result of [5] that connects the Volterra kernel of a Wiener-Hammerstein system to a canonical polyadic decomposition with circulant-structured factor matrices. Afterwards, we will generalize this to the parallel case and then we show that the entire problem leads to a joint and structured canonical polyadic decomposition.

3.1 Wiener-Hammerstein as Structured Tensor Decomposition

Let us try to understand how the canonical polyadic decomposition shows up in modeling a Wiener-Hammerstein system. Consider a (single-branch) Wiener-Hammerstein system as in Fig. 1 with FIR filters $P(z)$ and $Q(z)$ with memory lengths m_P and m_Q, respectively, and a static nonlinearity $f(x) = x^3$. The output $y(k)$ of the Wiener-Hammerstein model is obtained by passing the signal $w(k)$ through the filter $Q(z)$. We can write this as

$$\begin{aligned} y(k) &= \begin{bmatrix} w(k) \cdots w(k - m_Q) \end{bmatrix} \begin{bmatrix} 1 \ q_1 \cdots q_{m_Q} \end{bmatrix}^T \\ &= \mathbf{w}^T \mathbf{q}, \end{aligned} \tag{3}$$

where we fixed the first filter coefficient $q_0 = 1$ in order to ensure uniqueness of the identified model. The signal $w(k)$ is given by the expression $w(k) = f(v(k))$, or in this case $w(k) = v^3(k)$. To obtain $v(k), \ldots, v(k - m_Q)$ from $u(k)$, we require of $u(k)$ the samples k down to $k - m_Q - m_P$. This convolution operation can be expressed as a matrix equation as

$$\begin{bmatrix} v(k) \cdots v(k - m_Q) \end{bmatrix} = \begin{bmatrix} u(k) \cdots u(k - m_Q - m_P) \end{bmatrix} \begin{bmatrix} 1 & & \\ p_1 & \ddots & \\ \vdots & \ddots & 1 \\ p_{m_P} & & p_1 \\ & \ddots & \vdots \\ & & p_{m_P} \end{bmatrix}$$

$$\mathbf{v}^T = \mathbf{u}^T \mathbf{P},$$

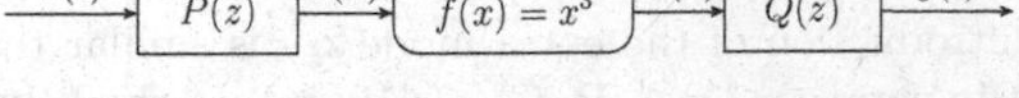

Fig. 1. A Wiener-Hammerstein system with input signal $u(k)$ and output signal $y(k)$ contains a static nonlinear function $f(\cdot)$ that is sandwiched between the FIR filters $P(z)$ and $Q(z)$.

with the circulant matrix $\mathbf{P}$ of size $m_P + m_Q + 1 \times m_Q + 1$. Notice that we fixed the first coefficient $p_0 = 1$ for uniqueness purposes. The matrix $\mathbf{P}$ will turn out to play a central role in the canonical polyadic decomposition of the Volterra kernels of a Wiener-Hammerstein system.

By fixing both $q_0 = 1$ and $p_0 = 1$, we are excluding the possibility that there is a pure delay present in the system. The presence of a delay in the system can be accounted for by setting $p_0 = 1$ and then performing a scaling on the nonlinearity, rather than on q_0. In case the system has a delay this will lead to an estimated $q_0 \approx 0$. However, for notational convenience we have chosen $p_0 = q_0 = 1$ in the remainder of this paper, but a more general scaling strategy to ensure uniqueness is possible and compatible with the presented method.

For the current Wiener-Hammerstein system we have $f(x) = x^3$, and hence $y(k) = \mathcal{H}_3 \times_1 \mathbf{u}^T \times_2 \mathbf{u}^T \times_3 \mathbf{u}^T$. In [5] it is shown that the Volterra kernel can be written as the canonical polyadic decomposition $\mathcal{H} = [\![\mathbf{P}, \mathbf{P}, \mathbf{P}\,\mathrm{diag}(\mathbf{q})]\!]$, which we can also write in a more symmetrical expression by extracting $\mathbf{q}^T$ into an extra mode as

$$\mathcal{H} = [\![\mathbf{P}, \mathbf{P}, \mathbf{P}, \mathbf{q}^T]\!]. \tag{4}$$

This fact can be appreciated by considering output $y(k)$ as

$$\begin{aligned}
y(k) &= \mathcal{H} \times_1 \mathbf{u}^T \times_2 \mathbf{u}^T \times_3 \mathbf{u}^T \\
&= [\![\mathbf{u}^T\mathbf{P}, \mathbf{u}^T\mathbf{P}, \mathbf{u}^T\mathbf{P}, \mathbf{q}^T]\!] \\
&= [\![\mathbf{v}^T, \mathbf{v}^T, \mathbf{v}^T, \mathbf{q}^T]\!] \\
&= \sum_{i=0}^{m_Q} q(i)v^3(k-i),
\end{aligned}$$

in which we recognize the convolution of the impulse response of $Q(z)$ with the time-shifted samples $v^3(k)$ as in (3).[1]

In case of a general polynomial function $f(x)$, the same reasoning can be developed for each degree d, which will lead to a structured canonical polyadic decomposition of the degree-d Volterra kernel as in (4). For instance, if $f(x) = ax^2 + bx^3$, we find the following expressions

$$\begin{aligned}
\mathbf{H}_2 &= a\,[\![\mathbf{P}, \mathbf{P}, \mathbf{q}^T]\!], \\
\mathcal{H}_3 &= b\,[\![\mathbf{P}, \mathbf{P}, \mathbf{P}, \mathbf{q}^T]\!].
\end{aligned}$$

In Sect. 3.3 we will discuss how this leads to a joint tensor decomposition.

3.2 Parallel Wiener-Hammerstein Structure

To understand how we can extend these results to the parallel case, let us consider a two-branch parallel Wiener-Hammerstein system where both branches have an identical nonlinearity $f_1(x) = f_2(x) = x^3$, as in Fig. 2. To avoid a

---

[1] Remark that the introduction of the extra mode $\mathbf{q}^T$ is similar to the extraction of the weights λ_i in the notation $[\![\lambda; \mathbf{A}, \mathbf{B}, \mathbf{C}]\!]$ of [10] where the columns of the factor matrices $\mathbf{A}$, $\mathbf{B}$ and $\mathbf{C}$ are scaled to have unit norm. Our notation is intentionally different in the sense that we have normalized the first elements of the columns of $\mathbf{P}$ and $\mathbf{q}$ equal to one, for practical purposes.

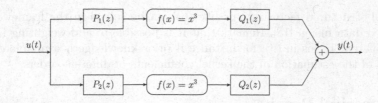

Fig. 2. An example of a two-branch parallel Wiener-Hammerstein system having an identical nonlinear function $f_1(x) = f_2(x) = x^3$.

notational overload, we will assume for the remainder of this paper that the memory lengths of all filters $P_i(z)$ are m_P, and likewise for the filters $Q_i(z)$ the lenghts are m_Q. The summation of the two branches leads to

$$\mathcal{H}_3 = [\![\mathbf{P}_1, \mathbf{P}_1, \mathbf{P}_1, \mathbf{q}_1^T]\!] + [\![\mathbf{P}_2, \mathbf{P}_2, \mathbf{P}_2, \mathbf{q}_2^T]\!]$$
$$= [\![\,[\mathbf{P}_1\ \mathbf{P}_2]\,, [\mathbf{P}_1\ \mathbf{P}_2]\,, [\mathbf{P}_1\ \mathbf{P}_2]\,, [\mathbf{q}_1^T\ \mathbf{q}_2^T]\,]\!],$$

with $\mathbf{P}_i$ and $\mathbf{q}_i$ defined similar as in the single-branch case. We may include a scaling of branch one by a scalar c_1 (i.e., $f_1(x) = c_1 x^3$) and branch two by a scalar c_2 (i.e., $f_2(x) = c_2 x^3$) by introducing an additional mode as

$$\mathcal{H}_3 = [\![\,[\mathbf{P}_1\ \mathbf{P}_2]\,, [\mathbf{P}_1\ \mathbf{P}_2]\,, [\mathbf{P}_1\ \mathbf{P}_2]\,, [\mathbf{q}_1^T\ \mathbf{q}_2^T]\,, [c_1 \mathbf{1}_{m_Q+1}^T\ c_2 \mathbf{1}_{m_Q+1}^T]\,]\!]. \quad (5)$$

Introducing the extra factor $[\mathbf{1}_{m_Q+1}^T c_1\ \mathbf{1}_{m_Q+1}^T c_2]$ does not change the size of the tensor, since it introduces a mode with dimension one. Formally, if we let $m = m_P + m_Q + 1$ denote the memory length of the Volterra model, the tensor in (5) has size $m \times m \times m \times 1 \times 1$ which is equivalent to $m \times m \times m$.

3.3   Coupled Tensor and Matrix Decompositions

The Volterra kernels of the parallel Wiener-Hammerstein model for a particular order d can be decomposed as a structured canonical polyadic decomposition. Hence, if the Volterra kernels of multiple orders are available, a joint decomposition of multiple Volterra kernels should be performed.

Ultimately, we find that the R-branch parallel Wiener-Hammerstein identification task is solved by minimizing the cost criterion

$$\underset{\mathbf{P},\mathbf{q},\mathbf{c}}{\text{minimize}} \|\mathbf{h}_1 - [\![\mathbf{P}, \mathbf{q}^T, \mathbf{c}_1^T]\!]\|_2^2 + \|\mathbf{H}_2 - [\![\mathbf{P}, \mathbf{P}, \mathbf{q}^T, \mathbf{c}_2^T]\!]\|_F^2$$
$$+ \|\mathcal{H}_3 - [\![\mathbf{P}, \mathbf{P}, \mathbf{P}, \mathbf{q}^T, \mathbf{c}_3^T]\!]\|_F^2 + \dots, \quad (6)$$

where

$$\mathbf{P} = [\mathbf{P}_1 \cdots \mathbf{P}_R],$$
$$\mathbf{q}^T = [\mathbf{q}_1^T \cdots \mathbf{q}_R^T],$$
$$\mathbf{c}_d^T = [c_{1d}\mathbf{1}_{m_Q+1}^T \cdots c_{Rd}\mathbf{1}_{m_Q+1}^T].$$

The factor matrices $\mathbf{P}$ and $\mathbf{q}^T$ are shared among all joint decompositions while the constants $\mathbf{c}_d^T$ depend on the order d of the considered Volterra kernel.

Joint and structured factorizations like (6) can be solved in the framework of structured data fusion [18]. Remark that it is possible to add weighting factors to the different terms in (6), for instance if prior knowledge is available on the accuracy of the estimation of the kernel coefficients of different orders.

4   Numerical Results

In this section we validate the proposed identification method on a simulation example. Numerical experiments were performed using MATLAB and structured data fusion [18] in Tensorlab 3.0 [21] (code available on request).

We consider a parallel Wiener-Hammerstein system having two branches with second and third-degree polynomial nonlinearities (Fig. 3). The finite impulse response coefficients of the filters $P_i(z)$ and $Q_i(z)$ are chosen as sums of decreasing exponentials with lengths $m_P = m_Q = 10$, such that the kth impulse response coefficient is given as $\sum_{i=1}^{S} \alpha_i^{k-1} / \sum_{i=1}^{S} \alpha_i$, with α_i drawn from a uniform distribution $U[-0.8, 0.8]$. For P_1 we have a sum of $S = 3$ exponentials, while P_2, Q_1 and Q_2 consist of a single decreasing exponential ($S = 1$). The coefficients c_{i2} and c_{i3} are drawn from a normal distribution $N(0, 0.1^2)$, and $c_{i0} = c_{i1} = 0$. The input signal is a Gaussian white noise sequence $u(k) \sim N(0, 0.7^2)$ and is applied without noise to the system, for $k = 1, \ldots, 10,000$. The outputs $y(k) = y_0(k) + n_y(k)$ are disturbed by additive Gaussian noise n_y with a signal-to-noise ratio of $10\,\mathrm{dB}$.

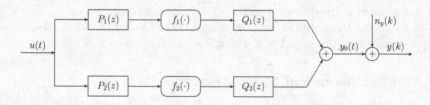

Fig. 3. A two-branch parallel Wiener-Hammerstein system with output noise.

The Volterra kernel coefficients $H_d(s_1, \ldots, s_2)$ with $s_i = 1, \ldots, m_P + m_Q + 1$ of degrees two and three are estimated using a standard linear least-squares method on the basis of the second and third degree time-shifted inputs $u(k - s_1) \cdots u(k - s_3)$ and the noisy outputs $y(k)$, for $k = 1, \ldots, 10,000$. The memory lengths $m_P = m_Q = 10$ give rise to 21×21 second-order Volterra kernel and a $21 \times 21 \times 21$ third-order Volterra kernel, having in total $231 + 1771 = 2002$ unique kernel elements.

The joint matrix and tensor decomposition with structured factors is then performed using Tensorlab 3.0 [21], returning the parameters of the parallel Wiener-Hammerstein system. We have performed a Monte Carlo experiment with 100 re-initializations of the optimization routine and were able to retrieve the true underlying system parameters in about 10% of the cases. In Fig. 4(a) we show a typical result of a successful identification that was able to retrieve the underlying system parameters accurately. A zoom of the reconstructed outputs of the identified model together with the true outputs and the noisy measurements from which the Volterra kernels were estimated is shown in Fig. 4(b). It is worth mentioning that in some cases, a relatively small output error was obtained, while the computed system parameters were very different from the underlying parameters. In other experiments we have observed that the success rate of the algorithm improves for lower noise levels and shorter filter lengths. If the method failed consistently on a system, this could almost always be understood from problem-specific system properties, such as similar impulse responses in the filters, dominance of one of the branches (i.e., unbalanced values of the coefficients of f_i), etc.

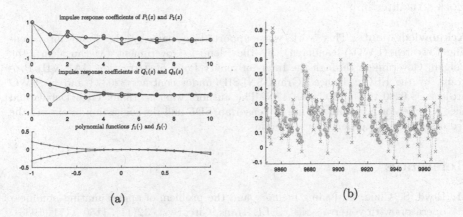

(a)   (b)

Fig. 4. (a) A typical result of the successful completion, where the method succeeds in retrieving the underlying parallel Wiener-Hammerstein system parameters. (b) The output of the identified parallel Wiener-Hammerstein system ($*$) reconstructs very accurately the noiseless output signal y_0 ($\circ$), starting from the Volterra kernel coefficients that were computed from the noisy data y with 10 dB SNR ($\times$).

Nevertheless, for extending the method to larger-scale problems, the issue of having good initial estimates becomes more relevant. Possibly the work of [19] which focuses on block-circulant structured canonical polyadic decomposition may provide good starting points for initializing our method.

Finally we would like to remark that in simulations, one can easily compare the retrieved system parameters (or simulated output data) with the underlying true system parameters or signals, which is obviously not possible in real experiments. However, it is worth mentioning that the tensor approximation error (6) was strongly correlated with the error between the simulated output and noiseless output data, which provides a proxy to select the best model among a set of candidates.

5   Conclusions

The joint decomposition of Volterra kernels with structured factors is able to retrieve up to a high degree of accuracy the underlying system parameters. The success rate of the method decreases as the noise level and number of parameters in the system grows, but even up to moderately long impulse responses (memory length of ten samples), the method was successful in about 10% of the re-initializations of the optimization routine. Ongoing work is concerned with obtaining good initializations [22], as this becomes an important issue when considering filters with longer memories and/or higher noise levels. In future work, the link should be investigated between rank properties and identifiability of the coupled and structured canonical polyadic decomposition and the identifiability of the parallel Wiener-Hammerstein structure, as was done for other block-structures in [9].

Acknowledgments. This work was supported in part by the Fund for Scientific Research (FWO-Vlaanderen), by the Flemish Government (Methusalem), the Belgian Government through the Inter-university Poles of Attraction (IAP VII) Program, by the ERC Advanced Grant SNLSID under contract 320378 and by FWO projects G.0280.15N and G.0901.17N. The authors want to thank Otto Debals and Nico Vervliet for help with the use of Tensorlab/SDF and the suggestion to extract the vector q into an additional tensor mode.

References

1. Boyd, S., Chua, L.: Fading memory and the problem of approximating nonlinear operators with volterra series. IEEE Trans. Circ. Syst. **32**(11), 1150–1171 (1985)
2. Carroll, J., Chang, J.: Analysis of individual differences in multidimensional scaling via an N-way generalization of "Eckart-Young" decomposition. Psychometrika **35**(3), 283–319 (1970)
3. Comon, P., Golub, G., Lim, L.H., Mourrain, B.: Symmetric tensors and symmetric tensor rank. SIAM J. Matrix Anal. Appl. **30**(3), 1254–1279 (2008)
4. de Goulart, J.H.M., Boizard, M., Boyer, R., Favier, G., Comon, P.: Tensor CP decomposition with structured factor matrices: algorithms and performance. IEEE J. Sel. Top. Signal Process. **10**(4), 757–769 (2016)
5. Favier, G., Kibangou, A.Y.: Tensor-based methods for system identification - Part 2: three examples of tensor-based system identification methods. Int. J. Sci. Tech. Autom. Control Comput. Eng. (IJ-STA) **3**(1), 870–889 (2006)
6. Giannakis, G.B., Serpedin, E.: A bibliography on nonlinear system identification. IEEE Trans. Signal Process. **81**(3), 533–580 (2001)
7. Giri, F., Bai, E.W.: Block-Oriented Nonlinear System Identification. Lecture Notes in Control and Information Sciences. Springer, London (2010)
8. Harshman, R.A.: Foundations of the PARAFAC procedure: model and conditions for an "explanatory" multi-mode factor analysis. UCLA Working Papers in Phonetics, vol. 16, no. 1, pp. 1–84 (1970)
9. Kibangou, A.Y., Favier, G.: Tensor analysis-based model structure determination and parameter estimation for block-oriented nonlinear systems. IEEE J. Sel. Top. Signal Process. **4**(3), 514–525 (2010)

10. Kolda, T.G., Bader, B.W.: Tensor decompositions and applications. SIAM Rev. **51**(3), 455–500 (2009)
11. Ljung, L.: System Identification: Theory for the User. Prentice Hall PTR, Upper Saddle River (1999)
12. Palm, G.: On representation and approximation of nonlinear systems. Biol. Cybern. **34**(1), 49–52 (1979)
13. Pintelon, R., Schoukens, J.: System Identification: A Frequency Domain Approach, 2nd edn. Wiley-IEEE Press, New York (2012)
14. Schetzen, M.: The Volterra and Wiener Theories of Nonlinear Systems. Wiley, New York (1980)
15. Schoukens, J., Pintelon, R., Dobrowiecki, T., Rolain, Y.: Identification of linear systems with nonlinear distortions. Automatica **41**, 491–504 (2005)
16. Schoukens, M., Tiels, K.: Identification of nonlinear block-oriented systems starting from linear approximations: a survey (2012). (Preprint arXiv:1607.01217)
17. Sjöberg, J., Zhang, Q., Ljung, L., Benveniste, A., Delyon, B., Glorennec, P.Y., Hjalmarsson, H., Juditsky, A.: Nonlinear black-box modeling in system identification: a unified overview. Automatica **31**(12), 1691–1724 (1995)
18. Sorber, L., Van Barel, M., De Lathauwer, L.: Structured data fusion. IEEE J. Sel. Top. Signal Process. **9**, 586–600 (2015)
19. Van Eeghem, F., De Lathauwer, L.: Algorithms for CPD with block-circulant factors. Technical report 16–100, KU Leuven ESAT/STADIUS (2016)
20. Van Overschee, P., De Moor, B.: Subspace Identification for Linear Systems: Theory, Implementation, Applications. Kluwer Academic Publishers, Dordrecht (1996)
21. Vervliet, N., Debals, O., Sorber, L., Van Barel, M., De Lathauwer, L.: Tensorlab 3.0, March 2016. http://www.tensorlab.net/
22. Westwick, D.T., Ishteva, M., Dreesen, P., Schoukens, J.: Tensor factorization based estimates of parallel Wiener-Hammerstein models. Technical report, University of Calgary, Calgary, Canada and Vrije Universiteit Brussel, Brussels, Belgium. IFAC World Congress (2016, submitted)

Fast Nonnegative Matrix Factorization and Completion Using Nesterov Iterations

Clément Dorffer, Matthieu Puigt$^{(\boxtimes)}$, Gilles Delmaire, and Gilles Roussel

Univ. Littoral Côte d'Opale, LISIC – EA 4491, 62228 Calais, France
{clement.dorffer,matthieu.puigt,gilles.delmaire,
gilles.roussel}@univ-littoral.fr

Abstract. In this paper, we aim to extend Nonnegative Matrix Factorization with Nesterov iterations (Ne-NMF)—well-suited to large-scale problems—to the situation when some entries are missing in the observed matrix. In particular, we investigate the Weighted and Expectation-Maximization strategies which both provide a way to process missing data. We derive their associated extensions named W-NeNMF and EM-W-NeNMF, respectively. The proposed approaches are then tested on simulated nonnegative low-rank matrix completion problems where the EM-W-NeNMF is shown to outperform state-of-the-art methods and the W-NeNMF technique.

Keywords: Low-rank matrix completion · Nonnegative matrix factorization · Nesterov iterations · Gradient descent

1 Introduction

Estimating missing entries in a low-rank matrix—known as Matrix Completion (MC)—received a lot of attention since some pioneering work in [2,8]. Indeed, such a problem finds many applications [1] including environmental monitoring [17,26] for example. Mathematically, it consists of estimating a matrix $M \in \mathbb{R}^{m \times n}$—partially known on a subset Ω of entries—by a low-rank matrix X, i.e.,

$$\min_{X} \operatorname{rank}(X) \quad \text{s.t.} \quad \mathcal{P}_{\Omega}(X) = \mathcal{P}_{\Omega}(M), \tag{1}$$

where $\mathcal{P}_{\Omega}(.)$ is the sampling operator of M.

Among the many approaches recently proposed to solve MC, matrix factorization techniques emerged as a powerful tool, especially when the rank of the data matrix M is known [11,31]. Moreover in some applications—e.g., blind sensor calibration [5,6]—M is nonnegative, thus yielding a Nonnegative MC (NMC) problem which can be tackled by Nonnegative Matrix Factorization (NMF). NMF is widely used in, e.g., signal processing or machine learning, and consists of finding two nonnegative matrices $G \in \mathbb{R}_{+}^{m \times r}$ and $F \in \mathbb{R}_{+}^{r \times n}$ such that

$$X \approx G \cdot F. \tag{2}$$

© Springer International Publishing AG 2017
P. Tichavský et al. (Eds.): LVA/ICA 2017, LNCS 10169, pp. 26–35, 2017.
DOI: 10.1007/978-3-319-53547-0_3

Most NMF algorithms consider two separate convex sub-problems that respectively and alternately estimate G for a fixed F and F for a fixed G. Such estimates are obtained using, e.g., Multiplicative Updates (MU) [19], Alternating Nonnegative Least Squares (ANLS) [15], or Projected Gradient (PG) [22]. Additionally, some authors incorporated some extra-information in the NMF model [30], e.g., weights [10,13], sum-to-one constraints [4,18,21], sparsity assumptions [6,12,14], or known and/or bounded values [3,5,6,12,20–22]. With the Big Data era, computational time reduction of NMF is particularly investigated, e.g., through optimal solvers [9], distributed strategies [23], online estimation [24], or randomization [29,33].

When applied to a data matrix M which is partially known on a subset Ω of entries, NMF consists of estimating both matrices G and F which satisfy

$$\min_{G,F \geq 0} ||\mathcal{P}_\Omega(M) - \mathcal{P}_\Omega(G \cdot F)||_{\mathcal{F}}, \qquad (3)$$

where $||.||_{\mathcal{F}}$ is the Frobenius norm. Solving Eq. (3) is usually done either by applying Weighted NMF (WNMF) [10,13] to M or by using an Expectation-Maximization (EM) strategy [16,32]. However, these approaches use MU [13,32] or ANLS iterations [16], and are limited to relatively small-sized matrices. As NMC may be applied to very large-scale problems, efficient WNMF methods must be proposed. As a consequence, in this paper we investigate NMF with Nesterov iterations (Ne-NMF) [9] that we propose to extend to the NMC problem. Indeed, NeNMF is improving PG-NMF [22] by replacing the line search of the optimal step in the update rules—known to be time consuming—by Nesterov iterations [25]. NeNMF is thus much faster than MU-NMF or PG-NMF[1]. Moreover, NeNMF can be easily extended to semi-NMF, which is of particular interest in some sensor calibration problems [7]. We thus propose to extend the NeNMF algorithm to deal with missing entries in a nonnegative data matrix M. In particular, we discuss the respective advantages and drawbacks of extensions using the above weighted and EM strategies. We experimentally show that the latter outperforms both the state-of-the-art methods and the proposed weighted extension in simulations of NMC problems.

The remainder of the paper is organized as follows. We recall the principles of the NeNMF method in Sect. 2. We then extend such an approach to the case of an observed matrix with missing entries in Sect. 3. The enhancement of the proposed methods is investigated in Sect. 4 while we conclude and discuss about future work in Sect. 5.

2   Standard NeNMF

As explained above, for a fixed $m \times n$ nonnegative data matrix X, NMF consists of finding both the $m \times r$ and $r \times n$ matrices G and F which provide the best

---

[1] See for example the CPU-time consumptions of [27] at https://github.com/andrewssobral/lrslibrary.

low-rank approximation of X, as shown in Eq. (2). The NeNMF method [9] iteratively and alternately updates G and F. For that, it solves

$$F = \arg\min_{\tilde{F} \geq 0} \mathcal{J}(G, \tilde{F}) = \arg\min_{\tilde{F} \geq 0} \frac{1}{2} \cdot \left\| X - G \cdot \tilde{F} \right\|_{\mathcal{F}}^2, \tag{4}$$

and

$$G = \arg\min_{\tilde{G} \geq 0} \mathcal{J}(\tilde{G}, F) = \arg\min_{\tilde{G} \geq 0} \frac{1}{2} \cdot \left\| X - \tilde{G} \cdot F \right\|_{\mathcal{F}}^2, \tag{5}$$

respectively, by applying in an inner loop the Nesterov accelerated gradient descent [25]. In order to update a factor, say F, the latter initializes $Y_0 \triangleq F^t$—where t is an NeNMF outer iteration—and considers a series α_k defined as

$$\alpha_0 = 1, \quad \alpha_{k+1} = \frac{1 + \sqrt{4\alpha_k^2 + 1}}{2}, \forall k \in \mathbb{N}. \tag{6}$$

For each inner loop index k, the Nesterov gradient descent then computes

$$F_k = \left(Y_k - \frac{1}{L} \nabla_F \mathcal{J}(G, Y_k) \right)^+, \tag{7}$$

and

$$Y_{k+1} = F_k + \frac{\alpha_k - 1}{\alpha_{k+1}}(F_k - F_{k-1}), \tag{8}$$

where $(.)^+$ is the operator which projects any negative entries to zero, $\nabla_F \mathcal{J}(G, .)$ is the gradient of $\mathcal{J}(G, .)$ in Eq. (4), and L is a Lipschitz constant equal to

$$L = \left\| G \cdot G^T \right\|_2 = \|G\|_2^2, \tag{9}$$

where $\|.\|_2$ is the spectral norm. Nesterov gradient descent thus performs a single step of gradient to go to F_k and then slides it in the direction of F_{k-1} to derive Y_{k+1}. Using the Karush-Kuhn-Tucker (KKT) conditions, a stopping criterion—considering both a maximum number Max_{iter} of iterations and a gradient bound—is proposed in [9], thus yielding $F^{t+1} = Y_K$, where Y_K is the last iterate of the above inner iterative gradient descent. In practice, the gradient descent stops if $K = \text{Max}_{\text{iter}}$ or if

$$\left\| \nabla_F^+ \mathcal{J}(G^t, Y_K) \right\|_{\mathcal{F}} \leq \epsilon \cdot \left\| [\nabla_G^+ \mathcal{J}(G^1, F^1), \nabla_F^+ \mathcal{J}(G^1, F^1)] \right\|_{\mathcal{F}}, \tag{10}$$

where ϵ is a user-defined threshold, $\nabla_F^+ \mathcal{J}(G^t, .)$ is the projected gradient of $\mathcal{J}(G^t, .)$, and $\mathcal{J}(G^1, F^1)$ is the value of the cost function computed at the outer loop initialization. The same strategy is applied to G.

3 Extending NeNMF to Missing Entries

3.1 Weigthed Extension of NeNMF

As explained in Sect. 1, two main strategies exist to process missing data in matrix factorization, i.e., using weights or an EM framework. In the weighted strategy, a binary matrix W is included in the NMF formalism. The (i, j)-th entry in W is defined as

$$W(i,j) \triangleq \begin{cases} 1 \text{ if } (i,j) \in \Omega, \\ 0 \text{ otherwise.} \end{cases} \tag{11}$$

Combining Eqs. (3) and (11), WNMF for NMC considers the following problem:

$$\min_{\tilde{G}, \tilde{F} \geq 0} \left\| W \circ \left(X - \tilde{G} \cdot \tilde{F} \right) \right\|_{\mathcal{F}}^2, \tag{12}$$

which is solved using the same splitting strategy as for NMF, e.g., for F:

$$F = \arg\min_{\tilde{F} \geq 0} \mathcal{J}_W(G, \tilde{F}) = \arg\min_{\tilde{F} \geq 0} \frac{1}{2} \cdot \left\| W \circ \left(X - G \cdot \tilde{F} \right) \right\|_{\mathcal{F}}^2. \tag{13}$$

Lemma 1. *The cost function $\mathcal{J}_W(G, .)$ is convex.*

Lemma 2. *In the case of binary weights—and assuming that any row and any column of W contains at least one nonzero element—the gradient of $\mathcal{J}_W(G, .)$ is L-Lipschitz continuous, where L is defined in Eq. (9).*

The proofs are provided in Appendix A.

Lemmas 1 and 2 allow the use of Nesterov iterations, which alternately compute updates (7) and (8), except that the gradient $\nabla_F \mathcal{J}(G, Y_k)$ is here replaced by

$$\nabla_F \mathcal{J}_W(G, Y_k) = G^T \cdot \left(W^2 \circ (G \cdot Y_k - X) \right), \tag{14}$$

where $W^2 = W \circ W$ and $\circ$ is the Hadamard product. The stopping criterion for the inner loop is the same as in Sect. 2 for the standard approach, except that $\mathcal{J}(.,.)$ is here replaced by $\mathcal{J}_W(.,.)$.

It should be noticed that weighted gradient optimization was also investigated in matrix factorization with no sign constraint [28]. In particular, the authors explain that many local minima of $\mathcal{J}_W(G, F)$ exist but can be avoided in practice by gradient descent methods with a well-chosen step size. As a consequence, the major risk of the proposed W-NeNMF might be to converge to a local minimum.

3.2 EM Extension of NeNMF

As an alternative to the above weighted framework, the authors in [32] proposed an EM strategy which consists of an unweighted low-rank approximation of a weighted combination of a previously computed approximation. It turns out that for a given NMF iteration t, the EM problem aims to solve

$$\min_{G^{t+1}, F^{t+1} \geq 0} \left\| W \circ M + (\mathbb{1}_{m,n} - W) \circ (G^t \cdot F^t) - G^{t+1} \cdot F^{t+1} \right\|_{\mathcal{F}}^2, \tag{15}$$

where $\mathbb{1}_{m,n}$ is the $m \times n$ matrix of ones. Indeed, the term $(\mathbb{1}_{m,n} - W) \circ (G^t \cdot F^t)$ in Eq. (15) provides the expected missing values of M. The latter are derived from the complete-data log-likelihood expression, with respect to the unknown data X and given the observed data matrix M and the current estimate of X, i.e., $G^t \cdot F^t$. Denoting $X^t \triangleq W \circ M + (\mathbb{1}_{m,n} - W) \circ (G^t \cdot F^t)$, Eq. (15) reads as a classical NMF problem, i.e.,

$$\min_{G^{t+1}, F^{t+1} \geq 0} \left|\left| X^t - G^{t+1} \cdot F^{t+1} \right|\right|_{\mathcal{F}}^2, \tag{16}$$

whose solution maximizes the previously computed likelihood [32]. As a consequence, the whole approach consists of:

E-step: Estimate the completed version X of M using

$$X \leftarrow W \circ M + (\mathbb{1}_{m,n} - W) \circ (G \cdot F). \tag{17}$$

M-step: Apply a standard NMF to update G and F using X instead of M.

The resulting EM extension of NeNMF then follows the same structure, except that the M-step runs Nesterov iterations instead of MU [32] or ANLS [16].

It should be noticed that the E-step is a key point in the method. Indeed, if the estimation of X is not accurate enough—which might happen in early NMF iterations or in noisy conditions—this might propagate inaccuracies in the estimation of G and F along iterations. We experimentally investigate such potential issues in Sect. 4.

Moreover, the authors in [32] state that the EM strategy combined with MU-NMF is less sensitive to initialization than MU-WNMF. However, they notice that MU-WNMF is faster than EM-NMF in their tests, because they run many MUs at each M-step. When applied to Nesterov iterations, such a strategy has to be rethought. As a consequence, we investigated in preliminary tests the number of Nesterov outer loops processed at each M-stage. Actually, this number is linked to the proportion of missing data in M. Indeed, when the proportion of missing data is low, the E-step is accurate enough to run many Nesterov outer loops and provide a good performance. On the contrary, when this proportion is high, we faced that running relatively few outer loops provides a better performance. As a trade-off, we run 2 outer loops per M-step.

4 Performance of the Weighted NeNMF Extensions

In this section, we investigate the completion performance of the proposed W-NeNMF and EM-W-NeNMF, with respect to the state-of-the-art WNMF [13] and EM-WNMF[2] [32] techniques[3]. For that purpose, we generate simulations

---

[2] As explained above, the standard approach was shown to be slow to converge in [32], because many MUs were applied at each M-step. To prevent such an issue, at each M-stage, we here run one NMF update per matrix factor, i.e., one MU.

[3] In some preliminary tests, we noticed that the EM-W-ANLS method [16] provided a performance similar to EM-WNMF [32]. We thus do not include it in these tests.

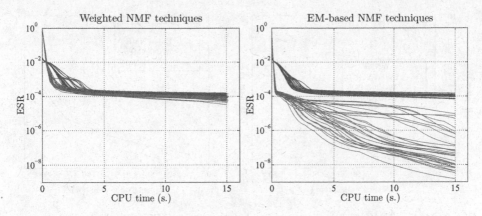

Fig. 1. Evolution of the NMC ESRs vs the CPU time, achieved with MU (in blue), and Nesterov iterations (in red). (Color figure online)

which each consist of the generation of 500×10 and 10×500 random matrices G and F, respectively. For each pair of matrices, we derive the rank-10 500×500 nonnegative matrix X and we randomly sample a percentage of data through an operator $\mathcal{P}_\Omega$, where this percentage ranges from 10 to 90 % (with a step-size of 20 %). The number Max_{iter} of passes in the Nesterov inner loops is set to 100. Indeed, such a value was shown to be a good trade-off in preliminary tests where Max_{iter} was set to 1, 10, 100, and 1000. In particular, we noticed a very low performance when $\text{Max}_{\text{iter}} = 1$—i.e., when the proposed extensions turn out to use a classical gradient descend instead of Nesterov iterations—while the achieved performance was close in the other cases, with a slightly better enhancement with 100 passes per inner loop for each extension. The user-defined threshold ϵ is initialized to 10^{-3} and is divided by 10 each time the stopping criterion (10) is reached within a very low number of inner iterations that we set to $\text{Min}_{\text{iter}} = 5$ in our tests.

All the methods are run using Matlab R2014a on a laptop with an Intel Core i7-4800MQ Quad Core processor, and 32 GB RAM memory. In particular, they are initialized with the same matrices and are stopped after 15 s. In order to compare their performance, we finally compute an Error Signal Ratio (ESR) defined as

$$\text{ESR}(X, G \cdot F) = \frac{\|X - G \cdot F\|_{\mathcal{F}}^2}{\|X\|_{\mathcal{F}}^2}. \tag{18}$$

Fig. 1 shows the ESRs—reached by the four tested methods on 30 simulations made with 50 % of missing entries in M—with respect to the CPU time expressed in seconds. Interestingly, the MU-based methods and the W-NeNMF method provide the same global ESR after 15 s. However, the W-NeNMF ESRs decrease faster in the early iterations. This shows the acceleration due to the Nesterov iterations. All these methods are outperformed by the EM-W-NeNMF approach whose much lower ESRs are still significantly decreasing. This implies that running the

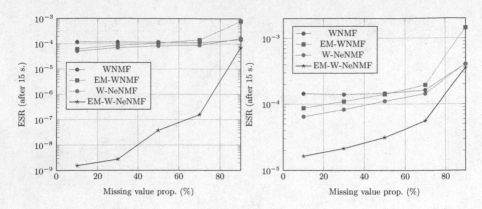

Fig. 2. Median NMC performance reached by the tested NMF methods: (left) noiseless data, (right) noisy data.

EM-W-NeNMF method more than 15 s should provide a much better enhancement. On the contrary, the other methods provide slightly decreasing ESRs, which implies that they will need a lot of time to converge to a limit matrix.

The left plot in Fig. 2 shows the median ESRs achieved by the tested methods after 15 s CPU time, with respect to the missing value proportion. Again, we notice that the EM-W-NeNMF outperforms all the other approaches. However, when only 10 % of the data in M are available, the achieved performance is almost similar to the one obtained with both the WNMF and W-NeNMF methods. Actually, the weighted extensions of MU-NMF and NeNMF are not very sensitive to the sampling rate, in contrast to the EM methods which look very sensitive to this parameter. Indeed, when a lot of data is missing, the E-step does not yield an accurate estimate of X, thus providing a little enhancement in the M-step, which is consistent with [32]. This results in a slower strategy, which is particularly visible with the EM-WNMF performance and which is partly compensated by the speed of Nesterov gradient for the EM-W-NeNMF method.

The right plot in Fig. 2 shows the achieved ESRs in the same simulations when an additive noise is added to X, with an input SNR around 34 dB. We first notice that the median performance achieved by the WNMF, EM-WNMF, and W-NeNMF methods is almost not affected by the presence of noise. However, the obtained ESRs were already quite high in the noiseless case and thus remain high. Despite the fact that the achieved ESRs are higher in the noisy configuration than in the previous noiseless case, the EM-W-NeNMF still outperforms the other methods. The performance loss is probably due to the fact that the E-step is less efficient, which prevents accurate estimation of G and F in the M-step.

5   Conclusion

In this paper, we proposed to extend a fast NMF method using Nesterov gradient descent to solve large-scale low-rank nonnegative matrix completion. Indeed, contrary to classical gradient methods, Nesterov gradient descent does not need the line search of the optimal step in the update rules, which makes the approach really fast. In order to process missing data, we proposed two extensions using the weighted and the EM strategies, respectively. The latter—which consists of an unweighted optimization of a weighted problem—was found to be better suited than the weighted strategy, in noiseless and noisy NMC simulations, hence showing the relevance of the method.

In future work, we aim to extend such approaches to an informed framework [5,6,20]—where some entries of G and/or F are provided—and to apply such extensions to large-scale blind calibration problems [5–7].

Acknowledgments. This work was funded by the "OSCAR" project within the Région Hauts-de-France "Chercheurs Citoyens" Program.

A   Proof of Lemmas 1 and 2

Following the structure of the proof in [9], the proof of Lemma 1 is shown by noticing that

$$\mathcal{J}_W(G, \lambda F_1 + (1 - \lambda)F_2) - \lambda \mathcal{J}_W(G, F_1) - (1 - \lambda)\mathcal{J}_W(G, F_2) = \tfrac{-\lambda(1-\lambda)}{2}\mathcal{J}_W(G, F_1 - F_2) \leq 0. \tag{19}$$

Let us now focus on the proof of Lemma 2. The Lipschitz continuity of $\nabla_F \mathcal{J}_W$ can be shown by extending the proof in [9] to the weighted situation considered in this paper. However, the key point lies in the estimation of the "best" Lipschitz constant. Indeed, let us first recall that Q is a Lipschitz constant of $\nabla_F \mathcal{J}_W(G, .)$ if, for any matrices F_1 and F_2,

$$||\nabla_F \mathcal{J}_W(G, F_1) - \nabla_F \mathcal{J}_W(G, F_2)||_{\mathcal{F}} \leq Q \cdot ||F_1 - F_2||_{\mathcal{F}}. \tag{20}$$

From Eq. (7), it is obvious that the larger is the majoring constant Q, the smaller is the associated gradient step size and thus the convergence speed.

Considering the gradient expression (14), we derive

$$||\nabla_F \mathcal{J}_W(G, F_1) - \nabla_F \mathcal{J}_W(G, F_2)||_{\mathcal{F}} = ||G^T \cdot (W^2 \circ (G \cdot (F_1 - F_2)))||_{\mathcal{F}}. \tag{21}$$

The singular value decomposition of G yields

$$||\nabla_F \mathcal{J}_W(G, F_1) - \nabla_F \mathcal{J}_W(G, F_2)||_{\mathcal{F}} \leq ||G||_2 ||W^2 \circ (G \cdot (F_1 - F_2))||_{\mathcal{F}}. \tag{22}$$

At this stage, we need to put W and G out of the norm to obtain a Lipschitz constant. This can be done by assuming that any column and any row of M contains at least one element—i.e., any column and row of W contains at least one 1—and thus noticing that

$$||W^2 \circ (G \cdot (F_1 - F_2))||_{\mathcal{F}} \leq ||G \cdot (F_1 - F_2)||_{\mathcal{F}}, \tag{23}$$

which provides, using another singular value decomposition of G:

$$||\nabla_F \mathcal{J}_W(G, F_1) - \nabla_F \mathcal{J}_W(G, F_2)||_{\mathcal{F}} \leq ||G||_2^2 ||F_1 - F_2||_{\mathcal{F}}, \tag{24}$$

i.e., the Lipschitz constant for W-NeNMF is L.

References

1. Candès, E.J., Plan, Y.: Matrix completion with noise. Proc. IEEE **98**(6), 925–936 (2010)
2. Candès, E.J., Recht, B.: Exact matrix completion via convex optimization. Found. Comput. Math. **9**(6), 717–772 (2009)
3. Choo, J., Lee, C., Reddy, C.K., Park, H.: Weakly supervised nonnegative matrix factorization for user-driven clustering. Data Min. Knowl. Disc. **29**(6), 1598–1621 (2015)
4. Chreiky, R., Delmaire, G., Puigt, M., Roussel, G., Courcot, D., Abche, A.: Split gradient method for informed non-negative matrix factorization. In: Vincent, E., Yeredor, A., Koldovský, Z., Tichavský, P. (eds.) Proceedings of LVA/ICA, pp. 376–383 (2015)
5. Dorffer, C., Puigt, M., Delmaire, G., Roussel, G.: Blind calibration of mobile sensors using informed nonnegative matrix factorization. In: Vincent, E., Yeredor, A., Koldovský, Z., Tichavský, P. (eds.) Proceedings of LVA/ICA, pp. 497–505 (2015)
6. Dorffer, C., Puigt, M., Delmaire, G., Roussel, G.: Blind mobile sensor calibration using an informed nonnegative matrix factorization with a relaxed rendezvous model. In: Proceedings of ICASSP, pp. 2941–2945 (2016)
7. Dorffer, C., Puigt, M., Delmaire, G., Roussel, G.: Nonlinear mobile sensor calibration using informed semi-nonnegative matrix factorization with a Vandermonde factor. In: Proceedings of SAM (2016)
8. Fazel, M.: Matrix rank minimization with applications. Ph.D. thesis, Stanford University (2002)
9. Guan, N., Tao, D., Luo, Z., Yuan, B.: NeNMF: an optimal gradient method for nonnegative matrix factorization. IEEE Trans. Signal Proc. **60**(6), 2882–2898 (2012)
10. Guillamet, D., Vitrià, J., Schiele, B.: Introducing a weighted non-negative matrix factorization for image classification. Pattern Recogn. Lett. **24**(14), 2447–2454 (2003)
11. Haldar, J.P., Hernando, D.: Rank-constrained solutions to linear matrix equations using powerfactorization. IEEE Signal Process. Lett. **16**(7), 584–587 (2009)
12. Hamon, R., Emiya, V., Févotte, C.: Convex nonnegative matrix factorization with missing data. In: Procceedings of MLSP (2016)
13. Ho, N.D.: Nonnegative matrix factorizations algorithms and applications. Ph.D. thesis, Université Catholique de Louvain (2008)
14. Hoyer, P.O.: Non-negative matrix factorization with sparseness constraints. J. Mach. Learn. Res. **5**, 1457–1469 (2004)
15. Kim, H., Park, H.: Nonnegative matrix factorization based on alternating nonnegativity constrained least squares and active set method. SIAM J. Matrix Anal. Appl. **30**(2), 713–730 (2008)
16. Kim, Y.D., Choi, S.: Weighted nonnegative matrix factorization. In: Proceedings of ICASSP, pp. 1541–1544, April 2009
17. Kumar, R., Da Silva, C., Akalin, O., Aravkin, A.Y., Mansour, H., Recht, B., Herrmann, F.J.: Efficient matrix completion for seismic data reconstruction. Geophysics **80**(5), V97–V114 (2015)
18. Lantéri, H., Theys, C., Richard, C., Févotte, C.: Split gradient method for nonnegative matrix factorization. In: Proceedings of EUSIPCO (2010)
19. Lee, D.D., Seung, H.S.: Algorithms for non-negative matrix factorization. In: NIPS, pp. 556–562 (2001)

20. Limem, A., Delmaire, G., Puigt, M., Roussel, G., Courcot, D.: Non-negative matrix factorization under equality constraints–a study of industrial source identification. Appl. Numer. Math. **85**, 1–15 (2014)
21. Limem, A., Puigt, M., Delmaire, G., Roussel, G., Courcot, D.: Bound constrained weighted NMF for industrial source apportionment. In: Proceedings of MLSP (2014)
22. Lin, C.: Projected gradient methods for nonnegative matrix factorization. Neural Comput. **19**(10), 2756–2779 (2007)
23. Liu, C., Yang, H.C., Fan, J., He, L.W., Wang, Y.M.: Distributed nonnegative matrix factorization for web-scale dyadic data analysis on MapReduce. In: Proceedings of WWW Conference, April 2010
24. Mairal, J., Bach, F., Ponce, J., Sapiro, G.: Online learning for matrix factorization and sparse coding. J. Mach. Learn. Res. **11**, 19–60 (2010)
25. Nesterov, Y.: A method of solving a convex programming problem with convergence rate O(1/k2). Sov. Math. Doklady **27**, 372–376 (1983)
26. Savvaki, S., Tsagkatakis, G., Panousopoulou, A., Tsakalides, P.: Application of matrix completion on water treatment data. In: Proceedings of CySWater, pp. 3:1–3:6 (2015)
27. Sobral, A., Bouwmans, T., Zahzah, E.: LRSLibrary: low-rank and sparse tools for background modeling and subtraction in videos. In: Robust Low-Rank and Sparse Matrix Decomposition: Applications in Image and Video Processing. CRC Press, Taylor and Francis Group
28. Srebro, N., Jaakkola, T.: Weighted low-rank approximations. In: Proceedings of ICML (2003)
29. Tepper, M., Sapiro, G.: Compressed nonnegative matrix factorization is fast and accurate. IEEE Trans. Signal Process. **64**(9), 2269–2283 (2016)
30. Wang, Y.X., Zhang, Y.J.: Nonnegative matrix factorization: a comprehensive review. IEEE Trans. Knowl. Data Eng. **25**(6), 1336–1353 (2013)
31. Wen, Z., Yin, W., Zhang, Y.: Solving a low-rank factorization model for matrix completion by a nonlinear successive over-relaxation algorithm. Math. Program. Comput. **4**(4), 333–361 (2012)
32. Zhang, S., Wang, W., Ford, J., Makedon, F.: Learning from incomplete ratings using non-negative matrix factorization, chap. 58, pp. 549–553 (2006)
33. Zhou, G., Cichocki, A., Xie, S.: Fast nonnegative matrix/tensor factorization based on low-rank approximation. IEEE Trans. Signal Process. **60**(6), 2928–2940 (2012)

Blind Source Separation of Single Channel Mixture Using Tensorization and Tensor Diagonalization

Anh-Huy Phan[1], Petr Tichavský[2(✉)], and Andrzej Cichocki[1,3]

[1] Lab for Advanced Brain Signal Processing,
Brain Science Institute - RIKEN, Wako, Japan
[2] Institute of Information Theory and Automation, Prague, Czech Republic
tichavsk@utia.cas.cz
[3] Skolkovo Institute of Science and Technology (SKOLTECH), Moscow, Russia

Abstract. This paper deals with estimation of structured signals such as damped sinusoids, exponentials, polynomials, and their products from single channel data. It is shown that building tensors from this kind of data results in tensors with hidden block structure which can be recovered through the tensor diagonalization. The tensor diagonalization means multiplying tensors by several matrices along its modes so that the outcome is approximately diagonal or block-diagonal of 3-rd order tensors. The proposed method can be applied to estimation of parameters of multiple damped sinusoids, and their products with polynomial.

Keywords: Blind source separation · Block tensor decomposition · Tensor diagonalization · Three-way folding · Three-way toeplitzation · Damped sinusoids

1 Introduction

Separation of hidden components or sources from mixtures with one or a few sensors appears in many real problems in signal processing. So far, there is no straightforward method for separation of arbitrary signals. However, the problem can be tackled in some cases when strong assumptions have been imposed onto the components or the data, e.g., non-negativity together with non-overlapping (i.e., orthogonality) as in nonnegative matrix factorisation [1,2], and statistical independence as in independent component analysis [3]. In some cases, the signal components of interest have some specific structures. For example, separation of the damped sinusoids appears in wide range of applications, e.g., electrical, mechanical, electromechanical, geophysical, chemical. The traditional methods for estimating damped sinusoid parameters are based on the linear self-prediction (auto regression) using signal samples, where the Prony least squares

---

The work of P. Tichavsky was supported by The Czech Science Foundation through Projects No. 14-13713S and 17-00902S.

P. Tichavský et al. (Eds.): LVA/ICA 2017, LNCS 10169, pp. 36–46, 2017.
DOI: 10.1007/978-3-319-53547-0_4

autoregressive model fitting, and the Pade approximation procedure are two well-known methods [4]. The other algorithmic families are based the signal subspace method, which computes self-linear prediction coefficients as eigenvectors of autocorrelation matrix of the measurement, e.g., Pisarenko, MUSIC, ESPRIT, or the singular value decomposition (SVD) of the Hankel-type matrix, e.g., the Kumaresan-Tufts and Matrix Pencil methods [5,6]. Extension of the methods to tensor decomposition can be found in [7–10].

Alternatively, another type of methods has been developed for this BSS problem, based on exploiting the structure of the components rather based on their statistical properties. More specifically, when the extracted components can be expressed in some low rank formats, e.g., low rank matrix or tensor, low multilinear rank tensor, by tensorization of the mixture, one can apply appropriate tensor decomposition methods to retrieve hidden factor matrices, which are then used to reconstruct the original sources [11,12]. De Lathauwer proposed to use the Hankelization to convert signals to be low-rank matrices, and resort the BSS problem to the canonical tensor decomposition or the $(L, L, 1)$-block term decomposition [11,12]. An advantage of this approach is that it allows to separate signals with relatively short length samples, even with dozens of samples.

In the same direction with the latter method, the method proposed in this paper first tensorizes the mixtures, and converts the BSS problem of R sources into block tensor decompositions (BTD) of R block terms, each term corresponds to a source. Tensorization for single channel data can be the simple re-shaping (folding), Toeplitzation or Hankelization. The blocks structure can be revealed through the tensor diagonalization technique, which is explained in an accompanying papers [13]. The proposed method not only separate sinusoids as, e.g., [14], but also allows to estimate a wider class of signals, e.g., products of polynomials and damped sinusoids. In addition, the proposed algorithm works even when the number of samples is small, or when the number of sources becomes significantly large.

In the simulation section, we test the proposed method and compare its performance with state-of-the-art methods of estimating parameters of damped sinusoids. We also show one example to which the damping estimation methods are not applicable.

2 General Framework for BSS of Single Channel Mixture

Consider a single mixture $y(t)$ which is composed of R component signals $x_r(t)$, $r = 1, \ldots, R$, and corrupted by additive Gaussian noise $e(t)$

$$y(t) = a_1 x_1(t) + a_2 x_2(t) + \ldots + a_R x_R(t) + e(t). \tag{1}$$

The aim is to extract the unknown sources (components) $x_r(t)$ from the observed signal $y(t)$. For this problem, we first apply some linear tensorization to the source vectors $\boldsymbol{x}_r = [x_r(1), \ldots, x_r(t), \ldots]$, in order to obtain 3-rd order tensors $\boldsymbol{\mathcal{X}}_r$, which are assumed to be low rank or low multilinear rank tensor, that is,

$$\boldsymbol{\mathcal{X}}_r = \boldsymbol{\mathcal{G}}_r \times_1 \mathbf{U}_r \times_2 \mathbf{V}_r \times_3 \mathbf{W}_r,$$

where $\mathcal{G}_r$ are core tensors of size $R_1 \times R_2 \times R_3$. Such tensorizations of interest can be reshaping, which is also known as segmentation [15,16], Hankelization [11], and Toeplization. From the mixing model in (1), and due to linearity of the tensorization, we have the following relation between tensorization $\mathcal{Y}$ of the mixture and those of the hidden components

$$\mathcal{Y} = a_1\mathcal{X}_1 + a_2\mathcal{X}_2 + \ldots + a_R\mathcal{X}_R + \mathcal{E}$$
$$= \sum_{r=1}^{R}(a_r\mathcal{G}_r) \times_1 \mathbf{U}_r \times_2 \mathbf{V}_r \times_3 \mathbf{W}_r + \mathcal{E} \tag{2}$$

where $\mathcal{E}$ is tensorization of the noise $e(t)$.

Now, by decomposition of $\mathcal{Y}$ into R blocks, i.e., BTD with R block terms, we can find approximations of $\mathcal{X}_r$ up to scaling. Finally, reconstruction of signals $\hat{x}_r(t)$ from $\mathcal{X}_r$ can be done straightforwardly because the tensorization is linear. Instead of applying the block term decomposition, we address the above tensor model as a tensor diagonalization, where each block of the core tensor corresponds to a hidden source.

For the multi-channel BSS, $y_m(t) = \sum_r a_{kr}x_r(t)$, channel mixtures $\boldsymbol{y}_m = [y_m(1), \ldots, y_m(t), \ldots]$, are tensorized separately, then all together they construct a tensor of order-4 which admits the BTD-$(R_1, R_2, R_3, 1)$

$$\mathcal{Y} = \sum_{r=1}^{R} \mathcal{G}_r \times_1 \mathbf{U}_r \times_2 \mathbf{V}_r \times_3 \mathbf{W}_r \times_4 \boldsymbol{a}_r + \mathcal{E} \tag{3}$$

where $\boldsymbol{a}_r$ are column vectors of the mixing matrix.

We are particular interested in the sinusoid signals and its modulated variant, e.g., the exponentially decaying signals

$$x(t) = \exp(-\gamma t)\sin(\omega t + \phi), \tag{4}$$
$$x(t) = t^n \sin(\omega t + \phi), \qquad x(t) = t^n \exp(-\gamma t), \tag{5}$$

for $t = 1, 1, \ldots, L$, $\omega \neq 0$, $n = 1, 2, \ldots$.

In the next sections, we present tensorizations to yield low-rank tensors from the above signals, and confirm their efficiency through examples for BSS of single mixture.

3 Tensorization of Sinusoid Signals

The tensorizations presented in this section are for the sinusoids but can also be applied to the other signals in (4)–(5) to yield tensors of multilinear rank-$(2, 2, 2)$, $(2(n + 1), 2(n + 1), 2(n + 1))$ or $(n + 1, n + 1, n + 1)$.

3.1   Two-Way and Three-Way Foldings

The simplest tensorization is reshaping (folding), which rearranges a vector to a matrix or tensor. This type of tensorization preserves the number of original data entries and their sequential ordering. It can be shown that reshaping of a sinusoid results a rank-2 matrix, or a multilinear rank-$(2, 2, 2)$ tensor.

Lemma 1 (Two-way folding). *A matrix of size $I \times J$ which is reshaped from a sinusoid signal $x(t)$ of length $L = IJ$ is of rank-2 and can be decomposed as*

$$\mathbf{Y} = \begin{bmatrix} y(1) & y(I+1) & \cdots & y(K-I+1) \\ y(2) & y(I+2) & \cdots & y(K-I+2) \\ \vdots & & \ddots & \vdots \\ y(I) & y(2I) & \cdots & y(K) \end{bmatrix} = \mathbf{U}_{\omega, I}\, \mathbf{S}\, \mathbf{U}_{\omega I, J}^T \tag{6}$$

where $\mathbf{S}$ is invariant to the folding size I, and depends only on the phase ϕ and takes the form

$$\mathbf{S} = \begin{bmatrix} \sin(\phi) & \cos(\phi) \\ \cos(\phi) & -\sin(\phi) \end{bmatrix}, \qquad \mathbf{U}_{\omega, I} = \begin{bmatrix} 1 & 0 \\ \vdots & \vdots \\ \cos(k\omega) & \sin(k\omega) \\ \vdots & \vdots \\ \cos((I-1)\omega) & \sin((I-1)\omega) \end{bmatrix}. \tag{7}$$

Lemma 2 (Three-way folding). *An order-3 tensor of size $I \times J \times K$, where $I, J, K > 2$, reshaped from a sinusoid signal of length $L = IJK$, can take a form of a multilinear rank-$(2, 2, 2)$ or rank-3 tensor*

$$\mathcal{Y} = \mathcal{H} \times_1 \mathbf{U}_{\omega, I} \times_2 \mathbf{U}_{\omega I, J} \times_3 \mathbf{U}_{\omega IJ, K} \tag{8}$$

where $\mathcal{H} = \mathcal{G} \times_3 \mathbf{S}$ is a small-scale tensor of size $2 \times 2 \times 2$, and

$$\mathcal{G}(:, :, 1) = \begin{bmatrix} 1 & 0 \\ 0 & -1 \end{bmatrix}, \qquad \mathcal{G}(:, :, 2) = \begin{bmatrix} 0 & 1 \\ 1 & 0 \end{bmatrix}. \tag{9}$$

Proof of Lemma 1 is obvious, while Lemma 2 can be deduced from Lemma 1 because the three-way folding can be performed over two foldings [17]. Note that in the complex field, the above tensors are of rank-2. An advantage of this tensorization is that it does not increase the number of samples; therefore, the method does not need extra space. However, when the signals are of short duration, the reshaped tensors are relatively small, and decomposition of these tensors does not give good approximation of the original sources. For such a case, tensorizations, which can increase the number of entries, e.g., the Toeplitzation and Hankelization, are recommended.

3.2   Toeplitzation

Definition 1 (Three-way Toeplitzation). *Tensorization of the signal $x(t)$ of length L to an order-3 tensor $\mathcal{X}$ of size $I \times J \times K$, where $I + J + K = L + 2$,*

$$\mathcal{X}(i, j, k) = x(I + J + k - i - j). \tag{10}$$

For this tensorization, each horizontal slice $\mathcal{X}(i, :, :)$ is a Toeplitz matrix composed of vectors $[x(I + 1 - i), \ldots, x(I + J - i)]$ and $[x(I + J - i), \ldots, y(I + J + K - 1 - i)]$.

Lemma 3. *Order-3 Toeplitz tensors tensorized from a sinusoid signal is of multilinear rank-$(2, 2, 2)$, and can be represented as*

$$\mathcal{X} = \frac{1}{\sin^3(\omega)}\, \mathcal{G} \times_1 \mathbf{U}_1 \times_2 \mathbf{U}_2 \times_3 \mathbf{U}_3 \tag{11}$$

where $\mathcal{G}$ is a tensor of size $2 \times 2 \times 2$

$$\mathcal{G}(:,:,1) = \begin{bmatrix} \sin(\omega(I_1 + 2) + 2\phi) & -\sin(\omega(I_1 + 1) + 2\phi) \\ -\sin(\omega(I_1 + 1) + 2\phi) & \sin(\omega I_1 + 2\phi) \end{bmatrix},$$

$$\mathcal{G}(:,:,2) = \begin{bmatrix} -\sin(\omega(I_1 + 1) + 2\phi) & \sin(\omega I_1 + 2\phi) \\ \sin(\omega I_1 + 2\phi) & -\sin(\omega(I_1 - 1) + 2\phi) \end{bmatrix},$$

and the three factor matrices are given by

$$\mathbf{U}_1 = \begin{bmatrix} x(1) & x(2) \\ \vdots & \vdots \\ x(I) & x(I+1) \end{bmatrix}, \ \mathbf{U}_2 = \begin{bmatrix} x(I) & x(I+1) \\ \vdots & \vdots \\ x(I+J-1) & x(I+J) \end{bmatrix}, \ \mathbf{U}_3 = \begin{bmatrix} x(I+J-1) & x(I+J-2) \\ \vdots & \vdots \\ x(L) & x(L-1) \end{bmatrix}.$$

Proof. The proof can be seen from the fact that

$$\mathcal{G} \times_1 \begin{bmatrix} x(i) & x(i+1) \end{bmatrix} = \sin(\omega) \begin{bmatrix} -x(I-i+3) & x(I-i+2) \\ x(I-i+2) & -x(I-i+1) \end{bmatrix} \tag{12}$$

and

$$[x(I+j-1), x(I+j)] \begin{bmatrix} -x(I-i+3) & x(I-i+2) \\ x(I-i+2) & x(I-i+1) \end{bmatrix} \begin{bmatrix} x(I+J+k-2) \\ x(I+J+k-3) \end{bmatrix}$$
$$= \sin^2(\omega)\, x(I+J+k-i-j). \qquad \square$$

In Appendix, we show that the three-way folding and Toeplitzation of $x(t) = t$ also yield multilinear rank-$(2, 2, 2)$ tensors or tensors of rank-3. A more general result is that the tensors of $x(t) = t^n$ have multilinear rank of $(n + 1, n + 1, n + 1)$. The closed-form expressions of the low multilinear representations are given in Lemmas 5–8. In addition, we note that the above three-way tensorizations to sinusoids yield tensors $\mathcal{X}_{\mathrm{sin}}$ of multilinear rank-$(2, 2, 2)$, and tensorization of exponentially decaying signal $\exp(-\omega t)$ yields a rank-1 tensor $\mathcal{X}_{\mathrm{exp}}$. Following Lemma 4, the Hadamard product $\mathcal{X}_{\mathrm{sin}} \circledast \mathcal{X}_{\mathrm{exp}}$ remains a tensor of multilinear rank-$(2, 2, 2)$, implying that the damped sinusoids can be represented by multilinear rank-$(2, 2, 2)$ tensors. Similarly, the signal $t \exp(-\gamma t)$ in (5) has multilinear rank-$(2, 2, 2)$, and $t \sin(\omega t + \phi)$ in (5) has multilinear rank-$(4, 4, 4)$.

Lemma 4 (Hadamard product of two tensors of low multilinear ranks). *Given two tensors of the same size, $\mathcal{X}$ of multilinear rank-(R, S, T) and $\mathcal{Y}$ of multilinear rank-(R', S', T'), the Hadamard product of them can be represented by a tensor which has multilinear rank at most (RR', SS', TT').*

Proof. We represent $\mathcal{X} = \mathcal{G} \times_1 \mathbf{A} \times_2 \mathbf{B} \times_3 \mathbf{C}$, and $\mathcal{Y} = \mathcal{H} \times_1 \mathbf{U} \times_2 \mathbf{V} \times_3 \mathbf{W}$, where $\mathcal{G}$ is a tensor of size $R \times S \times T$, and $\mathcal{H}$ is of size $R' \times S' \times T'$. Then the Hadamard product $\mathcal{X} \circledast \mathcal{Y}$ yields a tensor, for which entries are defined as

$$
\mathcal{X}(i,j,k)\mathcal{Y}(i,j,k) = \left(\sum_{r,s,t} g_{r,s,t}\, a_{ir} b_{js} c_{kt} \right) \left(\sum_{r',s',t'} h_{r',s',t'}\, u_{ir'} v_{js'} w_{kt'} \right)
$$

$$
= \sum_{r,r'} \sum_{s,s'} \sum_{t,t'} (g_{r,s,t} h_{r',s',t'})\,(a_{ir} u_{ir'})(b_{js} v_{js'})(c_{kt} w_{kt'})
$$

$$
= \sum_{\bar{r}}^{RR'} \sum_{\bar{s}}^{SS'} \sum_{\bar{t}}^{TT'} z_{\bar{r},\bar{s},\bar{t}}\, d_{i,\bar{r}}\, e_{j,\bar{s}}\, f_{j,\bar{t}},
$$

where $\mathcal{Z} = \mathcal{G} \otimes \mathcal{H}$ is of size $RR' \times SS' \times TT'$, and $\mathbf{D}$, $\mathbf{E}$ and $\mathbf{F}$ are row-wise Khatri-Rao products of the two corresponding factor matrices of $\mathcal{X}$ and $\mathcal{Y}$, i.e., $\boldsymbol{d}_i = \boldsymbol{a}_i \otimes \boldsymbol{u}_i$, $\boldsymbol{e}_j = \boldsymbol{b}_j \otimes \boldsymbol{v}_j$, and $\boldsymbol{f}_j = \boldsymbol{c}_k \otimes \boldsymbol{w}_k$. In summary, we obtain

$$
\mathcal{X} \circledast \mathcal{Y} = \mathcal{Z} \times_1 \mathbf{D} \times_2 \mathbf{E} \times_3 \mathbf{F}. \tag{13}
$$

$\square$

4   Simulations

Example 1. In this first example, we considered a signal of length $L = 414$, which were composed of two source signals $x_1(t)$ and $x_2(t)$ and corrupted by additive Gaussian noise $e(t)$

$$
y(t) = a_1 x_1(t) + a_2 x_2(t) + e(t),
$$

$$
x_1(t) = \exp(\frac{-2t}{L}) \sin(\frac{2\pi f_1}{f_s} t), \qquad x_2(t) = t \exp(\frac{-4t}{L}),
$$

where $f_1 = 5\,\mathrm{Hz}$, and $f_s = 135\,\mathrm{Hz}$, and the mixing coefficient $a_r = \frac{1}{\|\boldsymbol{x}_r\|_2}$ ($r = 1, 2$). We performed the three-way Toeplitzation for the mixture $y(t)$ to give an order-3 tensor of size $192 \times 32 \times 192$. Since the signal $x_2(t)$ can be represented by a multilinear rank-$(2, 2, 2)$ tensor (see Lemma 6), we can extract the two source signals $x_1(t)$ and $x_2(t)$ through block tensor decompositions. We verified the separation for various noise levels, SNR $= 0, 10, \ldots, 40\,\mathrm{dB}$, and assessed the performance over 100 independent runs for each noise level. In Fig. 1(a), we compared the mean and median Squared Angular Errors (SAE) achieved using TEDIA and the non-linear least squares (NLS) algorithm which utilises the

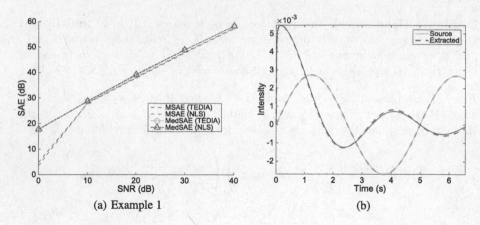

Fig. 1. (a) Mean and median squared angular errors (MSAE and MedSAE) of the considered algorithms at various SNRs in Example 1. (b) Sources and estimated signals in Example 2.

Gauss-Newton algorithm with dogleg trust region for the block tensor decomposition [18][1]. The performances were relatively stable and linearly decreased with the signal-noise-ratios in logarithmic scale.

Example 2. This example aims at showing that the proposed method can separate slowly time-varying signals. We considered two mixture signals of length $L = 262144$, $y_r(t) = a_{r1}x_1(t) + a_{r2}x_2(t)$, $r = 1, 2$, composed of two source signals

$$x_1(t) = \exp(\frac{-5t}{L})\sin(\frac{2\pi f_1}{f_s}t), \qquad x_2(t) = \frac{\sin(2\pi f_2/f_s t)}{t + 200}$$

by a mixing matrix $\mathbf{A} = \begin{bmatrix} -1.1896 & 0.1072 \\ 0.0946 & -1.3018 \end{bmatrix}$, where $f_1 = 0.2$ Hz, $f_1 = 0.3$ Hz, and $f_s = 40000$ Hz. We reshaped the mixtures to give an order-4 tensor of size $64 \times 64 \times 64 \times 2$. For this tensorization, the signal $x_1(t)$ yields a tensor of multilinear rank-$(2, 2, 2)$, while the signal $x_2(t)$ is well approximated by a tensor of multilinear rank-$(5, 5, 5)$, because it is elementwise product of a sinusoid and the rational function $1/(t + 200)$, which can be well approximated by a second-order polynomial. Hence we can apply the BTD with two blocks of rank-$(2, 2, 2)$ and rank-$(5, 5, 5)$ to retrieve the source signals. The obtained squared angular errors for the two sources are respective 118.7 and 54.1 dB.

Example 3 (BSS from a short length mixture). In this example, we used a signal of the length $L = 414$ composed of $R = 3$ component signals $x_r(t)$, $r = 1, 2, 3$, and corrupted by additive Gaussian noise $e(t)$

$$y(t) = a_1 x_1(t) + a_2 x_2(t) + a_3 x_3(t) + e(t)$$

---

[1] The NLS algorithm is available in the Tensorlab toolbox at www.tensorlab.net.

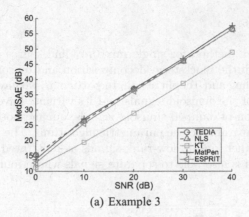

(a) Example 3

Fig. 2. Performance comparison of the considered algorithms at various SNRs in Example 3.

where

$$x_r(t) = \exp(\frac{-5t}{rL}) \sin(\frac{2\pi f_r}{f_s} t + \frac{(r-1)\pi}{R})$$

where the frequencies $f_r = 5, 7, 9$ Hz, and the sampling frequency $f_s = 135$ Hz. The mixing coefficients a_r were simply set to $a_r = \frac{1}{\|x_r\|_2}$ so that contributions of x_r to $y(t)$ are equivalent.

Tensorization using the three-way folding did not work for such short length signal. Instead, we applied the three-way Toeplitzation, and constructed a tensor of size $192 \times 32 \times 192$. It means that we increased the number of tensor entries to 1,179,648.

Similar to the previous example, after the tensorization, we compressed tensor of the measurement to one of size $6 \times 6 \times 6$ using the HOOI algorithm, and then applied the tensor diagonalization. In Fig. 2(a), we compare performance of the separation via the squared angular error SAE when SNR = 0, 10, ... 40 dB. The results were assessed over 100 independent runs for each noise level. An important observation is that performances achieved using TEDIA and the NLS algorithm for BTD are almost at the same level. Even when the signal-noise-ratio SNR = 0 dB, we were able to retrieve the sources with sufficiently good performance.

Performance of the parametric methods including the Kumaresan-Tufts KT algorithm [5], the Matrix Pencil [6], and ESPRIT methods [19] is also compared in Fig. 2(a). The results indicate that our non-parametric method achieved higher performance than the KT algorithm, and was comparable to the ESPRIT algorithm, and slightly worse than the Matrix Pencil algorithm, which performed best.

5   Conclusions

We have presented a method for single mixture blind source separation of low rank signals through the block term decomposition and tensor diagonalization, using three-way folding and Toeplitzation. In particular, we have also shown that the tensorizations of the sinusoid signals and its variants have low multi-linear ranks. For separation of damped sinusoid signals, our method achieved performance which is comparable to the parametric algorithms. The proposed method can also separate other kind of low-rank signals as illustrated in Example 1. In general, the method is also able to separate signals whose multilinear ranks are different.

A   Appendix: Low-Rank Representation of the Sequence $x(t) = t^n$

Lemma 5 (Three-way folding of $x(t) = t$). *An order-3 tensor of size $I \times J \times K$, reshaped (folded) from the sequence $1, 2, \ldots, IJK$, where $I, J, K > 2$, has multilinear rank-$(2, 2, 2)$ and rank-3, and can be represented as*

$$\mathcal{Y} = \mathcal{G} \times_1 \mathbf{U}_1 \times_2 \mathbf{U}_2 \times_3 \mathbf{U}_3 \tag{14}$$

where $\mathcal{G}$ is a tensor of size $2 \times 2 \times 2$

$$\mathcal{G}(:, 1, :) = \begin{bmatrix} -2 & -1 \\ 1 & 0 \end{bmatrix}, \qquad \mathcal{G}(:, 2, :) = \begin{bmatrix} 1 & 0 \\ 0 & 1 \end{bmatrix},$$

and the three factor matrices are given by

$$\mathbf{U}_1 = \begin{bmatrix} 0 & 1 \\ \vdots & \vdots \\ i & i+1 \\ \vdots & \vdots \\ (I-1) & I \end{bmatrix}, \ \mathbf{U}_2 = \begin{bmatrix} 0 & 1 \\ \vdots & \vdots \\ jI & jI+1 \\ \vdots & \vdots \\ (J-1)I & (J-1)I+1 \end{bmatrix}, \ \mathbf{U}_3 = \begin{bmatrix} 0 & 1 \\ \vdots & \vdots \\ kIJ & kIJ+1 \\ \vdots & \vdots \\ (K-1)IJ & (K-1)IJ+1 \end{bmatrix}.$$

Lemma 6 (Toeplitzation of $x(t) = t$). *An order-3 Toeplitz tensor of size $I \times J \times K$, tensorized from the sequence $1, 2, \ldots, L$, where $L = I + J + K - 2$, has multilinear rank-$(2, 2, 2)$ and rank-3, and can be represented as*

$$\mathcal{Y} = \mathcal{G} \times_1 \mathbf{U}_1 \times_2 \mathbf{U}_2 \times_3 \mathbf{U}_3 \tag{15}$$

where $\mathcal{G}$ is a tensor of size $2 \times 2 \times 2$

$$\mathcal{G}(:, :, 1) = \begin{bmatrix} I+4 & -(I+3) \\ -(I+3) & I+2 \end{bmatrix}, \qquad \mathcal{G}(:, :, 2) = \begin{bmatrix} -(I+3) & I+2 \\ I+2 & -(I+1) \end{bmatrix},$$

and the three factor matrices are given by

$$\mathbf{U}_1 = \begin{bmatrix} 1 & 2 \\ \vdots & \vdots \\ i & i+1 \\ \vdots & \vdots \\ I & I+1 \end{bmatrix}, \quad \mathbf{U}_2 = \begin{bmatrix} I & I+1 \\ \vdots & \vdots \\ j & j+1 \\ \vdots & \vdots \\ I+J-1 & I+J \end{bmatrix}, \quad \mathbf{U}_3 = \begin{bmatrix} I+J-1 & I+J-2 \\ \vdots & \vdots \\ k & k-1 \\ \vdots & \vdots \\ L & L-1 \end{bmatrix}.$$

Lemma 7 (Three-way folding of $x(t) = t^n$). *An order-3 tensor of size $I \times J \times K$, reshaped (folded) from the sequence $x(t) = t^n$, where $n = 1, 2, \dots$ and $I, J, K > 2$, has multilinear rank-$(n+1, n+1, n+1)$.*

Proof. By exploiting the closed-form expression of $x(t) = t$ in Lemma 5, and the property of the Hadamard product stated in Lemma 4 or in (13), we can prove that the tensor reshaped from $x(t) = t^n$ can be fully explained by three factor matrices which have $n + 1$ columns, and are defined as $\mathbf{U}_1 = \mathbf{F}(I, n)$, $\mathbf{U}_2 = \mathbf{F}(IJ, n)$ and $\mathbf{U}_3 = \mathbf{F}(IJK, n)$, where

$$\mathbf{F}(I, n) = \begin{bmatrix} 0 & \cdots & 0 & \cdots & 1 \\ \vdots & & \vdots & & \\ i^n & \cdots & i^k(i+1)^{n-k} & \cdots & (i+1)^n \\ \vdots & & \vdots & & \vdots \\ (I-1) & \cdots & (I-1)^k I^{n-k} & \cdots & I^n \end{bmatrix}.$$

□

Lemma 8 (Toeplitzation of $x(t) = t^n$). *An order-3 Toeplitz tensor of size $I \times J \times K$ of the sequence $x(t) = t^n$, has multilinear rank-$(n+1, n+1, n+1)$.*

Proof. Skipped for lack of space.

□

References

1. Cichocki, A., Zdunek, R., Phan, A.-H., Amari, S.: Nonnegative Matrix and Tensor Factorizations: Applications to Exploratory Multi-way Data Analysis and Blind Source Separation. Wiley, Chichester (2009)
2. Cichocki, A., Mandic, D., Caiafa, C., Phan, A.-H., Zhou, G., Zhao, Q., Lathauwer, L.D.: Multiway component analysis: tensor decompositions for signal processing applications. IEEE Sig. Process. Mag. **32**(2), 145–163 (2015)
3. Comon, P., Jutten, C. (eds.): Handbook of Blind Source Separation: Independent Component Analysis and Applications. Academic Press, Cambridge (2010)
4. Proakis, J., Manolakis, D.: Digital Signal Processing: Principles, Algorithms, Applications. Macmillan, London (1992)
5. Kumaresan, R., Tufts, D.: Estimating the parameters of exponentially damped sinusoids and pole-zero modelling in noise. IEEE Trans. Acoust. Speech Sig. Proces. **30**(7), 837–840 (1982)

6. Hua, Y., Sarkar, T.: Matrix pencil method for estimating parameters of exponentially damped/undamped sinusoid in noise. IEEE Trans. Acoust. Speech Sig. Process. **38**(5), 814–824 (1990)
7. Papy, J.M., De Lathauwer, L., Van Huffel, S.: Exponential data fitting using multilinear algebra: the single-channel and the multichannel case. Numer. Linear Algebra Appl. **12**(9), 809–826 (2005)
8. Nion, D., Sidiropoulos, N.D.: Tensor algebra and multidimensional harmonic retrieval in signal processing for MIMO radar. IEEE Trans. Sig. Process. **58**, 5693–5705 (2010)
9. Wen, F., So, H.C.: Robust multi-dimensional harmonic retrieval using iteratively reweighted HOSVD. IEEE Sig. Process. Lett. **22**, 2464–2468 (2015)
10. Sørensen, M., De Lathauwer, L.: Multidimensional harmonic retrieval via coupled canonical polyadic decomposition - Part 1: model and identifiability. IEEE Trans. Sig. Process. **65**(2), 517–527 (2017)
11. De Lathauwer, L.: Blind separation of exponential polynomials and the decomposition of a tensor in rank-$(L_r,L_r,1)$ terms. SIAM J. Matrix Anal. Appl. **32**(4), 1451–1474 (2011)
12. De Lathauwer, L.: Block component analysis, a new concept for blind source separation. In: Proceedings of the 10th International Conference on LVA/ICA, Tel Aviv, 12–15 March, pp. 1–8 (2012)
13. Tichavský, P., Phan, A.-H., Cichocki, A.: Tensor diagonalization-a new tool for PARAFAC and block-term decomposition (2014). arXiv preprint arXiv:1402.1673
14. Kouchaki, S., Sanei, S.: Tensor based singular spectrum analysis for nonstationary source separation. In: Proceedings of the 2013 IEEE International Workshop on Machine Learning for Signal Processing, Southampton, UK (2013)
15. Debals, O., Lathauwer, L.: Stochastic and deterministic tensorization for blind signal separation. In: Vincent, E., Yeredor, A., Koldovský, Z., Tichavský, P. (eds.) LVA/ICA 2015. LNCS, vol. 9237, pp. 3–13. Springer, Heidelberg (2015). doi:10.1007/978-3-319-22482-4_1
16. Boussé, M., Debals, O., De Lathauwer, L.: A tensor-based method for large-scale blind source separation using segmentation. IEEE Trans. Sig. Process. **65**(2), 346–358 (2017)
17. Cichocki, A., Namgil, L., Oseledets, I., Phan, A.-H., Zhao, Q., Mandic, D.P.: Tensor networks for dimensionality reduction and large-scale optimization. Found. Trends® Mach. Learn. **9**(4–5) (2017)
18. Sorber, L., Van Barel, M., De Lathauwer, L.: Structured data fusion. IEEE J. Sel. Top. Sign. Proces. **9**(4), 586–600 (2015)
19. Hayes, M.: Statistical Digital Signal Processing and Modeling. Wiley, Hoboken (1996)

High-Resolution Subspace-Based Methods: Eigenvalue- or Eigenvector-Based Estimation?

Konstantin Usevich$^{(\boxtimes)}$, Souleymen Sahnoun, and Pierre Comon

GIPSA-lab, CNRS and Univ. Grenoble Alpes, 38000 Grenoble, France
{konstantin.usevich,souleymen.sahnoun,pierre.comon}@gipsa-lab.fr

Abstract. In subspace-based methods for mulditimensional harmonic retrieval, the modes can be estimated either from eigenvalues or eigenvectors. The purpose of this study is to find out which way is the best. We compare the state-of-the art methods N-D ESPRIT and IMDF, propose a modification of IMDF based on least-squares criterion, and derive expressions of the first-order perturbations for these methods. The theoretical expressions are confirmed by the computer experiments.

Keywords: Frequency estimation · Multidimensional harmonic retrieval · Multilevel Hankel matrix · N-D ESPRIT · IMDF · Perturbation analysis

1 Introduction

Parameter estimation from bidimensional (2-D) and multidimensional (N-D) signals finds many applications in signal processing and communications such as magnetic resonance (NMR) spectroscopy [5], wireless communication channel estimation, antenna array processing, radar and medical imaging [1]. In these applications, signals are modeled by a superposition of damped or undamped N-D complex exponentials.

Signal model. Denote N the number of dimensions and M_n, $n = 1, \ldots, N$, the size of the sampling grid in each dimension. In this paper we consider the following model, for $m_n = 0, \ldots, M_n - 1$:

$$\tilde{y}(m_1, \ldots, m_N) = y(m_1, \ldots, m_N) + \varepsilon(m_1, \ldots, m_N), \tag{1}$$

where $\varepsilon(\cdot)$ is random noise (we leave the assumptions on the noise for later), and the signal y is a superposition of R N-D damped complex sinusoids:

$$y(m_1, \ldots, m_N) = \sum_{r=1}^{R} c_r \prod_{n=1}^{N} (a_{r,n})^{m_n}, \tag{2}$$

This work is funded by the European Research Council under the Seventh Framework Programme FP7/2007–2013 Grant Agreement no. 320594.

© Springer International Publishing AG 2017
P. Tichavský et al. (Eds.): LVA/ICA 2017, LNCS 10169, pp. 47–56, 2017.
DOI: 10.1007/978-3-319-53547-0_5

where

- c_r are complex amplitudes,
- $a_{r,n} = e^{-\alpha_{r,n}+\jmath\omega_{r,n}}$ are modes in the n-th dimension,
- $\{\alpha_{r,n}\}_{r=1,n=1}^{R,N}$ are (real and positive) damping factors,
- $\{\omega_{r,n} = 2\pi\nu_{r,n}\}_{r=1,n=1}^{R,N}$ are angular frequencies.

The problem is to estimate $\{a_{r,n}\}_{r=1}^{R}$ and $\{c_r\}_{r=1}^{R}$ from the observed signal $\tilde{y}(m_1,\ldots,m_N)$.

State of art. To deal with this problem, several methods have been proposed. They include linear prediction-based methods such as 2-D TLS-Prony [10], and subspace approaches such as matrix enhancement and matrix pencil (MEMP) [3], 2-D ESPRIT [8], improved multidimensional folding (IMDF) [6,7], Tensor-ESPRIT [2], principal-singular-vector utilization for modal analysis (PUMA) [13,14]. Among the most promising are N-D ESPRIT [8,12] and IMDF [6,7]. Both methods use the eigenvalue decomposition (EVD) of a so-called shift matrix constructed from the estimated basis of the signal subspace, but the modes are extracted differently: from eigenvalues in ND-ESPRIT and from eigenvectors in IMDF. Which method is the best? To our knowledge, there is no satisfactory answer to this question in the literature. The purpose of this paper is to shed some light on this question.

Contributions. In this paper, we perform a study to compare between methods that are based on eigenvalues and those based on eigenvectors to extract N-D modes. Our main contributions are:

- We derive simple expressions of first-order perturbations of IMDF that do not need to calculate the SVD of the MH matrix as it is needed in expressions given in [6].
- We propose a variation of IMDF in which the modes are estimated by minimizing the least squares criterion. It is shown through perturbation analysis and simulations that the proposed technique outperforms the original average-base technique.

Organisation of the paper. In Sect. 2, we recall the definition of multilevel Hankel (MH) matrices and their main properties. In Sect. 3, we recall the algorithms N-D ESPRIT and IMDF, and describe a proposed modification of IMDF (IMDF LS). In Sect. 4, we recall known results on first order perturbations and derive new expressions for IMDF and IMDF LS. Section 5 contains numerical experiments.

2   Multilevel Hankel Matrices and Their Subspaces

2.1   Definition and Factorization

Assume that the set of parameters $(L_n)_{n=1}^{N}$ is chosen such that $1 \leq L_n \leq M_n$ and define $K_n \stackrel{\text{def}}{=} M_n - L_n + 1$. The multilevel Hankel (MH) matrix $\mathbf{H} \in \mathbb{C}^{(L_1\cdots L_N)\times(K_1\cdots K_N)}$ is defined as

$$\mathbf{H} = \begin{bmatrix} \mathbf{H}_0 & \mathbf{H}_1 & \cdots & \mathbf{H}_{K_1-1} \\ \mathbf{H}_1 & \mathbf{H}_2 & \cdots & \mathbf{H}_{K_1} \\ \vdots & \vdots & & \vdots \\ \mathbf{H}_{L_1-1} & \mathbf{H}_{L_1} & \cdots & \mathbf{H}_{M_1-1} \end{bmatrix}, \tag{3}$$

where for $n = 1, \ldots, N-1$ the block matrices $\mathbf{H}_{m_1,\ldots,m_n}$ are defined recursively

$$\mathbf{H}_{m_1,\ldots,m_n} = \begin{bmatrix} \mathbf{H}_{m_1,\ldots,m_n,0} & \mathbf{H}_{m_1,\ldots,m_n,1} & \cdots & \mathbf{H}_{m_1,\ldots,m_n,K_{r+1}-1} \\ \mathbf{H}_{m_1,\ldots,m_n,1} & \mathbf{H}_{m_1,\ldots,m_n,2} & \cdots & \mathbf{H}_{m_1,\ldots,m_n,K_{r+1}} \\ \vdots & \vdots & & \vdots \\ \mathbf{H}_{m_1,\ldots,m_n,L_{r+1}-1} & \mathbf{H}_{m_1,\ldots,m_n,L_{r+1}} & \cdots & \mathbf{H}_{m_1,\ldots,m_n,M_{r+1}-1} \end{bmatrix} \tag{4}$$

and the blocks of the first level are scalars (1×1 matrices)

$$\mathbf{H}_{m_1,\ldots,m_N} = y(m_1,\ldots,m_N).$$

By $\widetilde{\mathbf{H}}$ we denote the MH matrix constructed from noisy observations $\widetilde{y}$. There are alternative equivalent ways to construct the MH matrix: using selection matrices [7] or using operations with tensors [12].

It can be verified (see [7, Lemma 2] or [12, Sect. 3.A]) that for the noiseless signal the MH matrix admits a factorization of the form

$$\mathbf{H} = \mathbf{P}\,\mathrm{diag}(\mathbf{c})\mathbf{Q}^\mathsf{T} \tag{5}$$

where

$$\mathbf{P} = \mathbf{A}_1^{(L_1)} \odot \mathbf{A}_2^{(L_2)} \odot \cdots \odot \mathbf{A}_N^{(L_N)}, \quad \mathbf{Q} = \mathbf{A}_1^{(K_1)} \odot \mathbf{A}_2^{(K_2)} \odot \cdots \odot \mathbf{A}_N^{(K_N)},$$

$\odot$ denotes the Khatri-Rao product, $\mathbf{A}_n^{(L_n)} \in \mathbb{C}^{L_n \times R}$ are Vandermonde matrices (with $(\mathbf{A}_n^{(L_n)})_{j,r} = a_{r,n}^{j-1}$), and $\mathbf{c} = [c_1, \ldots, c_R]^\mathsf{T}$ is the vector of amplitudes.

2.2 Shift Properties of Subspaces

Let us define the selection matrices

$$\overline{\,}_n^{\,}\mathbf{I} \stackrel{\text{def}}{=} \mathbf{I}_{L_1} \boxtimes \mathbf{I}_{L_2} \boxtimes \cdots \boxtimes \overline{\mathbf{I}}_{L_n} \boxtimes \cdots \boxtimes \mathbf{I}_{L_N} = \mathbf{I}_{\prod_{i=1}^{n-1} L_i} \boxtimes \overline{\mathbf{I}}_{L_n} \boxtimes \mathbf{I}_{\prod_{i=n+1}^{N} L_i} \tag{6}$$

$$\underline{\,}_n\mathbf{I} \stackrel{\text{def}}{=} \mathbf{I}_{L_1} \boxtimes \mathbf{I}_{L_2} \boxtimes \cdots \boxtimes \underline{\mathbf{I}}_{L_n} \boxtimes \cdots \boxtimes \mathbf{I}_{L_N} = \mathbf{I}_{\prod_{i=1}^{n-1} L_i} \boxtimes \underline{\mathbf{I}}_{L_n} \boxtimes \mathbf{I}_{\prod_{i=n+1}^{N} L_i} \tag{7}$$

and

$$\underline{\underline{\mathbf{J}}} = \underline{\mathbf{I}}_{L_1} \boxtimes \underline{\mathbf{I}}_{L_2} \boxtimes \cdots \boxtimes \underline{\mathbf{I}}_{L_N}, \tag{8}$$

$$\mathbf{J}_n = \underline{\mathbf{I}}_{L_1} \boxtimes \underline{\mathbf{I}}_{L_2} \boxtimes \cdots \boxtimes \overline{\mathbf{I}}_{L_n} \boxtimes \cdots \boxtimes \underline{\mathbf{I}}_{L_N}, \quad \overline{\overline{\mathbf{J}}} = \sum_{n=1}^{N} \beta_n \mathbf{J}_n \tag{9}$$

where $\underline{\mathbf{X}}$ (resp. $\overline{\mathbf{X}}$) represents $\mathbf{X}$ without the last (resp. first) row, $\boxtimes$ denotes the Kronecker product, and $\mathbf{I}_L$ is an $L \times L$ identity matrix.

Next, for a matrix $\mathbf{X}$ we define $\overset{n-}{\mathbf{X}} = \overset{n-}{\mathbf{I}}\mathbf{X}$ and $\underset{n-}{\mathbf{X}} = \underset{n-}{\mathbf{I}}\mathbf{X}$. Then the shifted versions of $\mathbf{P}$ satisfy the following equation:

$$\underset{n-}{\mathbf{P}}\boldsymbol{\Psi}_n = \overset{n-}{\mathbf{P}}, \tag{10}$$

where $\boldsymbol{\Psi}_n = \mathrm{diag}(\mathbf{a}_{(n)})$, $\mathbf{a}_{(n)} = [a_{1,n}, \ldots, a_{R,n}]^\mathsf{T}$.

Now consider the matrix $\mathbf{U}_s$ of the leading R left singular vectors of the noiseless matrix $\mathbf{H}$. Since the ranges of $\mathbf{U}_s$ and $\mathbf{P}$ coincide, they are linked by a nonsingular transformation:

$$\mathbf{P} = \mathbf{U}_s\mathbf{T}.$$

Hence, we have that the matrix $\mathbf{F}_n \overset{\mathrm{def}}{=} \mathbf{T}\boldsymbol{\Psi}_n\mathbf{T}^{-1}$ satisfies the equation

$$\underset{n-}{\mathbf{U}_s}\mathbf{F}_n = \overset{n-}{\mathbf{U}_s}. \tag{11}$$

If $\underset{n-}{\mathbf{U}_s}$ is full-column rank, then $\mathbf{F}_n$ can be obtained as:

$$\mathbf{F}_n = \left(\underset{n-}{\mathbf{I}}\,\mathbf{U}_s\right)^\dagger \left(\overset{n-}{\mathbf{I}}\,\mathbf{U}_s\right) := \left(\underset{n-}{\mathbf{U}_s}\right)^\dagger \left(\overset{n-}{\mathbf{U}_s}\right) \tag{12}$$

Hence, the matrices $\mathbf{F}_n$ can be computed from the signal subspace $\mathbf{U}_s$, and the modes of each dimension n can be estimated by the eigenvalues of $\mathbf{F}_n$.

On the other side it was shown in [6] that

$$\mathbf{G} = \underline{\mathbf{J}}\mathbf{U}_s\mathbf{T}, \quad \mathbf{G}\,\mathrm{Diag}(\boldsymbol{\eta}) = \overline{\overline{\mathbf{J}}}\mathbf{U}_s\mathbf{T}, \tag{13}$$

with $\mathbf{G} = \mathbf{A}_1^{(L_1-1)} \odot \cdots \odot \mathbf{A}_N^{(L_N-1)}$, $\boldsymbol{\eta} = [\eta_1, \ldots, \eta_R]^\mathsf{T}$ and $\eta_r = \sum_{n=1}^N \beta_n a_{r,n}$ where β_n are user parameters such that $\eta_r \neq \eta_i$ for $r \neq i$.

From (13) it follows that

$$\mathbf{T}\,\mathrm{Diag}(\boldsymbol{\eta})\mathbf{T}^{-1} = (\underline{\mathbf{J}}\mathbf{U}_s)^\dagger(\overline{\overline{\mathbf{J}}}\mathbf{U}_s). \tag{14}$$

Hence, the modes can be estimated from the elements of $\mathbf{G}$.

3   ESPRIT-Type Algorithms for MH Matrices

3.1   N-D ESPRIT Algorithm

The N-D ESPRIT algorithm [12] is an extension of the 2-D ESPRIT [8] and ESPRIT [9] algorithms. The algorithm consists of the following steps:

1. Choose $L_1, \ldots, L_N$.
2. Construct the MH matrix $\widetilde{\mathbf{H}}$ from the noisy signal.
3. Perform the SVD of $\widetilde{\mathbf{H}}$, and form the matrix $\widetilde{\mathbf{U}}_s \in \mathbb{C}^{(L_1\cdots L_N)\times R}$ of the R dominant singular vectors.
4. Compute the matrices $\widetilde{\mathbf{F}}_n$ such that:

$$\widetilde{\mathbf{F}}_n := \left(\underset{n-}{\widetilde{\mathbf{U}}_s}\right)^\dagger \left(\overset{n-}{\widetilde{\mathbf{U}}_s}\right) \tag{15}$$

5. For given parameters $\beta_1, \ldots, \beta_n$, compute a linear combination of $\widetilde{\mathbf{F}}_n$:

$$\widetilde{\mathbf{K}} = \sum_{n=1}^{N} \beta_n \widetilde{\mathbf{F}}_n \tag{16}$$

6. Compute a diagonalizing matrix $\mathbf{T}$ of $\widetilde{\mathbf{K}}$ from its EVD:

$$\widetilde{\mathbf{K}} = \mathbf{T} \operatorname{Diag}(\boldsymbol{\eta}) \mathbf{T}^{-1}. \tag{17}$$

7. Apply the transformation $\mathbf{T}$ to $\mathbf{F}_n$:

$$\widetilde{\mathbf{D}}_n = \mathbf{T}^{-1} \widetilde{\mathbf{F}}_n \mathbf{T}, \quad \text{for} \quad n = 1, \ldots, N. \tag{18}$$

8. Extract $\{[\widehat{a}_{1,n}, \ldots, \widehat{a}_{R,n}]\}_{n=1}^{N}$ from $\operatorname{diag}(\widetilde{\mathbf{D}}_n)$, $n = 1, \ldots, N$.

3.2 IMDF Algorithm

The IMDF algorithm consists of the following steps [6]:

1. Choose $L_1, \ldots, L_N$.
2. Construct the MH matrix $\widetilde{\mathbf{H}}$ from the noisy signal.
3. Perform the SVD of $\widetilde{\mathbf{H}}$, and form $\widetilde{\mathbf{U}}_s \in \mathbb{C}^{(L_1 \cdots L_N) \times R}$, as in N-D ESPRIT.
4. Compute the matrix $\widetilde{\mathbf{K}}_{\text{IMDF}} = (\underline{\mathbf{J}} \widetilde{\mathbf{U}}_s)^\dagger (\overline{\mathbf{J}} \widetilde{\mathbf{U}}_s)$.
5. Compute a diagonalizing matrix $\mathbf{T}$ of $\widetilde{\mathbf{K}}_{\text{IMDF}}$ from its EVD:

$$\widetilde{\mathbf{K}}_{\text{IMDF}} = \mathbf{T} \operatorname{Diag}(\boldsymbol{\eta}) \mathbf{T}^{-1}. \tag{19}$$

6. Estimate a scaled and permuted matrix $\mathbf{G}$:

$$\widetilde{\mathbf{G}} = \underline{\mathbf{J}} \widetilde{\mathbf{U}}_s \mathbf{T} \tag{20}$$

7. Extract $\{[\widehat{a}_{1,n}, \ldots, \widehat{a}_{R,n}]\}_{n=1}^{N}$ from $\widetilde{\mathbf{G}}$ by

$$\widehat{a}_{r,n} = \frac{1}{\mu_n} \sum_{\substack{k=1 \\ \bmod (k-1, L'_{n-1}) \geq L'_n}}^{L'_0} \frac{\widetilde{G}_{k,r}}{\widetilde{G}_{k-L'_n, r}}, \tag{21}$$

where $\mu_n = \frac{L'_0 (L_n - 2)}{L_n - 1}$ and $L'_n = \begin{cases} \prod_{i=n+1}^{N} (L_i - 1), & 0 \leq n \leq N-1, \\ 1, & n = N. \end{cases}$

3.3 IMDF Based on Least Squares (IMDF LS)

The averaging (21) may not be optimal, if some elements of $\widetilde{\mathbf{G}}$ take small values. To tackle this problem, we propose a modification of IMDF. The algorithm is the same as IMDF, except the last two steps, which are replaced by the following:

6. Estimate the scaled and permuted matrix $\mathbf{P}$

$$\widetilde{\mathbf{P}} = [\widetilde{\mathbf{p}}_1, \ldots, \widetilde{\mathbf{p}}_R] = \widetilde{\mathbf{U}}_s \mathbf{T}.$$

7. Extract $\{[\widetilde{a}_{1,n}, \ldots, \widetilde{a}_{R,n}]\}_{n=1}^{N}$ from $\widetilde{\mathbf{P}}$ as

$$\widehat{a}_{r,n} = \frac{(\underset{\underline{\mathbf{n}}}{\widetilde{\mathbf{p}}}_r)^{\mathsf{H}} \overset{\overset{\mathbf{n}}{-}}{\widetilde{\mathbf{p}}}_r}{\|\underset{\underline{\mathbf{n}}}{\widetilde{\mathbf{p}}}_r\|_2^2} = (\underset{\underline{\mathbf{n}}}{\widetilde{\mathbf{p}}}_r)^{\dagger} \overset{\overset{\mathbf{n}}{-}}{\widetilde{\mathbf{p}}}_r.$$

4 Perturbation Analysis

4.1 Basic Expressions

The SVD of the noiseless MH matrix $\mathbf{H}$ is given by:

$$\mathbf{H} = \mathbf{U}_s \boldsymbol{\Sigma}_s \mathbf{V}_s^{\mathsf{H}} + \mathbf{U}_n \boldsymbol{\Sigma}_n \mathbf{V}_n^{\mathsf{H}}, \tag{22}$$

where $\boldsymbol{\Sigma}_n = \mathbf{0}$. The subspace decomposition of the perturbed matrix $\tilde{\mathbf{H}} = \mathbf{H} + \Delta\mathbf{H}$ is given by

$$\tilde{\mathbf{H}} = \tilde{\mathbf{U}}_s \tilde{\boldsymbol{\Sigma}}_s \tilde{\mathbf{V}}_s^{\mathsf{H}} + \tilde{\mathbf{U}}_n \tilde{\boldsymbol{\Sigma}}_n \tilde{\mathbf{V}}_n^{\mathsf{H}} \tag{23}$$

We use the following lemma on the first-order approximation.

Lemma 1 ([4,15]). *The perturbed signal subspace is* $\tilde{\mathbf{U}}_s = \mathbf{U}_s + \Delta\mathbf{U}_s$, $\tilde{\mathbf{V}}_s = \mathbf{V}_s + \Delta\mathbf{V}_s$ *and* $\tilde{\boldsymbol{\Sigma}}_s = \boldsymbol{\Sigma}_s + \Delta\boldsymbol{\Sigma}_s$. *A first order perturbation is given by*

$$\Delta\mathbf{U}_s = \mathbf{U}_n \mathbf{U}_n^{\mathsf{H}} \Delta\mathbf{H} \mathbf{V}_s \boldsymbol{\Sigma}_s^{-1} \tag{24}$$

$$\Delta\mathbf{V}_s^{\mathsf{H}} = \boldsymbol{\Sigma}_s^{-1} \mathbf{U}_s^{\mathsf{H}} \Delta\mathbf{H} \mathbf{V}_n \mathbf{V}_n^{\mathsf{H}}, \quad \Delta\boldsymbol{\Sigma}_s = \mathbf{U}_s^{\mathsf{H}} \Delta\mathbf{H} \mathbf{V}_s \tag{25}$$

For N-D ESPRIT, an expression for first-order perturbation was derived in [12].

Proposition 1 ([12]). *Denote by* $\mathbf{b}_r \in \mathbb{C}^R$ *the r-th unit vector. Then first order perturbations of the modes obtained by the N-D ESPRIT admit an expansion*

$$\Delta a_{r,n} = \frac{1}{c_r} \mathbf{b}_r^{\mathsf{T}} \underline{\mathbf{P}}^\dagger (\overline{\underline{\mathbf{I}}} - a_{r,n}\underline{\mathbf{I}}) \Delta\mathbf{H} (\mathbf{Q}^{\mathsf{T}})^\dagger \mathbf{b}_r. \tag{26}$$

4.2 IMDF Perturbations

Perturbation analysis of IMDF have been done in [6]. However, the obtained expressions require the calculation of the SVD of the MH matrix $\mathbf{H}$. To get simplified perturbation expressions we use the following fact: from (8), $\mathbf{K}_{\text{IMDF}}$ can be written as a linear combination of $\mathbf{F}_n^{\text{IMDF}}$

$$\mathbf{K}_{\text{IMDF}} = \sum_{n=1}^{N} \beta_n \mathbf{F}_n^{\text{IMDF}}, \tag{27}$$

where

$$\mathbf{F}_n^{\text{IMDF}} = (\underline{\mathbf{J}}\mathbf{U}_s)^\dagger (\mathbf{J}_n \mathbf{U}_s). \tag{28}$$

We use the following lemma, which is a slight modification of [12, Lemma 4].

Lemma 2 ([12, Lemma 4, a modification]). *The first-order perturbation of* $\mathbf{F}_n^{\text{IMDF}}$ *is given by*

$$\Delta\mathbf{F}_n^{IMDF} = (\underline{\mathbf{J}}\mathbf{U}_s)^\dagger (\mathbf{J}_n \Delta\mathbf{U}_s - \underline{\mathbf{J}}\Delta\mathbf{U}_s \mathbf{F}_n). \tag{29}$$

Next, we derive perturbation expression of $a_{r,n}$ with respect to $\Delta \mathbf{g}_r$ from (21)

$$\Delta a_{r,n} = \frac{a_{r,n} G_{1,r}^{-1}}{\mu_n} \mathbf{v}_{r,n}^{\mathsf{T}} \Delta \mathbf{g}_r, \tag{30}$$

where

$$\mathbf{v}_{r,n} = (\boldsymbol{\phi}_{r,1} \boxtimes \cdots \boxtimes \boldsymbol{\phi}_{r,n-1}) \boxtimes \boldsymbol{\psi}_{r,n} \boxtimes (\boldsymbol{\phi}_{r,n+1} \boxtimes \cdots \boxtimes \boldsymbol{\phi}_{r,N}), \tag{31}$$

$$\boldsymbol{\phi}_{r,1} = \begin{bmatrix} 1 \ a_{r,n}^{-1} \cdots a_{r,n}^{-(L_n-2)} \end{bmatrix}^{\mathsf{T}} \in \mathbb{C}^{(L_n-1) \times 1}, \tag{32}$$

$$\boldsymbol{\psi}_{r,1} = \begin{bmatrix} -1 \ 0 \cdots 0 \ a_{r,n}^{-(L_n-2)} \end{bmatrix}^{\mathsf{T}} \in \mathbb{C}^{(L_n-1) \times 1}, \tag{33}$$

the vectors $\mathbf{g}_r$ are

$$\mathbf{g}_r = \underline{\mathbf{J}} \mathbf{U}_s \mathbf{t}_r,$$

and $\mathbf{t}_r$ are the eigenvectors of $\mathbf{K}_{\mathrm{IMDF}}$ (the columns of $\mathbf{T}$):

$$\mathbf{T} = [\mathbf{t}_1, \ldots, \mathbf{t}_R].$$

The perturbation of $\mathbf{g}_r$ can be expressed as:

$$\Delta \mathbf{g}_r = \underline{\mathbf{J}} \Delta \mathbf{U}_s \mathbf{t}_r + \underline{\mathbf{J}} \mathbf{U}_s \Delta \mathbf{t}_r, \tag{34}$$

where:

- $\Delta \mathbf{U}_s$ can be found from Eq. (24);
- $\Delta \mathbf{t}_r$ can be found as

$$\Delta \mathbf{t}_r = \sum_{i=1, i \neq r}^{R} \frac{1}{\eta_r - \eta_i} \mathbf{t}_i \boldsymbol{\tau}_i^{\mathsf{T}} \Delta \mathbf{K} \mathbf{t}_r \tag{35}$$

$$= \mathbf{T} \, \boldsymbol{\varXi}(r) \, \mathbf{T}^{-1} \Delta \mathbf{K} \, \mathbf{t}_r. \tag{36}$$

where $\boldsymbol{\tau}_r^{\mathsf{T}}$ denote the rows of $\mathbf{T}^{-1}$

$$\mathbf{T}^{-1} = [\boldsymbol{\tau}_1, \ldots, \boldsymbol{\tau}_R]^{\mathsf{T}},$$

and $\boldsymbol{\varXi}(r)$ is a diagonal matrix with $\boldsymbol{\varXi}_{ii}(r) = \frac{1}{\eta_r - \eta_i}$, for $i \neq r$ and $\boldsymbol{\varXi}_{rr}(r) = 0$.

From Eqs. (27), (28), and (35) we get, after some simplifications

$$\Delta \mathbf{t}_r = \sum_{n=1}^{N} \beta_n \left\{ \mathbf{T} \, \boldsymbol{\varXi}(r) \underline{\mathbf{P}}^{\dagger} (\mathbf{J}_n - a_{r,n} \underline{\mathbf{J}}) + \right.$$
$$\left. \sum_{i=1, i \neq r}^{R} \frac{(a_{r,n} - a_{i,n})}{\eta_r - \eta_i} \mathbf{t}_i \boldsymbol{\tau}_i^{\mathsf{T}} (\mathbf{T}^{-1})^{\mathsf{H}} \mathbf{P}^{\mathsf{H}} \right\} \frac{1}{c_r} \Delta \mathbf{H} (\mathbf{Q}^{\mathsf{T}})^{\dagger} \mathbf{b}_r. \tag{37}$$

Then, by combining (24) and (37), we get

$$\Delta \mathbf{g}_r = \underline{\mathbf{J}} \left\{ (\mathbf{I} - \mathbf{P} \mathbf{b}_r \mathbf{b}_r^{\mathsf{T}} \mathbf{P}^{\dagger}) + \mathbf{P} \boldsymbol{\varXi}(r) \underline{\mathbf{P}}^{\dagger} \left(\overline{\overline{\mathbf{J}}} - \eta_r \underline{\mathbf{J}} \right) \right\} \cdot \frac{1}{c_r} \Delta \mathbf{H} (\mathbf{Q}^{\mathsf{T}})^{\dagger} \mathbf{b}_r. \tag{38}$$

Here the SVD of $\mathbf{H}$ is not required.

4.3   IMDF LS Perturbations

For establishing the perturbations of $a_{r,n}$ for IMDF LS we can use Eq. (29):

$$\Delta a_{r,n} = (\underline{\mathbf{p}}_r)^\dagger(\Delta\underline{\overline{\mathbf{p}}}_r - \Delta\underline{\mathbf{p}}_r a_{r,n}) = \frac{1}{\|\underline{\mathbf{p}}_r\|_2^2}\mathbf{p}_r{}^{\mathsf{H}}(\mathbf{J}_n - a_{r,n}\underline{\mathbf{J}})\Delta\mathbf{p}_r$$

The perturbation $\Delta\mathbf{p}_r$ is, in fact, given in (34). Finally, since

$$(\mathbf{J}_n - a_{r,n}\underline{\mathbf{J}})\mathbf{P}\mathbf{b}_r = 0,$$

we have

$$\Delta a_{r,n} = \frac{1}{c_r\|\underline{\mathbf{p}}_r\|_2^2}\mathbf{p}_r{}^{\mathsf{H}}(\mathbf{J}_n - a_{r,n}\underline{\mathbf{J}})\left(\mathbf{I} + \mathbf{P}\boldsymbol{\Xi}(r)\underline{\mathbf{P}}^\dagger\left(\overline{\overline{\mathbf{J}}} - \eta_r\underline{\mathbf{J}}\right)\right)\Delta\mathbf{H}\,(\mathbf{Q}^{\mathsf{T}})^\dagger\mathbf{b}_r. \quad (39)$$

4.4   Computing the First-Order Perturbation and Its Moments

Similarly to [12, Sect. V.C], the perturbations in Eqs. (26), (38), (39) have the common form:

$$\Delta a_{r,n} = \mathbf{v}_{r,n}^{\mathsf{T}}\Delta\mathbf{H}\mathbf{x}_r,$$

where $\mathbf{x}_r = (\mathbf{Q}^{\mathsf{T}})^\dagger\mathbf{b}_r$, and the vector $\mathbf{v}_{r,n}^{\mathsf{T}}$ depends on the method. Since the MH matrix $\Delta\mathbf{H}$ depends linearly on the elements of $\mathbf{e}$ (vectorization of the noise term), the perturbation can be expressed as

$$\Delta a_{r,n} = \mathbf{z}_{r,n}^{\mathsf{H}}\mathbf{e}. \quad (40)$$

where $\mathbf{z}_{r,n}$ can be computed from $\mathbf{v}_{r,n}^{\mathsf{T}}$ and $\mathbf{x}_r$ efficiently using the N-D convolution, as shown in [12].

Therefore, we have the following:

1. $\mathbb{E}\left\{\Delta a_{r,n}\right\} = 0$ if $\mathbf{e}$ is zero-mean.
2. $\mathbb{E}\left\{\Delta a_{r,n}^2\right\} = 0$ if $\mathbf{e}$ is circular.
3. $\mathbb{E}\left\{|\Delta a_{r,n}|^2\right\} = \mathbf{z}_{r,n}^{\mathsf{H}}\Gamma\mathbf{z}_{r,n}$ if $\mathbf{e}$ has covariance matrix $\Gamma = \mathbb{E}\left\{\mathbf{e}\mathbf{e}^{\mathsf{H}}\right\}$.
4. $\mathbb{E}\left\{|\Delta a_{r,n}|^2\right\} = \sigma_e^2\|\mathbf{z}_{r,n}\|_2^2$ if $\mathbf{e}$ is white with variance σ_e^2.
5. Finally,

$$\mathrm{var}(\Delta\omega_{r,n}) = \mathrm{var}(\Delta\alpha_{r,n}) = \frac{\mathbb{E}\left\{|\Delta a_{r,n}|^2\right\}}{2|a_{r,n}|^2}$$

if $\mathbf{e}$ is complex circular Gaussian.

5   Simulations

Numerical simulations have been carried out to verify theoretical expressions and compare the performances of N-D ESPRIT, IMDF and IMDF LS algorithms in the presence of white Gaussian noise. The performances are measured by the total mean squared error (tMSE) on estimated parameters. The total MSE is

Table 1. 3-D signal with two modes

r	$\omega_{r,1}$	$\alpha_{r,1}$	$\omega_{r,2}$	$\alpha_{r,2}$	$\omega_{r,3}$	$\alpha_{r,3}$	c_r
1	0.2π	0.01	0.3π	0.01	0.26π	0.01	1
2	0.6π	0.01	0.8π	0.015	0.2π	0.01	1

defined as $\text{tMSE}_{\text{total}} = \frac{1}{RF}\mathbb{E}_p\left\{\sum_{r=1}^{R}\sum_{f=1}^{F}(\xi_{f,r} - \hat{\xi}_{f,r})^2\right\}$ where $\hat{\xi}_{f,r}$ is an estimate of $\xi_{f,r}$, and $\mathbb{E}_p$ is the average over p Monte-Carlo trials. In our simulations,
$\xi_{f,r}$ can be either a frequency or a damping factor.

In the following experiments we plot theoretical expressions of the variances
and compare them with empirical results of N-D ESPRIT, IMDF and IMDF LS.
Cramér-Rao bounds are also reported [11]. In all experiments, $L_n = \lceil\frac{M_n}{3}\rceil$.

Experiment 1. In this experiment, we simulate a 3-D signal of size $10 \times 10 \times 10$
containing two modes whose parameters are given in Table 1. Figure 1(a) shows
the obtained results. We can see that N-D ESPRIT and IMDF LS have the
similar results, which are almost equal to theoretical ones beyond -10 dB. We
can also remark that N-D ESPRIT and IMDF LS outperforms slightly IMDF.

Experiment 2. In this experiment, we simulate a 3-D signal of size $10 \times 10 \times 10$
containing nine modes. Figure 1(b) shows the obtained results. First, we remark
that theoretical variances match the empirical ones beyond thresholds. Then, we
can see that N-D ESPRIT outperforms IMDF and IMDF LS.

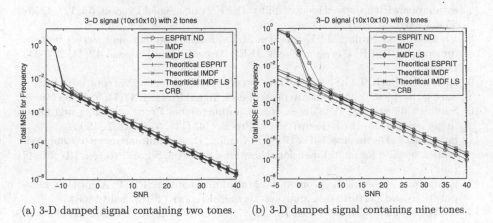

(a) 3-D damped signal containing two tones.   (b) 3-D damped signal containing nine tones.

Fig. 1. Theoretical and empirical tMSEs versus SNR. $(M_1, M_2, M_3) = (10, 10, 10)$.

6 Conclusions

Our study suggests that the N-D ESPRIT outperforms IMDF when the number of tones increases. The same conclusion holds for the improvement of IMDF proposed in this paper. We conjecture that the eigenvalue-based estimation should be preferred over the eigenvector-based ones, since they do not contain an additional estimation step. An extensive study (for different noise scenarios and parameter values of the methods) is needed to confirm our conjecture.

References

1. Gershman, A.B., Sidiropoulos, N.D.: Space-Time Processing for MIMO Communications. Wiley Online Library, Hoboken (2005)
2. Haardt, M., Roemer, F., Del Galdo, G.: Higher-order SVD-based subspace estimation to improve the parameter estimation accuracy in multidimensional harmonic retrieval problems. IEEE Trans. Signal Process. **56**(7), 3198–3213 (2008)
3. Hua, Y.: Estimating two-dimensional frequencies by matrix enhancement and matrix pencil. IEEE Trans. Signal Process. **40**(9), 2267–2280 (1992)
4. Li, F., Liu, H., Vaccaro, R.J.: Performance analysis for DOA estimation algorithms: unification, simplification, and observations. IEEE Trans. Aerosp. Electron Syst. **29**(4), 1170–1184 (1993)
5. Li, Y., Razavilar, J., Liu, K.: A high-resolution technique for multidimensional NMR spectroscopy. IEEE Trans. Biomed. Eng. **45**(1), 78–86 (1998)
6. Liu, J., Liu, X., Ma, X.: Multidimensional frequency estimation with finite snapshots in the presence of identical frequencies. IEEE Trans. Signal Process. **55**, 5179–5194 (2007)
7. Liu, J., Liu, X.: An eigenvector-based approach for multidimensional frequency estimation with improved identifiability. IEEE Trans. Signal Process. **54**(12), 4543–4556 (2006)
8. Rouquette, S., Najim, M.: Estimation of frequencies and damping factors by two-dimensional ESPRIT type methods. IEEE Trans. Signal Process. **49**(1), 237–245 (2001)
9. Roy, R., Kailath, T.: ESPRIT-estimation of signal parameters via rotational invariance techniques. IEEE Trans. Acoust. Speech Signal Process. **37**(7), 984–995 (1989)
10. Sacchini, J., Steedly, W., Moses, R.: Two-dimensional Prony modeling and parameter estimation. IEEE Trans. Signal Process. **41**(11), 3127–3137 (1993)
11. Sahnoun, S., Djermoune, E.H., Brie, D., Comon, P.: A simultaneous sparse approximation method for multidimensional harmonic retrieval. Signal Process. **137**, 36–48 (2017)
12. Sahnoun, S., Usevich, K., Comon, P.: Multidimensional ESPRIT: Algorithm, computations and perturbation analysis. Technical report (2016). hal-01360438
13. So, H.C., Chan, F., Lau, W., Chan, C.F.: An efficient approach for two-dimensional parameter estimation of a single-tone. IEEE Trans. Signal Process. **58**(4), 1999–2009 (2010)
14. Sun, W., So, H.C.: Accurate and computationally efficient tensor-based subspace approach for multidimensional harmonic retrieval. IEEE Trans. Signal Process. **60**(10), 5077–5088 (2012)
15. Xu, Z.: Perturbation analysis for subspace decomposition with applications in subspace-based algorithms. IEEE Trans. Signal Process. **50**(11), 2820–2830 (2002)

From Source Positions to Room Properties: Learning Methods for Audio Scene Geometry Estimation

Speaker Tracking on Multiple-Manifolds
with Distributed Microphones

Bracha Laufer-Goldshtein[1]([⊠]), Ronen Talmon[2], and Sharon Gannot[1]

[1] Bar-Ilan University, 5290002 Ramat-Gan, Israel
{bracha.laufer,sharon.gannot}@biu.ac.il
[2] Technion – Israel Institute of Technology, Technion City, 3200003 Haifa, Israel
ronen@ee.technion.ac.il

Abstract. Speaker tracking in a reverberant enclosure with an ad hoc network of multiple distributed microphones is addressed in this paper. A set of prerecorded measurements in the enclosure of interest is used to construct a data-driven statistical model. The function mapping the measurement-based features to the corresponding source position represents complex unknown relations, hence it is modelled as a random Gaussian process. The process is defined by a covariance function which encapsulates the relations among the available measurements and the different views presented by the distributed microphones. This model is intertwined with a Kalman filter to capture both the smoothness of the source movement in the time-domain and the smoothness with respect to patterns identified in the set of available prerecorded measurements. Simulation results demonstrate the ability of the proposed method to localize a moving source in reverberant conditions.

Keywords: Speaker tracking · Distributed microphones · Gaussian process · Acoustic manifold · Kalman filter

1 Introduction

Speaker localization and tracking in reverberant enclosures plays an important role in many applications, including: automatic camera steering, teleconferencing and beamforming. Conventional localization methods can be roughly divided into single- and dual-step approaches. In single-step approaches, a grid search is performed to find the position that maximizes a certain optimization criterion [5,13]. In dual-step approaches, the time difference of arrivals (TDOAs) of several microphone pairs are first estimated and then combined to perform the actual localization [2,9].

In dynamic scenarios, the measurements can be divided into short time frames, during which the source position is approximately static. Hence, in each time step the information available for the localization task is limited. However, the smoothness of the movement implies dependence across time. The temporal consistency across successive frames can be exploited by either Bayesian or

© Springer International Publishing AG 2017
P. Tichavský et al. (Eds.): LVA/ICA 2017, LNCS 10169, pp. 59–67, 2017.
DOI: 10.1007/978-3-319-53547-0_6

non-Bayesian models. Bayesian state-space models, which are usually nonlinear and non-Gaussian, are implemented using unscented Kalman filter or extended Kalman filter [7] and particle filter [16]. In non-Bayesian approaches, the trajectory is considered as a deterministic and time-varying parameter, and a maximum likelihood criterion can be applied [14].

In realistic environments, the presence of noise or reverberation often yields spurious observations which may lead to poor localization performance. In addition, traditional localization and tracking schemes are based on approximated physical and statistical assumptions which do not always meet the practical conditions in complex real-world scenarios. Recently, there is an attempt to overcome these limitations by applying supervised or unsupervised learning-based approaches [3,4,12,15]. The idea is to form a data-driven model for the spatial characteristics of an acoustic environment, rather than using a predefined statical model.

In this paper, we derive a semi-supervised tracking algorithm based on measurements from distributed pairs of microphones. The algorithm exploits a training set of prerecorded measurements from various locations in the enclosure of interest. Capitalizing this prior information, we identify the geometrical patterns, namely the underlying manifold to which the measurement-based features are confined, and relate it to the position of the source. Recently [11], we have presented a semi-supervised localization approach, which explores the acoustic manifold associated with each microphone pair, and composes these models in the definition of a multiple-manifold Gaussian process (MMGP). Here, this data-driven statistical model is integrated into a Kalman filter scheme to impose dual-domain smoothness, both with respect to the acoustic manifold and the time domain. The algorithm performance is examined using simulated trajectories of a moving source in a reverberant room with spatially-distributed microphone pairs.

2   Problem Formulation

We consider a reverberant enclosure consisting of M nodes, where each node comprises a pair of microphones. A single source is moving in the enclosure, generating an unknown speech signal $s(n)$, which is measured by all the microphones. The received signals are contaminated by additive stationary noise sources and are given by:

$$y^{mi}(n) = \sum_k a_n^{mi}(k)s(n-k) + u^{mi}(n), \quad m = 1, \ldots, M \qquad (1)$$

where n is the time index, a_n^{mi}, $i = \{1, 2\}$ is the time-varying acoustic impulse response (AIR) relating the source and the ith microphone in the mth node at time n, and $u^{mi}(n)$ is the corresponding noise signal. The measured signals are partitioned into short segments of a few hundred milliseconds, which are assigned with frame index t. From each segment we constitute a feature vector $\mathbf{h}^m(t)$ that preserves the relevant information for localization which is hidden in the AIRs, and is invariant to other irrelevant factors, namely the non-stationary source signal. More specifically, we use a feature vector based on relative transfer function (RTF) estimates in a certain frequency band, which is commonly used in acoustic array processing [6]. The RTF is typically represented in high dimension with a large number of coefficients to account for the room reverberation. The observation that the RTF is controlled by a small set of parameters, such as room dimensions, reverberation time, location of the source and the sensors etc., gives rise to the assumption that it is confined to a low dimensional manifold $\mathcal{M}_m$, as was demonstrated in [10].

To track a moving source, we consider the function f^m which attaches to an RTF sample $\mathbf{h}^m(t)$ from the mth node its corresponding x, y or z coordinate of the source position $p^m(t) \equiv f^m(\mathbf{h}(t))$, for frame t. Note that although the position of the source does not depend on the specific node, the notation $p^m(t)$ is used to express that the mapping is obtained from the measurement of the mth node. The different nodes represent different views of the same acoustic scene, hence incorporating the information from the different nodes in a unifying mapping denoted by f may enrich the spatial information utilized for localization and tracking. Let $\mathbf{h}(t) = \left[[\mathbf{h}(t)^1]^T, \ldots, [\mathbf{h}(t)^M]^T \right]^T$ denote the aggregated RTF (aRTF), which is a concatenation of the RTF vectors from all the nodes. The function f associates the corresponding source position to an aRTF sample $\mathbf{h}(t)$, namely $p(t) \equiv f(\mathbf{h}(t))$. The function f, which defines an instantaneous mapping, is used here to evaluate the position of the source along its track. In the dynamic scenario, the function is used to transform the observed propagation in the RTFs domain to the physical domain of the source positions, i.e. it assists the development of a simple Markovian relation between successive positions.

To estimate the function f, we assume the availability of a training set consisting of a limited number of labelled measurements from multiple nodes, attached with corresponding source positions, and a larger amount of unlabelled measurements with unknown source locations. All the training measurements apply to static sources. The labelled set consists of n_L pairs $\{\mathbf{h}_i, \bar{p}_i\}_{i=1}^{n_L}$, and the unlabelled set consists of n_U samples $\{\mathbf{h}_i\}_{i=n_L+1}^{n_D}$, where $n_D = n_L + n_U$. Note that the microphone positions may be unknown, since they are not required for the estimation. In the test phase, we receive the measurements of a moving source, partition them into n_T short segments, and compute the corresponding RTF separately for each segment. The set $\{\mathbf{h}(t)\}_{t=1}^{n_T}$ consists of the aRTFs of all the segments, where the index t denotes their chronological order. The goal is to estimate the corresponding n_T temporary positions $\{p(t)\}_{t=1}^{n_T}$ of each sample in the set $\{\mathbf{h}(t)\}_{t=1}^{n_T}$.

3 Multiple-Manifold Gaussian Process

We first define a statistical model for each node separately and the relation between the different nodes, and then combine them in a unified model [11]. We assume that the position p^m, which is associated with the measurements of the mth node, follows a zero-mean Gaussian process, i.e. the set of all possible positions mapped from the samples of the mth node, are joint Gaussian variables. The Gaussian process is completely defined by its covariance function, which is a pairwise affinity measure between two RTF samples. We use a manifold-based covariance function, defined by:

$$\text{cov}(p_r^m, p_l^m) \equiv \tilde{k}_m(\mathbf{h}_r^m, \mathbf{h}_l^m) = \sum_{i=1}^{n_D} k_m(\mathbf{h}_r^m, \mathbf{h}_i^m) k_m(\mathbf{h}_l^m, \mathbf{h}_i^m) \qquad (2)$$

where l and r represent ascription to certain positions, and k_m is a standard "kernel" function $k_m : \mathcal{M}_m \times \mathcal{M}_m \longrightarrow \mathbb{R}$. A common choice is to use a Gaussian kernel.

Considering multiple nodes, we similarly define the correlation between two source positions p_r^q and p_l^w associated with nodes q and w, respectively. We assume that p_r^q and p_l^w are jointly Gaussian and that their covariance is defined by:

$$\text{cov}(p_r^q, p_l^w) \equiv \tilde{k}_{qw}(\mathbf{h}_r^q, \mathbf{h}_l^w) = \sum_{i=1}^{n_D} k_q(\mathbf{h}_r^q, \mathbf{h}_i^q) k_w(\mathbf{h}_l^w, \mathbf{h}_i^w). \qquad (3)$$

Note that in both (2) and (3), the covariance is constituted by an average over all the available training samples. This averaging implies that the similarity between two samples from the manifold can be determined according to the way they are viewed by other samples residing on the same manifold. When two samples convey similar connections (i.e. proximity or remoteness) to other samples, it indicates that they are closely related with respect to the manifold. In (3), we cannot directly compute the distance between the corresponding RTF samples since they present different views of two nodes. Thus, we choose another sample $\mathbf{h}_i$, and compare the distances with respect to $\mathbf{h}_i$ as it is viewed by the different nodes. The inter-relations in the qth and wth manifolds are computed separately, and then they are composed by multiplying the corresponding kernels.

To fuse the different perspectives presented by the different nodes, we define the multiple-manifold Gaussian process (MMGP) p as the mean of the Gaussian processes of all the nodes:

$$p = \frac{1}{M}(p^1 + p^2 + \ldots + p^M). \qquad (4)$$

Due to the assumption that the processes are jointly Gaussian, the process p is also Gaussian with zero-mean and a covariance function given by:

$$\text{cov}(p_r, p_l) = \frac{1}{M^2} \text{cov}\left(\sum_{q=1}^{M} p_r^q, \sum_{w=1}^{M} p_l^w\right) = \frac{1}{M^2} \sum_{q,w=1}^{M} \text{cov}(p_r^q, p_l^w). \qquad (5)$$

Using the definitions of (2) and (3) we get the covariance for p_r and p_l:

$$\text{cov}(p_r, p_l) \equiv \tilde{k}(\mathbf{h}_r, \mathbf{h}_l) = \frac{1}{M^2} \sum_{i=1}^{n_D} \sum_{q,w=1}^{M} k_q(\mathbf{h}_r^q, \mathbf{h}_i^q) k_w(\mathbf{h}_l^w, \mathbf{h}_i^w). \qquad (6)$$

Here, the covariance is defined by averaging over all the available training samples as well as over all pairs of nodes. The induced kernel $\tilde{k}(\mathbf{h}_r, \mathbf{h}_l)$, can be considered as a *composition of kernels*, which, in addition to connections acquired in each node separately, incorporates the extra spatial information manifested in the mutual relationship between RTFs from different nodes.

4   Multiple-Manifold Speaker Tracking

The tracking is performed by a state-space representation, formulated according to the statistical relations implied by the MMGP. In the dynamic scenario, the test aRTF samples $\{\mathbf{h}(t)\}_{t=1}^{n_T}$, and their associated unknown source positions $\{p(t)\}_{t=1}^{n_T}$ are treated as two time-series, which are mutually-related through the mapping f. The propagation model, specifying the relation between the source positions in successive time steps, is defined according to similarities between the corresponding aRTFs, as induced by the covariance of the MMGP. This way the movement of the source is constrained to vary smoothly with respect to the manifolds of the different nodes. The measurement model relates the current sample $\mathbf{h}(t)$ to all the other available training samples. The resulting state-space representation is solved by a Kalman-filter, in which the source position, predicted through the local interpolation devised by successive samples, is updated by a global interpolation formed by all the training information.

We first define the propagation model. The position $p(t-1)$ at time $t-1$ and the current position $p(t)$ are two samples from the MMGP defined in the previous section. Hence, the random variables $p(t)$ and $p(t-1)$ have a joint normal distribution, and their conditional probability is given by:

$$p(t)|p(t-1) \sim \mathcal{N}\left(\frac{\tilde{\Sigma}_{t,t-1}}{\tilde{\Sigma}_{t-1}}p(t-1), \tilde{\Sigma}_t - \frac{\tilde{\Sigma}_{t,t-1}^2}{\tilde{\Sigma}_{t-1}}\right) \tag{7}$$

where $\tilde{\Sigma}_t$ and $\tilde{\Sigma}_{t-1}$ are the variances of $p(t)$ and $p(t-1)$ respectively, and $\tilde{\Sigma}_{t,t-1}$ is the covariance of $p(t)$ and $p(t-1)$. For a Gaussian process, the propagated probabilities in (7) can be equivalently represented by a linear propagation equation with an additive Gaussian noise ξ_t

$$p(t) = g_t \cdot p(t-1) + \xi_t \tag{8}$$

where $g_t = \frac{\tilde{\Sigma}_{t,t-1}}{\tilde{\Sigma}_{t-1}}$ and $\xi_t \sim \mathcal{N}\left(0, \sigma_\xi^2\right)$ with $\sigma_\xi^2 = \tilde{\Sigma}_t - \frac{\tilde{\Sigma}_{t,t-1}^2}{\tilde{\Sigma}_{t-1}}$. Since there is no prior information on the actual trajectory of the speaker, it is reasonable to use a simplified random walk model as in (8). However, it should be noted that the proposed model is data-driven, in the sense that both the transition factor g_t and the driving noise variance σ_ξ^2 are determined based on the relation between the current aRTF sample $\mathbf{h}(t)$ and the preceding one $\mathbf{h}(t-1)$. When the aRTF samples are close to each other, namely that the acoustic characteristics have hardly changed, it is assumed that only a slight movement of the source has occurred. In this case, we receive $\tilde{\Sigma}_{t,t-1} \approx \tilde{\Sigma}_t \approx \tilde{\Sigma}_{t-1}$, yielding $g_t \approx 1$ and $\sigma_\xi^2 \approx 0$, which implies that $p(t) \approx p(t-1)$ as desired. Overall, the proposed propagation model imposes a smooth variation of the position with respect to the manifolds associated with the different nodes, and reflects the strong relation between the physical domain and the aRTFs domain.

As for the measurement model, we can form an observation q_t that represents the estimated position based on the available training samples. Let $\bar{\mathbf{p}}_L = [\bar{p}_1, \ldots, \bar{p}_{n_L}]^T$ be a concatenation of the measured positions of the labelled set. We assume that the measured positions $\bar{p}_i = p_i + \eta_i$, are noisy versions of the actual position p_i, due to imperfections in the measurements while acquiring the labelled set. Assuming that η_i is an independent Gaussian noise with variance σ^2 yields that $p(t)$ and $\bar{\mathbf{p}}_L$ are jointly Gaussian, and their conditional distribution is given by:

$$p(t)|\bar{\mathbf{p}}_L \sim \mathcal{N}\left(\tilde{\Sigma}_{Lt}^H \left(\tilde{\Sigma}_L + \sigma^2 \mathbf{I}_{n_L}\right)^{-1} \bar{\mathbf{p}}_L, \tilde{\Sigma}_{t,t} - \tilde{\Sigma}_{Lt}^H \left(\tilde{\Sigma}_L + \sigma^2 \mathbf{I}_{n_L}\right)^{-1} \tilde{\Sigma}_{Lt}\right) \quad (9)$$

where $\tilde{\Sigma}_L$ is an $n_L \times n_L$ covariance matrix defined over the function values at the labelled samples, $\tilde{\Sigma}_{Lt}$ is an $n_L \times 1$ covariance vector between the function values at the labelled samples and $p(t)$, and $\mathbf{I}_{n_L}$ is the $n_L \times n_L$ identity matrix. Accordingly, we define the observation as $q_t = \mathbf{Q}_t \bar{\mathbf{p}}_L$, where $\mathbf{Q}_t = \tilde{\Sigma}_{Lt}^H \left(\tilde{\Sigma}_L + \sigma^2 \mathbf{I}_{n_L}\right)^{-1}$. The corresponding measurement model can be expressed as:

$$q_t = p(t) + \zeta_t \quad (10)$$

where $\zeta_t \sim \mathcal{N}\left(0, \sigma_\zeta^2\right)$ with $\sigma_\zeta^2 = \tilde{\Sigma}_{t,t} - \tilde{\Sigma}_{Lt}^H \left(\tilde{\Sigma}_L + \mathbf{I}_{n_L}\right)^{-1} \tilde{\Sigma}_{Lt}$. Here as well, the covariance terms are calculated using the kernel $\tilde{k}$ defined in the previous section. Since an *actual* noisy measurement of the position does not exist, we use instead an *artificial* data-driven measurement q_t, formed by the current sample $\mathbf{h}(t)$ and the entire training set. This measurement is, in fact, an estimated position, which is obtained by a global interpolation of the labelled samples, based on the learnt manifold-based model. The aRTF sample $\mathbf{h}(t)$ allows the calculation of the covariance $\tilde{\Sigma}_{Lt}$ that is essential for the evaluation of the artificial position measurement. Our confidence in the artificial measurement is determined according to the variance of the estimator, and is expressed in the model by the variance of the measurement noise ζ_t.

To summarize, the proposed state-space model is given by:

$$p(t) = g_t \cdot p(t-1) + \xi_t$$
$$q_t = p(t) + \zeta_t. \quad (11)$$

Since both the process and the observation models are linear, a standard Kalman filter can be applied for recursively solving (11). The Kalman filter recursion takes the following form:

$$\hat{p}(t|t-1) = g_t \cdot \hat{p}(t-1|t-1)$$
$$\gamma(t|t-1) = g_t^2 \gamma(t-1|t-1) + \sigma_\xi^2$$
$$\hat{p}(t|t) = \hat{p}(t|t-1) + \kappa(t)\left(q_t - \hat{p}(t|t-1)\right)$$
$$\gamma(t|t) = (1 - \kappa(t))\,\gamma(t|t-1) \tag{12}$$

where $\gamma(t|t-1)$ is the predicted covariance, $\gamma(t|t)$ is the posteriori covariance, and $\kappa(t)$ is the Kalman gain, defined as:

$$\kappa(t) = \frac{\gamma(t|t-1)}{\gamma(t|t-1) + \sigma_\xi^2}. \tag{13}$$

Note that the measurements of the moving source, accumulated through run-time, can be considered as additional unlabelled data. Thus, the current measurements can be used to update the manifold-based covariance terms in (7) and (9). An efficient recursive adaptation for the MMGP was presented in [11].

5   Experimental Study

We conducted a simulation of a 2-D tracking of a moving source. We simulated a $5.2 \times 6.2 \times 3$ room with 4 pairs of microphones mounted next to the room walls, using an efficient implementation [8] of the image method [1]. All the measurements were confined to a 2×2 m rectangular region, at a fixed height of 2 m (the same height of all the microphones). We generated a training set with $n_L = 36$ labelled samples, without additional unlabelled samples ($n_U = 0$). The labelled samples form a fixed grid with resolution of 0.4 m, and were generated using 10 s long speech signals. The room setup is presented in Fig. 1.

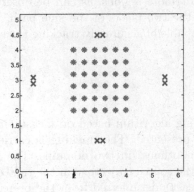

Fig. 1. Room setup: the blue x-marks denote the microphones and the red asterisks denote the labelled samples. (Color figure online)

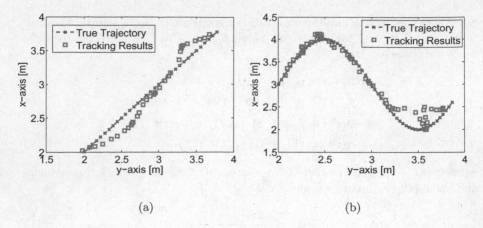

Fig. 2. True path and estimated path for (a) straight line movement and for (b) sinusoidal movement.

We examined two types of trajectories: a straight line along the diagonal of the rectangular region and a sinusoidal trajectory. The duration of the entire movement of the source was 3 s and 5 s for the straight line movement and for the sinusoidal movement, respectively. For both movement types, the source average velocity was approximately 1 m/s, and the measured signals were divided into segments of 330 ms with 75% overlap. For each segment, the corresponding RTF was estimated in 2048 frequency bins. In Fig. 2, we plot the two movements and the tracking results received for 300 ms reverberation time in noiseless conditions.

It can be observed that the proposed method is able to track the source for both types of trajectories. The root mean square errors (RMSEs) were 13 cm and 17 cm for the straight line movement and for the sinusoidal movement, respectively. The error is larger for the sine path compared to the straight path, since it is more complicated and neither the velocity nor the acceleration are fixed. In addition, for regions closer to the microphone positions we receive lower error compared to remote regions, as can be observed by comparing the tracking results around the two peaks of the sine path. We conclude that the proposed algorithm is capable of accurately tracking the source in a reverberant environment.

6   Conclusions

A semi-supervised tracking algorithm based on measurements from distributed pairs of microphones is presented. The tracking is carried out by Kalman filtering which exploits smoothness in two domains. The first is the commonly assumed smoothness of the source trajectory in the time domain. The second is related to the data-driven model inferred from the prerecorded measurements. The source position is assumed to vary smoothly with respect to the multiple acoustic manifolds associated with the different nodes. The resulting tracker is shown to accurately track a moving source in a simulated reverberant room. In future work, we intend to examine a more sophisticated modelling of the source movement.

References

1. Allen, J., Berkley, D.: Image method for efficiently simulating small-room acoustics. J. Acoust. Soc. Am. **65**(4), 943–950 (1979)
2. Benesty, J.: Adaptive eigenvalue decomposition algorithm for passive acoustic source localization. J. Acoust. Soc. Am. **107**(1), 384–391 (2000)
3. Bertin, N., Kitić, S., Gribonval, R.: Joint estimation of sound source location and boundary impedance with physics-driven cosparse regularization. In: IEEE International Conference on Acoustics, Speech and Signal Processing (ICASSP), pp. 6340–6344 (2016)
4. Deleforge, A., Forbes, F., Horaud, R.: Acoustic space learning for sound-source separation and localization on binaural manifolds. Int. J. Neural Syst. **25**(1), 1440003 (2015)
5. Dmochowski, J.P., Benesty, J.: Steered beamforming approaches for acoustic source localization. In: Cohen, I., Benesty, J., Gannot, S. (eds.) Speech Processing in Modern Communication, pp. 307–337. Springer, Heidelberg (2010)
6. Gannot, S., Burshtein, D., Weinstein, E.: Signal enhancement using beamforming and nonstationarity with applications to speech. IEEE Trans. Sign. Process. **49**(8), 1614–1626 (2001)
7. Gannot, S., Dvorkind, T.G.: Microphone array speaker localizers using spatial-temporal information. EURASIP J. Adv. Sig. Process. **2006**(1), 1–17 (2006)
8. Habets, E.A.P.: Room impulse response (RIR) generator, July 2006. http://home.tiscali.nl/ehabets/rir_generator.html
9. Knapp, C., Carter, G.: The generalized correlation method for estimation of time delay. IEEE Trans. Acoust. Speech Sig. Process. **24**(4), 320–327 (1976)
10. Laufer-Goldshtein, B., Talmon, R., Gannot, S.: A study on manifolds of acoustic responses. In: Vincent, E., Yeredor, A., Koldovský, Z., Tichavský, P. (eds.) LVA/ICA 2015. LNCS, vol. 9237, pp. 203–210. Springer, Heidelberg (2015). doi:10.1007/978-3-319-22482-4_23
11. Laufer-Goldshtein, B., Talmon, R., Gannot, S.: Semi-supervised source localization on multiple-manifolds with distributed microphones. pre-print arXiv:1610.04770v1, September 2016
12. Salvati, D., Drioli, C., Foresti, G.L.: A weighted MVDR beamformer based on SVM learning for sound source localization. Pattern Recogn. Lett. **84**, 15–21 (2016)
13. Schmidt, R.O.: Multiple emitter location and signal parameter estimation. IEEE Trans. Antennas Propag. **34**(3), 276–280 (1986)
14. Schwartz, O., Gannot, S.: Speaker tracking using recursive EM algorithms. IEEE/ACM Trans. Audio Speech Lang. Process. **22**(2), 392–402 (2014)
15. Smaragdis, P., Boufounos, P.: Position and trajectory learning for microphone arrays. IEEE Trans. Audio Speech Lang. Process. **15**(1), 358–368 (2007)
16. Ward, D.B., Lehmann, E.A., Williamson, R.C.: Particle filtering algorithms for tracking an acoustic source in a reverberant environment. IEEE Trans. Speech Audio Process. **11**(6), 826–836 (2003)

VAST: The Virtual Acoustic Space
Traveler Dataset

Clément Gaultier[1(✉)], Saurabh Kataria[2], and Antoine Deleforge[1]

[1] Inria Rennes - Bretagne Atlantique, Rennes, France
{clement.gaultier,antoine.deleforge}@inria.fr
[2] Indian Institute of Technology Kanpur, Kanpur, India
saurabhk@iitk.ac.in

Abstract. This paper introduces a new paradigm for sound source localization referred to as virtual acoustic space traveling (VAST) and presents a first dataset designed for this purpose. Existing sound source localization methods are either based on an approximate physical model (physics-driven) or on a specific-purpose calibration set (data-driven). With VAST, the idea is to learn a mapping from audio features to desired audio properties using a massive dataset of simulated room impulse responses. This virtual dataset is designed to be maximally representative of the potential audio scenes that the considered system may be evolving in, while remaining reasonably compact. We show that virtually-learned mappings on this dataset generalize to real data, overcoming some intrinsic limitations of traditional binaural sound localization methods based on time differences of arrival.

Keywords: Sound localization · Binaural hearing · Room simulation · Machine learning

1 Introduction

Human listeners have the stunning ability to understand complex auditory scenes using only two ears, *i.e.*, with binaural hearing. Advanced tasks such as sound source direction and distance estimation or speech deciphering in multi-source, noisy and reverberant environments are performed daily by humans, while they are still a challenge for artificial (two-microphone) binaural systems. The main line of research in machine binaural source localization along the past decades has been to estimate the time-difference of arrival (TDOA) of the signal of interest at the two microphones. An estimated TDOA can be approximately mapped to the azimuth angle of a frontal source if the distance between microphones is known, assuming free-field[1] and far-field[2] conditions. Two important limits

---

[1] Free-field means that the sound propagates from the source to the microphones through a single direct path, without interfering objects or reverberations.
[2] Far-field means that the source is placed far enough (*e.g.* >1.8 m [12]) from the receiver so that the effect of distance on recorded audio features is negligible.

© Springer International Publishing AG 2017
P. Tichavský et al. (Eds.): LVA/ICA 2017, LNCS 10169, pp. 68–79, 2017.
DOI: 10.1007/978-3-319-53547-0_7

of these assumptions can be identified. First, they are both violated in most practical scenarios. In the example of an indoor binaural hearing robot, users are typically likely to engage interaction in both far- and near-fields and non-direct sound paths exist due to reflection and diffusion on walls, ceiling, floor, other objects in the room and the robot itself. Second, the intrinsic symmetries of a free-field/far-field binaural system restrict any geometrical estimation to that of a frontal azimuth angle. Hence, 3D source position (azimuth, elevation, distance) is out of reach in this scope, let alone additional properties such as source orientation, receiver position or room shape.

To overcome intrinsic limitations of TDOA, richer binaural features have been investigated. These include frequency-dependent phase and level differences [4,20], spectral notches [7,14] or the direct-to-reverberant ratio [10]. To overcome the free-field/far-field assumptions, advanced mapping techniques from these features to audio scene properties have been considered. These mapping techniques divide in two categories. The first one is *physics-driven*, *i.e.*, the mapping is inferred from an approximate sound propagation model such as the Woodworth's spherical head formula [20] or the full wave-propagation equation [9]. The second category of mapping is *data-driven*. This approach is sometimes referred to as *supervised sound source localization* [19], or more generally *acoustic space learning* [2]. These methods bypass the use of an explicit, approximate physical model by directly learning a mapping from audio features to audio properties using manually recorded training data [4,19]. They generally yield excellent results, but because obtaining sufficient training data is very time consuming, they only work for a specific room and setup and are hard to generalize in practice. Unlike artificial systems, human listeners benefit from years of adaptive auditory learning in a multitude of acoustic environments. While machine learning recently showed tremendous success in the field of speech recognition using massive amounts of annotated data, equivalent training sets do not exist for audio scene geometry estimation, with only a few specialized manually annotated ones [2,4]. Interestingly, a recent data-driven method [13] used both real and simulated data to estimate room acoustic parameters and improve speech recognition performance, although it was not designed for sound localization.

We propose here a new paradigm that aims at making the best of physics-driven and data-driven approaches, referred to as *virtual acoustic space traveling*. The idea is to use a physics-based room-acoustic simulator to generate arbitrary large datasets of room-impulse responses corresponding to various acoustic environments, adapted to the physical audio system considered. Such impulse responses can be easily convolved with natural sounds to generate a wide variety of audio scenes including *cocktail-party* like scenarios. The obtained corpus can be used to learn a mapping from audio features to various audio scene properties using, *e.g.*, deep learning or other non-linear regression methods [3]. The *virtually-learned* mapping can then be used to efficiently perform real-world auditory scene analysis tasks with the corresponding physical system. Inspired by the idea of an artificial system learning to hear by exploring virtual acoustic environments, we name this proposal the *Virtual Acoustic Space Traveler* (VAST)

project. We initiate it by publicly releasing a dedicated project page: http://
theVASTproject.inria.fr and a first example of VAST dataset. This paper details
the guidelines and methodology that were used in the process of building this
training set. It then demonstrates that virtually-learned mappings can generalize
to real-world test sets, overcoming intrinsic limitations of TDOA-based sound
source localization methods.

2 Dataset Design

2.1 General Principles

The space of all possible acoustic scenes is vast. Therefore, some trade-offs
between the size and the representativity of the dataset must be made when
building a training corpus for audio scene geometry estimation. During the
process of designing the dataset, we imposed on ourselves the following guidelines:

- The dataset should consist of room impulse responses (RIR). This is a more
  generic representation than, *e.g.*, specific audio features or audio scenes involving specific sounds. Each RIR should be annotated by all the source, receiver
  and room properties defining it.
- Virtual acoustic space traveling aims at building a dataset for a **specific
  audio system** in a variety of environments. Following this idea, some intrinsic
  properties of the receiver such as its distance to the ground and its head-related
  transfer functions are kept fixed throughout the simulations. For this first
  dataset, called *VAST_KEMAR_0*, we chose the emblematic KEMAR acoustic
  dummy-head, whose measured HRTFs are publicly available. It was placed at
  1.70 from the ground, the average human's height.
- We are interested in modeling acoustic environments which are typically
  encountered in an office building, a university, a hotel or a modern habita-
  tion. Acoustics of the type encountered in a cathedral, a massive hangar, a
  recording studio or outdoor are deliberately left aside here. Surface materials
  and diffusion profiles are chosen accordingly.
- To make the dataset easily manipulable on a simple laptop, we aimed at
  keeping its total size under 10 GigaBytes. To handle datasets of larger order
  of magnitudes would require users to have access to specific hardware and
  software which is not desired here. *VAST_KEMAR_0* measures 6.4 GB.

2.2 Room Simulation and Data Generation

The efficient C++/MATLAB "shoebox" 3D acoustic room simulator ROOMSIM
developed by Schimmel et al. is selected for simulations [16]. This software takes
as input a room dimension (width, depth and height), a source and receiver position and orientation, a receiver's head-related-transfer function (HRTF) model,
and frequency-dependent absorption and diffusion coefficients for each surface.
It outputs a corresponding pair of room impulse responses (RIR) at each ear of

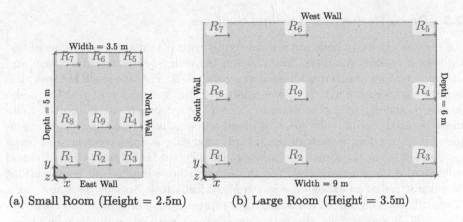

(a) Small Room (Height = 2.5m)   (b) Large Room (Height = 3.5m)

Fig. 1. Top views of training rooms with receiver positions and orientations.

the binaural receiver. Specular reflections are modeled using the image-source method [1], while diffusion is modeled using the so-called *rain-diffusion* algorithm. In the latter, sound rays uniformly sampled on the sphere are sent from the emitter and bounced on the walls according to specular laws, taking into account surface absorption. At each impact, each ray is also randomly bounced towards the receiver with a specified probability (the frequency-dependent *diffusion coefficient* of the surface). The total received energy at each frequency is then aggregated using histograms. This model was notably showed to realistically account for sound scattering due to the presence of objects, by comparing simulated RIRs with measured ones in [22]. The study [8] suggests that such diffusion effects play an important role in sound source localization. *VAST_KEMAR_0* contains over 110,000 RIR, which required about 700 CPU-hours of computation. This was done using a massively parallelized implementation on a large computer grid.

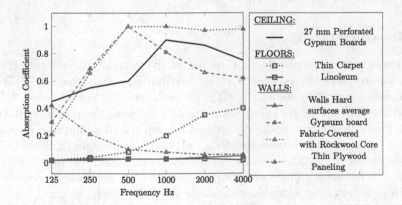

Fig. 2. Absorption profiles

2.3   Room Properties: Size and Surfaces

An obvious choice to generate virtual rooms with maximal variability would be to draw a random room size and random frequency-dependent absorption and diffusion profiles of surfaces for each generated RIR. This approach however, has several drawbacks. First, it makes impossible the generation of realistic audio scenes containing several sources, for which the receiver position and the room must be fixed. Second, the space of possible rooms is so vast that reliably sampling it at random is unrealistic. Third, changing source, receiver and room parameters all at the same time prevents from getting insights on the individual influence of these parameters. On the other hand, sampling all combinations of parameters in an exhaustive way quickly leads to enormous data size. As a trade-off, we designed 16 realistic rooms representative of typical reverberation time (RT_{60}) and surface absorption profiles encountered in modern buildings. Two room sizes were considered: a small one corresponding to a typical office or bed room (Fig. 1(a)), and a larger one corresponding to a lecture or entrance hall (Fig. 1(b)). For each room, floor, ceiling and wall materials which are representative in terms of absorption profile and are commonly encountered in nowadays buildings were chosen from [21]. The graph on Fig. 2 displays the absorption profiles of the selected materials, namely, 4 for the walls, 2 for the floor and 1 for the ceiling. The gypsum board material chosen for the ceiling was kept fixed throughout the dataset, as it represents well typical ceiling absorption profiles [21]. "Walls hard surface average" is in fact an average profile over many surfaces such as brick or plaster [21]. Combining all possible floors, walls and room sizes yielded the 16 rooms listed in Table 1.

Importantly, typical rooms also contain furniture and other objects responsible for random sound scattering effects, *i.e.*, diffusion. Following the acoustic study in [5], a unique frequency-dependent diffusion profile was used for all surfaces. The chosen profile is the average of the 8 configurations measured in [5], corresponding to varying numbers of chairs, table, computers and people in a room. Both absorption and diffusion profiles are piecewise-linearly interpolated from 8 Octave bands from 125 Hz to 4 kHz.

2.4   Reverberation Time

A common acoustic descriptor for rooms is the reverberation time (RT_{60}). Figure 3(a) displays the estimated RT_{60} distribution across the VAST Training Dataset. Figure 3(b) shows the RT_{60} for each room by octave band. RT_{60}'s were estimated from the room impulse responses following the recommendations in [17]. From these estimations, we decided to crop the room impulse responses provided in the datasets above the RT_{60}, with a 30 ms margin. This technique allows to shrink the dataset while keeping data points of interest and discarding the rest. To further complies with memory limitations, we chose to encode the room impulse response samples with single floats (16 bit). As can be seen in Fig. 3 the 16 chosen rooms present a quite good variability in terms of reverberation times in the range 100 ms–400 ms. Larger RT_{60} of the order of 1 s could

Table 1. Description of simulated training rooms in VAST

Room number	Floor	Ceiling	Walls	Width [m]	Depth [m]	Height [m]
1	Thin Carpet	Perforated 27 mm gypsum board	Walls Hard Surfaces Average	9	6	3.5
2	Thin Carpet	Perforated 27 mm gypsum board	Gypsum Board with Mineral Filling	9	6	3.5
3	Thin Carpet	Perforated 27 mm gypsum board	Fabric-Covered Panel with Rockwool Core	9	6	3.5
4	Thin Carpet	Perforated 27 mm gypsum board	Thin Plywood Paneling	9	6	3.5
5	Linoleum	Perforated 27 mm gypsum board	Walls Hard Surfaces Average	9	6	3.5
6	Linoleum	Perforated 27 mm gypsum board	Gypsum Board with Mineral Filling	9	6	3.5
7	Linoleum	Perforated 27 mm gypsum board	Fabric-Covered Panel with Rockwool Core	9	6	3.5
8	Linoleum	Perforated 27 mm gypsum board	Thin Plywood Paneling	9	6	3.5
9	Thin Carpet	Perforated 27 mm gypsum board	Walls Hard Surfaces Average	3.5	5	2.5
10	Thin Carpet	Perforated 27 mm gypsum board	Gypsum Board with Mineral Filling	3.5	5	2.5
11	Thin Carpet	Perforated 27 mm gypsum board	Fabric-Covered Panel with Rockwool Core	3.5	5	2.5
12	Thin Carpet	Perforated 27 mm gypsum board	Thin Plywood Paneling	3.5	5	2.5
13	Linoleum	Perforated 27 mm gypsum board	Walls Hard Surfaces Average	3.5	5	2.5
14	Linoleum	Perforated 27 mm gypsum board	Gypsum Board with Mineral Filling	3.5	5	2.5
15	Linoleum	Perforated 27 mm gypsum board	Fabric-Covered Panel with Rockwool Core	3.5	5	2.5
16	Linoleum	Perforated 27 mm gypsum board	Thin Plywood Paneling	3.5	5	2.5
0	Anechoic room					

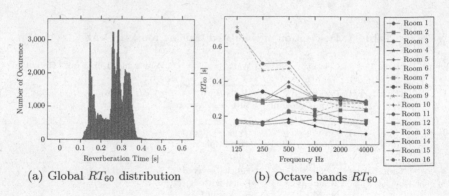

(a) Global RT_{60} distribution   (b) Octave bands RT_{60}

Fig. 3. Reverberation time.

be obtain by using highly reflective materials on all surfaces, creating an echo chamber. However, this rarely occurs in realistic buildings.

2.5   Source and Receiver Positions

A relatively poorly-studied though important effect in sound source localization is the influence of the receiver's position in the room, especially its distance to the nearest surface. In order to accurately capture this effect, 9 receiver positions are used for each of the 16 rooms, while the height of the receiver is fixed at 1.7 m. Figure 1 shows top views of the rooms with receiver positions. Positions from R_1 to R_8 are set 50 cm from the nearest wall(s) whereas R_9 is approximately placed in the middle of the room. Perfectly symmetrical configurations are avoided to make the dataset as generic as possible, without singularities. The receiver is always facing the north wall as a convention. For each of the 9 receiver positions, sources are placed on spherical grids centered on the receiver. Each sphere consists of regularly-spaced elevation lines each containing sources at regularly-spaced azimuths, with a spacing of $9°$. The equator elevation line and the first azimuth angle of each line are randomly offset by $-4.5°$ to $+4.5°$ in order to obtain a dense sphere sampling throughout the dataset. Six spherical grid radii are considered, yielding source distances of 1, 1.5, 2, 3, 4 and 6 m. Sources falling outside of the room or less than 20 cm from a surface are removed.

2.6   Test Sets

To test the generalizability of mappings learned on the *VAST_KEMAR_0* dataset, we built four simulated test sets differing from the training dataset on various levels. A first challenge is to test robustness to random positioning, since the training set is built with regular spherical source grids and fixed listener positions. Hence, the 4 testing sets contain completely random source and receiver positions in the room. Only the receiver's height is fixed to 1.7 m, and both receiver and source are set within a 20 cm safety margin within the room boundaries. Test sets 2 and 4 feature random receiver orientation (yaw angle), as opposed to the receiver facing north in the training set. Test 1 and 2 contain 1,000 binaural RIRs (BRIRs) for each of the 16 rooms of Table 1. Finally, test sets 3 and 4 contain 10,000 BRIRs, each corresponding to a random room size (walls from $3\,m \times 2\,m$ to $10\,m \times 4\,m$) and random absorption properties of walls and floor picked from Fig. 2. Different surfaces for all 4 walls are allowed.

In addition to these simulated test sets, three binaural RIR datasets recorded with the KEMAR dummy head in real rooms have been selected, as listed below:

- **Auditorium 3** [11] was recorded at TU Berlin in 2014 in a trapezium-shaped lecture room of dimensions $9.3\,m \times 9\,m$ and $RT_{60} \approx 0.7\,s$. 3 individual sources placed $1.5\,m$ from the receiver at different azimuth and $0°$ elevation were recorded. For each source, one pair of binaural RIR is recorded for each receivers' head yaw angle from $-90°$ to $+90°$, with $1°$ steps.
- **Spirit** [11] was recorded at TU Berlin in 2014 in a small rectangular office room of size $4.3\,m' \times 5\,m$, $RT_{60} \approx 0.5\,s$, containing various objects, surfaces and furniture near the receiver. The protocol is the same as Auditorium 3 except sources are placed $2\,m$ from the receiver.
- **Classroom** [18] was recorded at Boston University in 2005 in a $5\,m \times 9\,m \times 3.5\,m$ carpeted classroom with 3 concrete walls and one sound-absorptive wall ($RT_{60} = 565\,ms$). The receiver is placed in 4 locations of the room including 3 with at least one nearby wall.

Note that the KEMAR HRTF measurements used to simulate the VAST dataset was recorded by yet another team, in MIT's anechoic chamber in 1994, as described in [6].

3 Virtually Supervised Sound Source Localization

For all experiments in this section, all training and test sets used are reduced to contain only frontal sources (azimuth in $[-90°, +90°]$) with elevation in $[-45°, +45°]$ and distances between 1 and $3\,m$. As mentioned in the introduction, sound source localization consists in two steps: calculating auditory features from binaural signals followed by mapping these features to a source position. Robustly estimating features can be difficult when dealing with additive noise, sources with sparse spectra such as speech or music, and source mixtures. We leave this problematic aside in this paper, and focus on mapping clean features to source positions. Hence, we use *ideal* features directly calculated from the clean room impulse responses in all experiments.

We first make an experiment to put forward some intrinsic limitations of TDOA-based azimuth estimation. Figure 4 plots TDOAs against the source's azimuth angle for different subsets of VAST. TDOAs (in samples) were computed as the integer delay in $[-15, +15]$ maximizing the correlation between the first 500 samples of the left and the right impulse responses. As can be seen in Fig. 4(a), a near-linear relationship between frontal azimuth and TDOA exists in the anechoic case, regardless of the elevation. This matches previously observed

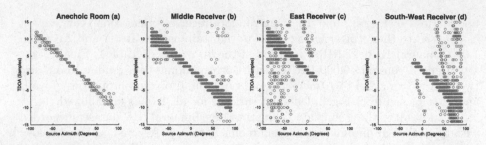

Fig. 4. TDOA as a function of source azimuth in various settings

results in binaural sound localization [4,15,20]. When the receiver is placed in the middle of the 16 reverberant rooms, (Fig. 4(b)), some outliers appear due to reflections. This effect is dramatically increased when the receiver is placed 50 cm from a wall (Fig. 4(c) and (d)), where stronger early reflections are present. This suggests that the TDOA, even when ideally estimated, is not adapted to binaural sound source localization in realistic indoor environments.

Table 2. Azimuth absolute estimation errors in degrees with 3 different methods, showed in the form $avg \pm std(out\%)$, where avg and std denote the mean and standard deviation of *inlying* absolute errors ($<30°$) while out denotes the percentage of outliers.

Test data ↓	TDOA	GLLiM (Anech. train.)	GLLiM (VAST train.)
VAST Testing Set 1	$5.49 \pm 4.6\,(5.6\%)$	$8.63 \pm 7.6\,(12\%)$	$4.38 \pm 4.9\,(1.8\%)$
VAST Testing Set 2	$5.37 \pm 4.4\,(6.0\%)$	$8.09 \pm 7.5\,(12\%)$	$4.32 \pm 4.7\,(1.6\%)$
VAST Testing Set 3	$5.21 \pm 4.5\,(4.6\%)$	$8.46 \pm 7.5\,(5.2\%)$	$4.23 \pm 4.4\,(1.8\%)$
VAST Testing Set 4	$5.14 \pm 4.4\,(3.3\%)$	$8.21 \pm 7.2\,(4.8\%)$	$4.25 \pm 4.4\,(0.6\%)$
Auditorium 3 [11]	$7.02 \pm 4.7\,(1.4\%)$	$8.01 \pm 7.0\,(5.9\%)$	$5.03 \pm 4.5\,(0.0\%)$
Spirit [11]	$5.19 \pm 3.4\,(0.0\%)$	$12.2 \pm 8.3\,(15\%)$	$4.50 \pm 5.6\,(0.4\%)$
Classroom [18]	$5.71 \pm 3.7\,(3.7\%)$	$9.47 \pm 7.3\,(5.2\%)$	$6.50 \pm 5.9\,(0.0\%)$

We then compare azimuth estimation errors obtained with the TDOA-based method described above, a learning-based method trained on anechoic HRTF measurements (Room 0), and a learning-based method trained on VAST, using the 4 simulated and 3 real test sets described in Sect. 2.6. TDOAs were mapped to azimuth values using the affine regression coefficients corresponding to the red line in Fig. 4(a). The chosen learning-based sound source localization method is the one described in [4]. It uses Gaussian Locally Linear Regression (GLLiM, [3]) to map high-dimensional feature vectors containing frequency-dependent inter-aural level and phase differences from 0 to 8000 Hz to low-dimensional source positions. In our case, the GLLiM model with K locally-linear components was trained on N interaural feature vectors of dimension $D = 1537$ associated to

Table 3. Elevation and distance absolute estimation errors obtained with GLLiM trained on VAST. Outliers correspond to errors larger than 15° or 1 m.

Test data ↓	Elevation (°)	Distance (m)
VAST Testing Set 1	5.91 ± 4.1 (23%)	0.43 ± 0.3 (19%)
VAST Testing Set 2	6.05 ± 4.2 (27%)	0.44 ± 0.3 (20%)
VAST Testing Set 3	6.05 ± 4.1 (27%)	0.43 ± 0.3 (21%)
VAST Testing Set 4	6.03 ± 4.2 (26%)	0.44 ± 0.3 (21%)
Auditorium 3 [11]	7.92 ± 4.4 (44%)	0.45 ± 0.3 (23%)
Spirit [11]	7.44 ± 4.3 (30%)	0.52 ± 0.3 (25%)
Classroom [18]	8.40 ± 4.1 (45%)	0.41 ± 0.3 (6.5%)

3-dimensional source positions in spherical coordinate (azimuth, elevation and distance). $K = 8$ components were used for the anechoic training set ($N = 181$) and $K = 100$ for the (reduced) VAST dataset ($N \approx 41,000$). All 3 methods showed comparably low testing computational times, in the order of 10 ms for 1 s of input signal. Table 2 summarizes obtained azimuth estimation errors. As can be seen, the learning method trained on VAST outperforms the two others on all datasets, with significantly less outliers and a globally reduced average error of inliers. This is encouraging considering the variety of testing data used. In addition, Table 3 shows that GLLiM trained on VAST is capable of approximately estimating the elevation and distance of the source, which is known to be particularly difficult from binaural data. While elevation estimation on real data remains a challenge, results obtained on simulated sets are promising.

4   Conclusion

We introduced the new concept of virtual acoustic space traveling and released a first dataset dedicated to it. A methodology to efficiently design such a dataset was provided, making extensions and improvements of the current version easily implementable in the future. Results show that a learning-based sound source localization method trained on this dataset yields better localization results than when trained on anechoic HRTF measurements, and performs better than a TDOA-based approach in azimuth estimation while being able to estimate source elevation and distance. To the best of the authors' knowledge, this is the first time a sound localization method trained on simulated data is successfully used on real data, validating the new concept of virtual acoustic space traveling. The learning approach could still be significantly improved by considering other auditory features, by better adapting the mapping technique to spherical coordinates and by annotating training data with further acoustic information. Other learning methods such as deep neural networks may also be investigated.

References

1. Allen, J.B., Berkley, D.A.: Image method for efficiently simulating small-room acoustics. J. Acoust. Soc. Am. **65**(4), 943–950 (1979)
2. Deleforge, A., Forbes, F., Horaud, R.: Acoustic space learning for sound-source separation and localization on binaural manifolds. Int. J. Neural Syst. **25**(01), 1440003 (2015)
3. Deleforge, A., Forbes, F., Horaud, R.: High-dimensional regression with gaussian mixtures and partially-latent response variables. Stat. Comput. **25**(5), 893–911 (2015)
4. Deleforge, A., Horaud, R., Schechner, Y.Y., Girin, L.: Co-localization of audio sources in images using binaural features and locally-linear regression. IEEE Trans. Audio Speech Lang. Process. **23**(4), 718–731 (2015)
5. Faiz, A., Ducourneau, J., Khanfir, A., Chatillon, J.: Measurement of sound diffusion coefficients of scattering furnishing volumes present in workplaces. In: Acoustics 2012 (2012)
6. Gardner, W.G., Martin, K.D.: HRTF measurements of a kemar. J. Acoust. Soc. Am. **97**(6), 3907–3908 (1995)
7. Hornstein, J., Lopes, M., Santos-Victor, J., Lacerda, F.: Sound localization for humanoid robots-building audio-motor maps based on the HRTF. In: 2006 IEEE/RSJ International Conference on Intelligent Robots and Systems, pp. 1170–1176. IEEE (2006)
8. Kataria, S., Gaultier, C., Deleforge, A.: Hearing in a shoe-box: binaural source position and wall absorption estimation using virtually supervised learning. In: 2017 IEEE International Conference on Acoustics, Speech and Signal Processing (ICASSP). IEEE (2017)
9. Kitić, S., Bertin, N., Gribonval, R.: Hearing behind walls: localizing sources in the room next door with cosparsity. In: 2014 IEEE International Conference on Acoustics, Speech and Signal Processing (ICASSP), pp. 3087–3091. IEEE (2014)
10. Lu, Y.C., Cooke, M.: Binaural estimation of sound source distance via the direct-to-reverberant energy ratio for static and moving sources. IEEE Trans. Audio Speech Lang. Process. **18**(7), 1793–1805 (2010)
11. Ma, N., May, T., Wierstorf, H., Brown, G.J.: A machine-hearing system exploiting head movements for binaural sound localisation in reverberant conditions. In: 2015 IEEE International Conference on Acoustics, Speech and Signal Processing (ICASSP), pp. 2699–2703. IEEE (2015)
12. Otani, M., Hirahara, T., Ise, S.: Numerical study on source-distance dependency of head-related transfer functions. J. Acoust. Soc. Am. **125**(5), 3253–3261 (2009)
13. Parada, P.P., Sharma, D., Lainez, J., Barreda, D., van Waterschoot, T., Naylor, P.A.: A single-channel non-intrusive c50 estimator correlated with speech recognition performance. IEEE/ACM Trans. Audio Speech Lang. Process. **24**(4), 719–732 (2016)
14. Raykar, V.C., Duraiswami, R., Yegnanarayana, B.: Extracting the frequencies of the pinna spectral notches in measured head related impulse responses. J. Acoust. Soc. Am. **118**(1), 364–374 (2005)
15. Sanchez-Riera, J., Alameda-Pineda, X., Wienke, J., Deleforge, A., Arias, S., Čech, J., Wrede, S., Horaud, R.: Online multimodal speaker detection for humanoid robots. In: 2012 12th IEEE-RAS International Conference on Humanoid Robots (Humanoids 2012), pp. 126–133. IEEE (2012)

16. Schimmel, S.M., Muller, M.F., Dillier, N.: A fast and accurate "shoebox" room acoustics simulator. In: 2009 IEEE International Conference on Acoustics, Speech and Signal Processing, pp. 241–244. IEEE (2009)
17. Schroeder, M.R.: New method of measuring reverberation time. J. Acoust. Soc. Am. **37**(3), 409–412 (1965)
18. Shinn-Cunningham, B.G., Kopco, N., Martin, T.J.: Localizing nearby sound sources in a classroom: binaural room impulse responses. J. Acoust. Soc. Am. **117**(5), 3100–3115 (2005)
19. Talmon, R., Cohen, I., Gannot, S.: Supervised source localization using diffusion kernels. In: 2011 IEEE Workshop on Applications of Signal Processing to Audio and Acoustics (WASPAA), pp. 245–248. IEEE (2011)
20. Viste, H., Evangelista, G.: On the use of spatial cues to improve binaural source separation. In: Proceedings of 6th International Conference on Digital Audio Effects (DAFx-03), pp. 209–213. No. LCAV-CONF-2003-026 (2003)
21. Vorländer, M.: Auralization: Fundamentals of Acoustics, Modeling, Simulation, Algorithms and Acoustic Virtual Reality. Springer, Heidelberg (2007)
22. Wabnitz, A., Epain, N., Jin, C., Van Schaik, A.: Room acoustics simulation for multichannel microphone arrays. In: Proceedings of the International Symposium on Room Acoustics, pp. 1–6 (2010)

Sketching for Nearfield Acoustic Imaging of Heavy-Tailed Sources

Mathieu Fontaine[1]([⊠]), Charles Vanwynsberghe[2], Antoine Liutkus[1], and Roland Badeau[3]

[1] Inria, Speech Processing Team, Nancy Grand-Est, Nancy, France
mathieu.fontaine@inria.fr
[2] Institut Jean le Rond d'Alembert, Saint-Cyr l'École, France
[3] LTCI, CNRS, Télécom ParisTech, Université Paris-Saclay, Paris, France

Abstract. We propose a probabilistic model for acoustic source localization with known but arbitrary geometry of the microphone array. The approach has several features. First, it relies on a simple nearfield acoustic model for wave propagation. Second, it does not require the number of active sources. On the contrary, it produces a heat map representing the energy of a large set of candidate locations, thus imaging the acoustic field. Second, it relies on a heavy-tail α-stable probabilistic model, whose most important feature is to yield an estimation strategy where the multichannel signals need to be processed only once in a simple online procedure, called *sketching*. This sketching produces a fixed-sized representation of the data that is then analyzed for localization. The resulting algorithm has a small computational complexity and in this paper, we demonstrate that it compares favorably with state of the art for localization in realistic simulations of reverberant environments.

1 Introduction

Source localization has attracted a lot of research interest, notably in acoustics [5] and wireless communications [15]. It aims at identifying the position or *direction of arrival* (DoA) of *sources* that are captured by an array of sensors. It has many applications, notably for isolating the target signals. In this paper, we are focused on the acoustic application.

Popular approaches for localization largely exploit the geometry of the sensor array. When the positions of the sensors are known, we can indeed predict and exploit the time difference of arrival (TDOA) to all sensors. In a more realistic environment with echoes and reverberation, localization becomes a much more challenging inverse problem composed of two classical parts. First, the knowledge of the geometry of the sensor array along with physics provides us with a *direct model*. Then, localization tries to invert this direct path so as to estimate the most likely location of the sources based on the observations. As in any challenging inverse problem, the difficulties come from having less observations than unknowns, and/or from uncertainties in the direct model. Furthermore, localization should ideally work regardless of the particular source signals considered, which brings an additional difficulty.

© Springer International Publishing AG 2017
P. Tichavský et al. (Eds.): LVA/ICA 2017, LNCS 10169, pp. 80–88, 2017.
DOI: 10.1007/978-3-319-53547-0_8

Many methods for source localization have already been proposed in the past. Since we usually have a huge number of candidate locations for only a limited amount of sensors, they all attempt to reduce the number of parameters. One approach is to fix the number of sources to look for, yielding for instance high resolution methods [16] such as MUSIC [10,14] that provide good performance when the microphone array is not too massive and obeys some geometry assumptions. Another approach for exploiting this relative *sparsity* of active sources' locations is to use greedy methods [17] that iteratively detect the most predominant source and then remove its influence from the observation using the direct model. Provided the amount of reverberation is not too large and the direct model is sufficiently good, these methods yield good performance. Another direction is grounded on a probabilistic setting [3,11,12] where a prior distribution such as a multivariate Gaussian is assigned to both the unknown source signals and the mixing model.

Apart from raw performance, one important issue of source localization methods is their computational complexity. For the purpose of imaging, the *steering response power* method (SRP) simply averages the power of beamformed outputs targeted at all candidate directions. Although very simple computationally, it yields a very poor contrast. See however [19] for an improvement involving hierarchical search. On the same topic of computational complexity, localization under the Gaussian model [12] involves a demanding Expectation-Maximization algorithm (EM) that requires going through the data many times and inverting many covariance matrices. To a lesser extent, the same goes for RELAX and CLEAN [17].

In this paper, we propose a new imaging technique, conceptually close to SRP because it only requires going through the recordings once. However, it is also grounded in a probabilistic setting but the source signals are no longer assumed Gaussian as in [12] but rather α-stable, which is a heavy-tailed distribution permitting to describe audio signals with very large dynamics using only a very small amount of parameters [7,13]. Departing from the costly EM, estimation in this model is based on moment-fitting, appearing as one instance of the recently popularized sketching methodology [4]. We use a near-field acoustic model here and simulate challenging reverberant environments.

2   Mixture Model and α-Stable Theory

2.1   Notation and Convolutive Model

Let $x \in \mathbb{C}^{F \times T \times K}$ be the Short-Term Fourier Transforms (STFT) of the observations, where F is the number of frequency bins, T the number of time frames and K the number of microphones. $x(f,t) \in \mathbb{C}^K$ gathers its entries for Time-Frequency (TF) bin (f,t). Now, we assume this recording is the superposition of signals originating from L potential locations, corresponding to a grid in the 3D-space. Let $s \in \mathbb{C}^{F \times T \times L}$ denote the STFT of the L corresponding sources, with entries $s_l(f,t) \in \mathbb{C}$. Our objective becomes to estimate the power of the sources at all these L locations. Of course, we expect most of them to be inactive.

Then, the acoustic model defines the mixture as a superposition of filtered versions of the sources. In the frequency domain, this convolution may be approximated as a simple multiplication with steering vectors $A_l(f) \in \mathbb{C}^K$:

$$\forall (f,t), \, x(f,t) = \sum_{l=1}^{L} A_l(f) \, s_l(f,t). \tag{1}$$

A particular direct model then consists in a specific choice for the steering vectors. In this study, we adopt the near field region assumption, thus taking the steering vectors $A_l(f)$ as:

$$\forall l, f \, A_l(f) = \left[\frac{1}{r_{1l}} \exp\left(-i\frac{\omega_f r_{1l}}{c_0} \right), \ldots, \frac{1}{r_{Kl}} \exp\left(-i\frac{\omega_f r_{Kl}}{c_0} \right) \right]^\top, \tag{2}$$

where $\cdot^\top$ stands for transposition, c_0 is the speed of sound in the air, r_{kl} the distance between the kth microphone and the lth source and ω_f is the angular frequency at frequency band f. Note that if applicable, actual measurements may be used instead of the model (2) to provide numerical values for $A_l(f)$ at every candidate location l.

2.2 Independent Isotropic α-Stable Model for the Sources

We assume that all the L sources are independent α-stable harmonizable processes as defined in [7]. In practice, it means that all $s_l(f,t)$ are independent and distributed w.r.t. a complex symmetric α-stable ($S\alpha S_c$) distribution:

$$s_l(f,t) \sim S\alpha S_c(\Upsilon_l), \tag{3}$$

where $\alpha \in (0,2]$ is called the *characteristic exponent*, controlling the tail of the distribution: the closer it is to 0, the heavier the tails. The nonnegative scale parameters $\Upsilon_l \in \mathbb{R}_+$ are the central quantity of interest in our study. Gathering them together in the $L \times 1$ vector $\Upsilon = [\Upsilon_1, \ldots, \Upsilon_L]^\top$, we call it the *discrete spatial measure*. Our objective is to estimate this measure, since it gives the scale of the signal present at each location.

A remarkable fact of the model (3) is that the entries of s_l are modeled as having the same distribution for all f and t. This is made possible thanks to the heavy-tailed nature of the $S\alpha S_c$ distribution. In contrast, the classical Gaussian model [8] requires variances to depend on (f,t) to fit well the data.

2.3 The Levy Exponent and the Spatial Measure

Since the distributions (3) and the acoustic model (1) do not depend on time, neither does the distribution of $x(f,t)$. For a given f, let φ_f be the characteristic function (chf.) of $x(f,t)$ and let I_f be the *Levy exponent*, i.e. the logarithm of its opposite:

$$\forall \theta \in \mathbb{C}^K, \, \varphi_f(\theta) \triangleq \mathbb{E}\left[\exp\left(i\Re\langle \theta, x(f,t)\rangle\right)\right] \text{ and } I_f(\theta) = -\log\varphi_f(\theta), \tag{4}$$

with $\langle .,. \rangle$ the inner product on $\mathbb{C}^K$. In this study, the argument $\boldsymbol{\theta} \in \mathbb{C}^K$ of the chf. is called a *sketching frequency*. Combining the $S\alpha S_c$ model for the sources and the propagation model (1), it can be shown that we have:

$$\forall \boldsymbol{\theta}, \in \mathbb{C}^K, \ I_f(\boldsymbol{\theta}) = \sum_{l=1}^{L} |\langle \boldsymbol{\theta}, \boldsymbol{a}_l(f) \rangle|^\alpha \Upsilon_l, \tag{5}$$

where $\boldsymbol{a}_l(f) = \boldsymbol{A}_l(f) / \|\boldsymbol{A}_l(f)\|_2 \in \mathbb{C}^K$ are the normalized steering vectors.

Now, the approach undertaken in this study is to pick a set of L sketching frequencies and exploit relation (5). Even if we could pick any complex vector for $\boldsymbol{\theta}$, informal experiments shows that taking the normalized steering vectors $\boldsymbol{\theta} = \boldsymbol{a}_l(f)$ gives good performance. This yields L relations of the form (5), that can be expressed in compressed form as $\boldsymbol{I}_f = \boldsymbol{\Psi}_f \boldsymbol{\Upsilon}$, where:

$$\boldsymbol{I}_f \triangleq [I_f(\boldsymbol{a}_1(f)), \ldots, I_f(\boldsymbol{a}_L(f))]^\top \ \text{and} \ \forall l, l' \ [\boldsymbol{\Psi}_f]_{ll'} = |\langle \boldsymbol{a}_l(f), \boldsymbol{a}_{l'}(f) \rangle|^\alpha. \tag{6}$$

Finally by gathering all $\boldsymbol{I}_f$ and $\boldsymbol{\Psi}_f$ into $\boldsymbol{I} \in \mathbb{R}^{FL}$ and $\boldsymbol{\Psi} \in \mathbb{R}_+^{FL \times L}$, respectively, we get:

$$\boldsymbol{I} = \boldsymbol{\Psi} \boldsymbol{\Upsilon}, \tag{7}$$

which is our main tool for estimating $\boldsymbol{\Upsilon}$. Indeed, $\boldsymbol{I}$ is estimated from the data and $\boldsymbol{\Psi}$ is given by combining our acoustic model for $\boldsymbol{a}_l(f)$ and (6).

3   Parameter Estimation

3.1   Sketching for the Levy Exponent

As noted above in (4), the Levy exponent is defined as the logarithm of the negative chf. A naive idea would be to simply replace $\varphi_f(\boldsymbol{\theta})$ in (4) by its empirical counterpart averaged over the different time frames. However, this may lead to numerical instability in case of negative empirical chf. To address this issue, a new unbiased estimator for the chf. specific to symmetric α-stable random vectors is proposed here:

$$\forall \boldsymbol{\theta} \in \mathbb{C}^K, \ \widehat{\varphi}_f(\boldsymbol{\theta}) = \left| \frac{1}{T} \sum_{t=1}^{T} \exp\left(i \frac{\Re \langle \boldsymbol{\theta}, \boldsymbol{x}(f,t) \rangle}{2^{1/\alpha}} \right) \right|^2. \tag{8}$$

As can be seen, this estimate is guaranteed to be nonnegative. Hence, no numerical instability is to be expected when considering the empirical Levy exponent $\widehat{\boldsymbol{I}}_f \in \mathbb{R}_+^L$, defined as:

$$\forall f, \ \widehat{\boldsymbol{I}}_f = [-\ln(\widehat{\varphi}_f(\boldsymbol{a}_1(f))), \ldots, -\ln(\widehat{\varphi}_f(\boldsymbol{a}_L(f)))]^\top. \tag{9}$$

Gathering them as $\widehat{\boldsymbol{I}} = \left[\widehat{\boldsymbol{I}}_1^\top, \ldots, \widehat{\boldsymbol{I}}_F^\top \right]^\top \in \mathbb{R}_+^{FL}$, we obtain a relation similar to (7):

$$\widehat{\boldsymbol{I}} \approx \boldsymbol{\Psi} \boldsymbol{\Upsilon}. \tag{10}$$

Interestingly enough, relation (10) provides us with a linear model where all factors except the desired spatial measure $\boldsymbol{\Upsilon}$ are either empirically estimated from the data ($\widehat{\boldsymbol{I}}$) or provided by the acoustic model ($\boldsymbol{\Psi}$). The fundamental fact here is that the observed data is only used once for estimating the Levy exponent in (8), through a very simple procedure, producing the $LF \times 1$ fixed-sized vector $\widehat{\boldsymbol{I}}$. This is reminiscent of the sketching strategy recently described, e.g. in [4].

3.2   A Proposed NMF Algorithm to Determine $\boldsymbol{\Upsilon}$

The estimation method for $\boldsymbol{\Upsilon}$ is undertaken by a classical minimization of the divergence between the two terms of (10):

$$\widehat{\boldsymbol{\Upsilon}} \leftarrow \arg\min_{\boldsymbol{\Upsilon} \geq 0} d_\beta \left(\widehat{\boldsymbol{I}} | \boldsymbol{\Psi}\boldsymbol{\Upsilon} \right) + \lambda \|\boldsymbol{\Upsilon}\|_1, \tag{11}$$

where d_β depicts a data-fit cost function such as the β-divergence [1], and $\lambda\|\boldsymbol{\Upsilon}\|_1$ is an ℓ_1-regularization penalty term to enforce sparsity of $\boldsymbol{\Upsilon}$. Following classical multiplicative updates strategy, we can straightforwardly estimate $\boldsymbol{\Upsilon}$. The algorithm box below summarizes the whole process, which is of total complexity $\mathcal{O}\left(FTL^2\right)$.

Algorithm 1. Estimation of the spatial measure $\boldsymbol{\Upsilon}$

---

1. Input
   - Number L of possible locations, distances r_{lk} with the microphones.
   - Characteristic exponent α
   - β-divergence to use, number of iterations, regularization parameter λ.
2. Compute steering vectors $\boldsymbol{A}_l(f)$ as in (2).
3. Sketching: $\forall f,\ \widehat{\boldsymbol{I}}_f \leftarrow$(9)
   (the mixture $\boldsymbol{x}$ may only be streamed and not stored)
4. Analysis
   - Gather all $\boldsymbol{\Psi}_f$ and $\widehat{\boldsymbol{I}}_f$ to form $\boldsymbol{\Psi}$ and $\widehat{\boldsymbol{I}}$ (6)
   - Estimation of $\boldsymbol{\Upsilon}$: iterate $\widehat{\boldsymbol{\Upsilon}} \leftarrow \widehat{\boldsymbol{\Upsilon}} \cdot \dfrac{\boldsymbol{\Psi}^\top\left((\boldsymbol{\Psi}\widehat{\boldsymbol{\Upsilon}})^{\beta-2} \cdot \widehat{\boldsymbol{I}}\right)}{\boldsymbol{\Psi}^\top\left((\boldsymbol{\Psi}\widehat{\boldsymbol{\Upsilon}})^{\beta-1}\right)+\lambda}$.

---

4   Evaluation

We now compare the proposed approach with several baseline methods for wide-band source localization. We consider $J = 5$ speech signals lasting $10\,\text{s}$ and taken for the CMU[1] dataset. They are sampled at $16\,\text{kHz}$ and placed randomly in a simulated room of dimensions $5 \times 4 \times 3\,\text{m}$, featuring up to $K = 50$ omnidirectional

---

[1] Carnegie Mellon University dataset: http://www.festvox.org/cmu_faf/.

microphones at random positions. The room impulse responses are obtained with the RIR[2] generator toolbox [2] by simulated a 0.4 s reverberation time. Because of computational cost, the sources' positions are restrained to lie in a flat 2D surface that is 1.5 m high. All source localization methods operate with a grid of 10 cm step-size, located on the source plane, but which does not contain the exact sources' locations. This results in $L = 2091$ candidate locations. To optimize computational cost, the frequency range considered was reduced from 1 kHz to 3 kHz, since it proved sufficient for speech signals. The different techniques compared are the following ones:

DSM The Discrete Spatial Measure (proposed). We take $\alpha = 1$, corresponding to the Cauchy distribution [9], $\lambda = 1$ for sparsity regularization and we pick the Itakura-Saito divergence $\beta = 0$ as the NMF cost function.

SRP The Steering Response Power, also called delay-and-sum [18], is the most classical source localization approach. It is based on the nearfield propagation model (2) and projects the STFT of observations on the steering vectors: $\forall l,\ SRP = \frac{1}{FT} \sum_{f,t} \frac{|A_l^\star(f)x(f,t)|}{\|A_l(f)\|}$. We use the same frequency range for SRP as for the proposed method.

CLEAN is a greedy algorithm [17]: at one iteration, it successively identifies the strongest source in the grid with SRP, and removes its contribution. The algorithm is repeated until all sources are identified.

RELAX is an enhanced variation of CLEAN [6] presented in [17].

A Monte Carlo simulation is carried out with arrays of $K = 5, 10, 20$ and 50 microphones. For each array configuration, we perform 50 trials with random positions of the sources on the 5×4 m source plane. One trial for $K = 50$ is illustrated in Fig. 1, showing the estimated heat-maps. It first demonstrates that DSM is more accurate with better contrast than SRP, with only a slight increase in computational cost[3]. Indeed, the energy is focused on the ground

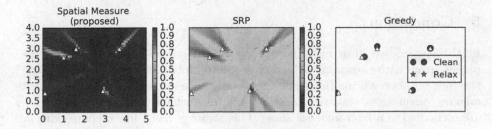

Fig. 1. Heat maps of spatial measure, SRP and both greedy algorithms.

[2] https://www.audiolabs-erlangen.de/fau/professor/habets/software/rir-generator.

[3] In the specific case where $J = 5$, the computation time of each method are 5.2 s for SRP, 54 s for DSM (comprising 24 s for computing Ψ, which only needs to be done once). CLEAN and RELAX are implemented in GPU, and respectively need 0.45 s and 55 s. Note that the complexity of these two last methods depends on the a priori number of sources J and that our implementation for DSM did not exploit its highly parallelisable capabilities.

truth positions and is close to 0 elsewhere, whereas the SRP map is noisier because of side lobes. Since CLEAN and RELAX exactly look for $J = 5$ sources, they result in the sparsest representations.

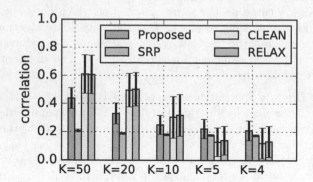

Fig. 2. Correlation with ground truth. Deviation is depicted with black whiskers.

The Monte Carlo experiment is evaluated by correlations between the estimated maps and the ground truth map. The latter is built by setting ones at ground truth positions, and zeros elsewhere, followed by a Gaussian smoothing with a 10 cm length-scale. Correlation means and standard deviations along the 50 trials are depicted in Fig. 2. First, it shows that DSM outperforms SRP in all cases. For $K \geq 10$ microphones, CLEAN and RELAX have the highest correlation, notably thanks to the a priori on the source number J. However their performance decreases rapidly when K decreases. On the contrary, DSM performance appears more robust to a decrease of K. Lastly, the standard deviation of DSM is smaller than that of CLEAN/RELAX, showing that it also has a more stable behavior at different configurations.

5   Conclusion

In this paper, we have introduced an acoustic imaging method for microphone arrays with known but arbitrary geometry. Interestingly, it requires going through the observed multichannel signals only once in order to compute a fixed amount of sufficient statistics called *sketch* from which the model parameters are estimated in a later analysis stage. This strategy has a linear complexity in terms of signal duration.

A fundamental feature of the probabilistic α-stable model we use is to describe the source emitting at each spatial location using a single scale parameter. This is possible because α-stable distributions correctly account for the marginal distribution of an acoustic signal in the Time-Frequency plane. Gathering all these location-specific scale parameters, we defined the Discrete Spatial Measure (DSM) and showed how it can be very easily estimated based on the sketch with a simple matrix factorization procedure.

In a very challenging simulation of heavily reverberant environments, the DSM method proved competitive with state-of-the-art methods, particularly when the number of microphones is comparable with the number of sources. Open directions include incorporating time-varying scale parameters and experimentally validating robustness to noise.

Acknowledgments. This work was partly supported by the research programme KAMoulox (ANR-15-CE38-0003-01) and EDiSon3D (ANR-13-CORD-0008-01) funded by ANR, the French State agency for research.

References

1. Févotte, C., Idier, J.: Algorithms for nonnegative matrix factorization with the β-divergence. Neural Comput. **23**(9), 2421–2456 (2011)
2. Habets, E.A.P.: Room impulse response (RIR) generator (2008)
3. Ito, N., Araki, S., Nakatani, T.: Complex angular central Gaussian mixture model for directional statistics in mask-based microphone array signal processing. In: 2016 24th European Signal Processing Conference (EUSIPCO), pp. 1153–1157. IEEE (2016)
4. Keriven, N., Bourrier, A., Gribonval, R., Pérez, P.: Sketching for large-scale learning of mixture models. In: Proceedings of IEEE International Conference on Acoustics, Speech and Signal Processing (ICASSP), pp. 6190–6194, May 2016
5. Kundu, T.: Acoustic source localization. Ultrasonics **54**(1), 25–38 (2014)
6. Li, J., Stoica, P.: Efficient mixed-spectrum estimation with applications to target feature extraction. IEEE Trans. Signal Process. **44**(2), 281–295 (1996)
7. Liutkus, A., Badeau, R.: Generalized Wiener filtering with fractional power spectrograms. In: Proceedings of IEEE International Conference on Acoustics, Speech and Signal Processing (ICASSP), pp. 266–270, April 2015
8. Liutkus, A., Badeau, R., Richard, G.: Gaussian processes for underdetermined source separation. IEEE Trans. Signal Process. **59**(7), 3155–3167 (2011)
9. Liutkus, A., Fitzgerald, D., Badeau, R.: Cauchy nonnegative matrix factorization. In: Proceedings of IEEE Workshop on Applications of Signal Processing to Audio and Acoustics (WASPAA), pp. 1–5, October 2015
10. Ma, N., Goh, J.: Ambiguity-function-based techniques to estimate DOA of broadband chirp signals. IEEE Trans. Signal Process. **54**(5), 1826–1839 (2006)
11. Mandel, M.I., Weiss, R.J., Ellis, D.P.W.: Model-based expectation-maximization source separation and localization. IEEE Trans. Audio Speech Lang. Process. **18**(2), 382–394 (2010)
12. Nikunen, J., Virtanen, T.: Direction of arrival based spatial covariance model for blind sound source separation. IEEE/ACM Trans. Audio Speech Lang. Process. **22**(3), 727–739 (2014)
13. Samoradnitsky, G., Taqqu, M.: Stable Non-Gaussian Random Processes: Stochastic Models with Infinite Variance, vol. 1. CRC Press, Boca Raton (1994)
14. Schmidt, R.: Multiple emitter location and signal parameter estimation. IEEE Trans. Antennas Propag. **34**(3), 276–280 (1986)
15. Sheng, X., Hu, Y.: Maximum likelihood multiple-source localization using acoustic energy measurements with wireless sensor networks. IEEE Trans. Signal Process. **53**(1), 44–53 (2005)

16. Stoica, P., Moses, R.: Introduction to spectral analysis, vol. 1. Prentice Hall, Upper Saddle River (1997)
17. Wang, Y., Li, J., Stoica, P., Sheplak, M., Nishida, T.: Wideband RELAX and wideband CLEAN for aeroacoustic imaging. J. Acoust. Soc. Am. **115**(2), 757–767 (2004)
18. Williams, E.: Fourier Acoustics: Sound Radiation and Nearfield Acoustical Holography. Academic Press, London (1999)
19. Zotkin, D.N., Duraiswami, R.: Accelerated speech source localization via a hierarchical search of steered response power. IEEE Trans. Speech Audio Process. **12**(5), 499–508 (2004)

Acoustic DoA Estimation by One Unsophisticated Sensor

Dalia El Badawy[1(✉)], Ivan Dokmanić[2], and Martin Vetterli[1]

[1] École Polytechnique Fédérale de Lausanne, Lausanne, Switzerland
{dalia.elbadawy,martin.vetterli}@epfl.ch
[2] University of Illinois at Urbana-Champaign, Champaign, USA
dokmanic@illinois.edu

Abstract. We show how introducing known scattering can be used in direction of arrival estimation by a single sensor. We first present an analysis of the geometry of the underlying measurement space and show how it enables localizing white sources. Then, we extend the solution to more challenging non-white sources like speech by including a source model and considering convex relaxations with group sparsity penalties. We conclude with numerical simulations using an unsophisticated sensing device to validate the theory.

Keywords: Monaural localization · Compressed sensing · Direction of arrival · Group sparsity · Scattering · Sound source localization

1 Introduction

Walking down a street, we (or a cat) are able to tell where a bird song is coming from. Perhaps it helps that we know birds live in trees, but it is the auditory scene analysis performed by the brain that enables us to almost instantaneously determine the direction of arrival (DoA), even for multiple sound sources [10]. In this paper, we study computational DoA estimation with a single sensor, a task usually referred to as monaural sound source localization. We begin by a brief review of the biological mechanisms from which we draw some inspiration.

First of all, we have two ears. Sound reaches each ear at a slightly different time and loudness providing us with so-called binaural cues. The shape of the outer ear as well as the shape of the head and torso additionally modify the sound as it reaches our ears, and thus provide us with *monaural* cues. These cues are encoded by the head-related transfer function (HRTF) [1]. Both types of cues are necessary for accurate localization. Indeed, obstructing one ear hurts the localization accuracy [7]. However, monaural localization is still possible, though it is known that monaurally deaf people usually require certain prior knowledge about the source to be localized [11]. We will see that (unless the sources are white), the same is true of algorithms we propose.

Consider a generalized *ear*, a sensor with the directional frequency response $a(\omega; \theta)$ for sounds arriving from direction θ at a frequency ω. For J sources

© Springer International Publishing AG 2017
P. Tichavský et al. (Eds.): LVA/ICA 2017, LNCS 10169, pp. 89–98, 2017.
DOI: 10.1007/978-3-319-53547-0_9

emitting from directions $\Theta = \{\theta_1, \ldots, \theta_J\}$, what we measure at the single sensor is

$$y(\omega) = \sum_{j=1}^{J} a(\omega; \theta_j) s_j(\omega) + n(\omega), \qquad (1)$$

where s_j are the source spectra and n is the measurement noise which will be ignored in the large part of the ensuing discussion.

DoA estimation is then an inverse problem concerned with mapping the measurement back to the directions Θ. The properties of the directional response a are key in determining whether it can be successful. For instance, if our sensor is omnidirectional, then $a(\omega, \theta) = 1$ for all frequencies ω and directions θ, and no directional information is present. That is, the measurement remains the same even if the sources are rotated to different directions. Thus, we would prefer that a are diverse and act as distinguishable spectral signatures for their corresponding directions. Still, as can be seen from (1), the inverse problem is ill-posed since decomposing y back into a sum of products has infinitely many solutions. This ill-posedness can be resolved by a combination of scattering and proper source modeling.

Requiring that the responses a be diverse is similar to the HRTF case where for each ear, the frequency response differs with the angle of arrival. An especially interesting HRTF is that of a cat: it features prominent notches at frequencies that depend on the direction of arrival [10].[1] In fact, notches are one of two possibilities to get strong diversity, the other being resonances. They both enable localization of wideband sources, but while in enclosures such as rooms, they are easy to obtain and have been successfully used for localization [2], they otherwise require special design. For example, resonances were obtained in recent work [14] with a metamaterial-coated device which was then used to localize noise. Similarly, diversity of a was achieved in [9] using several microphone enclosures which were designed and tested for localizing a single sound source. In our prior work [3], we used a randomly shaped device to introduce random scattering and showed that noise can be localized without a source model. While all the latter work relies on the idea of a directional spectral signature, it was not made precise why or how such spectral signatures are good for DoA estimation. As we will show, whereas any *incoherence* of a is sufficient to localize noise sources, in order to compensate for the lack of diversity and to handle complex sound sources, an adequate source model is required, for example, a Hidden Markov Model [9] or a dictionary [14].

In this paper, we achieve the desired a by scattering by a very simple, haphazard structure. Unlike prior work, we show in Sect. 2 that the underlying principle that makes scattering useful requires neither a sophisticated sensing device nor a source model to localize noise. The geometry of the problem suggests a matched field processing approach [13] to DoA estimation, which has reasonable complexity for few sources. Then in Sect. 3, we turn to sparse reconstruction techniques with group sparsity penalties that can be optimized efficiently. Beyond having

---

[1] This is specific for localization in elevation.

controlled complexity for larger numbers of sources, this more sophisticated formulation also allows us to include more general source models like dictionaries. Finally, we present numerical results in Sect. 4.

2   Localization of Noise Sources

We assume having a set of D possible directions in the azimuth interval $[0, 2\pi)$ for which we know the sensing functions $a_1, a_2, \ldots, a_D$. Further, we choose a set of F frequencies at which we examine the recorded signal; this can be done through a filterbank of F narrowband filters. We can then re-write (1) as

$$y = \sum_{j \in \Theta} a_j \odot s_j + n, \tag{2}$$

where $y \in \mathbb{C}^F$, $a_j \in \mathbb{C}^F$, $s_j \in \mathbb{C}^F$, and $\odot$ denotes the Hadamard product. We think of y as corresponding to one audio frame.

2.1   Geometrical Structure

White. In the presence of $J \geq 1$ independent white sources at locations Θ, the expected power of the frame y from (2) is

$$\mathbb{E}[|y|^2] = \sum_{j \in \Theta} \sigma_j^2 |a_j|^2, \tag{3}$$

where σ_j^2 is the power of the jth source and we again set $n = 0$. Thus, even if σ_j^2 are unknown, we see that the measured power spectrum is, in expectation, a positive linear combination of the power spectra of the sensing vectors, with coefficients being the source powers. Put differently, all power measurements arising from a certain configuration Θ lie in a cone characterized by the corresponding sensing vectors:

$$\mathbb{E}[|y|^2] \in C_\Theta = \{ w \mid w = \sum_{j \in \Theta} p_j |a_j|^2, \ p_j \geq 0 \}$$

as shown in Fig. 1. The entire space of measurements, for all possible configurations, is a union of those cones. It follows that if we can find the right cone, we will have identified the source locations. More precisely, the source localization task amounts to identifying which of the cones $\{ C_\Theta \mid \Theta \text{ a set of } J \text{ directions} \}$ contains $\mathbb{E}[|y|^2]$ or its empirical estimate. Without scattering, the measurement space is collapsed into a single cone.

Color. Unlike white, colored sources will modulate the sensing functions and move them about in space as seen in (2):

$$\mathbb{E}[|y|^2] = \sum_{j \in \Theta} \mathbb{E}[|s_j|^2] \odot |a_j|^2 = \sum_{j \in \Theta} \sigma_j^2 |b_j|^2 \odot |a_j|^2, \tag{4}$$

where $|b_j|^2$ are the prototype power spectra for each source. Consequently, without knowing the modulation, we cannot identify the cones for localization. So if we know the sources' power spectral density or simply the time-varying power spectra, we are again in similar situation as for white sources except that the number of cones increases due to ambiguities in assigning $|b_j|$ to directions.

2.2 Structure Quality

Not all unions of cones are created equal. To ensure that we correctly identify the cone and hence solve the localization problem, we require adequate separation between the different cones. Thus, we examine the angles between every pair of cones (for a certain number of sources J) as illustrated in Fig. 1a. Consider two cones C_Θ and C_Φ for two sets of J directions Θ and Φ. The largest angle between them is

$$\check{\alpha} = \max_{\substack{\mathbf{p}\in C_\Theta, \mathbf{q}\in C_\Phi \\ \|\mathbf{p}\|=\|\mathbf{q}\|=1}} \cos^{-1}\langle \mathbf{p}, \mathbf{q}\rangle. \tag{5}$$

For simplicity, instead of the inter-cone angle $\check{\alpha}$, we will in the following look at the maximal angle between the smallest subspaces that contain the cones. For this to make sense, we need to assume that $J < F$ since otherwise cones will lie in the same subspace. We note that this relaxation will then give us sufficient conditions for localization.

Denote the orthonormal bases for the smallest subspaces containing C_Θ and C_Φ by $\mathbf{B}_\Theta$ and $\mathbf{B}_\Phi$, and define the largest angle as

$$\alpha = \cos^{-1}\sigma_{\min}(\mathbf{B}_\Theta^{\mathrm{T}}\mathbf{B}_\Phi), \tag{6}$$

where $\sigma_{\min}(.)$ denotes the smallest singular value. We do not consider the smaller angles because what matters is that the two cones are distinct i.e., the largest angle is non-zero. If the smaller angles include zero, it means that the cones intersect; by definition the cones here indeed intersect at exactly the sensing

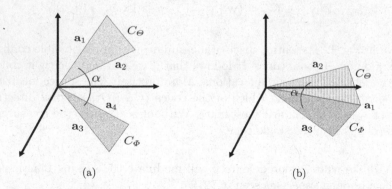

(a)   (b)

Fig. 1. Cones in the measurement space. (a) The angle between two cones formed by a pair of different sensing vectors. (b) Two distinct cones that share a sensing vector.

vectors. For example as shown in Fig. 1b, consider the following two cones $C_\Theta = \{\mathbf{w} = p\mathbf{a}_1 + q\mathbf{a}_2,\ p, q \geq 0\}$ and $C_\Phi = \{\mathbf{w} = p\mathbf{a}_1 + q\mathbf{a}_3,\ p, q \geq 0\}$, then $C_\Theta \cap C_\Phi = \{p\mathbf{a}_1,\ p \geq 0\}$.

The smaller the angle α, the more sensitive the sensing device is to noise. Hence, a good set of sensing functions are ones that result in large angles between every pair of cones. Thus, we are interested in the worst-case angle or alternatively the worst-case coherence between the cones which we define as

$$\mu_J = \max_{\Theta \neq \Phi} \sigma_{\min}(\mathbf{B}_\Theta^\mathsf{T} \mathbf{B}_\Phi), \tag{7}$$

where $\sigma_{\min}(.)$ denotes the smallest singular value. For the case of a single white source $J = 1$, (7) reduces to conventional coherence in the power domain

$$\mu_1 = \max_{i \neq j} \frac{\langle |\mathbf{a}_i|^2, |\mathbf{a}_j|^2 \rangle}{\||\mathbf{a}_i|^2\| \||\mathbf{a}_j|^2\|}. \tag{8}$$

The lower the coherence, the better. Nevertheless as we will see next, in the noiseless case, a sufficient condition for the accurate localization of any number of sources J is to simply have the corresponding coherence $\mu_J < 1$.

2.3   Conditions for Localization

We now turn our attention to the actual localization problem. Let $\mathbf{y} := |y|^2$ denote the power spectrum of y. Based on the analysis in Sect. 2.1, we have $\mathbb{E}[\mathbf{y}] \in C_\Theta$ and accordingly $\mathbb{E}[\mathbf{y}] = \mathbf{B}_\Theta \mathbf{w}$. Then, we can write $\mathbf{y} = \mathbf{P}_\Theta \mathbf{y} + (\mathbf{I} - \mathbf{P}_\Theta)\mathbf{y}$ where $\mathbf{P}_\Theta$ denotes the projection onto range($\mathbf{B}_\Theta$). For a particular realization, the error vector $\mathbf{z}_\Theta = (\mathbf{I} - \mathbf{P}_\Theta)\mathbf{y}$ will be non-zero, but by the law of large numbers, its average over many frames will converge to zero.

Thus, a straightforward akin to matched field processing is to calculate the sample mean of N power frames and test it against every cone: perform an exhaustive search for the right match as determined by the minimum distance to the cone (more precisely, the corresponding subspace) of the empirical power mean. This procedure is summarized in Algorithm 1.

Algorithm 1. DoA estimation of J sources

Input: Number of sources J, bases $\mathbf{B}_\Theta\ \forall \Theta$, $|\Theta| = J$, N power frames $\mathbf{y}_n$ for $n = 1, \ldots, N$.
Output: Directions of arrival $\Theta^* = \{\theta_1^*, \ldots, \theta_J^*\}$.
  Compute $\widetilde{\mathbf{y}} = \frac{1}{N} \sum_{n=1}^N \mathbf{y}_n$
  $\Theta^* = \arg\min_\Theta \|(\mathbf{I} - \mathbf{B}_\Theta \mathbf{B}_\Theta^T)\widetilde{\mathbf{y}}\| = \arg\max_\Theta \|\mathbf{B}_\Theta^T \widetilde{\mathbf{y}}\|$

Algorithm 1 relies on the law of large numbers to justify using the empirical mean in lieu of the expectation, but the whole discussion has made no mention of noise. The following proposition suggests that the localization will be correct even with measurement noise as long as a certain relationship holds between the signal-to-noise ratio (SNR) in the power domain and the worst-case coherence.

Proposition 1 (Correct localization). *Assuming J independent sources, let the source configuration be specified by the cone C_Θ and denote by $\widetilde{\mathbf{y}}$ the sample mean of N independent power frames. Consider further a zero-mean noise term* n *independent of the source signals, with power $\mathbf{n} := |\mathsf{n}|^2$. Then as long as the SNR in the power domain exceeds $\|\mathbb{E}[\sum_{j\in\Theta}\mathbf{a}_j \odot \mathbf{s}_j]\|/\|\mathbb{E}[\mathbf{n}]\| > \sqrt{2/(1-\mu_J)}$, localization by Algorithm 1 with input $\widetilde{\mathbf{y}}$ is correct with arbitrarily high probability for a sufficiently large N.*

Proof (sketch). Suppose first that we can measure the expected value of the power measurements $\mathbb{E}[\mathbf{y}] = \mathbb{E}[\mathbf{x}] + \mathbb{E}[\mathbf{n}]$, where $\mathbf{x} := \sum_j \mathbf{a}_j \odot \mathbf{s}_j$. In Algorithm 1, we take subspace membership as a proxy to cone membership, meaning that we will have localized correctly as long as $\mathbb{E}[\mathbf{y}]$ is closer to the true subspace range$(\mathbf{B}_\Theta)$ than to any other range$(\mathbf{B}_\Phi)$; equivalently, we ask that $\langle \mathbb{E}[\mathbf{y}], \widehat{\mathbf{P}_\Theta \mathbb{E}[\mathbf{y}]}\rangle > \langle \mathbb{E}[\mathbf{y}], \widehat{\mathbf{P}_\Phi \mathbb{E}[\mathbf{y}]}\rangle$, where $\widehat{\mathbf{u}} := \frac{\mathbf{u}}{\|\mathbf{u}\|}$. This can be rewritten as (denoting $\overline{\mathbf{u}} := \mathbb{E}[\mathbf{u}]$ and setting $\widetilde{\mu} := \langle \widehat{\mathbf{P}_\Theta \overline{\mathbf{y}}}, \widehat{\mathbf{P}_\Phi \overline{\mathbf{y}}}\rangle$):

$$\langle \mathbf{P}_\Theta \overline{\mathbf{y}} + \overline{\mathbf{n}}, \widehat{\mathbf{P}_\Theta \overline{\mathbf{y}}}\rangle > \langle \mathbf{P}_\Theta \overline{\mathbf{y}} + \overline{\mathbf{n}}, \widehat{\mathbf{P}_\Phi \overline{\mathbf{y}}}\rangle$$

$$\Leftrightarrow \|\mathbf{P}_\Theta \overline{\mathbf{y}}\| + \langle \widehat{\mathbf{P}_\Theta \overline{\mathbf{y}}}, \overline{\mathbf{n}}\rangle > \|\mathbf{P}_\Theta \overline{\mathbf{y}}\|\widetilde{\mu} + \langle \widehat{\mathbf{P}_\Phi \overline{\mathbf{y}}}, \overline{\mathbf{n}}\rangle$$

$$\Leftrightarrow \|\mathbf{P}_\Theta \overline{\mathbf{y}}\|(1-\widetilde{\mu}) > \langle \widehat{\mathbf{P}_\Phi \overline{\mathbf{y}}} - \widehat{\mathbf{P}_\Theta \overline{\mathbf{y}}}, \overline{\mathbf{n}}\rangle$$

$$\Leftarrow \|\mathbf{P}_\Theta \overline{\mathbf{y}}\|(1-\widetilde{\mu}) \overset{(a)}{>} \|\widehat{\mathbf{P}_\Phi \overline{\mathbf{y}}} - \widehat{\mathbf{P}_\Theta \overline{\mathbf{y}}}\|\|\overline{\mathbf{n}}\|$$

$$\Leftrightarrow \|\mathbf{P}_\Theta \overline{\mathbf{y}}\|(1-\widetilde{\mu}) \overset{(b)}{>} \sqrt{2}\sqrt{1-\widetilde{\mu}}\|\overline{\mathbf{n}}\| \Leftarrow \frac{\|\mathbf{P}_\Theta \overline{\mathbf{y}}\|}{\|\overline{\mathbf{n}}\|} \overset{(c)}{>} \sqrt{\frac{2}{1-\mu_J}}$$

where in (a) we used the Cauchy-Schwarz inequality to upper bound the right-hand side, (b) we used the law of cosines and (c) we used the fact that the function $\frac{1}{\sqrt{1-t}}$ is increasing with $t \in [0,1)$ to replace $\widetilde{\mu}$ with the worst-case μ_J. Convergence in probability then follows from replacing expectations by empirical means and invoking the weak law of large numbers. □

Note that Proposition 1 does not quantify the number of frames N required to guarantee correct localization. However, the concentration of measure phenomenon for the Lipschitz $\|\cdot\|$ suggests that N is tightly controlled [6].

3   Algorithms

With the described matched field processing approach, it is not straightforward to use more complex source models such as overcomplete dictionaries. Moreover, for large D and J, computational complexity makes the search unfavorable. Therefore, in this section, we resort to convex relaxations for sparse recovery which can be optimized efficiently.

Let $\mathbf{a}_j \in \mathbb{R}_+^F$ denote the power spectrum of the jth sensing function. Let $\mathbf{V} \in \mathbb{R}_+^{F\times K}$ be the source model such that $\mathbf{s}_j = \mathbf{V}\mathbf{x}_j$, e.g. a subspace basis or an overcomplete dictionary. Then we can write

$$\mathbf{y} = \mathbf{A}\mathbf{x} + \mathbf{z}, \tag{9}$$

where $\mathbf{A} \in \mathbb{R}_+^{F \times KD} = [\text{diag}(\mathbf{a}_1)\mathbf{V}, \ldots, \text{diag}(\mathbf{a}_D)\mathbf{V}]$, $\mathbf{x} \in \mathbb{R}_+^{KD}$ is a vector of concatenated source coefficients $\mathbf{x}_j \in \mathbb{R}_+^K$ and $\mathbf{z} \in \mathbb{R}_+^F$ is a term grouping all the cross-terms which arise when calculating the power of y (2).

Since the system of equations in (9) is underdetermined, we consider the solution of the following optimization problem

$$\min_{x \geq 0} \quad \frac{1}{2}\|\mathbf{y} - \mathbf{Ax}\|_2^2 + \Psi(\mathbf{x}), \tag{10}$$

where the first term is the data fidelity and Ψ is an appropriate regularization. The choice of Ψ is inspired by the underlying geometrical structure and is discussed in the following sections.

Once we solve for $\mathbf{x}$, localization amounts to finding the J direction indices corresponding to the $\mathbf{x}_j$ with the highest norms $\|\mathbf{x}_j\|_2$.

3.1 Subspace Model

The appropriate regularization for signals from a union of cones is to enforce group sparsity, i.e., only few $\mathbf{x}_j$ are non-zero. The ℓ_1/ℓ_2 penalty known to promote group sparsity [15] is defined as

$$\Psi(\mathbf{x}) = \lambda \sum_{j=1}^{D} \|\mathbf{x}_j\|_2, \tag{11}$$

where $\lambda > 0$ determines the weight of the penalty.

The source model for white sources is one-dimensional i.e., $\mathbf{V} = \mathbf{1}$ and thus Ψ reduces to the ℓ_1 penalty. We emphasize that in that case we do not need an explicit source model and only require knowledge of the sensing vectors where $\mathbf{A} = [\mathbf{a}_1, \ldots, \mathbf{a}_D]$.

3.2 Dictionary Model

For colored sources, we consider using an overcomplete dictionary (i.e., $K > F$) to represent their time-varying power spectra. The dictionary is chosen such that every source admits a sparse representation and while we still have a union of cones structure, the elements in the union depend dynamically on the sources being localized and are not known a priori. Thus, to appropriately select the right subset, we add the ℓ_1 penalty

$$\Psi(\mathbf{x}) = \lambda \sum_{j=1}^{D} \|\mathbf{x}_j\|_2 + \gamma \|\mathbf{x}\|_1, \tag{12}$$

where $\lambda > 0$ and $\gamma > 0$ are the trade-off parameters determining the weights of their respective terms. This penalty (12) promotes sparsity across groups and within active groups. The corresponding objective is known as the sparse-group lasso [4] which we augment by the non-negativity constraint.

4   Numerical Results

In this section, we present numerical results for 2D DoA estimation in a 3D environment using a simulated 3D model of a randomly shaped sensing device. The sensing device consists of an omnidirectional sensor surrounded by 7 cubes of randomly chosen sizes (side lengths $\in [10, 14]$ cm) and orientations, spread over an area $60 \times 60\,\text{cm}^2$ as shown in Fig. 2a. The mesh was generated using Gmsh [5] and the directional frequency responses were calculated using the boundary element method package BEM++ [12]. Taking into consideration the sizes of the cubes, we use 193 frequencies between 2000 Hz and 8000 Hz which are most affected by the scattering. The power spectra of the sensing vectors for 36 directions equally spaced in the interval $[0°, 360°)$ are shown in Fig. 2b.

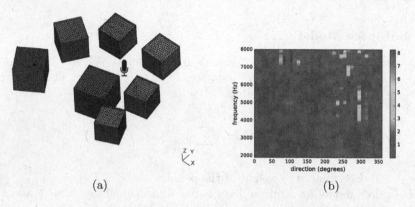

(a)   (b)

Fig. 2. The sensing device. (a) Illustration of the sensing device consisting of 7 cubes surrounding a microphone. (b) The corresponding transfer functions per direction.

To add modeling mismatch in the simulations, the sources are randomly placed at a $\pm 1°$ shift from the assumed model. We implemented consensus ADMM [8] to solve (10). Finally, we consider the localization successful when the estimate is the closest shift for all J sources.

4.1   White Sources

First we show that we can localize white sources without having an explicit dictionary or knowing the distribution parameters. The corresponding coherences of our sensing device are $\mu_1 = 0.88$ and $\mu_2 = 0.93$ so Proposition 1 guarantees perfect localization of one and two sources. We simulate one and two white Gaussian (zero-mean unit-variance) and Bernoulli ($p_{\text{success}} = 0.5$) sources at all 36 directions. We solve the non-negative lasso (11) with $\lambda = 10$. The rate of successful localization averaged over 10 runs is shown in Table 1. We conjecture that any error is strictly due to the modeling mismatch.

4.2   Speech Sources

Next we show preliminary results for localization of one or two speakers with the help of a dictionary. Four speakers (two female, two male) were randomly chosen from the TIMIT speech corpus; the maximum amplitudes were normalized to 1 before computing the frequency representation. Every speaker emits 100 frames. As discussed in Sect. 2.1, colored sources require some prior knowledge. Thus, we assume knowing the power spectra for each speaker's frames where $\mathbf{V} = [\mathbf{V}_1, \mathbf{V}_2, \mathbf{V}_3, \mathbf{V}_4]$ and $\mathbf{V}_i \in \mathbb{R}_+^{F \times 100}$; this is similar to what was done in [14] and we leave for future work incorporating a more general learned speech dictionary. We solve the non-negative sparse group lasso (12) with $\lambda = 0.1$ and $\gamma = 0.1$. The average success rates are shown in Table 1. First, we note how it still possible to perfectly localize one speaker even at a very high coherence. In two-source cases, one source was almost always localized accurately (in 99 % of all cases). Second, the lower performance for localizing two sources compared to the white case is likely due to the higher coherence μ_2. In particular, the lower performance in localization of male speakers can probably be attributed to unfavorable interplay between the structure response and the source spectrum. It remains, however, to be completely explained.

Table 1. Success rates for DoA estimation of one or two sources

Type	Success rate
One Gaussian source	100%
One Bernoulli source	100%
Two Gaussian sources	86.7%
Two Bernoulli sources	86.7%
One female speaker	100%
One male speaker	100%
Two speakers (female)	75.9%
Two speakers (male)	41.7%
Two speakers (female & male)	41%

5   Conclusion

In conclusion, we demonstrated the potential of using a sensing device that introduces known scattering in the measurements for DoA estimation. In particular, we showed that the scattering induces a union of cones structure which allows us to localize any number of white sources in the noiseless case granted the coherence is strictly less than 1. We then showed that with the proper modeling, in the form of an overcomplete dictionary and group sparsity penalties, we are able to localize more challenging sources like speech, all while using a single sensor with what may be considered a rather poor response, corrupted by scattering off of random clutter. Future work includes running a real-world experiment and using a general learned dictionary as well as extending the approach to handle reverberation.

References

1. Blauert, J.: Spatial Hearing: The Psychophysics of Human Sound Localization. The MIT Press, Cambridge (1997)
2. Dokmanić, I.: Listening to Distances and Hearing Shapes: Inverse Problems in Room Acoustics and Beyond. Ph.D. thesis, École polytechnique fédérale de Lausanne (2015)
3. El Badawy, D.: Acoustic sensing using scattering microphones. Master's thesis, École polytechnique fédérale de Lausanne, July 2015
4. Friedman, J., Hastie, T., Tibshirani, R.: A note on the group lasso and a sparse group lasso. arXiv (2010)
5. Geuzaine, C., Remacle, J.F.: Gmsh: A 3-D finite element mesh generator with built-in pre- and post-processing facilities. Int. J. Numer. Meth. Eng. **79**(11), 1309–1331 (2009)
6. Ledoux, M.: The concentration of measure phenomenon. Mathematical Surveys and Monographs. American Mathematical Society, Providence (R.I.) (2001)
7. Lessard, N., Pare, M., Lepore, F., Lassonde, M.: Early-blind human subjects localize sound sources better than sighted subjects. Nature **395**(6699), 278–280 (1998)
8. Parikh, N., Boyd, S.: Proximal algorithms. Found. Trends Optim. **1**(3), 123–231 (2014)
9. Saxena, A., Ng, A.: Learning sound location from a single microphone. In: IEEE International Conference on Robotics and Automation (ICRA), pp. 1737–1742 (2009)
10. Schnupp, J., Nelken, I., King, A.: Auditory Neuroscience: Making Sense of Sound. The MIT Press, Cambridge (2010)
11. Slattery, W.H.I., Middlebrooks, J.C.: Monaural sound localization: acute versus chronic unilateral impairment. Hear. Res. **139**(6), 38–46 (1994)
12. Śmigaj, W., Betcke, T., Arridge, S., Phillips, J., Schweiger, M.: Solving boundary integral problems with BEM++. ACM Trans. Math. Softw. **41**(2), 6:1–6:40 (2015)
13. Tolstoy, A.: Applications of matched-field processing to inverse problems in underwater acoustics. Inverse Prob. **16**(6), 1655 (2000)
14. Xie, Y., Tsai, T., Konneker, A., Popa, B., Brady, D.J., Cummer, S.A.: Single-sensor multispeaker listening with acoustic metamaterials. Proc. Natl. Acad. Sci. **112**(34), 10595–10598 (2015)
15. Yuan, M., Lin, Y.: Model selection and estimation in regression with grouped variables. J. R. Stat. Soc. Ser. B (Statistical Methodology) **68**(1), 49–67 (2006)

Acoustic Source Localization by Combination of Supervised Direction-of-Arrival Estimation with Disjoint Component Analysis

Jörn Anemüller[✉] and Hendrik Kayser

Medical Physics Unit and Cluster of Excellence Hearing4all, Computational Audition Group, Carl von Ossietzky Universität Oldenburg, 26111 Oldenburg, Germany
joern.anemueller@uol.de

Abstract. Analysis and processing in reverberant, multi-source acoustic environments encompasses a multitude of techniques that estimate from sensor signals a spatially resolved "image" of acoustic space, a high-level representation of physical sources that consolidates several source components into a single sound object, and the estimation of filter parameters that would permit enhancement of target and attenuation of interfering signal components.

The contribution of the present manuscript is the introduction of a combination of different algorithms from the field of supervised learning, unsupervised subspace decomposition and multi-channel signal enhancement to accomplish these goals.

Specifically, we propose a system that (1) uses a bank of trained support vector machine classifiers to estimate source activity probability for each spatial position and (2) employs disjoint component analysis (DCA) to obtain from this probabilistic spatial source activity map those components that pertain to individual sound objects. We conclude with a brief outline for (3) estimation of multi-channel filter parameters based on DCA components in order to perform target source enhancement.

We illustrate the proposed method with decomposition results obtained with a four-channel hearing aid geometry setup that comprises two localized sources plus isotropic background noise in an anechoic environment.

1 Introduction

Acoustic source localization is a common task in everyday acoustic communication and an important component of computational auditory scene analysis systems, e.g., [8,11,12]. It distinguishes spatially localized acoustic sources from each other and from diffuse, non-localized sounds such as background noise.

Estimation of the time difference of arrival (TDOA) between multiple microphone signals is the dominant approach in acoustic localization. The generalized cross-correlation (GCC) function, introduced in [10], together with the phase transform frequency weighting permits robust estimation of inter-microphone

© Springer International Publishing AG 2017
P. Tichavský et al. (Eds.): LVA/ICA 2017, LNCS 10169, pp. 99–108, 2017.
DOI: 10.1007/978-3-319-53547-0_10

propagation differences. In previous work, we adopted a data-driven, training-based approach to source localization where optimal decision functions, indicating presence probability of a source at a given position, are learned from real measurement data without the need to compute a precise acoustic propagation model [6].

Individual physical sources in reverberant conditions, however, are in general not characterized by sound impinging from a single direction (with a single TDOA), but may also include components from, e.g., reverberation or reflections. Thus, a "grouping" of TDOA component pertaining to a single physical source is a key requirement for proper analysis of acoustic scenes. It forms the major part of the present work and is accomplished by disjoint component analysis (DCA, [2]). In addition, we propose to leverage knowledge of DCA component activations to better estimate parameters of the spatial sound field that are required in order to perform signal enhancement.

2   Methods

The method proposed here is constituted of three different building blocks, cf. Fig. 1. Multi-channel microphone signals are processed by a support vector machine-based speech source localization algorithm that transforms observed inter-channel cross-correlation features into a spatio-temporal probabilistic map. This provides an estimate of the a-posteriori probability for localized speech activity at each spatial position and time, cf. Sect. 2.1.

The main contribution of the present manuscript is the subsequent step of decomposing the probabilistic map into components that each pertain to an individual acoustic source, cf. Sect. 2.2. As the probabilistic map derived in the first step is agnostic as to whether, e.g., high a-posteriori estimates at different spatial positions belong to the same physical source, e.g., through reverberation added to a point source or by means of a spatially distributed ambient sound field, this second step is crucial for forming spatial "images" pertaining to a discrete number of acoustic sources.

Subsequent signal enhancement (Sect. 2.4) could operate using the component activations identified in the second step and use these as probabilistic weights during estimation of system parameters for standard multi-channel microphone enhancement algorithms, e.g., minimum-variance distortionless response (MVDR) filters.

2.1   Probabilistic Source Localization

Reliable estimation of spatially localized speech source probability is the first step in the proposed method which the subsequent steps build upon. We here employ the discriminative classification approach to probabilistic sound source localization described in [6]. It estimates the a-posteriori probability of speech for

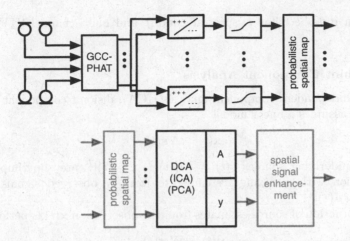

Fig. 1. Schematic overview of the proposed system. From 4-channel microphone signals, generalized cross-correlation with phase-transform (GCC-PHAT) features are computed. The short-time GCC-PHAT vectors form the input for a bank of linear SVM classifiers, one for each azimuth angle, that are trained to optimally discriminate speech vs. non-speech at each respective location. A component-wise non-linear transformation obtained with a generalized linear model transforms SVM decision values to a-posteriori probability estimates. These form a probabilistic spatial map that indicates probability of speech source presence for each azimuth angle and time-frame. Operating on this probability map, spatial source patterns are computed by disjoint component analysis (DCA), that each represent the spatial image of an individual sound object. Independent (ICA) and principal (PCA) component analysis are used for comparison. In a subsequent step (covered here through an outline of the approach, though without experimental validation), derived spatial maps (**A**) and source activations (**y**) could be used to steer multi-channel signal enhancement algorithms.

a defined set of source locations θ using short-term generalized cross-correlation [10] with phase transform (GCC-PHAT) input features

$$\rho_{ij}(t,\tau) = \mathcal{IFFT} \left\{ \frac{X_i(t,f)}{|X_i(t,f)|} \frac{X_j^*(t,f)}{|X_j(t,f)|} \right\} \tag{1}$$

of channels i and j, computed from a time-frame centered at time t, from short-term spectral transforms $X_i(t,f)$ and $X_j(t,f)$. Cross-correlation lag is denoted as τ and inverse Fourier transformation as $\mathcal{IFFT}$.

The set of coefficients $\rho_{ij}(t,\tau)$ is used to train a bank of discriminative linear support-vector machine (SVM) classifiers, with presence and absence of a speech source for a given position serving as the training class label. Each SVM is followed by a generalized linear model (GLM) classifier, that converts SVM decision values into the estimated spatial source probability map $p^S(\theta,t)$. Let $\mathcal{G}_\theta(\cdot)$ denote the combined localizer for direction θ as described above, then the source probability map is given by

$$p^S(\theta,t) = \mathcal{G}_\theta(\mathbf{X}(t,f)) \tag{2}$$

for location index θ, time t, spectral band f and multi-channel STFT input vector $\mathbf{X}(t, f)$.

2.2 Disjoint Component Analysis

Similar to independent component analysis (ICA), disjoint component analysis (DCA, [2]) assumes a linear model

$$\mathbf{x}(t) = \mathbf{A}\,\mathbf{s}(t) \tag{3}$$

where N underlying sources $\mathbf{s}(t) = [s_1(t), \ldots, s_N(t)]^T$ are superimposed by multiplication with a mixing system $\mathbf{A}$ to form N observed signals $\mathbf{x}(t) = [x_1(t), \ldots, x_N(t)]^T$.

Reconstruction of source estimates from the observation $\mathbf{x}(t)$ is performed as

$$\mathbf{y}(t) = \mathbf{W}\,\mathbf{x}(t) \tag{4}$$

with components $\mathbf{y}(t) = [y_1(t), \ldots, y_N(t)]^T$ and some estimated separation system $\mathbf{W}$.

While ICA maximizes mutual statistical independence of the source estimates $\mathbf{y}$, DCA's objective function is defined as

$$H = \frac{1}{2} \sum_{m \neq n} o_{mn} = \frac{1}{2} \sum_{m \neq n} E(|y_m|\,|y_n|) \tag{5}$$

with the overlap

$$o_{mn} = E(|y_m|\,|y_n|) \tag{6}$$

of output signals y_m and y_n. Two signals are mutually *disjoint* if and only if $y_m(t)\,y_n(t) = 0$ for all t and $m \neq n$, i.e., at most one of the signals is non-zero at any time. We note that disjoint signals are not mutually *independent* but exhibit statistical *dependencies* through the negative correlations of their signal envelopes or signal power time-courses. Note also that by *not* assuming zero-means signals, DCA retains particular physical information assigned to amplitude zero of a signal, namely the absence of energy.

The minimum of H is obtained by gradient descent optimization on the gradient with respect to matrix $\mathbf{W}$,

$$\nabla_{\mathbf{W}} H = E\left(-\mathbf{y}\mathbf{x}^H + ||\mathbf{y}||_1 \mathrm{sign}(\mathbf{y})\mathbf{x}^H\right), \tag{7}$$

where $||\mathbf{y}||_1 = \sum_i |y_i|$ denotes the 1-norm of $\mathbf{y}$.

Right-multiplication with $\mathbf{W}^T\mathbf{W}$ yields an expression in analogy to the natural gradient approach of [1],

$$\tilde{\nabla}_{\mathbf{W}} H = E\left(-\mathbf{y}\mathbf{y}^H + ||\mathbf{y}||_1 \mathrm{sign}(\mathbf{y})\mathbf{y}^H\right) \mathbf{W}. \tag{8}$$

Minimization of H is carried out under the unit-norm constraint $||\mathbf{w}_i||_2 = 1$ for the rows $\mathbf{w}_i$ of matrix $\mathbf{W}$. Gradients (7) and (8) are similar to the corresponding gradients derived from infomax or maximum-likelihood ICA with a sparse prior.

2.3   Decomposition of Source Probability Map

The source probability map $p^S(\theta, t) = \mathcal{G}_\theta(\mathbf{X}(t, f))$ is decomposed into a discrete number of components, each of which may represent a physical sound source. To this end, the input vector $\mathbf{x}(t)$ of the DCA decomposition is defined as the vector of spatial source probabilities, i.e.,

$$x_n(t) \equiv p^S(\theta_n, t) \tag{9}$$

with $n = 1, \ldots, N$. Since the DCA decomposition does not include spatial priors or spatial continuity assumptions, the resulting source images may correspond to an arbitrary spatial arrangement of coherently modulated source probabilities that may represent spatially localized sources as well as spatially distributed sound fields. Note that a physically localized source may be represented by a non-delta-shaped probability map component since, e.g., reverberation may result in broadened source peaks and also spurious peaks due to reflections.

2.4   Multi-channel Signal Enhancement

Knowledge of localizer $\mathcal{G}_\theta(\cdot)$ and DCA parameters $\mathbf{W}$ and $\mathbf{y}$ corresponds to implicit knowledge of a spatial source model. However, a model that is appropriate for source localization does not necessarily imply knowledge of spatial filter parameters that would permit to optimally enhance a target speech source and attenuate interference from other sound sources.

To this end, we extend our previous work [9] to estimation of spatial filters from the source activations $\mathbf{y}(t)$ obtained through source localization and DCA decomposition. These are used as weights to compute a speech covariance matrix $\mathbf{\Phi}_k(f)$ corresponding to the k-th DCA component in spectral band f, with ij-element

$$[\mathbf{\Phi}_k(f)]_{ij} \equiv E[y_k(t)\, X_i^*(t, f)\, X_j(t, f)]. \tag{10}$$

Knowledge of $\mathbf{\Phi}_k(f)$ is sufficient to steer multi-channel signal enhancement algorithms, such as minimum variance distortionless response (MVDR) filter.

3   Experiments and Results

We evaluated the proposed source localization and decomposition scheme on data from a six-channel binaural hearing aid geometry setup of which four channels (front and rear microphone pairs) were employed for estimation of the spatial source probability map as described in [6] with discrete azimuth angles $\theta = 0°, \ldots, 355°$ in steps of $5°$. STFT frame length was 10 ms with 25% shift. All acoustic signals used in the experiments were generated by filtering single-channel speech signals with head-related impulse responses (HRIR) captured with the binaural hearing aid setup as described in detail in [7] and in the accompanying publicly available dataset.

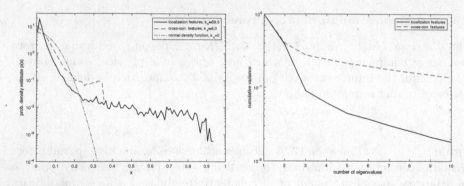

Fig. 2. Statistical properties of the probabilistic spatial map features of the SVM-based localization model ("localization features") and the inter-microphone GCC-PHAT cross-correlation features ("cross-corr. features"). Left panel: Binned probability density estimates and empirical kurtosis values. The normal density function is shown for comparison ("normal density function"). Right panel: Cumulative eigenvalue spectra of the covariance matrix of the localization features and the cross-correlation features, respectively. Both have the dominant contribution from two sources (target and spatially localized interference) in the first two eigenvectors. In case of the localization features, however, the noise subspace (corresponding predominantly to the isotropic noise field) has a higher degree of separation from the signal subspace as implied by the steeper drop-off from second to third eigenvalue.

Three-seconds-long speech signals, each from the same (female or male) speaker, were randomly sampled from the TIMIT speech database [5]. A head-related isotropic noise field was obtained by convolution of speech shaped noise [4] with anechoic HRIRs from the whole horizontal plane. The resulting signals were combined to a set of test scenarios containing a target speech source, an interfering speaker from a different position and isotropic noise. Thereby, the energy ratio between target and interferer, signal-to-interference ratio (SIR), was varied between $-10\,$dB, $10\,$dB, $10\,$dB, $20\,$dB and $\infty\,$dB, as well as the energy ratio between target and noise field, signal-to-noise-ratio (SNR). The resulting overall acoustic complexity is then represented by the signal-to-noise-plus-interferer-ratio (SINR). The target in the anechoic environment was located in the left hemisphere at DOAs ranging from $-180°$ (back) to $0°$ (front) in steps of $30°$. The interfering speaker impinged from positions in the entire sphere around the head in the range from $-165°$ to $+165°$ in steps of $30°$.

Resulting signals were processed with the localization method (Sect. 2.1) and the probabilistic localization map was decomposed with DCA (Sect. 2.2). For comparison, we also computed infomax independent component analysis [3] and principal component analysis decompositions of the probabilistic map.

In a first analysis step, we computed bin-wise probability density estimates for the GCC-PHAT cross-correlation feature vector that forms the input of the localization algorithm, and of the probabilistic map coefficients, i.e., the localization algorithm's output. As shown in Fig. 2, left panel, both distributions are highly super-kurtotic with normalized kurtosis values of about 6.0 and 59.5, respectively. The localization algorithm results in a further increase of sparsity compared to the already sparse cross-correlation coefficients, thus, fulfilling a necessary requirement for the applicability of sparsity-based decomposition techniques.

A subspace analysis of cross-correlation features and probabilistic map features (Fig. 2, right panel) shows that, as expected, the two localized sources (target and interferer) result in a dominant contribution of the first two eigenvectors of each feature covariance matrix. The variance explained by the remaining eigenvectors (EV 3 and beyond) for the cross-correlation features amounts

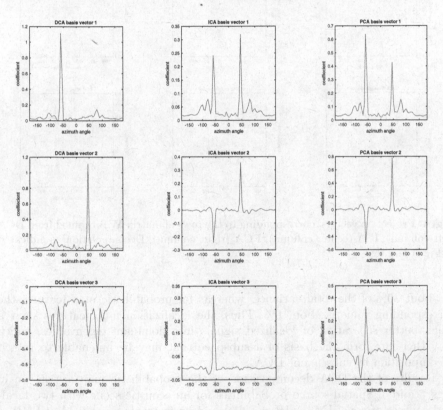

Fig. 3. Decomposition of the probabilistic spatial map that was computed by the SVM localization algorithm into separate components. The 72-dimensional map was projected with PCA onto the dominant three-dimensional subspace, which was further decomposed by disjoint component analysis (DCA, left panel), independent component analysis (ICA, center panel) and principal component analysis (PCA, right panel). Rows: Estimated spatial pattern pertaining to decomposition source 1 (top row), source 2 (second row), source 3 (bottom row), i.e., columns of matrix **A**. Physical source locations of the two separated sources coincide with DCA basis vectors 1 and 2. ICA extracts a plausible source in component 3. PCA does not result in components that correspond to separated sources.

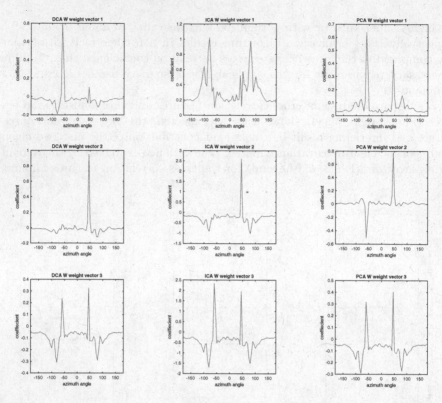

Fig. 4. Projection vectors, corresponding to the rows of matrix **W**, obtained from DCA (left column), ICA (center column), PCA (right column). Decomposition as indicated in Fig. 3 and text.

to about 30% of the total variance, whereas for probabilistic map features the corresponding value is about 9%. Thus, the localization map features show a much better separation of localized signals into dominant eigenvalues, which may therefore form the basis of a subspace that may be amenable to robust decomposition by subsequent DCA.

Results of the DCA decomposition of the probabilistic map are shown in Figs. 3 and 4. Spatial source probabilities for an acoustic scene with two localized sources of equal level (SIR = 0 dB) and an isotropic noise field at SNR = 20 dB were projected to the dominant three-dimensional subspace by PCA, thus capturing in excess of 93% of total variance (cf. Fig. 2). This subspace was further decomposed by DCA and the resulting source basis vectors (i.e., columns of estimate of matrix **A**) and separation vectors (i.e., rows of **W**) are displayed. The same data were decomposed by Infomax ICA [3] and PCA. Figure 3 shows that DCA successfully identifies the two source positions in the first two basis vectors, accounting for in excess of 90% of the total variance. ICA and PCA, in contrast, extract components that contain contributions from the true sources' positions, albeit in most cases not in full separation but as source superpositions. The third (lower variance) ICA component contains a single peak at a true source position, while the two remaining ICA components correspond to mixed sources.

4   Summary and Discussion

In this contribution, we presented an approach for acoustic source localization and sound object formation that combines a spatially-resolved source probability estimator that was trained discriminatively on source/no-source signal examples with an unsupervised decomposition stage that identifies a small number of sound objects and their corresponding spatial activity patterns. A concept for a multi-channel filter approach that aims at enhancement of identified (target) obejct(s) has been outlined, the experimental validation of which, however, would be beyond the scope of the present work.

Our approach, thus, combines prior-learned localization models that have been trained with labeled examples, and unsupervised decomposition on a lower-dimensional subspace. As experiments in several acoustic environments have shown [6], the trained localization models generalizes robustly from the anechoic conditions under which it was learned to new, reverberant environments without retraining. The unsupervised decomposition is expected to yield basis vectors that reflect acoustic conditions including reverberation and therefore needs to be adapted to each environment. Its adaptation, however, is computationally efficient since it operates on a lower-dimensional subspace which permits fast and robust optimization. Future work will focus on the proposed multi-channel signal enhancement scheme and its validation in reverberant and multi-source acoustic scenes.

Acknowledgments. Supported by DFG grants SFB/TRR 31 "The Active Auditory System" and FOR 1732 "Individualized Hearing Acoustics".

References

1. Amari, S.I.: Natural gradient works efficiently in learning. Neural Comput. **10**, 251–276 (1998)
2. Anemüller, J.: Maximization of component disjointness: a criterion for blind source separation. In: Davies, M.E., James, C.J., Abdallah, S.A., Plumbley, M.D. (eds.) ICA 2007. LNCS, vol. 4666, pp. 325–332. Springer, Heidelberg (2007). doi:10.1007/978-3-540-74494-8_41
3. Bell, A., Sejnowski, T.: An information-maximization approach to blind separation and blind deconvolution. Neural Comput. **7**, 1129–1159 (1995)
4. Dreschler, W.a., Verschuure, H., Ludvigsen, C., Westermann, S.: ICRA noises: artificial noise signals with speech-like spectral and temporal properties for hearing instrument assessment. Audiology **40**(3), 148–157 (2001)
5. Garofolo, J.S., Lamel, L.F., Fisher, W.M., Fiscus, J.G., Pallett, D.S., Dahlgren, N.L., Zue, V.: TIMIT Acoustic-Phonetic Continuous Speech Corpus. CDROM (1993)

6. Kayser, H., Anemüller, J.: A discriminative learning approach to probabilistic acoustic source localization. In: Proceedings of IWAENC 2014 - International Workshop on Acoustic Echo and Noise Control, pp. 100–104 (2014)
7. Kayser, H., Ewert, S.D., Anemüller, J., Rohdenburg, T., Hohmann, V., Kollmeier, B.: Database of multichannel in-ear and behind-the-ear head-related and binaural room impulse responses. EURASIP J. Adv. Sig. Process. **2009**(1), 1–10 (2009). ID 298605
8. Kayser, H., Hohmann, V., Ewert, S.D., Kollmeier, B., Anemüller, J.: Robust auditory localization using probabilistic inference and coherence-based weighting of interaural cues. J. Acoust. Soc. Am. **138**(5), 2635–2648 (2015)
9. Kayser, H., Moritz, N., Anemüller, J.: Probabilistic spatial filter estimation for signal enhancement in multi-channel automatic speech recognition. In: Proceedings of INTERSPEECH 2016 (2016)
10. Knapp, C., Carter, G.: The generalized correlation method for estimation of time delay. IEEE Trans. Acoust. Speech Sig. Process. **24**(4), 320–327 (1976)
11. May, T., van de Par, S., Kohlrausch, A.: A probabilistic model for robust localization based on a binaural auditory front-end. IEEE Trans. Audio Speech Lang. Process. **19**, 1–13 (2011)
12. Woodruff, J., Wang, D.: Binaural localization of multiple sources in reverberant and noisy environments. IEEE Trans. Audio Speech Lang. Process. **20**, 1913–1928 (2012)

Tensors and Audio

An Initialization Method for Nonlinear Model Reduction Using the CP Decomposition

Gabriel Hollander[✉], Philippe Dreesen, Mariya Ishteva, and Johan Schoukens

Vrije Universiteit Brussel, Brussels, Belgium
gabriel.hollander@vub.ac.be

Abstract. Every parametric model lies on the trade-off line between accuracy and interpretability. Increasing the interpretability of a model, while keeping the accuracy as good as possible, is of great importance for every existing model today. Currently, some nonlinear models in the field of block-oriented modeling are hard to interpret, and need to be simplified. Therefore, we designed a model-reduction technique based on the Canonical Polyadic tensor Decomposition, which can be used for a special type of static nonlinear multiple-input-multiple-output models. We analyzed how the quality of the model varies as the model order is reduced. This paper introduces a special initialization and compares it with a randomly chosen initialization point.

Using the method based on tensor decompositions ensures smaller errors than when using the brute-force optimization method. The resulting simplified model is thus able to keep its accuracy as high as possible.

Keywords: Tensor decomposition · CP decomposition · Model order reduction · Multiple-input-multiple-output model

1 Introduction

The process of simplifying complex parametric models by reducing the number of parameters is called *model reduction* (for a general introduction, see for example [1]). In order to understand the trade-off between accuracy and interpretability, model reduction techniques must be developed for different types of models. For this, we have selected one special type of multiple-input-multiple-output models, and have developed a model reduction technique using tensor decompositions.

The models that we have studied form a central part of block-oriented system identification (see [2,3]), and the results apply also to nonlinear state space models, see [4]. For instance, in the block-oriented framework, the so-called parallel Wiener-Hammerstein models, which consist of a certain combination of linear time-invariant blocks and nonlinear static blocks, is a universal approximation and was intensively researched during the last few years, see [5–7].

In this paper, we will give an overview of how to find a good initialization for an iterative optimization, and will compare this to a random initizaliation.

© Springer International Publishing AG 2017
P. Tichavský et al. (Eds.): LVA/ICA 2017, LNCS 10169, pp. 111–120, 2017.
DOI: 10.1007/978-3-319-53547-0_11

We will also analyze one case study in detail, and show how the accuracy and complexity of this model varies with different reduced orders.

This article is organized as follows: Sect. 2 states the problem discussed in this article, and sets some notation conventions. Next, Sect. 3 gives an overview of the method using tensor decompositions, while Sect. 4 shows the results of this method based on simulations. Finally, Sect. 5 examines a test case in detail and Sect. 6 concludes this paper.

2   Notations and Problem Statement

In the domain of nonlinear system identification, recent research has been focused on special types of models and their properties, used, amongst others, in the identification of parallel Wiener-Hammerstein models, see [3,7,8]. These are the models considered in this paper and they are defined as follows. A model $\mathbf{f} \colon \mathbb{R}^m \to \mathbb{R}^n$ has m inputs and n outputs and is defined internally with the following parts, given from an earlier identification procedure, or chosen:

– a set of r nonlinear single-input-single-output functions $g_1(x_1), \ldots, g_r(x_r)$, also called *branches*, surrounded by
– two transformation matrices $\mathbf{V} \in \mathbb{R}^{m \times r}$ and $\mathbf{W} \in \mathbb{R}^{n \times r}$.

In this paper, we consider polynomials for the function $g_i(x_i)$, so they have the form $g_i(x_i) = \sum_{k=0}^{N_i} c_{ij} x_i^k$, where c_{ij} are the coefficients of the polynomial $g_i(x_i)$. If we denote $\mathbf{u} = (u_1, \ldots, u_m)$ for the input of $\mathbf{f}$, then the output $\mathbf{y} = (y_1, \ldots, y_n)$ of the model is given by (Fig. 1)

$$\mathbf{y} = \mathbf{f}(\mathbf{u}) = \mathbf{W}\mathbf{g}(\mathbf{V}^T\mathbf{u}). \tag{1}$$

In Eq. (1), $\mathbf{g}$ represents the set of single-input-single-output functions, so $\mathbf{g} = (g_1(x_1), \ldots, g_r(x_r))$. The nonlinear parts g_i of the models analyzed in this article are chosen to be univariate polynomials of degree less than or equal to $d \in \mathbb{N}\backslash\{0\}$.

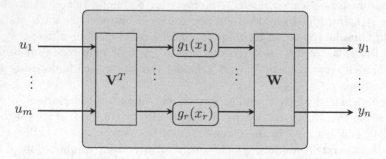

Fig. 1. The nonlinear models in this article have the same structure, given by $\mathbf{y} = \mathbf{W}\mathbf{g}(\mathbf{V}^T\mathbf{u})$.

In this paper, we are concerned with the following problem: given a model as described above, find a reduced description involving $\tilde{r} < r$ branches that approximates the given model well. That is, find appropriate transformation matrices $\tilde{\mathbf{V}} \in \mathbb{R}^{m \times \tilde{r}}$ and $\tilde{\mathbf{W}} \in \mathbb{R}^{n \times \tilde{r}}$ and nonlinear single-input-single-output functions $\tilde{g}_1(x_1), \ldots, \tilde{g}_r(x_{\tilde{r}})$—and thus a multiple-input-multiple-output function $\tilde{\mathbf{f}}$, such that the reduced modeled output approximates the original output:

$$\mathbf{f}(\mathbf{u}) = \mathbf{W}\mathbf{g}(\mathbf{V}^T\mathbf{u}) \approx \tilde{\mathbf{W}}\tilde{\mathbf{g}}(\tilde{\mathbf{V}}^T\mathbf{u}) = \tilde{\mathbf{f}}(\mathbf{u}),$$

for a given set of inputs $\mathbf{u}$. Formally, the following cost function should be minimized, over the parameters of $\tilde{\mathbf{V}}$, $\tilde{\mathbf{W}}$ and $\tilde{\mathbf{g}}$:

$$\min_{\tilde{\mathbf{V}}, \tilde{\mathbf{W}}, \tilde{\mathbf{g}}} \frac{\sum_{i=1}^{N} \left\| \mathbf{W}\mathbf{g}(\mathbf{V}^T\mathbf{u}^{(i)}) - \tilde{\mathbf{W}}\tilde{\mathbf{g}}(\tilde{\mathbf{V}}^T\mathbf{u}^{(i)}) \right\|^2}{\sum_{i=1}^{N} \left\| \mathbf{W}\mathbf{g}(\mathbf{V}^T\mathbf{u}^{(i)}) \right\|^2}. \tag{2}$$

The search space during this optimization are the matrix elements of $\tilde{\mathbf{V}}$ and $\tilde{\mathbf{W}}$, together with the coefficients of the function $\tilde{\mathbf{g}}$, which also have degree d. Also, the used norm is the Euclidean norm of vectors. Finally, in (2), the inputs $\mathbf{u}^{(1)}, \ldots, \mathbf{u}^{(N)}$ are N sampling points; in our experiments, these are uniform randomly chosen in a fixed interval. Because of the nonlinear function $\tilde{\mathbf{g}}$, this is a nonlinear and non-convex optimization problem. Any iterative method, see [9], might thus end in local minima, and it is very sensitive to the starting point.

Finding an appropriate initialization for the optimization problem (2) is crucial in order to not obtain poor local minima. Using a tensor decomposition approach, it is possible to guarantee good initialization points, as described in the next section.

3 Finding an Appropriate Initialization

In order to optimize the cost function (2), a good initial starting point is needed for any iterative optimization method. The starting point suggested in this article is created using tensor decompositions. A general overview of tensors, multidimensional arrays of numbers, can be found in [10], and we will review a few general definitions, needed for the proposed technique. Because we will solely use tensors up to the third order (also called three-way tensors), we will limit the definitions to this case.

When considering a tensor, it is always possible to write it as a sum of so-called rank-one tensors: these are tensors which can be written as outer products of non-zero vectors. So, using the same notation as in [11], if we denote by $\mathcal{X}$ a third-order tensor, we say that

$$\mathcal{X} = \mathbf{a} \circ \mathbf{b} \circ \mathbf{c} \tag{3}$$

is the outer product of the vectors $\mathbf{a}$, $\mathbf{b}$ and $\mathbf{c}$ if each tensor element can be written as the product of the corresponding vector elements,

$$x_{ijk} = \mathbf{a}(i)\,\mathbf{b}(j)\,\mathbf{c}(k),$$

for all indices i, j, k in the appropriate bounds.

More generally, if the following equation holds,

$$\mathcal{X} = \sum_{r=1}^{R} \mathbf{a}_r \circ \mathbf{b}_r \circ \mathbf{c}_r, \tag{4}$$

and R is minimal, then we call R the rank of $\mathcal{X}$. In this case, we call (4) the Canonical Polyadic (CP) decomposition and $\mathbf{A} = [\mathbf{a}_1 \cdots \mathbf{a}_R]$ and analogous for $\mathbf{B}$ and $\mathbf{C}$ the factors. The CP decomposition thus factors a given tensor into its rank-one terms. This is represented graphically in Fig. 2.

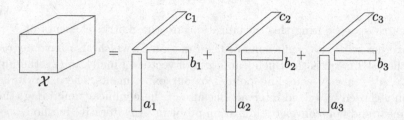

Fig. 2. The Canonical Polyadic decomposition can be represented graphically as a factorization into rank-one tensors.

Several methods exist in the literature for finding the CP decomposition of a given tensor, for example [12,13]. Also, MATLAB implementations exist in order to compute the CP decomposition, for example [14–16].

In order to find an appropriate initialization point for the optimization problem (2), we will base the solution on the decoupling algorithm proposed in [7]. This algorithm was originally used and designed to find the decomposition (1), see Fig. 1. We will use it here to approximate $\mathbf{f}$ using fewer branches. The summary of this method is described as follows (Sect. 2.5 of [7]):

– Starting from the nonlinear multiple-input-multiple-output function $\mathbf{f}$, evaluate its Jacobian matrix $\mathbf{J}(\mathbf{u}) \in \mathbb{R}^{n \times m}$ with respect to the input $\mathbf{u}$ in N sampling points. These sampling points are chosen in a random way, and the number N is chosen sufficiently large in order to avoid ill-posed problems in the following steps;
– Stack these Jacobian matrices into a three-dimensional tensor $\mathcal{J}$ of dimensions $n \times m \times N$;
– Find the CP decomposition of $\mathcal{J}$ for a given rank $\tilde{r}$. This will be an approximate decoupling, until a given stopping criterion is met, and will be dependent on a maximal number of iterations, as well as a threshold for the approximated

error. The output of this step can be represented as three matrices: $\tilde{\mathbf{V}}$, $\tilde{\mathbf{W}}$ and $\tilde{\mathbf{H}}$;

– Reconstruct the univariate functions $g_1(x_1)$, ..., $g_{\tilde{r}}(x_{\tilde{r}})$ by integrating an interpolation of the matrix $\tilde{\mathbf{H}}$, see [17]. This step is necessary in order to reconstruct the univariate functions, because of the first-order derivative step to obtain the Jacobian matrices.

We note that this CP decomposition technique passes through the level of first-order derivatives of $\mathbf{f}$ before integrating the result and arriving back on the original level for $\tilde{\mathbf{f}}$. This process may seem like an unnecessary step in order to solve the optimization problem (2), but this extra step offers a solution for so-called decoupling problem, see [7,18]. The major advantage is that a three dimensional tensor needs to be decoupled, which is independent of the polynomial degree of $\mathbf{f}$.

We also note that the technique discussed above was originally designed to work in the exact case (4) and $\tilde{r} = R$. But here, the equality should be replaced by an approximation

$$\mathcal{X} \approx \sum_{r=1}^{R} \mathbf{a}_r \circ \mathbf{b}_r \circ \mathbf{c}_r,$$

and the following cost function is used for this approximation:

$$\min_{\tilde{\mathbf{V}}, \tilde{\mathbf{W}}, \tilde{\mathbf{H}}} \left\| \mathcal{J} - \sum_{r=1}^{R} \tilde{\mathbf{w}}_r \circ \tilde{\mathbf{v}}_r \circ \tilde{\mathbf{h}}_r \right\|_F^2, \tag{5}$$

see [18]. The cost function (5) is defined on the level of the Jacobian elements and uses the Frobenius norm.

Finally, we note that the optimization problem (2) with $\tilde{r} < r$ may yield a different result than when using the algorithm discussed in [7]. This is because the latter uses the optimization (5) defined on the level on the Jacobian elements, while the former is defined on the level of the function $\mathbf{f}$. The extra step using the proposed method of this paper may be able to better handle local optima, by starting with a good initial point, described in [7].

4   Simulations and Results

In order to compare the proposed initialization using the CP decomposition with a random initialization, we experimented with a computer-based simulation. In this section, we give an overview of this experiment and analyze the results.

An experiment is defined by the following set of variables:

– A number of inputs m ($3 \le m \le 20$);
– A number of outputs n ($3 \le n \le 20$);
– A number r of "true" internal branches, before the model reduction ($3 \le r \le 20$);

- A number $\tilde{r}$ of "reduced" internal branches, after the model reduction ($3 \leq \tilde{r} \leq r$);
- The maximal degree d of the polynomials in the internal branches ($d = 3$);
- Two uniform randomly chosen transformation matrices $\mathbf{V}$ and $\mathbf{W}$;
- r uniform randomly chosen polynomials $g_1(x_1), \ldots, g_r(x_r)$;
- Uniform randomly chosen input samples $\mathbf{u}^{(1)}, \ldots, \mathbf{u}^{(N)}$ ($N = 500$).

Once these parameters are set, twenty random initialization points are generated. Each initialization point represent all the parameters in $\tilde{\mathbf{V}}$, $\tilde{\mathbf{W}}$ and $>$. The first half of those (initialization 1 in Fig. 3) will serve as starting points for the CP decomposition technique outlined in Sect. 3, yielding the starting point for the optimization problem (2). The other half of the random initial points (initialization 2 in Fig. 3) are immediately used to start the same optimization problem (2), see Fig. 3. We note that while other initialization methods of the CP decomposition exist, we chose the initialization points described here in order to compare both methods with the same initialization.

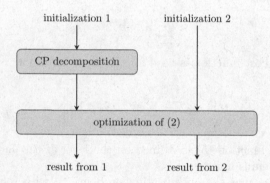

Fig. 3. One experiment examines the optimization with twenty initial points. The first half serves as a starting point for the CP decomposition technique, while the other half is immediately used in order to optimize (2).

Each of these twenty results of such an experiment are defined by

- Two transformation matrices $\tilde{\mathbf{V}}$ and $\tilde{\mathbf{W}}$;
- A set of $\tilde{r}$ internal branches, polynomials $g_1(x_1), \ldots, g_{\tilde{r}}(x_{\tilde{r}})$, of degree d;
- The error given by the cost function (2).

In order to compare the two sets of initialization points, we repeated this for 192 different experiments, each chosen uniform randomly. The specific values are given at the start of this section. All of these experiments are chosen such that, no identifiability issues arise, see for example [19].

To analyze the results, we have plotted both errors against each other (Fig. 4). For this, we have chosen the smallest error for every set of ten initialization points.

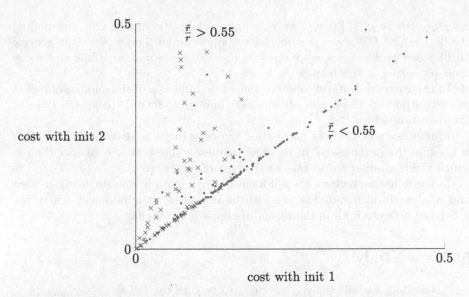

Fig. 4. All 192 experiments show a better (or equal) error after optimization using the CPD as initialization point instead of a random initialization. This difference is most significant when $\frac{\tilde{r}}{r} > 0.55$ (blue crosses), because these lie in the upper left part of the plot. This happens whenever m or n is quite large, as there is then more reduction possible. In case $\frac{\tilde{r}}{r} \leq 0.55$ (red points), the difference is less significant. The number 0.55 is chosen such that it divides the points of this plot in two parts: most experiments in the upper-half portion of this plot satisfy $\frac{\tilde{r}}{r} > 0.55$. If the search space of (2) is not too large, then it is highly probable that the difference in initializations is high.

From the experiments, we observe that adding the extra step with the CP decomposition only makes the resulting model error smaller. Also, adding this CP decomposition does not add any significant computational time to the experiment, as CP decomposition algorithms exist with very little overhead time (decomposing a $20 \times 20 \times 1000$-tensor using [16] takes under 3 seconds to compute, while the optimization problem (2) may be much slower). In these experiments, no noise error was added, so the errors are solely due to model approximations. As a first step in this research, this ensures a sound analysis of the proposed method. What happens in general when noise is considered, is outside the scope of this paper.

Also, we note that solely adding the CP decomposition step without optimizing the problem (2) is not sufficient, as the model errors still remain larger than the discussed method. Furthermore, existing CP decomposition implementations as [15,16] offer predefined initialization points, but these do not guarantee better results as using a sample of ten random initializations, as other simulations have shown. These experiments show that the choice of the initialization method for the CPD in the context of the optimization problem (2) does not seem to much influence the results. Finally, we note that, while is it known [20] that finding the best rank $\tilde{r}$ CP decomposition for tensors of order 3 or higher, is in general an ill-posed problem, we are interested in workable approximations and study their properties in this context.

In these experiments, we observe almost no influence of the number of inputs m and outputs n. These were chosen large enough to be able to see the effect of the reduction with $\tilde{r} \leq r$.

Finally, we note that the normalized errors in Fig. 4 increase up to 50%. This is because the parameter $\tilde{r}$ in these experiments may be much smaller than r, which yields a larger reductions, at the cost of larger errors.

In the following section, we will focus our attention to one single experiment and analyze it in detail. For this, we have chosen an experiment where the difference between both initialization methods is significant.

5   Case Study

In this section, we will analyze the case study with the following parameters:

- Number of inputs $m = 20$;
- Number of outputs $n = 20$;
- Number of internal branches, before the model reduction $r = 20$. The coefficients of the polynomials $g_1(x_1), \ldots, g_{20}(x_{20})$ are uniform randomly chosen in the intervals $[-1, 0.5]$ or $[0.5, 1]$;
- Number of internal branches, after the model reduction $\tilde{r} = 1, \ldots, 20$;
- The maximal degree of the polynomials in the internal branches $d = 3$;
- Two transformation matrices $\mathbf{V} \in \mathbb{R}^{20 \times 20}$ and $\mathbf{W} \in \mathbb{R}^{20 \times 20}$ whose elements are uniformly chosen in the interval $[-1, 1]$;
- Uniform randomly chosen input samples $\mathbf{u}^{(1)}, \ldots, \mathbf{u}^{(N)}$ ($N = 500$), chosen in the ball $[-3, 3]^m$.

We conclude from these experiments that using the initial point coming from the CP decomposition yields a smaller cost function from a certain threshold of $\tilde{r}$ (Fig. 5). Also, using these initial points, the original model is found back when $\tilde{r} = r = 20$ (the cost function there is in the order of 10^{-11} in magnitude—this is not exactly machine precision due to the CP iteration algorithm, which may terminate the iterations at a predefined threshold level), which is not the case for a random initial point.

Furthermore, using the random initial point, the error increases above $\tilde{r} = 5$. This seems to be the threshold when the search space for the optimization problem (2) becomes too large. Also, this is why the errors using random initialization points when $\tilde{r} \approx 20$ are so large: bad local minima are found using because of the bad starting point of the optimization procedure.

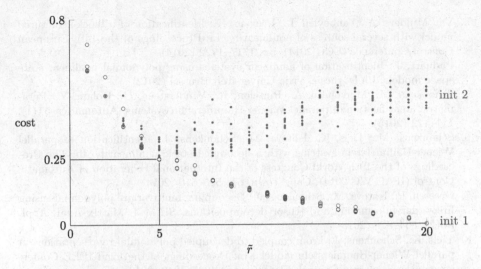

Fig. 5. Starting from $\tilde{r} = 5$, the error using the CP-based initialization point is smaller than when using a randomly chosen initialization point. Moreover, the variations using ten different CP-based initializations are smaller than using ten random points.

6 Conclusion

To reduce the model order of certain nonlinear model structures, we have to minimize a nonlinear and non-convex cost function. For this, we can use an iterative method, which needs an initialization point. A randomly chosen initialization point may yield bad local minima, while using a Canonical Polyadic decomposition-based initialization point results in smaller model errors. The models thus reduced better approximate the original models, and thus could be used in several engineering applications, for example parallel Wiener-Hammerstein system identification. For future work, we would like to investigate the influence of the number of inputs and outputs of the coupled function, with respect to [19]. Finally, we would like to analyze if the tensor decomposition method could also be applied to other types of model reduction problems.

Acknowledgments. This work was supported in part by the Fund for Scientific Research (FWO-Vlaanderen), the Flemish Government (Methusalem), the Belgian Government through the Interuniversity Poles of Attraction (IAP VII) Program, the ERC advanced grant SNLSID under contract 320378, the ERC starting grant SLRA under contract 258581, and FWO project G028015N.

References

1. Johnson, K., Kuhn, M.: Applied Predictive Modeling. Springer, New York (2013)
2. Giri, F., Bai, E.W.: Block-Oriented Nonlinear System Identification. Springer, London (2010)

3. Van Mulders, A., Vanbeylen, L., Usevich, K.: Identification of a block-structured model with several sources of nonlinearity. In: Proceedings of the 14th European Control Conference (ECC 2014), pp. 1717–1722 (2014)
4. Paduart, J.: Identification of nonlinear systems using polynomial nonlinear state space models. Ph.D. thesis, Vrije Universiteit Brussel (2012)
5. Schoukens, M., Marconato, A., Pintelon, R., Vandersteen, G., Rolain, Y.: Parametric identification of parallel Wiener-Hammerstein systems. Automatica 51(1), 111–122 (2015)
6. Schoukens, M., Tiels, K., Ishteva, M., Schoukens, J.: Identification of parallel Wiener-Hammerstein systems with a decoupled static nonlinearity. In: The Proceedings of the 19th World Congress of the International Federation of Automatic Control (IFAC WC 2014), Cape Town, pp. 505–510 (2014)
7. Dreesen, P., Ishteva, M., Schoukens, J.: Decoupling multivariate polynomials using first-order information and tensor decompositions. SIAM J. Matrix Anal. Appl. 36(2), 864–879 (2015)
8. Tiels, K., Schoukens, J.: From coupled to decoupled polynomial representations in parallel Wiener-Hammerstein models. In: Proceedings of the 52nd IEEE Conference on Decision and Control, Florence, pp. 4937–4942 (2013)
9. Nocedal, J., Wright, S.: Numerical Optimization, 2nd edn. Springer, New York (2006)
10. Cichocki, A., Mandic, D., Phan, A.-H., Caiafa, C., Zhou, G., Zhao, Q., De Lathauwer, L.: Tensor decompositions for signal processing applications, from two-way to multiway component analysis. IEEE Siganl Process. Mag. 32(2), 145–163 (2015)
11. Kolda, T.G., Bader, B.W.: Tensor decomposition and applications. SIAM Rev. 51(3), 455–500 (2009)
12. Comon, P.: Tensor decomposition, state of the art and applications. In: Mathematics in Signal Processing V, pp. 1–24 (2002)
13. Sorber, L., Van Barel, M., De Lathauwer, L.: Optimization-based algorithms for tensor decompositions: canonical polyadic decomposition, decomposition in rank-$(L_r, L_r, 1)$ terms, and a new generalization. SIAM J. Optim. 23(2), 695–720 (2013)
14. Bader, B.W., Kolda, T.G., et al.: Matlab tensor toolbox version 2.6, February 2015
15. Andersson, C., Bro, R.: The N-way toolbox for MATLAB. Chemom. Intell. Lab. Syst. 52(1), 1–4 (2000)
16. Vervliet, N., Debals, O., Sorber, L., Van Barel, M., De Lathauwer, L.: Tensorlab 3.0, March 2016
17. Dreesen, P., Schoukens, M., Tiels, K., Schoukens, J.: Decoupling static nonlinearities in a parallel Wiener-Hammerstein system: a first-order approach, pp. 987–992 (2015)
18. Hollander, G., Dreesen, P., Ishteva, M., Schoukens, J.: Weighted tensor decomposition for approximate decoupling of multivariate polynomials. Technical report (2016)
19. Comon, P., Qi, Y., Usevich, K.: X-rank and identifiability for a polynomial decomposition model. arXiv:1603.01566 (2016)
20. Lim, L., Comon, P.: Nonnegative approximations of nonnegative tensors. J. Chemom. 23, 432–441 (2009)

Audio Zoom for Smartphones Based on Multiple Adaptive Beamformers

Ngoc Q.K. Duong[1(✉)], Pierre Berthet[2], Sidkièta Zabre[3], Michel Kerdranvat[1],
Alexey Ozerov[1], and Louis Chevallier[1]

[1] Technicolor, 975 avenue des Champs Blancs, 35576 Cesson Sévigné, France
{quang-khanh-ngoc.duong,michel.kerdranvat,alexey.ozerov,
louis.chevallier}@technicolor.com
[2] 3D Sound Labs, 22 rue de la Rigourdière, 35510 Cesson Sévigné, France
p.berthet@3dsoundlabs.com
[3] Altran Technologies, 3 Rue Louis Braille, 35136 Saint-Jacques-de-la-Lande, France
sidkieta.zabre@altran.com

Abstract. Some recent smartphones have offered the so-called *audio
zoom* feature which allows to focus sound capture in the front direction while attenuating progressively surrounding sounds along with video
zoom. This paper proposes a complete implementation of such function
involving two major steps. First, targeted sound source is extracted by a
novel approach that combines multiple adaptive beamformers having different look directions with a post-processing algorithm. Second, spatial
zooming effect is created by leveraging the microphone signals and the
enhanced target source. Subjective test with real-world audio recordings
using a mock-up simulating an usual shape of the smartphone confirms
the rich user experience obtained by the proposed system.

Keywords: Audio zoom on smartphone · Sound capture · Robust adaptive beamformer · Post-processing

1 Introduction

Mobile devices such as smartphones and tablets have become very popular
nowadays for many users. Their hardware and processing power has also been
improved day by day, that makes them able to offer more enhanced applications
with richer user experience. This paper considers a so-called *audio zoom* application [1,2] where mobile devices can focus the sound capture on a desired direction
while attenuating progressively surrounding sounds[1]. Audio zoom has been com-
mercialized in recent smartphones (*e.g.,* Samsung Galaxy S5 and LG G2), and
it would be even more powerful in future products owning larger microphone
array.

This work has been done while the Pierre Berthet and the Sidkièta Zabre were with
Technicolor.
[1] https://www.youtube.com/watch?v=7DEyuapmRCs.

© Springer International Publishing AG 2017
P. Tichavský et al. (Eds.): LVA/ICA 2017, LNCS 10169, pp. 121–130, 2017.
DOI: 10.1007/978-3-319-53547-0_12

In order to perform audio zoom, the target sound source needs to be isolated from other surrounding sounds (*i.e.*, interferences originated from unwanted spatial directions) first. Thanks to the hardware improvement where most smartphone nowadays possesses two or more microphones (*e.g.*, Apple iPhone 5 showed up with not one or two, but even three microphones[2]), research in microphone array processing field can be well-applied to this considered problem. Specifically, beamforming [3,4] and (blind) audio source separation (BSS) [5,6] can be considered as the most appropriate approaches. BSS may require higher computation cost than beamforming algorithms since it usually involves, in addition, the advanced spectral modelling of the audio sources [7,8]. Thus, by considering the critical constraint of limited processing power in mobile devices, and also the interest here is to enhance a spatial region but not a particular source, we design our signal enhancement algorithm for the target source grounded on beamforming technique[3]. However, since beamforming usually requires a large microphone array in order to create a narrow beam capturing sound from a desired direction, we propose in this paper a novel approach that combines multiple robust adaptive beamformers [9,10] with a derived post-processing algorithm taking into account outputs of the beamformers so as to greatly enhance the targeted sound source. Once the target sound source is extracted, we further propose the creation of zooming effect as second step of the audio zoom system. Note that in the considered beamforming implementation, one beamformer has directivity pattern that emphasizes the target source while, on the contrary, the other beamformers suppress the target source. Similar strategy has been presented in [11] with the use of two fixed null beamformers, instead of multiple adaptive beamformers as considered in this paper, and spectral substraction as post-processing algorithm. Some other related work concerning the use of multiple beamformers for audio enhancement can be found also in *e.g.*, [12–15].

The paper aims to design a complete audio zoom system which can be implemented in mobile devices as an emerging application with reasonable processing cost. Yet, to the best of our knowledge, non of the scientific publications has been described such a similar system. It is also worth noting that the proposed approach has been implemented as part of a MediaPlayer running real-time on Android smartphones[4].

The rest of the paper is organized as follows. In Sect. 2 we present the global workflow as well as the detail steps of the proposed audio zoom system. We conduct experiment with subjective test on real-world sound scene recordings to validate the effectiveness of the proposed approach in Sect. 3. Finally we conclude in Sect. 4.

---

[2] http://www.idownloadblog.com/2012/09/12/iphone-5-three-mics/.

[3] Note that, preliminary study in [8] did not show remarkable advantage of BSS compared to beamforming in some specific setups such as a single target source in noise field.

[4] The demonstration has been presented at the Show and Tell session of the 41st IEEE International Conference on Acoustics, Speech and Signal Processing (ICASSP 2016).

2   Proposed Audio Zoom System

General workflow of the proposed audio zoom approach is shown in Fig. 1. It consists in two major steps: (1) target sound source extraction and (2) zooming effect creation. These steps will be described in detail in Sects. 2.1 and 2.2, respectively.

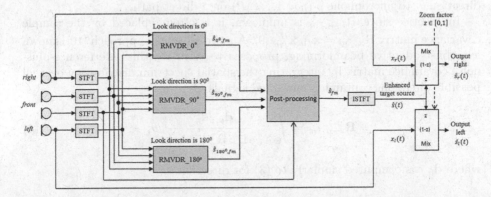

Fig. 1. General workflow of the proposed audio zoom implementation.

2.1   Target Sound Source Enhancement

Robust Adaptive Beamforming. Let us denote by $\mathbf{x}_{fm} \in \mathbb{C}^{P \times 1}$ the complex-valued STFT coefficients in time frame m and frequency bin f of the mixture signal recorded by P microphones. Beamforming isolates sound coming from a target spatial direction θ_t by deriving a frequency dependent weight vector $\mathbf{w}_{\theta_t,f}$ such that its output is given by

$$\hat{s}_{\theta_t,fm} = \mathbf{w}_{\theta_t,f}^H \mathbf{x}_{fm}, \tag{1}$$

where $(.)^H$ denotes Hermitian transpose. As relevant to the considered audio zoom application, where users usually focus sound capture in the front direction that is perpendicular to the device's surface, in the rest of the paper we consider $\theta_t = 90^0$.

The optimal weight vector $\mathbf{w}_{\theta_t,f}$ can be obtained by minimizing the energy of the interfering sources and noise under the constraint to keep unit response in the target direction. In this derivation, the beamformer is known as Minimum Variance Distortionless Response (MVDR) [9] - a well-known one in the literature and the resulting weight vector is given by

$$\mathbf{w}_{\theta_t,f} = \frac{\mathbf{R}_{i+n,fm}^{-1} \mathbf{d}_{\theta_t,f}}{\mathbf{d}_{\theta_t,f}^H \mathbf{R}_{i+n,fm}^{-1} \mathbf{d}_{\theta_t,f}}, \tag{2}$$

where $\mathbf{R}_{i+n,fm}$ is the interference-plus-noise covariance matrix and $\mathbf{d}_{\theta_t,f}$ is the steering vector which accounts for the time differences of arrival from the target source to the microphones and is computed as

$$\mathbf{d}_{\theta_t,f} = [1, e^{i2\pi f(\tau_{\theta_t 2} - \tau_{\theta_t 1})}, \ldots, e^{i2\pi f(\tau_{\theta_t P} - \tau_{\theta_t 1})}], \tag{3}$$

where $\tau_{\theta_t p}$ is the the time it takes for the sound to travel from target source at direction θ_t to microphone p $(p = 1, \ldots, P)$ on a direct path.

In practice since $\mathbf{R}_{i+n,fm}$ is unknown, it is often replaced by the sample covariance matrix $\widehat{\mathbf{R}}_{\mathbf{x},fm} = \mathbf{x}_{fm}\mathbf{x}_{fm}^H$ [9]. A more advanced approach [10], known as robust adaptive beamforming, proposes to estimate this interference-plus-noise covariance matrix by integrating the spatial spectrum distribution over all possible directions containing unwanted signals Θ_{i+n} as

$$\widehat{\mathbf{R}}_{i+n,fm} = \int_{\Theta_{i+n}} \frac{\mathbf{d}_{\theta,f}\mathbf{d}_{\theta,f}^H}{\mathbf{d}_{\theta,f}^H \widehat{\mathbf{R}}_{\mathbf{x},fm}^{-1}\mathbf{d}_{\theta,f}} d\theta, \tag{4}$$

where $\mathbf{d}_{\theta,f}$ is computed similarly to (3) for direction θ.

Proposed Implementation of Multiple Beamformers. In our implementation, in order to reduce the computation cost for smartphone application we first replace the integration (4) by the sum over several unwanted directions ($e.g.$, $\widehat{\Theta}_{i+n} = 0^0, 45^0, 135^0, 180^0$ when the target direction is $\theta_t = 90^0$). Additionally, we incorporate the diagonal loading technique investigated in [16] to enhance the directivity pattern design. The resulting weight of the proposed beamforming implementation, named robust MVDR (RMVDR), is estimated as

$$\widehat{\mathbf{w}}_{\theta_t,f} = \frac{(\widehat{\mathbf{R}}_{i+n,fm} + \gamma\mathbf{I})^{-1}\mathbf{d}_{\theta_t,f}}{\mathbf{d}_{\theta_t,f}^H(\widehat{\mathbf{R}}_{i+n,fm} + \gamma\mathbf{I})^{-1}\mathbf{d}_{\theta_t,f}}, \tag{5}$$

where $\mathbf{I}$ is the $P \times P$ identity matrix, γ is a loading factor preventing instability [16], and the interference-plus-noise covariance matrix is computed by

$$\widehat{\mathbf{R}}_{i+n,fm} = \sum_{\theta \in \widehat{\Theta}_{i+n}} \frac{\mathbf{d}_{\theta,f}\mathbf{d}_{\theta,f}^H}{\mathbf{d}_{\theta,f}^H \widehat{\mathbf{R}}_{\mathbf{x},fm}^{-1}\mathbf{d}_{\theta,f}}. \tag{6}$$

Since equipping a large microphone array for a smartphone so that beamforming can isolate well the target source is not feasible in practice, we further propose to use multiple RMVDR where one of them enhances the target source ($i.e.$, RMVDR_90^0) and the others enhance unwanted sound coming from other directions ($e.g.$, RMVDR_0^0 and RMVDR_180^0, these beamformers have look directions perpendicular to the desired one). This implementation is depicted in Fig. 1 for the case when three RMVDRs are used. Note that the overall computational cost does not increase linearly with respect to the number of RMVDRs used since the sample covariance matrix $\widehat{\mathbf{R}}_{\mathbf{x},fm}$ needs to be computed once, and similarly steering vectors $\mathbf{d}_{\theta,f}$ needed in (5) and (6) can be shared between RMV-DRs. We will discuss the post-processing of the outputs of these beamformers so as to further isolate the target sound source, compared to the conventional case where only RMVDR_90^0 is used, in the following section.

Proposed Post-processing Algorithm. Denoting by $\hat{s}_{\theta_t,fm}$ and $\hat{s}_{\theta,fm}$ the output of RMVDRs looking at the target direction θ_t and other directions $\theta \neq \theta_t$, respectively. As example in our setting shown in Fig. 1, $\theta_t = 90^0$ while $\theta = 0^0$ or 180^0. However, one can easily extend the algorithm with the use of more RMVDRs and any desired direction than the 90^0. We propose to compute the STFT coefficients of the post-processed output signal for audio zoom as

$$\hat{s}_{fm} = \begin{cases} x_{p,fm} & \text{if } |\hat{s}_{\theta_t,fm}| > \alpha \max\{|\hat{s}_{\theta,fm}|, \forall\theta \neq \theta_t\} \\ \beta_{fm}\hat{s}_{\theta_t,fm} & \text{otherwise} \end{cases} \tag{7}$$

where $|.|$ denotes the absolute value, p denotes a reference microphone signal such as $p = 2$ for a front microphone in our setting, $\alpha > 1$ is a tuning constant, and

$$\beta_{fm} = \frac{1}{\epsilon + \frac{\max\{|\hat{s}_{\theta,fm}|, \forall\theta \neq \theta_t\}}{|\hat{s}_{\theta_t,fm}|}} \tag{8}$$

where ϵ is a constant (e.g., $\epsilon = 1$).

Our derivation to Eqs. (7) and (8) is motivated by the well-known observation that the sound sources are usually non-overlapped in the time-frequency (T-F) domain. As can be seen from the first line of (7), for time-frequency (T-F) points where the estimated target source is really dominant than the others, we take signal from a front microphone $x_{p,fm}$ as the final output so as to maximize the sound quality[5]. In this case, a reference microphone signal is a good estimate of the target source since other sources are considered to be inactive. Otherwise, the estimated target STFT coefficients $\hat{s}_{\theta_t,fm}$ will be considered. The derivation to Eq. (8) can be explained by the fact that in T-F points where sound from non-desired directions is really dominant (i.e., $\max\{|\hat{s}_{\theta,fm}|, \forall\theta \neq \theta_t\} \gg \hat{s}_{\theta_t,fm}$), the target source $\hat{s}_{fm}$ should be considered as inactive. Thus its value should close to 0 as β_{fm} will be very small. In neutral case where none of the estimated sources is really dominant, the smaller $\hat{s}_{\theta_t,fm}$ compared to the other sources, the more amplification it should be, as β_{fm} increase, in order to further improve the designed zooming effect as presented in Sect. 2.2. Finally, the time domain signal $\hat{s}(t)$ of the enhanced target source is obtained by the inverse STFT of $\hat{s}_{fm}$.

2.2 Proposed Audio Zoom Effect Creation

Let us denote by $z \in [0, 1]$ the zooming factor where the higher value of z the more target sound source is focused, and $z = 1$ corresponds to the maximum

---

[5] Note that in the output of RMVDR there is usually some artifact due to the nonlinear processing, and the signal distortion is more severe at high frequencies where the array's geometry error has more impact.

zoom (*i.e.*, 100 %). In order to maintain spatial effect of the perceived stereo output signal, we propose to mix the estimated target source after the post-processing $\hat{s}(t)$ with the original signals recorded by left and right microphones, denoted as $x_l(t)$ and $x_r(t)$, respectively. The final left and right channels of the output signal, denoted by $\tilde{s}_l(t)$, and $\tilde{s}_l(t)$, respectively, are computed as

$$\tilde{s}_l(t) = z * \hat{s}(t) + (1 - z) * x_l(t), \tag{9}$$
$$\tilde{s}_r(t) = z * \hat{s}(t) + (1 - z) * x_r(t). \tag{10}$$

It can be seen that there is no zooming effect when $z = 0$, and when z increases the estimated target source $\hat{s}(t)$ contributes more to the output signal as it should be more progressively focused. In case of maximum zoom with $z = 1$, both output channels take the same value (*i.e.*, $\tilde{s}_l(t) = \tilde{s}_r(t) = \hat{s}(t)$) so that the user can experience spatial effect of the isolated sound as if it comes from the front direction ($\theta_t = 90^0$) and the target sound source is most focused.

3   Experiments

We fist describe the recording setup in Sect. 3.1. We then present the algorithm implementation and result of the subjective test where different users experienced audio zooming effect created by the proposed approach in Sect. 3.2.

3.1   Experiment Setup

In order to make a test close to the real situation, we built a mock-up containing four microphones mimicking a smartphone as shown in Fig. 2. In this setting, two microphones are located at the top and bottom of the mock-up as usual with most available smartphones, two other microphones are located at the back side so as to ease sound capture during the video recording. The detail (x, y, z) coordinates of these microphones, measured in centimeter, are (6.5, 2, 0.5); (3.3, 0, 0); (−0.033, 0, 0); (−6.5, 2, 0.5), respectively.

We performed two 40 s length indoor audio recordings without video capture. The setups are shown in Fig. 3(a) and (b), respectively, where M_1 and M_2 are two musical instruments while S_1 and S_2 are two speeches. In both cases, audio zoom algorithm aims to enhance two sound sources located near the center while progressively attenuating two other unwanted sources. For a more realistic evaluation of the user perception when audio zoom is performed together with video zoom, we made an additional outdoor recording in a park as shown in Fig. 3(c) where audio and video is captured together. The recording duration is 90 s and audio zoom algorithm aims to focus on the bird song while canceling surrounding sounds including human walking, speech, environmental wind, etc.

3.2   Result with Subjective Test

We developed an application with a friendly graphical user interface (GUI) so as user can perform audio zoom and experience the audio quality obtained by

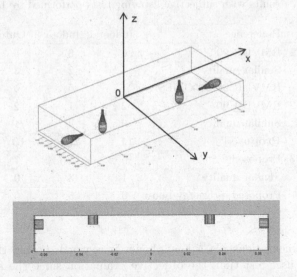

Fig. 2. Mock-up with 4 microphones for the experiment.

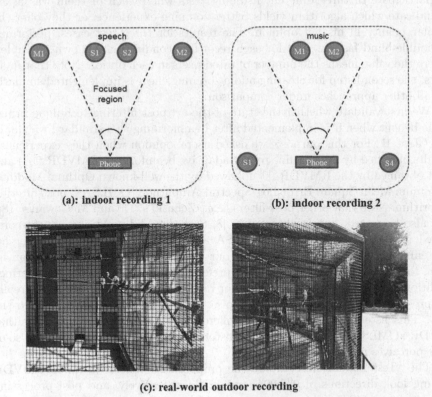

(a): indoor recording 1

(b): indoor recording 2

(c): real-world outdoor recording

Fig. 3. Experiment setup for user test on audio zoom feature.

Table 1. Results with subjective listening tests performed by 13 users.

Setup	Preference	Indoor 1	Indoor 2	Outdoor
Test 1	RMVDR_90^0	8	6	5
	Similar quality	4	2	3
	RMVDR_90^0 + OMLSA	1	5	5
Test 2	RMVDR_90^0	2	1	2
	Similar quality	0	2	1
	Proposed	**11**	**10**	**10**
Test 3	Proposed	6	5	7
	Similar quality	1	3	0
	Proposed + Energy boost	6	5	6

different implementations via a headphone. It is worth noting that we prefer the real-world user test than the objective evaluation since the former case is more relevant to the target application. We invited 13 people at different ages to participate in three different listening tests where each of them was asked to indicate which algorithm yields better zooming experience, or they offer the similar quality in his/her opinion. The results for three test cases, performed in double-blind fashion, and for each recording condition are shown in Table 1 where the value means the number of rated users in each option. Note that in all tests, the second step for creating audio zooming effect is implemented similarly for all other approaches under comparison.

We first validate whether the state-of-the-art post filtering technique brings some benefit when it is implemented after beamforming as a standard way [8] in the "Test 1". For this purpose, we asked users' opinion when they experienced results obtained by the baseline robust adaptive beamformer (RMVDR_90^0) and that obtained by the RMVDR_90^0 followed by the well-known Optimal Modified Minimum Mean-Square Error Log-Spectral Amplitude (OMLSA)[6] post-filtering algorithm. Note that other post-filters (*e.g.*, Zelinski's [17] and McCowan's [18]) can also be tested, but as observed in [8] that they did not bring benefit compared to OMLSA, we consider OMLSA as the state-of-the-art post filter for the enhanced single-channel signal in our implementation and test. As can be seen, for indoor recording more users prefer not to use OMLSA since it brings additional signal distortion. For outdoor recording, even though OMLSA really suppresses more background diffuse noise, it still does not bring benefit in the test. This listening test is actually coherent with the observation in [8] that MVDR+OMLSA adds further signal distortion compared to MVDR alone so as user perceives more artifact.

The "Test 2" aims to compare the proposed approach, *i.e.* three RMVDR having look directions of 0^0, 90^0, and 180^0, respectively, and post-processing

---

[6] Matlab code is available at: http://webee.technion.ac.il/Sites/People/IsraelCohen/Download/omlsa.m.

as shown in Fig. 1, (named "Proposed"), with the state-of-the-art beamforming approach using one RMVDR_90^0. Note that we did not compare to the case where RMVDR_90^0 is followed by OMLSA here since it has been shown in the Test 1 that users prefer RMVDR_90^0 alone. We also implemented a method using null-beamformers and post-processing algorithm described in [11], but subjectively observed that it performs poorer than the two considered algorithms, so we did not formally perform user test with it in order to avoid too much listening for users. As can be seen in Table 1, most users prefer the audio zoom quality obtained by the proposed approach in all three recording conditions. As example, for the real-world outdoor recording where audio zoom was performed together with video zoom to maximize the user experience, 10 users prefer the result of the proposed approach while only 2 users prefer the result of the baseline. It is also worth noting that our informal listening test in case of using two microphones, instead of four, also shares the same experience that the proposed approach performs better than the others.

The final test was devoted to the zooming effect only where we want to validate if increasing the volume of the enhanced signal can improve overall user experience. Thus we compare the "proposed" with a case where the enhanced signal after beamforming and post processing $\hat{s}(t)$ is boosted by 6 dB energy before mixing with the original microphone signals in the zooming creation step. The result is shown in "Test 3". Surprisingly, overall performance for three recording conditions shows that user experience is generally not improved as expected when increasing volume of the target sound. This can be explained by the fact that $\hat{s}(t)$ still contains noticeable distortion so that when its volume increases users also perceive more artifacts.

4 Conclusion

In this paper, we have presented a practical approach for performing audio zoom, an emerging application, in mobile devices with low computation cost. The proposed implementation combines several robust adaptive beamformers with a derived post-processing algorithm to further enhance the targeted sound source. We also describe the design of zooming effect so as to improve the user perceptual experience. Subjective tests with both real-world indoor and outdoor recordings confirm the effectiveness of the derived approach. Future research would be devoted to perform a formal objective evaluation where ground truth is available. Additionally, the investigation of audio source separation based approach where the target direction can be taken into account as prior information [19] would be potential.

References

1. Avendano, C., Solbach, L.: Audio zoom. US Patent Submitted 20 110 129 095A1 (2011). http://www.google.com/patents/US20110129095
2. Lee, K., Song, H., Lee, Y., Son, Y., Kim, J.: Mobile terminal and audio zooming method thereof. US Patent Submitted 20 130 342 730A1 (2013). http://www.google.com/patents/US20130342730

3. Veen, B.V., Buckley, K.: Beamforming: a versatile approach to spatial filtering. IEEE ASSP Mag. **5**(2), 4–24 (1988)
4. Li, J., Stoica, P.: Robust Adaptive Beamforming. Wiley, New York (2005)
5. Makino, S., Lee, T.-W., Sawada, H.: Blind Speech Separation. Springer, New York (2007)
6. Vincent, E., Araki, S., Theis, F., Nolte, G., Bofill, P., Sawada, H., Ozerov, A., Gowreesunker, V., Lutter, D., Duong, N.Q.K.: The signal separation campaign (2007–2010): achievements and remaining challenges. Sig. Process. **92**, 1928–1936 (2012)
7. Duong, N.Q.K., Vincent, E., Gribonval, R.: Under-determined reverberant audio source separation using a full-rank spatial covariance model. IEEE Trans. Audio Speech Lang. Process. **18**(7), 1830–1840 (2010)
8. Thiemann, J., Vincent, E.: An experimental comparison of source separation and beamforming techniques for microphone array signal enhancement. In: Proceedings of International Workshop on Machine Learning for Signal Processing (MLSP), pp. 1–5 (2013)
9. Bitzer, J., Simmer, K.U.: Superdirective microphone arrays. In: Brandstein, M., Ward, D. (eds.) Microphone Arrays. Digital Signal Processing, pp. 19–38. Springer, Heidelberg (2010)
10. Gu, Y., Leshem, A.: Robust adaptive beamforming based on interference covariance matrix reconstruction and steering vector estimation. IEEE Trans. Sig. Process. **60**(7), 3881–3885 (2012)
11. Takada, S., Kanba, S., Ogawa, T., Akagiri, K., Kobayashi, T.: Sound source separation using null-beamforming and spectral subtraction for mobile devices. In: Proceedings of IEEE Workshop on Applications of Signal Processing to Audio and Acoustics (WASPAA), pp. 30–33 (2007)
12. Bianchi, L., D'Amelio, F., Antonacci, F., Sarti, A., Tubaro, S.: A plenacoustic approach to acoustic signal extraction. In: Proceedings of IEEE Workshop on Applications of Signal Processing to Audio and Acoustics (WASPAA) (2005)
13. Markovich, S., Gannot, S., Cohen, I.: Multichannel eigenspace beamforming in a reverberant noisy environment with multiple interfereing speech signals. IEEE Trans. Audio Speech Lang. Process. **17**(6), 1071–1086 (2009)
14. Loesch, B., Yang, B.: Online blind source separation based on time-frequency sparseness. In: Proceedings of IEEE International Conference on Acoustics, Speech, and Signal Processing (ICASSP), pp. 117–120 (2009)
15. Masnadi-Shirazi, A., Rao, B.D.: Separation and tracking of multiple speakers in a reverberant environment using a multiple model particle filter glimpsing method. In: Proceedings of IEEE International Conference on Acoustics, Speech, and Signal Processing (ICASSP), pp. 2516–2519 (2011)
16. Mestre, X., Lagunas, M.: On diagonal loading for minimum variance beamformers. In: Proceedings of IEEE International Symposium on Signal Processing and Information Technology (ISSPIT), pp. 459–462 (2003)
17. Zelinski, R.: A microphone array with adaptive post-filtering for noise reduction in reverberant rooms. In: Proceedings of IEEE International Conference on Acoustics, Speech, and Signal Processing (ICASSP), pp. 2578–2581 (1998)
18. McCowan, I.A., Bourlard, H.: Microphone array post-filter for diffuse noise field. In: Proceedings of IEEE International Conference on Acoustics, Speech, and Signal Processing (ICASSP), pp. 1905–1908 (2002)
19. Duong, N.Q.K., Vincent, E., Gribonval, R.: Spatial location priors for gaussian model based reverberant audio source separation. EURASIP J. Adv. Sig. Process. **1**, 1–11 (2013)

Complex Valued Robust Multidimensional SOBI

Niko Lietzén[1]([✉]), Klaus Nordhausen[2,3], and Pauliina Ilmonen[1]

[1] Department of Mathematics and Systems Analysis,
Aalto University School of Science, P.O. Box 11100, 00076 Aalto, Finland
niko.lietzen@aalto.fi
[2] Department of Mathematics and Statistics, University of Turku,
20014 Turku, Finland
[3] School of Health Sciences, University of Tampere, 33014 Tampere, Finland

Abstract. Complex valued random variables and time series are common in various applications, for example in wireless communications, radar applications and magnetic resonance imaging. These applications often involve the famous blind source separation problem. However, the observations rarely fully follow specific models and robust methods that allow deviations from the model assumptions and endure outliers are required. We propose a new algorithm, robust multidimensional eSAM-SOBI, for complex valued blind source separation. The algorithm takes into account possible multidimensional spatial or temporal dependencies, whereas traditional SOBI-like procedures only consider dependencies in a single direction. In applications like functional magnetic resonance imaging, the dependencies are indeed not only one-dimensional. We provide a simulation study with complex valued data to illustrate the better performance of the methods that utilize multidimensional autocovariance in the presence of two-dimensional dependency. Moreover, we also examine the performance of the multidimensional eSAM-SOBI in the presence of outliers.

Keywords: Complex valued BSS · SOBI · Time series · Multidimensional autocovariance

1 Introduction

In statistics, a procedure that endures observations that violate some model assumptions is called robust. In particular, robust methods have been developed for situations where outliers are expected. For an overview of robust methods, see [1].

There has been an increasing interest towards robust methods in blind source separation (BSS) problems. In BSS the objective is to reverse the effects of a mixing process to find underlying structures from an observed data set. Usually, little is known about the mixing process and the latent sources. However, the sources can be estimated by making some model specific assumptions related to the latent sources. Applications which involve BSS include wireless

© Springer International Publishing AG 2017
P. Tichavský et al. (Eds.): LVA/ICA 2017, LNCS 10169, pp. 131–140, 2017.
DOI: 10.1007/978-3-319-53547-0_13

communications, radar applications and magnetic resonance imaging. Additionally, in these applications complex valued data occurs frequently. The literature related to complex valued BSS has been relatively narrow compared to the real counterpart.

Algorithms for complex valued BSS based on autocovariance matrices with different lags include for example AMUSE and SOBI [2]. General statistical properties of SOBI for real data have been given in [3] and robust properties of different SOBI algorithms have been discussed in [4]. These classical procedures assume that the latent sources are vectors that have one dimensional dependency structures.

However, in many applications the dependency structure is not one-dimensional. In the real valued case, an extension to multidimensional SOBI was given in [5]. In [5], multidimensional autocovariance matrices are applied to utilize the information contained in the multidimensional dependency structure. This approach is particularly fruitful when the latent sources are for example images. The approach proposed in [5] only considers real valued variables and it is sensitive to outliers. In this paper, we extend the previous work to include complex valued BSS and provide a more robust version of the algorithm. We then compare the performance of the different methods in the presence of multidimensional temporal dependence and outliers.

2   Preliminaries

Let $x(t)$ be a p-variate stochastic process and let $X = [x(1) \ldots x(T)]$ denote a realization of the corresponding process. The classical formulation of the complex valued BSS model is the following

$$x(t) = \Omega z(t) + \mu \quad \text{and} \quad t = 1, 2, \ldots, T, \tag{1}$$

where $x(t)$ is a p-variate complex valued stationary stochastic process, $z(t)$ is an unobservable p-variate complex valued stationary stochastic process, Ω is a $p \times p$ complex valued full-rank mixing matrix and μ is a p-variate location vector that is usually a nuisance parameter in the model. The objective in BSS is to find an unmixing matrix Γ, such that $\Gamma x(t)$ satisfies some model assumptions.

In this paper, we assume that x and z do not have dependence with respect to a single variable t. Instead, we assume that the dependencies are multidimensional i.e. $z = z(t_1, t_2, \ldots, t_m)$ and $x = x(t_1, t_2, \ldots, t_m)$. In practice, the observations are vectorized, but the multidimensional dependencies are considered in the analysis. We write $z(t_1, t_2)$ and $x(t_1, t_2)$ for variables with two-dimensional dependency structure. For the model, we fix a mapping from $z(t_1, t_2)$ to $z(t)$, which is achieved by simply vectorizing $z(t_1, t_2)$ according to the rows or the columns. Hereby, in the multidimensional case, we still have the BSS model from Eq. 1.

A p-vector valued functional $T(x(t))$ is said to be a location functional if it is affine equivariant in the sense that $T(Ax(t) + \mu) = AT(x(t)) + \mu$, for all full rank complex valued matrices A and complex valued p-vectors μ. Furthermore, a positive definite $p \times p$ matrix valued functional $S(x(t))$ is said to be a scatter functional if it is affine equivariant in the sense that $S(Ax(t) + \mu) = AS(x(t))A^*$, for all full rank complex valued matrices A and complex valued p-vectors μ. (Here, A^* denotes the conjugate transpose of A.) The corresponding sample statistics of location and scatter are obtained by replacing T and S by their sample counterparts, denoted by $\hat{T}$ and $\hat{S}$.

Note that many common estimators of centrality and spread are not affine equivariant and thus not true location and scatter measures, by definition. However, clearly the expected value and the regular covariance matrix are examples of location and scatter functionals.

The definition for regular autocovariance matrix with lag τ is

$$S_\tau(x(t)) = \mathbb{E}\left((x(t) - \mathbb{E}(x(t)))(x(t+\tau) - \mathbb{E}(x(t)))^*\right),$$

where the regular covariance matrix is obtained with $\tau = 0$. The autocovariance matrix is also, under mild model assumptions, an example of a scatter matrix functional. The sample version is given by

$$\hat{S}_\tau(X) = \frac{1}{T-\tau} \sum_{t=1}^{T-\tau} (x(t) - m_x)(x(t+\tau) - m_x)^*,$$

where m_x is the sample mean of X.

A complex valued second-order stationary time series $x(t)$ satisfies the following two conditions:

(S1) $\mathbb{E}(x(t)) = \mu$,   for every t,
(S2) $S_\tau(x(t)) = \Sigma_\tau$,   for every t and $\tau = 0, \pm 1, \pm 2, \ldots$,

where μ and every Σ_τ are finite constants.

In the literature, there exists many other location and scatter functionals that have different statistical properties. One example is the class of M-functionals. M-functionals of location, T, and scatter, V, satisfy the following implicit equations

$$T(x(t)) = \mathbb{E}(w_1(r))^{-1} \mathbb{E}(w_1(r)x(t)),$$
$$V(x(t)) = \mathbb{E}\left(w_2(r)(x - T(x(t)))(x - T(x(t)))^*\right),$$

where w_1 and w_2 are nonnegative continuous functions of the Mahalanobis distance $r = \|V(x(t))^{-\frac{1}{2}}(x(t) - T(x(t)))\|_F$ and $\|\cdot\|_F$ is the Frobenius norm. The Hettmansperger-Randels estimator [6] is obtained with weight functions $w_1(r) = 1/r$ and $w_2(r) = p/r^2$. This estimator is essentially a combination of the spatial median and Tyler's shape matrix. This estimator is considered to be one of the most robust M-estimators. Furthermore, the Hettmansperger-Randels estimators are affine equivariant.

Assume that $x(t)$ is centered. Multidimensional complex valued autocovariance matrix functionals with different lag sets $\mathcal{T} = \{\tau_1, \tau_2, \ldots, \tau_m\}$, where $\tau_i \in \{0, 1, 2, \ldots\}$, are defined as follows

$$S_{\mathcal{T}}(x(t_1, t_2, \ldots, t_m)) = \mathbb{E}\left(x(t_1, t_2, \ldots, t_m)x(t_1 + \tau_1, t_2 + \tau_2, \ldots, t_m + \tau_m)^*\right).$$

The sample version is given by replacing the expected values by averages. For example, consider three $h \times w$ matrices. (Matrices can represent three images, for example.) Assume that there exists both, row-wise and column-wise dependencies. Assume that the three matrices are vectorized and mixed. That is, the full data is given as a large $3 \times T$ matrix. We use the notation $x(k, j) = (x_1(k, j), x_2(k, j), x_3(k, j))$ for the column of the full large data matrix that contains the (k, j) elements of the original three $h \times w$ matrices (that is the (k, j) pixels of the three images, for example). The corresponding multidimensional complex valued autocovariance matrix estimate is then given by

$$\hat{S}_{\tau_1, \tau_2}\left(X\left(t_1, t_2\right)\right) = \frac{1}{N} \sum_{k=1}^{h-\tau_1} \sum_{j=1}^{w-\tau_2} x\left(k, j\right) x\left(k + \tau_1, j + \tau_2\right)^*,$$

where $N = (h - \tau_1)(w - \tau_2)$.

3   Algorithms

In this section we consider algorithms in the case of two-dimensional dependence structures (i.e. for example images). We assume that we have p observations that are $h \times w$ matrices. It is assumed that the p matrices are, if vectorized, linear combinations of p unobservable vectorized $h \times w$ matrices. The goal is to find the corresponding $p \times p$ mixing/unmixing matrix and reveal the unobservable hidden $h \times w$ matrices. The algorithms that we consider are SOBI, eSAM-SOBI, mdSOBI and the new robust multidimensional SOBI (RmdSOBI).

All the algorithms presented here begin by vectorizing the matrices and whitening the obtained p-variate data. Note that the whitening and robust whitening do not utilize the order of the observations. Thus, the way the matrices are vectorized (for example by columns or by rows) does not have an effect on the whitening step. However, for SOBI and eSAM-SOBI the choice of vectorization has an impact on the estimation of the autocovariance matrices. On the contrary, the methods utilizing multidimensional autocovariances are not affected by this choice.

3.1   SOBI

The SOBI algorithm was originally presented in [2]. The algorithm is based on first whitening the data and then jointly diagonalizing K autocovariance matrices.

1. Vectorizing the $h \times w$ matrices with respect to rows or columns. Series $x(t)$ is obtained.
2. Whitening of the series: $y(t) = \hat{S}_0^{-1/2}\left(x(t)\right)\left(x(t) - m_x\right)$, where m_x is the sample mean vector of $x(t)$, and where $A^{-1/2}$ denotes the conjugate symmetric square root of A.
3. Computing K autocovariance matrices: $\hat{S}_{\tau_i}\left(y(t)\right)$, with different lags τ_i.

4. Symmetrizing the autocovariance matrices: $\hat{S}_{\tau_i}^{S} = \left(\hat{S}_{\tau_i} + \hat{S}_{\tau_i}^{*} \right) / 2$.

5. Finding a unitary matrix U that jointly diagonalizes the K matrices $\hat{S}_{\tau_i}^{S}$ computed in step 4.

3.2 Affine Equivariant SAM-SOBI

An affine equivariant robust version of SOBI, eSAM-SOBI, was introduced in [4]. In eSAM-SOBI, the whitening is performed using the Hettmansperger-Randels estimators of location and scatter. The second scatter matrix is then the joint diagonalizer of spatial sign autocovariance matrices.

For centered time series, the spatial sign autocovariance matrix with lag τ, defined as

$$R_\tau = \mathbb{E} \left(\frac{x(t)}{||x(t)||_F} \frac{x(t+\tau)}{||x(t+\tau)||_F}^{*} \right),$$

provides a robust version of the regular autocovariance matrix. For centered data, the corresponding estimator is $\hat{R}_{\tau_i} = \frac{1}{T-\tau_i} \sum_{t=1}^{T-\tau_i} \left(\frac{x(t)}{||x(t)||_F} \frac{x(t+\tau_i)}{||x(t+\tau_i)||_F}^{*} \right)$.

1. Vectorizing the $h \times w$ matrices with respect to rows or columns: Series $x(t)$ is obtained.
2. Robust whitening of the series: $y(t) = \hat{V}^{-1/2} \left(x(t) \right) \left(x(t) - \hat{T} \left(x(t) \right) \right)$, where $\hat{T}$ and $\hat{V}$ are the Hettmansperger-Randels estimates of location and scatter.
3. Computing K spatial autocovariance matrices: $\hat{R}_{\tau_i}$.
4. Symmetrizing the robust autocovariance matrices: $\hat{R}_{\tau_i}^{S} = \left(\hat{R}_{\tau_i} + \hat{R}_{\tau_i}^{*} \right) / 2$
5. Finding a unitary matrix U that jointly diagonalizes the K matrices $\hat{R}_{\tau_i}^{S}$ computed in step 4.

3.3 Multidimensional SOBI

An extension to SOBI, named multidimensional SOBI (mdSOBI), was proposed in [7]. The main advantage compared to regular SOBI is that dependencies can be considered in m directions.

1. Vectorizing the $h \times w$ matrices with respect to rows or columns. Series $x(t)$ is obtained.
2. Whitening of the series: $y(t) = \hat{S}_0^{-1/2} \left(x(t) \right) \left(x(t) - m_x \right)$, where m_x is the sample mean vector of $x(t)$.
3. Computing K multidimensional autocovariance matrices: $\hat{S}_{\tau_{i,1}, \tau_{i,2}} \left(Y(t_1, t_2) \right)$, with different lags sets $\{ \tau_{i,1}, \tau_{i,2} \}$.
4. Symmetrizing the multidimensional autocovariance matrices:
$\hat{S}_{\tau_{i,1}, \tau_{i,2}}^{S} = \left(\hat{S}_{\tau_{i,1}, \tau_{i,2}} + \hat{S}_{\tau_{i,1}, \tau_{i,2}}^{*} \right) / 2$
5. Finding a unitary matrix U that jointly diagonalizes the K matrices $\hat{S}_{\tau_{i,1}, \tau_{i,2}}^{S}$ computed in step 4.

3.4   Robust Multidimensional SOBI

As a novel contribution we extend the mdSOBI method for complex valued data and at the same time we robustify the method. For this purpose we propose a new multidimensional complex valued spatial sign autocovariance matrix functional. Assume that $x(t)$ is centered. The novel multidimensional complex valued spatial sign autocovariance matrix functionals with different lag sets $\mathcal{T} = \{\tau_1, \tau_2, \ldots, \tau_m\}$, where $\tau_i \in \{0, 1, 2, \ldots\}$ is then defined as follows

$$R_{\mathcal{T}}\left(x\left(t_1, \ldots, t_m\right)\right) = \mathbb{E}\left(\frac{x\left(t_1, \ldots, t_m\right)}{\|x\left(t_1, \ldots, t_m\right)\|_F} \frac{x\left(t_1 + \tau_1, \ldots, t_m + \tau_m\right)}{\|x\left(t_1 + \tau_1, \ldots, t_m + \tau_m\right)\|_F}^*\right).$$

The sample version is given by replacing the expected values by averages. In the case of two-dimensional dependency structure that we consider in the algorithm of this section, the corresponding multidimensional complex valued spatial sign autocovariance matrix estimate is then given by

$$\hat{R}_{\tau_1, \tau_2}\left(X\left(t_1, t_2\right)\right) = \frac{1}{N} \sum_{k=1}^{h-\tau_1} \sum_{j=1}^{w-\tau_2} \frac{x\left(k, j\right)}{\|x\left(k, j\right)\|_F} \frac{x\left(k + \tau_1, j + \tau_2\right)}{\|x\left(k + \tau_1, j + \tau_2\right)\|_F}^*,$$

where $N = (h - \tau_1)(w - \tau_2)$, and where we again use the notation $x(k, j) = (x_1(k, j), \ldots x_p(k, j))$ for the column of the full large data matrix that contains the (k, j) elements of the original p matrices of size $h \times w$.

1. Vectorizing the $h \times w$ matrices with respect to rows or columns. Series $x(t)$ is obtained.
2. Robust whitening of the series: $y(t) = \hat{V}^{-1/2}\left(x(t)\right)\left(x(t) - \hat{T}\left(x(t)\right)\right)$, where $\hat{T}$ and $\hat{V}$ are the Hettmansperger-Randels estimates of location and scatter.
3. Computing K multidimensional spatial sign autocovariance matrices: $\hat{R}_{\tau_{i,1}, \tau_{i,2}}\left(Y(t_1, t_2)\right)$, with different lags sets $\{\tau_{i,1}, \tau_{i,2}\}$.
4. Symmetrizing the multidimensional autocovariance matrices: $\hat{R}^{S}_{\tau_{i,1}, \tau_{i,2}} = \left(\hat{R}_{\tau_{i,1}, \tau_{i,2}} + \hat{R}^{*}_{\tau_{i,1}, \tau_{i,2}}\right)/2$
5. Finding a unitary matrix U that jointly diagonalizes the K matrices $\hat{R}^{S}_{\tau_{i,1}, \tau_{i,2}}$ computed in step 4.

To ensure that all algorithms presented here estimate the same population quantity, we have to require that all vectorized sources are symmetrically distributed.

4   Simulation Study

We performed a simulation study to compare the four different versions of SOBI, discussed in Sect. 3. To evaluate the performance, we used the extension of the minimum distance (MD) index for complex valued ICA, see [8,9]. The MD index is defined as:

$$MD = \frac{1}{\sqrt{p - 1}} \inf_{PD} \|PD\hat{\Gamma}\Omega - I_p\|_F,$$

where Ω is the known mixing matrix, $\hat{\Gamma}$ is the unmixing matrix estimate, P is a permutation matrix and D is a complex valued diagonal matrix. The values of MD are between 0 and 1, where 0 corresponds to complete separation.

In the simulation study, the latent source signals were matrices with a two dimensional spatial dependency structure. Consider a so called neighboring matrix W, which is a $m \times m$ matrix with diagonal elements zero and $w_{ij} = 1$, if i and j are considered to be neighbors and $w_{ij} = 0$ otherwise. The neighboring matrix is utilized in the simultaneous autoregressive model (SAR). The SAR model is for example popular in spatial econometrics, see for example [10,11] for details. The model can be defined as

$$y = (I_n - \rho W)^{-1} \varepsilon,$$

where ε is a complex valued random vector with $\mathbb{E}(\varepsilon) = 0$, ρ is a real valued parameter from the following interval $\{\rho : \max(1/\lambda_{min}, 0) \le \rho < \min(1/\lambda_{max}, 1)\}$, λ_{min} and λ_{max} are the largest and smallest eigenvalues of W and y is then a complex valued random spatial process with a multivariate dependency structure.

Assume that we have an image of $n \times n$ pixels in a vectorized form. Then, if the ith and jth element of the vectorized image are considered to be neighbors in the original image, set $w_{ij} = 1$ and otherwise $w_{ij} = 0$. In the simulation study we used the three different neighborhood structures presented in Fig. 1. The figure presents the neighbors of every pixel in the original image, where the black cell is the pixel of interest and the gray cells are the neighbors of the pixel of interest. The matrices W were then formed by considering every pixel separately as the point of interest. Hereby, an $n \times n$ image generates an $n^2 \times n^2$ neighborhood matrix W.

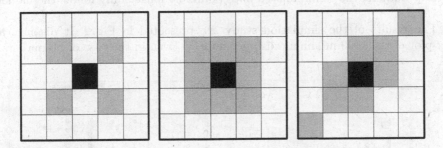

Fig. 1. Neighborhood structures used to generate the different W matrices. Black denotes the point of interest and gray denotes a neighboring cell. The left pattern is the corresponding pattern in W_1, the middle in W_2 and the right in W_3.

In the simulation study, we had 30×30 matrices with patterns presented in Fig. 1 yielding $30^2 \times 30^2$ neighborhood matrices W_1, W_2 and W_3. After the generation of the W matrices, we chose the following feasible ρ parameters, $\rho_1 = 0.2$, $\rho_2 = 0.1$ and $\rho_3 = 0.05$ and calculated the inverses $(I_{n^2} - \rho_i W_i)^{-1}$, $i = 1, 2, 3$. We then repeated the following steps 1000 times:

1. Generate $n^2 = 900$ observations $\varepsilon_1, \varepsilon_2, \varepsilon_3$ independently from univariate standard complex normal distribution.
2. Obtain z_i from $z_i = (I_{n^2} - \rho_i W_i)^{-1} \varepsilon_i$, $i = 1, 2, 3$.
3. Obtain $X = \Omega Z$, where $Z = (z_1, z_2, z_3)$ and Ω is a fixed mixing matrix shown below.
4. Find the estimate $\hat{\Gamma}$ using the four different algorithms.
5. Calculate MD index for the different estimates.

We used lags $\tau = 1, 2, \ldots, 16$ for SOBI and eSAM-SOBI. Vectorization was performed both by columns and by rows. The following lag pairs were used

$$\mathcal{T} = \{\{1,0\}, \{2,0\}, \{3,0\}, \{1,2\}, \{4,1\}, \{1,1\}, \{1,2\}, \{2,3\}, \{0,1\}, \{0,2\},$$
$$\{0,3\}, \{2,1\}, \{1,4\}, \{2,1\}, \{3,2\}\},$$

for mdSOBI and robust mdSOBI. The mixing matrix Ω was the following

$$\Omega = \begin{pmatrix} 0.1 + 0.9i & -0.4 + 0.15i & -0.2 + 0.6i \\ -0.3 - 0.2i & 0.9 - 0.2i & 0.1 + 0.2i \\ 0.2 + 0.1i & 0 - 0.5i & 0.1 + 0.1i \end{pmatrix}.$$

As all methods are affine equivariant the mixing matrix Ω has no effect on the simulations when no outliers are present but must be considered when designing outliers. To evaluate the robustness of the estimators, we then had two different outlier scenarios. Consider X in vectorized form. The two scenarios were, (i) isolated outlier: observation $X(100)$ was replaced with an isolated outlier $(100 + 100i, 100 + 100i, 100 + 100i)$; (ii) patchy outliers: observations $X(101), \ldots, X(116)$ were added with noise coming from a three-variate complex normal distribution with mean $20 + 20i$ and variance 1. With the presence of these outliers, the contaminated data points no longer fully follow the model $X = \Omega Z$.

The results of the simulation study are presented in Fig. 2. It displays a box-plot of the 1000 minimum distance indices. Even in the case of no outliers,

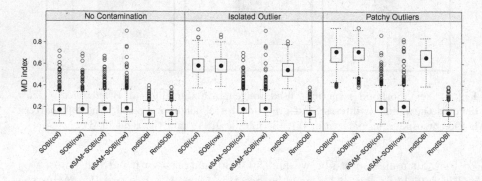

Fig. 2. Box-plots of the MD indices for SOBI, eSAM-SOBI, mdSOBI and robust mdSOBI (RmdSOBI). For SOBI and eSAM-SOBI vectorization was performed both, by rows and by columns.

the performance of SOBI and eSAM-SOBI is considerably worse compared to their multidimensional counterparts. Furthermore, SOBI fails completely in the two settings where outliers are present. However, eSAM-SOBI seems to be quite unaffected by the outliers. For SOBI and eSAM-SOBI vectorization was performed both by rows and by columns. However, since the neighboring patterns are symmetric, the results were similar.

The multidimensional methods perform both relatively well in the case of no outliers. However, the performance of mdSOBI is considerably worse in the case of outliers compared to robust mdSOBI. Robust mdSOBI performs well in all settings.

In the simulation study, the robust mdSOBI has the best performance in the case of multidimensional dependency structure and in the presence of outliers.

5   Conclusions

In this paper we proposed a new robust algorithm for complex valued blind source signal separation. The algorithm utilizes multidimensional dependency structures of the data. In the real valued case, the superior performance of multidimensional SOBI over the vectorized SOBI was already illustrated in [7].

In this paper, we demonstrate similar results in the complex valued case. Simulations show that in the presence of multidimensional dependency, the robust mdSOBI outperforms its one-dimensional counterpart and its non-robust counterparts. Investigating the performance of the robust mdSOBI with authentic data from some application might be worthwhile in the future. Note that mdSOBI requires that the spatial data comes from a regular grid. Irregularly spaced real valued data has been considered for example in [12]. It would be of interest to also consider irregularly spaced complex valued data.

Acknowledgements. The work of Klaus Nordhausen was supported by the Academy of Finland Grant 268703. The authors wish to thank the referees for their excellent comments that helped improve the paper.

References

1. Maronna, R., Martin, D., Yohai, V.: Robust Statistics: Theory and Methods. Wiley, Chichester (2006)
2. Belouchrani, A., Abed-Meraim, K., Cardoso, J.F., Moulines, E.: A blind source separation technique using second-order statistics. IEEE Trans. Signal Process. **45**(2), 434–444 (1997)
3. Miettinen, J., Illner, K., Nordhausen, K., Oja, H., Taskinen, S., Theis, F.J.: Separation of uncorrelated stationary time series using autocovariance matrices. J. Time Ser. Anal. **37**, 336–354 (2016)
4. Ilmonen, P., Nordhausen, K., Oja, H., Theis, F.: An affine equivariant robust second-order BSS method. In: Vincent, E., Yeredor, A., Koldovský, Z., Tichavský, P. (eds.) LVA/ICA 2015. LNCS, vol. 9237, pp. 328–335. Springer, Heidelberg (2015). doi:10.1007/978-3-319-22482-4_38

5. Theis, F.J., Meyer-Bäse, A., Lang, E.W.: Second-order blind source separation based on multi-dimensional autocovariances. In: Puntonet, C.G., Prieto, A. (eds.) ICA 2004. LNCS, vol. 3195, pp. 726–733. Springer, Heidelberg (2004). doi:10.1007/978-3-540-30110-3_92

6. Hettmansperger, T.P., Randles, R.H.: A practical affine equivariant multivariate median. Biometrika **89**(4), 851–860 (2002)

7. Theis, F.J., Müller, N.S., Plant, C., Böhm, C.: Robust second-order source separation identifies experimental responses in biomedical imaging. In: Vigneron, V., Zarzoso, V., Moreau, E., Gribonval, R., Vincent, E. (eds.) LVA/ICA 2010. LNCS, vol. 6365, pp. 466–473. Springer, Heidelberg (2010). doi:10.1007/978-3-642-15995-4_58

8. Ilmonen, P., Nordhausen, K., Oja, H., Ollila, E.: A new performance index for ICA: properties, computation and asymptotic analysis. In: Vigneron, V., Zarzoso, V., Moreau, E., Gribonval, R., Vincent, E. (eds.) LVA/ICA 2010. LNCS, vol. 6365, pp. 229–236. Springer, Heidelberg (2010). doi:10.1007/978-3-642-15995-4_29

9. Lietzén, N., Nordhausen, K., Ilmonen, P.: Minimum distance index for complex valued ICA. Stat. Prob. Lett. **118**, 100–106 (2016)

10. Haining, R.: Spatial Data Analysis in the Social and Environmental Sciences. Cambridge University Press, Cambridge (1990)

11. Schabenberger, O., Gotway, C.A.: Statistical Methods for Spatial Data Analysis. CRC Press, Boca Ranton (2005)

12. Nordhausen, K., Oja, H., Filzmoser, P., Reimann, C.: Blind source separation for spatially correlated compositional data. Math. Geosci. **47**, 753–770 (2015)

Ego Noise Reduction for Hose-Shaped Rescue Robot Combining Independent Low-Rank Matrix Analysis and Multichannel Noise Cancellation

Narumi Mae[1(✉)], Masaru Ishimura[1], Shoji Makino[1], Daichi Kitamura[2], Nobutaka Ono[2,3], Takeshi Yamada[1], and Hiroshi Saruwatari[4]

[1] University of Tsukuba, 1-1-1 Tennodai, Tsukuba, Ibaraki 305-8573, Japan
{mae,ishimura}@mmlab.cs.tsukuba.ac.jp, maki@tara.tsukuba.ac.jp,
takeshi@cs.tsukuba.ac.jp
[2] SOKENDAI (The Graduate University for Advanced Studies),
Shonan Village, Hayama, Kanagawa 240-0193, Japan
d-kitamura@nii.ac.jp
[3] National Institute of Informatics (NII),
2-1-2 Hitotsubashi, Chiyoda-ku, Tokyo 101-8430, Japan
onono@nii.ac.jp
[4] The University of Tokyo, 7-3-1 Hongo, Bunkyo-ku, Tokyo 113-8654, Japan
hiroshi_saruwatari@ipc.i.u-tokyo.ac.jp

Abstract. In this paper, we present an ego noise reduction method for a hose-shaped rescue robot, developed for search and rescue operations in large-scale disasters. It is used to search for victims in disaster sites by capturing their voices with its microphone array. However, ego noises are mixed with voices, and it is difficult to differentiate them from a call for help from a disaster victim. To solve this problem, we here propose a two-step noise reduction method involving the following: (1) the estimation of both speech and ego noise signals from observed multichannel signals by multichannel nonnegative matrix factorization (NMF) with the rank-1 spatial constraint, and (2) the application of multichannel noise cancellation to the estimated speech signal using reference signals. Our evaluations show that this approach is effective for suppressing ego noise.

Keywords: Rescue robot · Tough environment · Noise reduction · Nonnegative matrix factorization · Independent vector analysis · Multichannel noise cancellation

1 Introduction

It is important to develop robots for search and rescue operations during large-scale disasters such as earthquakes. Robots are required for emergency responses and for the restoration of disaster sites, which are difficult and dangerous tasks

© Springer International Publishing AG 2017
P. Tichavský et al. (Eds.): LVA/ICA 2017, LNCS 10169, pp. 141–151, 2017.
DOI: 10.1007/978-3-319-53547-0_14

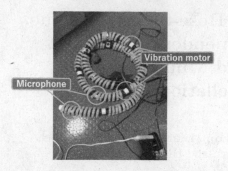

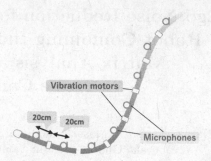

Fig. 1. Hose-shaped rescue robot.

Fig. 2. Structure of hose-shaped rescue robot.

for humans. The "Tough Robotics Challenge" is one of the research and development programs in the Impulsing Paradigm Change through Disruptive Technologies Program (ImPACT) [1]. One of the robots developed in this program is a hose-shaped rescue robot [2]. This robot is long and slim like a snake, allowing it to investigate narrow spaces into which conventional remotely operable robots cannot enter. This robot searches for a disaster victim by capturing his/her voice using a microphone array attached around itself at regular intervals. However, there is a serious problem of "ego noise". This noise is generated by the vibration motors used to move the robot via the vibrating cilia tape wrapped around the robot. In this study, we focus on reducing the ego noise from the sound recorded by the microphone array of the robot.

Recently, many ego noise reduction methods have been proposed [3–6]. In addition, the many microphones on the hose-shaped rescue robot enable the application of the overdetermined source separation method. However, the microphone arrangement changes as the robot moves, making it difficult to control the microphone array geometry. Hence, in [7], we proposed a noise reduction method for the hose-shaped rescue robot combining determined rank-1 multichannel nonnegative matrix factorization [8,9] proposed by Kitamura *et al.* which can be interpreted as an independent low-rank matrix analysis (hereafter referred to as ILRMA), and a noise canceller (NC). As a reference input of the NC, we used the sum of all the noise components of the ILRMA outputs. On the other hand, in this study, we use a multichannel NC and confirm the applicability of the proposed method for reducing ego noise.

2   Hose-Shaped Rescue Robot and Ego Noise

2.1   Hose-Shaped Rescue Robot

Figure 1 shows an image of the hose-shaped rescue robot and Fig. 2 shows its structure. The hose-shaped rescue robot basically consists of a hose as its axis

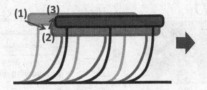

Fig. 3. Principle of movement of hose-shaped rescue robot [2].

with cilia tape wrapped around it; it moves forward slowly as a result of the reaction between the cilia and floor through the vibration of the cilia tape with the vibration motors. Figure 3 schematically shows the principle of movement of the hose-shaped rescue robot. When the motors vibrate, state (1) changes to state (2) through the friction between the cilia and the floor, then state (2) changes to state (3) as a result of the cilia slipping. The hose-shaped rescue robot moves by repeating such changes in its state. It performs various sensing functions using sensors such as microphones, cameras, an inertial measurement unit and light sensors.

2.2 Problem in Recording Speech

Recording speech using the hose-shaped rescue robot has a serious problem. During the operation of the robot, very loud ego noise is mixed in the input to the microphones. The main sources of the ego noise are the driving sound of the vibration motors, the fricative sound generated between the cilia and floor, and the noise generated by microphone vibration. In an actual disaster site, the voice of a person seeking help is not sufficiently loud to capture and it is smaller than the ego noise.

3 Overview of Independent Low-Rank Matrix Analysis

Recently, many ego noise reduction methods have been proposed [3–6]. In [3], noise reduction based on generalizations of K-singular value decomposition (K-SVD) was proposed, which can be used for an underdetermined multichannel situation. Also, the authors of [4,5] proposed a method of improving the performance of ego noise reduction using an adaptive microphone array geometry. On the other hand, the many microphones on the rescue robot enable the application of an overdetermined source separation method. In a determined situation, independent vector analysis (IVA) [10–12] is a commonly used method. IVA requires independence between the sources to estimate a demixing matrix. In general, in IVA, a spherical multivariate distribution is assumed as the source model to ensure a higher-order correlation between the frequency bins in all sources. However, this model does not include any particular information on the sources, that is, IVA cannot capture specific spectral structures of the sources. Thus, the utilization of nonnegative matrix factorization (NMF) [13–15] as the

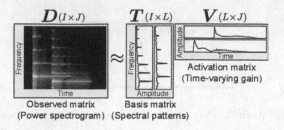

Fig. 4. Decomposition model of NMF.

source model has been proposed [8,9], which enables us to capture the spectral structures.

NMF decomposes a given spectrogram into several spectral bases T and temporal activations V, as shown in Fig. 4, then the decomposed components are clustered into each source. Multichannel NMF (MNMF) [16–18] is one of the techniques for clustering the NMF bases and activations using a sourcewise spatial model. MNMF separately models the mixing system and the nonnegative power spectra of sources. However, this method is strongly dependent on its initial values because there are no constraints in the spatial models.

To solve the problem of MNMF, ILRMA [8,9] was proposed, in which a rank-1 spatial model is introduced into MNMF [18]. This method estimates a demixing matrix while representing a source using NMF bases, and can be optimized by the update rules of IVA and conventional single-channel NMF. Therefore, ILRMA is a method that unifies IVA and NMF.

Since the hose-shaped rescue robot moves very slowly and the spatial locations of the sources and microphones barely change, we can assume a linear time-varying mixing system. In this case, ILRMA is effective for the separation because it does not require the locations of the sources and microphones. In particular, ILRMA can efficiently capture the time-frequency structure of the ego noise because it is the repetition of several types of similar spectra.

3.1   Formulation

We assume that M sources are observed using M microphones (determined case). The sources, the observed and separated signals in each time-frequency slot are as follows:

$$\boldsymbol{s}_{ij} = (s_{ij,1} \ \cdots \ s_{ij,M})^t, \tag{1}$$

$$\boldsymbol{x}_{ij} = (x_{ij,1} \ \cdots \ x_{ij,M})^t, \tag{2}$$

$$\boldsymbol{y}_{ij} = (y_{ij,1} \ \cdots \ y_{ij,M})^t, \tag{3}$$

where $1 \leq i \leq I$ and $1 \leq j \leq J$ are indexes of frequency and time, respectively, and t denotes the vector transpose. All the entries of these vectors are complex values. When the window size in an STFT is sufficiently longer than the impulse response between a source and microphone, we can approximately represent the observed signal as

$$x_{ij} = A_i s_{ij}. \tag{4}$$

Here, $A_i = (a_{i,1} \; \cdots \; a_{i,M})$ is an $M \times M$ mixing matrix of the observed signals. When $W_i = (w_{i,1} \; \cdots \; w_{i,M})^h$ denotes the demixing matrix, the separated signal y_{ij} is represented as

$$y_{ij} = W_i x_{ij}, \tag{5}$$

where h is the Hermitian transpose.

3.2 Independent Low-Rank Matrix Analysis

We use ILRMA [8,9] to impose a rank-1 spatial model on MNMF [18]. We explain the formulation and algorithm derived by Kitamura *et al.* [8,9] MNMF is an extension of simple NMF for multichannel signals. The observed signals are represented as

$$\mathsf{X}_{ij} = x_{ij} x_{ij}^h, \tag{6}$$

where X_{ij} of size $M \times M$ is the correlation matrix between channels. The diagonal elements of X_{ij} represent real-valued powers detected by the microphones, and the nondiagonal elements represent the complex-valued correlations between the microphones. The separation model of MNMF $\hat{\mathsf{X}}_{ij}$ used to approximate X_{ij} is represented as

$$\mathsf{X}_{ij} \approx \hat{\mathsf{X}}_{ij} = \sum_m \mathsf{H}_{i,m} \sum_l t_{il,m} v_{lj,m}, \tag{7}$$

where $m = 1 \cdots M$ is the index of the sound sources. $\mathsf{H}_{i,m}$ is an $M \times M$ spatial covariance matrix for each frequency i and source m, and $\mathsf{H}_{i,m} = a_{i,m} a_{i,m}^h$ is limited to a rank-1 matrix. This assumption corresponds to $t_{il,m} \in \mathbb{R}_+$ and $v_{lj,m} \in \mathbb{R}_+$ being the elements of the basis matrix T_m and activation matrix V_m, respectively. This rank-1 spatial constraint leads to the following cost function:

$$\mathcal{Q} = \sum_{i,j} \left[\sum_m \frac{|y_{ij,m}|^2}{\sum_l t_{il,m} v_{lj,m}} - 2 \log |\det W_i| + \sum_m \log \sum_l t_{il,m} v_{lj,m} \right], \tag{8}$$

namely, the estimation of $\mathsf{H}_{i,m}$ can be transformed to the estimation of the demixing matrix W_i. This cost function is equivalent to the Itakura-Saito divergence between X_{ij} and $\hat{\mathsf{X}}_{ij}$, and we can derive

$$t_{il,m} \leftarrow t_{il,m} \sqrt{\frac{\Sigma_j |y_{ij,m}|^2 v_{lj,m} \left(\Sigma_{l'} t_{il',m} v_{l'j,m}\right)^{-2}}{\Sigma_j v_{lj,m} \left(\Sigma_{l'} t_{il',m} v_{l'j,m}\right)^{-1}}}, \quad (9)$$

$$v_{lj,m} \leftarrow v_{il,m} \sqrt{\frac{\Sigma_i |y_{ij,m}|^2 t_{il,m} \left(\Sigma_{l'} t_{il',m} v_{l'j,m}\right)^{-2}}{\Sigma_i t_{il,m} \left(\Sigma_{l'} t_{il',m} v_{l'j,m}\right)^{-1}}}, \quad (10)$$

$$r_{ij,m} = \sum_l t_{il,m} v_{lj,m}, \quad (11)$$

$$V_{i,m} = \frac{1}{J} \sum_j \frac{1}{r_{ij,m}} \boldsymbol{x}_{ij} \boldsymbol{x}_{ij}^h, \quad (12)$$

$$\boldsymbol{w}_{i,m} \leftarrow \left(\boldsymbol{W}_i V_{i,m}\right)^{-1} \boldsymbol{e}_m, \quad (13)$$

where $\boldsymbol{e}_m$ is a unit vector whose mth element is one. We can simultaneously estimate both the sourcewise time-frequency model $r_{ij,m}$ and the demixing matrix $\boldsymbol{W}_i$ by iterating (9)–(13) alternately. After the cost function converges, the separated signal $\boldsymbol{y}_{ij}$ can be obtained as (5). Note that since the signal scale of $\boldsymbol{y}_{ij}$ cannot be determined, we apply a projection-back method [19] to $\boldsymbol{y}_{ij}$ to determine the scale.

The demixing filter in ILRMA is time-invariant over several seconds. To achieve time-variant noise reduction, in a previous study [7], we applied a single-channel NC for the postprocessing of ILRMA to reduce the remaining time-variant ego noise components. An NC usually requires a reference microphone to observe only the noise signal. Thus, we utilized the noise estimates obtained by ILRMA as the noise reference signals.

4   Multichannel Noise Canceller

4.1   Conventional Method

The NC proposed in [7,20] requires a reference microphone located near a noise source. The recorded noise reference signals $n_1(t), \ldots, n_k(t)$ are utilized to reduce

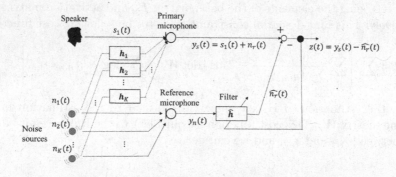

Fig. 5. Noise canceller.

the noise in the observed speech signal $s_1(t)$ as shown in Fig. 5. We here assume that both $s_1(t)$ and $n_1(t), \ldots, n_k(t)$ are simultaneously recorded. The observed signal contaminated with the noise source can be represented as

$$y_s(t) = s_1(t) + n_1(t) + \cdots + n_k(t). \tag{14}$$

We can consider that the noise signal $n_r(t)$ is strongly correlated with the reference noise signal $y_n(t)$ and that $n_r(t) = n_1(t) + \cdots + n_k(t)$ can be represented by a linear convolution model as

$$n_r(t) \simeq \hat{n}_r(t) = \hat{\boldsymbol{h}}(t)^t \boldsymbol{y}_n(t), \tag{15}$$

where $\boldsymbol{y}_n(t) = [y_n(t) \, y_n(t-1) \, \cdots \, y_n(t-N+1)]^t$ is the reference microphone input from the current time t to the past N samples and $\hat{\boldsymbol{h}}(t) = [\hat{\boldsymbol{h}}_1(t) \, \hat{\boldsymbol{h}}_2(t) \cdots \hat{\boldsymbol{h}}_N(t)]^t$ is the estimated impulse response. From (15), the speech signal $s_1(t)$ is extracted by subtracting the estimated noise $\hat{\boldsymbol{h}}(t)^t \boldsymbol{y}_n(t)$ from the observation as

$$z(t) = x(t) - \hat{\boldsymbol{h}}(t)^t \boldsymbol{y}_n(t), \tag{16}$$

where $z(t)$ is the estimated speech signal.

4.2 Proposed Method

In the conventional NC proposed in [7,20], we used the sum of all the noise components of the ILRMA outputs applied projection-back method to the same microphone. This means that the change in each mixing system $\boldsymbol{h}_1, \boldsymbol{h}_2, \ldots, \boldsymbol{h}_k$ in Fig. 5 is the same. However, the mixture systems may change differently according to the noise sources. Thus, in this study, we use a multichannel NC. Figure 6 shows the multichannel NC model. In this model, the filter of the multichannel NC is estimated for each noise source. Thereby, the filter and noise can be more precisely estimated. The filter $\hat{\boldsymbol{h}}(t)$ can be obtained by minimization of the mean square error. In this paper, we use the normalized least mean square (NLMS) algorithm [21] to estimate $\hat{\boldsymbol{h}}(t)$. From the NLMS algorithm, the update rule of the filter $\hat{\boldsymbol{h}}(t)$ is given as

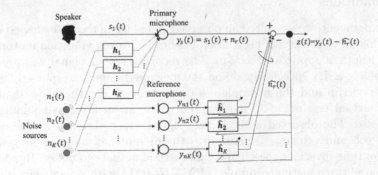

Fig. 6. Multichannel noise canceller.

$$\hat{\boldsymbol{h}}(t+1) = \hat{\boldsymbol{h}}(t) + \mu \frac{z(t)}{||\boldsymbol{y}_n(t)||^2} \boldsymbol{y}_n(t), \tag{17}$$

where

$$\hat{\boldsymbol{h}}(t) = [\hat{\boldsymbol{h}}_1(t)^t \ \hat{\boldsymbol{h}}_2(t)^t \cdots \ \hat{\boldsymbol{h}}_K(t)^t]^t, \tag{18}$$

$$\hat{\boldsymbol{h}}_k(t) = [\hat{h}_{k,0}(t) \ \hat{h}_{k,1}(t) \cdots \ \hat{h}_{k,N-1}(t)]^t, \tag{19}$$

$$\boldsymbol{y}_n(t) = [\boldsymbol{y}_{n1}(t)^t \ \boldsymbol{y}_{n2}(t)^t \cdots \ \boldsymbol{y}_{nK}(t)^t]^t, \tag{20}$$

$$\boldsymbol{y}_{nk}(t) = [y_{nk}(t), y_{nk}(t-1), \cdots, y_{nk}(t-N+1)]^t. \tag{21}$$

4.3 Flow of the Proposed Method

Figure 7 shows the flow of the proposed method. In Fig. 7, $y_{1(t)}$, ..., $y_{8(t)}$ are the ILRMA outputs, $y_s(t)$ is the speech signal estimated by ILRMA, and $y_{n1(t)}$, ..., $y_{n7(t)}$ are the residual outputs corresponding to the various components of ego noise. In the first step, the observed signals are separated into independent signals via ILRMA, where the number of separated signals is the same as the number of microphones ($M = 8$). Note that ILRMA cannot determine the order of the output signals. Therefore, we must find an estimated signal that includes most of the speech components to be used as $y_s(t)$. In this paper, we manually choose the speech estimate from the output signals, while such speech estimate detection may be possible by employing statistics or spectrograms of the output signals. Since a time-invariant spatial demixing matrix (demixing filter) is applied for the separation in the first step, the ego noise, which does not follow the time-invariant assumption, remains in the separated speech signal $y_s(t)$. In the second step, we apply the multichannel NC with the ego noise reference signals $y_{n1(t)}$, ..., $y_{n7(t)}$. In this step, we expect that the multichannel NC will reduce the residual noise component in $y_s(t)$ because the multichannel NC models the time-variant noise as $\hat{\boldsymbol{h}}_1(t)^t \boldsymbol{y}_{n1}(t)$, ..., $\hat{\boldsymbol{h}}_7(t)^t \boldsymbol{y}_{n7}(t)$, which can update the filter $\hat{\boldsymbol{h}}_1(t), \ldots, \hat{\boldsymbol{h}}_7(t)$ for each time sample.

5   Experiment

5.1 Conditions

In this experiment, we measured an observed signal using the hose-shaped rescue robot. This robot consists of eight microphones and seven vibration motors, and its total length is approximately 3 m. The recorded speech signal was produced by convolving a dry speech signal and the measured impulse responses between a disaster victim and the microphones on the robot. For the noise signal, we recorded actual ego noise by moving the robot in an area that simulated a disaster site. The observed multichannel signals were obtained as the sum of these speech and ego noise signals in each microphone, namely, they were a mixture of time-invariant speech and time-variant actual ego noises. In addition, we compared three methods: simple ILRMA, ILRMA with a single-channel NC,

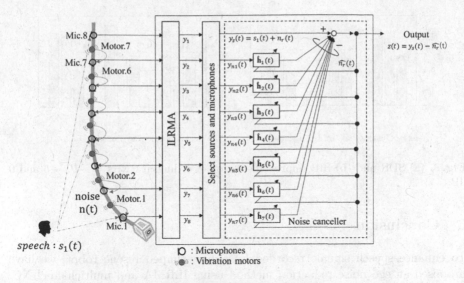

Fig. 7. Flow of proposed method.

Table 1. Experimental conditions

Sampling frequency	16 kHz
Window length	1024 samples
Window shift	STFT length/4
Number of bases	15
Number of iterations	100
Filter length of noise canceller	1600 taps
Step size of NLMS	0.1
Input SNR	0, −5, −10 dB

and the proposed method (ILRMA + multichannel NC). The signal-to-distortion ratio (SDR) and the signal-to-interference ratio (SIR) [22] were used to evaluate the separation performance. The other experimental conditions are shown in Table 1. The estimated signal that includes most of the speech components was projected back to microphone 1. Also, each estimated noise signal was projected back to microphone 2.

5.2   Results

Figure 8 shows the improvements in the SDR and SIR for each method. The results show that the multichannel NC improves the separation performance. This is because the multichannel NC efficiently estimates the changes in each filter from the estimation result of ILRMA.

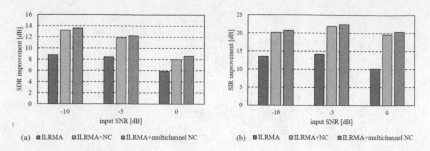

Fig. 8. (a) SDR and (b) SIR improvements for recording at SNRs = −10, −5 and 0 dB.

6   Conclusion

To enhance speech signals recorded by a hose-shaped rescue robot, we have proposed an ego noise reduction method using ILRMA and multichannel NC. We evaluated the proposed method by an experiment and compared ILRMA, ILRMA with a single-channel NC, and ILRMA with a multichannel NC in terms of the SDR and SIR. It was found that the proposed method exhibited the best performance under all conditions, thus confirming the effectiveness of combining ILRMA and the multichannel NC.

Acknowledgments. This work was supported by the Japan Science and Technology Agency and the Impulsing Paradigm Change through Disruptive Technologies Program (ImPACT) designed by the Council for Science, Technology and Innovation, and partly supported by SECOM Science and Technology Foundation. We would also like to express our gratitude to Prof. Hiroshi Okuno and Mr. Yoshiaki Bando for providing experimental data.

References

1. Impulsive Paradigm Change through Disruptive Technologies Program (ImPACT). http://www.jst.go.jp/impact/program07.html
2. Namari, H., Wakana, K., Ishikura, M., Konyo, M., Tadokoro, S.: Tube-type active scope camera with high mobility and practical functionality. In: Proceedings of IEEE/RSJ IROS, pp. 3679–3686 (2012)
3. Deleforge, A., Kellerman, W.: Phase-optimized K-SVD for signal extraction from underdetermined multichannel sparse mixtures. In: Proceedings of IEEE ICASSP, pp. 355–359 (2015)
4. Barfuss, H., Kellerman, W.: Improving blind source separation performance by adaptive array geometries for humanoid robots. In: Proceedings of HSCMA (2014)
5. Barfuss, H., Kellerman, W.: An adaptive microphone array topology for target signal extraction with humanoid robots. In: Proceedings of IWAENC, pp. 16–20 (2014)
6. Aichner, R., Zourub, M., Buchner, H., Kellerman, W.: Post-processing for convolutive blind source separation. In: Proceedings of ICASSP (2006)

7. Mae, N., Kitamura, D., Ishimura, M., Yamada, T., Makino, S.: Ego noise reduction for hose-shaped rescue robot combining independent low-rank matrix analysis and noise cancellation. In: Proceedings of APSIPA (2016, to be published)
8. Kitamura, D., Ono, N., Sawada, H., Kameoka, H., Saruwatari, H.: Efficient multichannel nonnegative matrix factorization exploiting rank-1 spatial model. In: Proceedings of ICASSP, pp. 276–280 (2015)
9. Kitamura, D., Ono, N., Sawada, H., Kameoka, H., Saruwatari, H.: Determined blind source separation unifying independent vector analysis and nonnegative matrix factorization. IEEE/ACM Trans. Audio Speech Lang. Process. 24(9), 1626–1641 (2016)
10. Kim, T., Eltoft, T., Lee, T.-W.: Independent vector analysis: an extension of ICA to multivariate components. In: Rosca, J., Erdogmus, D., Príncipe, J.C., Haykin, S. (eds.) ICA 2006. LNCS, vol. 3889, pp. 165–172. Springer, Heidelberg (2006). doi:10.1007/11679363_21
11. Hiroe, A.: Solution of permutation problem in frequency domain ICA, using multivariate probability density functions. In: Rosca, J., Erdogmus, D., Príncipe, J.C., Haykin, S. (eds.) ICA 2006. LNCS, vol. 3889, pp. 601–608. Springer, Heidelberg (2006). doi:10.1007/11679363_75
12. Kim, T., Attias, H.T., Lee, S.-Y., Lee, T.-W.: Blind source separation exploiting higher-order frequency dependencies. IEEE Trans. Speech Audio Process. 15(1), 70–79 (2007)
13. Lee, D.D., Seung, H.S.: Learning the parts of objects by nonnegative matrix factorization. Nature 401, 788–791 (1999)
14. Lee, D.D., Seung, H.S.: Algorithms for non-negative matrix factorization. Proc. NIPS 13, 556–562 (2001)
15. Cichocki, A., Zdunek, R., Phan, A.H., Amari, S.: Nonnegative Matrix and Tensor Factorizations: Applications to Exploratory Multi-way Data Analysis and Blind Source Separation. Wiley, New York (2009)
16. Ozerov, A., Févotte, C.: Multichannel nonnegative matrix factorization in convolutive mixtures for audio source separation. IEEE Trans. ASLP 18(3), 550–563 (2010)
17. Kameoka, H., Yoshioka, T., Hamamura, M., Roux, J., Kashino, K.: Statistical model of speech signals based on composite autoregressive system with application to blind source separation. In: Vigneron, V., Zarzoso, V., Moreau, E., Gribonval, R., Vincent, E. (eds.) LVA/ICA 2010. LNCS, vol. 6365, pp. 245–253. Springer, Heidelberg (2010). doi:10.1007/978-3-642-15995-4_31
18. Sawada, H., Kameoka, H., Araki, S., Ueda, N.: Multichannel extensions of nonnegative matrix factorization with complex-valued data. IEEE Trans. ASLP 21(5), 971–982 (2013)
19. Murata, N., Ikeda, S., Ziehe, A.: An approach to blind source separation based on temporal structure of speech signals. Neurocomputing 41(14), 1–24 (2001)
20. Ishimura, M., Makino, S., Yamada, T., Ono, N., Saruwatari, H.: Noise reduction using independent vector analysis and noise cancellation for a hose-shaped rescue robot. In: Proceedings of IWAENC (2016)
21. Hänsler, E., Schmidt, G.: Acoustic Echo and Noise Control: A Practical Approach. Wiley, New York (2004)
22. Vincent, E., Gribonval, R., Févotte, C.: Performance measurement in blind audio source separation. IEEE Trans. ASLP 14, 1462–1469 (2006)

Some Theory on Non-negative Tucker Decomposition

Jeremy E. Cohen[1(✉)], Pierre Comon[2], and Nicolas Gillis[1]

[1] Departement of Mathematics and Operational Research,
Faculté Polytechnique, Université de Mons, Rue de Houdain 9, Mons, Belgium
jeremy.cohen@umons.ac.be
[2] Gipsa-lab, 11 Rue des Mathematiques, 38210 St Martin d'Hères, France

Abstract. Some theoretical difficulties that arise from dimensionality reduction for tensors with non-negative coefficients is discussed in this paper. A necessary and sufficient condition is derived for a low non-negative rank tensor to admit a non-negative Tucker decomposition with a core of the same non-negative rank. Moreover, we provide evidence that the only algorithm operating mode-wise, minimizing the dimensions of the features spaces, and that can guarantee the non-negative core to have low non-negative rank requires identifying on each mode a cone with possibly a very large number of extreme rays. To illustrate our observations, some existing algorithms that compute the non-negative Tucker decomposition are described and tested on synthetic data.

Keywords: Non-negative Tucker Decomposition · Non-negative Canonical Polyadic Decomposition · Dimensionality reduction · Non-negative Matrix Factorization

Notation

The following notation will be used: bold calligraphic letters $\boldsymbol{T}$ for tensors, bold uppercase letters $\boldsymbol{U}$ for matrices or linear operators, and bold lowercase letters $\boldsymbol{a}$ for vectors. Here tensors are real-valued vectors in $\mathbb{R}^K \otimes \mathbb{R}^L \otimes \mathbb{R}^M$ or multilinear operators in $\mathbb{R}^{K \times R_1} \otimes \mathbb{R}^{L \times R_2} \otimes \mathbb{R}^{M \times R_3}$ with K, L, M, R_i integers and the product $\otimes$ is a tensor product [1], which implies $\lambda \boldsymbol{x} \otimes \boldsymbol{y} = \boldsymbol{x} \otimes \lambda \boldsymbol{y} = \lambda(\boldsymbol{x} \otimes \boldsymbol{y})$. Rank-one linear operators acting on tensors are denoted as $\boldsymbol{U} \otimes \boldsymbol{V} \otimes \boldsymbol{W}$, where the tensor product is the canonical tensor product for linear applications inherited from the tensor product of vectors, and by definition, $(\boldsymbol{U} \otimes \boldsymbol{V} \otimes \boldsymbol{W})(\boldsymbol{U}_2 \otimes \boldsymbol{V}_2 \otimes \boldsymbol{W}_2) = \boldsymbol{UU}_2 \otimes \boldsymbol{VV}_2 \otimes \boldsymbol{WW}_2$. Also, for two-way arrays, $(\boldsymbol{U} \otimes \boldsymbol{V})\boldsymbol{T} = \boldsymbol{UTV}^t$. The Kronecker product [2] is denoted by $\boxtimes$ and is one possible expression of a tensor product in $\mathbb{R}^{KLM}$. Further discussion on notations can be found in [3].

J.E. Cohen—Research funded by ERC advanced grant "DECODA" no. 320594, ERC starting grant "COLORAMAP" no. 679515, and F.R.S.-FNRS incentive grant for scientific research n° F.4501.16.

© Springer International Publishing AG 2017
P. Tichavský et al. (Eds.): LVA/ICA 2017, LNCS 10169, pp. 152–161, 2017.
DOI: 10.1007/978-3-319-53547-0_15

1   Tensor Decomposition Models

In this section we quickly survey two tensor decomposition models, namely the Tucker Decomposition (TD) and the Canonial Polyadic Decomposition (CPD) [4].

1.1   Tucker Decompositions

Given a tensor $\mathcal{T} \in \mathbb{R}^K \otimes \mathbb{R}^L \otimes \mathbb{R}^M$, the TD finds so-called factor matrices U, V, W of respective sizes $K \times R_1$, $L \times R_2$ and $M \times R_3$ defining bases onto which the tensor can be expressed mode-wise:

$$\mathcal{T} = (U \otimes V \otimes W)\,\mathcal{G}, \tag{1}$$

where $\mathcal{G}$ is a tensor of coefficient often called the core of the TD. In other words, the span of U contains all the columns in $\mathcal{T}$. This is interesting for dimensionality reduction if R_1 is strictly smaller than K. The same observation holds for the two other modes. TD has been first investigated by Hitchcock in 1927 [4], and is now a widely used data mining model [5]. The main drawback is that there are infinitely many solution to decompose $\mathcal{T}$, so that it may not be possible to recover the ground truth for $U, V, W, \mathcal{G}$ from the data $\mathcal{T}$ solely.

Similarly to the matrix factorization problem, in the hope to restore identifiability of the parameters, the Non-negative Tucker Decomposition (NTD) was introduced recently [6]:

$$\begin{cases} \mathcal{T} = (U \otimes V \otimes W)\,\mathcal{G}, \\ \mathcal{G} \geq 0,\ U \geq 0,\ V \geq 0,\ W \geq 0, \end{cases} \tag{2}$$

but NTD was later shown to be unique up to permutation and scaling ambiguities if and only if NMF of each unfolding is unique, which is a very strong assumption and may not be verified in practice [7]. Imposing non-negativity constraints however improves the interpretability of the results of the Tucker Decomposition in some applications, see for instance [7] for an application in neuroscience.

Again to reduce the set of solutions to (2), a Sparse Non-negative Tucker Decomposition (SNTD) was suggested by Morup *et al.* [6] in which the factors matrices and the core are also constrained to be sparse. As we will show below, imposing sparsity on the factors may not be sufficient to restore identifiability.

Note that other constraints have been imposed on the factors of TD in the literature, notably orthogonality constraints and slice-orthogonality on the core [8].

1.2   Canonical Polyadic Decomposition

Maybe the most widely used tensor decomposition model is the Canonical Polyadic Decomposition (CPD) also called PARAFAC. It is similar to TD in the sense that a basis is sought on each mode, but in CPD the core is required to be diagonal, which makes CPD a much more constrained model than TD:

$$\mathcal{T} = (A \otimes B \otimes C)\,\mathcal{I}_R, \tag{3}$$

where A, B, C are respectively of sizes $K \times R$, $L \times R$ and $L \times R$ and R is the minimal integer so that (3) holds. CPD is often unique (up to permutations and scaling ambiguities) under mild conditions on the factors often verified in practice. A very common assumption is that R is much smaller than the dimensions of the data, in which case T is said to be a low rank tensor. CPD has been used in many applications ranging from chemometrics to social sciences [9].

In those applications, it often makes sense to look for non-negative factors. The Non-negative CPD (NCPD) [10] can then be used instead as a decomposition model:

$$\begin{cases} T = (A \otimes B \otimes C)\mathcal{I}_R, \\ A \geq 0, \ B \geq 0, \ C \geq 0, \end{cases} \tag{4}$$

where R is now called the non-negative rank of T if it is the smallest integer so that (4) holds. It is denoted $\mathrm{rank}_+(T)$.

2   Propagating Non-negativity and Non-negative Rank Through NTD

In the following, we show that NTD may not propagate the low non-negative rank of the original tensor T, and that to ensure $\mathcal{G}$ has the same non-negative rank as T, it is sufficient to identify the rays of a particular cone. We also show that no mode-wise procedure with $R_1 = R_2 = R_3 = R$ can guarantee non-negative rank propagation.

2.1   Elements of Cone Theory

First let us define some basic tools of cone theory that we shall use later in this section, most of which can be found in [11]. We start with a possible definition of the cone generated by columns of a matrix U:

Definition 1. *The cone generated by the columns of a matrix $U \in \mathbb{R}^{K \times R_1}$ is the set $cone(U) = \{Ux, \ x \in \mathbb{R}_+^{R_1}\}$.*

Another important notion is the extreme rays of a cone, intuitively the generating set of all elements in the cone:

Definition 2. *A vector y in $cone(U)$ spans an extreme ray if there does not exists $x, z \in cone(U) \backslash cone(y)$ such that $y = x + z$.*

Moreover, a cone is said to be simplicial if and only if all the extreme rays are linearly independent. Clearly, given a full column rank matrix U in $\mathbb{R}^{K \times R_1}$ with R_1 strictly smaller than K, then $cone(U)$ is simplicial and the columns of U are the extreme rays.

A set of interest for what follows is $\mathcal{H}(U) = \mathrm{span}(U) \cap \mathbb{R}_+^K$, namely the intersection of the non-negative orthant with the span of the columns of matrix $U \geq 0$. It can be seen that $\mathcal{H}(U)$ is a cone [12], and its number of extreme rays is between R_1 and $O(C_K^{R_1})$ [11] (the upper bound is attained by cones whose slices are cyclic polytopes with many vertices). This means that $\mathcal{H}(U)$ may be a cone with a very large number of extreme rays. Note however that $\mathcal{H}(U) \subset cone(I)$ which has K rays and corresponds to a trivial factorization ($U = IU$).

2.2   Working Hypotheses

In this paper we wish to explore the properties of the NTD. In particular, we found the case of a low non-negative rank tensor $\mathcal{T}$ of particular interest; see below. These results are meant as a first step in the understanding of NTD so we allow ourselves to make restrictive hypotheses. Note however that these hypotheses are often verified in real-life applications, for instance in fluorescence spectroscopy or neuroimaging.

Here are the working hypotheses that we need in order to establish the results presented in the remainder of this section:

- **H1**: $\mathcal{T}$ is non-negative, i.e. all entries of $\mathcal{T}$ are greater or equal to 0.
- **H2**: $\mathcal{T}$ admits a unique NCPD with factors A, B, C and its non-negative rank R is smaller than all dimensions.
- **H3**: the factors of the NCPD of $\mathcal{T}$ have full column rank.

All three hypotheses are required for results presented in Subsect. 2.3 to hold, but only **H1** is used in Sect. 2.4.

2.3   Propagating the Non-negative Rank to the Core

Our goal in this subsection is to study the propagation of the non-negative rank of $\mathcal{T}$ to the core $\mathcal{G}$ in (2). A property enjoyed by Tucker Decomposition is that the rank of $\mathcal{T}$ and the rank of $\mathcal{G}$ are always equal in the exact setting provided factors U, V, W admit left inverses. This may not be the case however for the non-negative rank and the NTD. First, we give a necessary and sufficient condition for the two non-negative ranks to match:

Proposition 1. *Let $\mathcal{T}$ be a $K \times L \times M$ non-negative tensor of non-negative rank R satisfying **H1,H2,H3**. Let $\mathcal{T} = (U \otimes V \otimes W)\,\mathcal{G}$ be a NTD with $\mathcal{G}$ of size $R_1 \times R_2 \times R_3$ so that U, V, W admit left inverses. Then $R = \mathrm{rank}_+(\mathcal{G})$ if and only if A, B, C belongs respectively to $\mathrm{cone}(U)$, $\mathrm{cone}(V)$ and $\mathrm{cone}(W)$. Moreover, if U, V, W do not admit left inverses, then there exists a core $\mathcal{G}'$ of non-negative rank R such that $\mathcal{T} = (U \otimes V \otimes W)\,\mathcal{G}'$, where $\mathcal{G} - \mathcal{G}'$ belongs to the null space of $U \otimes V \otimes W$.*

Proof. First suppose that $\mathrm{rank}_+(\mathcal{G}) = R$. Then there exists A_c, B_c, C_c such that:

$$\mathcal{T} = (U \otimes V \otimes W)\,(A_c \otimes B_c \otimes C_c)\,\mathcal{I}_R \tag{5}$$

so that $\mathcal{T}$ admits a NCPD with factors UA_c, VB_c, WC_c. Because the NCPD of $\mathcal{T}$ is unique, we can conclude that $A = UA_c$ and similarly on the other modes. Conversely, first note that since the factors of the NCPD have full column rank, the factors of the NTD cannot span a smaller linear subspace, so that $\mathrm{rank}_+(\mathcal{G}) \geq R$. Moreover, because A, B, C belong to the cones spanned by U, V, W, there exist A_c, B_c, C_c non-negative $R_i \times R$ matrices so that $A = UA_c$, $B = VB_c$ and $C = WC_c$. These non-negative coefficient matrices are factors in a NCPD of $\mathcal{G}$ because U, V, W are left invertible. From this, we get that $\mathrm{rank}_+(\mathcal{G}) \leq R$ which concludes the proof. If the factors of the NTD are not invertible, then simply set $\mathcal{G}' = (A_c \otimes B_c \otimes C_c)\,\mathcal{I}_R$.

In the next section, some algorithms designed for NTD and SNTD will be tested to check whether this condition is verified or not in practice. But in a theoretical perspective, it is natural to wonder whether matrices U, V, W can be found solely from $\mathcal{T}$ so that the necessary and sufficient condition from Proposition 1 is always verified. Since this problem can be cast mode-wise, it is closely related to recent uniqueness results obtained for Non-negative Matrix Factorization [13]. In what follows, we study mode-wise approaches to this problem. These involve the unfoldings of the data tensor, which are the columns/rows/fibers stacked into matrices. Using the matricization suggested in [3], the unfoldings of $\mathcal{T}$ can be expressed as follows:

$$
\begin{aligned}
T_1 &= A\,(B \odot C)^T = U G_1\,(V \boxtimes W)^T, \\
T_2 &= B\,(A \odot C)^T = V G_2\,(U \boxtimes W)^T, \\
T_3 &= C\,(A \odot B)^T = W G_3\,(U \boxtimes V)^T,
\end{aligned}
\tag{6}
$$

where $\odot$ is the Khatri-Rao product, that is, the column-wise Kronecker product.

Now, how can we guarantee that, say on the first mode, cone(U) contains A? A first (non mode-wise) solution is to constrain the core to be diagonal and actually look for the NCPD instead of the NTD. In the following, we restrict our preliminary study to the case where span$(U) = $ span(A). In that case it is possible to choose U as extreme rays of $\mathcal{H}(A)$. By definition, $\mathcal{H}(A)$ is the largest cone in the intersection of the non-negative orthant and the column space of A containing T_1. It also contains A since A belongs to the non-negative orthant. This means that extreme rays U of $\mathcal{H}(A)$ can be used in the NTD to ensure that the non-negative rank is preserved using Proposition 1.

However finding the extreme rays of $\mathcal{H}(A)$ is likely to be of little interest in practice since the number of extreme rays needed can be larger than K. Yet a special and easy case is when the non-negative matrix factorization of each unfolding is unique, then any cone spanning the unfolding on one mode also spans the NCPD factor on that mode.

In the light of the previous paragraph, a more interesting question is the following: can we design a procedure to find a simplicial cone cone(U) with $R_1 = R$ extreme rays (i.e. of order R) which always contains A? If a solution to this problem is found, then in theory it would be possible to compress the non-negative tensor $\mathcal{T}$ into $\mathcal{G}$ and to only compute the NCPD on $\mathcal{G}$.

Such a procedure needs to compute a maximal volume cone. Indeed, suppose the procedure outputs a set U of extreme rays, and suppose there exists a larger cone U' also enclosing T_1, then because the only requirement for A in this problem is that T_1 belongs to cone(A), then possibly $U' = A$ and cone(U) may not contain the columns of A.

However, the largest simplicial one of order R may not be unique, and this provides a counter example to the idea that the largest cone of order R could always contain the columns of A (see Fig. 1).

Sketch of a counter example. Let us build a matrix A in $\mathcal{R}^{4 \times 3}$ and data T_1 so that there will be at least two largest cones $\mathcal{H}_3$ containing T_1 with three extreme rays in the cone $\mathcal{H}$ defined by intersection of the span of $\boldsymbol{A}$ and the non-negative orthant. We set $\boldsymbol{A}^T = \begin{bmatrix} 1 & 1 & 0 & 0 \\ 0 & 1 & 1 & 0 \\ 0 & 0 & 1 & 1 \end{bmatrix}$. Using Theorem 9.1.1 from [12], we know that a non-zero vector from $\mathcal{H}$ belongs to an extreme ray if and only if it is as sparse as possible, and that for each set of 0 indices there is only one extreme ray. Here this can be applied in a straightforward manner, since in the span of $\boldsymbol{A}$ there can be no vector with three zeros. This means that the extreme rays can only have two zeros among 4 coefficients, and we thus need to check which combinations among the 6 belong to the span of $\boldsymbol{A}$ and the non-negative orthant. Since $\boldsymbol{A}$ has a simple structure, it is easy to check that $\mathcal{H}$ has 4 extreme rays, containing the columns of $\boldsymbol{A}$ and $e = [0, 0, 1, 1]^T$. Finally, the problem admits a rotational symmetry and it is easy to build $\boldsymbol{T}_1$ as a smaller cone contained in both cone($[\boldsymbol{a}_1, \boldsymbol{a}_2, e]$) and cone($A$), see Fig. 1.

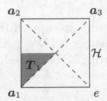

Fig. 1. A case where symmetry gives birth to two maximal volume cones with 3 extreme rays. The figure is the projection of the cones and data on the subspace $\{\boldsymbol{x} \in \mathbb{R}^4 | \sum\limits_{i=1}^{4} x_i = 1\}$.

This maximal volume cone $\mathcal{H}_R$ of order R is actually what SNTD computes since SNTD imposes minimal ℓ_1 norm on the factors, meaning they should be as close as possible to the border of $\mathcal{H}$. Whether SNTD actually manages to compute cones containing the factors or not is investigated in the simulation section.

As a partial conclusion here, the only procedure that computes factors independently on each mode that can guarantee the propagation of the non-negative rank and under the constraint span($\boldsymbol{A}$) = span($\boldsymbol{U}$) is the computation of $\mathcal{H}(\boldsymbol{A})$. This provides evidence that using NTD as a preprocessing step for NCPD is difficult, but we cannot conclude that it is impossible since there may exist procedures working globally on the tensor (not mode-wise) or increasing the dimension of the column space of $\boldsymbol{U}$ that can guarantee non-negative rank propagation other than NCPD.

2.4   Propagating Non-negativity to the Core

A very fundamental question to answer for computing the NTD is whether some conditions can be imposed on the factor matrices $\boldsymbol{U}, \boldsymbol{V}, \boldsymbol{W}$ to ensure that $\mathcal{G}$

has non-negative entries. At first glance, a natural condition to impose on the factors is that their cones contain the columns of the unfoldings of $\mathcal{T}$. Clearly this condition is necessary, otherwise from (6), the product of the core and the Kronecker product of two factors has to contain negative entries, which itself is possible only if either the core or the factors contain negative entries.

However contrary to what is trivial for NMF, finding cones containing the columns of the unfoldings in each mode does not guarantee a non-negative core. We do not provide in this communication a simple counter-example, but we made this observation after running some numerical experiments reported in the next section, and this was confirmed by simulations run by the reviewers for this communication. This means that computing NMF on each unfolding to obtain U, V, W and infer the core by inverting a linear system may not yield a non-negative core.

3   Simulations

In this section we run numerical tests to support the previous theoretical discussion and provide evidence that neither NTD nor SNTD propagates non-negative rank in practice, and that computing NMF on each mode does not ensure obtaining a non-negative core.

3.1   Some Algorithms for NTD and NMF

There has been a few algorithms reported in the literature to compute NTD. In the following simulations we make use of Hierarchical Alternating Least Squares (HALS) by Phan *et al.* [14].

HALS is based on coordinate descent, where the set of variables is alternatively each columns of U, V and W, because with respect to these columns the underlying constrained least squares optimization problem admits a closed-form solution. For computing NMF, we used an algorithm based on the same idea from Gillis *et al.* called accHALS for accelerated HALS [15]. Again, other algorithms exist for computing NMF and NTD that offer at least the same performances, but the goal here is not to compare state-of-the-art algorithms.

To compute SNTD, we used the algorithms from Morup *et al.* which was the first algorithm designed for SNTD in the literature [6]. It is based on multiplicative updates, which are known to be slow for least square problems.

3.2   Some Tests on the Outputs of Algorithms

Settings. For both experiments, the tensors are rank $R = 3$ non-negative tensors build using the NCPD from factors drawn from the uniform distribution over $[0, 1]$, with sizes $K = L = M = 20$. The NCPD factors are normalized column-wise using the ℓ_2 norm. No noise is added. We set $R_1 = R_2 = R_3 = R$ - we will study the case where the number of components used in NTD are larger than the true rank in a longer communication. The maximal number of iterations is set to

1000 for the HALS algorithm, and to 3000 for the multiplicative algorithm solving SNTD - which we will abusively denote by SNTD. For accHALS applied on each unfolding, the maximal number of iterations is set to 1000. We chose the number of maximal iterations large enough so that convergence is always reached. In SNTD, the sparsity coefficient on the core is set to 0, and set to 10^{-3} on factors.

HALS and accHALS compute exact NTD and NMF up to around 10^{-8} relative error on the reconstructed tensor when no noise is added on the data and a good initialization is provided. We chose to initialize with High Order Singular Value Decomposition [8] to start in the right subspace on each mode. For SNTD, relative error with respect to the norm of the original data is of order of magnitude 10^{-4} in the following simulations.

Experiment 1: Number of Negative Entries in the Core Computed by Mode-Wise AccHALS. In this first experiment, NTD is computed using NMF on each unfolding of a hundred tensors. We plot the number of negative entries in $\mathcal{G}$ obtained by an unconstrained linear system. We also plot the percentage of negative coefficients in $U^\dagger T_1$, $V^\dagger T_2$ and $W^\dagger T_3$, † denoting the left pseudo inverse. If the unfoldings are contained in the cones spanned by U, V, W, then there should be no negative coefficients in these products.

Results reported in Fig. 2 show that although the unfoldings are indeed almost contained in the cone of computed factor matrices, the core $\mathcal{G}$ obtained contains a high number of negative entries. Moreover, the negative entries have a non-negligible intensity. This observation supports the idea that spanning the columns of the unfoldings is not a sufficient condition to ensure non-negativity of the core.

Experiment 2: Estimation of the Span of Factors and Propagation of Non-negative Rank. In this second experiment, 10 different tensors are decomposed using the NTD model using HALS and mode-wise accHALS and the SNTD model using the multiplicative algorithm. We check that the span of factors from the known NCPD and the computed NTD or SNTD are the

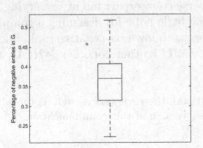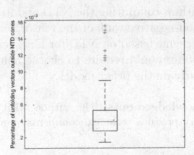

Fig. 2. Left: Percentage of negative entries in the core $\mathcal{G}$ estimated by three NMF. Right: Percentage of negative values in the coefficients of vectors of unfoldings of $\mathcal{T}$ in subspaces spanned by U, V, W.

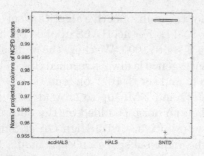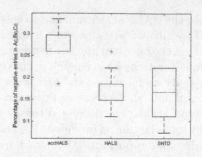

Fig. 3. Left: Average norm of the projected column of factors A, B, C onto the subspaces spanned by estimated U, V, W. Right: Percentage of negative entries in the coefficients A_c, B_c, C_c so that $A = U A_c$ among all coefficients.

same by comparing the norm of projected columns of A, B, C on the subspaces spanned by NTD factors. Moreover, we want to show that NCPD factors are not contained in the cones spanned by U, V, W. The latter is checked by computing the amount of negative entries in products of the form $A_c = U^\dagger A$. Results are presented in Fig. 3.

We observe that although the spans of factors from NCPD and NTD are the same (with a small variation for SNTD), the necessary and sufficient condition from Proposition 1 is not verified in this example. This means that neither NTD nor SNTD propagate the non-negative rank.

4   Conclusion

Non-negative Tucker Decomposition is a relatively unexplored research topic among constrained tensor decomposition models. We have shown in this paper that choosing the maximum volume cone generating the data does not necessarily restore identifiability of the factor matrices. We have also illustrated on some numerical experiments that choosing cones containing the unfoldings of the tensor on each mode does not necessarily yield a non-negative core, and that both algorithms computing the NTD and its sparse counterpart fail at preserving the low non-negative rank of the tensor, leaving little hope for designing a compression scheme based on NTD for large tensors with low non-negative rank. Such a procedure would require to choose non-trivial U so that $\mathrm{cone}(U) \supseteq \mathcal{H}(A)$, and similarly on the other modes.

Acknowledgements. The authors wish to thank the reviewers as well as the editor for very precious technical comments on a first version of this communication.

References

1. Schwartz, L., Bamberger, Y., Bourguignon, J.-P.: Les tenseurs (1977)
2. Brewer, J.: Kronecker products and matrix calculus in system theory. IEEE Trans. Circ. Syst. **25**(9), 772–781 (1978)
3. Cohen, J.E.: Environmental multiway data mining, Ph.D. dissertation, Universite Grenoble Alpes (2016)
4. Hitchcock, F.L.: The expression of a tensor or a polyadic as a sum of products. J. Math. Phys. **6**(1), 164–189 (1927)
5. Papalexakis, E.E., Faloutsos, C., Sidiropoulos, N.D.: Tensors for data mining and data fusion: models, applications, and scalable algorithms. ACM Trans. Intell. Syst. Technol. (TIST) **8**(2), 16 (2016)
6. Mørup, M., Hansen, L.K., Arnfred, S.M.: Algorithms for sparse nonnegative tucker decompositions. Neural Comput. **20**(8), 2112–2131 (2008)
7. Zhou, G., Cichocki, A., Zhao, Q., Xie, S.: Efficient nonnegative tucker decompositions: algorithms and uniqueness. IEEE Trans. Image Process. **24**(12), 4990–5003 (2015)
8. De Lathauwer, L., De Moor, B., Vandewalle, J.: A multilinear singular value decomposition. SIAM J. Matrix Anal. Appl. **21**(4), 1253–1278 (2000)
9. Acar, E., Yener, B.: Unsupervised multiway data analysis: a literature survey. IEEE Trans. Knowl. Data Eng. **21**(1), 6–20 (2009)
10. Bro, R.: Parafac. tutorial and applications. Chemometr. Intell. Lab. Syst. **38**(2), 149–171 (1997)
11. Ziegler, G.M.: Lectures on Polytopes, vol. 152. Springer Science & Business Media, New York (1995)
12. D'Alessandro, P.: Conical Approach to Linear Programming. CRC Press (1997)
13. Laurberg, H., Christensen, M.G., Plumbley, M.D., Hansen, L.K., Jensen, S.H.: Theorems on positive data: on the uniqueness of NMF. Comput. Intell. Neurosci. **2008** (2008)
14. Phan, A.H., Cichocki, A.: Extended HALS algorithm for nonnegative tucker decomposition and its applications for multiway analysis and classification. Neurocomputing **74**(11), 1956–1969 (2011)
15. Gillis, N., Glineur, F.: Accelerated multiplicative updates and hierarchical ALS algorithms for nonnegative matrix factorization. Neural Comput. **24**(4), 1085–1105 (2012)

A New Algorithm for Multimodal Soft Coupling

Farnaz Sedighin[1(✉)], Massoud Babaie-Zadeh[1], Bertrand Rivet[2],
and Christian Jutten[2]

[1] Department of Electrical Engineering, Sharif University of Technology, Tehran, Iran
f_sedighin@ee.sharif.edu, mbzadeh@yahoo.com
[2] GIPSA-lab, CNRS, Univ. Grenoble Alpes, Grenoble INP, Grenoble, France
{bertrand.rivet,christian.jutten}@gipsa-lab.grenoble-inp.fr

Abstract. In this paper, the problem of multimodal soft coupling under the Bayesian framework when variance of probabilistic model is unknown is investigated. Similarity of shared factors resulted from Nonnegative Matrix Factorization (NMF) of multimodal data sets is controlled in a soft manner by using a probabilistic model. In previous works, it is supposed that the probabilistic model and its parameters are known. However, this assumption does not always hold. In this paper it is supposed that the probabilistic model is already known but its variance is unknown. So the proposed algorithm estimates the variance of the probabilistic model along with the other parameters during the factorization procedure. Simulation results with synthetic data confirm the effectiveness of the proposed algorithm.

Keywords: Nonnegative matrix factorization · Bayesian framework · Soft coupling

1 Introduction

Multimodal signals are recorded by different sensors viewing a same physical phenomenon. These signals can be of the same type (different microphones recording a same speech) or different types (audio and video recordings of a speech). Since the physical origin of the multimodal signals are the same, some similarities and correlations are expected among them. Utilizing this similarity by the joint analysis of the multimodal signals is known as data fusion [1,2]. Coupled factorization of the multimodal data sets is a common approach for data fusion [3] and can be achieved by coupled matrix factorization [4], coupled matrix-tensor factorization [2] or coupled tensor factorization [5].

Factorization of matrix $\mathbf{V}^m$ (a 2-way array data set) can be achieved by using Nonnegative Matrix Factorization (NMF). NMF is decomposing a data

---

This work has been partly supported by the European project ERC-2012-AdG-320684-CHESS and also by the Center for International Scientific Studies and Collaboration (CISSC).

P. Tichavský et al. (Eds.): LVA/ICA 2017, LNCS 10169, pp. 162–171, 2017.
DOI: 10.1007/978-3-319-53547-0_16

matrix with nonnegative elements as a product of two matrices with nonnegative elements as [6]

$$\mathbf{V}^m = \mathbf{W}^m \mathbf{H}^m, \qquad m = 1, ..., M \qquad (1)$$

where $\mathbf{V}^m \in \mathbb{R}_+^{F \times N}$ is the m-th data set, $\mathbf{W}^m \in \mathbb{R}_+^{F \times K}$ and $\mathbf{H}^m \in \mathbb{R}_+^{K \times N}$ ($K < \min(F, N)$) are the factorization parameters of the m-th data set and M is the number of the data sets.

Due to the correlation among the multimodal data sets ($\mathbf{V}^m, m = 1, ..., M$), one or some of their factorization parameters is (are) similar which is (are) called shared factor(s). The other parameters which are different for each of the data sets are called unshared factors [5,7]. Since factorization of a data set is not unique, joint (coupled) factorization of multimodal data sets and utilizing the similarity of their shared factors can improve the quality of the factorization, and especially can reduce the indeterminacies.

In some algorithms such as [8] the shared factors are assumed to be equal among the data sets. These algorithms are usually named as the hard coupling algorithms. The "equality" constraint of the shared factors is relaxed to their "similarity" in algorithms such as [4]. These algorithms are known as the soft coupling algorithms and are exploited in different applications such as source separation [4] or speaker diarization [9]. The similarity of the shared factors is usually controlled by using penalty terms. The penalty terms can be in the form of ℓ_1 or ℓ_2 norms [4] or can be achieved in the Bayesian framework and based on the joint distribution of the shared factors [7].

The soft coupling in the Bayesian framework is studied in [7] and is based on the statistical dependence between the shared factors which is assumed to be known. But this assumption does not always hold. The statistical dependence between the shared factors can be unknown. Even if the kind of the statistical dependence is known, its parameters such as its variance can be unknown. In this paper, the soft coupling of the shared factors in the Bayesian framework when the variance of the statistical model is unknown is studied. Factorization parameters of a data set are computed by the help of the parameters of another data set using soft coupling. It is supposed that the kind of the statistical model between the shared factors (Gaussian) is known, but the variance of the model is unknown. So the variance is also estimated along with the other parameters. In this paper, the update rules for updating the parameters are derived by using majorization minimization algorithm and exploiting auxiliary functions and an stopping criterion for stopping the update of the variance is also defined.

The paper is organized as follows. Soft coupling for NMF is reviewed in Sect. 2. The proposed algorithm is presented in Sect. 3, and finally Sect. 4 is devoted to the experimental results.

2 Soft Coupling for NMF

2.1 NMF Model

As mentioned in the introduction, NMF is decomposing a matrix $\mathbf{V}$ with non-negative elements to the product of two matrices $\mathbf{W}$ and $\mathbf{H}$ with nonnegative elements. The decomposition is achieved by solving [6]

$$\min_{\mathbf{W}\geq 0,\mathbf{H}\geq 0} D(\mathbf{V}\|\mathbf{WH}), \tag{2}$$

where D measures the difference between $\mathbf{V}$ and $\mathbf{WH}$. Different functions are used for D such as the Kulback-Leibler divergence or the Itakura-Saito divergence [4,8]. The Itakura-Saito divergence is defined as [8]

$$D_{\mathrm{IS}}(\mathbf{V}\|\mathbf{WH}) = \sum_{f,n}\left\{\frac{v(f,n)}{\sum_k w(f,k)h(k,n)} - \log\frac{v(f,n)}{\sum_k w(f,k)h(k,n)} - 1\right\}, \tag{3}$$

where $v(f,n)$, $w(f,k)$ and $h(k,n)$ are the elements of $\mathbf{V}$, $\mathbf{W}$ and $\mathbf{H}$, respectively.

The parameters $\mathbf{W}$ and $\mathbf{H}$ in (2) are estimated during an update procedure. Multiplicative update rules with nonnegative initialization which preserve the nonnegativity of the elements of the final parameters are proposed for estimating $\mathbf{W}$ and $\mathbf{H}$ in (3) as [4,8,10]

$$w(f,k) \leftarrow w(f,k) \times \frac{\sum_n h(k,n)v(f,n)/\hat{v}^2(f,n)}{\sum_n h(k,n)/\hat{v}(f,n)}, \tag{4}$$

$$h(k,n) \leftarrow h(k,n) \times \frac{\sum_f w(f,k)v(f,n)/\hat{v}^2(f,n)}{\sum_f w(f,k)/\hat{v}(f,n)}, \tag{5}$$

where $\hat{v}(f,n)$ is the (f,n)-th element of $\hat{\mathbf{V}} = \mathbf{WH}$.

2.2 Coupled NMF

Coupled Factorization. As mentioned in the introduction, the coupled factorization of the multimodal data sets is a common approach for data fusion. Coupled factorization of two multimodal data sets in a hard manner (hard coupling) is modeled as [8]

$$\min_{\mathbf{W}_1,\mathbf{W}_2,\mathbf{H}} \lambda_1 D(\mathbf{V}_1\|\mathbf{W}_1\mathbf{H}) + \lambda_2 D(\mathbf{V}_2\|\mathbf{W}_2\mathbf{H}), \tag{6}$$

where $\mathbf{V}_1$ and $\mathbf{V}_2$ are the multimodal data sets, $\mathbf{H}$ is the shared factor, $\mathbf{W}_1$ and $\mathbf{W}_2$ are the unshared factors, and λ_1 and λ_2 are the weights of each term. For coupled factorization in a soft manner (soft coupling) the above cost function changes to [4]

$$\min_{\mathbf{W}_1,\mathbf{W}_2,\mathbf{H}_1,\mathbf{H}_2} \lambda_1 D(\mathbf{V}_1\|\mathbf{W}_1\mathbf{H}_1) + \lambda_2 D(\mathbf{V}_2\|\mathbf{W}_2\mathbf{H}_2) + \lambda_3\ell_p(\mathbf{H}_1,\mathbf{H}_2), \tag{7}$$

where $\mathbf{H}_1$ and $\mathbf{H}_2$ are the shared factors, $\ell_p(\mathbf{H}_1, \mathbf{H}_2)$ is the penalty term which controls the similarity of the shared factors, and λ_3 weights the penalty term. As mentioned before, the penalty term can be for example in the form of ℓ_1 or ℓ_2 norms or can be obtained in the Bayesian framework which will be discussed in the next subsection.

Soft Coupling in the Bayesian Framework. The problem of estimating $\mathbf{W}$ and $\mathbf{H}$ given $\mathbf{S}$ can be modeled as the Maximum A Posteriori (MAP) estimation of the parameters as [7,8]

$$\operatorname*{argmax}_{\boldsymbol{\theta}} p(\boldsymbol{\theta}, \mathbf{S}) = \operatorname*{argmin}_{\boldsymbol{\theta}} \{ -\log p(\mathbf{S}|\boldsymbol{\theta}) - \log p(\boldsymbol{\theta}) \}, \tag{8}$$

where $\boldsymbol{\theta} = \{\mathbf{W}, \mathbf{H}\}$ and p denotes the probability density function. The joint estimation of the parameters of the two multimodal data sets $\mathbf{S}_1$ and $\mathbf{S}_2$ can also be modeled as [7]

$$\operatorname*{argmax}_{\boldsymbol{\theta}} p(\boldsymbol{\theta}, \mathbf{S}_1, \mathbf{S}_2) = \operatorname*{argmin}_{\boldsymbol{\theta}} \{ -\log p(\mathbf{S}_1|\boldsymbol{\theta}_1) - \log p(\mathbf{S}_2|\boldsymbol{\theta}_2) - \log p(\boldsymbol{\theta}_1, \boldsymbol{\theta}_2) \},$$

$$\tag{9}$$

where $\boldsymbol{\theta} = \{\mathbf{W}_1, \mathbf{H}_1, \mathbf{W}_2, \mathbf{H}_2\}$, $\boldsymbol{\theta}_1 = \{\mathbf{W}_1, \mathbf{H}_1\}$ and $\boldsymbol{\theta}_2 = \{\mathbf{W}_2, \mathbf{H}_2\}$. The third term, $\log p(\boldsymbol{\theta}_1, \boldsymbol{\theta}_2)$, is the logarithm of the joint density of $\boldsymbol{\theta}_1$ and $\boldsymbol{\theta}_2$. In (9) it is assumed that the data sets $\mathbf{S}_1$ and $\mathbf{S}_2$ are conditionally independent given $\boldsymbol{\theta}_1$ and $\boldsymbol{\theta}_2$. $\mathbf{H}_1$ and $\mathbf{H}_2$ are the shared factors and $\mathbf{W}_1$ and $\mathbf{W}_2$ are the unshared factors.

Similar to [7], it is assumed that $\mathbf{H}_1$ is random but $\mathbf{H}_2$, $\mathbf{W}_1$ and $\mathbf{W}_2$ are deterministic, and $\mathbf{H}_1$ only depends on $\mathbf{H}_2$ (shared factors). So the last term of (9) can be written as

$$-\log p(\boldsymbol{\theta}_1, \boldsymbol{\theta}_2) = -\log p(\mathbf{H}_1|\mathbf{H}_2). \tag{10}$$

So, the joint estimation of the parameters in the Bayesian framework is modeled as [7]

$$\operatorname*{argmin}_{\boldsymbol{\theta}} \{ -\log p(\mathbf{S}_1|\boldsymbol{\theta}_1) - \log p(\mathbf{S}_2|\boldsymbol{\theta}_2) - \log p(\mathbf{H}_1|\mathbf{H}_2) \}, \tag{11}$$

where $-\log p(\mathbf{H}_1|\mathbf{H}_2)$ relates the shared factors and is the soft coupling term. For modeling $-\log p(\mathbf{S}_i|\boldsymbol{\theta}_i)(i = 1, 2)$, in [8], it is assumed that $\mathbf{S}_i$ is the Short Time Fourier Transform (STFT) matrix of a source ($F \times N$ matrix) whose elements at discrete time "n" and frequency "f", $(s_i(f, n))$, have the complex Gaussian distribution: $s_i(f, n) \sim \mathcal{N}_c(0, \sum_k w_i(f, k) h_i(k, n))$, where $w_i(f, k)$ and $h_i(k, n)$ are the elements of $\mathbf{W}_i$ and $\mathbf{H}_i$, respectively. Under this assumption, it is shown in [11] that:

$$-\log p(\mathbf{S}_i|\boldsymbol{\theta}_i) = -\log p(\mathbf{S}_i|\mathbf{W}_i\mathbf{H}_i) = D_{\mathrm{IS}}(\mathbf{V}_i^s \| \mathbf{W}_i\mathbf{H}_i) + cst, \tag{12}$$

where $\mathbf{V}_i^s \in \mathbb{R}_+^{F \times N}$ is a matrix whose elements are $v_i^s(f, n) = |s_i(f, n)|^2$. In [7], it is assumed that the coupling model $(-\log p(\mathbf{H}_1|\mathbf{H}_2))$ and its parameters are known. In this paper we assume that although the statistical model between the shared factors is known, the variance of the model is unknown. So the variance should also be estimated along with the other parameters. This will be discussed in the next section.

3   The Proposed Algorithm

In this paper, it is assumed that the second data set, $\mathbf{V}_2^s = |\mathbf{S}_2|^2$, is factorized beforehand and $\mathbf{H}_2$ has been computed and kept constant during the updating procedure. In addition, the variance of the model is unknown and should be estimated along with the other parameters. So the problem is computing the parameters $\mathbf{W}_1$, $\mathbf{H}_1$ and the variance of the statistical model given $\mathbf{S}_1$ and $\mathbf{H}_2$. The problem is formulated as the MAP estimation of the parameters as

$$\underset{\mathbf{W}_1,\mathbf{H}_1,\sigma}{\text{argmax}} \, p(\mathbf{S}_1,\mathbf{H}_2,\mathbf{W}_1,\mathbf{H}_1,\sigma)$$

$$= \underset{\mathbf{W}_1,\mathbf{H}_1,\sigma}{\text{argmin}} \left\{ -\log p(\mathbf{S}_1|\mathbf{W}_1\mathbf{H}_1) - \log p(\mathbf{H}_1|\mathbf{H}_2,\sigma) \right\}, \tag{13}$$

where σ^2 is the variance of the statistical model which is unknown. In the above model, σ is the same for all of the elements of $\mathbf{H}_1$ and $\mathbf{H}_2$, but the problem can also be investigated when each element has a particular variance. Recall that it is assumed that $\mathbf{S}_1$ only depends on $\mathbf{W}_1$ and $\mathbf{H}_1$ and $\mathbf{W}_1$ and σ are assumed to be deterministic. Supposing that p is the Gaussian probability density function and

$$(h_1(k,n)|h_2(k,n),\sigma) \perp\!\!\!\perp (h_1(k',n')|h_2(k',n'),\sigma), \quad (k,n) \neq (k',n')$$

where $\perp\!\!\!\perp$ shows the independence between two random variables, and $h_1(k,n)$ and $h_2(k,n)$ are the (k,n)-th elements of $\mathbf{H}_1$ and $\mathbf{H}_2$, respectively. So the soft coupling term can be written as $-\log p(\mathbf{H}_1|\mathbf{H}_2,\sigma) = \frac{\sum_{k,n} \|h_1(k,n)-h_2(k,n)\|^2}{2\sigma^2} + \sum_{k,n}\{\frac{1}{2}\log 2\pi + \log\sigma\}$. By considering (12), (13) can be written as

$$\underset{\mathbf{W}_1,\mathbf{H}_1,\sigma}{\text{argmin}} \left\{ D_{\text{IS}}(\mathbf{V}_1^s\|\mathbf{W}_1\mathbf{H}_1) + \frac{\sum_{k,n} \|h_1(k,n)-h_2(k,n)\|^2}{2\sigma^2} + \sum_{k,n}\log\sigma \right\}. \tag{14}$$

The parameters are updated sequentially in each iteration using update rules. We use (4) for updating $\mathbf{W}_1$, but new update rules are needed for updating $\mathbf{H}_1$ as well as σ which will be discussed in the following subsections.

3.1   Update Rule for Updating $\mathbf{H}_1$

The update rule for estimating $\mathbf{H}_1$ is derived using the majorization minimization approach and auxiliary functions [6,12]. For minimizing $F(h)$, an auxiliary function $G(h^t,h)$ is defined as

$$G(h^t,h) \geq F(h),$$
$$G(h^t,h^t) = F(h^t), \tag{15}$$

where $G(h^t,h)$ is an auxiliary function for $F(h)$ and h^t is the point that $G(h^t,h^t)$ is equal to $F(h^t)$. $G(h^t,h^t)$ has the property that $F(h)$ is nonincreasing under the following update [6]

$$h^{t+1} = \underset{h}{\text{argmin}} \, G(h^t,h).$$

It means that $F(h^{t+1}) \leq F(h^t)$. So an update rule for minimizing $F(h)$ can be achieved by using a proper auxiliary function (details can be found in [6]). An auxiliary function for minimizing the Itakura-Saito divergence of (3) with respect to $\mathbf{H}$ is proposed in [12] as

$$G(\mathbf{H}|\mathbf{H}^t) = \sum_{k,n} \left\{ \frac{h^{t^2}(k,n)}{h(k,n)} \sum_f w(f,k) \frac{v(f,n)}{\hat{v}^2(f,n)} + h(k,n) \sum_f \frac{w(f,k)}{\hat{v}(f,n)} \right\} + cst,$$

(16)

where $\hat{v}(f,n)$ is the (f,n)-th element of $\hat{\mathbf{V}} = \mathbf{W}\mathbf{H}^t$. Since the above auxiliary function is convex with respect to $\mathbf{H}$ (noting that $h(k,n) \geq 0 \quad \forall k,n$), its minimum can be found by finding the roots of its derivative. In (14), the Itakura-Saito divergence is coupled with the penalty term. So the convex auxiliary function for minimizing the cost function (14) with respect to $\mathbf{H}_1$ is

$$G_2(\mathbf{H}_1|\mathbf{H}_1^t) = G(\mathbf{H}_1|\mathbf{H}_1^t) + \frac{\sum_{k,n} \|h_1(k,n) - h_2(k,n)\|^2}{2\sigma^2}.$$

(17)

The derivative of the above auxiliary function with respect to $h_1(k,n)$ is

$$-\frac{h_1^{t^2}(k,n)}{h_1^2(k,n)} \left(\sum_f w_1(f,k) \frac{v_1^s(f,n)}{\hat{v}_1^2(f,n)} \right) + \left(\sum_f \frac{w_1(f,k)}{\hat{v}_1(f,n)} \right) + \frac{(h_1(k,n) - h_2(k,n))}{\sigma^2}.$$

(18)

The above equation should be solved with respect to $h_1(k,n)$. Denoting $a(k,n) = -h_1^{t^2}(k,n) \left(\sum_f w_1(f,k) \frac{v_1^s(f,n)}{\hat{v}_1^2(f,n)} \right)$, $b(k,n) = \left(\sum_f \frac{w_1(f,k)}{\hat{v}_1(f,n)} \right) - \frac{h_2(k,n)}{\sigma^2}$ and $c(k,n) = \frac{1}{\sigma^2}$, (18) changes to

$$\frac{a(k,n) + b(k,n) \times h_1^2(k,n) + c(k,n) \times h_1^3(k,n)}{h_1^2(k,n)},$$

(19)

where $\hat{v}_1(f,n)$ and $v_1^s(f,n)$ are the (f,n)-th elements of $\hat{\mathbf{V}}_1 = \mathbf{W}_1\mathbf{H}_1$ and $\mathbf{V}_1^s$, respectively, and $a(k,n) < 0$, $c(k,n) > 0$ and $b(k,n)$ can be positive or negative. One of the roots of the numerator of (19) is $\frac{1}{3}\left(z(k,n) + \frac{1}{z(k,n)} - 1\right)\frac{b(k,n)}{c(k,n)}$ where $z(k,n)$ is equal to (for the sake of simplicity, (k,n) is removed in the rest of the equations)

$$z = \frac{\sqrt[3]{3\sqrt{3}\sqrt{27a^2c^4 + 4ab^3c^2} - 27ac^2 - 2b^3}}{b\sqrt[3]{2}}.$$

(20)

For $\sqrt{27a^2c^4 + 4ab^3c^2}$ being real, the condition $b \leq \sqrt[3]{-\frac{27}{4}ac^2}$ (noting that $a < 0$) should be held. Simple calculation shows that if $b \leq \sqrt[3]{-\frac{27}{4}ac^2}$, then $-27ac^2 - 2b^3$ is also positive. So if $b \leq \sqrt[3]{-\frac{27}{4}ac^2}$, the numerator of (20) is positive and the sign of z is the same as the sign of b. The sign of $z + \frac{1}{z} - 1$ is the same as the sign of the z and the sign of z is the same as the sign of b, therefore if the constraint

$b \leq \sqrt[3]{-\frac{27}{4}ac^2}$ holds, $\frac{1}{3}\left(z + \frac{1}{z} - 1\right)\frac{b}{c}$ is positive. So $h_1(k,n) = \frac{1}{3}\left(z + \frac{1}{z} - 1\right)\frac{b}{c}$ is the positive root of (18). For when the condition $b \leq \sqrt[3]{-\frac{27}{4}ac^2}$ does not hold, for decreasing the auxiliary function and consequently the proposed cost function, if (18)> 0 then $h_1^t(k,n)$ decreases by dividing by $1 + \beta$. Otherwise $h_1^t(k,n)$ is increased by multiplying to $1 + \beta$ where β is a small positive constant. Based on this discussion, the update procedure of $\mathbf{H}_1$ is summarized below:

Algorithm 1. Update procedure for $\mathbf{H}_1$ (($t + 1$)-th iteration)

1: **if** $b \leq \sqrt[3]{-\frac{27}{4}ac^2}$ **then**
2:     $h_1^{t+1}(k,n) \leftarrow \frac{1}{3}\left(z + \frac{1}{z} - 1\right)\frac{b}{c}$
3: **else**
4:     **if** (18)> 0 **then**
5:         $h_1^{t+1}(k,n) \leftarrow h_1^t(k,n)/(1 + \beta)$
6:     **else**
7:         $h_1^{t+1}(k,n) \leftarrow h_1^t(k,n) \times (1 + \beta)$
8:     **end if**
9: **end if**

3.2   Update Rule for Updating σ

Similar to $\mathbf{H}_1$, we use auxiliary function for updating σ (at point σ^t) as

$$G(\sigma|\sigma^t) = \frac{\sum_{k,n} \|h_1(k,n) - h_2(k,n)\|^2}{2\sigma^2} + (\log \sigma^t + \frac{\sigma - \sigma^t}{\sigma^t})K \times N, \qquad (21)$$

where "log" function is replaced by its tangent [12] which is the same for all of the elements of $\mathbf{H}_1$. So the last summation in (14) changes to the product of $(\log \sigma^t + \frac{\sigma - \sigma^t}{\sigma^t})$ by $(K \times N)$, the entry number of $\mathbf{H}_1$. The auxiliary function (21) is convex with respect to σ and the root of its derivative with respect to σ is

$$\sigma = \sqrt[3]{\frac{\sum_{k,n} \|h_1(k,n) - h_2(k,n)\|^2 \sigma^t}{K \times N}}. \qquad (22)$$

So σ is updated using (22). Updating σ without any additional constraint results in the convergence of σ to zero (very small values) and $\mathbf{H}_1$ will become equal to $\mathbf{H}_2$ and finally the cost function (14) converges to $-\infty$. So updating σ should be stopped after some iterations. In this paper, σ is updated as long as $D_{\text{IS}}(\mathbf{V}_1^s\|\mathbf{W}_1^{t+1}\mathbf{H}_1^{t+1}) \leq D_{\text{IS}}(\mathbf{V}_1^s\|\mathbf{W}_1^t\mathbf{H}_1^t)$, where $\mathbf{W}_1^t$ and $\mathbf{H}_1^t$ are the parameters of the t-th iteration and $\mathbf{W}_1^{t+1}$ and $\mathbf{H}_1^{t+1}$ are the parameters of the ($t + 1$)-th iteration. $D_{\text{IS}}(\mathbf{V}_1^s\|\mathbf{W}_1^t\mathbf{H}_1^t)$ is the cost function of (14) without the coupling penalty term in the t-th iteration. Excessive reduction in σ gives a significant weight to the coupling term which results in too much similarity of $\mathbf{H}_1$ and $\mathbf{H}_2$. This makes $D_{\text{IS}}(\mathbf{V}_1^s\|\mathbf{W}_1\mathbf{H}_1)$ to increase (instead of decrease), especially when

$\mathbf{H}_1$ and $\mathbf{H}_2$ are not very similar. This can be used as a criterion for stopping the update of σ. So updating σ stops and σ is kept fixed in the rest of the updating procedure as soon as $D_{\text{IS}}(\mathbf{V}_1^s \| \mathbf{W}_1^{t+1}\mathbf{H}_1^{t+1}) \leq D_{\text{IS}}(\mathbf{V}_1^s \| \mathbf{W}_1^t\mathbf{H}_1^t)$ is violated.

4   Experimental Results

In this section, the effectiveness of the proposed algorithm is investigated. In the first simulation, the quality of the proposed algorithm in estimating the variance is investigated. The matrices $\mathbf{W}_1 \in \mathbb{R}_+^{100 \times 10}$ and $\mathbf{H}_2 \in \mathbb{R}_+^{10 \times 100}$ are produced with random nonnegative elements. $\mathbf{H}_1$ is produced by adding Gaussian noise to $\mathbf{H}_2$ as $p(\mathbf{H}_1|\mathbf{H}_2,\sigma) = \mathcal{N}(\mathbf{H}_2,\sigma^2)$ where $\mathcal{N}(\mathbf{H}_2,\sigma^2)$ is the Gaussian noise with the mean $= \mathbf{H}_2$ and the variance $= \sigma^2$. The negative elements of $\mathbf{H}_1$ are replaced with zero. The data matrix $(\mathbf{V}_1^s)$ is produced by multiplying $\mathbf{W}_1$ and $\mathbf{H}_1$. β is set to 0.1, the initial value for the estimation of σ is set to 10 and all of the other parameters are initialized randomly with positive values. The results for the estimation of σ are shown in Fig. 1. It is clear from the results that the algorithm has the ability to estimate σ.

The estimation error of $\mathbf{H}_1$ is calculated as $\frac{\|\mathbf{H}_1 - \hat{\mathbf{H}}_1\|_F^2}{K \times N}$ where $\hat{\mathbf{H}}_1$ is the estimation of $\mathbf{H}_1$. The estimation errors for the proposed algorithm and for the situation when the variance is known are shown in Fig. 2. The results show that except for some large values of the actual σ, the proposed algorithm and the situation in which the variance is known has nearly the same estimation errors. Note that when the actual variance (σ^2) is known, only $\mathbf{W}_1$ and $\mathbf{H}_1$ are updated using (4) and Algorithm 1.

The decrease of the cost function (14) during the iterations under the proposed update rules is shown in Fig. 3. The proposed algorithm is executed for the actual σ equal to 0.1 and $\beta = 10^{-3}$.

In Table 1, the estimation errors of the proposed algorithm are compared to the hard coupling situation in which $\hat{\mathbf{H}}_1 = \mathbf{H}_2$. It is clear from the results that the proposed algorithm has lower estimation errors comparing to the hard coupling situation, especially for the larger variances. But by decreasing the actual variance the estimation errors become closer to each other.

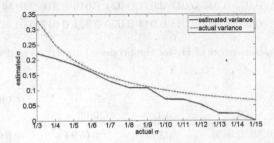

Fig. 1. The estimated σ (continuous line) using the proposed algorithm and the actual σ (dashed line).

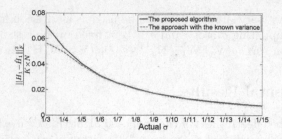

Fig. 2. Comparing the estimation errors using the proposed algorithm (continuous line) and the situation when the variance is known (dashed line).

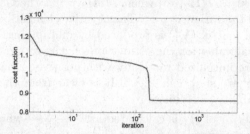

Fig. 3. The decrease of the proposed cost function during the iterations.

And finally, we have compared the proposed algorithm with the situation when the variance is not estimated but is chosen arbitrarily (not necessarily equal to the actual variance) for several values of the actual σ. The estimation errors are presented in Table 2 (the estimation errors of the proposed algorithm is presented in the last column). It is clear from the results that choosing an incorrect variance, especially when the actual $\sigma = 0$, can result in a significant estimation error. But this error is reduced using the proposed algorithm.

Table 1. The estimation errors of H_1 for the proposed algorithm and the hard coupling situation.

Actual σ	$\frac{1}{3}$	$\frac{1}{4}$	$\frac{1}{5}$	$\frac{1}{6}$	$\frac{1}{7}$	$\frac{1}{8}$	$\frac{1}{9}$	$\frac{1}{10}$	$\frac{1}{11}$	$\frac{1}{12}$
Proposed algorithm	0.073	0.055	0.041	0.031	0.024	0.019	0.016	0.0134	0.0111	0.0099
Hard coupling	0.087	0.062	0.045	0.034	0.026	0.021	0.017	0.0139	0.0116	0.0099

Table 2. The estimation errors of H_1 for the proposed algorithm and when σ is chosen arbitrarily.

Actual σ	Chosen σ					Proposed algorithm
	3	1	0.3	0.1	0.03	
0.3	0.0733	0.0308	0.0481	0.0714	0.0794	0.0471
0.1	0.0357	0.0027	0.0069	0.0087	0.0093	0.0087
0	0.0460	0.0019	5.477×10^{-6}	2.816×10^{-9}	3.715×10^{-12}	9.969×10^{-21}

5   Conclusion

In this paper, we have proposed an algorithm for the soft coupling of the shared factors in the Bayesian framework. As mentioned before, for the soft coupling of the shared factors in the Bayesian framework the statistical model between the shared factors should be known. But this assumption does not always hold. In this paper, it is assumed that the general statistical model between the shared factors is known but the variance of the model is unknown. So the proposed algorithm estimates the variance of the model along with the estimation of the factorization parameters. The presented results show the ability of the proposed algorithm in the estimation of the model variance.

References

1. Lahat, D., Adalı, T., Jutten, C.: Challenges in multimodal data fusion. In: 2014 22nd European Signal Processing Conference (EUSIPCO), pp. 101–105 September 2014
2. Acar, E., Kolda, T.G., Dunlavy, D.M.: All-at-once optimization for coupled matrix and tensor factorizations. arXiv preprint arXiv:1105.3422 (2011)
3. Acar, E., Rasmussen, M.A., Savorani, F., Næs, T., Bro, R.: Understanding data fusion within the framework of coupled matrix and tensor factorizations. Chemometr. Intell. Lab. Syst. 129, 53–63 (2013)
4. Seichepine, N., Essid, S., Févotte, C., Cappé, O.: Soft nonnegative matrix co-factorization. IEEE Trans. Signal Process. 62(22), 5940–5949 (2014)
5. Rivet, B., Duda, M., Guérin-Dugué, A., Jutten, C., Comon, P.: Multimodal approach to estimate the ocular movements during EEG recordings: a coupled tensor factorization method. In: 2015 37th Annual International Conference of the IEEE Engineering in Medicine and Biology Society (EMBC), pp. 6983–6986 (2015)
6. Lee, D.D., Seung, H.S.: Algorithms for non-negative matrix factorization. In: Advances in Neural Information Processing Systems, pp. 556–562 (2001)
7. Farias, R.C., Cohen, J.E., Comon, P.: Exploring multimodal data fusion through joint decompositions with flexible couplings. IEEE Trans. Signal Process. 64(18), 4830–4844 (2016)
8. Ozerov, A., Févotte, C.: Multichannel nonnegative matrix factorization in convolutive mixtures for audio source separation. IEEE Trans. Audio Speech Lang. Process. 18(3), 550–563 (2010)
9. Seichepine, N., Essid, S., Févotte, C., Cappé, O.: Soft nonnegative matrix co-factorizationwith application to multimodal speaker diarization. In: 2013 IEEE International Conference on Acoustics, Speech and Signal Processing, pp. 3537–3541, May 2013
10. Sawada, H., Kameoka, H., Araki, S., Ueda, N.: Multichannel extensions of nonnegative matrix factorization with complex-valued data. IEEE Trans. Audio Speech Lang. Process. 21(5), 971–982 (2013)
11. Févotte, C., Bertin, N., Durrieu, J.L.: Nonnegative matrix factorization with the itakura-saito divergence: with application to music analysis. Neural Comput. 21(3), 793–830 (2009)
12. Févotte, C.: Majorization-minimization algorithm for smooth itakura-saito nonnegative matrix factorization. In: 2011 IEEE International Conference on Acoustics, Speech and Signal Processing (ICASSP), pp. 1980–1983, May 2011

Adaptive Blind Separation of Instantaneous Linear Mixtures of Independent Sources

Ondřej Šembera[1(✉)], Petr Tichavský[1], and Zbyněk Koldovský[2]

[1] Institute of Information Theory and Automation of the CAS,
Prague, Czech Republic
{sembera,tichavsk}@utia.cas.cz
[2] Faculty of Mechatronics, Informatics and Interdisciplinary Studies,
Technical University Liberec, Liberec, Czech Republic

Abstract. In many applications, there is a need to blindly separate independent sources from their linear instantaneous mixtures while the mixing matrix or source properties are slowly or abruptly changing in time. The easiest way to separate the data is to consider off-line estimation of the model parameters repeatedly in time shifting window. Another popular method is the stochastic natural gradient algorithm, which relies on non-Gaussianity of the separated signals and is adaptive by its nature. In this paper, we propose an adaptive version of two blind source separation algorithms which exploit non-stationarity of the original signals. The results indicate that the proposed algorithms slightly outperform the natural gradient in the trade-off between the algorithm's ability to quickly adapt to changes in the mixing matrix and the variance of the estimate when the mixing is stationary.

1 Introduction

Blind separation of instantaneous mixtures of independent signals or independent component analysis (ICA) usually assumes that a mixing matrix and source signals are stationary. In practice, however, the mixing matrix may vary in time - for example in audio signal separation, the audio scene may change in time, speakers may move, or there are some other changes in the environment.

Traditional methods of the blind source separation (BSS) can be adapted to such cases by applying them to time-shifting windows. There is always a trade-off between adaptability of the algorithms to changes of the mixing systems and accuracy (stability) of the estimation when the mixing matrix is constant. Such trade-off can be controlled through one or more tuning parameters, often called step size or forgetting factor. Some of the first BSS methods were adaptive [1–3].

In this paper, we design algorithms to blindly and adaptively separate linear instantaneous mixtures of signals that are non-stationary or, more precisely, piecewise stationary with varying variances in different blocks (also called

---

This work was supported by California Community Foundation through Project No. DA-15-114599 and by the Czech Science Foundation through Project No. 17-00902S.

P. Tichavský et al. (Eds.): LVA/ICA 2017, LNCS 10169, pp. 172–181, 2017.
DOI: 10.1007/978-3-319-53547-0_17

epochs) of data. We compare their performance to the widely used stochastic natural gradient algorithm (NG) [4], which is adaptive by its nature and separates the independent signals based on the assumption of their non-Gaussianity. In fact, NG is the most popular method applied in the frequency domain BSS algorithms [5]. The algorithms proposed in this paper are adaptive versions of BGSEP (Block Gaussian SEParation) [6] and of BARBI (Block AutoRegressive Blind Identification) [7]. Both BGSEP and BARBI are based on approximate joint diagonalization of matrices [8]. BARBI is more complex and works with covariance matrices of lag 1, also.

The cause of the gain in performance is that the piecewise stationary modeling of the speech signals is more appropriate for the blind separation than the pure non-Gaussianity. We support the above empirical observation by a theoretical analysis. We compare expressions characterizing the best achievable separation accuracy (Cramer-Rao-induced bounds) obtained trough separation based on non-Gaussianity, and similar expressions for separation based on non-stationarity. However, the performance depends on the degree of non-stationarity of the separated signals.

Next, we compare the performance of a non-Gaussianity based EFICA [9] and non-stationarity-based BGSEP and BARBI when they are applied to mixtures of short speech signals. Then, in Sect. 3, we describe the stochastic natural gradient algorithm and present details of the proposed algorithms, adaptive BGSEP and adaptive BARBI. Section 4 contains simulation results and Sect. 5 concludes the paper.

2   Signal Model and Separation Performance Limits

In this paper, we consider for simplicity squared instantaneous mixtures of independent signals

$$\mathbf{x}_t = \mathbf{A}\mathbf{z}_t, \quad t = 1 \ldots T, \tag{1}$$

where $\mathbf{A}$ is an $N \times N$ mixing matrix, which may be constant or slowly varying in time, and $\mathbf{z}_t = [\mathbf{z}_{1t}, \ldots, \mathbf{z}_{Nt}]^T$ is the vector of the separated signals.

Below we consider three models of the separated signals:

1. **Non-Gaussianity:** $\mathbf{z}_{nt}$ are i.i.d with zero mean, unit variance. We shall assume that mean square score function of the probability density exists and is finite,

$$\kappa_n = \mathrm{E}\left[((\partial \log(p_n(x))/\partial x)^2)\right] < \infty, \tag{2}$$

where $p_n(x)$ is the probability density function of the distribution of $\mathbf{z}_{nt}$.
2. **Non-Stationarity:** The observation period $t = [1, \ldots, T]$ can be divided into M epochs of equal size, T/M, such that on each epoch m, the $\mathbf{z}_{nt}$ is Gaussian-distributed with zero mean and variance s_{nm}, $m = 1, \ldots, M$ for $t = (m-1)T/M + 1, \ldots, mT/M$.
3. **Piecewise AR(1) modeling:** The observation period is divided into M epochs, and within each the signal is Gaussian AR(1) process with zero mean, variance s_{nm} and an autoregressive coefficient $\rho_{nm} = \mathrm{E}[z_{nt}z_{n,t+1}]/s_{nm}$ for $n = 1, \ldots, N$ and $t = (m-1)T/M + 1, \ldots, mT/M$.

There are many methods of the independent component analysis relying on the source non-Gaussianity, see, e.g., [10–13] and references therein. A few BSS methods relying on the source non-stationarity exist, see e.g. [14].

The separation performance can be measured in terms of the estimated interference-to-signal ratio (ISR) matrix, which tells how much energy of the jth original signal is contained in the kth estimated signal.

The ISR matrix can be estimated by examining statistical properties of the separated signals. In particular, for the non-Gaussianity model it was shown in [15] that the ISR matrix elements are lower bounded by the Cramer-Rao-induced bound as

$$\text{ISR}_{jk} \geq \frac{1}{N} \frac{\kappa_k}{\kappa_j \kappa_k - 1}. \tag{3}$$

Note that it was shown that $\kappa_j \geq 1$ for all distribution functions $p_j(x)$, and the equality holds if and only if (iff) $p_j(x)$ is Gaussian. This observation is in accord with the well known fact that the mixture of two random signals is separable (ISR is finite) iff at least one of the probability distributions is non-Gaussian.

For the non-stationarity model it can be shown in a similar way as in [16] that the ISR matrix elements are lower bounded by the Cramer-Rao-induced bound as

$$\text{ISR}_{jk} \geq \frac{1}{N} \frac{\phi_{kj}}{\phi_{jk} \phi_{kj} - 1} \frac{\sum_{m=1}^{M} s_{mj}}{\sum_{m=1}^{M} s_{mk}}, \tag{4}$$

where

$$\phi_{jk} = \frac{1}{M} \sum_{m=1}^{M} \frac{s_{mj}}{s_{mk}}. \tag{5}$$

It can be easily shown that the product $\phi_{jk} \phi_{kj}$ is always greater or equal to one, and it is equal to one if the variances of the separated signals are multiples each of the other, $s_{mj} = \alpha s_{mk}$ for some α and all $m = 1, \ldots, M$. The last fraction in (4) is the ratio of average powers of the jth and kth signal.

Similarly, for the piecewise AR(1) models, the bound on ISR_{jk} has the same form as in (4). The difference resides in the definition of ϕ_{jk}, which is

$$\phi_{jk} = \frac{1}{M} \sum_{m=1}^{M} \frac{s_{mj}}{s_{mk}} \frac{1 + \rho_{km}^2 - 2\rho_{km}\rho_{jm}}{1 - \rho_{jm}^2}. \tag{6}$$

Note that both the models 2 and 3 (non-stationarity and block AR(1) modeling) lead effectively to non-Gaussian signals, so that the principle of non-Gaussianity is a valid approach to decompose the signals. The overall probability distribution function of the data in model 2 and 3 is a mixture of Gaussian, and therefore it is non-Gaussian unless the variances in the blocks are the same. The statistical dependence of the signal in different times is ignored in this model. Parameter κ for the mixture of Gaussians is hard to handle analytically, but we can compute it by numerical integration. Assume that the signal can be divided into 100 epochs and that variances of a signal in the epochs are uniformly distributed in the interval $[1 - \Delta, 1 + \Delta]$, where Δ is a free parameter from the interval $[0, 1]$. We can consider a mixture of two signals of the same type. For Δ close to zero, the signals are nearly stationary and hard to separate for both methods. For Δ close to 1, the separation is more accurate, as we can see in Fig. 1. We can observe the difference in performance about 10 dB.

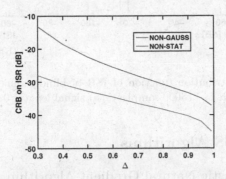

Fig. 1. Cramer-Rao bound on ISR for separation of two piecewise stationary signals with variances uniformly distributed in the interval $[1 - \Delta, 1 + \Delta]$ versus Δ for the signal length $N = 10000$.

Next, we compare performance of one non-Gaussianity-based method and two non-stationarity-based methods in the following experiment. We consider the set of 16 speech signals from [17]. Assuming that the mixing can be considered to be stationary for a second, we take pairs of one second long pieces of different signals, mix them together with a fixed mixing matrix $\mathbf{A} = [1, -0.5; 0.5, 1]$ and demix them blindly with three BSS algorithms: EFICA, as a representative of non-Gaussianity based algorithms, BGSEP and BARBI(1), both with the block length of 200 samples. In total, we did 8 trials (with different beginnings) of all $16.15/2 = 120$ pairs of signals. In Fig. 2(a) we plot the cumulative distribution functions of the achieved ISR for the three methods. We can see that the ISR varies in the range -20 dB to -100 dB, and statistically, there are gaps between the ISR of EFICA, BGSEP and BARBI(1) of 5 dB and another 5 dB, respectively. The BARBI(1) achieves the best separation with BGSEP following and EFICA performing the worst.

Next, we have repeated the same experiment with shorter signals, of the half length, 0.5 s. The difference in performance becomes smaller, as we can see in Fig. 2(b).

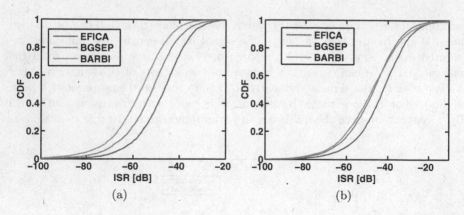

Fig. 2. Cumulative distribution function of ISR of blindly demixed pairs of speech signals: (a) signal length 1 s (16000 samples), (b) signal length: 0.5 s (8000 samples).

3   Adaptive BSS Algorithms

3.1   Scaled Stochastic Natural Gradient Algorithm

Given the current sample of the mixtures $\mathbf{x}_t$ and an estimate of the demixing matrix $\mathbf{W}_t$, the natural gradient updates $\mathbf{W}_t$ as

$$\mathbf{W}_{t+1} = c\left(\mathbf{W}_t + \mu(\mathbf{I} - cf(\mathbf{W}_t\mathbf{x}_t)(\mathbf{W}_t\mathbf{x}_t)^T)\mathbf{W}_t\right),$$

where $f(\,.\,)$ is an appropriately chosen nonlinear function, μ is the step length parameter and c is a scaling parameter

$$c = \frac{N}{\sum_{i,j=1}^{N}|(f(\mathbf{W}_t\mathbf{x}_t)(\mathbf{W}_t\mathbf{x}_t)^T)_{ij}|}.$$

The function $f(\,.\,)$ is applied elementwise. In our simulations, we use the commonly used nonlinear function

$$f(\mathbf{W}_t\mathbf{x}_t) = \tanh(10\mathbf{W}_t\mathbf{x}_t).$$

3.2   Adaptive BGSEP

The adaptive BGSEP algorithm is initialized by an estimate of the demixing matrix $\mathbf{W}_0$ and by M sample covariance matrices computed in the past $M - 1$ epochs of the given mixture, each of the length L, as

$$\mathbf{R}_{M-m} = \frac{1}{L}\sum_{\ell=1}^{L}\mathbf{W}_0\mathbf{x}_{\ell-mL}(\mathbf{W}_0\mathbf{x}_{\ell-mL})^T \tag{7}$$

$m = 1,\ldots,M-1$. $\mathbf{R}_M$ is set to zero matrix.

Given a t-th new sample $\mathbf{x}_t$, BGSEP updates the M-th covariance as

$$\mathbf{R}_M = ((s-1)\mathbf{R}_M + \mathbf{W}_t\mathbf{x}_t(\mathbf{W}_s\mathbf{x}_t)^T)/s \tag{8}$$

where $s = \mathrm{rem}(t,L)$ is the remainder in the division of t by L, and the demixing matrix as

$$\mathbf{W}_{t+1} = (\mathbf{I} - \mu\mathbf{B})\mathbf{W}_t. \tag{9}$$

Here, $\mathbf{B}$ is matrix with a zero diagonal, such that each pair of non-diagonal elements B_{ij}, B_{ji} is computed separately as a solution to a 2×2 set of linear equations

$$\begin{bmatrix} B_{ij} \\ B_{ji} \end{bmatrix} = \begin{bmatrix} \sum_{m=1}^{M} \frac{r_{jjm}}{r_{iim}} & M \\ M & \sum_{m=1}^{M} \frac{r_{iim}}{r_{jjm}} \end{bmatrix}^{-1} \begin{bmatrix} \sum_{m=1}^{M} \frac{r_{ijm}}{r_{iim}} \\ \sum_{m=1}^{M} \frac{r_{ijm}}{r_{jjm}} \end{bmatrix}, \tag{10}$$

where r_{ijm} is the (i,j)th element of $\mathbf{R}_m$. The update formula (9) was obtained by modifying the off-line BGSEP, see [6,7,18]. The step length parameter μ is chosen so that the estimate varies smoothly while still follows the changes of the demixing matrix. If t equals a multiple of the length of the blocks L, we discard $\mathbf{R}_1$ and set $\mathbf{R}_i \leftarrow \mathbf{R}_{i+1}$ for $i = 1\ldots M-1$ and $\mathbf{R}_M = \mathbf{0}$, thus resetting the algorithm. Each update of the demixing matrix depends not only on the actual sample of the mixtures but also on previous samples, number of which is given by the block length L and the number of blocks M. Therefore we can expect the increase in L and/or in M will result in the decreased variance of the estimate and increased reaction time for the change in the true mixing matrix and vice versa. The algorithm is summarized in Algorithm 1.

Algorithm 1. Adaptive BGSEP update

> **Input:** $\mathbf{x}_t$, $\mathbf{W}$, $\mathbf{R}_1 \ldots \mathbf{R}_M$, t, L;
> $t = t + 1$, $s \leftarrow \mathrm{rem}(t-1, L) + 1$;
> $\mathbf{R}_M \leftarrow ((s-1)\mathbf{R}_M + \mathbf{W}\mathbf{x}_t(\mathbf{W}\mathbf{x}_t)^T)/s$;
> Compute elements of $\mathbf{B}$ via (10);
> $\mathbf{W} \leftarrow (\mathbf{I} - \mu\mathbf{B})\mathbf{W}$;
> if $s = L$ then
>   for $m = 1 : M - 1$ do
> $\mathbf{R}_m \leftarrow \mathbf{R}_{m+1}$;
>   end for
> $\mathbf{R}_M \leftarrow \mathbf{0}$;
> end if
> **Output:** $\mathbf{W}$, $\mathbf{R}_1 \ldots \mathbf{R}_M$, t;

3.3 Adaptive BARBI

The online BARBI algorithm works similarly to BGSEP, but in addition to covariances $\mathbf{R}_m$, its initialization requires also a set of symmetrized sample lag one covariances

$$S_m = \frac{1}{2L} \sum_{\ell=1}^{L} [W_0 x_{\ell-mL}(W x_{\ell-mL-1})^T + W_0 x_{\ell-mL-1}(W_0 x_{\ell-mL})^T] \quad (11)$$

for $m = 1 \ldots M - 1$, $S_M = 0$. Given a k-th new sample x_k, BARBI updates the M-th lag one covariance as

$$S_M = \frac{s-1}{s} S_M + \frac{W x_t (W x_{t-1})^T}{2s} + \frac{W x_{t-1}(W x_t)^T}{2s}.$$

The updates of lag zero covariance and the demixing matrix are the same as in BGSEP, except the equations for the non-diagonal elements of B take the form

$$\begin{bmatrix} B_{ij} \\ B_{ji} \end{bmatrix} = \begin{bmatrix} \sum_{m=1}^{M} q_{im}^T p_{jjm} & M \\ M & \sum_{m=1}^{M} q_{jm}^T p_{iim} \end{bmatrix}^{-1} \cdot \begin{bmatrix} \sum_{m=1}^{M} q_{im}^T p_{ijm} \\ \sum_{m=1}^{M} q_{jm}^T p_{ijm} \end{bmatrix}, \quad (12)$$

where

$$q_{im} = \frac{1}{r_{iim}(r_{iim}^2 - |s_{iim}|^2)} \begin{bmatrix} r_{iim}^2 + |s_{iim}|^2 \\ -2 s_{iim} r_{iim} \end{bmatrix}, \quad (13)$$

$$p_{ijm} = \begin{bmatrix} r_{ijm} \\ s_{ijm} \end{bmatrix}. \quad (14)$$

Here s_{ijm} is the (i,j)th element of S_m. The update formula (12) was obtained by modifying the off-line BARBI, see [7,18]. After L iterations the algorithm is reset as in BGSEP. The update step of adaptive BARBI is summarized in Algorithm 2.

Algorithm 2. Adaptive BARBI update

Input: x_t, x_{t-1}, W, $R_1 \ldots R_M$, $S_1 \ldots S_M$, t, L;
$t = t + 1$, $s \leftarrow \mathrm{rem}(t - 1, L) + 1$;
$R_M \leftarrow ((s-1)R_M + W x_t (W x_t)^T)/s$;
$S_M \leftarrow ((s-1)S_M + 1/2(W x_t (W x_{t-1})^T + W x_{t-1}(W x)^T))/s$;
Compute elements B via (12);
$W \leftarrow (I - \mu B)W$;
if $s = L$ **then**
    for $m = 1 : M - 1$ **do**
        $R_m \leftarrow R_{m+1}, S_m \leftarrow S_{m+1}$;
    end for
    $R_M \leftarrow 0$, $S_M \leftarrow 0$;
end if
Output: W, $R_1 \ldots R_M$, $S_1 \ldots S_M$, t, L;

---

4   Experiments

We examine tracking properties of the natural gradient, adaptive BGSEP and adaptive BARBI in the following example. Consider a pair of natural speech signals taken from the same database as in Sect. 2. We mix them by the mixing matrix $\mathbf{A}_1 = \begin{bmatrix} 1 & 2 \\ 2 & -1 \end{bmatrix}$, and change it abruptly to another matrix $\mathbf{A}_2 = \begin{bmatrix} -1 & 2 \\ 2 & 1 \end{bmatrix}$ at time instant $t = 4.1875\,\text{s}$. For NG the step length parameter μ was set to 0.001. For BGSEP the block length was set to $L = 200$ samples, the number of blocks was set to $M = 10$, and the step length parameter was set $\mu_{bg} = 0.01$. The parameters were manually selected as such that the methods yield best performances. For adaptive BARBI we have chosen the same block length and the same number of blocks, and the step length $\mu_{barbi} = 0.001$. The ability of the algorithms to adapt to the change of the mixing matrix is studied in terms of the estimated gain matrix $\mathbf{G}_t = \widehat{\mathbf{W}}_t \mathbf{A}_t$, where $\mathbf{A}_t$ is the mixing matrix at time t and $\widehat{\mathbf{W}}_t$ is the estimated demixing matrix. In the ideal case, $\mathbf{G}_t$ should be a diagonal or counter-diagonal matrix. The results for all three algorithms are plotted in Fig. 3. Gain matrices for all three algorithms switch from near diagonal to near counter-diagonal following the abrupt change of the true mixing matrix.

Next, we have computed the instantaneous interference to signal ratio ISR of the separated signals in moving time window of the length of 10000 samples (0.625 s). The results are plotted in Fig. 4. The BGSEP achieves the same separation as NG in the first half of the signal, attaining average ISR of $-32.84\,\text{dB}$ and $-33.41\,\text{dB}$ respectively, both outperforming BARBI with $-27.13\,\text{dB}$. In the

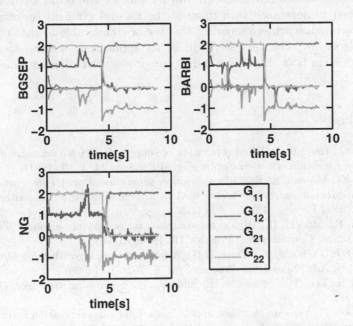

Fig. 3. Evolution of the elements of the gain matrix $\mathbf{G}$ for BGSEP, BARBI and NG algorithms in the case of an abrupt change in the mixing matrix. The gain matrices switch between diagonal and counterdiagonal in reaction to the change in the true mixing matrix.

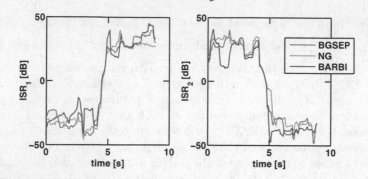

Fig. 4. Instantaneous SIR for BGSEP, NG and BARBI algorithms in the case of an abrupt change in the mixing matrix.

second half of the signal, BARBI attains the lowest average ISR of -38.17 dB, BGSEP being second best with -35.51 dB and NG achieving -33.09 dB.

5   Conclusion

We have proposed two novel adaptive algorithms for the blind separation and compare their performance with those of the natural gradient technique. The proposed techniques achieve separation better or comparable to that of natural gradient algorithm. The next step will be an application of these methods in frequency domain BSS algorithms and a comparison with adaptive time domain BSS [19].

References

1. Jutten, C., Herault, J.: Blind separation of sources, Part I: an adaptive algorithm based on neuromimetic architecture. Signal Process. **24**, 1–10 (1991)
2. Macchi, O., Moreau, E.: Self-adaptive source separation, Part I: convergence analysis of a direct linear network controled by the Herault-Jutten algorithm. IEEE Trans. Signal Process. **45**, 918–926 (1997)
3. Moreau, E., Macchi, O.: High order contrasts for self-adaptive source separation. Int. J. Adapt. Control Signal Process. **10**, 19–46 (1996)
4. Amari, S.I., Cichocki, A., Yang, H.H.: A new learning algorithm for blind signal separation. Adv. Neural Inf. Process. Syst. **8**, 752–763 (1996)
5. Makino, S., Lee, T.W., Sawada, H.: Blind Speech Separation. Springer, Dordrecht (2007)
6. Tichavský, P., Yeredor, A.: Fast approximate joint diagonalization incorporating weight matrices. IEEE Trans. Signal Process. **57**, 878–891 (2009)
7. Tichavský, P., Yeredor, A., Koldovský, Z.: A fast asymptotically efficient algorithm for blind separation of a linear mixture of block-wise stationary autoregressive processes. In: ICASSP 2009, Taipei, pp. 3133–3136 (2009)

8. Chabriel, G., Kleinsteuber, M., Moreau, E., Shen, H., Tichavský, P., Yeredor, A.: Joint matrices decompositions and blind source separation. IEEE Signal Process. Mag. **31**, 34–43 (2014)
9. Koldovský, Z., Tichavský, P., Oja, E.: Efficient variant of algorithm fastICA for independent component analysis attaining the Cramér-Rao lower bound. IEEE Trans. Neural Netw. **17**, 1265–1277 (2006)
10. Comon, P., Jutten, C.: Handbook of Blind Source Separation, Independent Component Analysis and Applications. Academic Press/Elsevier, Amsterdam (2010)
11. Hyvärinen, A., Karhunen, J., Oja, E.: Independent Component Analysis. Wiley-Interscience, New York (2001)
12. Tichavský, P., Koldovský, Z.: Fast and accurate methods of independent component analysis: a survey. Kybernetika **47**, 426–438 (2011)
13. Koldovský, Z., Tichavský, P.: A comparison of independent component and independent subspace analysis algorithms. In: Proceedings of the European Signal Processing Conference (EUSIPCO), Glasgow, pp. 1447–1451 (2009)
14. Cardoso, J.-F., Pham, D.T.: Separation of non stationary sources. Algorithms and performance. In: Roberts, S.J., Everson, R.M. (eds.) Independent Components Analysis: Principles and Practice, pp. 158–180. Cambridge University Press, Cambridge (2001)
15. Tichavský, P., Koldovský, Z., Oja, E.: Performance analysis of the fastICA algorithm and Cramér-Rao bounds for linear independent component analysis. IEEE Trans. Signal Process. **54**, 1189–1203 (2006)
16. Doron, E., Yeredor, A., Tichavský, P.: Cramér-Rao-induced bound for blind separation of stationary parametric Gaussian sources. IEEE Signal Process. Lett. **14**, 417–420 (2007)
17. Tichavský, P., Koldovský, Z.: Weight adjusted tensor method for blind separation of underdetermined mixtures of nonstationary sources. IEEE Trans. Signal Process. **59**, 1037–1047 (2011)
18. Tichavský, P., Šembera, O., Koldovský, Z.: Blind separation of mixtures of piecewise AR(1) processes and model mismatch. In: Vincent, E., Yeredor, A., Koldovský, Z., Tichavský, P. (eds.) LVA/ICA 2015. LNCS, vol. 9237, pp. 304–311. Springer, Heidelberg (2015). doi:10.1007/978-3-319-22482-4_35
19. Málek, J., Koldovský, Z., Tichavský, P.: Adaptive time-domain blind separation of speech signals. In: Vigneron, V., Zarzoso, V., Moreau, E., Gribonval, R., Vincent, E. (eds.) LVA/ICA 2010. LNCS, vol. 6365, pp. 9–16. Springer, Heidelberg (2010). doi:10.1007/978-3-642-15995-4_2

Source Separation, Dereverberation and Noise Reduction Using LCMV Beamformer and Postfilter

Ofer Schwartz[1], Sebastian Braun[2], Sharon Gannot[1],
and Emanuël A.P. Habets[2]

[1] Faculty of Engineering, Bar-Ilan University, Ramat-Gan, Israel
[2] International Audio Laboratories Erlangen, Erlangen, Germany

Abstract. The problem of source separation, dereverberation and noise reduction using a microphone array is addressed in this paper. The observed speech is modeled by two components, namely the early speech (including the direct path and some early reflections) and the late reverberation. The minimum mean square error (MMSE) estimator of the early speech components of the various speakers is derived, which jointly suppresses the noise and the overall reverberation from all speakers. The overall time-varying level of the reverberation is estimated using two different estimators, an estimator based on a temporal model and an estimator based on a spatial model. The experimental study consists of measured acoustic transfer functions (ATFs) and directional noise with various signal-to-noise ratio levels. The separation, dereverberation and noise reduction performance is examined in terms of perceptual evaluation of speech quality (PESQ) and signal-to-interference plus noise ratio improvement.

1 Introduction

Speech enhancement techniques, utilizing microphone arrays, have attracted the attention of many researchers during the last three decades, especially in the context of hands-free communication tasks. Usually, the received speech signals are corrupted by interfering sources, such as competing speakers, reverberation and noise sources.

A generalization of the minimum variance distortionless response (MVDR) beamformer (BF), which deals with multiple linear constraints, is the linearly constrained minimum variance (LCMV) BF [1,2]. The LCMV BF can be applied to construct a beam-pattern, satisfying multiple constraints for a set of directions, while minimizing the output noise power.

In [3], it was shown that the minimum mean square error (MMSE) estimator of a single speech signal can be decomposed into two stages, an MVDR-BF and a postfilter. In [4], the authors proved that the output of the MVDR-BF is a sufficient statistic for estimating a single speech signal from multichannel inputs in the presence of additive Gaussian noise. Hence, any MMSE estimator necessarily

© Springer International Publishing AG 2017
P. Tichavský et al. (Eds.): LVA/ICA 2017, LNCS 10169, pp. 182–191, 2017.
DOI: 10.1007/978-3-319-53547-0_18

includes an initial MVDR-BF stage. In our previous work [5], the aforementioned decomposition was utilized to jointly suppress reverberation and noise for single-speaker scenarios. The reverberation and the noise were suppressed both in the BF stage and in the postfiltering stage.

In a recent work [6], we analyzed the multichannel MMSE estimator for the case of multiple desired speakers in a noisy and low reverberant environment. The estimator was decomposed into a multi-speaker LCMV BF, followed by a multi-speaker Wiener postfilter, inspired by the decomposition for the single-speaker case presented in [3]. The output of the first stage is a vector with dimensions equal to the number of desired speakers with each element dominated by one source plus residual noise. The multi-speaker Wiener postfilter is a square matrix that minimizes the residual noise at the output signals of the BF stage. Moreover, it was proved, by using the Fisher-Neyman factorization, that the output signals of the multi-speaker LCMV BF is the sufficient statistic for estimating the individual speech signals. In [7], an MMSE estimator for dereverberation of multiple sources was proposed, where the aim was to extract the sum of all direct sound components while reducing reverberation and noise.

In this paper, the MMSE estimator analyzed in [6] is exploited to separate highly reverberant speech signals in a noisy environment. The acoustic impulse response (AIR) is modeled by two components (that are assumed to be uncorrelated), namely the early speech (including the direct path and some early reflections) and the late reverberation [8,9]. The early speech is characterized by discrete reflections of sound waves on the walls and other rigid objects. In the short-time Fourier transform (STFT) domain, the early speech component can be modeled as a multiplication of the transformed sound source signal frame by the frequency response of the early component of the AIR. The late reflections are usually dense, since they are a summation of many reflections arriving from all directions. Therefore, the late reverberation and an ideal diffuse sound-field have very similar spatial properties [8]. Accordingly, in the STFT domain the late reverberation can be modeled as a diffuse sound field with a time-varying level. In a multi speaker-environment, the overall reverberation of the speakers (the sum of the reverberant signals from all speakers) can be modeled as a diffuse field due to the multiple overlapping reflections. The overall time-varying level of the reverberation is estimated using two alternative estimators: (1) an estimator based on a temporal exponentially decaying model [9] and (2) an estimator based on a spatial model [7]. In the experimental section the proposed source separation, dereverberation and noise reduction system is examined using the two latter reverberation PSD level estimators in terms of output segmental signal-to-interference plus noise ratio (SINR) results and perceptual evaluation of speech quality (PESQ) scores.

2   Problem Formulation

The source separation, dereverberation and noise reduction problem is formulated in the STFT domain with ℓ denoting the frame index and k denoting the frequency index. Assume that J coherent sources are captured by an array of N microphones and the i-th microphone signal can be expressed as

$$Y_i(\ell, k) = \sum_{j=1}^{J} X_{i,j}(\ell, k) + V_i(\ell, k), \ i = 1, 2, \ldots, N, \tag{1}$$

where $X_{i,j}(\ell, k)$ denotes the reverberant speech signal of the j-th speaker as received by the i-th microphone and $V_i(\ell, k)$ denotes the ambient noise. The reverberant speech signal can be separated into two components

$$X_{i,j}(\ell, k) = E_{i,j}(\ell, k) + R_{i,j}(\ell, k), \tag{2}$$

where $E_{i,j}(\ell, k)$ denotes the early speech signal of the j-th speaker and $R_{i,j}(\ell, k)$ denotes the reverberation originating from the j-th speaker. The observed early speech component at the i-th microphone is modeled in the STFT domain as a multiplication of the direct speech at the first microphone with a time-invariant relative early transfer function (RETF) $G_{i,j}(k)$ (i.e., assuming a static scenario) relating the source position and the i-th microphone as

$$E_{i,j}(\ell, k) = G_{i,j}(k)E_{1,j}(\ell, k). \tag{3}$$

The N microphone signals are stacked in a vector form

$$\mathbf{y}(\ell, k) = \sum_{j=1}^{J} \left(\mathbf{g}_j(k)E_{1,j}(\ell, k) + \mathbf{r}_j(\ell, k) \right) + \mathbf{v}(\ell, k) = \mathbf{G}(k)\mathbf{e}(\ell, k) + \mathbf{r}(\ell, k) + \mathbf{v}(\ell, k),$$
$$\tag{4}$$

where

$$\mathbf{y}(\ell, k) = \left[Y_1(\ell, k) \ Y_2(\ell, k) \ldots Y_N(\ell, k) \right]^{\mathrm{T}}, \tag{5}$$

$$\mathbf{g}_j(k) = \left[G_{1,j}(k) \ G_{2,j}(k) \cdots G_{N,j}(k) \right]^{\mathrm{T}}, \tag{6}$$

$$\mathbf{r}_j(\ell, k) = \left[R_{1,j}(\ell, k) \ R_{2,j}(\ell, k) \cdots R_{N,j}(\ell, k) \right]^{\mathrm{T}}, \tag{7}$$

$$\mathbf{v}(\ell, k) = \left[V_1(\ell, k) \ V_2(\ell, k) \cdots V_N(\ell, k) \right]^{\mathrm{T}}, \tag{8}$$

$$\mathbf{G}(k) = \left[\mathbf{g}_1(k) \ \mathbf{g}_2(k) \cdots \mathbf{g}_J(k) \right], \tag{9}$$

$$\mathbf{e}(\ell, k) = \left[E_{1,1}(\ell, k) \ E_{1,2}(\ell, k) \cdots E_{1,J}(\ell, k) \right]^{\mathrm{T}}, \tag{10}$$

and $\mathbf{r}(\ell, k) = \sum_{j=1}^{J} \mathbf{r}_j(\ell, k)$.

The probability density function (p.d.f.) of the observed data given the early speech is modeled as a complex-Gaussian

$$f(\mathbf{y}(\ell, k)|\mathbf{e}(\ell, k); \mathbf{G}(k), \boldsymbol{\Phi}_{\mathrm{in}}(\ell, k)) = \mathcal{N}_{\mathrm{C}}\left(\mathbf{y}(\ell, k); \mathbf{G}(k)\mathbf{e}(\ell, k), \boldsymbol{\Phi}_{\mathrm{in}}(\ell, k) \right), \tag{11}$$

where $\boldsymbol{\Phi}_{\mathrm{in}}(\ell, k) = \boldsymbol{\Phi}_{\mathbf{v}}(k) + \boldsymbol{\Phi}_{\mathbf{r}}(\ell, k)$ is the interference PSD matrix, $\boldsymbol{\Phi}_{\mathbf{v}}(k)$ is the power spectral density (PSD) matrix of the ambient noise and $\boldsymbol{\Phi}_{\mathbf{r}}(\ell, k)$ is the aggregated reverberation PSD matrix, i.e.,

$$\boldsymbol{\Phi}_{\mathbf{v}}(k) = E\{\mathbf{v}(\ell, k)\mathbf{v}^{\mathrm{H}}(\ell, k)\} \tag{12}$$

$$\boldsymbol{\Phi}_{\mathbf{r}}(\ell, k) = E\{\mathbf{r}(\ell, k)\mathbf{r}^{\mathrm{H}}(\ell, k)\}. \tag{13}$$

The function $\mathcal{N}_C(\cdot;\cdot,\cdot)$ denotes the complex Gaussian probability

$$\mathcal{N}_C(\mathbf{z};\boldsymbol{\mu},\boldsymbol{\Phi}) = \frac{1}{\pi^N \det(\boldsymbol{\Phi})} \exp\left((\mathbf{z}-\boldsymbol{\mu})^H \boldsymbol{\Phi}^{-1}(\mathbf{z}-\boldsymbol{\mu})\right), \qquad (14)$$

where $\mathbf{z}$ is a Gaussian random vector, $\boldsymbol{\mu}$ is its mean and $\boldsymbol{\Phi}$ is its PSD matrix. The noise PSD matrix $\boldsymbol{\Phi}_\mathbf{v}(k)$ can be estimated during speech-absence. Estimating the noise PSD matrix is beyond the scope of this contribution. The p.d.f. of the early speech signals is given by:

$$f(\mathbf{e}(\ell,k);\boldsymbol{\Phi}_\mathbf{e}(\ell,k)) = \mathcal{N}_C(\mathbf{e}(\ell,k);\mathbf{0},\boldsymbol{\Phi}_\mathbf{e}(\ell,k)) \qquad (15)$$

where $\boldsymbol{\Phi}_\mathbf{e}(\ell,k) = E\{\mathbf{e}(\ell,k)\mathbf{e}^H(\ell,k)\}$ is the PSD matrix of the early speech signals of the multiple speakers.

The late reverberation PSD matrix of each speaker $\boldsymbol{\Phi}_{\mathbf{r}_j}(\ell,k)$ is time-varying, since the reverberation originates from the speech sources. However, the spatial characteristic of the reverberation is assumed to be time-invariant, as long as the speaker and microphone constellation remains fixed. Therefore, it is reasonable to model the PSD matrix of the reverberation as a time-invariant matrix with time-varying levels

$$\boldsymbol{\Phi}_{\mathbf{r}_j}(\ell,k) = \phi_{R,j}(\ell,k)\,\boldsymbol{\Gamma}(k), \qquad (16)$$

where $\boldsymbol{\Gamma}(k)$ is the time-invariant spatial coherence matrix of the reverberation and $\phi_{R,j}(\ell,k)$ is the temporal level of the marginal reverberation of speaker j. It should be emphasized that $\boldsymbol{\Gamma}(k)$ is identical for all sources. Assuming statistical independence between the speakers, the total reverberation PSD matrix is the summation of the marginal reverberation PSD matrix of the individual speakers

$$\boldsymbol{\Phi}_\mathbf{r}(\ell,k) = \sum_{j=1}^J \boldsymbol{\Phi}_{\mathbf{r}_j}(\ell,k) = \phi_R(\ell,k)\,\boldsymbol{\Gamma}(k), \qquad (17)$$

where $\phi_R(\ell,k) = \sum_j \phi_{R,j}(\ell,k)$ is the overall level of the reverberation from all speakers per time-frequency bin.

In the current work, we assume that the reverberation can be modeled using a spatially homogenous and spherically isotropic sound field (as in many previous works [5]), and determine $\boldsymbol{\Gamma}(k)$ accordingly $\boldsymbol{\Gamma}_{m,n}(k) = \text{sinc}\left(\frac{2\pi f_s k d_{m,n}}{Kc}\right)$, where $\text{sinc}(x) = \sin(x)/x$, K is the number of frequency bins, $d_{m,n}$ is the distance between microphones m and n, and c is the sound velocity.

The aim of this work is to provide an MMSE-optimal multichannel estimate of $\mathbf{e}$. The well-known MMSE estimate of $\mathbf{e}(\ell,k)$ given the microphone signals is given by

$$\underset{\hat{\mathbf{e}}}{\arg\min} \, E\left\{\|\hat{\mathbf{e}}-\mathbf{e}(\ell,k)\|^2 \,|\, \mathbf{y}(\ell,k)\right\} = E\left\{\mathbf{e}(\ell,k)\,|\,\mathbf{y}(\ell,k)\right\}. \qquad (18)$$

3   Optimal Multichannel Speaker Separation, Dereverberation and Noise Reduction

In this section, we first describe the optimal MMSE estimator of the signal $\mathbf{e}(\ell, k)$. In the following, whenever applicable, the frequency index k and the time index ℓ are omitted for brevity. Since $\mathbf{e}$ and $\mathbf{y}$ are assumed to be zero-mean complex-Gaussian random variables, the MMSE estimator of $\mathbf{e}$ is given by the optimal linear estimator, i.e. the multichannel Wiener filter (MCWF)

$$\widehat{\mathbf{e}}_{\mathrm{MCWF}} = E\{\mathbf{e}\mathbf{y}^{\mathrm{H}}\} \times E\{\mathbf{y}\mathbf{y}^{\mathrm{H}}\}^{-1} \mathbf{y} = \mathbf{\Phi}_{\mathbf{e}}\mathbf{G}^{\mathrm{H}} \times \left[\mathbf{G}\mathbf{\Phi}_{\mathbf{e}}\mathbf{G}^{\mathrm{H}} + \mathbf{\Phi}_{\mathrm{in}}\right]^{-1} \mathbf{y}. \quad (19)$$

Using the Woodbury identity and some straightforward algebraic steps (as shown in [6]), $\widehat{\mathbf{e}}_{\mathrm{MCWF}}$ can be expressed as

$$\widehat{\mathbf{e}}_{\mathrm{MCWF}} = \underbrace{\mathbf{\Phi}_{\mathbf{e}}\left(\mathbf{\Phi}_{\mathbf{e}} + \mathbf{\Phi}_{\mathrm{RE}}\right)^{-1}}_{\mathbf{H}_{\mathrm{WF}}^{\mathrm{H}}}\widehat{\mathbf{e}}_{\mathrm{LCMV}}, \quad (20)$$

where

$$\widehat{\mathbf{e}}_{\mathrm{LCMV}} = \underbrace{\left(\mathbf{G}^{\mathrm{H}}\mathbf{\Phi}_{\mathrm{in}}^{-1}\mathbf{G}\right)^{-1}\mathbf{G}^{\mathrm{H}}\mathbf{\Phi}_{\mathrm{in}}^{-1}}_{\mathbf{H}_{\mathrm{LCMV}}^{\mathrm{H}}}\mathbf{y} \quad (21)$$

and $\mathbf{\Phi}_{\mathrm{RE}} = \left(\mathbf{G}^{\mathrm{H}}\mathbf{\Phi}_{\mathrm{in}}^{-1}\mathbf{G}\right)^{-1}$. The filtering matrix $\mathbf{H}_{\mathrm{LCMV}}$ is an $N \times J$ matrix which denotes the multi-speaker LCMV BF, and $\mathbf{H}_{\mathrm{WF}}$ is a $J \times J$ symmetric matrix which represents a multi-speaker Wiener postfilter.

4   Estimation of the Late Reverberation PSD Matrix

In the following, two estimators of the total late reverberation PSD ϕ_R are described. One estimator is based on a temporal model and the second is based on a spatial model.

4.1   Estimator Based on a Temporal Model

An estimate of the marginal reverberation PSD level in each microphone and for each speaker, $\phi_{R,i,j}$, can be obtained using Polack's model [10] (c.f. [9]):

$$\widehat{\phi}_{R,i,j}(\ell, k) = \exp(-2\alpha R L)\phi_{X_{i,j}}(\ell - L, k), \quad (22)$$

where $\phi_{X_{i,j}}(\ell, k) = E\left\{|X_{i,j}(\ell, k)|^2\right\}$, $\alpha = \frac{3\log(10)}{T_{60}f_s}$, L is the time in frames between the arrival of the direct sound and the start of the late reverberation, R is the number of samples between two subsequent STFT frames, T_{60} is the reverberation time, and f_s is the sampling frequency in Hz. Following (17), an estimate of the overall late reverberation level of all speakers at each microphone is obtained by

$$\widehat{\phi}_{R,i}(\ell, k) = \exp(-2\alpha R L)\sum_{j=1}^{J}\phi_{X,i,j}(\ell - L, k). \quad (23)$$

Assuming statistical independence between the speakers and using the signal model in (1), we can compute $\sum_{j=1}^{J} \phi_{X,i,j}(\ell,k) = [\phi_{Y,i}(\ell,k) - \phi_{V,i}(k)]$, where $\phi_{Y_i}(\ell,k) = E\left\{|Y_i(\ell,k)|^2\right\}$ and $\phi_{V_i}(k) = E\left\{|V_{i,j}(\ell,k)|^2\right\}$. We assume that the speakers are sufficiently far from the microphones and that the late reverberant sound field is homogenous such that the reverberation level is approximately equal at all microphones, i.e., $\phi_{R,i}(\ell,k) = \phi_R(\ell,k)$ for all $i \in \{1,2,\ldots,N\}$. This assumption might be violated if the distance between the speakers and the microphones is small. When this assumption holds, an estimate of the total late reverberation level is obtained by averaging the PSD estimates across all channels

$$\widehat{\phi}_R(\ell,k) = \frac{1}{N}\sum_{i=1}^{N}\widehat{\phi}_{R,i}(\ell,k). \tag{24}$$

The PSD of $Y_i(\ell,k)$ can be directly estimated from the microphone signals using

$$\widehat{\phi}_{Y,i}(\ell,k) = \beta_y\widehat{\phi}_{Y,i}(\ell-1,k) + (1-\beta_y)|Y_i(\ell,k)|^2 \tag{25}$$

where β_y is a forgetting factor.

4.2 Estimator Based on a Spatial Model

To simplify the estimator, a blocking matrix may be used to cancel out the direct sound. Let $\mathbf{u}$ be the $(N-J)$-dimensional output of the blocking matrix

$$\mathbf{u} = \mathbf{B}^{\mathrm{H}}\mathbf{y}, \tag{26}$$

where the blocking matrix $\mathbf{B}$ of size $N \times (N-J)$ has to satisfy the constraint $\mathbf{B}^{\mathrm{H}}\mathbf{G} = \mathbf{0}_{(N-J)\times J}$. Possible alternatives for designing the blocking matrix are discussed and analyzed in [7]. As a consequence of using (4) and (11), the PSD matrix of the blocking matrix outputs $\mathbf{u}$ is given by

$$\mathbf{\Phi}_\mathbf{u} = \mathbf{B}^{\mathrm{H}}\mathbf{\Phi}_\mathbf{y}\mathbf{B} = \phi_R\underbrace{\mathbf{B}^{\mathrm{H}}\mathbf{\Gamma}\mathbf{B}}_{\widetilde{\mathbf{\Gamma}}} + \underbrace{\mathbf{B}^{\mathrm{H}}\mathbf{\Phi}_\mathbf{v}\mathbf{B}}_{\widetilde{\mathbf{\Phi}}_\mathbf{v}}. \tag{27}$$

Denote $\widehat{\mathbf{\Phi}}_\mathbf{u}$, the estimated PSD matrix of $\mathbf{u}$, which can be directly computed from (26). Matching $\widehat{\mathbf{\Phi}}_\mathbf{u}$ and its model in (27), the problem at hand may be recast as a system of $(N-J)^2$ equations in one variable ϕ_R. Since there are more equations than variables, the best fitting ϕ_R that minimizes the total squared error may be found by minimizing the Frobenius norm between $\widehat{\mathbf{\Phi}}_\mathbf{u}$ and its model in (27). Accordingly, ϕ_R is the minimizer of the following cost-function:

$$\widehat{\phi}_R = \operatorname*{argmin}_{\phi_R} ||\widehat{\mathbf{\Phi}}_\mathbf{u} - \mathbf{\Phi}_\mathbf{u}||_{\mathrm{F}}^2, \tag{28}$$

where $||\cdot||_{\mathrm{F}}^2$ denotes the Frobenius norm. The solution to the minimization problem (28) is given by [7]

$$\widehat{\phi}_R = \frac{\Re\left\{\mathrm{Tr}\left[\widetilde{\mathbf{\Gamma}}^{\mathrm{H}}\left(\widehat{\mathbf{\Phi}}_\mathbf{u} - \widetilde{\mathbf{\Phi}}_\mathbf{v}\right)\right]\right\}}{\mathrm{Tr}\left[\widetilde{\mathbf{\Gamma}}^{\mathrm{H}}\widetilde{\mathbf{\Gamma}}\right]}. \tag{29}$$

5   Performance Evaluation

In this section, we evaluate the performance of the proposed estimators. In Sect. 5.1, the setup of the experiments is described. The separation, dereverberation and noise reduction performance using the two late reverberation estimators is reported in Sect. 5.2.

5.1   Setup

The received microphone signals were generated by convolving clean speech signals and measured room impulse responses (RIRs) recorded in our acoustic lab [11]. The lab with dimensions $6 \times 6 \times 2.4$ m is equipped with dedicated panels to control the reverberation level. In our setup, the lab reverberation time was $T_{60} = 0.61$ s. We used $J = 2$ speaker positions each at a distance of 2 m from an eight-microphone array at various angles. The inter-microphone distances were $\{3, 3, 3, 8, 3, 3, 3\}$ cm. Each experiment consists of two sentences with equal level, each 4–8 s long, one by a male and one by a female speaker. The speakers were positioned at $0°, -45°$ relative to the array. The noise signal $\mathbf{v}$ consists of two components: (1) a stationary noise signal from the NOISEX-92 database [12] convolved with RIRs for a source located at $45°$ relative to the array at a distance of 2 m and (2) a sensor noise with a level 10 dB below the level of the directional noise source. The noise signal $\mathbf{v}$ was added to the speech signals with various input signal-to-noise ratio (SNR) levels

$$\text{iSNR} = 10 \log_{10} \frac{\sum_{k,\ell,j} ||\mathbf{x}_j(\ell, k)||^2}{\sum_{k,\ell} ||\mathbf{v}(\ell, k)||^2}, \tag{30}$$

where $|| \cdot ||$ is the Euclidean norm. The sampling frequency of the speech signals was set to 16 kHz. The frame length of the STFT was 32 ms with 8 ms between successive time frames (i.e., 1024 frequency bins and 75% overlap). We assumed that the late reverberation starts 32 ms after the arrival of the direct-path by using $L = 4$. The direct-to-reverberation ratio (DRR) and the early-speech-to-reverberation ratio (ERR) were measured as approximately -3 dB and 6 dB. The noise PSD matrix $\mathbf{\Phi_v}$, which is non-diagonal, was estimated using time-segments where all speakers are inactive.

Each RETF $\mathbf{g}_j$ of the various speakers was estimated by the least squares (LS) technique described in [5], using time-segments where one speaker was exclusively active (the noise was always present). Further details about the system identification procedure can be found in [5]. Assuming statistically independent speakers, $\mathbf{\Phi_e}$ can be modeled as a diagonal matrix. To maintain this structure, only the diagonal elements of $\mathbf{\Phi_e}$ were estimated and the off-diagonal elements are set to zero. Similarly to [6], we adopt the decision-directed approach proposed in [13] for estimating the diagonal elements of $\mathbf{\Phi_e}$. Further details about the implementation of the various component of the proposed MMSE estimator can be found in [6].

The aforementioned estimators were compared in terms of output segmental SINR and PESQ. The output segmental SINR was aggregated for all speakers and for all time-frequency bins and is defined as

$$\text{oSINR}_{\text{BF}} = \sum_{\ell} 10 \log_{10} \frac{\sum_k \|\mathbf{e}(\ell, k)\|^2}{\sum_k \|\widehat{\mathbf{e}}_{\text{BF}}(\ell, k) - \mathbf{e}(\ell, k)\|^2}, \tag{31}$$

where $\text{BF} \in \{\text{LCMV}, \text{MCWF}\}$. In addition, the input SINR was calculated, i.e., the SINR in (31) where $\widehat{\mathbf{e}}_{\text{BF}}(\ell, k)$ is substituted by $Y_1(\ell, k)$. The PESQ scores were calculated for the estimated early speech of the speakers w.r.t. the actual early speech and were averaged for the two speakers. Note, that the output segmental SINR and PESQ actually measure the total performance of the source separation, the dereverberation and the noise reduction. All measures were computed by averaging the output segmented SINR results and PESQ scores obtained using 2×50 sentences, i.e., 50 experiments for each scenario where each experiment consists of 2 concurrent speakers (captured by $N = 8$ microphones).

5.2   Results

In Table 1, the output segmental SINR results and PESQ scores are presented for the LCMV BF estimator presented in (21) and for the complete MCWF presented in (20), using the late reverberation PSD estimators in either (24) or (29). The best results are depicted in boldface type. It can be verified that the MCWF outperforms the LCMV BF. According to our results, it is hard to determine which reverberation PSD estimator is better, as they both achieve comparable performance measures. It should be noted that the estimator based

Table 1. Performance measures for the LCMV in (21) and the MCWF in (20) using the PSD estimators in (22)–(24) (without brackets) and (29) (in brackets).

	iSINR			
Method	10 dB	20 dB	30 dB	∞ dB
Unprocessed	−8.50	−5.80	−4.20	−2.57
LCMV	−1.15 (−0.84)	0.86 (1.03)	1.93 (2.01)	1.16 (1.11)
MCWF	3.05 (**3.30**)	**3.86** (3.75)	**4.04** (3.80)	**3.67** (3.59)
	(a) Output SNR in dB			
	iSNR			
Method	10 dB	20 dB	30 dB	∞ dB
Unprocessed	1.40	1.56	1.59	1.60
LCMV	1.88 (1.90)	2.01 (2.01)	2.03 (2.03)	1.97 (1.97)
MCWF	**1.99** (1.97)	**2.07** (2.04)	**2.14** (2.07)	2.07 (**2.22**)
	(b) PESQ scores			

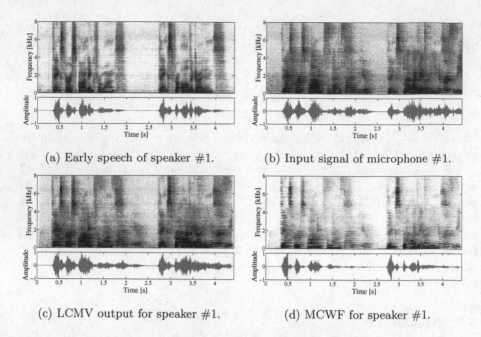

(a) Early speech of speaker #1.   (b) Input signal of microphone #1.

(c) LCMV output for speaker #1.   (d) MCWF for speaker #1.

Fig. 1. Spectrograms with $T_{60} = 0.61$ s and input iSNR of 30 dB.

on the temporal model requires an estimate of the reverberation time while the estimator based on the spatial model only requires an estimate of the spatial coherence matrix $\mathbf{\Gamma}$. Audio examples are available on our website[1]. By listening to these examples, it is evident that the proposed multi-speaker estimators produce satisfactory speaker separation, dereverberation and noise reduction.

Example spectrograms of the various output signals for iSNR of 30 dB and $T_{60} = 0.61$ s are depicted in Fig. 1. Figure 1a depicts $E_{1,1}$, the early speech signal of speaker #1 (positioned at $-45°$). Figure 1b depicts Y_1, the total received signal (including the noise component). Figure 1c depicts $\widehat{X}_{\text{LCMV},1}$, the first component of the multi-speaker LCMV output, corresponding to speaker #1. Likewise, Fig. 1d depicts $\widehat{X}_{\text{MCWF},1}$. Using careful examination, it can be seen that the MCWF output in Fig. 1d, exhibits better separation, dereverberation and noise reduction than the LCMV output in Fig. 1c.

6 Conclusions

In this paper, the MMSE estimator of the individual early speech of multiple concurrent speakers in reverberant and noisy environments was proposed. The overall reverberation was spatially modeled as diffuse noise with time varying level. The overall time-varying level of the reverberation was estimated using

---

[1] http://www.eng.biu.ac.il/gannot/speech-enhancement/.

two estimators, one based on a temporal model and the other based on a spatial model. The MMSE estimator was tested in a room with a reverberation time of 0.61 s for several signal-to-noise levels of directional noise. In terms of output SINR and PESQ scores, the proposed MMSE estimator well separates and enhances the early components of the various speakers.

References

1. Van Veen, B.D., Buckley, K.M.: Beamforming: a versatile approach to spatial filtering. IEEE AASP Mag. **5**(2), 4–24 (1988)
2. Markovich, S., Gannot, S., Cohen, I.: Multichannel eigenspace beamforming in a reverberant noisy environment with multiple interfering speech signals. IEEE Trans. Audio, Speech Lang. Process. **17**(6), 1071–1086 (2009)
3. Simmer, K.U., Bitzer, J., Marro, C.: Post-filtering techniques. Microphone Arrays: Signal Process. Tech. Appl. **3**, 39–60 (2001)
4. Balan, R., Rosca, J.: Microphone array speech enhancement by Bayesian estimation of spectral amplitude and phase. In: IEEE Sensor Array and Multichannel Signal Processing Workshop, pp. 209–213 (2002)
5. Schwartz, O., Gannot, S., Habets, E.A.P.: Multi-microphone speech dereverberation and noise reduction using relative early transfer functions. IEEE/ACM Trans. Audio, Speech, Lang. Process. **23**(2), 240–251 (2015)
6. Schwarz, O., Gannot, S., Habets, E.A.P.: Multi-speaker LCMV beamformer and postfilter for source separation and noise reduction. IEEE/ACM Trans. Audio, Speech Lang. Process., December 2016. (Accepted for publication)
7. Braun, S., Habets, E.A.P.: A multichannel diffuse power estimator for dereverberation in the presence of multiple sources. EURASIP J. Audio Speech, Music Process **2015**, 34 December 2015
8. Allen, J., Berkley, D.A., Blauert, J.: Multimicrophone signal-processing technique to remove room reverberation from speech signals. J. Acoust. Soc. Am. **62**(4), 912–915 (1977)
9. Lebart, K., Boucher, J.-M., Denbigh, P.: A new method based on spectral subtraction for speech dereverberation. Acta Acustica united with Acustica **87**(3), 359–366 (2001)
10. Polack, J.D.: La transmission de l'énergie sonore dans les salles, Ph.D. dissertation, Université du Maine, Le Mans, France (1988)
11. Hadad, E., Heese, F., Vary, P., Gannot, S.: Multichannel audio database in various acoustic environments. In: 14th International Workshop on Acoustic Signal Enhancement (IWAENC), pp. 313–317 (2014)
12. Varga, A., Steeneken, H.J.: Assessment for automatic speech recognition: II. NOISEX-92: a database and an experiment to study the effect of additive noise on speech recognition systems. Speech Commun. **12**(3), 247–251 (1993)
13. Ephraim, Y., Malah, D.: Speech enhancement using a minimum-mean square error short-time spectral amplitude estimator. IEEE Trans. Acoust. Speech Signal Process. **32**(6), 1109–1121 (1984)

Toward Rank Disaggregation: An Approach Based on Linear Programming and Latent Variable Analysis

Vincent Vigneron[1] and Leonardo Tomazeli Duarte[2]([envelope])

[1] IBISC, Université d'Evry, 40 Rue du Pelvoux, 91020 Courcouronnes, France
vincent.vigneron@ibisc.univ-evry.fr
[2] School of Applied Sciences (FCA), University of Campinas (UNICAMP),
Rua Pedro Zaccaria, 1300, Limeira, Brazil
leonardo.duarte@fca.unicamp.br

Abstract. This work presents an unsupervised approach to the problem of rank disaggregation, which can be defined as the task of decomposing a set of rankings provided by different people (or entities). To accomplish this task, we first discuss the problem of rank aggregation and how it can be solved via linear programming. Then, we introduce a disaggregation method based on rank aggregation and inspired by decomposition methods such as principal component analysis (PCA). The results are preliminary but may pave the way for a better understating of relevant features found in applications such as group decision.

Keywords: Rank aggregation · Rank disaggregation · Linear programming · Latent variable analysis · Condorcet distance

1 Introduction

In many applications, from data mining to computational social choice theory [3], a fundamental task is to obtain a global ranking of a set of alternatives with respect to a given criterion and by considering distinct ranking models [7,9]—these rankings may represent preference models from different people. This problem of combining a set of rankings in such a way to optimize the joint ranking is known as *rank aggregation*.

Conversely, in *rank disaggregation*, which is the focus of the present paper, the main idea is to decompose the observed rankings into a new set of rankings.

This research was supported by the program *Cátedras Franco-Brasileiras no Estado de São Paulo*, an initiative of the French consulate and the state of São Paulo (Brazil). We thank our colleagues Prof. João M. T. Romano, Dr. Kenji Nose and Dr. Michele Costa, who provided insights that greatly assisted this work. L.T. Duarte thanks the São Paulo Research Foundation (FAPESP) (Grant 2015/16325-1) and the National Council for Scientific and Technological Development (CNPq) for funding his research.

© Springer International Publishing AG 2017
P. Tichavský et al. (Eds.): LVA/ICA 2017, LNCS 10169, pp. 192–200, 2017.
DOI: 10.1007/978-3-319-53547-0_19

In a certain sense, one may draw a parallel between rank disaggregation and blind source separation (BSS) [5]. Indeed, in both problems, the observed signals are transformed into latent variables which may provide useful information on the problem at hand. The main difference, however, is that in rank disaggregation the observed signals correspond to vectors of orders and, therefore, it becomes difficult to obtain decompositions by relying on statistics such as those considered in principal and independent component analysis (PCA and ICA).

Motivated by the above-mentioned parallel, the present paper aims at providing an initial study on the decomposition of a set of rankings. Although we do not focus on any particular application, ranking decomposition may be useful in applications in which one is interested in understanding the behavior of individuals in a group decision problem. In such a context, there are several questions that must be addressed. The first one is how to decompose a set of rankings. Besides, it becomes necessary to interpret the obtained decomposition. The focus of the present work shall be on the first question, which will be tackled by formulating linear optimization problems based on ranking distances.

The paper is organized as follows. We first discuss in Sect. 2 the problem of rank aggregation. We show how to address this problem via linear programming for different types of ranking distances. Then, in Sect. 3, we present a disaggregation method. Our final remarks are stated in Sect. 4.

2   Rank Aggregation

2.1   Rank-Aggregation as an Optimization Problem

Let $A = \{a_1, a_2, \ldots, a_n\}$ be a set of n alternatives that may represent different strategies, objects, candidates, etc., and let $V = \{v_1, v_2, \ldots, v_m\}$ be a set of m attributes related to A. These attributes may correspond to different criteria, in a context of multicriteria decision aiding, or different people, in applications such as group decision. In order to keep a consistent nomenclature, V will be related to m voters and A to n alternatives—we shall assume that m is an odd number.

The observed data in the rank aggregation problem is a $n \times m$ matrix R whose element $\{r_i^{(k)}\}$ refers to the ranking of alternative i with respect to the voter k—we consider both the case of *total order*, i.e., $r_{i'}^{(k)} \neq r_i^{(k)}$, $\forall i' \neq i$ [4] and the situation in which a voter can provide *ex-aequo* positions. In the following, we denote $r_{ik} = r_i^{(k)}$. Moreover, the ranking provided by each voter will be represented by a vector of integers $\mathbf{r}^{(k)} = [r_1^{(k)}, r_2^{(k)}, \ldots, r_n^{(k)}]^T$ — these vectors are the columns of R.

Among the different approaches to rank aggregation is the formulation of an optimization problem in which the cost function is related to a measure of dissensus between the voters. In mathematical terms, this problem can be formulated as follows:

$$\min_{\mathbf{r}} \sum_{k=1}^{m} d_\alpha(\mathbf{r}, \mathbf{r}^{(k)}), \quad \text{s.t.} \quad \mathbf{r} \in \mathcal{S}_n. \tag{1}$$

where $\mathcal{S}_n$ is the symmetric group of the $n!$ permutations [6] and the metric $d_\alpha : \mathcal{S}_n \times \mathcal{S}_n \to \mathbb{R}^+$ is a distance function, i.e. a function with metric properties on $\mathcal{S}_n$. Solving Eq. (1) is more complex than the classical scale aggregation problem [7] due to the constraint $\mathbf{r} \in \mathcal{S}_n$.

An alternative representation of $\mathbf{r}^{(k)}$ is to consider a permutation matrix $X^{(k)} = \{x_{ij}^{(k)}\}$, where $x_{ij}^{(k)} \in \{0, 1\}$. If $x_{ij}^{(k)} = 1$, then the i-th alternative is placed at the j-th position. By considering the representation $X^{(k)}$, one can rewrite the constraint $\mathbf{r} \in \mathcal{S}_n$ in Eq. (1) as follows

$$\sum_{j=1}^{n} x_{ij}^{(k)} = \sum_{i=1}^{n} x_{ij}^{(k)} = 1, \forall i, j. \tag{2}$$

As will be discussed in the sequel the search for a global ranking, which can also be represented by a permutation matrix, will lead to a linear optimization problem.

2.2   Criteria for Rank Aggregation

Having defined the rank aggregation problem as an optimization problem, a central question to be addressed refers to the choice of the distance $d_\alpha(\mathbf{r}, \mathbf{r}^{(k)})$ in Eq. (1). In the present work, we consider two distances: the disagreement distance [8] and the Condorcet distance [2, 10]. The choice of these metrics are motivated by the fact that they have an intuitive and plausible interpretation and also because of their nice mathematical properties.

Disagreement distance d_D. In this case, the distance between the ranking of voter k and the ranking of voter k' is given by

$$d_D(\mathbf{r}^{(k)}, \mathbf{r}^{(k')}) = \sum_{i=1}^{n} \operatorname{sgn} |r_{ik} - r_{ik'}|.$$

By considering this distance in (1), one obtains the following optimization problem:

$$\min_{\mathbf{r}} F_D(\mathbf{r}) = \min_{\mathbf{r} \in \mathcal{S}_n} \sum_{k=1}^{m} d_D(\mathbf{r}, \mathbf{r}^{(k)}) = \min_{\mathbf{r} \in \mathcal{S}_n} \sum_{k=1}^{m} \sum_{i=1}^{n} \operatorname{sgn} |r_i - r_{ik}| \tag{3}$$

where r_i denotes the rank of the i-th alternative in the global (aggregated) rank represented by $\mathbf{r}$.

By representing the ranking vectors $\mathbf{r}$ and $\mathbf{r}^{(k)}$ by the permutation matrices X and $X^{(k)}$, respectively, it is possible to rewrite (3). Indeed, since $r_i = \sum_{j=1}^{n} j x_{ij}$, the cost function of Eq. (3) is given by

$$F_D(\mathbf{r}) = \sum_{k=1}^{m} \sum_{i=1}^{n} \operatorname{sgn} \left| \sum_{j=1}^{n} j x_{ij} - r_{ik} \right|. \tag{4}$$

or $F_D(\mathbf{r}) = \sum_{k=1}^m \sum_{i=1}^n \operatorname{sgn}|\sum_j (j - r_{ik})x_{ij}|$. After some manipulation, Eq. (4) can be rewritten as

$$F_D(\mathbf{r}) = \sum_{k=1}^m \sum_{i=1}^n \operatorname{sgn}\left(\sum_j^n |j - r_{ik}|x_{ij}\right) = \sum_{k=1}^m \sum_{i=1}^n \sum_{j=1}^n (\operatorname{sgn}|j - r_{ik}|)x_{ij}. \qquad (5)$$

Let us define by $\phi_{ij}(\mathbf{r}) = \sum_{k=1}^m \operatorname{sgn}|j - r_{ik}| = \sum_{k=1}^m \left| x_{ij} - x_{ij}^{(k)}\right|$ the cost of attributing the alternative i in the position j, or the number of voters who do not place the alternative i in the position j. $\phi_{ij}(\mathbf{r})$ is equivalent to

$$m - \pi_{ij},$$

where

$$\pi_{ij} = \sum_{k=1}^m x_{ij}^{(k)}$$

is the number of voters who place the alternative i in the position j. Given that $|x_{ij} - x_{ij}^{(k)}| = (x_{ij} - x_{ij}^{(k)})^2$ (since $|x_{ij} - x_{ij}^{(k)}| \in \{0,1\}$) it asserts that $F_D(\mathbf{r}) = \frac{1}{2}\sum_k \sum_i \sum_j (x_{ij} - x_{ij}^{(k)})^2 = \sum_{k=1}^m (m - \sum_{i=1}^n \sum_{i=1}^n x_{ij}x_{ij}^{(k)})$ and, thus,

$$F_D(\mathbf{r}) = \sum_{i=1}^n \sum_{i=1}^n \left(m - \sum_{k=1}^m x_{ij}^{(k)} \right) x_{ij}. \qquad (6)$$

Therefore, the linear programming model associated with Eq. (3) is given by

$$\min_X \sum_{i=1}^n \sum_{j=1}^n (m - \pi_{ij})x_{ij} \quad \text{s.t.} \sum_{i=1}^n x_{ij} = \sum_{j=1}^n x_{ij} = 1, \text{ and } x_{ij} \in \{0,1\}. \qquad (7)$$

Condorcet distance d_C. In order to define this distance, let us introduce a new set of matrices $\{Y^{(1)}, \dots, Y^{(m)}\}$. These matrices provide an alternative representation of R in the following manner: for a given voter k, $Y_{ij}^{(k)} = \mathbf{1}_{i<j}$ denotes the indicator matrix for which $y_{ij}^{(k)} = 1$ if the rank of the alternative a_i is lower than the ranking of alternative a_j and 0 otherwise.

By considering the matrices $Y^{(k)}$, the Condorcet distance between $\mathbf{r}^{(k)}$ and $\mathbf{r}^{(k')}$ is defined by

$$d_C(\mathbf{r}^{(k)}, \mathbf{r}^{(k')}) = d_C(Y^{(k)}, Y^{(k')}) = \frac{1}{2}\sum_{i=1}^n \sum_{j=1}^n |y_{ij}^{(k)} - y_{ij}^{(k')}| = \frac{1}{2}\sum_{i=1}^n \sum_{j=1}^n (y_{ik}^{(k)} - y_{ik}^{(k')})^2.$$

In the case of total order, this distance can be written as $d_C(\mathbf{r}^{(k)}, \mathbf{r}^{(k')}) = d_C(Y^{(k)}, Y^{(k')}) = \sum_i \sum_j y_{ij}^{(k)} y_{ji}^{(k')}$.

Since $y_{ij}^2 = y_{ij} = {y_{ij}^{(k)}}^2 = y_{ij}^{(k)} = 0$ or 1, the dissensus function associated with the Condorcet distance d_C is given by

$$F_C(\mathbf{r}) = \frac{1}{2}\left[\sum_{i=1}^n \sum_{j=1}^n m y_{ij} + \sum_{i=1}^n \sum_{j=1}^n \left(\sum_{k=1}^m y_{ij}\right) - 2\sum_{i=1}^n \sum_{j=1}^n y_{ij} \sum_{k=1}^m y_{ij}^{(k)} \right]. \qquad (8)$$

Let $\alpha_{ij} = \sum_{k=1}^{m} y_{ij}^{(k)}$ be the number of voters preferring alternative a_i to a_j. The criterion $F_C(\mathbf{r})$ can be rewritten as a function of α_{ij}, as follows:

$$F_C(\mathbf{r}) = \frac{1}{2} \left[\sum_{i=1}^{n} \sum_{j=1}^{n} m y_{ij} + \sum_{i=1}^{n} \sum_{j=1}^{n} \alpha_{ij} - 2 \sum_{i=1}^{n} \sum_{j=1}^{n} \alpha_{ij} y_{ij} \right]. \tag{9}$$

In the case of a total order,

$$\sum_{i=1}^{n} \sum_{j=1}^{n} y_{ij} = \frac{n(n-1)}{2}$$

and

$$\sum_{i=1}^{n} \sum_{j=1}^{n} \alpha_{ij} < \frac{n(n-1)}{2}.$$

Let $K = \frac{1}{2} \left(\frac{n(n-1)}{2} + \sum_{i=1}^{n} \sum_{j=1}^{n} \alpha_{ij} \right) = $ constant, then $F_C(\mathbf{r})$ is given by:

$$F_C(\mathbf{r}) = K - \sum_{i=1}^{n} \sum_{j=1}^{n} \alpha_{ij} y_{ij}. \tag{10}$$

Finally the search of a total order given by a matrix Y is the optimal solution of the following linear program

$$\max_X \sum_{i=1}^{n} \sum_{j=1}^{n} \alpha_{ij} y_{ij} \quad \text{s.t.} \qquad\qquad y_{ij} + y_{ji} = 1, i < j,$$

$$y_{ii} = 0 \;\; y_{ij} + y_{ji} - y_{ik} \leq 1, i \neq j \neq k, \; y_{ij} \in \{0, 1\}. \tag{11}$$

From an optimization perspective, both the programs expressed by Eqs. (7) and (11) are remarkably simple and provide an exact solution through the application of a simple linear programming solver [1].

2.3 Numerical Example: CAC 40 Ranking of the Top 10 French Companies

In order to illustrate the use of the disagreement and Cordorcet distances according to the optimization problems expressed in (7) and (11), let us consider the problem of ranking the top 10 CAC 40 French companies. The voters in this case are related to the following criteria: funds, sales, cash flow and reported earnings. In Table 1, we provide the matrix R. We also provide in this table the aggregated rankings obtained via the optimization of (7) and (11). Note that in this example, the aggregated rankings were the same.

Table 1. Rank aggregation: numerical example.

Company	$r^{(1)}$	$r^{(2)}$	$r^{(3)}$	$r^{(4)}$	r with d_D	r with d_C
TOTAL	1	4	3	10	3	3
LOREAL	2	2	1	2	2	2
SANOFI	3	1	2	1	1	1
LVMH	4	6	4	3	4	4
BNP PARIBAS	5	5	7	5	5	6
DANONE	6	3	6	4	6	5
AXA	7	8	5	6	7	7
VINCI	8	7	8	9	8	8
AIRBUS	9	10	10	8	1	10
ORANGE	10	9	9	7	9	9

3   Towards Rank Disaggregation

3.1   Rank Disaggregation via a Multivariate Decomposition Approach

Many strategies were proposed in the literature to extract important variables or develop parsimonious models to deal with dimensionality issues. Moreover, since the dimension of the observed data is usually higher than their intrinsic dimension, it is theoretically possible to reduce the dimension without losing relevant information. Among the unsupervised tools, principal component analysis (PCA) and factor analysis (FA) are traditional and certainly the most used ones to extract relevant features from a multivariate data set.

In the present work, the goal is to perform rank disaggregation by performing multivariate decomposition, as in PCA. The question is thus simple to formulate: Can one extract a set of ranking-order variables that are analogous to principal components?

In order to provide an initial answer to this question, we propose an algorithm that is based on rank aggregation. For instance, given a set of rankings represented by $R = [r^{(1)}, r^{(2)}, \ldots, r^{(m)}]$, which corresponds to the observed matrix, we propose to extract a first rank-order component a_1 by solving, for instance, the linear program expressed in (11). Then, the second component a_2 is again obtained by minimizing (11) and by simultaneously maximizing the distance to the rank component a_1. More generally, the component a_ℓ is obtained via joint minimization of (11) and maximization of its distance to the set $\{a_1, \ldots, a_{\ell-1}\}$. Such a procedure is described in Algorithm 1, which stops when $\ell = m$.

In mathematical terms, the search of the ℓ-th component, a_ℓ, can be simplified by considering a representation given by the matrix $Z^{(\ell)}$, which is defined in the same way as the matrix $Y^{(k)}$ considered in the minimization of the Condorcet distance (see Sect. 2.2). By considering the representation $Z^{(\ell)}$, the problem of finding $\mathbf{a}_\ell$ is given by

$$\max_{Z^{(\ell)}} \sum_{i=1}^{n}\sum_{j=1}^{n} \alpha_{ij} z_{ij}^{(\ell)} - \sum_{i=1}^{n}\sum_{j=1}^{n} \beta_{ij} z_{ij}^{(\ell)} \quad (12)$$

$$\text{s.t.} \quad z_{ij}^{(\ell)} + z_{ji}^{(\ell)} = 1, i < j,$$

$$z_{ii} = 0 \quad z_{ij}^{(\ell)} + z_{ji}^{(\ell)} - z_{ik}^{(\ell)} \le 1, i \ne j \ne k, \ z_{ij}^{(\ell)} \in \{0,1\},$$

where

$$\alpha_{ij} = \sum_{k=1}^{m} y_{ij}^{(k)},$$

and

$$\beta_{ij} = \sum_{k=1}^{\ell-1} z_{ij}^{(k)}.$$

Algorithm 1. Rank disaggregation algorithm.

Require: $Y^{(1)}, \ldots, Y^{(m)} \ \leftarrow \ \{\mathbf{r}^{(1)}, \mathbf{r}^{(2)}, \ldots, \mathbf{r}^{(m)}\}$ {Condorcet matrices $Y^{(k)} = \{y_{ij}^{(k)}\}\} \vee$ stack $A = \emptyset$ {contains ranking components}
Ensure: $\{\mathbf{a}_1, \ldots, \mathbf{a}_m\}$ {Outputs}
1: **for** $\ell = 1$ **to** m **do**
2:     Compute $\alpha_{ij} = \sum_{k=1}^{m} y_{ij}^{(k)}, \ \beta_{ij} = \sum_{k=1}^{\ell-1} z_{ij}^{(k)}$
3:     $\alpha = \{\alpha_{ij}\}, \beta = \{\beta_{ij}\}$
4:     $\mathbf{LP}(\alpha, \beta, Z^{(\ell)})$ under constraints (12) {solve linear program}
5:     $\mathbf{a}_\ell \leftarrow Z^{(\ell)}$
6: **end for**
7: **return** $\{\mathbf{a}_1, \mathbf{a}_2, \ldots, \mathbf{a}_m\}$

3.2 Numerical Experiment

Let us consider again the problem of ranking the top 10 CAC 40 French companies. By applying Algorithm 1 to the observed rankings $\mathbf{r}^{(1)}, \mathbf{r}^{(2)}, \mathbf{r}^{(3)}, \mathbf{r}^{(4)}$, the obtained components $\mathbf{a}_1, \mathbf{a}_2, \mathbf{a}_3, \mathbf{a}_4$ were those shown in Table 2.

The pairwise Kendall's rank correlation coefficients between the obtained components were given by

$$\rho_Z = \begin{bmatrix} 1.0 & 0.5556 & 0.5556 & 0.3333 \\ & 1.0 & 0.1111 & 0.4478 \\ & & 1.0 & -0.1111 \\ & & & 1.0 \end{bmatrix}$$

Table 2. Rank disaggregation: numerical example.

Company	a_1	a_2	a_3	a_4
TOTAL	4	5	6	5
LOREAL	1	3	1	4
SANOFI	2	2	2	2
LVMH	3	6	3	7
BNP PARIBAS	6	1	10	1
DANONE	5	4	7	3
AXA	7	7	5	9
VINCI	8	8	8	8
AIRBUS	10	9	9	6
ORANGE	9	10	4	10

whereas the correlation coefficients for the observed rankings were given by

$$\rho_X = \begin{bmatrix} 1.0 & 0.6000 & 0.7333 & 0.3778 \\ & 1.0 & 0.6889 & 0.5111 \\ & & 1.0 & 0.4667 \\ & & & 1.0 \end{bmatrix}.$$

Therefore, the ranking components a_1, a_2, a_3, a_4 are less correlated than the observed rankings.

4 Conclusion

In this paper, an initial study on rank disaggregation was provided. In this spirit, we firstly considered a matrix representation of rankings that allowed us to write the problem of rank aggregation as a linear program. Then, we provided an algorithm to perform rank disaggregation. The algorithm was based on rank aggregation and was conceived under the same spirit of decomposition techniques such as PCA.

References

1. Baeza-Yates, R., Ribeiro-Neto, B.: Modern Information Retrieval. ACM Press/Addison-Wesley, New York (1999)
2. Médianes, B.J., Kendall, C.: Mathématiques et Sciences Humaines, pp. 5–13, note SEMA (1980)
3. Brandt, F., Conitzer, V., Endriss, U., Lang, J., Procaccia, A.D.: Handbook of Computational Social Choice. Cambridge University Press, Cambridge (2016)
4. Brüggemann, R., Patil, G.: Ranking and Prioritization for Multi-indicator Systems: Introduction to Partial Order Applications. Environmental and Ecological Statistics, vol. 5. Springer, New York (2011)

5. Comon, P., Jutten, C. (eds.): Handbook of Blind Source Separation: Independent Component Analysis and Applications. Academic Press, New York (2010)
6. Diaconis, P.: Group Representation in Probability and Statistics. Institute of Mathematical Statistics, IMS Lecture Series, vol. 11, Harvard, USA (1988)
7. Klementiev, A., Roth, D., Small, K.: An unsupervised learning algorithm for rank aggregation. In: Kok, J.N., Koronacki, J., Mantaras, R.L., Matwin, S., Mladenič, D., Skowron, A. (eds.) ECML 2007. LNCS (LNAI), vol. 4701, pp. 616–623. Springer, Heidelberg (2007). doi:10.1007/978-3-540-74958-5_60
8. Truchon, M.: An extension of the Condorcet criterion and Kemeny orders, cahier 9813, Universite Laval, Quebec, Canada, October 1998
9. Vogt, C., Cottrell, G.: Fusion via a linear combination of scores. Inf. Retrieval **1**(3), 151–173 (1999)
10. Young, H.: Condorcet's theory of voting. Am. Polit. Sci. Rev. **82**, 1231–1244 (1988)

A Proximal Approach for Nonnegative Tensor Decomposition

Xuan Vu[1,2,3], Caroline Chaux[3(✉)], Nadège Thirion-Moreau[1,2],
and Sylvain Maire[1,2]

[1] Aix-Marseille Université, CNRS, ENSAM, LSIS, UMR 7296,
13397 Marseille, France
thi-thanh-xuan.vu@lsis.org
[2] Université de Toulon, CNRS, LSIS, UMR 7296, 83957 La Garde, France
{thirion,maire}@univ-tln.fr
[3] Aix Marseille Univ, CNRS, Centrale Marseille, I2M, Marseille, France
caroline.chaux@univ-amu.fr

Abstract. This communication deals with N-th order tensor decompositions. More precisely, we are interested in the (Canonical) Polyadic Decomposition. In our case, this problem is formulated under a variational approach where the considered criterion to be minimized is composed of several terms: one accounting for the fidelity to data and others that can represent not only regularization (such as sparsity prior) but also hard constraints (such as nonnegativity). The resulting optimization problem is solved by using the Block-Coordinate Variable Metric Forward-Backward (BC-VMFB) algorithm. The robustness and efficiency of the suggested approach is illustrated on realistic synthetic data such as those encountered in the context of environmental data analysis and fluorescence spectroscopy. Our simulations are performed on 4-th order tensors.

Keywords: Constrained optimization · Proximal algorithm · Block alternating minimization · Nonnegative tensor factorization (NTF)

1 Introduction

In numerous applications, the data sets that are collected can be organized into multi-way (or N-way with $N \geq 3$) arrays of numerical values. Consequently, a growing interest has been dedicated to the development of efficient methods and derived algorithms, capable of both processing such multi-way arrays and extracting as much relevant information as possible. The most famous tensor decomposition certainly remains the (Canonical) Polyadic Decomposition (CPD) since it has been proven effective in many application fields (see [6,13] for an overview of applications). Another of its main advantages is its uniqueness under mild conditions [14,20]. In some leading applications of CPD particularly those linked to the image processing field (examples include 3D fluorescence spectroscopy and functional magnetic resonance imaging (fMRI) for brain mapping),

© Springer International Publishing AG 2017
P. Tichavský et al. (Eds.): LVA/ICA 2017, LNCS 10169, pp. 201–210, 2017.
DOI: 10.1007/978-3-319-53547-0_20

some specific properties are generally known about the latent variables due to their physical meaning. Standing for concentrations, percentage, fractional abundance, spectra, and so on, these latent variables should be nonnegative and/or smooth and/or sparse quantities and imposing these physical constraints can "help" the algorithms to recover more "relevant" pure constituent compounds. As a consequence, our main aim, here, is to properly tackle the relatively general problem of the CPD of N-way tensors subject to a certain number of constraints linked to *a priori knowledge* we may have about the involved latent variables. To that purpose, this general problem is formulated under the general framework of variational approaches where the cost function to be minimized is composed of several terms: the classical one accounting for the fidelity to data and additional ones that can either stand for regularization (such as sparsity prior) or represent hard constraints such as nonnegativity. The resulting optimization problem can become numerically difficult; the adopted algorithm is based upon a "Block Coordinate Variable Metric Forward-Backward" (BC-VMFB) approach [5] that gathers four main stages: (1) a gradient step involved in the forward stage, (2) a proximal step involved in the backward stage, (3) a preconditioning step ("variable metric") and finally (4) a block arrangement ("Block Coordinate") of the unknown (latent) variables that will be swept according to a random (or cyclic or other [21]) rule. Such an approach but without preconditioning has been used to deal with third order tensor decompositions (CPD and Tucker) [23]. Alternatively, an alternating optimization approach based on an alternating direction method of multipliers has been recently proposed in [12]. Finally, this algorithm recently proved its effectiveness in third order tensor decomposition for 3D fluorescence spectroscopy [22].

The remaining of this communication is organized as follows. Section 2 is devoted to the presentation of the considered multilinear model and the objective to be reached. Section 3 describes the proposed approach which consists of two steps: after formulating the problem under a variational approach and introducing the resulting criterion to be minimized, the proximal algorithm based on the Block coordinate Variable Metric Forward-Backward algorithm is presented. The efficiency of the proposed approach is emphasized through numerical experiments conducted in Sect. 4. A complicated ill-posed scenario (noisy overestimated model) is considered for the decomposition of a synthetic, yet, realistic 4-th order tensor. Finally, a conclusion is drawn and perspectives are delineated.

2   Canonical Polyadic Decomposition of N-th Order Tensors

2.1   Model

The Canonical Polyadic Decomposition (CPD) of tensors, also known as Parafac (PARAllel FACtor analysis [10]), CanDecomp (Canonical Decomposition [2]) and CP (for CanDecomp/Parafac [8]), constitutes a compact and informative model. It consists of decomposing an original tensor $\overline{T}$ into a minimal sum of rank-1 terms:

$$\overline{\mathcal{T}} = \sum_{r=1}^{\overline{R}} \bar{\mathbf{a}}_r^{(1)} \circ \bar{\mathbf{a}}_r^{(2)} \circ \ldots \circ \bar{\mathbf{a}}_r^{(N)} = [\![\bar{\mathbf{A}}^{(1)}, \bar{\mathbf{A}}^{(2)}, \ldots, \bar{\mathbf{A}}^{(N)}]\!], \qquad (1)$$

where $N \in \mathbb{N}$ is the tensor order and $\circ$ is the outer product of vectors. The minimal $\overline{R} \in \mathbb{N}$ such that Eq. (1) holds is called the tensor rank. For every $n \in \{1, \ldots, N\}, r \in \{1, \ldots, \overline{R}\}$, the real column vector $\bar{\mathbf{a}}_r^{(n)} = (\bar{a}_{1r}^{(n)}, \bar{a}_{2r}^{(n)}, \ldots, \bar{a}_{I_n r}^{(n)})^\top \in \mathbb{R}^{I_n}$ is called a loading factor (where $(\cdot)^\top$ stands for the transpose operator) and the unknown latent matrices $\bar{\mathbf{A}}^{(n)} = [\bar{\mathbf{a}}_1^{(n)}, \bar{\mathbf{a}}_2^{(n)}, \ldots, \bar{\mathbf{a}}_{\overline{R}}^{(n)}] = (\bar{a}_{i_n r}^{(n)})_{i_n, r} \in \mathbb{R}^{I_n \times \overline{R}}$ are called the loading matrices.

2.2 Objective

Given a tensor $\mathcal{T}$ (which can be an observation, possibly noisy, of an original tensor $\overline{\mathcal{T}}$), we aim at approximating it using the CP model *i.e.* we intend to determine for all $n \in \{1, \ldots, N\}$ an estimation of the loading matrices $\bar{\mathbf{A}}^{(n)}$.

To estimate the loading matrices $\bar{\mathbf{A}}^{(n)}$ for all $n \in \{1, \ldots, N\}$, it can be more convenient to rewrite Eq. (1) under a matrix form by using flattening. Indeed, let $\overline{\mathbf{T}}_{I_n, I_{-n}}^{(n)} \in \mathbb{R}^{I_n \times I_{-n}}$ be the matrix obtained by unfolding tensor $\overline{\mathcal{T}}$ along mode n, where $n \in \{1, \ldots, N\}$ and $I_{-n} = I_1 \ldots I_{n-1} I_{n+1} \ldots I_N$, then the model given in Eq. (1) can be written in a compact matrix form [6, p. 352] as follows

$$\overline{\mathbf{T}}_{I_n, I_{-n}}^{(n)} = \bar{\mathbf{A}}^{(n)} (\overline{\mathbf{Z}}^{(-n)})^\top, \qquad \forall n \in \{1, \ldots, N\} \qquad (2)$$

where

$$\overline{\mathbf{Z}}^{(-n)} = \bar{\mathbf{A}}^{(N)} \odot \ldots \odot \bar{\mathbf{A}}^{(n+1)} \odot \bar{\mathbf{A}}^{(n-1)} \odot \ldots \odot \bar{\mathbf{A}}^{(1)} \in \mathbb{R}^{I_{-n} \times \overline{R}}, \qquad (3)$$

and $\odot$ denotes the Khatri-Rao product.

3 Optimization Problem and Proximal Algorithm

We choose, here, to express the problem of estimating the loading matrices under a variational framework *i.e.* to solve an optimization problem whose solution constitutes an estimation of the initial loading matrices.

3.1 Criterion Formulation, Assumptions and Properties

In classical variational approaches, the criterion is divided into two main terms: a data fidelity term denoted by $\mathcal{F}$ and a regularization term which is here constituted of the sum of N regularization functions, each linked to one of the loading matrices.

Mathematically, this problem is formulated as

$$\underset{\mathbf{A}^{(n)} \in \mathbb{R}^{I_n \times R}, n \in \{1, \ldots, N\}}{\text{minimize}} \quad \mathcal{F}(\mathbf{A}^{(1)}, \ldots, \mathbf{A}^{(N)}) + \sum_{n=1}^{N} \mathcal{R}_n(\mathbf{A}^{(n)}) \qquad (4)$$

where $\mathcal{F}$ and $(\mathcal{R}_n)_{n \in \{1,...,N\}}$ are assumed to be proper lower semi-continuous functions such that $\mathcal{F}$ is differentiable with a β-Lipschitz gradient where $\beta \in]0, +\infty[$ and such that for all $n = 1,\ldots, N$, $\mathcal{R}_n : \mathbb{R}^{I_n \times R} \to]-\infty, +\infty[$ is bounded from below by an affine function, and its restriction to its domain is continuous.

The numerical method used to solve Eq. (4) is described in the next section.

3.2 Proposed Algorithm

Here, we suggest to use the Block Coordinate Variable Metric Forward Backward (BC-VMFB) algorithm [4,5] to solve the problem described by Eq. (4). The general principle of the resulting iterative method is detailed in the following paragraph and is summed up in Algorithm 1. Our different choices (of cost function, preconditioning matrix, etc.) to tackle specifically the CPD problem are discussed in Sect. 3.3. The approach chosen here mainly consists of two steps:

❶ a gradient step (linked to $\mathcal{F}$ which is assumed to be differentiable with a β-Lipschitz gradient). It requires to compute the partial gradient matrices of $\mathcal{F}$ with respect to $\mathbf{A}^{(n)}$ for all $n = 1,\ldots, N$. In the following, they are denoted by $\nabla_n \mathcal{F}(\mathbf{A}^{(1)}, \ldots, \mathbf{A}^{(N)})$.

❷ a proximal step (linked to $(\mathcal{R}_n)_{n \in \{1,...,N\}}$): for all $n = 1,\ldots, N$ it requires to compute the proximity operator of $\mathcal{R}_n$ associated to the metric $\mathbf{P}^{(n)}$. The definition of the proximity operator is recalled hereafter.

The proximity operator of a proper, lower semicontinuous function from $\mathbb{R}^I$ to $]-\infty, +\infty[$ associated with a Symmetric Positive Definite (SPD) matrix $\mathbf{P}$ is defined as [11]

$$\text{prox}_{\mathbf{P},\varphi} : \mathbb{R}^I \to \mathbb{R}^I : v \mapsto \arg \min_{u \in \mathbb{R}^I} \frac{1}{2} \|u - v\|_{\mathbf{P}}^2 + \varphi(u). \qquad (5)$$

where $\forall x \in \mathbb{R}^I$, $\|x\|_{\mathbf{P}}^2 = \langle x, \mathbf{P}x \rangle$, and $\langle \cdot, \cdot \rangle$ stands for the inner product. The original definition of the proximity operator [17] is recovered when $\mathbf{P}$ reduces to the identity matrix.

To simplify the notations, the partial gradient matrices $\nabla_n \mathcal{F}(\mathbf{A}^{(1)}[k], \ldots \mathbf{A}^{(N)}[k])$ associated to k-th iteration are simply denoted by $\nabla_n[k]$. The Hadamard division between two matrices is denoted by $\oslash$. Finally, we recall that under some technical assumptions [5, Sect. 2.2] (concerning the preconditioning matrices $\mathbf{P}^{(n)}$, the step-size γ, the block scanning rule, and the fact that $\mathcal{F} + \mathcal{R}$ satisfies the Kurdyka-Łojasiewicz inequality) the convergence of the algorithm to a critical point is guaranteed [5, Theorem 3.1].

3.3 Criterion Choice: Related Gradient and Proximity Operators

The algorithm described hereabove was presented in a very general way. We now introduce the objective that we have chosen to minimize, explain some of our choices and provide the resulting involved quantities (partial gradient matrices, preconditioning matrix, and so on).

Algorithm 1. BC-VMFB algorithm to minimize Eq. (4).

1: Let $\mathbf{A}^{(n)} \in \mathrm{dom}\mathcal{R}_n$, $n \in \{1, \dots, N\}$, $k \in \mathbb{N}$ and $\gamma[k] \in]0, +\infty[$ // *Initialization*
2: **for** $k = 0, 1, \dots, K$ **do**   // *k-th iteration*
3:       Choose a block $n \in \{1, \dots, N\}$ // *Quasi cyclic rule*
4:       Compute $\mathbf{P}^{(n)}[k] = \mathbf{P}^{(n)}(\mathbf{A}^{(1)}[k], \dots, \mathbf{A}^{(N)}[k])$ // *Preconditioner construction*
5:       Compute the Gradient Matrix $\nabla_n[k]$ // *Calculation of Gradient*
6:       $\mathbf{A}^{(n)}[k + \frac{1}{2}] = \mathbf{A}^{(n)}[k] - \gamma[k]\nabla_n[k] \oslash \mathbf{P}^{(n)}[k]$ // *Gradient step*
7:       $\mathbf{A}^{(n)}[k + 1] \in \mathrm{prox}_{\gamma[k]^{-1}\widetilde{\mathbf{P}}^{(n)}[k], \mathcal{R}_n}(\mathbf{A}^{(n)}[k + \frac{1}{2}])$ // *Proximal step*
8:       $\mathbf{A}^{(\bar{n})}[k + 1] = \mathbf{A}^{(\bar{n})}[k]$ where $\bar{n} = \{1, \dots, N\} \setminus \{n\}$ // *Other blocks kept unchanged*
9: **end for**
10: $\forall n \in \{1, \dots, N\}$, $\widehat{\mathbf{A}}^{(n)} = \mathbf{A}^{(n)}[K]$ // *Convergence reached at K-th iteration*

For the computer simulations provided in this communication, the data fidelity term $\mathcal{F}$ takes a quadratic form. It thus leads to the following definition

$$\mathcal{F}(\mathbf{A}^{(1)}, \dots, \mathbf{A}^{(N)}) = \frac{1}{2}\|\mathcal{T} - [\![\mathbf{A}^{(1)}, \dots, \mathbf{A}^{(N)}]\!]\|_F^2 = \frac{1}{2}\|\mathbf{T}_{I_n, I_{-n}}^{(n)} - \mathbf{A}^{(n)}\mathbf{Z}^{(-n)^\top}\|_F^2,$$
(6)

where $\|\cdot\|_F$ stands for the Frobenius norm. As a consequence

❶ the associated partial gradient matrices are given by [9]

$$\nabla_n \mathcal{F}(\mathbf{A}^{(1)}, \dots, \mathbf{A}^{(N)}) = -(\mathbf{T}_{I_n, I_{-n}}^{(n)} - \mathbf{A}^{(n)}\mathbf{Z}^{(-n)^\top})\mathbf{Z}^{(-n)}.$$
(7)

❷ in the same spirit as in [16,19], the preconditioning matrix $\mathbf{P}^{(n)}$, $\forall n \in \{1, \dots, N\}$, can be defined as follows

$$\mathbf{P}^{(n)}(\mathbf{A}^{(1)}, \dots, \mathbf{A}^{(N)}) = \left(\mathbf{A}^{(n)}(\mathbf{Z}^{(-n)^\top}\mathbf{Z}^{(-n)})\right) \oslash \mathbf{A}^{(n)}.$$
(8)

It is based on the n-th mode unfolding of the tensor (see (2)) and on the definition of a majorant function of the restriction of $\mathcal{F}$ to the n-th loading matrix on the domain of $\mathcal{R}_n$. Additional details about preconditioning matrix construction can be found in [22].

Concerning the regularization terms, they may account at the same time for the nonnegativity constraint we want to impose on the solution and to the sparsity of the data (possible overfactoring). For all $\mathbf{A}^{(n)} = (a_{i_n r}^{(n)})_{(i_n, r) \in \{1, \dots, I_n\} \times \{1, \dots, R\}}$, we thus choose [5, pp. 18–20],

$$\mathcal{R}_n(\mathbf{A}^{(n)}) = \sum_{i_n=1}^{I_n} \sum_{r=1}^{R} \rho_n(a_{i_n r}^{(n)})$$
(9)

where $\forall n \in \{1, \dots, N\}$

$$\rho_n(\omega) = \begin{cases} \alpha^{(n)}|\omega|^{\pi^{(n)}} & \text{if } \eta_{\min}^{(n)} \leq \omega \leq \eta_{\max}^{(n)} \\ +\infty & \text{otherwise} \end{cases}$$
(10)

and $\alpha^{(n)} \in]0, +\infty[$, $\pi^{(n)} \in \mathbb{N}^*$, $\eta_{\min}^{(n)} \in [-\infty, +\infty[$, and $\eta_{\max}^{(n)} \in [\eta_{\min}^{(n)}, +\infty]$ (block dependent regularization parameters[1]). This choice enables to ensure nonnegativity by taking for example $\eta_{\min}^{(n)} = 10^{-10}$ and $\eta_{\max}^{(n)} = +\infty$ and to promote sparsity by choosing the exponent $\pi^{(n)} = 1$ (hence performing an ℓ_1-norm regularization).

To be properly computed, the associated proximity operator requires first to define[2] (i) $\mathbf{a}^{(n)} = \mathsf{vec}(\mathbf{A}^{(n)}) \in \mathbb{R}^{RI_n}$ (vectorization of loading matrices and operator $\mathsf{vec}(\cdot)$ stacks the columns of the matrix given in argument into a vector) and (ii) $\widetilde{\mathbf{P}}^{(n)} = \mathsf{Diag}(\mathsf{vec}(\mathbf{P}^{(n)})) \in \mathbb{R}^{RI_n \times RI_n}$ (vectorization and diagonalization of preconditioning matrices, the $\mathsf{Diag}(\cdot)$ operator builds a diagonal matrix whose diagonal elements are the elements of the vector passed as a parameter).

By using definition in Eq. (5), we can derive the expression of $\mathrm{prox}_{\gamma^{-1}\widetilde{\mathbf{P}}^{(n)},\mathcal{R}_n}(\mathbf{a}^{(n)})$ as

$$\left(\forall \mathbf{A}^{(n)} = (a_{i_n r}^{(n)})_{(i_n,r) \in \{1,...,I_n\} \times \{1,...,R\}}\right)$$

$$\mathrm{prox}_{\gamma^{-1}\widetilde{\mathbf{P}}^{(n)},\mathcal{R}_n}(\mathbf{A}^{(n)}) = \left(\mathrm{prox}_{\gamma^{-1}p_{i_n r}^{(n)},\rho_n}(a_{i_n r}^{(n)})\right)_{(i_n,r) \in \{1,...,I_n\} \times \{1,...,R\}} \tag{11}$$

where $\forall(i_n, r) \in \{1, ..., I_n\} \times \{1, ..., R\}$, we have [7] $(\forall \upsilon \in \mathbb{R})$

$$\mathrm{prox}_{\gamma^{-1}p_{i_n r}^{(n)},\rho_n}(\upsilon) = \min\left\{\eta_{\max}^{(n)}, \max\left\{\eta_{\min}^{(n)}, \mathrm{prox}_{\gamma\alpha^{(n)}(p_{i_n r}^{(n)})^{-1}|\cdot|^{\pi^{(n)}}}(\upsilon)\right\}\right\}. \tag{12}$$

A closed form expression of the proximity operator presented in Eq. (12) can be found in [3]. Note that in Algorithm 1, at iteration k, the proximity operator is associated with metric $\mathbf{P}^{(n)}[k]$ and is computed at $\mathbf{A}^{(n)}[k + \frac{1}{2}]$ with stepsize $\gamma[k]$.

4   Numerical Simulations: Application to 4-th Order CPD

We consider here a tensor of order $N = 4$. It has been constructed synthetically but following realistic guidelines. Inspired by 3D fluorescence spectroscopy, we build this tensor simulating: (uni or bimodal type) emission and excitation spectra, smooth (either linear or unimodal) concentrations (the 3 classical components of 3D fluorescence spectroscopy) and an additional 4-th dimension modelling the lifetime (exponential decay) of compounds (such as those observed when time resolved spectroscopy is performed [15]).

The tensor rank has been fixed here to $\overline{R} = 5$. The resulting tensor $\overline{\mathcal{T}}$ is of size $I_1 = I_2 = I_3 = I_4 = 100$ and $\overline{\mathcal{T}} \in \mathbb{R}_+^{100 \times 100 \times 100 \times 100}$. Original spectra, concentrations, lifetimes are displayed in Fig. 1 (black curves).

---

[1] In our case, the easiest way to proceed is to consider that each block matches a loading matrix, but other choices could have been made.

[2] In practice, elementwise operations are performed instead making it possible to avoid memory issues.

The following scenario has been considered: the observed tensor $\mathcal{T}$ is assumed to be a perturbed version of the original tensor, that is $\mathcal{T} = \overline{\mathcal{T}} + \mathcal{B}$ where $\mathcal{B}$ stands for an additive white Gaussian noise with $\sigma = 0.001$ resulting in an initial SNR of $18.46\,$dB. Furthermore, the tensor rank $\overline{R}$ is assumed to be unknown and the decomposition is performed assuming a tensor rank $\widehat{R} = 7$ (which corresponds to an overestimation of a factor 2).

We compare our algorithm (with $\pi^{(n)} \equiv \pi = 1$, $\alpha^{(n)} \equiv \alpha = 0.05$, $\eta_{\min}^{(n)} \equiv \eta_{\min} = 10^{-16}$ and $\eta_{\max}^{(n)} \equiv \eta_{\max} = 10^2$) performances to two state-of-the-art methods: (1) fast HALS algorithm [18] and (2) Bro's N-way algorithm [1] for which, to be fair, we used the non-negativity constrained versions. Algorithm initialization is random.

In addition to visual results, we compute three error measures. Let $\widehat{\mathbf{A}}^{(n)} = [\widehat{\mathbf{a}}_1^{(n)}, \widehat{\mathbf{a}}_2^{(n)}, \ldots, \widehat{\mathbf{a}}_{\widehat{R}}^{(n)}]$ denote the normalized permuted estimate of $\bar{\mathbf{A}}^{(n)} = [\bar{\mathbf{a}}_1^{(n)}, \bar{\mathbf{a}}_2^{(n)}, \ldots, \bar{\mathbf{a}}_{\overline{R}}^{(n)}]$. The considered error measures are given by

1. Signal to Noise Ratio (SNR) defined as $\mathrm{SNR} = 20 \log_{10} \frac{\|\overline{\mathcal{T}}\|_F}{\|\widehat{\mathcal{T}} - \overline{\mathcal{T}}\|_F}$
2. Estimation error

$$\mathbf{E}_1 = 10 \log_{10} \left(\frac{\sum_{n=1}^4 \|\widehat{\mathbf{A}}^{(n)}(1:\overline{R}) - \bar{\mathbf{A}}^{(n)}\|_1}{\sum_{n=1}^4 \|\bar{\mathbf{A}}^{(n)}\|_1} \right) \tag{13}$$

3. Over-factoring error $\mathbf{E}_2$:

$$\mathbf{E}_2 = 10 \log_{10} \left(\| \sum_{r=\overline{R}+1}^{\widehat{R}} \widehat{\mathbf{a}}_r^{(1)} \circ \widehat{\mathbf{a}}_r^{(2)} \circ \widehat{\mathbf{a}}_r^{(3)} \circ \widehat{\mathbf{a}}_r^{(4)} \|_1 \right) \tag{14}$$

All the considered approaches being iterative, the following stopping conditions were used: either the maximum number of iterations fixed to $K = 10^5$ has been reached or the relative diminishing rate of the quadratic criterion reads $\frac{\|\mathcal{F}[\cdot+1] - \mathcal{F}[\cdot]\|}{\mathcal{F}[\cdot]} < 10^{-8}$.

The estimated spectra are displayed in Fig. 1. We can see that despite the overestimation factor and the noise, the proposed algorithm, contrary to fHALS or N-way approaches, allows to accurately recover the original data without creating phantoms in the artificially added compounds. This is confirmed by the numerical results given in Table 1 where we can see that the estimation error is equal or higher for fHALS and N-way methods and that the over-factoring error is much more smaller for the proposed method. Concerning the algorithm computation times, we can see that the proposed approach is very competitive but requires more iterations to reach the stopping criterion.

Table 1. Computation times and numerical performances of fHALS, N-way with non negativity constraints and BC-VMFB algorithms. Simulations were performed on a 8 cores Intel i7 @3.40 GHz.

	fHALS	N-way	BC-VMFB
Time (in s) for 50 iterations	15.4	251	9.7
Time (in s) to reach stopping criterion	139	554	1689
Iteration number to reach stopping criterion	506	176	17000
Associated (SNR, $\mathbf{E}_1$, $\mathbf{E}_2$) in dB	$(29.61, 0.21, 47.74)$	$(29.61, 0.11, 46.61)$	$(34.15, 0.11, -543)$

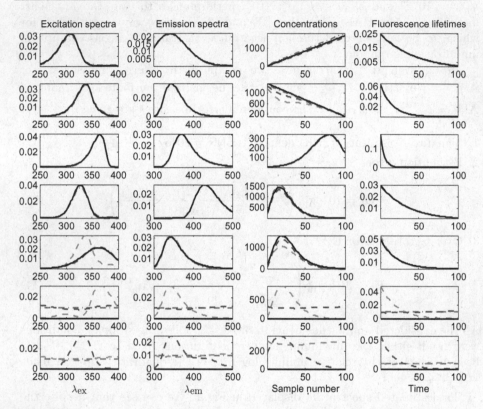

Fig. 1. Estimated scaled spectra using N-way (blue) with non negativity constraints, fHALS (green) and BC-VMFB (red). Black curves: ground truth. (Color figure online)

5   Conclusion

In this communication, we addressed the problem of the CP decomposition of N-th order tensors subject to given constraints such as nonnegativity, sparsity, regularity, etc. We tackled this problem within the very general framework of Block Coordinate Variable Metric Forward-Backward (BC-VMFB) approaches. An algorithm was provided and its robustness and efficiency was demonstrated on synthetic yet realistic 4-th order data inspired by those encountered in the context of environmental data analysis and more precisely fluorescence spectroscopy. The obtained results are encouraging, and future developments could be to apply this algorithm on raw data sets, to test other cost functions, other preconditioning matrices, etc. It could be also interesting to test other kind of regularization functions and to better understand their impact on the performance of the algorithm.

References

1. Bro, R.: Parafac: tutorial and applications. Chemom. Intell. Lab. Syst. **38**(2), 149–171 (1997)
2. Carroll, P., Chang, J.J.: Analysis of individual differences in multi-dimensional scaling via n-way generalization of Eckart-Young decomposition. Psychometrika **35**(3), 283–319 (1970)
3. Chaux, C., Combettes, P.L., Pesquet, J.C., Wajs, V.R.: A variational formulation for frame based inverse problems. Inverse Probl. **23**(4), 1495–1518 (2007)
4. Chouzenoux, E., Pesquet, J.C., Repetti, A.: Variable metric forward-backward algorithm for minimizing the sum of a differentiable function and a convex function. J. Optim. Theory Appl. **162**(1), 107–132 (2014)
5. Chouzenoux, E., Pesquet, J.C., Repetti, A.: A block coordinate variable metric forward-backward algorithm. J. Global Optim. **66**(3), 457–485 November 2016
6. Cichocki, A., Zdunek, R., Phan, A.H., Amari, S.I.: Non Negative Matrix and Tensor Factorizations: Application to Exploratory Multi-way Data Analysis and Blind Separation. Wiley, Chichester (2009)
7. Combettes, P.L., Pesquet, J.C.: Proximal splitting methods in signal processing. In: Bauschke, H.H., Burachik, R., Combettes, P.L., Elser, V., Luke, D.R., Wolkowicz, H. (eds.) Fixed-point Algorithms for Inverse Problems in Science and Engineering, pp. 185–212. Springer, New York (2010)
8. Comon, P., Jutten, C.: Handbook of Blind Source Separation, Independent Component Analysis and Applications. Academic Press, Oxford (2010). ISBN: 978-0-12-374726-6
9. Franc, A.: Etude algébrique des multi-tableaux: apport de l'algèbre tensorielle. Ph.D. thesis, University of Montepellier II, Montpellier, France (1992)
10. Harshman, R.A.: Foundation of the Parafac procedure: models and conditions for an explanatory multimodal factor analysis. UCLA Working Papers in Phonetics, vol. 16, pp. 1–84 (1970)
11. Hiriart-Urruty, J.B., Lemarechal, C.: Convex Analysis and Minimization Algorithms. Springer, Heidelberg (1993)
12. Huang, K., Sidiropoulos, N.D., Liavas, A.P.: A flexible and efficient algorithmic framework for constrained matrix and tensor factorization. IEEE Trans. Sig. Process. **64**(19), 5052–5065 (2016)
13. Kolda, T.G., Bader, B.W.: Tensor decompositions and applications. SIAM Rev. **51**(3), 455–500 (2009)
14. Kruskal, J.B.: Rank, decomposition and uniqueness for 3-way and n-way arrays. In: Coppi, R., Bolasco, S. (eds.) Multiway Data Analysis, pp. 7–18. North-Holland Publishing Co., Amsterdam (1989)
15. Lakowicz, J.R., Szmacinski, H., Nowaczyk, K., Berndt, K.W., Johnson, M.: Fluorescence lifetime imaging. Anal. Biochem. **202**(2), 316–330 (1992)

16. Lee, D.D., Seung, H.S.: Algorithms for non-negative matrix factorization. In: Leen, T.K., Dietterich, T.G., Tresp, V. (eds.) Advances in Neural Information Processing Systems, vol. 13, pp. 556–562. MIT Press (2001). http://papers.nips.cc/paper/1861-algorithms-for-non-negative-matrix-factorization.pdf

17. Moreau, J.J.: Proximité et dualité dans un espace hilbertien. Bull. Soc. Math. France **93**, 273–299 (1965)

18. Phan, A.H., Tichavskỳ, P., Cichocki, A.: Fast alternating is algorithms for high order CANDECOMP/PARAFAC tensor factorizations. IEEE Trans. Sig. Proc. **61**(19), 4834–4846 (2013)

19. Repetti, A., Chouzenoux, E., Pesquet, J.C.: A preconditioned Forward-Backward approach with application to large-scale nonconvex spectral unmixing problems. In: 39th IEEE International Conference on Acoustics, Speech, and Signal Processing (ICASSP 2014), Florence, Italie, May 2014

20. Sidiropoulos, N., Bro, R.: On the uniqueness of multilinear decomposition of N-way arrays. J. Chemom. **14**(3), 229–239 (2000)

21. Vervliet, N., Lathauwer, L.D.: A randomized block sampling approach to canonical polyadic decomposition of large-scale tensors. IEEE J. Sel. Top. Sig. Proces. **10**(2), 284–295 (2016)

22. Vu, X.T., Chaux, C., Thirion-Moreau, N., Maire, S.: A new penalized nonnegative third order tensor decomposition using a block coordinate proximal gradient approach: application to 3D fluorescence spectroscopy. J. Chemometr., special issue on penalty methods **3** (2017, to appear). CEM. doi:10.1002/cem.2859

23. Xu, Y., Yin, W.: A block coordinate descent method for regularized multiconvex optimization with applications to nonnegative tensor factorization and completion. SIAM J. Imaging Sci. **6**(3), 1758–1789 (2013)

Psychophysical Evaluation of Audio Source Separation Methods

Andrew J.R. Simpson[1(✉)], Gerard Roma[1], Emad M. Grais[1],
Russell D. Mason[2], Christopher Hummersone[2],
and Mark D. Plumbley[1]

[1] Centre for Vision, Speech and Signal Processing, Guildford, UK
{andrew.simpson,m.plumbley}@surrey.ac.uk
[2] Institute of Sound Recording, University of Surrey, Guildford, UK

Abstract. Source separation evaluation is typically a *top-down* process, starting with perceptual measures which capture fitness-for-purpose and followed by attempts to find physical (objective) measures that are predictive of the perceptual measures. In this paper, we take a contrasting *bottom-up* approach. We begin with the physical measures provided by the Blind Source Separation Evaluation Toolkit (BSS Eval) and we then look for corresponding perceptual correlates. This approach is known as *psychophysics* and has the distinct advantage of leading to interpretable, *psychophysical* models. We obtained perceptual similarity judgments from listeners in two experiments featuring vocal sources within musical mixtures. In the first experiment, listeners compared the overall quality of vocal signals estimated from musical mixtures using a range of competing source separation methods. In a loudness experiment, listeners compared the loudness balance of the competing musical accompaniment and vocal. Our preliminary results provide provisional validation of the psychophysical approach.

Keywords: Deep learning · Source separation · Perceptual evaluation

1 Introduction

Audio source separation methods typically attempt to recover or estimate signals, known as 'sources', that have been mixed. The success of this "unmixing" process is evaluated both objectively, using BSS Eval [1], and subjectively by asking listeners to rate the perceived quality using methods such as MUSHRA [2, 3]. Unfortunately, observed correlation between the BSS Eval measures and the subjective evaluations has proved sufficiently poor [4–9] that the community has rejected the BSS Eval measures as invalid.

It is difficult to relate physical measures (such as BSS Eval) to subjective evaluations of audio quality, as the latter is affected by a wide range of perceptual attributes, dependencies on suitability for purpose, as well as individual opinion and preference. This approach could be considered as being *top-down*, where we begin with perceptual (subjective) ratings that we find are important, and then seek physical (objective) measures which are correlated. The downside of the top-down approach is that we may

© Springer International Publishing AG 2017
P. Tichavský et al. (Eds.): LVA/ICA 2017, LNCS 10169, pp. 211–221, 2017.
DOI: 10.1007/978-3-319-53547-0_21

not find such physical measures because we do not fully understand the process of audition.

An alternative method would be to approach the problem *bottom-up*, where we instead begin with physical measures which are descriptive of the audio signals and we look for perceptual measures which correlate to the physical measures. This approach is known as *psychophysics* and has the advantage of being able to produce interpretable, psychophysical models relating the known physical measures to corresponding perceptual measures. This paper examines whether bottom-up psychophysical evaluation principles can be applied in source separation research to obtain perceptual measures with better correlation to the objective measures.

A. *Psychophysics*

The main principle of psychophysics is that there exist psychological correlates of physical parameters [10, 11]. A psychological correlate is a measurable behaviour, relating to body or mind, that is a (usually monotonic) function of some controllable, physical parameter of the stimulus. These psychophysical correlates are typically identified or substantiated using perceptual data. Perhaps the most well established example of a psychophysical correlate is loudness [12]. Loudness theory defines the loudness of a given sound as the auditory perceptual correlate of acoustic intensity. Hence, for a given signal, variations in intensity correspond to perceptual variations in loudness. Subjective judgments of loudness are elicited from listeners for stimuli with varying degrees of intensity. Although acoustic intensity may be considered as an absolute physical measurement (e.g., sound pressure level), psychophysical paradigms for the study of loudness typically involve relative judgments relating the loudness of one sound to that of another sound [13]. The most common methods for studying loudness involve comparisons of pairs of signals. The most direct method for loudness comparison is known as magnitude ratio estimation [13], where listeners provide a numerical ratio estimate that captures the loudness ratio between a given pair of stimuli.

B. *Application of Psychophysics to Source Separation*

A psychophysical approach could be applied to source separation. When undertaking objective measurements of an arbitrary mixture of known signals, the separated signal from a given separation algorithm (the signal estimate) may be compared to the known separate signals for evaluation. Any difference between the signal estimate and the corresponding known signal is often described as 'distortion' [1]. By subtracting the known signal from the signal estimate, a difference (i.e. distortion) signal may be obtained. The ratio of the known signal energy to the energy of this difference signal is known as the signal-to-distortion ratio (SDR) [1]. The difference signal is decomposed into two parts: an interference signal, considered to be due to the influence of other sources on the target source estimate, and an artefacts signal, considered to be that part of the estimate that is not due to either the target signal or any of the other signals. The ratio of the known target signal energy to an estimate of the energy remaining from the interfering signals is known as the source-to-interference ratio (SIR), and the ratio of the known target signal to the artefacts signal energy is known as the source-to-artefacts ratio (SAR). If SAR and SIR are employed as the physical measures, a magnitude ratio estimation methodology of subjective attributes related to these factors could result in a close correlation between the subjective and objective metrics. As for the loudness example above, the magnitude ratio estimation methodology would involve comparative judgements, in this case judgements of the similarity of certain aspects of the signals.

The subjective attribute selected to compare to SAR was the similarity of the vocal; this is a judgement of the effect of artefacts on the perceived similarity of the target signal. This can be analysed such that judgements that are similar to the known target signal are positioned at one end of the scale, and judgements that are dissimilar to the known target signal are positioned towards the opposite end of the scale. The subjective attribute selected to compare to SIR was the loudness-balance-similarity; this is a judgement of the similarity of the perceived loudness balance between the target and interferer signals. This can be analysed such that judgements that are similar to the known target signal are positioned at one end of the scale, and judgements that are dissimilar to the known target signal, or similar to the unseparated mixture signal, are positioned at the towards the opposite end of the scale.

C. Overview

In this paper, we introduce a psychophysical evaluation method based on magnitude ratio estimation corresponding as closely as possible to SAR and SIR. In the following section we describe two preliminary listening tests featuring real-world audio examples obtained from a range of state-of-the-art audio source separation methods (see [14]). Next, we describe the resulting perceptual data and correlations with the physical measures to evaluate the match between the perceptual and physical data. Then, we analyse the data from the point of view of model comparison in order to evaluate whether the subjective judgements produce meaningful results. Finally, we provide some brief discussion.

2   Method

The prevailing MUSHRA perceptual evaluation methods [2–9] are focused on obtaining interpretable perceptual measures and, hence, any attempt at modelling is a secondary consideration. In contrast, the sole aim of our psychophysical study is to establish perceptual correlates of the physical parameters. Focussing on separation of vocals from musical mixtures, we conducted two listening tests using stimuli generated using competing methods for source separation. Listeners were asked to locate each of the respective versions of each given vocal signal on a perceptual line such that the placement on the line captured the perceptual similarity relationships between the respective sounds. In the first experiment, as a perceptual correlate of the physical measure SAR, we evaluated *similarity* of the vocal. In the second experiment, to capture the loudness balance (between the vocal and accompaniments) as a perceptual correlate of the physical measure SIR, we evaluated *loudness-balance-similarity*. Critically, in both cases the stimuli placed on each perceptual line included the mixture and original (pre-mixture) vocal signal.

We consider five competing source separation methods [14] featuring deep neural networks (DNN). One of the five methods comprises a baseline DNN model (*M_baseline*) and the remaining four methods comprise multi-stage architectures which extend the baseline DNN model with various parameterisations (*M_DNN*_1-3) and/or non-negative matrix factorisation (*M_NMF*) – see [14]. Critically, the DNN architectures in [14] are designed as augmentations of the baseline DNN architecture which attempt to improve the method.

In an experiment where multiple stimuli were compared directly, six listeners were asked to provide judgements on the perceptual similarity between 30-second musical excerpts. The excerpts, chosen as representative of typical musical mixtures featuring typical vocal and accompaniment, were taken from the mixes of 10 songs selected from the *test set* of the 'MUS' task of the SiSEC challenge [15]. The mixture for each song was a summation of the available stems, where the stems comprised vocals, bass, drums and other (accompaniment). Listeners reported normal hearing and were naïve to the purpose of the test. Most listeners had some prior experience of listening tests and all were familiar with music listening, audio production and/or recording technologies.

Stimuli. We consider the separation of the vocal (stem) signal from the accompaniment signal within a mixture. All mixtures were collapsed (summed) to mono (single channel) and the various, competing source separation methods were each independently applied to each of the 10 mixtures. The voice separation output of each of the five source separation methods was used. In addition, the mixture and original vocal stem signals were also used, providing 7 alternate stimuli for each song. From the point of view of the source separation methods employed, we note that the mixtures separated were not used in training the deep neural networks (i.e., we are concerned with evaluation of the models on the *test set*). Closed (isolating) headphones were used to present the stimuli. Presentation was monaural and diotic (same in both ears). Listeners were instructed to set the volume control on the amplifier for a comfortable listening level at the beginning of the test and did not adjust it further during the test. Listeners were unpaid volunteers.

Procedure. In the case of the first experiment, similarity judgements about the vocal sources were solicited. Listeners were instructed to compare only the vocal component of the mixture in this experiment. Even in the case of the full mixture, this means that the listeners had to isolate their perception of the vocal component for comparison. The listeners declared that they were able to do this to their satisfaction. In the second experiment, similarity judgements were solicited comparing the loudness balance (see [16]) between the vocal and accompaniment in each presented stimulus.

Listeners were presented with an interface featuring seven play buttons and seven respective sliders (on a computer screen). Each play button and slider represented either one of the five voice separation outputs from the respective models or the mixture or the original vocal source. Using each individual 'play' button, listeners were able to listen to each of the respective alternate versions of the vocal at will and could repeat an unlimited number of times. Listeners arranged the vertical placement of the seven sliders to capture the similarity relationships between the various stimuli, such that sliders for very similar stimuli were placed closely (on the vertical axis) and sliders for dissimilar stimuli were placed with greater distance. Note that the absolute placements of the sliders is not informative. Listeners were briefed to maximise their use of the scale for each song but were not briefed to attempt to make consistent judgements across songs.

Listeners evaluated the stimuli in sessions of 10 songs. There was no explicit time constraint on the sessions. Most sessions were completed in under 25 min. The presentation order (both the order of the songs and the order of stimuli for each song) was randomised so that each slider corresponded to a different stimulus each time. Sliders were reset before each new song. When the listener had completed the arrangement of the sliders for a given song a 'next' button was pushed for the next song.

Analysis I: Correlation. Listeners did not make comparisons between songs but only within each song, hence we must first test correlations within the context of each song, before summarising the correlation over the 10 songs. For the perceptual data resulting from each listening test, the slider placements corresponding to the original vocal were subtracted (by way of reference) from the other placements on a song-by-song basis. The slider placement data for the original vocal were subsequently discarded. For each of the stimuli used in the experiments (except the original vocal signals which are not suitable for analysis), the corresponding physical measures (SAR/SIR) were computed using the toolbox associated with [1].

For the data of each listening test, a separate correlation analysis was conducted on a song-by-song basis. The medians of the subjective data were calculated for each song and each stimulus (therefore averaging across the results of the listeners). This resulted in 6 perceptual measures per song, per listening test. Next, the corresponding SAR and SIR values for each song (6 measures per song) were used to compute linear (Pearson) correlation coefficients with the respective perceptual data for that song. The correlation was calculated between the data from the first (vocal similarity) experiment and SAR, and the data from the second (loudness balance similarity) experiment and SIR. This provided, for each listening test, a set of 10 (song-wise) correlation coefficients. To summarise (over songs) each of the two respective distributions we take the median correlation coefficient.

Next, in order to provide a measure of statistical significance for the respective median correlation coefficients, permutation tests [17] were conducted. The above procedure for computing the song-wise distribution of correlation coefficients was repeated 10,000 times. Each of these 10,000 times, prior to the correlation computation, the order of the data were randomly shuffled. The median of this distribution was then taken and, over the 10,000 replications, an empirical null distribution (of null across-song medians) was accumulated. Finally, the number of median correlation coefficients that was greater than or equal to (and with the same sign as) the actual median correlation coefficient was counted and the resulting count divided by 10,000. This provides an empirical estimate of the probability of the respective across-song median correlation coefficient occurring by chance (a *P* value).

Analysis II: Model comparison. In contrast to the correlation analysis described above, in this analysis we are interested in overall performance comparisons of the models in terms of perception. Across-song means were computed for each listener and collated.

The resulting data were analysed using non-parametric statistical methods (see [18]). Initially, a main-effects analysis was conducted for the data of each test using a Friedman test. Next, post-hoc analyses were conducted on a pair-wise basis in order to determine which pairs of models showed evidence of being significantly different. The post-hoc pairwise analyses can be considered 'planned tests' and so contrasts were limited (in advance) to comparisons between the baseline model and the respective, competing multi-stage models. We do not provide correction for multiple comparisons, primarily because of the limited number of listeners involved and because of the minimal number of planned contrasts.

3   Results: Analysis I – Psychophysical Correlation

In the correlation analysis, we are not concerned with the question of which model is best, but rather we are concerned with the question of whether the physical measures correlate with the perceptual measures. For the perceptual data of the first listening test (*similarity*), Fig. 1 plots, on a song-by-song basis, the across-listener medians as a function of the respective SAR measures. Note that, for illustrative purposes only, in these plots we limited the upper SAR (for the original mixtures) to 10 dB (because these numbers would otherwise be at the limits of precision). Figure 2 plots the equivalent for the perceptual data of the second (*loudness balance similarity*) listening test. Linear regression lines are shown in grey for illustration. Qualitatively, the scatter plots of Fig. 1 show some evidence of monotonic trends but are somewhat noisy and appear to be dominated by the extremes of slider placement. The scatter plots of Fig. 2 show more obvious monotonic trends.

Figure 3 shows box-plots capturing the respective correlation coefficient distributions (each over the 10 songs) relating to the plots of Figs. 1 and 2. The median

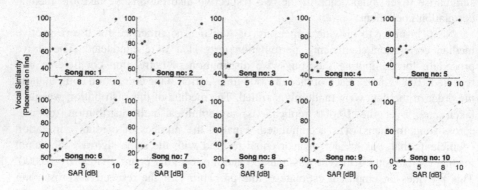

Fig. 1. Perceptual similarity versus SAR. Listeners organized sliders representing the seven respective stimuli along perceptual lines which depict the perceived similarity of the vocal component. These scatter plots show, on a song-by-song basis, across-listener median perceptual slider placement as a function of SAR. Dashed grey lines indicate linear regression lines shown for illustration only. Note: axis scale and range vary.

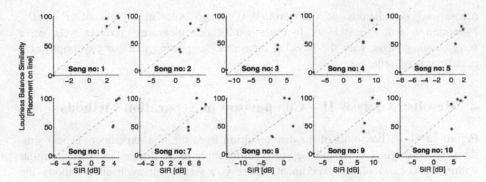

Fig. 2. Loudness balance similarity versus SIR. Listeners organized sliders representing the seven respective stimuli along perceptual lines which depict the perceived similarity of the loudness balance between the vocal and the accompaniment. These scatter plots show, on a song-by-song basis, across-listener median perceptual slider placement as a function of SIR. Dashed grey lines indicate linear regression lines shown for illustration only. Note: axis scale and range vary.

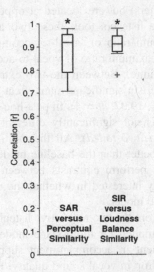

Fig. 3. Song-wise correlation coefficients. For each listening test, linear (Pearson) correlation coefficients were computed, on a song-by-song basis, for the listener-wise medians of the perceptual data with the respective BSS Eval measures. The above box-plots show median (in red), inter-quartile range (box) and 1.5 × IQR respectively (whiskers). Outliers are given as red crosses. Asterisks (above box-plots) denote significant median correlation coefficients ($P < 0.01$, *Permutation test*, $n = 10,000$). (Color figure online)

across-song correlations are both around 0.91 and both are significant ($P < 0.01$, *Permutation test*, $n = 10,000$). In other words, our measures correlate well, on a song-by-song basis, with the physical measures, suggesting that our psychophysical paradigm is reliable.

4   Results: Analysis II – Comparison of Separation Methods

Figure 4a shows box-plots of the data resulting from the first experiment (vocal similarity). The original vocal and the vocal-within-mixture are deemed to be very similar within the context of the experiment. There is a significant main effect among the different models ($P < 0.05$, Friedman Test, $\chi^2 = 10.93$, $df = 4$). In the pairwise post-hoc analysis, we find no evidence that the results for multi-stage model M_DNN1 or the M_NMF model are significantly different from the baseline ($P > 0.05$, paired Wilcox tests, two tailed). However, multi-stage models M_DNN2-3 are significantly less similar to the original vocal than the baseline ($P < 0.05$, *paired Wilcox tests, two tailed*).

Figure 4b shows the respective box-plots of the data resulting from the second (loudness balance) experiment. In this case, the mixture and original voice are not located together (by the listeners) but are located at opposite ends of the perceptual space. This indicates that the listeners took these two as bounding the space; the mixture providing the minimum ratio of vocal-to-accompaniment loudness and the original vocal providing the maximum ratio of vocal-to-accompaniment loudness (i.e., the ratio was theoretically infinite). Between the two extremes, there is a reasonable spread over the models. There is a significant main effect among the different models ($P < 0.05$, *Friedman Test*, $\chi^2 = 19.07$, $df = 4$). In post-hoc analysis, the baseline model is the worst performer and is not significantly different from the M_NMF model ($P > 0.05$, *paired Wilcox Test, two tailed*). All the multi-stage models suppress the accompaniment significantly better than the baseline model ($P < 0.05$, *paired Wilcox test, two-tailed*). We did not perform contrasts between the respective multi-stage models because we are chiefly interested in whether the multi-stage models offer an improvement over the baseline model.

Combining the evidence from the two respective listening tests, with respect to the performance of the baseline model, for multi-stage model M_DNN1˙there is a significant perceptual improvement in accompaniment suppression and no associated evidence of a corresponding drop in vocal sound quality. However, for the alternative multi-stage models (M_DNN2, M_DNN3, M_NMF) although there is evidence of significantly better suppression than baseline, there is also evidence of correspondingly significantly worse distortion than baseline. Therefore, in these cases, it would appear that there has been a trade-off [19], with improved accompaniment suppression coming at the expense of vocal sound quality.

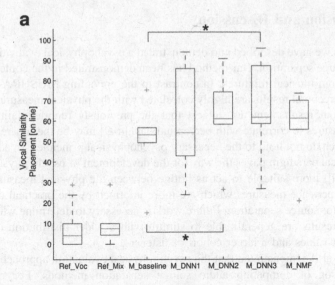

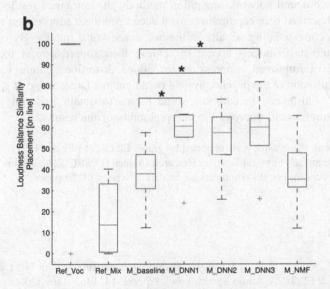

Fig. 4. **Model Comparison:** *Perceived Similarity.* Listeners organized sliders representing the seven respective stimuli along perceptual lines which depict the perceived similarity. **a** plots the data of experiment 1, capturing the similarity of the vocal component of the respective stimuli. **b** plots the respective data of experiment 2, capturing the loudness balance similarity. *Ref_Voc* refers to the original vocal signal, *Ref_Mix* refers to the mixture, *M_baseline* refers to the baseline DNN, *M_DNN1-3* refer to the respective multi-stage DNNs and *M_NMF* refers to the NMF model. Medians are shown in red. Boxes describe inter-quartile range and 'whiskers' indicate 95% confidence intervals. Bars with asterisks denote significant differences (*P* < 0.05, *paired Wilcox Test*). All other contrasts are not significant (*P* > 0.05, *paired Wilcox Test*). (Color figure online)

5 Conclusion and Discussion

In this paper, we have described and demonstrated a psychophysical evaluation method for audio source separation. Our method has been demonstrated in the context of vocal separation from musical mixtures. In contrast to the prevailing MUSHRA paradigms [4–9], our perceptual results are highly correlated with the physical measures SAR and SIR. Thus, our results tend to suggest that the previously reported failures of the physical measures to correlate with perceptual data [4–9] may be the inherent result of methods which do not hold to the necessary psychophysical principles. In addition, our psychophysical paradigm paves the way for the development of psychophysical models (e.g., see [19]) more suitable to act as bridge between the physical measures and the quality-of-experience measures which are more informed by the practical uses of and motivations for source separation. Future work is necessary to determine whether these preliminary results are generalizable to stimuli with a wider distribution of physical measurement values and a larger cohort of listeners.

We have also demonstrated that the psychophysical evaluation approach is suitable for comparison of competing audio source separation methods. For one of the multi-stage deep neural network separation methods, the combined results of the two experiments described here capture improved accompaniment suppression without any evidence of a corresponding penalty in the associated vocal quality. By contrast, the alternative multi-stage models appear to achieve their suppression at the cost of a trade-off [19] of improved suppression for added distortion. Future work should include generalisation of the psychophysical paradigm to a larger range of stimuli and a larger cohort of listeners. In addition, some means to obtain uniformly distributed physical measures would improve the interpretability of the results.

Acknowledgment. This work was supported by grants EP/L027119/1 and EP/L027119/2 from the UK Engineering and Physical Sciences Research Council (EPSRC). The authors also wish to thank the reviewers for helpful comments on an earlier version of the paper.

References

1. Vincent, E., Gribonval, R., Févotte, C.: Performance measurement in blind audio source separation. IEEE Trans. Audio Speech Lang. Process. **14**, 1462–1469 (2006)
2. Vincent, E., Jafari, M.G., Plumbley. M.D.: Preliminary guidelines for subjective evaluation of audio source separation algorithms. In: Nandi, A.K., Zhu, X., (eds.) Proceedings of ICA Research Network International Workshop, Liverpool, UK, pp. 93–96 (2006)
3. ITU. Recommendation ITU-R BS.1534-3: Method for the subjective assessment of intermediate quality level of audio systems (2014)
4. Emiya, V., Vincent, E., Harlander, N., Hohmann, V.: Subjective and objective quality assessment of audio source separation. IEEE Trans. Audio Speech Lang. Process. **19**, 2046–2057 (2011)
5. Cartwright, M., Pardo, B., Mysore, G.J., Hoffman, M.: Fast and easy crowdsourced perceptual audio evaluation. In: 2016 IEEE International Conference on Acoustics, Speech and Signal Processing (ICASSP), pp. 619–623 (2016)

6. Kornycky, J., Gunel, B., Kondoz, A.: Comparison of subjective and objective evaluation methods for audio source separation. In: Meetings on Acoustics, Paris, France, vol. 123, no. 5, p. 3569 (2008)
7. Langjahr, P., Mowlaee, P.: Objective quality assessment of target speaker separation performance in multisource reverberant environment. In: 4th International Workshop on Perceptual Quality of Systems, Vienna, Austria, pp. 89–94 (2013)
8. Gupta, U., Moore, E., Lerch, A.: On the perceptual relevance of objective source separation measures for singing voice separation. In: IEEE Workshop on Applications of Signal Processing to Audio and Acoustics (WASPAA 2015) (2015)
9. Cano, E., FitzGerald, D., Brandenburg, K.: Evaluation of quality of sound source separation algorithms: human perception vs quantitative metrics. In: EUSIPCO 2016, pp. 1758–1762 (2016)
10. Fechner, G.T.: Elemente der Psychophysik. Breitkopf und Härtel, Leipzig (1860)
11. Gescheider, G.: Psychophysics: The Fundamentals, 3rd edn. Lawrence Erlbaum Associates, Mahwah (1997)
12. Fletcher, H., Munson, W.A.: Loudness, its definition, measurement and calculation. J. Acoust. Soc. Am. **5**, 82–108 (1933)
13. Moore, B.C.J.: An Introduction to the Psychology of Hearing, 6th edn. Brill, Leiden (2012)
14. Grais, E.M., Roma, G., Simpson, A.J.R., Plumbley, M.D.: Discriminative enhancement for single channel audio source separation using deep neural networks. In: 13th International Conference on Latent Variable Analysis and Signal Separation (LVA/ICA) (2017)
15. Ono, N., Rafii, Z., Kitamura, D., Ito, N., Liutkus, A.: The 2015 signal separation evaluation campaign. In: Vincent, E., Yeredor, A., Koldovský, Z., Tichavský, P. (eds.) LVA/ICA 2015. LNCS, vol. 9237, pp. 387–395. Springer, Heidelberg (2015). doi:10.1007/978-3-319-22482-4_45
16. Terrell, M.J., Simpson, A.J.R., Sandler, M.: The mathematics of mixing. J. Audio Eng. Soc. **62**(1/2), 4–13 (2014)
17. Dwass, M.: Modified randomization tests for nonparametric hypotheses. Ann. Math. Stat. **28**, 181–187 (1957)
18. Simpson, A.J.R., Roma, G., Grais, E.M., Mason, R.D., Hummersone, C., Liutkus, A., Plumbley, M.D.: Evaluation of audio source separation models using hypothesis-driven non-parametric statistical methods. In: European Signal Processing Conference (EUSIPCO) (2016)
19. Simpson, A.J.R., Roma, G., Plumbley, M.D.: Deep karaoke: Extracting vocals from musical mixtures using a convolutional deep neural network. In: Proceedings of International Conference on Latent Variable Analysis and Signal Separation, pp. 429–436 (2015)

Audio Signal Processing

On the Use of Latent Mixing Filters in Audio Source Separation

Laurent Girin[1(✉)] and Roland Badeau[2]

[1] GIPSA-lab / Grenoble Alpes Univ. / INRIA Rhône-Alpes, Grenoble, France
`laurent.girin@gipsa-lab.grenoble-inp.fr`
[2] LTCI, CNRS, Télécom ParisTech, Université Paris-Saclay, Paris, France
`roland.badeau@telecom-paristech.fr`

Abstract. In this paper, we consider the underdetermined convolutive audio source separation (UCASS) problem. In the STFT domain, we consider both source signals and mixing filters as latent random variables, and we propose to estimate each source image, i.e. each individual source-filter product, by its posterior mean. Although, this is a quite straightforward application of the Bayesian estimation theory, to our knowledge, there exist no similar study in the UCASS context. In this paper, we discuss the interest of this estimator in this context and compare it with the conventional Wiener filter in a semi-oracle configuration.

Keywords: Audio source separation · Source image · Latent mixing filters · MMSE estimator · MCMC sampling

1 Introduction

To address the difficult problem of underdetermined audio source separation, probabilistic methods working in the Short-Term Fourier Transform (STFT) domain have been developed, e.g. [1–4]. These methods combine a physical mixture model, including source-to-microphone channel, with a source prior model. The mixture is often considered as convolutive, while using a (complex-valued) local Gaussian model (LGM) for the sources is now very popular. The convolutive mixture is generally approximated in the STFT domain as an instantaneous mixture at each frequency [2], even though this approximation can be questioned when the impulse response of the mixing filters is longer than the STFT window. A more general channel model has been proposed in [3], and combined with source LGM: the covariance matrix of the source image[1] is modeled as the product of the source power spectral density (PSD) with a spatial covariance matrix (SCM). A full-rank SCM is claimed to appropriately model diffuse sources and

This work is partly supported by the French National Research Agency (ANR) as a part of the EDISON 3D project (ANR-13-CORD-0008-02), and by the European Research Council (ERC) Advanced Grant VHIA 340113.

[1] A source image is defined as the multichannel version of the source signal, as recorded at the microphones [5].

P. Tichavský et al. (Eds.): LVA/ICA 2017, LNCS 10169, pp. 225–235, 2017.
DOI: 10.1007/978-3-319-53547-0_22

to overcome to some extent the limitation of the convolutive-to-multiplicative approximation [3], whereas the SCM model reduces to the convolutive model when the SCM is rank-1.

In all these papers, the coefficients of the channel model are considered as parameters of the overall probabilistic model. The source signals are considered as latent variables. The inference of sources and the estimation of (source and channel) parameters are made using an Expectation-Maximization (EM) algorithm or a similar two-step iterative procedure. Considering the channel coefficients as random variables or random processes, hence additional latent variables, has been recently proposed in a few audio source separation studies [6–9]. In [6,8], a prior distribution is assigned to the channel coefficients. This enables to introduce prior information about the mixture and acoustic environment in a principled manner, e.g. dependencies of the channel with source location or knowledge on room acoustics. In [7,9] a time-varying channel is considered as a random process and is estimated using a hidden Markov model with states corresponding to source direction of arrival (DoA) [7] or using a more general Kalman smoother [9]. The whole model solution is obtained following the variational EM methodology, which relies on the approximation of the joint posterior distribution of hidden variables (for instance source and channel) into a factorized form [10].

In a general manner, in all the above-mentioned studies, the extraction of the source signals from the mixture signal is made by some kind of Wiener filtering, in the E-step of the EM. Wiener filters are built from the current value of source parameters and from the current value of channel coefficients, be these latter considered as parameters or random variables. In turn, the new source estimates are used to update the channel coefficients (in the M-step or in some other part of the E-step). Therefore, *channel estimation and source signal estimation are two separate sequential processes.* Yet, in a fully Bayesian approach, where source and channel coefficients are considered as random variables, the posterior distribution of the source, and the associated source MMSE estimator, take a more general form: a stochastic integral that is generally not tractable [1,11]. Therefore, the (standard or variational) EM methodology can be seen as a way to break this intractability into an iterative sequential process that is suboptimal at each iteration but that is globally efficient.

In the present study, we consider the convolutive case, and we consider the mixing matrix in the STFT domain as a latent variable affected with a prior distribution. Instead of the sequential channel estimation and Wiener filtering inherent to the EM, we propose to *directly estimate the source image, i.e. the product of a monochannel source and the corresponding mixing vector, by its posterior expectation,* i.e. the "fully Bayesian" MMSE estimator applied to the source image. Hence, in contrast to the EM, the mixing filters are considered here as a latent random variable *during the source inference step.* This may sound quite trivial at first sight, but the inference of the product of two random variables is not easy. In particular, we assume that the posterior probability of the filter-source product does *not* factorize, as opposed to what is done in approximate variational methods [7,9]. In Sect. 2, we discuss this important point in more details and explain how the source image estimator contrasts with the conventional (convolutive) Wiener estimator. Actually, we put in evidence theoretical links between latent mixing vectors and the SCM model of [3].

Unfortunately, just like in the general case, the source image MMSE estimator takes the form of an intractable stochastic integral. Nevertheless, we further derive an advanced formulation of that stochastic integral in the case where the mixing filters follow a complex Gaussian distribution. The resulting expression depends on the source distribution, and though we use the LGM source model in the present study, the formulation is valid for any other distribution. We then turn to numerical approximation techniques to compute values of the source image estimator. We conduct experiments using a very basic sampling technique, for instance the Metropolis algorithm [12, Chap. 3]. We validate this approach in a "semi-oracle" configuration, where the source and channel parameters are estimated "offline" from the individual source images. In the present study, we only implement and discuss the inference step (in this semi-oracle configuration). The design of a complete blind separation process based on the proposed inference scheme and most likely of iterative nature, is out of the scope of the present paper. This paper must be considered as a prospective paper that discusses the use of a direct source image inference scheme in the UCASS framework, and positions this approach w.r.t. Wiener filtering.

Note that although the principle of the direct source image estimator is simple in essence, we could not find any paper exploring this idea in the present UCASS framework and reporting associated experiments. Probably the need to resort to computationally heavy sampling schemes can explain it. For example, a sampling process was applied to source separation in [11], but this study only dealt with instantaneous mixtures with mixing parameters assumed to be known. The present study however echoes [13], in which a *joint system and signal Kalman filter* was proposed and applied to single-channel speech enhancement and speech dereverberation. Interestingly, this joint scheme was opposed to a *dual* scheme, with sequential system and signal estimation, which can be seen, according to the authors of [13], "as a sequential variant of the EM procedure." The distribution of the source-system product within the joint Kalman filter was sampled using the Unscented Transform. In short, [13] considered a unique speech signal and a dynamic system, and the present paper considers a source separation problem, with stationary filters that can be easily extended to non-stationary filters.

2 Latent Mixing Filters and Estimation of Source Image

2.1 Principle

As in many source separation methods, the mixture signal is modeled as a convolutive noisy mixture of the source signals. Relying on the so-called narrow-band assumption, i.e. the impulse responses of the mixing filters are shorter than the time-frequency (TF) analysis window, the $I \times 1$ mixture signal is expressed in the short-time Fourier transform (STFT) domain as:

$$\mathbf{x}_{f\ell} = \mathbf{A}_{f\ell}\mathbf{s}_{f\ell} + \mathbf{b}_{f\ell} = \sum_{j=1}^{J} \mathbf{a}_{j,f\ell}s_{j,f\ell} + \mathbf{b}_{f\ell}, \tag{1}$$

where $f \in [1, F]$ is the frequency bin index, $\ell \in [1, L]$ is the frame index, $\mathbf{s}_{f\ell} = [s_{1,f\ell}, \ldots, s_{J,f\ell}]^\top \in \mathbb{C}^J$ (where symbol $.^\top$ denotes the transpose operator) is the vector of source coefficients, considered as a latent variable, $\mathbf{A}_{f\ell} = [\mathbf{a}_{1,f\ell}, \ldots, \mathbf{a}_{J,f\ell}] \in \mathbb{C}^{I \times J}$ is the mixing matrix ($\mathbf{a}_{j,f\ell} \in \mathbb{C}^I$ is the mixing vector for source j), and $\mathbf{b}_{f\ell} = [b_{1,f\ell}, \ldots, b_{I,f\ell}]^\top \in \mathbb{C}^I$ is a residual noise.

In the present study, we consider the mixing filter matrix $\mathbf{A}_{f\ell}$ as a latent variable, as opposed to a parameter as done in most audio source separation studies[2]. Moreover, in contrast to the classical use of a Wiener filter, we propose to estimate a source image signal $\mathbf{y}_{j,f\ell} = \mathbf{a}_{j,f\ell}s_{j,f\ell}$ directly by its posterior expectation, i.e. the MMSE estimator:

$$\hat{\mathbf{y}}_{j,f\ell} = \mathbb{E}_{q(\mathcal{H})}[\mathbf{a}_{j,f\ell}s_{j,f\ell}] = \mathbb{E}_{q(\mathbf{a}_{j,f\ell}, s_{j,f\ell})}[\mathbf{a}_{j,f\ell}s_{j,f\ell}], \tag{2}$$

where $\mathbb{E}_q$ denotes the mathematical expectation w.r.t. the probability density function (PDF) q, $q(.)$ denotes the posterior probability of a variable, i.e. $q(.) = p(.|\mathbf{x})$, and $\mathcal{H}$ denotes the complete set of hidden variables, i.e. $\mathcal{H} = \{\mathbf{A}_{f\ell}, \mathbf{s}_{f\ell}\}_{f,\ell=1}^{F,L} = \{\mathbf{a}_{j,f\ell}, s_{j,f\ell}\}_{f,\ell,j=1}^{F,L,J}$. Note that we assume for simplicity that all distributions factorize over f and ℓ. We also naturally assume that sources and filters are independent in the prior sense, i.e. $p(\mathbf{a}_{j,f\ell}, s_{j,f\ell}) = p(\mathbf{a}_{j,f\ell})p(s_{j,f\ell})$. However, and very importantly, we do *not* want here $q(\mathbf{a}_{j,f\ell}, s_{j,f\ell})$ to factorize over $\mathbf{a}_{j,f\ell}$ and $s_{j,f\ell}$, as opposed to what was done in the variational approximation approach, e.g. [7,9]. This is for two reasons: (i) In a general manner, a joint process is optimal compared to a combination of subprocesses. For instance, we want to take benefit from a possible posterior correlation between source and mixing filter. (ii) We want the proposed inference process to account for a diffuse source, seen as the "sum" of (possibly many) punctual sources with identical PSD and filtered with slightly different filters. Here the expectation in (2) takes the role of such summation. In contrast, factorizing $q(\mathbf{a}_{j,f\ell}, s_{j,f\ell})$ over $\mathbf{a}_{j,f\ell}$ and $s_{j,f\ell}$ would lead to $\hat{\mathbf{y}}_{j,f\ell} = \mathbb{E}_{q(\mathbf{a}_{j,f\ell})}[\mathbf{a}_{j,f\ell}]\mathbb{E}_{q(s_{j,f\ell})}[s_{j,f\ell}] = \hat{\mathbf{a}}_{j,f\ell}\hat{s}_{j,f\ell}$, i.e. a "unique" filtered source estimate, loosing the ability to represent diffuse sources. Note that the EM/Wiener approach within the convolutive mixture

---

[2] Considering the filters as latent variables enables us to make them depend on the time frame ℓ at no additional cost, compared to frame-independent latent filters $\mathbf{A}_f$, given that both models have the same set of parameters. This also comes at a much lower cost than the parametric case. However this does not necessarily mean that we have "trajectories" of filters, as for the moving sources or moving sensors in [7,9]. This simply allows the realization of the filters to be different for each frame, e.g. modeling slight movements of sources around their mean position. In the following, $\mathbf{a}_{j,f\ell}$ is assumed wide-sense stationary (WSS) along ℓ, hence its mean and covariance matrix do not depend on ℓ.

model is also problematic in this regard: a single mixing vector estimate $\hat{\mathbf{a}}_{j,f\ell}$ is used to build a single Wiener filter, whose ability to filter out diffuse sources is questionable.

Before entering into the technical derivation of (2), we now want to mention that considering the mixing filters as (WSS) latent variables also has a very interesting interpretation in terms of spatial properties of the sources from the prior distribution point of view. Indeed, let us define $\boldsymbol{\mu}_{\mathbf{a},j,f} = \mathbb{E}_{p(\mathbf{a}_{j,f\ell})}[\mathbf{a}_{j,f\ell}]$ the (prior) mean vector of $\mathbf{a}_{j,f\ell}$, and $\boldsymbol{\Sigma}_{\mathbf{a},j,f} = \mathbb{E}_{p(\mathbf{a}_{j,f\ell})}[(\mathbf{a}_{j,f\ell} - \boldsymbol{\mu}_{\mathbf{a},j,f})(\mathbf{a}_{j,f\ell} - \boldsymbol{\mu}_{\mathbf{a},j,f})^{\mathrm{H}}]$ its (prior) covariance matrix ($.^{\mathrm{H}}$ denotes the conjugate transpose operator). Then, assuming prior uncorrelation between source and filter, the prior covariance matrix of a source image is given by:

$$\mathbf{R}_{\mathbf{y},j,f\ell} = \mathbb{E}_{p(\mathcal{H})}[\mathbf{y}_{j,f\ell}\mathbf{y}_{j,f\ell}^{\mathrm{H}}] = \mathbb{E}_{p(s_{j,f\ell})}[|s_{j,f\ell}|^2]\,\mathbb{E}_{p(\mathbf{a}_{j,f\ell})}[\mathbf{a}_{j,f\ell}\mathbf{a}_{j,f\ell}^{\mathrm{H}}], \qquad (3)$$

hence

$$\mathbf{R}_{\mathbf{y},j,f\ell} = v_{j,f\ell}\mathbf{R}_{\mathbf{a},j,f}, \qquad (4)$$

where $v_{j,f\ell} = \mathbb{E}_{p(s_{j,f\ell})}[|s_{j,f\ell}|^2]$ is the PSD of source j at TF-bin (f,ℓ), and

$$\mathbf{R}_{\mathbf{a},j,f} = \mathbb{E}_{p(\mathbf{a}_{j,f\ell})}[\mathbf{a}_{j,f\ell}\mathbf{a}_{j,f\ell}^{\mathrm{H}}] = \boldsymbol{\mu}_{\mathbf{a},j,f}\boldsymbol{\mu}_{\mathbf{a},j,f}^{\mathrm{H}} + \boldsymbol{\Sigma}_{\mathbf{a},j,f} \qquad (5)$$

is the 2nd-order moment of the corresponding mixing filter. In conventional studies using the (time-invariant) convolutive model with $\mathbf{a}_{j,f}$ considered as a parameter, (4) holds with $\mathbf{R}_{\mathbf{a},j,f}$ being defined as $\mathbf{R}_{\mathbf{a},j,f} = \mathbf{a}_{j,f}\mathbf{a}_{j,f}^{\mathrm{H}}$ and thus limited to be rank-1. In the parametric context $\mathbf{R}_{\mathbf{a},j,f}$ is referred to as the spatial covariance matrix (SCM) of source j, and an extension to a full-rank SCM has been proposed in [3]. This full-rank matrix is assumed to model well a diffuse source, though interpreting this model in terms of the process generating the source image is not easy. An interpretation was given in [4,6] in the form of a finite summation of punctual sources filtered by different filters, all considered as parameters during the source inference step. *In contrast, considering the mixing filter as a latent variable as proposed in the present study enables to directly define $\mathbf{R}_{\mathbf{a},j,f}$ as a full-rank matrix with (5), while keeping the mixture comfortably described by a simple convolutive model (i.e. one source-filter product per image source signal).* Obviously, the proposed filter model reduces to the parametric convolutive case when $\boldsymbol{\Sigma}_{\mathbf{a},j,f}$ tends to zero. Hence, we believe that the fully probabilistic model presented in the present paper generalizes —or at least provides an elegant interpretation of—, the "parametric" definition of the SCM. It actually provides an elegant probabilistic interpretation of both the generation of diffuse source signals, as a "probabilistic convolution" (a probabilistic source-filter product in the TF domain), and their estimation, as a continuous summation of source-filter products.

2.2   General Expression of the Source Image MMSE Estimator

Let us now provide some technical derivations, starting with a general formulation of the source image MMSE estimator. Equation (2) writes:

$$\hat{\mathbf{y}}_{j,f\ell} = \int \int \mathbf{a}_{j,f\ell} s_{j,f\ell} p(\mathbf{A}_{f\ell}, \mathbf{s}_{f\ell} | \mathbf{x}_{f\ell}) d\mathbf{A}_{f\ell} d\mathbf{s}_{f\ell}. \tag{6}$$

Since we have $p(\mathbf{A}_{f\ell}, \mathbf{s}_{f\ell} | \mathbf{x}_{f\ell}) = \frac{p(\mathbf{x}_{f\ell} | \mathbf{A}_{f\ell}, \mathbf{s}_{f\ell}) p(\mathbf{A}_{f\ell}) p(\mathbf{s}_{f\ell})}{p(\mathbf{x}_{f\ell})}$, (6) rewrites:

$$\hat{\mathbf{y}}_{j,f\ell} = \frac{1}{p(\mathbf{x}_{f\ell})} \int s_{j,f\ell} \Big(\int \mathbf{a}_{j,f\ell} p(\mathbf{x}_{f\ell} | \mathbf{A}_{f\ell}, \mathbf{s}_{f\ell}) p(\mathbf{A}_{f\ell}) d\mathbf{A}_{f\ell} \Big) p(\mathbf{s}_{f\ell}) d\mathbf{s}_{f\ell}. \tag{7}$$

Note that this expression is completely independent of the form of all densities. It only relies on definition (6) and the Bayes product rule. Obviously, (7) can be extended in the Bayesian sense by including priors on the parameters of the different distributions. In the present work, we stick to the above form.

2.3   The Gaussian Case

In this section, we go a bit further and derive a "simplified" or "advanced" form of the source image MMSE estimator in the case where the mixing filters are assumed to follow a complex Gaussian distribution. For this aim, let us first specify and reshape $p(\mathbf{x}_{f\ell} | \mathbf{A}_{f\ell}, \mathbf{s}_{f\ell}) p(\mathbf{A}_{f\ell})$. As in several other studies, $\mathbf{b}_{f\ell}$ is assumed to be a zero-mean circular stationary complex Gaussian noise, i.e. $p(\mathbf{b}_{f\ell}) = \mathcal{N}_c(\mathbf{b}_{f\ell}; \mathbf{0}, \boldsymbol{\Sigma}_{\mathbf{b},f})$, where $\boldsymbol{\Sigma}_{\mathbf{b},f}$ is the noise covariance matrix to be estimated[3]. In addition, $\mathbf{b}_{f\ell}$ may be assumed to be isotropic, i.e. $\boldsymbol{\Sigma}_{\mathbf{b},f} = v_{\mathbf{b},f} \mathbf{I}_I$ with $v_{\mathbf{b},f} \in \mathbb{R}^+$ and $\mathbf{I}_I$ denoting the identity matrix of size I. We thus have $p(\mathbf{x}_{f\ell} | \mathbf{A}_{f\ell}, \mathbf{s}_{f\ell}) = \mathcal{N}_c(\mathbf{x}_{f\ell}; \mathbf{A}_{f\ell} \mathbf{s}_{f\ell}, \boldsymbol{\Sigma}_{\mathbf{b},f})$. Now it is natural to assume that the mixing filters $\mathbf{A}_{f\ell}$ follow a complex Gaussian prior distribution, since the latter is the conjugate prior of the Gaussian distribution for the mean parameter. For the sake of technical derivation, $\mathbf{A}_{f\ell}$ is first vectorized by vertically concatenating its J columns $\mathbf{a}_{j,f\ell}$ into a single column vector $\mathbf{a}_{:,f\ell}$, i.e. $\mathbf{a}_{:,f\ell} = \text{vec}(\mathbf{A}_{f\ell}) = [\mathbf{a}_{1,f\ell}^\top, \dots, \mathbf{a}_{J,f\ell}^\top]^\top \in \mathbb{C}^{IJ}$. Then we assume:

$$p(\mathbf{A}_{f\ell}) = p(\mathbf{a}_{:,f\ell}) = \mathcal{N}_c(\mathbf{a}_{:,f\ell}; \boldsymbol{\mu}_{\mathbf{a},f}, \boldsymbol{\Sigma}_{\mathbf{a},f}), \tag{8}$$

where the mean vector $\boldsymbol{\mu}_{\mathbf{a},f} \in \mathbb{C}^{IJ}$ and the covariance matrix $\boldsymbol{\Sigma}_{\mathbf{a},f} \in \mathbb{C}^{IJ \times IJ}$ are parameters to be estimated. $\boldsymbol{\mu}_{\mathbf{a},f}$ is the concatenation of the individual mean mixing vectors $\boldsymbol{\mu}_{\mathbf{a},j,f}, j \in [1, J]$, defined for each source. $\boldsymbol{\Sigma}_{\mathbf{a},f}$ is block diagonal, assuming prior decorrelation of filters corresponding to different sources.

---

[3] The proper complex Gaussian distribution is defined as $\mathcal{N}_c(\mathbf{x}; \boldsymbol{\mu}, \boldsymbol{\Sigma}) = |\pi \boldsymbol{\Sigma}|^{-1} \exp \big(-[\mathbf{x} - \boldsymbol{\mu}]^H \boldsymbol{\Sigma}^{-1} [\mathbf{x} - \boldsymbol{\mu}] \big)$, where $|.|$ denotes the matrix determinant [14].

Let us then rewrite $\mathbf{A}_{f\ell}\mathbf{s}_{f\ell} = \sum_{j=1}^{J} \mathbf{a}_{j,f\ell}s_{j,f\ell} = (\mathbf{s}_{f\ell}^{\top} \otimes \mathbf{I}_I)\mathbf{a}_{:,f\ell} = \mathbf{U}_{f\ell}\mathbf{a}_{:,f\ell}$, with $\mathbf{U}_{f\ell} = \mathbf{s}_{f\ell}^{\top} \otimes \mathbf{I}_I$ ($\otimes$ denotes the Kronecker matrix product). Then, we can write:

$$p(\mathbf{x}_{f\ell}|\mathbf{a}_{:,f\ell}, \mathbf{s}_{f\ell})p(\mathbf{a}_{:,f\ell}) = p(\mathbf{a}_{:,f\ell}|\mathbf{x}_{f\ell}, \mathbf{s}_{f\ell})p(\mathbf{x}_{f\ell}|\mathbf{s}_{f\ell}), \tag{9}$$

since both sides are equal to $p(\mathbf{x}_{f\ell}, \mathbf{a}_{:,f\ell}|\mathbf{s}_{f\ell})$. Because $p(\mathbf{a}_{:,f\ell})$ is the conjugate prior of $p(\mathbf{x}_{f\ell}|\mathbf{a}_{:,f\ell}, \mathbf{s}_{f\ell})$, $p(\mathbf{a}_{:,f\ell}|\mathbf{x}_{f\ell}, \mathbf{s}_{f\ell})$ is a complex-Gaussian distribution that can be written $p(\mathbf{a}_{:,f\ell}|\mathbf{x}_{f\ell}, \mathbf{s}_{f\ell}) = \mathcal{N}_c(\mathbf{a}_{:,f\ell}; \boldsymbol{\mu}_{\mathbf{d},f\ell}, \boldsymbol{\Sigma}_{\mathbf{d},f\ell})$. Then, since $\mathbf{a}_{:,f\ell}$ is Gaussian and $\mathbf{b}_{f\ell}$ is Gaussian, it follows that $p(\mathbf{x}_{f\ell}|\mathbf{s}_{f\ell})$ is a Gaussian distribution that can be written $p(\mathbf{x}_{f\ell}|\mathbf{s}_{f\ell}) = \mathcal{N}_c(\mathbf{x}_{f\ell}; \boldsymbol{\mu}_{\mathbf{e},f\ell}, \boldsymbol{\Sigma}_{\mathbf{e},f\ell})$. Identifying the quadratic terms in $\mathbf{a}_{:,f\ell}$ in (9), we get:

$$\boldsymbol{\Sigma}_{\mathbf{d},f\ell}^{-1} = \mathbf{U}_{f\ell}^{\mathrm{H}}\boldsymbol{\Sigma}_{\mathbf{b},f}^{-1}\mathbf{U}_{f\ell} + \boldsymbol{\Sigma}_{\mathbf{a},f}^{-1}. \tag{10}$$

Then, identifying the linear terms in $\mathbf{a}_{:,f\ell}$ in (9), we get:

$$\boldsymbol{\mu}_{\mathbf{d},f\ell} = \boldsymbol{\Sigma}_{\mathbf{d},f\ell}(\mathbf{U}_{f\ell}^{\mathrm{H}}\boldsymbol{\Sigma}_{\mathbf{b},f}^{-1}\mathbf{x}_{f\ell} + \boldsymbol{\Sigma}_{\mathbf{a},f}^{-1}\boldsymbol{\mu}_{\mathbf{a},f}). \tag{11}$$

Then, identifying the quadratic terms in $\mathbf{x}_{f\ell}$ in (9) and applying the matrix inversion lemma [15, pp. 18–19], we get:

$$\boldsymbol{\Sigma}_{\mathbf{e},f\ell}^{-1} = \boldsymbol{\Sigma}_{\mathbf{b},f}^{-1} - \boldsymbol{\Sigma}_{\mathbf{b},f}^{-1}\mathbf{U}_{f\ell}\boldsymbol{\Sigma}_{\mathbf{d},f\ell}\mathbf{U}_{f\ell}^{\mathrm{H}}\boldsymbol{\Sigma}_{\mathbf{b},f}^{-1}$$
$$\Leftrightarrow \boldsymbol{\Sigma}_{\mathbf{e},f\ell} = \boldsymbol{\Sigma}_{\mathbf{b},f} + \mathbf{U}_{f\ell}\boldsymbol{\Sigma}_{\mathbf{a},f}\mathbf{U}_{f\ell}^{\mathrm{H}}. \tag{12}$$

Finally, identifying the remaining linear terms in $\mathbf{x}_{f\ell}$, we get:

$$\boldsymbol{\mu}_{\mathbf{e},f\ell} = \boldsymbol{\Sigma}_{\mathbf{e},f\ell}(\boldsymbol{\Sigma}_{\mathbf{b},f}^{-1}\mathbf{U}_{f\ell}\boldsymbol{\Sigma}_{\mathbf{d},f\ell}\boldsymbol{\Sigma}_{\mathbf{a},f}^{-1}\boldsymbol{\mu}_{\mathbf{a},f}). \tag{13}$$

Now we can inject (9) into (7), and we get:

$$\hat{\mathbf{y}}_{j,f\ell} = \frac{1}{p(\mathbf{x}_{f\ell})} \int s_{j,f\ell}\boldsymbol{\mu}_{\mathbf{d},j,f\ell}\mathcal{N}_c(\mathbf{x}_{f\ell}; \boldsymbol{\mu}_{\mathbf{e},f\ell}, \boldsymbol{\Sigma}_{\mathbf{e},f\ell})p(\mathbf{s}_{f\ell})d\mathbf{s}_{f\ell}, \tag{14}$$

with $\boldsymbol{\mu}_{\mathbf{d},j,f\ell}$ being the sub-vector of $\boldsymbol{\mu}_{\mathbf{d},f\ell}$ that corresponds to source j. If we concatenate the source images as $\mathbf{y}_{f\ell} = [\mathbf{y}_{1,f\ell}^{\top}, \ldots, \mathbf{y}_{J,f\ell}^{\top}]^{\top} \in \mathbb{C}^{IJ}$, we can rewrite (14) for all sources in compact form:

$$\hat{\mathbf{y}}_{f\ell} = \frac{1}{p(\mathbf{x}_{f\ell})} \int (\mathbf{s}_{f\ell} \otimes \mathbf{I}_I)\boldsymbol{\mu}_{\mathbf{d},f\ell}\mathcal{N}_c(\mathbf{x}_{f\ell}; \boldsymbol{\mu}_{\mathbf{e},f\ell}, \boldsymbol{\Sigma}_{\mathbf{e},f\ell})p(\mathbf{s}_{f\ell})d\mathbf{s}_{f\ell}. \tag{15}$$

Note that (14) and (15) are valid for any source distribution. In the following, we use the LGM with diagonal covariance matrix: $p(\mathbf{s}_{f\ell}) = \mathcal{N}_c(\mathbf{s}_{f\ell}; \mathbf{0}, \mathbf{v}_{f\ell} = \mathrm{diag}_J(v_{j,f\ell}))$. Even for such a classical source distribution, the integral in (14) or (15) has no closed-form expression since $\boldsymbol{\mu}_{\mathbf{d},f\ell}$ is a non-linear function of $\mathbf{s}_{f\ell}$ implying the inversion of a quadratic form (which is also present in $\boldsymbol{\mu}_{\mathbf{e},f\ell}$). Also, (14) or (15) requires the calculation of the observation marginal density $p(\mathbf{x}_{f\ell})$, which is a classical obstacle in inference problems. Therefore we have to turn towards sampling techniques.

2.4   Inference of Source Image Using Metropolis Algorithm

For the computation of values of the source image estimator (15), in the present study, we propose to use the Metropolis algorithm. Because this algorithm is very classical and quite basic, and because of room limitation, we will not present it into details. The reader is referred to [12] for a general overview of sampling techniques, and to [12, chap. 3] for a tutorial on the Metropolis algorithm.

3   Experiments

In this section, we report experiments conducted with three different stereo ($I = 2$) mixtures of $J = 3$ speech signals[4]. In Mix 1 and Mix 2, the source signals were monochannel 16 kHz signals randomly taken from the TIMIT database [16]. The source images $\mathbf{y}_j(t)$ were individually generated using the room impulse response (RIR) simulator of AudioLabs Erlangen[5]. The setting was the following: room size 7 m × 5 m × 2.5 m, sensor array placed at (3.5 m , 1.5 m , 1.5 m), distance between microphones $d = 0.15$ m, reverberation time $T_{60} = 150$ ms, source-to-sensor distance 1.2 m. In Mix 1, sources s_1, s_2 and s_3 are initially located at azimuths $-45°$, $0°$, $45°$, respectively, and they all move by $20°$ around the microphone array, within the signal duration of 2 s. In Mix 2, they start at azimuths $-75°$, $-25°$, $25°$ and they all move by $50°$. Finally, for Mix 3, three speakers were (separately) recorded in an office ($T_{60} \approx 0.6$ s). They were initially located at azimuths $-45°$, $0°$, $45°$, at 1.5 m from a two-microphone array (omnidirectional), and moved by about $45°$ in 2 s.

The STFT window was a 1024-point sine window with 50 % overlap. The parameters $\boldsymbol{\mu}_{\mathbf{a},f}$ and $\boldsymbol{\Sigma}_{\mathbf{a},f}$ were set to "semi-oracle" values calculated from the individual source images. More precisely, for each $j \in [1, J]$, $\boldsymbol{\mu}_{\mathbf{a},j,f}$ and $\boldsymbol{\Sigma}_{\mathbf{a},j,f}$ were calculated from $\mathbf{y}_{j,f\ell}$, the STFT of $\mathbf{y}_j(t)$, following the spirit of the full-rank SCM initialization in [3]: $\mathbf{y}_{j,f\ell}$ was first normalized in phase, i.e. we calculated $\tilde{\mathbf{y}}_{j,f\ell} = \mathbf{y}_{j,f\ell}e^{-i\arg(y_{1,f\ell})}$; then $\boldsymbol{\mu}_{\mathbf{a},j,f}$ and $\boldsymbol{\Sigma}_{\mathbf{a},j,f}$ were calculated as the empirical mean and empirical covariance matrix of $\tilde{\mathbf{y}}_{j,f\ell}$, $\ell \in [1, L]$; finally, $v_{j,f\ell}$ was calculated for each frame by $v_{j,f\ell} = \frac{1}{I}\mathrm{trace}(\mathbf{R}_{\mathbf{a},j,f}^{-1}\mathbf{y}_{j,f\ell}\mathbf{y}_{j,f\ell}^{H})$. The noise variance $v_{\mathbf{b},f}$ was set to 10^{-6} times the average PSD of the mixture signal. The semi-oracle setting of the parameters is of course an artificial close-to-optimal configuration that ensures very good separation performance (as verified in Table 1).

The computation of the Metropolis source image estimator was made using the semi-oracle values of the parameters and the mixture signal, with the PDF in the integral of (15) used as the target distribution and a complex-Gaussian distribution used as the candidate distribution. 15, 000 samples were drawn at each TF bin (1, 000 for burn-in). The separation of each 2s-mixture required about 4 hours on a 4-core 2.3 GHz Intel Core i7 using the Matlab Parallel Toolbox. For comparison, the rank-1 Wiener estimator (R1W; as used in [2]) and the

---

[4] Matlab code and data are available at: http://www.gipsa-lab.grenoble-inp.fr/~laurent.girin/demo/lva2017.zip.

[5] www.audiolabs-erlangen.de/fau/professor/habets/software/rir-generator.

Table 1. Separation performance (in dB). Best scores across methods are in bold (when the difference is larger than 0.1 dB).

Method	Meas	Mix 1			Mix 2			Mix 3		
		s_1	s_2	s_3	s_1	s_2	s_3	s_1	s_2	s_3
Rank-1 Wiener	SDR	14.28	12.49	8.83	12.52	10.01	7.47	−3.39	−1.28	−2.55
	SAR	16.28	15.12	8.78	14.95	13.12	8.70	2.21	1.62	1.92
	SIR	17.18	14.25	7.77	15.52	12.71	7.05	0.56	−0.96	0.93
	ISR	16.30	18.70	14.56	14.59	14.88	14.14	1.35	3.73	3.96
Full-Rank Wiener	SDR	19.88	15.54	13.98	18.88	14.56	**13.67**	7.68	8.96	8.22
	SAR	22.43	17.25	**16.44**	20.89	**16.79**	12.05	**10.74**	**6.71**	**7.13**
	SIR	24.94	19.31	18.14	23.47	**19.46**	12.44	**11.56**	**6.14**	**7.27**
	ISR	24.04	20.29	19.74	23.33	18.59	19.38	11.82	12.64	12.26
Proposed	SDR	**19.99**	**15.82**	14.02	18.86	14.64	13.56	7.62	8.94	8.29
	SAR	**22.62**	**18.27**	16.24	**21.30**	16.46	**12.41**	9.70	6.14	6.85
	SIR	**26.24**	**21.93**	**18.38**	**25.16**	19.21	**13.05**	10.88	5.54	7.10
	ISR	**24.48**	**20.57**	**20.12**	23.32	**18.88**	**19.77**	**12.10**	**12.89**	**12.84**

full-rank Wiener estimator (FRW; as used in [3]), using the same semi-oracle values of the parameters, were calculated as:

$$\hat{\mathbf{y}}_{j,f\ell} = v_{j,f\ell}\boldsymbol{\mu}_{\mathbf{a},j,f}\boldsymbol{\mu}_{\mathbf{a},j,f}^{H}\left(\sum_{k=1}^{J}v_{k,f\ell}\boldsymbol{\mu}_{\mathbf{a},k,f}\boldsymbol{\mu}_{\mathbf{a},k,f}^{H} + v_{\mathbf{b},f}\mathbf{I}_{I}\right)^{-1}\mathbf{x}_{f\ell}, \qquad (16)$$

$$\hat{\mathbf{y}}_{j,f\ell} = v_{j,f\ell}\mathbf{R}_{\mathbf{a},j,f}\left(\sum_{k=1}^{J}v_{k,f\ell}\mathbf{R}_{\mathbf{a},k,f} + v_{\mathbf{b},f}\mathbf{I}_{I}\right)^{-1}\mathbf{x}_{f\ell}. \qquad (17)$$

Four standard audio source separation objective measures were calculated between the estimated and ground truth source images, namely: signal-to-distortion ratio (SDR), signal-to-interference ratio (SIR) signal-to-artifact ratio (SAR) and image-to-spatial distortion ratio (ISR) [17]. The results are presented in Table 1. We can see that both the FRW and the proposed estimator provide separation measures that are notably larger than the R1W. This confirms that both are able to efficiently exploit the spatial information on the mixture encoded in the SCM (remember that for each source, the SCM is equivalent to the second-order moment of the mixing filter, see (5)). For Mix 1 (sources moving relatively slowly), the proposed estimator performs globally better than the FRW. For Mix 2 (sources moving more rapidly), the results of the proposed estimator and FRW are more similar. Finally, the results for the real recordings tend to slightly favor FRW, even if the difference in SDR is especially small[6].

---

[6] So far, no statistical test could be performed on a large set of mixtures to test the significativity of the results because of the huge computational cost of the Metropolis.

4   Conclusion

Altogether, these results show the potential of the proposed method to overcome the state-of-the-art. As opposed to the Wiener filter build from the full-rank spatial covariance matrix of [3], the proposed source image estimator has the freedom to use the latter to independently estimate (an infinite set of) filter values at every frame and use it for image source estimation. In contrast, the Wiener filter of [3] directly uses the same spatial information at every frame. Yet, the results for real recordings are mitigated. The proposed estimator may be more sensible than the full-rank Wiener filter to the convolutive-to-multiplicative approximation for long mixing filters, for reasons that must be investigated. We will also work on improving the sampling scheme, and integrating the proposed estimator in a fully blind (iterative) separation process.

References

1. Vincent, E., Jafari, M., Abdallah, S., Plumbley, M., Davies, M.: Probabilistic modeling paradigms for audio source separation. In: Machine Audition: Principles, Algorithms and Systems, pp. 162–185 (2010)
2. Ozerov, A., Févotte, C.: Multichannel nonnegative matrix factorization in convolutive mixtures for audio source separation. IEEE Trans. Audio Speech Lang. Process. **18**(3), 550–563 (2010)
3. Duong, N., Vincent, E., Gribonval, R.: Under-determined reverberant audio source separation using a full-rank spatial covariance model. IEEE Trans. Audio Speech Lang. Process. **18**(7), 1830–1840 (2010)
4. Ozerov, A., Vincent, E., Bimbot, F.: A general flexible framework for the handling of prior information in audio source separation. IEEE Trans. Audio Speech Lang. Process. **20**(4), 1118–1133 (2012)
5. Sturmel, N., Liutkus, A., Pinel, J., Girin, L., Marchand, S., Richard, G., Badeau, R., Daudet, L.: Linear mixing models for active listening of music productions in realistic studio conditions. In: Convention of the Audio Engineering Society (AES). Budapest, Hungary (2012)
6. Duong, N., Vincent, E., Gribonval, R.: Spatial location priors for Gaussian model based reverberant audio source separation. EURASIP J. Adv. Signal Process. **149**, 2013 (2013)
7. Higuchi, T., Takamune, N., Tomohiko, N., Kameoka, H.: Underdetermined blind separation and tracking of moving sources based on DOA-HMM. In: Proceedings of the International Conference on Acoustics, Speech and Signal Proceedings (ICASSP) (2014)
8. Leglaive, S., Badeau, R., Richard, G.: Multichannel audio source separation with probabilistic reverberant modeling. In IEEE Workshop Applications of Signal Processing to Audio and Acoustics (WASPAA) (2015)
9. Kounades-Bastian, D., Girin, L., Alameda-Pineda, X., Gannot, S., Horaud, R.: A variational EM algorithm for the separation of moving sound sources. In: IEEE Workshop Application Signal Process. to Audio and Acoustics (WASPAA) (2015)
10. Smidl, V., Quinn, A.: The Variational Bayes Method in Signal Processing. Springer, Berlin (2006)

11. Cemgil, A., Févotte, C., Godsill, S.: Variational and stochastic inference for Bayesian source separation. Digit. Signal Proc. **2007**(17), 891–913 (2007)
12. Liang, F., Liu, C., Carroll, R.: Advanced Markov Chain Monte Carlo Methods: Learning from Past Samples. Wiley, New York (2010)
13. Gannot, S., Moonen, M.: On the application of the unscented Kalman filter to speech processing. In: IEEE International Workshop on Acoustic Echo and Noise Control (IWAENC), Japan, Kyoto, p. 811(2003)
14. Neeser, F., Massey, J.: Proper complex random processes with applications to information theory. IEEE Trans. Info. Theory **39**(4), 1293–1302 (1993)
15. Horn, R., Johnson, C.: Matrix Analysis. Cambridge University Press, Cambridge (1985)
16. Garofolo, J.S., Lamel, L.F., Fisher, W.M., Fiscus, J.G., Pallett, D.S., Dahlgren, N.L., Zue, V.: TIMIT acoustic-phonetic continuous speech corpus. In: Linguistic Data Consortium, Philadelphia (1993)
17. Vincent, E., Sawada, H., Bofill, P., Makino, S., Rosca, J.P.: First stereo audio source separation evaluation campaign: data, algorithms and results. In: Davies, M.E., James, C.J., Abdallah, S.A., Plumbley, M.D. (eds.) ICA 2007. LNCS, vol. 4666, pp. 552–559. Springer, Heidelberg (2007). doi:10.1007/978-3-540-74494-8_69

Discriminative Enhancement for Single Channel Audio Source Separation Using Deep Neural Networks

Emad M. Grais$^{(\boxtimes)}$, Gerard Roma, Andrew J.R. Simpson,
and Mark D. Plumbley

Centre for Vision, Speech and Signal Processing, University of Surrey, Guildford, UK
{grais,g.roma,andrew.simpson,m.plumbley}@surrey.ac.uk

Abstract. The sources separated by most single channel audio source separation techniques are usually distorted and each separated source contains residual signals from the other sources. To tackle this problem, we propose to enhance the separated sources to decrease the distortion and interference between the separated sources using deep neural networks (DNNs). Two different DNNs are used in this work. The first DNN is used to separate the sources from the mixed signal. The second DNN is used to enhance the separated signals. To consider the interactions between the separated sources, we propose to use a single DNN to enhance all the separated sources together. To reduce the residual signals of one source from the other separated sources (interference), we train the DNN for enhancement discriminatively to maximize the dissimilarity between the predicted sources. The experimental results show that using discriminative enhancement decreases the distortion and interference between the separated sources.

Keywords: Single channel audio source separation · Deep neural networks · Audio enhancement · Discriminative training

1 Introduction

Audio single channel source separation (SCSS) aims to separate sources from their single mixture [3,17]. Deep neural networks (DNNs) have recently been used to tackle the SCSS problem [2,6,18,20]. DNNs have achieved better separation results than nonnegative matrix factorization which is considered as one of the most common approaches for the SCSS problem [2,5,14,18]. DNNs are used for SCSS to either predict the sources from the observed mixed signal [5,6], or to predict time-frequency masks that are able to describe the contribution of each source in the mixed signal [2,14,18]. The masks usually take bounded values between zero and one. It is normally preferred to train the DNNs to predict masks that take bounded values to avoid training them on the full dynamic ranges of the sources [2,18].

© Springer International Publishing AG 2017
P. Tichavský et al. (Eds.): LVA/ICA 2017, LNCS 10169, pp. 236–246, 2017.
DOI: 10.1007/978-3-319-53547-0_23

Most SCSS techniques produce separated sources accompanied by distortion and interference from other sources [2,3,12,17]. To improve the quality of the separated sources, Williamson et al. [20] proposed to enhance the separated sources using nonnegative matrix factorization (NMF). The training data for each source is modelled separately, and each separated source is enhanced individually by its own trained model. However, enhancing each separated source individually does not consider the interaction between the sources in the mixed signal [4,20]. Furthermore, the residuals of each source that appear in the other separated sources are not available to enhance their corresponding separated sources.

In this paper, to consider the interaction between the separated sources, we propose to enhance all the separated sources together using a single DNN. Using a single model to enhance all the separated sources together allows each separated source to be enhanced using its remaining parts that appear in the other separated sources. This means most of the available information of each source in the mixed signal can be used to enhance its corresponding separated source. DNNs have shown better performance than NMF in many audio signal enhancement applications [2,18]. Thus, in this work we use a DNN to enhance the separated sources rather than using NMF [20]. We train the DNN for enhancement discriminatively to maximize the differences between the estimated sources [6,7]. A new cost function to discriminatively train the DNN for enhancement is introduced in this work. Discriminative training for the DNN aims to decrease the interference of each source in the other estimated sources and has also been found to decrease distortions [6]. Unlike other enhancement approaches such as NMF [20] and denoising deep autoencoders [16,21] that aim to only enhance the quality of an individual signal, our new discriminative enhancement approach in this work aims to both enhance the quality and achieve good separation for the estimated sources.

The main contributions of this paper are: (1) the use of a single DNN to enhance all the separated signals together; (2) discriminative training of a DNN for enhancing the separated sources to maximize the dissimilarity of the predicted sources; (3) a new cost function for discriminatively training the DNN.

This paper is organized as follows. In Sect. 2 a mathematical formulation of the SCSS problem is given. Section 3 presents our proposed approach for using DNNs for source separation and enhancement. The experimental results and the conclusion of this paper are presented in Sects. 4 and 5.

2   Problem Formulation of Audio SCSS

Given a mixture of I sources as $y(t) = \sum_{i=1}^{I} s_i(t)$, the aim of audio SCSS is to estimate the sources $s_i(t)$, $\forall i$, from the mixed signal $y(t)$. The estimate $\hat{S}_i(n, f)$ for source i in the short time Fourier transform (STFT) domain can be found by predicting a time-frequency mask $M_i(n, f)$ that scales the mixed signal according to the contribution of source i in the mixed signal as follows [2,14,18]:

$$\hat{S}_i(n, f) = M_i(n, f) \times Y(n, f) \tag{1}$$

where $Y(n, f)$ is the STFT of the observed mixed signal $y(t)$, while n and f are the time and frequency indices respectively. The mask $M_i(n, f)$ takes real values between zero and one. The main goal here is to predict masks $M_i(n, f)$, $\forall i$, that separate the sources from the mixed signal. In this framework, the magnitude spectrogram of the mixed signal is approximated as a sum of the magnitude spectra of the estimated sources [12,17] as follows:

$$|Y(n, f)| \approx \sum_{i=1}^{I} \left| \hat{S}_i(n, f) \right|. \tag{2}$$

For the rest of this paper, we denote the magnitude spectrograms and the masks in a matrix form as Y, $\hat{S}_i$, and M_i.

3   DNNs for source separation and enhancement

In this paper, we use two deep neural networks (DNNs) to perform source separation and enhancement. The first DNN (DNN-A) is used to separate the sources from the mixed signal. The separated sources are then enhanced by the second DNN (DNN-B) as shown in Fig. 1. DNN-A is trained to map the mixed signal in its input into reference masks in its output. DNN-B is trained to map the separated sources (distorted signals) from DNN-A into their reference/clean signals. As in many machine learning tasks, the data used to train the DNNs is usually different than the data used for testing [2,6,14,18]. The performance of the trained DNN on the test data is often worse than the performance on the training set. The trained DNN-A is used to separate data that is different than the training data, and since the main goal of using DNN-B is to enhance the separated signals by DNN-A, then DNN-B should be trained on a different set of data than the set of data that was used to train DNN-A. Thus, in this work we divide the available training data into two sets. The first set of the training data is used to train DNN-A for separation and the second set is for training DNN-B for enhancement.

3.1   Training DNN-A for Source Separation

Given the magnitude spectrograms of the sources in the first set of the training data $S_{tri}^{(1)}$, $\forall i$, DNN-A is trained to predict a reference mask $M_{tri}^{(1)}$. The subscript tri indicates the training data for source i, and the superscript "(1)" indicates the first set of the training data is used for training. Different types of masks have been proposed in [9,18]. We chose to use the ratio mask from [18], which gives separated sources with reasonable distortion and interference. The reference ratio mask in [18] is defined as follows:

$$M_{tri}^{(1)} = \frac{S_{tri}^{(1)}}{\sum_{i=1}^{I} S_{tri}^{(1)}} \tag{3}$$

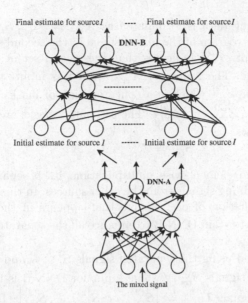

Fig. 1. The overview of the proposed approach of using DNNs for source separation and enhancement. DNN-A is used for separation. DNN-B is used for enhancement.

where the division is done element-wise, $S_{tri}^{(1)}$ is the magnitude spectrogram of reference source i, and $M_{tri}^{(1)}$ is the mask which defines the contribution of source i in every time-frequency bin (n, f) in the mixed signal. The input of DNN-A is the magnitude spectrogram $X_{tr}^{(1)}$ of the mixed signal of the first set of the training data which is formulated as $X_{tr}^{(1)} = \sum_{i=1}^{I} S_{tri}^{(1)}$. The reference/target output of DNN-A for all sources is formed by concatenating the reference masks for all sources as

$$M_{tr}^{(1)} = \left[M_{tr1}^{(1)}, \ldots, M_{tri}^{(1)}, \ldots, M_{trI}^{(1)} \right]. \tag{4}$$

DNN-A is trained to minimize the following cost function as in [10,18]:

$$C_1 = \sum_{n,f} \left(Z_{tr}^{(1)}(n, f) - M_{tr}^{(1)}(n, f) \right)^2 \tag{5}$$

where $Z_{tr}^{(1)}$ is the actual output of the final layer of DNN-A and $M_{tr}^{(1)} \in [0, 1]$ is computed from Eqs. (3) and (4). The activation functions of the output layer for DNN-A are sigmoid functions, thus $Z_{tr}^{(1)} \in [0, 1]$.

3.2   Training DNN-B for Discriminative Enhancement

To generate the training data to train DNN-B, the trained DNN-A is used to separate mixed signals from the second set of the training data. The mixed signal of this set of training data is formulated as $X_{\text{tr}}^{(2)} = \sum_{i=1}^{I} S_{\text{tr}i}^{(2)}$, where $X_{\text{tr}}^{(2)}$ is the magnitude spectrogram of the mixed signal in the second set of the training data, the superscript "(2)" indicates that the second set of the training data is used in this stage. The frames of $X_{\text{tr}}^{(2)}$ are fed as inputs to DNN-A, which then produces mask $Z_{\text{tr}}^{(2)}$ which is a concatenation of masks for many sources as $Z_{\text{tr}}^{(2)} = \left[Z_{\text{tr}1}^{(2)}, \ldots, Z_{\text{tr}i}^{(2)}, \ldots, Z_{\text{tr}I}^{(2)} \right]$. The estimated masks are used to estimate the sources as follows:

$$\tilde{S}_{\text{tr}i}^{(2)} = Z_{\text{tr}i}^{(2)} \odot X_{\text{tr}}^{(2)}, \forall i \tag{6}$$

where $\odot$ denotes an element-wise multiplication. Each separated source $\tilde{S}_{\text{tr}i}^{(2)}$ often contains remaining signals from the other sources. In this work, to consider the available information of each source that appears in the other separated sources, we propose to train DNN-B to enhance all the separated sources $\tilde{S}_{\text{tr}i}^{(2)}, \forall i$ together.

DNN-B is trained using the separated signals $\tilde{S}_{\text{tr}i}^{(2)}, \forall i$ and their corresponding reference/clean signals $S_{\text{tr}i}^{(2)}, \forall i$. The input for DNN-B is the concatenation of the separated signals $U_{\text{tr}}^{(2)} = \left[\tilde{S}_{\text{tr}1}^{(2)}, \ldots, \tilde{S}_{\text{tr}i}^{(2)}, \ldots, \tilde{S}_{\text{tr}I}^{(2)} \right]$. DNN-B is trained to produce in its output layer the concatenation of the reference signals as $V_{\text{tr}}^{(2)} = \left[S_{\text{tr}1}^{(2)}, \ldots, S_{\text{tr}i}^{(2)}, \ldots, S_{\text{tr}I}^{(2)} \right]$. Each frame in $S_{\text{tr}i}^{(2)}, \forall i$ is normalized to have a unit Euclidean norm. This normalization allows us to train DNN-B to produce bounded values in its output layer without any need to train DNN-B over a wide range of values that the sources can have. Since the reference normalized signals have values between zero and one, we choose the activation functions of the output layer of DNN-B to be sigmoid functions.

DNN-B is trained to minimize the following proposed cost function:

$$C_2 = \sum_{n,f} \left(Q_{\text{tr}}^{(2)}(n,f) - V_{\text{tr}}^{(2)}(n,f) \right)^2 - \lambda \sum_{j \neq i}^{I} \sum_{n,f} \left(Q_{\text{tr}i}^{(2)}(n,f) - S_{\text{tr}j}^{(2)}(n,f) \right)^2 \tag{7}$$

where λ is a regularization parameter, $Q_{\text{tr}}^{(2)}$ is the actual output of DNN-B which is a concatenation of estimates for all sources as $Q_{\text{tr}}^{(2)} = \left[Q_{\text{tr}1}^{(2)}, \ldots, Q_{\text{tr}i}^{(2)}, \ldots, Q_{\text{tr}I}^{(2)} \right]$. The output $Q_{\text{tr}i}^{(2)}$ is the set of DNN-B output nodes that correspond to the normalized reference output $S_{\text{tr}i}^{(2)}$. The first term in the cost function in Eq. (7) minimizes the difference between the outputs of DNN-B and their corresponding reference signals. The second term of the cost function maximizes the dissimilarity/differences between DNN-B outputs of different sources, which is considered as "discriminative learning" [6,7]. The cost function in Eq. (7) aims to decrease the possibility of each set of the outputs of DNN-B from representing the other set, which helps in achieving better separation for the estimated sources. Note that, DNN-A is trained to predict masks in its output layer, while DNN-B is trained to predict normalized magnitude spectrograms for the sources. Both DNNs are trained to produce bounded values between zero and one.

3.3 Testing DNN-A and DNN-B

In the separation stage, we aim to use the trained DNNs (DNN-A and DNN-B) to separate the sources from the mixed signal. Given the magnitude spectrogram Y of the mixed signal $y(t)$. The frames of Y are fed to DNN-A to predict concatenated masks in its output layer as $\tilde{Z}_{ts} = \left[\tilde{Z}_{ts1}, \ldots, \tilde{Z}_{tsi}, \ldots, \tilde{Z}_{tsI} \right]$. The output masks are then used to compute initial estimates for the magnitude spectra of the sources as follows:

$$\tilde{S}_{tsi} = Z_{tsi} \odot Y, \forall i. \tag{8}$$

The initial estimates for the sources $\tilde{S}_{tsi}$ are usually distorted [4,20], and need to be enhanced by DNN-B. The sources can have any values but the output nodes of DNN-B are composed of sigmoid activation functions that take values between zero and one. To retain the scale information between the sources, the Euclidean norm (gain) of each frame in the spectrograms of the estimated source signals $\tilde{S}_{tsi}, \forall i$ are computed as $\alpha_{tsi} = [\alpha_{1,i}, .., \alpha_{n,i}, .., \alpha_{N,i}]$ and saved to be used later, where N is the number of frames in each source. The estimated sources are concatenated as $\tilde{S}_{ts} = \left[\tilde{S}_{ts1}, \ldots, \tilde{S}_{tsi}, \ldots, \tilde{S}_{tsI} \right]$, and then fed to DNN-B to produce a concatenation of estimates for all sources $\hat{S}_{ts} = \left[\hat{S}_{ts1}, \ldots, \hat{S}_{tsi}, \ldots, \hat{S}_{tsI} \right]$. The values of the outputs of DNN-B are between zero and one. The output of DNN-B is then used with the gains in $\alpha_{tsi}, \forall i$ to build a final mask as follows:

$$M_{tsi} = \frac{\alpha_{tsi} \otimes \hat{S}_{tsi}}{\sum_{i=1}^{I} \alpha_{tsi} \otimes \hat{S}_{tsi}} \tag{9}$$

where the division here is also element-wise and the multiplication $\alpha_{tsi} \otimes \hat{S}_{tsi}$ means that each frame n in $\hat{S}_{tsi}$ is multiplied (scaled) with its corresponding gain entry $\alpha_{n,i}$ in α_{tsi}. The scaling using α_{tsi} here helps in using DNN-B with bounded outputs between zero and one without the need to train DNN-B over all possible values of the source signals. Each $\alpha_{n,i}$ here is considered as an estimate for the scale of its corresponding frame n in source i. The final enhanced estimate for the magnitude spectrogram of each source i is computed as

$$\hat{S}_i = M_{tsi} \odot Y. \tag{10}$$

The time domain estimate for source $\hat{s}_i(t)$ is computed using the inverse STFT of $\hat{S}_i$ with the phase angle of the STFT of the mixed signal.

4   Experiments and Discussion

We applied the proposed separation and enhancement approaches to separate vocal and music signals from various songs in the dataset of SiSEC-2015-MUS-task [11]. The dataset has 100 stereo songs with different genres and instrumentations. To use the data for the proposed SCSS approach, we converted the stereo songs into mono by computing the average of the two channels for all songs and sources in the data set. We consider to separate each song into vocal signals and accompaniment signals. The accompaniment signals tend to have higher energy than the vocal signals in most of the songs in this dataset [11]. The first 35 songs were used to train DNN-A for separation as shown in Sect. 3.1. The next 35 songs were used to train DNN-B for enhancement as shown in Sect. 3.2. The remaining 30 songs were used for testing. The data was sampled at 44.1kHz. The magnitude spectrograms for the data were calculated using the STFT: a Hanning window with 2048 points length and overlap interval of 512 was used and the FFT was taken at 2048 points, the first 1025 FFT points only were used as features for the data.

For the parameters of the DNNs: For DNN-A, the number of nodes in each hidden layer was 1025 with three hidden layers. Since we separate two sources, DNN-A is trained to produce a single mask for the vocal signals $M_{\text{voc}}^{(1)}$ in its output layer and the second mask that separates the accompaniment source is computed as $M_{\text{acc}}^{(1)} = 1 - M_{\text{voc}}^{(1)}$, where 1 is a matrix of ones. Thus, the dimension of the output layer of DNN-A is 1025. For DNN-B, the number of nodes in the input and output layers is 2050 which is the length of the concatenation of the two sources 2×1025. For DNN-B, we used three hidden layers with 4100 nodes in each hidden layer. Sigmoid nonlinearity was used at each node including the output nodes for both DNNs. The parameters for the DNNs were initialized randomly. We used 200 epochs for backpropagation training for each DNN. Stochastic gradient descent was used with batch size 100 frames and learning rate 0.1. We implemented our proposed algorithms using Theano [1]. For the regularization parameter λ in Eq. (7), we tested with different values as shown in Fig. 2 below. We also show the results of using enhancement without discriminative learning where $\lambda = 0$.

We compared our proposed discriminative enhancement approach using DNN with using NMF to enhance the separated signals similar to [20]. In [20], a DNN was used to separate speech signals from different background noise signals and then NMF was used to improve the quality of the separated speech signals only. Here we modified the method in [20] to suit the application of enhancing all the separated sources. NMF uses the magnitude spectrograms of the training data in Sect. 3.2 to train basis matrices W_{tr1} and W_{tr2} for both sources as follows:

$$S_{\text{tr1}}^{(2)} \approx W_{\text{tr1}} H_{\text{tr1}} \quad \text{and} \quad S_{\text{tr2}}^{(2)} \approx W_{\text{tr2}} H_{\text{tr2}} \tag{11}$$

where H_{tr1} and H_{tr2} contain the gains of the basis vectors in W_{tr1} and W_{tr2} respectively. As in [20], we trained 80 basis vectors for each source and the generalized Kullback-Leibler divergence [8] was used as a cost function for NMF. NMF was then used to decompose the separated spectrograms $\tilde{S}_{\text{ts}i}, \forall i = 1, 2$ in Eq. (8) with the trained basis matrices W_{tr1} and W_{tr2} as follows:

$$\tilde{S}_{\text{ts1}} \approx W_{\text{tr1}} H_{\text{tst1}} \quad \text{and} \quad \tilde{S}_{\text{ts2}} \approx W_{\text{tr2}} H_{\text{tst2}} \tag{12}$$

where the gain matrices H_{tst1} and H_{tst2} contain the contribution of each trained basis vector of W_{tr1} and W_{tr2} in the mixed signal. In [20], the product $W_{\text{tr1}} H_{\text{tst1}}$

was used directly as an enhanced-separated speech signal. Here we used the product $W_{tr1}H_{tst1}$ and $W_{tr2}H_{tst2}$ to build a mask equivalent to Eq. (9) as follows:

$$M_{1_{nmf}} = \frac{W_{tr1}H_{tst1}}{W_{tr1}H_{tst1} + W_{tr2}H_{tst2}}, \text{ and } M_{2_{nmf}} = 1 - M_{1_{nmf}}. \tag{13}$$

These masks are then used to find the final estimates for the source signals as in Eq. (10).

Performance of the separation and enhancement algorithms was measured using the signal to distortion ratio (SDR), signal to interference ratio (SIR), and signal to artefact ratio (SAR) [15]. SIR indicates how well the sources are separated based on the remaining interference between the sources after separation. SAR indicates the artefacts caused by the separation algorithm to the estimated separated sources. SDR measures how distorted the separated sources are. The SDR values are usually considered as the overall performance evaluation for any source separation approach [15]. Achieving high SDR, SIR, and SAR indicates good separation performance.

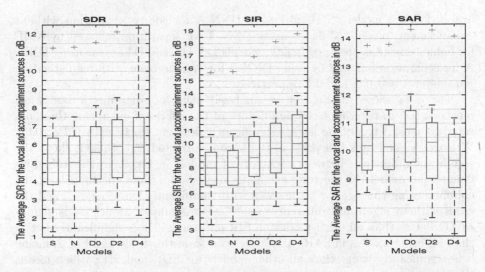

Fig. 2. The box-plot of the average SDR, SIR, and SAR of the vocal and accompaniment signals for the test set. Model "S" is for using DNN-A for source separation without enhancement. Model "N" is for using DNN-A for separation and NMF for enhancement. Models D0, D2, and D4 are for using DNN-A for separation followed by using DNN-B for enhancement with regularization parameter $\lambda = 0.0, 0.2,$ and 0.4 receptively.

The average SDR, SIR, and SAR values of the separated vocal and accompaniment signals for the 30 test songs are reported in Fig. 2. To plot this figure, the average of the vocal and accompaniment for each song was calculated as $(SDR_{voc} + SDR_{acc})/2$ for each model. The definitions of the models in Fig. 2

Table 1. The significant differences between each pair of models in Fig. 2. The signs + and − in each cell at a certain row and column mean that the model in this row is significantly better or worse respectively than the model in this column, the sign "0" means no evidence for significant differences between the models. Model S is for separation only using DNN-A without enhancement. Models N, D0, D2, and D4 are for enhancing the separated sources using NMF, DNN-B with $\lambda = 0$, DNN-B with $\lambda = 0.2$, and DNN-B with $\lambda = 0.4$ respectively.

	S	N	D0	D2	D4
D4	+	+	0	0	
D2	+	+	+		0
D0	+	+		−	0
N	+		−	−	−
S		−	−	−	−

SDR

	S	N	D0	D2	D4
D4	+	+	+	+	
D2	+	+	+		−
D0	+	+		−	−
N	+		−	−	−
S		−	−	−	−

SIR

	S	N	D0	D2	D4
D4	−	−	−	−	
D2	0	0	−		+
D0	+	+		+	+
N	0		−	0	+
S		0	−	0	+

SAR

are as follows: model "S" is for using DNN-A for source separation without enhancement; model "N" is for using DNN-A for separation followed by NMF for enhancement as proposed in [20]; models D0, D2, and D4 are for using DNN-A for separation followed by using DNN-B for enhancement with regularization parameter $\lambda = 0, 0.2$, and 0.4 respectively.

The data shown in Fig. 2 were analysed using non-parametric statistical methods [13] to determine the significance of the effects of enhancing the separated sources. A pair of models are significantly different statistically if $P < 0.05$, Wilcoxon signed-rank test [19] and Bonferroni corrected [22]. Table 1, shows the significant differences between each pair of models in Fig. 2. In this table, we denote the models in the rows as significantly better than the models in the columns using the sign "+", the cases with significantly worse as "−" and the cases without significant differences as "0". For example, Model D4 is significantly better than all other models in SIR and model D0 is significantly better than all other models in SAR. As can be seen from this table and Fig. 2, model S is significantly worse than all other models for SDR and SIR, which means there is significant improvements due to using the second stage of enhancement compared to using DNN-A only for separation without enhancement (model S). Also, we can see significant improvements in SDR, SIR and some SAR values between the proposed enhancement methods using DNNs (models D0 to D4) compared to the enhancement method in [20] using NMF (model N). This means that the proposed enhancement methods using DNN-B is significantly better than using NMF for enhancement. Model D0 achieves the highest SAR values and it is also significantly better in SDR and SIR than models S and N, which means that using DNN-B for enhancement even without discriminative learning ($\lambda = 0$) still achieves good results compared with no enhancement (S) or using NMF for enhancement (N). The regularization parameter λ in models

D0 to D4 has significant impact on the results, and can be used as a trade-off parameter between achieving high SIR values verses SAR and vice versa.

From the above analysis we can conclude that using DNN-B for enhancement improves the quality of the separated sources by decreasing the distortion (high SDR values) and interference (high SIR values) between the separated sources. Using discriminative learning for DNN-B improves the SDR and SIR results. Using DNN-B for enhancement gives better results than using NMF for most SDR, SIR, and SAR values.

The implementation of the separation and enhancement approaches in this paper is available at: http://cvssp.org/projects/maruss/discriminative/.

5   Conclusion

In this work, we proposed a new discriminative enhancement approach to enhance the separated sources after applying source separation. Discriminative enhancement was done using a deep neural network (DNN) to decrease the distortion and interference between the separated sources. To consider the interaction between the sources in the mixed signal, we proposed to enhance all the separated sources together using a single DNN. We enhanced the separated sources discriminatively by introducing a new cost function that decreases the interference between the separated sources. Our experimental results show that the proposed discriminative enhancement approach using DNN decreases the distortion and interference of the separated sources. In our future work, we will investigate the possibilities of using many stages of enhancement (multi-stages of enhancement).

Acknowledgment. This work is supported by grants EP/L027119/1 and EP/L027119/2 from the UK Engineering and Physical Sciences Research Council (EPSRC).

References

1. Bergstra, J., Breuleux, O., Bastien, F., Lamblin, P., Pascanu, R., Desjardins, G., Turian, J., Warde-Farley, D., Bengio, Y.: Theano: a CPU and GPU math expression compiler. In: Proceedings of the Python for Scientific Computing Conference (SciPy) (2010)
2. Erdogan, H., Hershey, J., Watanabe, S., Roux, J.L.: Phase-sensitive and recognition-boosted speech separation using deep recurrent neural networks. In: Proceedings of the ICASSP, pp. 708–712 (2015)
3. Grais, E.M., Erdogan, H.: Hidden Markov models as priors for regularized non-negative matrix factorization in single-channel source separation. In: Proceedings of the InterSpeech (2012)
4. Grais, E.M., Erdogan, H.: Spectro-temporal post-enhancement using MMSE estimation in NMF based single-channel source separation. In: Proceedings of the InterSpeech (2013)

5. Grais, E.M., Sen, M.U., Erdogan, H.: Deep neural networks for single channel source separation. In: Proceedings of the ICASSP, pp. 3734–3738 (2014)
6. Huang, P.S., Kim, M., Hasegawa-Johnson, M., Smaragdis, P.: Singing-Voice separation from monaural recordings using deep recurrent neural networks. In: Proceedings of the ISMIR, pp. 477–482 (2014)
7. Huang, P.S., Kim, M., Hasegawa-Johnson, M., Smaragdis, P.: Joint optimization of masks and deep recurrent neural networks for monaural source separation. IEEE/ACM Trans. Audio Speech Lang. Process. **23**(12), 2136–2147 (2015)
8. Lee, D.D., Seung, H.S.: Algorithms for non-negative matrix factorization. Adv. Neural Inf. Process. Syst. (NIPS) **13**, 556–562 (2001)
9. Narayanan, A., Wang, D.: Ideal ratio mask estimation using deep neural networks for robust speech recognition. In: Proceedings of the ICASSP, pp. 7092–7096 (2013)
10. Nugraha, A.A., Liutkus, A., Vincent, E.: Multichannel audio source separation with deep neural networks. IEEE/ACM Trans. Audio Speech Lang. Process. **24**(9), 1652–1664 (2016)
11. Ono, N., Rafii, Z., Kitamura, D., Ito, N., Liutkus, A.: The 2015 signal separation evaluation campaign. In: Proceedings of the LVA/ICA, pp. 387–395 (2015)
12. Ozerov, A., Fevotte, C., Charbit, M.: Factorial scaled hidden Markov model for polyphonic audio representation and source separation. In: Proceedings of the WASPAA, pp. 121–124 (2009)
13. Simpson, A.J.R., Roma, G., Grais, E.M., Mason, R., Hummersone, C., Liutkus, A., Plumbley, M.D.: Evaluation of audio source separation models using hypothesis-driven non-parametric statistical methods. In: Proceedings of the EUSIPCO (2016)
14. Simpson, A.J.R., Roma, G., Plumbley, M.D.: Deep Karaoke: extracting vocals from musical mixtures using a convolutional deep neural network. In: Proceedings of the LVA/ICA, pp. 429–436 (2015)
15. Vincent, E., Gribonval, R., Fevotte, C.: Performance measurement in blind audio source separation. IEEE Trans. Audio Speech Lang. Process. **14**(4), 1462–1469 (2006)
16. Vincent, P., Larochelle, H., Lajoie, I., Bengio, Y., Manzagol, P.A.: Stacked denoising autoencoders: learning useful representations in a deep network with a local denoising criterion. J. Mach. Learn. Res. **11**, 3371–3408 (2010)
17. Virtanen, T.: Monaural sound source separation by non-negative matrix factorization with temporal continuity and sparseness criteria. IEEE Trans. Audio Speech Lang. Process. **15**, 1066–1074 (2007)
18. Weninger, F., Hershey, J.R., Roux, J.L., Schuller, B.: Discriminatively trained recurrent neural networks for single-channel speech separation. In: Proceedings of the GlobalSIP, pp. 577–581 (2014)
19. Wilcoxon, F.: Individual comparisons by ranking methods. Biometrics Bull. **1**(6), 80–83 (1945)
20. Williamson, D., Wang, Y., Wang, D.: A two-stage approach for improving the perceptual quality of separated speech. In: Proceedings of the ICASSP, pp. 7034–7038 (2014)
21. Xie, J., Xu, L., Chen, E.: Image denoising and inpainting with deep neural networks. In: Advances in Neural Information Processing Systems (NIPS) (2012)
22. Hochberg, Y., Tamhane, A.C.: Multiple Comparison Procedures. Wiley, New York (1987)

Audiovisual Speech Separation Based on Independent Vector Analysis Using a Visual Voice Activity Detector

Pierre Narvor[1,2(✉)], Bertrand Rivet[1,2], and Christian Jutten[1,2]

[1] GIPSA-Lab, University of Grenoble Alpes, 38000 Grenoble, France
{pierre.narvor,bertrand.rivet,christian.jutten}@gipsa-lab.fr
[2] GIPSA-Lab, CNRS, 38000 Grenoble, France

Abstract. In this paper, we present a way of improving the Independent Vector Analysis in the context of blind separation of convolutive mixtures of speech signals. The periods of activity and inactivity of one or more speech signals are first detected using a binary visual voice activity detector based on lip movements and then fed into a modified Independent Vector Analysis algorithm to achieve the separation. Presented results show that this approach improves separation and identification of sources in a determined case with a higher convergence rate, and is also able to enhance a specific source in an underdetermined mixture.

Keywords: Audiovisual speech separation · Convolutive mixture · Blind source separation · Visual voice activity detector · Independent vector analysis · Multimodality

1 Introduction

The problem of extracting a speech signal of interest from a mixture of sounds in a natural reverberant environment is still a difficult task. This problem, well known as the cocktail-party problem [3], has been heavily investigated within the field of Convolutive Blind Source Separation (CBSS) in the past decades [7]. The Independent Vector Analysis (IVA) framework introduced in [5] and similarly in [4] has been proposed as a possible way of achieving such a separation. Indeed, the CBSS can be performed in the frequency space. Each frequency bin of the Discrete Short-Term Fourier Transform (D-STFT) of the observed signal is a linear instantaneous mixture of the D-STFT of the source signals. Therefore, the separation can be carried out at each frequency bin using Independent Component Analysis (ICA). However, because of the permutation ambiguity inherent to blind separation of signals, a random permutation between frequency bins

This work have been partly supported by the ERC project CHESS: 2012-ERC-AdG-320684.

P. Tichavský et al. (Eds.): LVA/ICA 2017, LNCS 10169, pp. 247–257, 2017.
DOI: 10.1007/978-3-319-53547-0_24

occurs during the separation process. A post-processing step is thus needed to reassociate the frequency bins to the proper sources, as in [9]. On the contrary, the IVA is able to perform a joint ICA of frequency bins, allowing to keep a coherence between those. Unfortunately, even if each estimated frequency bin is associated to the right estimated source, a global permutation indeterminacy still remains between the sources. This could be a problem in a case where there are less sources of interest than the total number of sources. Identifying the right ones might be a real challenge without further information.

In the context of separation or extraction of speech signals, these further information can be given by a video of the speaker's face. Indeed, the information carried by a video of the speaker's face is strongly related to the speech signal itself [11], but usually independent from the remaining sounds of the scene. For a recent overview of the field of audiovisual speech source separation, see [8]. In this paper, we propose to use a Visual Voice Activity Detector (V-VAD) to get the periods when the speech signal is actually active. Then, this binary information is included into an IVA algorithm to perform the separation. Presented results show that the estimated sources are associated to the right estimated activity after separation. The extraction is also faster and the quality is higher than when the estimated activity is not used. Moreover, we show that this method can also be used to enhance a specific source of interest in an underdetermined mixture. The method, designated as AV-IVA in the following, is compared to a reference IVA algorithm in which no other information than the audio is used. The method is also compared to an IVA where the actual information of speech activity is given by an oracle (O-IVA). The O-IVA gives us the highest performance bound that can be expected from this method.

Mathematical notations and mixing and separation models are defined in Sect. 2. The IVA algorithm is shortly described in Sect. 3, before a detailed presentation of our contribution. Experiments descriptions, numerical results and a discussion can be found in Sect. 4.

2   Mathematical Preliminaries

2.1   Notations

For now, only the determined case is considered. The number of sources N is the same than the number of microphones. Since the separation is processed in the frequency domain, the audio signals are represented by their D-STFT. The D-STFTs are processed over T frames of size $2(K-1)$ time samples. The signals are real in the time domain, so the first K points of the Discrete Fourier Transform (DFT) are sufficient to represent a frame in the frequency domain. Finally, a complete set of audio data is represented by a 3D array of $N \times T \times K$ complex numbers. The arrays associated to the sources, the observations, and the estimated sources are denoted $S \in \mathbb{C}^{N \times T \times K}$, $X \in \mathbb{C}^{N \times T \times K}$ and $Y \in \mathbb{C}^{N \times T \times K}$, respectively. In the latter, we will work on subsets of these arrays. An element of one array is denoted x_{ntk}. Vectors taken along the first, second and third dimensions are denoted $x_{:tk}$, $x_{n:k}$ and $x_{nt:}$, respectively. Unless specified otherwise, these vectors are considered as column vectors. The matrices $X_{n::} \in \mathbb{C}^{T \times K}$ and $X_{::k} \in \mathbb{C}^{N \times T}$ are slices of a 3D array respectively orthogonal to the first and third dimensions. $X_{n::}$ is the D-STFT of the n-th observation and $X_{::k}$ is the k-th set of frequency bins where each line is the k-th bin of the D-STFT of an observation. The determinant of a matrix is denoted det, the Hermitian transposition operator is denoted $'$ and the identity matrix is denoted I. To select the k-th element in a vector, we use a dot product with the k-th column vector of the canonical basis, denoted e_k. The probability density function (pdf) of a uni/multivariate random variable is denoted $P_{y_n}(y_{nt:}; \theta)$, with θ a set of parameters defining the pdf.

2.2 · Mixing and Separation Models

The audio mixing model is the classical mixing model of CBSS expressed in the frequency domain [7]. Each microphone picks a sum of several source contributions with an eventual additive noise. The relationship between the i-th source and its contribution to the j-th microphone is modeled as a convolution. This convolution becomes a product in the frequency domain if the analysis window is long compared to the impulse response of the filter. Under these assumptions the mixing model can be written as follows:

$$X_{::k} = A_{::k} S_{::k} \qquad \forall k = 1 \ldots K. \tag{1}$$

$A \in \mathbb{C}^{N \times N \times K}$ is a 3D array in which the DFTs of the mixing filters are stored along the third dimension. These impulse responses are $2(K-1)$ points long but their DFTs are represented by K points since those are real. The separation process which estimates the sources is represented by $W \in \mathbb{C}^{N \times N \times K}$ and is defined in a similar fashion to (1): $Y_{::k} = W_{::k} X_{::k}, \forall k = 1 \ldots K.$

3   Method

Our contribution is based on the IVA algorithm described in [2]. In this section, we first recall this method. For the sake of simplicity we shall refer to it in the following as the reference IVA algorithm. The proposed improvement is presented in the Subsect. 3.3.

3.1   Cost Function and Learning Algorithm

This algorithm is based on a Maximum Likelihood Estimation. The time samples are considered independent and identically distributed, and the sources are considered mutually independent. The cost function to minimize is then:

$$C_{IVA} = -\frac{1}{T} \log P_{X}(X) = -\frac{1}{T} \sum_{n=1}^{N} \sum_{t=1}^{T} \log P_{y_n}(y_{nt:}; \theta) - \sum_{k=1}^{K} \log |\det W_{::k}|. \tag{2}$$

The optimization is carried out with a natural gradient learning rule [1]. The derivation of the cost function is straightforward and starts with the computation of the regular gradient:

$$\frac{\partial C_{IVA}}{\partial w_{n_1 n_2 k}} = -\frac{1}{T} \sum_{t=1}^{T} \frac{\partial \log P_{y_{n_1}}(y_{n_1 t:}; \theta)'}{\partial y_{n_1 t:}} \frac{\partial x_{n_2 t:}}{\partial w_{n_1 n_2 k}} - e'_{n_2}(W_{::k}^{-1})e_{n_1}. \quad (3)$$

Denoting $\Phi \in \mathbb{C}^{N \times T \times K}$ by $\phi_{ntk} = (\partial \log P_{y_n}(y_{nt:}; \theta)/\partial y_{nt:})e_k$, the gradient can be expressed in a more compact way: $\nabla_{W_{::k}} C_{IVA} = -\frac{1}{T}\Phi_{::k}X'_{::k} - (W_{::k}^{-1})'$. The natural gradient, defined as $\tilde{\nabla}_{W_{::k}} C_{IVA} = \nabla_{W_{::k}} C_{IVA} \cdot W'_{::k}$, can then be computed as follows:

$$\tilde{\nabla}_{W_{::k}} C_{IVA} = -\frac{1}{T}\Phi_{::k} Y'_{::k} - I. \quad (4)$$

Please note that now in (4) only Φ depends on the choice of the pdf prior on the sources. The natural gradient update rule is $W_{::k}^+ = (I + \mu \tilde{\nabla}_{W_{::k}} C_{IVA}) W_{::k}$. The step size μ is fixed to 0.01.

3.2 Choice of the Source Prior

To end the derivation of this cost function, the joint pdf expressing the relationship between the frequency bins must be defined. We chose to use the zero mean multivariate Student-t distribution (5) since it has been shown to be well suited to speech signals [6].

$$P_{y_n}(y_{nt:}; \Sigma_n) \propto (1 + \frac{1}{v}y'_{nt:}\Sigma_n^{-1}y_{nt:})^{-\frac{v+K}{2}},$$

$$(5)$$

$$\frac{\partial \log P_{y_n}(y_{nt:}; \Sigma_n)}{\partial y_{nt:}} = -\frac{v+K}{2}\frac{\Sigma_n^{-1}y_{nt:}}{1 + \frac{1}{v}y'_{nt:}\Sigma_n^{-1}y_{nt:}}.$$

The covariance matrix $\Sigma_n \in \mathbb{C}^{K \times K}$ is usually taken as the identity matrix, which implies that the time samples are assumed identically distributed. In (5), K is the number of frequency bins and v is the degrees of freedom for the Multivariate Student-t distribution. In this study, v is taken equal to T, the number of time samples in the D-STFTs.

3.3 Integration of Activity Information

In the AV-IVA algorithm, we make $\Sigma_{n,t}^{AV}$ dependent on the activity of the n-th source, through the use of a visual voice activity detector described in the next paragraph. Therefore, time samples are no longer considered identically distributed. $\Sigma_{n,t}^{AV}$ is defined as $\Sigma_{n,t}^{AV} = \frac{1}{K}diag(\sigma_{nt}^2, ..., \sigma_{nt}^2)$ with:

$$\sigma_{nt}^2 = \begin{cases} 1, & \text{if the V-VAD of the } n-\text{th source is active at time } t \\ \epsilon, & \text{else} \end{cases}$$

The prior on the n-th source now models the non-stationarity of this source:

$$\frac{\partial \log P_{\boldsymbol{y}_n}(\boldsymbol{y}_{nt:}; \boldsymbol{\Sigma}^{AV}_{n,t})}{\partial \boldsymbol{y}_{nt:}} = -\frac{v+K}{2} \frac{(\boldsymbol{\Sigma}^{AV}_{n,t})^{-1} \boldsymbol{y}_{nt:}}{1 + \frac{1}{v} \boldsymbol{y}'_{nt:} (\boldsymbol{\Sigma}^{AV}_{n,t})^{-1} \boldsymbol{y}_{nt:}} \tag{6}$$

The matrix $\boldsymbol{\Sigma}^O_{n,t}$ is defined in a similar fashion, but the activity detector is ideal and given by an oracle. The matrix $\boldsymbol{\Sigma}^{IVA}_n = I$ is also defined for the reference IVA algorithm, in which all the sources are considered stationary (active at all time).

The variable $\epsilon \in\ ]0, 1]$ allows to weight the contribution of the activity detector. Its value can be chosen accordingly to the accuracy of the activity detector. The more accurate the activity detector is, the closer to zero ϵ should be. Note that if $\epsilon = 1$, the AV-IVA is behaving exactly like the reference IVA. In the evaluation, ϵ was set to 0.05.

It must be pointed out that either for the IVA, the AV-IVA or the O-IVA, the proposed $\boldsymbol{\Sigma}_{n,t}$ definition assumes that all of the frequency bins carries the same power at a given time, since all the diagonal coefficients are equal. This is not a realistic model for a speech signal. To overcome this problem, each frequency bins of the observation is normalized before processing the separation. However, the total power carried by all observations in a frequency band is saved, and after separation, the total power carried by a frequency band across by all estimated source is restored to its original value. This is done because the mean power of a specific frequency band across all observations should not be modified by the separation step.

Visual Voice Activity Detector: The audiovisual data that we used for this study are those that were used in [10]. These are composed of three time vectors. One is the speech signal itself, and width and height of the lips of the speakers during the locution. These visual features have been extracted from a video sampled at 50 Hz. The V-VAD itself is based only on the evolution of the height of the lips. The data are first smoothed using a 80 ms long mean filter and its derivative was computed. In a first step, the source activity is set to true when the absolute value of the derivative of the lips height is above a threshold. In a second step, all the detected silences shorter than 300 ms are suppressed of the estimated activity to reduce the number of miss-detection.

The performances of this V-VAD can be evaluated by comparison with an oracle activity detector (O-VAD). In this case, this oracle activity detector is simply based on audio of the signal before mixing and is a thresholded version of the power profile of the audio signal. By taking the oracle as reference, we can define a false positive error rate (the V-VAD gives true and the O-VAD gives false) and a false negative error rate (the V-VAD gives false but the O-VAD gives true). In the results presented in Sect. 4, the false negative rate is 3% and the false positive rate is 24%. Also, since the V-VAD inherently cannot detect silences shorter than 300 ms, it may be more meaningful to not take into account these short silences. In this case, the false positive rate is 16% (Fig. 1).

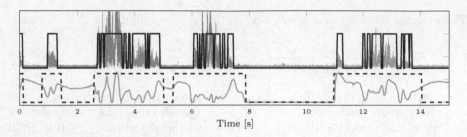

Fig. 1. Example of typical data used in this study. Top plot, Gray line: absolute value of speech signal. Black line: Oracle activity detector. Bottom plot: Gray line, height of lips. Black dashed line: Visual Voice Activity Detector

4   Experimental Results

The proposed method was evaluated over 39 trials. All of the figures presented below are averaged over these trials. At each trial, three different sources were randomly selected over a database of 77 fifteen seconds length speech samples. The mixing filters where generated at each trial from measured impulse responses found in the MARDY database [13]. These impulse responses where measured between 3 speakers and 24 different microphone positions and were truncated to 150 ms in our study. Also, only the 8 microphones the farthest of the speakers were used. The mixtures were then generated by convolution in the time domain. The audio sampling rate was 8 kHz and the STFTs were processed using 150 ms frames with an overlap of 90%. The unmixing filters were initialized to identity (i.e. $\forall k \in [1, K], \boldsymbol{W}_{::k} = I$).

The behaviour of the proposed method where evaluated over three different scenarios: a separation in a determined case where the activity of each source is estimated, an extraction in a determined case where only one source activity is estimated, and a extraction case where only one activity is estimated but in an underdetermined case where only two observation are available for three sources.

4.1   Performance Measure

The separation performance where evaluated with two criteria. The first one is the Signal to Interference Ratio (SIR) [12]. It measures the quality of extraction of a source. It is a signal to noise ratio where the signal is an estimated source, and the noise is the remaining signals of the other sources in the estimation of the first one. The higher the SIR is, the better the extraction is. In a simulation, the computation of the SIR is straight forward because we have access the separate contribution of each source to each estimated source after the separation stage. We denote $\tilde{s}_{m,n}(t)$ the contribution of the n-th source to the m-th estimated source. The definition of SIR is then:

$$SIR_m = 10 \log \frac{\sum_t \tilde{\boldsymbol{s}}_{m,m}^2(t)}{\sum_t (\sum_{n \neq m} \tilde{\boldsymbol{s}}_{m,n}(t))^2} \tag{7}$$

The second criterion is the Identification Failure Rate (IFR). It measures how often each estimated source has been associated to the rightful activity after the separation stage. In the separation case it means that the output permutation should be the same than the input permutation of the sources. The permutation computed at the performance evaluation stage is the one maximizing the mean SIR of the estimated sources. In the extraction case, only the target source has to be rightfully aligned. The IFR is the failure rate on these criteria, so the lesser it is, the better is the identification performance.

4.2 Experiments

In this experiment, the proposed method, designated AV-IVA, is compared to a standard IVA algorithm, where no visual information is used. It is also compared to an oracle IVA designated O-IVA where the ideal VAD presented in Sect. 3.3 is used. This allows to compute an upper limit on the performance of the proposed method. In the three methods, the same implementation of the IVA algorithm was used. Only the activity data carried by $\Sigma_{n,t}$ was changed.

Determined Separation Case: In this experiment the goal is to recover all of the sources. So, the displayed SIR are the SIR averaged over the three estimated sources. To compute the IFR, we consider that the output order of sources is valid only if it is the same than the order of the source before the mixing step. Results in Fig. 2a and Table 1a show that the proposed method improves both

Table 1. SIR and IFR after convergence in the determined case

Prior	SIR start (dB)	SIR end (dB)	IFR		Prior	SIR start (dB)	SIR end (dB)	IFR
AV-IVA	-7.0	14	0%		AV-IVA	-8.2	11	3%
O-IVA	-7.0	15	0%		O-IVA	-8.2	12	3%
IVA	-7.0	9.5	92%		AV-IVA_n	-6.4	8.2	43%
					O-IVA_n	-6.4	7.8	48%
(a) Separation case					(b) Extraction case			

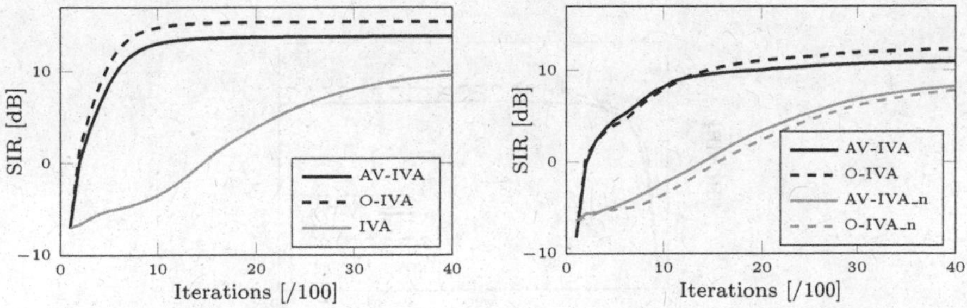

(a) Separation case.   (b) Extraction case.

Fig. 2. Determined case: Evolution of performance index during optimisation (SIR vs. number of iterations)

the separation quality and the convergence speed. The permutation error is also dramatically reduced, but such a comparison might be unfair to the IVA. The output permutation of the IVA depends only on the mixing process because no other information is given before the separation step. Also as expected, the AV-IVA performances are below the O-IVA case.

Determined Extraction Case: In this experiment, the goal is to extract a specific speech from the determined mixture. Only the activity about this source is used. The other ones are considered active at all time (i.e. $\Sigma_{n,t} = I \quad \forall n \neq 1$). To compute the IFR, we consider that only the target source should be aligned with its associated activity. Results in Fig. 2b and Table 1b show that for the target source, SIR and IFR improvements are a bit inferior to the one in the separation case, and the extraction converges slower, but the results are still better than those of the reference IVA. The SIR and IFR of the non-target sources are behaving like the ones of the reference IVA in the separation case (The IFR are inferior to the ones in the separation case, but still correspond to the one given by a random permutation between sources).

Underdetermined Extraction Case: In this section, the goal is to extract a specific speech from the mixture and only the activity information about this source is known. However, only two observations are used to perform this extraction while the mixture is composed of three sources. As shown in Fig. 3 and Table 2, the target source is better enhanced than the other sources but the overall performances are much below the ones in the determined case. This is due to the fact that the mixing matrix are not invertible in an underdetermined

Table 2. Separation results in the under-determined case.

Prior	SIR start (dB)	SIR end (dB)	IFR
AV-IVA	-8.7	3.5	8%
O-IVA	-8.7	5.7	8%
IVA	-8.7	0.6	60%

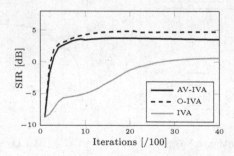

Fig. 3. Underdetermined case: Evolution of performance index during optimisation: (SIR vs. number of iterations)

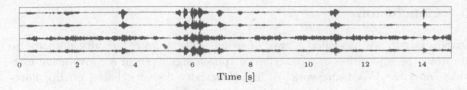

Fig. 4. Example of a speech signal enhancement using AV-IVA in an underdetermined mixture. From top to bottom: Original speech, Recording of speech alone in the reverberant environment, Mixture, Enhanced Speech.

mixture. While the target source can be enhanced, a complete extraction cannot be reached.

An example of enhancement of a speech signal in an underdetermined mixture with real reverberation in presented in Fig. 4. The mixture was composed of three sources recorded one after another by two microphones. The speaker were relocated in the room for each recording and the recordings were added afterwards to compose the mixture. For the target source, the SIR was $-1\,\text{dB}$ at initialisation, and $2.5\,\text{dB}$ after enhancement.

4.3   Discussion

Several points need to be discussed about the proposed method. The first point is about the underdetermined case. While this technique cannot be used to achieve an exact extraction of the target speech in an underdetermined mixture, it still can be used as a preprocessing step to enhance the desired signal before using another extraction technique, like frequency masking (see [7]), or before a voice recognition software.

The second point is that the V-VAD proposed in this paper is based on segmented data of the face: the lips. These might be challenging to detect in a non-controlled environment. However, only the absolute value of the derivative of the mouth height is used. Using the norm of the optical flow computed below the eyes of the speaker should gives the same results, as long as the face of the speaker is detected. An accurate segmentation of the lips is therefore not needed.

A third point is that in this study, the activity information is used in the same way across all frequency bands because the activity detector gives only a binary information about the presence of the source at a certain time. However, this methods is based on making the covariance matrix of the frequency bins of a source dependent on time. This matrix can carry much more information about the frequency structure of the signal. If the available activity detector can give more information about the frequency structure of the desired source, it can also be used by the proposed method. An example would be the musical score of an instrument we want to enhance in a stereo recording containing several instruments. However, more studies shall be necessary to confirm that last point.

5   Conclusion

In this paper, we presented a way of using a visual voice activity detector to improve the separation or extraction of speech signals from a convolutive mixture using the IVA framework. A simple activity detector based on lip movements was presented and its output was included into a natural-gradient based IVA algorithm. Presented results show that the proposed method improves the separation of speech signals in a determined mixture, accelerate the convergence of the algorithm and allows to identify the target sources. The proposed method is also able to enhance a specific source in an underdetermined mixture.

References

1. Amari, S.I., Cichocki, A., Yang, H.H.: A new learning algorithm for blind signal separation. In: Advances in Neural Information Processing Systems, pp. 757–763 (1996)
2. Anderson, M.: Independent vector analysis: theory, algorithms, and applications. Ph.D. thesis (2013)
3. Cherry, E.C.: Some experiments on the recognition of speech, with one and with two ears. J. Acoust. Soc. Am. 25(5), 975–979 (1953)
4. Hiroe, A.: Solution of permutation problem in frequency domain ICA, using multivariate probability density functions. In: Rosca, J., Erdogmus, D., Príncipe, J.C., Haykin, S. (eds.) ICA 2006. LNCS, vol. 3889, pp. 601–608. Springer, Heidelberg (2006). doi:10.1007/11679363_75
5. Kim, T., Eltoft, T., Lee, T.-W.: Independent vector analysis: an extension of ICA to multivariate components. In: Rosca, J., Erdogmus, D., Príncipe, J.C., Haykin, S. (eds.) ICA 2006. LNCS, vol. 3889, pp. 165–172. Springer, Heidelberg (2006). doi:10.1007/11679363_21
6. Liang, Y., Chen, G., Naqvi, S., Chambers, J.: Independent Vector Analysis with multivariate student's t-distribution source prior for speech separation. Electron. Lett. 49(16), 1035–1036 (2013)
7. Pedersen, M.S., Larsen, J., Kjems, U., Parra, L.C.: A survey of convolutive blind source separation methods. In: Multichannel Speech Processing Handbook, pp. 1065–1084 (2007)
8. Rivet, B., Wang, W., Naqvi, S.M., Chambers, J.A.: Audiovisual speech source separation: an overview of key methodologies. IEEE Signal Process. Mag. 31(3), 125–134 (2014)
9. Sawada, H., Mukai, R., Araki, S., Makino, S.: A robust and precise method for solving the permutation problem of frequency-domain blind source separation. IEEE Trans. Speech Audio Process. 12(5), 530–538 (2004)
10. Sodoyer, D., Rivet, B., Girin, L., Schwartz, J.L., Jutten, C.: An analysis of visual speech information applied to voice activity detection. In: 2006 IEEE International Conference on Acoustics Speech and Signal Processing Proceedings. IEEE (2006)
11. Summerfield, Q.: Use of visual information for phonetic perception. Phonetica 36(4–5), 314–331 (1979)

12. Vincent, E., Gribonval, R., Févotte, C.: Performance measurement in blind audio source separation. IEEE Trans. Audio Speech Lang. Process. **14**(4), 1462–1469 (2006)
13. Wen, J.Y., Gaubitch, N.D., Habets, E.A., Myatt, T., Naylor, P.A.: Evaluation of speech dereverberation algorithms using the MARDY database. In: Proceedings of the International Workshop on Acoustics Echo Noise Control (IWAENC). Citeseer (2006)

Monoaural Audio Source Separation Using Deep Convolutional Neural Networks

Pritish Chandna, Marius Miron, Jordi Janer, and Emilia Gómez[(⊠)]

Music Technology Group, Universitat Pompeu Fabra, Barcelona, Spain
{pritish.chandna,marius.miron,jordi.janer,emilia.gomez}@upf.edu
http://mtg.upf.edu

Abstract. In this paper we introduce a low-latency monaural source separation framework using a Convolutional Neural Network (CNN). We use a CNN to estimate time-frequency soft masks which are applied for source separation. We evaluate the performance of the neural network on a database comprising of musical mixtures of three instruments: voice, drums, bass as well as other instruments which vary from song to song. The proposed architecture is compared to a Multilayer Perceptron (MLP), achieving on-par results and a significant improvement in processing time. The algorithm was submitted to source separation evaluation campaigns to test efficiency, and achieved competitive results.

Keywords: Convolutional autoencoder · Music source separation · Deep learning · Convolutional Neural Networks · Low-latency

1 Introduction

Monoaural audio source separation has drawn the attention of many researchers in the past few years, with approaches varying from using timbre models such as those proposed by [4], to those exploiting the repetitive nature of music such as [13]. While being an interesting problem in itself, the separation of sources from a mixture can serve as a intermediary step for other tasks such as automatic speech recognition, [9] and fundamental frequency estimation, [5]. Some applications, such as speech enhancement for cochlear implant users, [7,9], require low-latency processing, which we will focus on in this paper.

Techniques using Non-Negative Matrix Factorization (NMF) have historically been the most prominent in this field, as seen in [4]. While effective, these approaches have a high processing time and are difficult to adapt for real-time applications.

Approaches directly using deep neural networks for separation have been proposed recently. A deep architecture for estimating Ideal Binary Masks (IBMs) to separate speech signals from a noisy mixture was proposed by [18]. Nugraha et al. [12] adapt deep neural networks for multichannel source separation, using both phase and magnitude information. With respect to monaural source separation, Huang et al. [8] propose a method using deep neural networks, which takes a single frame of the magnitude spectrogram of a mixture as an input feature to learn

© Springer International Publishing AG 2017
P. Tichavský et al. (Eds.): LVA/ICA 2017, LNCS 10169, pp. 258–266, 2017.
DOI: 10.1007/978-3-319-53547-0_25

single-frame timbre features for each source. Temporal evolution is then modeled using a recurrent layer. Uhlich et al. [16] propose another method which takes multiple frames of the magnitude spectrogram of a mixture as input and consists of only fully connected layers. This method models timbre features across multiple time frames. While these approaches work well, they do not exploit completely local time-frequency features. Instead, they rely on global features across the entire frequency spectrum, over a longer period of time. Convolutional neural networks (CNNs), as seen in [10], take advantage of small scale features present in data. CNNs require less memory and resources than regular fully connected neural networks, allowing for a faster, more efficient model. CNNs have recently been used by [15] for extracting vocals from a musical mixture.

CNNs have proved to be successful in image processing for tasks such as image super-resolution [3] and semantic segmentation of images as proposed by [11]. In the image processing field, CNNs take as input a two-dimensional vector of pixel intensities across the spatial dimension and exploit the local spatial correlation among input neurons to learn localized features. A similar two-dimensional representation is used in our model for audio mixtures, using the Short-Time Fourier Transform (STFT), which has frequency and time dimensions. Unlike 2D images, the STFT does not have symmetry across both axis, but a local symmetry can be found along each single axes. Therefore, the filters used in CNNs need to be adapted to the STFT representation of audio. To this end, a network architecture is proposed in Sect. 2. In Sect. 3 we evaluate the proposed model on the DSD100 dataset for the separation of four sources from a mix and compare the results with a Multilayer Perceptron architecture.

2 Proposed Framework

Figure 1 shows the block diagram for the proposed source separation framework. The STFT is computed on a segment of time context T of the mixture audio. The resulting magnitude spectrogram is then passed through the network, which outputs an estimate for each of the separated sources. The estimate is used to compute time-frequency soft masks, which are applied to the magnitude spectrogram of the mixture to compute final magnitude estimates for the separated sources. These estimates, along with the phase of the mixture, are used to obtain the audio signals corresponding to the separated sources.

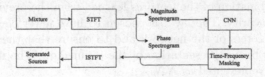

Fig. 1. Data flow

2.1   Model Architecture

State-of-the-art deep learning frameworks model source separation as a regression problem, where the network yields full resolution output for all the sources. This elicits a high numbers of parameters, which increase the processing time of the network. We take advantage of the parameter reduction property of a CNN architecture to alleviate this problem. In order to keep the multi-scale reasoning used in classification problems, we use a CNN which functions as a variation of an autoencoder architecture, as used by [14]. The network is able to learn an end-to-end model for the separated sources by finding a compressed representation for the training data. The model proposed in this paper is shown in Fig. 2. It uses a CNN with two stages, a convolution or encoding stage and the inverse operation, the deconvolution or decoding stage. We use vertical and horizontal convolutions, which have been successfully used in automatic speech recognition [1,6].

Encoding Stage. This part of the network consists of two convolution layers and a fully connected dense layer, which acts as a bottleneck to compress information.

1. Vertical Convolution Layer: This convolution layer has the shape (t_1, f_1), spanning across t_1 time frame and taking into account f_1 frequency bins. This layer tries to capture local timbre information, allowing the model to learn timbre features, similar to the approach used in NMF algorithms for source separation. These features are shared among the sources to be separated, contrary to the NMF approach, where specific basis and activation gains are derived for each source. Therefore, the timbre features learned by this layer need to be robust enough to separate the required source across songs of different genres, where the type of instruments and singers might vary. N_1 filters were used in this layer.
2. Horizontal Convolution layer: This layer models temporal evolution for different instruments from the features learned in the *Vertical Convolution Layer*. This is particularly useful for modeling time-frequency characteristics of the different instruments present in the sources to be separated. The filter shape of this layer is (t_2, f_2) and N_2 filters were used.
3. Fully Connected Layer: The output of the *Horizontal Convolution Layer* is connected to a fully connected Rectified Linear Unit (ReLU) layer which acts as a bottleneck, achieving dimensional reduction [14]. This layer consists of a non-linear combination of the features learned from the previous layers, with a ReLU non-linearity. The layer is chosen to have fewer elements to reduce the total parameters of the network and to ensure that the network is able to produce a robust representation of the input data. The number of nodes in this layer is represented as NN.

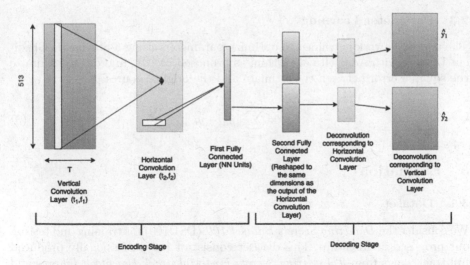

Fig. 2. Network architecture for source separation, using vertical and horizontal convolutions, showing the encoding and decoding stages

Decoding Stage. The output of the first fully connected layer is passed to another fully connected layer, with a ReLU non-linearity and the same size as the output of the second convolution layer. Thereafter, this layer is reshaped to the same dimensions as the horizontal convolution layer and passed through successive deconvolution layers, the inverse operations to the convolution stage. This approach is similar to the one proposed by [11] and is repeated to compute estimates, $\hat{y}_n$, for each of the sources, y_n.

2.2   Time-Frequency Masking

As advocated in [8,18], it is desirable to integrate the computation of a soft mask for each of the sources into the network. From the output of the network $\hat{y}_n(f)$, we can compute a soft mask, $m_n(f)$ as follows:

$$m_n(f) = \frac{|\hat{y}_n(f)|}{\sum_{n=1}^{N} |\hat{y}_n(f)|} \tag{1}$$

where $\hat{y}_n(f)$ represents the output of the network for the nth source and N is the total number of sources to be estimated.

The estimated mask is then applied to the input mixture signal to estimate the sources $\tilde{y}_n$.

$$\tilde{y}_n(f) = m_n(f)x(f) \tag{2}$$

where $x(f)$ is the spectrogram of the input mixture signal.

2.3   Parameter Learning

The neural·network is trained to optimize parameters using a Stochastic Gradient Descent with AdaDelta algorithm, as proposed by [19], in order to minimize the squared error between the estimate and the original source y_n.

$$L_{sq} := \sum_{i=1}^{N} \|\tilde{y}_n - y_n\|^2 \tag{3}$$

3   Evaluation

3.1   Dataset

We consider the *Demixing Secrets Dataset 100* (DSD100) for training and testing our proposed architecture. This dataset consist of 100 professionally produced full track songs from *The Mixing Secrets Free Multitrack Download Library* and is designed to evaluate signal source separation methods from music recordings. The dataset contains separate tracks for drums, bass, vocals and other instruments for each song in the set, present as stereo wav files with a sample rate of 44.1 KHz. The four source tracks are mixed using a professional Digital Audio Workstation to create the mixture track for each song. The dataset is divided into a dev set, used for training the network and a test set, which is used for testing the network. Both of these sets consist of 50 songs each.

3.2   Adjustments to Learning Objective

After some initial experimentation, we observed that an additional loss term, L_{diff}, representing the difference between the estimated sources, as used by [8], improved the performance of the system. In addition, we observed that, while voice, bass and drums were consistently present across songs, the other instruments varied a lot. Thus, the network was not able to efficiently learn a representation for this category as it tries to learn a general timbre class instead of particularities of the different instruments to be separated. We overcame this by modifying L_{sq} to $L_{sq'}$ and incorporating an additional loss term, L_{other}. $L_{sq'}$ excludes the source corresponding to other instruments, while L_{other} encourages differences between sources such as 'vocals' and 'other', 'bass' and 'other', and 'drums' and 'other'.

Also, we noted that the 'other' source comprised of harmonic instruments such as guitars and synths, which were similar to the 'vocals' source. To emphasize the difference between these two sources in the separation stage, a $L_{othervocals}$ loss element, which represents the difference between the estimated vocals and the other stem, was introduced.

$$L_{sq'} = \sum_{i=1}^{N-1} \|\tilde{y}_n - y_n\|^2 \tag{4}$$

$$L_{diff} = \sum_{n=1}^{N-1} \|\tilde{y}_n - \tilde{y}_{\hat{n} \neq n}\|^2 \qquad L_{othervocals} = \|\tilde{y}_1 - y_N\|^2 \qquad L_{other} = \sum_{n=2}^{N-1} \|\tilde{y}_n - y_N\|^2 \quad (5)$$

The total cost is then written as:

$$L_{total} = L_{sq'} - \alpha L_{diff} - \beta L_{other} - \beta_{vocals} L_{othervocals} \qquad (6)$$

y_1 represents the source corresponding to vocals and y_N represents that corresponding to other instruments.

3.3 Evaluation Setup

During the training phase, the input mixture and the individual sources comprising the mixture were split into 20 s segments, and the STFT for each of these segments was computed. We used a Hanning window of length 1024 samples, which, at a sampling rate of 44.1 KHz corresponds to 23 milliseconds (ms), and a hopsize of 256 samples (5.8 ms), leading to an overlap of 75 % across frames.

The frames generated from this procedure were grouped into batches of T frames, representing the maximum time context that the network tries to model. Batches were also generated using a 50 % overlap to generate more data for training. These batches were shuffled to avoid over-fitting and fed to the model for training, with 30 batches being fed at each round. Thus, each batch consists of T frames of 513 frequency bins. A complete pass over the entire set is considered as one training epoch and the network is trained for 30 epochs, an experimentally determined variable. Lasagne, a framework for neural networks built on top of Theano[1], was used for data flow and network training on a computer with GeForce GTX TITAN X GPU, Intel Core i7-5820K 3.3 GHz 6-Core Processor, X99 gaming 5 x99 ATX DDR44 motherboard.

For evaluation, the measures proposed by [17] were used. These include: *Source to Distortion Ratio* (SDR), *Source to Interference Ratio* (SIR), *Source to Artifacts Ratio* (SAR), and *Image to Spatial distortion Ratio* (ISR). These measures are averaged for overlapping 30 s frames of each song in both the development and the test set.

3.4 Experiments

The number of parameters of the network is directly proportional to the processing time required by the network. Since our aim was to design a low-latency source separation algorithm, we tried to minimize the parameters of the network, by adjusting the variables, Time context T in frames, Filter shapes (t_1, f_1) and (t_2, f_2), the number of filters, N_1 and N_2 and the number of nodes in the bottleneck, NN, while not compromising on performance. These variables were determined to be 25 (290 ms), (1, 513), (12, 1), 50, 30 and 128 respectively. For more details on these experiments, please refer to [2]. The parameters α, β and β_{vocals} were also experimentally determined to be 0.001, 0.01 and 0.03 respectively [2].

---

[1] http://lasagne.readthedocs.io/en/latest/Lasagne   and   http://deeplearning.net/ software/theano/Theano.

Table 1. Evaluation Results. Values in the table are presented in Decibels: Mean ± Standard Deviation. N represents the number of sources to be separated.

Model Details	Measure	Bass	Drums	Vocals	Others	Acc.
Multilayer Perceptron (MLP)	SDR	1.2 ± 2.7	2.0 ± 2.1	1.5 ± 2.7	1.4 ± 1.4	4.1 ± 0.9
	SIR	3.7 ± 4.2	6.5 ± 4.1	6.6 ± 3.7	4.7 ± 4.2	15.0 ± 3.7
Processing Time = 654.3 ms	SAR	7.2 ± 2.3	7.7 ± 2.6	6.8 ± 2.5	2.9 ± 2.2	14.4 ± 3.0
6617704+N × 1654426 params	ISR	11.4 ± 3.8	8.5 ± 2.4	7.8 ± 3.0	3.7 ± 1.5	6.2 ± 1.3
Convolution With Horizontal	SDR	0.9 ± 2.7	2.4 ± 2.0	1.3 ± 2.4	0.8 ± 1.5	3.7 ± 0.8
And Vertical Filters (CONV)	SIR	4.6 ± 4.4	9.1 ± 4.3	7.2 ± 3.6	3.8 ± 4.0	14.7 ± 3.5
Processing Time = 160.8 ms	SAR	6.9 ± 2.3	7.0 ± 2.8	5.3 ± 2.9	2.8 ± 2.4	14.0 ± 3.4
97698+N × 54181 params	ISR	11.5 ± 3.4	8.5 ± 2.2	7.3 ± 3.0	4.4 ± 1.7	6.1 ± 1.3

The evaluation of the CNN model and an MLP with similar training criterion is shown in Table 1, for each of the four aforementioned sources plus the accompaniment (Acc.), which refers to the entire mix minus the vocals. Moreover, in order to asses low-latency capabilities, the total number of parameters to be optimized and the processing time for a batch of T time frames for each model are also reported. The processing time reported was calculated on the CPU, without the use of the GPU.

Table 1 and Fig. 3 show that the performance of the MLP and CNN architectures was similar. However, a significant increase in the number of parameters, up to $26x$, involved in the network was observed when using an MLP architecture.

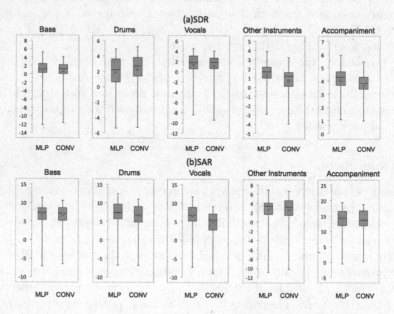

Fig. 3. Comparison between the output of the Multilayer Perceptron (MLP) and the Convolutional Neural Network (CONV) in terms of (a) SDR and (b) SAR

The processing time required by the MLP on the computer system used was $4x$ higher than the processing time required for the CNN. For an input time from of 290 ms, the CNN took just 161 ms to process, whereas the MLP took an average of 654 ms. This shows that for low-latency requirements, it is preferable to use convolution networks with horizontal and vertical layers over a simple MLP architecture.

Comparison with State-of-the-Art. The algorithm was submitted to the MIREX2016 singing voice separation task and achieved results on-par with the best algorithms in the challenge, both in terms of runtime and evaluation metrics. These results can be found at http://www.music-ir.org/mirex/wiki/2016:Singing _Voice_Separation_Results. The framework was also submitted to SiSEC 2016, for comparison with other state-of-the-art algorithms for source separation. The evaluation results for this campaign can be found at https://sisec.inria.fr/.

Sound examples of applying the model to real-world mp3 songs can be found at https://www.youtube.com/watch?v=71WwHyNaDfE, demonstrating the robustness of the model. Source code for the framework can be found on GitHub at https://github.com/MTG/DeepConvSep.

4   Conclusions and Future Work

We designed a low-latency monoaural audio source separation algorithm using a deep convolutional neural network. It was noted that the use of convolutional filters specifically designed for audio data allowed a significant gain in processing time over a simple multilayer perceptron. Dimensional reduction in the fully connected layer allows the model the learn a more compact representation of the input data from which the sources can be separated. Contrary to other approaches, which try to model both the target instrument and other background instruments, the presented algorithm solely models the target sources while the stems of the other instruments are used primarily to increase their dissimilarity with the targets.

We plan to explore the potential use of the algorithm for low-latency applications such as remixing for cochlear implants. We believe that the performance of the framework can be improved by providing additional input information such as fundamental frequency of the harmonic sources to be separated or indeed, midi information related to the various sources.

Acknowledgments. The TITANX used for this research was donated by the NVIDIA Corporation. This work is partially supported by the Spanish Ministry of Economy and Competitiveness under CASAS project (TIN2015-70816-R).

References

1. Abdel-Hamid, O., Mohamed, A.R., Jiang, H., Deng, L., Penn, G., Yu, D.: Convolutional neural networks for speech recognition. IEEE/ACM Trans. Audio Speech Lang. Process. **22**(10), 1533–1545 (2014)

2. Chandna, P.: Audio source separation using deep neural networks, Master Thesis, Universitat Pompeu Fabra (2016)
3. Dong, C., Loy, C.C., He, K., Tang, X.: Image super-resolution using deep convolutional networks. CoRR, abs/1501.00092 (2015)
4. Durrieu, J., Ozerov, A., Févotte, C.: Main instrument separation from stereophonic audio signals using a source/filter model. In: 17th European Signal Processing Conference (2009)
5. Gómez, E., Cañadas, F., Salamon, J., Bonada, J., Vera, P., Cabañas, P.: Predominant fundamental frequency estimation vs singing voice separation for the automatic transcription of accompanied flamenco singing. In: 13th International Society for Music Information Retrieval Conference (ISMIR 2012) (2012)
6. Han, Y., Lee, K.: Acoustic scene classification using convolutional neural network and multiple-width frequency-delta data augmentation (2016)
7. Hidalgo, J.: Low latency audio source separation for speech enhancement in cochlear implants, Master Thesis, Universitat Pompeu Fabra (2012)
8. Huang, P.-S., Kim, M., Hasegawa-Johnson, M., Smaragdis, P.: Deep learning for monaural speech separation. In: Acoustics, Speech and Signal Processing (ICASSP), pp. 1562–1566 (2014)
9. Kokkinakis, K., Loizou, P.C.: Using blind source separation techniques to improve speech recognition in bilateral cochlear implant patients. J. Acoust. Soc. Am. 123(4), 2379–2390 (2008)
10. Krizhevsky, A., Sutskever, I., Hinton, G.E.: ImageNet classification with deep convolutional neural networks. In: Advances in Neural Information Processing Systems, pp. 1097–1105 (2012)
11. Noh, H., Hong, S., Han, B.: Learning deconvolution network for semantic segmentation. CoRR, abs/1505.04366 (2015)
12. Nugraha, A.A., Liutkus, A., Vincent, E.: Multichannel audio source separation with deep neural networks. Technical report (2016)
13. Rafii, Z., Liutkus, A., Pardo, B.: REPET for background/foreground separation in audio. In: Naik, G.R., Wang, W. (eds.) Blind Source Separation, pp. 395–411. Springer, Heidelberg (2014)
14. Sainath, T.N., Kingsbury, B., Ramabhadran, B.: Auto-encoder bottleneck features using deep belief networks. In: 2012 IEEE International Conference on Acoustics, Speech and Signal Processing (ICASSP), pp. 4153–4156. IEEE (2012)
15. Simpson, A.J.R.: Probabilistic Binary-Mask Cocktail-Party Source Separation in a Convolutional Deep Neural Network (2015)
16. Uhlich, S., Giron, F., Mitsufuji, Y.: Deep neural network based instrument extraction from music. In: 2015 IEEE International Conference on Acoustics, Speech and Signal Processing (ICASSP), pp. 2135–2139. IEEE (2015)
17. Vincent, E., Gribonval, R., Fevotte, C.: Performance measurement in blind audio source separation. IEEE Trans. Audio Speech Lang. Process. 14(4), 1462–1469 (2006)
18. Wang, Y., Narayanan, A., Wang, D.: On training targets for supervised speech separation. IEEE/ACM Trans. Audio Speech Lang. Process. 22(12), 1849–1858 (2014)
19. Zeiler, M.D.: Adadelta: an adaptive learning rate method. arXiv preprint arXiv:1212.5701

Theoretical Developments

On the Behaviour of the Estimated Fourth-Order Cumulants Matrix of a High-Dimensional Gaussian White Noise

Pierre Gouédard and Philippe Loubaton[✉]

Université Paris-Est Marne la Vallée, Laboratoire d'Informatique Gaspard Monge,
UMR CNRS 8049, 5 Bd. Descartes, Cité Descartes,
77454 Marne la Vallée Cedex 2, France
pierre.mgouedard@gmail.com, loubaton@univ-mlv.fr

Abstract. This paper is devoted to the study of the traditional estimator of the fourth-order cumulants matrix of a high-dimensional multivariate Gaussian white noise. If M represents the dimension of the noise and N the number of available observations, it is first established that this $M^2 \times M^2$ matrix converges towards 0 in the spectral norm sens provided $\frac{M^2 \log N}{N} \to 0$. The behaviour of the estimated fourth-order cumulants matrix is then evaluated in the asymptotic regime where M and N converge towards $+\infty$ in such a way that $\frac{M^2}{N}$ converges towards a constant. In this context, it is proved that the matrix does not converge towards 0 in the spectral norm sense, and that its empirical eigenvalue distribution converges towards a shifted Marcenko-Pastur distribution. It is finally claimed that the largest and the smallest eigenvalue of the cumulant matrix converges almost surely towards the rightend and the leftend points of the support of the Marcenko-Pastur distribution.

Keywords: Estimated joint fourth-order cumulants matrices · Large random matrices · Blind source separation in the high-dimensional context

1 Introduction

It is now well understood that the statistical signal processing of high-dimensional signals poses a number of new problems which stimulated the development of appropriate new tools, e.g. large random matrices or approaches exploiting sparsity. Statistical methods based on the use of the empirical covariance matrix of a M-dimensional time series $(\mathbf{y}_n)_{n \in \mathbb{Z}}$ (e.g. detection of noisy low rank signals, estimation of direction of arrival using subspace methods,...) provide a number of convincing examples illustrating this point. In effect, when the dimension M of the observation is large, it is very often difficult to collect a number N of observations much larger than M, so that in practice M

This work was supported by project ANR-12-MONU-OOO3 DIONISOS.

P. Tichavský et al. (Eds.): LVA/ICA 2017, LNCS 10169, pp. 269–278, 2017.
DOI: 10.1007/978-3-319-53547-0_26

and N appear to be of the same order of magnitude. In this context, it is well established that the empirical covariance matrix $\hat{\mathbf{R}}_N = \frac{1}{N}\sum_{n=1}^{N} \mathbf{y}_n\mathbf{y}_n^*$ is a poor estimator of $\mathbf{R} = \mathbb{E}(\mathbf{y}_n\mathbf{y}_n^*)$. When no a priori information on $\mathbf{R}$ (e.g. various kinds of sparsity, see e.g. [3]) is available, it appears that large random matrix theory provides useful informations on the behaviour of $\hat{\mathbf{R}}_N$ in the asymptotic regime where M and N converge towards $+\infty$ at the same rate. This allows to analyse the behaviour of the standard statistical inference algorithms based on the assumption that $\hat{\mathbf{R}}_N \simeq \mathbf{R}$, and more importantly to propose modifications which allow to improve the performance (see e.g. [2, 6–8, 14]).

The present paper is motivated by the blind source separation of an instantaneous mixture of K independent sources in the case where the observed signal $(\mathbf{y}_n)_{n\in\mathbb{Z}}$ is high-dimensional and where the additive noise is Gaussian with unknown statistics. Popular approaches developed and evaluated in the low dimensional observation context use the particular structure of the fourth-order cumulants tensor of the observed signal which appears as the sum of K rank 1 tensors generated by the columns of the mixing matrix. Under mild assumptions, the column vectors of the mixing matrix can be identified from the eigenvalue/eigenvector decomposition of the $M^2 \times M^2$ fourth-order cumulants matrix (see e.g. the algorithm ICAR in [1]). In practice, the fourth-order cumulants matrix has to be estimated from the N available M–dimensional observations $\mathbf{y}_1,\ldots,\mathbf{y}_N$, and the presence of the additive Gaussian noise has of course an influence on the eigenstructure of the estimated fourth-order cumulants matrix and thus on the statistical performance of the estimator of the mixing matrix. When the dimension of the observation M is much smaller than the sample size N, standard large sample analysis conducted in the regime M fixed and $N \to +\infty$ can be used in order to prove that the estimated fourth-order cumulants matrix converges towards the true cumulants matrix in the spectral norm sense, a property that immediately implies the consistency of the mixing matrix estimates. When M is large, the above regime may not be relevant, and asymptotic regimes for which both M and N converge towards $+\infty$ at possibly different rates may produce more reliable results. In this context, a crucial issue is to determine the rates of convergence of M and N towards $+\infty$ for which the estimated fourth-order cumulants matrix still converges towards the true cumulant matrix in the spectral norm sense. When these conditions are not met, the traditional estimates of the mixing matrix are non consistent, but it may be useful to characterize the properties of the estimated fourth-order cumulants matrix in order to be able to derive improved performance estimates. This research program appears highly non trivial, and needs to develop a number of new large random matrix tools. In this paper, we thus consider the preliminary problem of characterizing the behaviour of the estimated fourth-order cumulants matrix in the absence of source when both M and N converge towards $+\infty$. We do not claim that the results of this paper can be used as is in order to analyse the behaviour of blind source separation algorithms in the high-dimensional case. However, the study of this simpler problem will provide a number of useful insights to address the more complicated scenario in which sources are present.

This paper is organized as follows. In Sect. 2, we present more precisely the addressed problem. In Sect. 3, we study the conditions on M and N under which the estimated fourth-order cumulants matrix of the M-dimensional white Gaussian noise sequence $(\mathbf{y}_n)_{n \in \mathbb{Z}}$ converges towards 0 in the spectral norm sense. We show that this is the case as soon as $\frac{M^2 \log N}{N} \to 0$, a condition which is close from $\frac{M^2}{N} \to 0$. In Sect. 4, we consider the regime in which $\frac{M^2}{N}$ converges towards a non zero constant, and prove that the estimated fourth-order cumulants matrix does not converge towards 0 in the spectral norm sense. In particular, we establish that its empirical eigenvalue distribution converges towards a shifted Marcenko-Pastur (MP) distribution, and that its smallest and largest eigenvalues converge towards the leftend and rightend points of the support of the MP distribution. This result shows that parameter $\frac{M^2}{N}$ controls the spreading of the eigenvalues, thus confirming the evaluations of Sect. 3. Some numerical experiments illustrating these results are also provided.

General notations. $(\mathbf{e}_i)_{i=1,\ldots,M}$ represents the canonical basis of $\mathbb{C}^M$. If $\mathbf{x}$ is an element of $\mathbb{C}^{M^2}$, and if $(k, i) \in \{1, 2, \ldots, M\}$, we denote its $k + i(M - 1)$-th component as $\mathbf{x}_{i,k}$. If $\mathbf{A}$ and $\mathbf{B}$ are 2 matrices, $\mathbf{A} \otimes \mathbf{B}$ represents the block matrix whose blocks are the $\mathbf{A}_{i,j}\mathbf{B}$. If $\mathbf{A}$ is a $M^2 \times M^2$ matrix, we denote by $\mathbf{A}_{i,k,j,l}$ the element $k + i(M - 1), l + M(j - 1)$ of matrix $\mathbf{A}$. In other words, matrix $\mathbf{A}$ can be written as

$$\mathbf{A} = \sum_{i,j,k,l=1,\ldots,M} \mathbf{A}_{i,k,j,l} \ (\mathbf{e}_i \otimes \mathbf{e}_k) \, (\mathbf{e}_j \otimes \mathbf{e}_l)^*$$

In the following, we denote by $\boldsymbol{\Pi}$ the $M^2 \times M^2$ matrix defined by $(\boldsymbol{\Pi}\mathbf{x})_{i,j} = \mathbf{x}_{j,i}$ for each element $\mathbf{x}$ of $\mathbb{C}^{M^2}$ and for each pair (i, j). It is clear that for each pair $(\mathbf{x}_1, \mathbf{x}_2)$ of $\mathbb{C}^M$, then it holds that $\boldsymbol{\Pi}(\mathbf{x}_1 \otimes \mathbf{x}_2) = \mathbf{x}_2 \otimes \mathbf{x}_1$. It is easily seen that matrix $\frac{1}{2}(\mathbf{I} + \boldsymbol{\Pi})$ can be expressed as

$$\frac{1}{2}(\mathbf{I} + \boldsymbol{\Pi}) = \boldsymbol{\Gamma}\boldsymbol{\Gamma}^*$$

where $\boldsymbol{\Gamma}$ is the $M^2 \times M(M + 1)/2$ matrix whose columns $\boldsymbol{\Gamma}_{i,j}$, $1 \leq i \leq j \leq M$ are $\boldsymbol{\Gamma}_{i,i} = \mathbf{e}_i \otimes \mathbf{e}_i$ and, for $i < j$,

$$\boldsymbol{\Gamma}_{i,j} = \frac{1}{\sqrt{2}} \left(\mathbf{e}_i \otimes \mathbf{e}_j + \mathbf{e}_i \otimes \mathbf{e}_j \right)$$

Matrix $\boldsymbol{\Gamma}$ verifies $\boldsymbol{\Gamma}^*\boldsymbol{\Gamma} = \mathbf{I}_{M(M+1)/2}$. Therefore, for each pair $(\mathbf{x}_1, \mathbf{x}_2)$ of $\mathbb{C}^M$, it holds that

$$\boldsymbol{\Gamma}\boldsymbol{\Gamma}^*(\mathbf{x}_1 \otimes \mathbf{x}_2) = \frac{1}{2} (\mathbf{x}_1 \otimes \mathbf{x}_2 + \mathbf{x}_2 \otimes \mathbf{x}_1) \tag{1}$$

2   Presentation of the Problem

We consider a sequence $(\mathbf{y}_n)_{n=1,\ldots,N}$ of complex Gaussian i.i.d. $\mathcal{N}_c(0, \sigma^2 \mathbf{I}_M)$ random vectors. The fourth order joint cumulants of vectors $\mathbf{y}_n$ are of course identically 0, and we study in this paper the behaviour of $M^2 \times M^2$ matrix $\hat{\mathbf{C}}_N$ whose entries $(\hat{\mathbf{C}}_{i,k,j,l})_{1 \leq i,k,j,l \leq M}$ coincide with the traditional empirical estimate of the joint cumulant $c_4(\mathbf{y}_{i,n}, \mathbf{y}_{k,n}, \mathbf{y}_{j,n}^*, \mathbf{y}_{l,n}^*)$, i.e.

$$\hat{\mathbf{C}}_{i,k,j,l} = \frac{1}{N} \sum_{n=1}^{N} \mathbf{y}_{i,n} \mathbf{y}_{k,n} \mathbf{y}_{j,n}^* \mathbf{y}_{l,n}^* - \hat{\mathbf{R}}_{i,j} \hat{\mathbf{R}}_{k,l} - \hat{\mathbf{R}}_{i,l} \hat{\mathbf{R}}_{k,j}$$

where we recall that $M \times M$ matrix $\hat{\mathbf{R}}_N$ represents the empirical covariance matrix $\hat{\mathbf{R}}_N = \frac{1}{N} \sum_{n=1}^{N} \mathbf{y}_n \mathbf{y}_n^*$. Using Eq. (1), it is easily seen that

$$\hat{\mathbf{R}}_{i,j} \hat{\mathbf{R}}_{k,l} + \hat{\mathbf{R}}_{i,l} \hat{\mathbf{R}}_{k,j} = \left(2\, \boldsymbol{\Gamma} \boldsymbol{\Gamma}^* \left(\hat{\mathbf{R}}_N \otimes \hat{\mathbf{R}}_N \right) \boldsymbol{\Gamma} \boldsymbol{\Gamma}^* \right)_{i,k,j,l}$$

Therefore, matrix $\hat{\mathbf{C}}_N$ can be written as

$$\hat{\mathbf{C}}_N = \hat{\mathbf{D}}_N - 2\, \boldsymbol{\Gamma} \boldsymbol{\Gamma}^* \left(\hat{\mathbf{R}}_N \otimes \hat{\mathbf{R}}_N \right) \boldsymbol{\Gamma} \boldsymbol{\Gamma}^* \tag{2}$$

where matrix $\hat{\mathbf{D}}_N$ is defined by

$$\hat{\mathbf{D}}_N = \frac{1}{N} \sum_{n=1}^{N} (\mathbf{y}_n \otimes \mathbf{y}_n)(\mathbf{y}_n \otimes \mathbf{y}_n)^* \tag{3}$$

We remark that for each n, $\boldsymbol{\Gamma} \boldsymbol{\Gamma}^*(\mathbf{y}_n \otimes \mathbf{y}_n)$ coincides with $\mathbf{y}_n \otimes \mathbf{y}_n$ so that the range of $\hat{\mathbf{D}}_N$ is included in the column space of $\boldsymbol{\Gamma} \boldsymbol{\Gamma}^*$. Therefore, matrix $\hat{\mathbf{C}}_N$ is rank deficient, and its range is included into the $M(M+1)/2$–dimensional space generated by the columns of $\boldsymbol{\Gamma}$. This point will be used in Sect. 4 below.

It is clear that if $N \to +\infty$ while M remains fixed, the law of large number implies that each element of $\hat{\mathbf{C}}_N$ converges towards 0. This implies $\|\hat{\mathbf{C}}_N\| \to 0$ because the size of matrix $\hat{\mathbf{C}}_N$ does not scale with N. When both M and N converge towards $+\infty$, the convergence of the individual entries of $\hat{\mathbf{C}}_N$ towards 0 no longer implies the convergence of $\|\hat{\mathbf{C}}_N\|$ towards 0. In the two following sections, we provide some results concerning the asymptotic behaviour of $\hat{\mathbf{C}}_N$ and $\|\hat{\mathbf{C}}_N\|$ when M and N converge to $+\infty$.

3   Conditions Under Which $\|\hat{\mathbf{C}}_N\| \to 0$

In this section, we derive conditions on M and N under which $\|\hat{\mathbf{C}}_N\| \to 0$. For this, we prove the following proposition.

Proposition 1. *Assume that M and N both converge towards $+\infty$ in such a way that*

$$\frac{M^2 \log N}{N} \to 0 \tag{4}$$

Then it holds that

$$\left\| \hat{\mathbf{D}}_N - 2\sigma^4 \boldsymbol{\Gamma}\boldsymbol{\Gamma}^* \right\| \to 0 \ a.s. \tag{5}$$

and that

$$\|\hat{\mathbf{C}}_N\| \to 0 \tag{6}$$

Moreover, almost surely, for N large enough, $\left\| \hat{\mathbf{D}}_N - 2\sigma^4 \boldsymbol{\Gamma}\boldsymbol{\Gamma}^* \right\|$ and $\|\hat{\mathbf{C}}_N\|$ are less that $\mu \left(\frac{\log N}{N} \right)^{1/2} (M + \log N)$ for some positive constant μ.

We just provide a sketch of the proof of Proposition 1. We first mention that $\mathbb{E}\left((\mathbf{y}_n \otimes \mathbf{y}_n)(\mathbf{y}_n \otimes \mathbf{y}_n)^* \right) = \mathbb{E}\left(\hat{\mathbf{D}}_N \right) = 2\sigma^4 \boldsymbol{\Gamma}\boldsymbol{\Gamma}^*$. We now provide some insights on (5). For this, we recall the matrix Bernstein inequality (see Theorem 6.6.1 in [13]).

Theorem 1. *Let $(\mathbf{S}_n)_{n=1,\dots,N}$ be a sequence of Hermitian zero mean i.i.d. $m \times m$ random matrices satisfying $\sup_{n=1,\dots,N} \|\mathbf{S}_n\| \le \kappa_N$ for some deterministic constant κ_N. If $\overline{\mathbf{S}}_N$ denotes matrix $\sum_{n=1}^N \mathbf{S}_n$, and if $v_N = \mathbb{E}\left(\overline{\mathbf{S}}_N \right)^2$, then, for each $\epsilon > 0$, it holds that*

$$\mathbb{P}\left(\|\overline{\mathbf{S}}_N\| > \epsilon \right) \le m \exp\left(\frac{-\epsilon^2}{2v_N + \kappa_N \epsilon/3} \right). \tag{7}$$

We denote by $\mathbf{R}_n$ matrix

$$\mathbf{R}_n = \frac{1}{N} \left((\mathbf{y}_n \otimes \mathbf{y}_n)(\mathbf{y}_n \otimes \mathbf{y}_n)^* - \mathbb{E}(\mathbf{y}_n \otimes \mathbf{y}_n)(\mathbf{y}_n \otimes \mathbf{y}_n)^* \right) \tag{8}$$

We have thus to establish that if $\overline{\mathbf{R}}_N = \sum_{n=1}^N \mathbf{R}_n$, then $\|\overline{\mathbf{R}}_N\| \to 0$ almost surely. For this, we prove that if $\epsilon = \mu \left(\frac{\log N}{N} \right)^{1/2} (M + \log N)$, it exists a constant μ for which

$$\sum_{N=1}^{+\infty} \mathbb{P}(\|\overline{\mathbf{R}}_N\| > \epsilon_N) < +\infty \tag{9}$$

If (9) holds, Borel-Cantelli's lemma implies that almost surely, for N large enough, $\|\overline{\mathbf{R}}_N\| \le \epsilon_N$, which leads to the conclusion that $\|\overline{\mathbf{R}}_N\|$ converges towards 0 at rate ϵ_N. Bernstein inequality could be used to evaluate an upper bound of $\mathbb{P}(\|\overline{\mathbf{R}}_N\| > \epsilon_N)$ if the norms of matrices $(\mathbf{R}_n)_{n=1,\dots,N}$ were bounded everywhere by a constant κ_N. As this property does not hold, it is possible to use a classical truncation argument (see e.g. [12]), and consider matrices $(\mathbf{S}_n)_{n=1,\dots,N}$ (resp. $(\mathbf{T}_n)_{n=1,\dots,N}$) defined in the same way than $(\mathbf{R}_n)_{n=1,\dots,N}$ in (8), but when $\mathbf{y}_n$ is replaced by $\mathbf{y}_n \mathbb{1}_{(\|\mathbf{y}_n\|^2 \le \alpha_N)}$ (resp. by $\mathbf{y}_n \mathbb{1}_{(\|\mathbf{y}_n\|^2 > \alpha_N)}$) where α_N is a well chosen deterministic constant. Due to the lack of space, we do not provide more details.

It remains to justify that (5) implies (6). For this, we recall that $\boldsymbol{\Gamma}^*\boldsymbol{\Gamma} = \mathbf{I}$, and express $\hat{\mathbf{C}}_N$ as

$$\hat{\mathbf{C}}_N = \hat{\mathbf{D}}_N - 2\sigma^4\boldsymbol{\Gamma}\boldsymbol{\Gamma}^* + 2\boldsymbol{\Gamma}\boldsymbol{\Gamma}^*\left(\sigma^4\mathbf{I} - \hat{\mathbf{R}}_N \otimes \hat{\mathbf{R}}_N\right)\boldsymbol{\Gamma}\boldsymbol{\Gamma}^* \tag{10}$$

We remark that (4) implies that $\frac{M\log N}{N} \to 0$. Using the same approach as above, it is easily seen that $\frac{M\log N}{N} \to 0$ implies that $\|\hat{\mathbf{R}}_N - \sigma^2\mathbf{I}_M\| \to 0$, and that, almost surely, for N large enough, $\|\hat{\mathbf{R}}_N - \sigma^2\mathbf{I}_M\|$ is less $\mu\left(\frac{\log N}{N}\right)^{1/2}(M + \log N)^{1/2}$ for some positive constant μ. Using this, it is easy to check that $\|\hat{\mathbf{R}}_N \otimes \hat{\mathbf{R}}_N - \sigma^4\mathbf{I}\|$ converges towards 0 at the same rate that $\|\hat{\mathbf{R}}_N - \sigma^2\mathbf{I}_M\|$. (10) and the triangular inequality completes the proof.

We now comment Proposition (1). We first remark that, up to the term $\log N$, condition (4) is optimal because, as shown below in Sect. 4, if $\frac{M^2}{N}$ converges towards a non zero constant, then (6) does not hold. Next, it is easy to verify that Proposition 1 holds if $(\mathbf{y}_n)_{n=1,\dots,N}$ is a temporally white, but spatially correlated Gaussian noise: in this case, $\mathbf{y}_n$ can be written as $\mathbf{y}_n = \mathbf{R}^{1/2}\mathbf{v}_n$ where $\mathbf{R}$ represents the covariance matrix of vectors $(\mathbf{y}_n)_{n=1,\dots,N}$, and where $(\mathbf{v}_n)_{n=1,\dots,N}$ is a temporally and spatially white Gaussian noise. Using Proposition 1 to sequence $(\mathbf{v}_n)_{n=1,\dots,N}$ leads immediately to the conclusion that the estimated fourth-order cumulants matrix of $(\mathbf{y}_n)_{n=1,\dots,N}$ converges towards 0 provided the spectral norm of $\mathbf{R}$ remains bounded when $N \to +\infty$. It would of course be useful to evaluate the behaviour of the estimated fourth-order cumulants matrix in the presence of a linear mixing of K independent signals. We feel that the use of similar tools should lead to the conclusion that $\hat{\mathbf{C}}_N$ converges towards the true fourth-order cumulants matrix provided $\frac{M^2\log N}{N} \to 0$. This condition has to be compared with the condition $\frac{M\log N}{N} \to 0$ which ensures that the empirical covariance matrix converges towards the true covariance matrix in the spectral norm sense: estimating the true fourth-order cumulants matrix necessitates a much larger number of samples. We also conjecture that the consistent estimation of the $2p$-order cumulant matrix would need that $\frac{M^p\log N}{N} \to 0$.

4 Study of the Case Where N and M^2 Are of the Same Order of Magnitude

We now consider the asymptotic regime M, N converge towards $+\infty$ in such a way that $\frac{M^2}{N}$ converges towards a constant. In the following, we denote by L the integer $L = \frac{M(M+1)}{2}$, and define c_N as $c_N = \frac{L}{N}$. In the present asymptotic regime, c_N converges towards $c_* > 0$. In order to simplify the presentation of the following results, we assume that $c_* \leq 1$.

We first recall that the empirical eigenvalue distribution of an hermitian $m \times m$ random matrix $\mathbf{B}$ is the random probability distribution $\frac{1}{m}\sum_{k=1}^{m}\delta_{\lambda_k}$ where $(\lambda_k)_{k=1,\dots,m}$ are the eigenvalues of $\mathbf{B}$, and where δ_λ represents the Dirac distribution at point λ. It is well known that under certain assumptions on $\mathbf{B}$, its empirical eigenvalue distribution converges weakly almost surely towards a deterministic probability distribution when $m \to +\infty$. Unformally, this means that the histogram of the eigenvalues of a realization of $\mathbf{B}$ tends to accumulate around the graph of a deterministic probability distribution when m increases. For example, if $\mathbf{B} = \frac{1}{p}\mathbf{A}\mathbf{A}^*$ where $\mathbf{A}$ is a $m \times p$ random matrix with zero mean and variance σ^2 i.i.d. entries, when m and p both converge towards $+\infty$ in a such a way that $d_m = \frac{m}{p}$ converges towards $d_* > 0$, the empirical eigenvalue distribution of $\mathbf{B}$ converges towards the so-called Marcenko-Pastur distribution with parameter (σ^2, d_*) denoted $\mathrm{MP}(\sigma^2, d_*)$ in the following (see e.g. [10] or the tutorial [4]). If $d_* \leq 1$, distribution $\mathrm{MP}(\sigma^2, d_*)$ is absolutely continuous, and is supported by interval $[\sigma^2(1-\sqrt{d_*})^2, \sigma^2(1+\sqrt{d_*})^2]$. It is important to remark that in the context of this particular random matrix model, the asymptotic behaviour of the measure $\frac{1}{m}\sum_{k=1}^{m}\delta_{\lambda_k}$ does not depend on the probability distribution of the entries of $\mathbf{A}$.

In the regime considered in this section, we will see that the norm of $\hat{\mathbf{C}}_N$ does not converge towards 0. Therefore, the eigenvalues of $\hat{\mathbf{C}}_N$ do not concentrate around 0, and we propose to evaluate the behaviour of the empirical eigenvalue distribution of $\hat{\mathbf{C}}_N$ to understand how $\hat{\mathbf{C}}_N$ deviates from the null matrix. For this, we notice that in the present asymptotic regime, it holds that $\frac{M\log N}{N} \to 0$. Therefore, matrix $\hat{\mathbf{R}}_N$ converges towards $\sigma^2\mathbf{I}_M$ in the spectral norm sense. Consequently, matrix $\hat{\mathbf{C}}_N$ has the same behaviour than matrix $\hat{\mathbf{D}}_N - 2\sigma^4\boldsymbol{\Gamma}\boldsymbol{\Gamma}^*$ in the sense that

$$\left\| \hat{\mathbf{C}}_N - (\hat{\mathbf{D}}_N - 2\sigma^4\boldsymbol{\Gamma}\boldsymbol{\Gamma}^*) \right\| \to 0 \, a.s. \tag{11}$$

Therefore, the eigenvalues of $\hat{\mathbf{C}}_N$ behave like the eigenvalues of $\hat{\mathbf{D}}_N - 2\sigma^4\boldsymbol{\Gamma}\boldsymbol{\Gamma}^*$. We thus study the empirical eigenvalue distribution of the latter matrix. For this, we recall that the range of $\hat{\mathbf{D}}_N$ coincides with the range of $\boldsymbol{\Gamma}$. Hence, it holds that

$$\boldsymbol{\Gamma}\boldsymbol{\Gamma}^*\hat{\mathbf{D}}_N\boldsymbol{\Gamma}\boldsymbol{\Gamma}^* = \hat{\mathbf{D}}_N$$

and matrix $\hat{\mathbf{D}}_N - 2\sigma^4\boldsymbol{\Gamma}\boldsymbol{\Gamma}^*$ can be written as

$$\hat{\mathbf{D}}_N - 2\sigma^4\boldsymbol{\Gamma}\boldsymbol{\Gamma}^* = \boldsymbol{\Gamma}\left(\boldsymbol{\Sigma}_N\boldsymbol{\Sigma}_N^* - 2\sigma^4\mathbf{I}_L\right)\boldsymbol{\Gamma}^* \tag{12}$$

where $\boldsymbol{\Sigma}_N$ is the $L \times N$ random matrix defined by

$$\boldsymbol{\Sigma}_N = \frac{1}{\sqrt{N}}\boldsymbol{\Gamma}^* (\mathbf{y}_1 \otimes \mathbf{y}_1, \ldots, \mathbf{y}_N \otimes \mathbf{y}_N)$$

The eigenvalues of $\hat{\mathbf{C}}_N$ are thus 0 with multiplicity $M^2 - L = M(M-1)/2$ as well as the eigenvalues of matrix $\boldsymbol{\Sigma}_N\boldsymbol{\Sigma}_N^* - 2\sigma^4\mathbf{I}_L$. In order to evaluate the asymptotic behaviour of the eigenvalue distribution of $\hat{\mathbf{C}}_N$, it is thus sufficient to study the empirical eigenvalue distribution of matrix $\boldsymbol{\Sigma}_N\boldsymbol{\Sigma}_N^*$. We denote by $\xi_1, \ldots, \xi_N$ the columns of $\boldsymbol{\Sigma}_N$. It is clear that vectors $(\xi_n)_{n=1,\ldots,N}$ are independent and identically distributed, that $\mathbb{E}(\xi_n\xi_n^*) = \frac{2\sigma^4}{N}\mathbf{I}_L$, but that for each n, the entries of vector ξ_n are of course not independent. However, the behaviour of the eigenvalues of $\boldsymbol{\Sigma}_N \boldsymbol{\Sigma}_N^*$ behave as if the entries of $\boldsymbol{\Sigma}_N$ were i.i.d. More precisely, the following result holds.

Theorem 2. *The empirical eigenvalue distribution of $\boldsymbol{\Sigma}_N \boldsymbol{\Sigma}_N^*$ converges almost surely towards $\mathrm{MP}(2\sigma^4, c_*)$. Moreover, the largest eigenvalue and the smallest eigenvalue of $\boldsymbol{\Sigma}_N \boldsymbol{\Sigma}_N^*$ converge almost surely towards $2\sigma^4(1 + \sqrt{c_*})^2$ and $2\sigma^4(1 - \sqrt{c_*})^2$ respectively, and this implies that for each $\epsilon > 0$, almost surely, all the eigenvalues of $\boldsymbol{\Sigma}_N \boldsymbol{\Sigma}_N^*$ lie in the interval $[2\sigma^4(1 - \sqrt{c_N})^2 - \epsilon, 2\sigma^4(1 + \sqrt{c_N})^2 + \epsilon]$ for N large enough. Finally, 0 is eigenvalue of $\hat{\mathbf{C}}_N$ with multiplicity $M^2 - L = M(M - 1)/2$. For each $\epsilon > 0$, almost surely, the L remaining eigenvalues of $\hat{\mathbf{C}}_N$ are located in the interval $[-2\sigma^4(2\sqrt{c_N} - c_N) - \epsilon, 2\sigma^4(2\sqrt{c_N} + c_N) + \epsilon]$ for N large enough, and the distribution of these remaining eigenvalues converge towards a translated version of $\mathrm{MP}(2\sigma^4, c_*)$.*

The convergence of the eigenvalue distribution of $\boldsymbol{\Sigma}_N \boldsymbol{\Sigma}_N^*$ towards $\mathrm{MP}(2\sigma^4, c_*)$ follows immediately from the general results of [9]. The convergence of the extreme eigenvalues of $\boldsymbol{\Sigma}_N \boldsymbol{\Sigma}_N^*$ is more demanding, and follows an approach developed in a different context in [5, 11]. The behaviour of the eigenvalues of $\hat{\mathbf{C}}_N$ is a direct consequence of (11) and (12).

Theorem 2 implies that the non zero eigenvalues of $\hat{\mathbf{C}}_N$ lie in the neighbourhood of an interval whose lenght δ_N is equal to $\delta_N = 8\sigma^4 \sqrt{c_N}$. δ_N is thus proportional to $2\sigma^4$ which corresponds to the fourth-order moment of the components of vectors $(\mathbf{y}_n)_{n=1,\ldots,N}$. δ_N also depends on M and N through $\sqrt{c_N} = \left(\frac{M(M+1)}{2N}\right)^{1/2}$ which is nearly equal to $\frac{1}{\sqrt{2}}\left(\frac{M^2}{N}\right)^{1/2}$. The spreading of the eigenvalues thus depends on the ratio $\frac{M^2}{N}$, which, in order to be a small factor, needs N to be very large. We notice that the spreading of the eigenvalues of the empirical matrix $\hat{\mathbf{R}}_N$ is equal to $4\sigma^2 \left(\frac{M}{N}\right)^{1/2}$. This tends to indicate that to estimate $\mathbf{C}_N$ with the same accuracy than the covariance matrix $\mathbf{R}_N$, the sample size should be increased by a factor M.

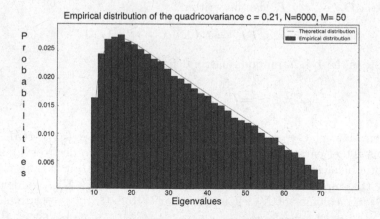

Fig. 1. Histogram of the eigenvalues of $\boldsymbol{\Sigma}_N \boldsymbol{\Sigma}_N^*$

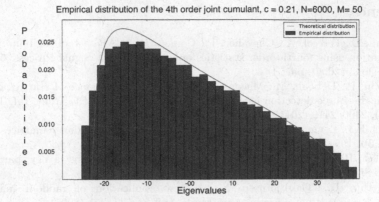

Fig. 2. Histogram of the non zero eigenvalues of $\hat{\mathbf{C}}_N$

We finally illustrate Theorem 2. In our numerical experiments, $\sigma^2 = 4$, $N = 6000$ and $M = 50$ so that $c_N = 0.21$. In Fig. 1, we represent the histogram of the eigenvalues of a realization of matrix $\boldsymbol{\Sigma}_N \boldsymbol{\Sigma}_N^*$. It appears that the histogram fits quite well with the graph of the probability density of $MP(32, 0.21)$, and that all the eigenvalues lie in the support of the MP distribution, thus confirming the practical reliability of the first statements of Theorem 2. In Fig. 2, we represent the histogram of the non zero eigenvalues of $\hat{\mathbf{C}}_N$. This time, we can observe a larger gap between the histogram and the limit distribution. This extra gap follows from the errors due to the approximation $\hat{\mathbf{R}}_N \simeq \sigma^2 \mathbf{I}_M$.

5   Concluding Remarks

In this paper, we have shown that the estimated fourth-order cumulant matrix $\hat{\mathbf{C}}_N$ of a temporally and spatially white Gaussian noise converges towards 0 in the spectral norm sense if $\frac{M^2}{N} \log N$ converges towards 0. When $\frac{M^2}{N}$ converges towards a non zero constant, the empirical eigenvalue distribution of $\hat{\mathbf{C}}_N$ converges towards a translated Marcenko-Pastur distribution, and all the eigenvalues of $\hat{\mathbf{C}}_N$ lie for N large enough in an interval whose lenght is a $\mathcal{O}\left(\sigma^4(\frac{M^2}{N})^{1/2}\right)$ term. In the presence of sources, this suggests that in order to estimate accurately $\hat{\mathbf{C}}_N$ when M is large, the number of observations N should be much larger than M^2. In the next future, we will study the behaviour of the largest eigenvalues and corresponding eigenvectors of $\hat{\mathbf{C}}_N$ in the presence of sources when $\frac{M^2}{N}$ does not converge towards 0. Hopefully, this will allow to propose improved performance estimation algorithms of the mixing matrix when $\frac{M^2}{N}$ is not small enough.

References

1. Albéra, L., Ferréol, A., Chevalier, P., Comon, P.: ICAR, a tool for blind source separation using fourth order statistics only. IEEE Trans. Signal Process. **53**(10), 3633–3643 (2005). part 1
2. Bianchi, P., Debbah, M., Maëda, M., Najim, J.: Performance of statistical tests for single source detection using random matrix theory. IEEE Trans. Inf. Theory **57**(4), 2400–2419 (2011)
3. Bickel, P.J., Levina, E.: Regularized estimation of large covariance matrices. Ann. Stat. **36**, 199–227 (2008)
4. Couillet, R., Debbah, M.: Signal processing in large systems: a new paradigm. IEEE Signal Process. Mag. **30**(1), 24–39 (2013)
5. Haagerup, U., Thorbjornsen, S.: A new application of random matrices: $\mathrm{Ext}(C^*_{red}(F_2))$ is not a group. Ann. Math. **162**(2), 711–775 (2005)
6. Hiltunen, S., Loubaton, P., Chevalier, P.: Large system analysis of a GLRT for detection with large sensor arrays in temporally white noise. IEEE Trans. Signal Process. **63**(20), 5409–5423 (2015)
7. Kritchman, S., Nadler, B.: Non-parametric detection of the number of signals, hypothesis testing and random matrix theory. IEEE Trans. Signal Process. **57**(10), 3930–3941 (2009)
8. Nadakuditi, R.R., Silverstein, J.W.: Fundamental limit of sample generalized eigenvalue based detection of signals in noise using relatively few signal-bearing and noise-only samples. IEEE J. Sel. Topics Signal Process. **4**(3), 468–480 (2010)
9. Pajor, A., Pastur, L.A.: On the limiting empirical measure of the sum of rank one matrices with log-concave distribution. Stud. Math. **195**, 11–29 (2009)
10. Pastur, L.A., Shcherbina, M.: Mathematical surveys and monographs. In: Eigenvalue Distribution of Large Random Matrices. American Mathematical Society, Providence (2011)
11. Schultz, H.: Non commutative polynomials of independent Gaussian random matrices. Probab. Theory Relat. Fields **131**, 261–309 (2005)
12. Tao, T.: Graduate studies in mathematics. In: Topics in Random Matrix Theory, vol. 132. American Mathematical Society, Providence, Rhosde Island (2012)
13. Tropp, J.: An introduction to matrix concentration inequalities. Found. Trends Mach. Learn. **8**(1–2), 1–230 (2015)
14. Vallet, P., Mestre, X., Loubaton, P.: Performance analysis of an improved MUSIC DoA estimator. IEEE Trans. Signal Process. **63**(23), 6407–6422 (2015)

Caveats with Stochastic Gradient and Maximum Likelihood Based ICA for EEG

Jair Montoya-Martínez[(⊠)], Jean-François Cardoso, and Alexandre Gramfort

LTCI, CNRS, Télécom ParisTech, Université Paris-Saclay, 75013 Paris, France
jmontoya@telecom-paristech.fr

Abstract. Stochastic gradient (SG) is the most commonly used optimization technique for maximum likelihood based approaches to independent component analysis (ICA). It is in particular the default solver in public implementations of Infomax and variants. Motivated by experimental findings on electroencephalography (EEG) data, we report some caveats which can impact the results and interpretation of neuroscience findings. We investigate issues raised by controlling the step size in gradient updates combined with early stopping conditions, as well as initialization choices which can artificially generate biologically plausible brain sources, so called *dipolar* sources. We provide experimental evidence that pushing the convergence of Infomax using non stochastic solvers can *reduce* the number of highly dipolar components and provide a mathematical explanation of this fact. Results are presented on public EEG data.

Keywords: Independent component analysis (ICA) · Maximum likelihood · Stochastic gradient method · Infomax · Electroencephalography (EEG) · Neuroscience

1 Introduction

Independent Component Analysis (ICA) is a multidimensional statistical method that seeks to uncover hidden latent variables in multivariate and potentially high-dimensional data. In the ICA model we consider here, the observations $\mathbf{x}$ satisfy $\mathbf{x} = \mathbf{As}$, where $\mathbf{s}$ are referred to as the sources or independent components, and $\mathbf{A}$ is the mixing matrix considered unknown [12]. In the following, we assume as many sources as sensors: $\mathbf{A}$ is a square matrix. This model is usually described as a latent linear stochastic model, where $\mathbf{x}$ and $\mathbf{s}$ are random variables (r.v.) in $\mathbb{R}^N$, and $\mathbf{A} \in \mathbb{R}^{N \times N}$ is a nonsingular matrix. The goal of ICA is, given a set of observations of the r.v. $\mathbf{x}$, to estimate the hidden sources $\mathbf{s}$ and the unknown mixing matrix $\mathbf{A}$. In order to accomplish this task, the key assumption in ICA is that the components $s_1, s_2, \ldots, s_N$ are mutually statistically independent [6], a plausible assumption if each individual source signal is thought to be generated by a process unrelated to any other source signal.

In neuroscience, and in particular when working with Electroencephalography (EEG) data, ICA is extremely popular. It is used for artifact removal as well as

© Springer International Publishing AG 2017
P. Tichavský et al. (Eds.): LVA/ICA 2017, LNCS 10169, pp. 279–289, 2017.
DOI: 10.1007/978-3-319-53547-0_27

estimation of brain sources. Linear ICA is justified by the fact that EEG data are linear mixtures of volume-conducted neural activities [13]. Each brain source is thought to represent near-synchronous local field activity across a small cortical patch [14], which can be modeled as an electrical current dipole (ECD) located within the brain [18].

To solve the ICA estimation problem, we need to estimate a linear operator $\widehat{\mathbf{W}} \in \mathbb{R}^{N \times N}$, such that $\hat{\mathbf{s}} = \widehat{\mathbf{W}}\mathbf{x} = \widehat{\mathbf{W}}\mathbf{A}\mathbf{s} \approx \mathbf{s}$, where $\hat{\mathbf{s}}$ is an estimation of the sources. In the context of EEG, the estimated mixing matrix $\widehat{\mathbf{A}} = \widehat{\mathbf{W}}^{-1}$ gives us information about how the estimated sources are seen at the sensor level. Indeed each column of $\widehat{\mathbf{A}}$ can be visualized as a scalp map (topography). This helps the EEG users to identify plausible brain sources which correspond to ECDs. For such sources, topographies are spatially smooth and exhibit a dipolar pattern.

A common approach to tackle the ICA problem is to cast it as a maximum likelihood estimation problem [16]: given a probability density function (p.d.f.) $p_{\mathbf{s}}(\mathbf{s})$, associated with the sources, and a set $\mathcal{X} = \{\mathbf{x}(1), \mathbf{x}(2), \ldots, \mathbf{x}(T)\} = \{\mathbf{x}_j\}_{j=1}^{j=T}$, containing independent and identically distributed (i.i.d) samples of the r.v. $\mathbf{x}$, one wants to find the unmixing matrix $\mathbf{W}$ that maximizes the log-likelihood function $\ell(\mathbf{W}, \mathcal{X})$ (for the sake of simplicity, we make the dependence on $\mathcal{X}$ implicit in ℓ):

$$\ell(\mathbf{W}) = \sum_{j=1}^{T} \left[\sum_{i=1}^{N} \log p_{s_i}(\mathbf{w}_i^\top \mathbf{x}_j) + \log|\det \mathbf{W}| \right] \tag{1}$$

where $\mathbf{w}_i^\top$ denotes the i-th row of $\mathbf{W}$. One of the most popular algorithm in the EEG community is Infomax [2] and it can be shown to follow this likelihood approach [5].

In order to maximize the log-likelihood function (1), we have at our disposal two different families of optimization methods: batch methods, such as gradient descent, which use at each iteration the entire set of observations $\mathcal{X}$, and stochastic methods which access at each iteration only one observation $\mathbf{x}_j$, or a small group of observations $\mathcal{B} = \{\mathbf{x}_j\} \subset \mathcal{X}$ (also known as mini-batch). When using stochastic gradient (SG) methods, the gradients used as update directions, with one sample or a mini-batch, are affected by 'noise' [4]. The consequence is that unless so-called step size annealing strategies are employed, SG will not reach a minimum of the minimized function [17]. On the contrary, gradient descent (GD), which is a non-stochastic batch method, does guarantee a decay of the minimized function at every iteration and does reach points with zero gradients (cf. Proposition 1.2.1 in [3]). Convergence rates can be up to linear for strongly convex functions with Lipschitz gradients. However, one update of the parameters by GD requires a full pass on the whole dataset while SG does already reduce the cost function after accessing a fraction of it. That is why when working with many samples, which is the case for EEG, SG exhibits a rapid convergence during the early stages of the optimization procedure, yet this convergence then slows down and the cost function reaches a plateau well before a point with zero gradient is reached. In other words, plain SG will stop too early if a high numerical precision solution is needed.

Infomax uses SG to maximize the log-likelihood function (1). In particular, it uses a mini-batch SG method in combination with a step size annealing policy, which is applied after one pass on the full data (Infomax considers one iteration as one pass on the full data). As in any stochastic method, the Infomax solver needs an initial step size.

In the first part of the paper, we explore the impact of algorithm initialization, the initial value of the step size, jointly with the annealing policy and the stopping criterion used by the standard Infomax implementation. We explain theoretically why the commonly used initialization of Infomax produces highly dipolar sources. We then explain the observation that Infomax can eventually waste a lot of computation time without converging, or worse can report convergence while the norm of the gradient is still high. Finally, using public EEG data and an alternative optimization strategy we investigate the impact of convergence on source dipolarity, highlighting specificities of EEG.

2   Infomax: Description of the Optimization Algorithm

The maximum likelihood problem tackled by Infomax can be written as the following minimization problem:

$$\widehat{\mathbf{W}} = \underset{\mathbf{W}}{\text{argmin}}\ L(\mathbf{W}) \tag{2}$$

where $L(\mathbf{W}) = -\ell(\mathbf{W})/T$ denotes the normalized negative log-likelihood function. In order to solve the problem (2), Infomax uses the relative gradient [1,7]

$$\widetilde{L}'(\mathbf{W}) = \frac{1}{T} \sum_{j=1}^{T} \left[\boldsymbol{\phi}(\mathbf{y}_j)\mathbf{y}_j^\top - \mathbf{I}_N \right] \tag{3}$$

where $\mathbf{y}_j = \mathbf{W}\mathbf{x}_j$, where $\boldsymbol{\phi}(\mathbf{y}_j) = [\phi_1(y_{j_1}), \ldots, \phi_N(y_{j_N})]^\top$, and where

$$\phi_k(y_{j_k}) = -p'_{s_k}(y_{j_k})/p_{s_k}(y_{j_k}) = \tanh\left(y_{j_k}/2\right)$$

In order to solve (2), the reference Infomax implementation, included for example in the EEGLAB software [8], uses a mini-batch stochastic gradient method, whose iterative expression can be written as follows:

$$\mathbf{W}_{k+1} = \mathbf{W}_k - \alpha \sum_{j \in \mathcal{B}_k} \left[\boldsymbol{\phi}(\mathbf{y}_j)\mathbf{y}_j^\top - \mathbf{I}_N \right] \mathbf{W}_k \tag{4}$$

where, before each pass on the full data, the set of samples $\{\mathbf{x}_j\}_{j=1}^{j=T}$ is randomly permuted, and then, during the full pass, each mini-batch $\mathcal{B}_k$ is created by taking, sequentially, a subset of samples $\{\mathbf{x}_j\}$ of size $|\mathcal{B}_k|$. Once the pass on the full data is completed, the stopping criterion is checked and the annealing policy is applied to determine whether or not the step size α should be decreased. In this policy, the step size is never increased. Finally, this process is repeated until the stopping criterion is fulfilled or until the maximum number of iterations is reached.

Let us denote $\Delta_k = \mathbf{W}_{k+1} - \mathbf{W}_k$. The stopping criterion used by the standard Infomax implementation is $\|\Delta_k\|_F^2 < \text{tol}$, where $\|\cdot\|_F$ is the Frobenius norm, tol is by default 10^{-6} if $N \leq 32$, and 10^{-7} otherwise. This implementation uses the following heuristic for its annealing policy: if the angle between matrices Δ_k and Δ_{k-1} is larger than $60°$, i.e., $\arccos\left(\text{Trace}(\Delta_k^\top \Delta_{k-1})/(\|\Delta_k\|_F \|\Delta_{k-1}\|_F)\right) > \pi/3$, it decreases the step size by 10% ($\alpha \leftarrow 0.9\alpha$), otherwise the step size remains the same.

Regarding the stopping criterion $\|\Delta_k\|_F^2 < \text{tol}$, it is important to notice that by Eq. (4), Δ_k is proportional to the step size α so that the algorithm will stop if the gradient *or* the step size is small. Even if the stopping criterion is not met, the step size may have become small enough (even using the standard default values) to prevent any significant update. Example of such behaviors on EEG data are given in Sect. 4.

3   Assessing the Performance of ICA Using Dipolarity

Dipolarity metric. ICA can be seen an unsupervised learning method, therefore, it is in general difficult to assess its performance in real life scenarios where the ground truth is unknown. In order to help to mitigate this issue, when using EEG data, Delorme et al. [9] proposed to use the physics underlying the propagation of the electromagnetic field throughout the head. Physics states that the signals measured on the scalp can be modeled as linear mixtures of the electrical activities generated by ECDs located inside the brain. To assess the biological plausibility of ICA sources, Delorme et al. [9] proposed to take each column of the estimated mixing matrix $\widehat{\mathbf{A}}$, which can be represented as a topography, and compute how well it can be modeled by a single ECD. The following metric is defined by:

$$\text{dipolarity}(\widehat{\mathbf{A}}_j) = (1 - \|\widehat{\mathbf{A}}_j - \bar{\mathbf{A}}_j\|_2^2 / \|\widehat{\mathbf{A}}_j\|_2^2) \times 100 \tag{5}$$

where $\widehat{\mathbf{A}}_j$, $\bar{\mathbf{A}}_j$ denote respectively the j-th column of the estimated mixing matrix and the corresponding topography obtained by fitting a single dipole. Taking into account (5), the "Near-Dipolar percentage (ND%)" of an ICA decomposition is defined in [9] as: $\text{ND}\%(\widehat{\mathbf{A}}) = \{\#j : \text{dipolarity}(\widehat{\mathbf{A}}_j) > \tau\}/N$, that is, the percentage of returned components whose topographies can be modeled by a single ECD with more than a specified dipolarity threshold τ (specified as percentage of explained/residual variance). Following [9], we will consider an ICA source to be biologically plausible when its dipolarity is larger than $\tau = 90$.

Relationship between initialization and dipolarity. ICA leads to nonconvex optimization problems: solutions found by algorithms necessarily depend on their initialization. In this section we discuss connections between initialization and dipolarity.

Learning of a separating matrix $\mathbf{W}$ starts with some initial value $\mathbf{W}_0$. While it is generally possible to start with the identity matrix $\mathbf{W}_0 = \mathbf{I}_N$, it is a sound and common practice to start with some whitening matrix, that is, with a matrix $\mathbf{W}_0$ such that $\mathbf{W}_0 \mathbf{\Sigma}_\mathbf{x} \mathbf{W}_0^\top = \mathbf{I}_N$ where $\mathbf{\Sigma}_\mathbf{x} = \mathrm{Cov}(\mathbf{x})$ denotes the covariance matrix of $\mathbf{x}$. There are infinitely many such matrices; two popular choices are 'PCA' and 'sphering', which can be defined in terms of the eigen-value decomposition $\mathbf{\Sigma}_\mathbf{x} = \mathbf{U}\mathbf{D}\mathbf{U}^\top$:

$$\mathbf{W}^{\mathrm{pca}} = \mathbf{D}^{-1/2}\mathbf{U}^\top, \ \mathbf{W}^{\mathrm{sph}} = \mathbf{U}\mathbf{D}^{-1/2}\mathbf{U}^\top.$$

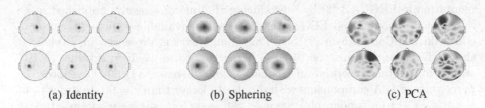

(a) Identity   (b) Sphering   (c) PCA

Fig. 1. Topographies associated with different initializations (actually a subset of 6 of them).

The topographies (the columns of matrix $\mathbf{A}_0 = \mathbf{W}_0^{-1}$) associated with the three aforementioned initializations $\mathbf{W}_0 = \mathbf{I}_N, \mathbf{W}^{\mathrm{sph}}, \mathbf{W}^{\mathrm{pca}}$, are displayed on Fig. 1. Of course, the topographies associated with $\mathbf{W}_0 = \mathbf{I}_N$ (Fig. 1(a)) are 'quasi-dipolar' in the sense that activating only one channel could be interpreted as the effect of a single source located just beneath the scalp. Much more striking is the fact that the sphering $\mathbf{W}_0 = \mathbf{W}^{\mathrm{sph}}$ produces topographies which all look dipolar. Figure 1(b) shows 6 of them, randomly selected. Nothing similar is observed in Fig. 1(c) after PCA $\mathbf{W}_0 = \mathbf{W}^{\mathrm{pca}}$. See also Fig. 4, which shows the dipolarity index for all components after sphering or PCA, sorted in decreasing order.

If the dipolarity criterion is to be used for assessing the biological plausibility of a source, one has to understand why a simple sphering would produce dipolar topographies. An explanation can be provided by the observation that somehow *sphering is the 'smallest' whitening transform*. Indeed, if a whitening matrix is close to the identity, then it should not modify much the 'quasi-dipolar' patterns of Fig. 1(a) and this is what seems to happen upon observation of Fig. 1(b). So, in which sense would sphering be the 'smallest whitening transform'? An answer is provided by Theorem 1 of Eldar *et al.* [10], which implies that, among all whitening matrices $\mathbf{W}$, the sphering matrix is the one with the minimal mean-squared difference $\mathrm{E}\|\mathbf{x} - \mathbf{W}\mathbf{x}\|^2$. In other words, among all white random vectors $\mathbf{W}\mathbf{x}$, $\mathbf{W}^{\mathrm{sph}}\mathbf{x}$ is the closest to $\mathbf{x}$ with closeness measured in the mean-squared sense. In other words, sphering is the whitening transform which moves the data the least. In terms of matrix norms, we can write the mean-squared difference $\mathrm{E}\|\mathbf{x} - \mathbf{W}\mathbf{x}\|^2$ as $\mathrm{E}\|(\mathbf{I}_N - \mathbf{W})\mathbf{x}\|^2 = \mathrm{Trace}\left[(\mathbf{I}_N - \mathbf{W})\mathbf{\Sigma}_\mathbf{x}(\mathbf{I}_N - \mathbf{W})^\top\right]$, so that, the sphering matrix is the closest to the identity in the matrix norm $\|\mathbf{M}\|_\Sigma^2 = \mathrm{Trace}\left[\mathbf{M}\mathbf{\Sigma}_\mathbf{x}\mathbf{M}^\top\right]$. For later reference, we note that sphering is the default initialization used by the current Infomax implementation in the EEGLAB package [8].

4   Numerical Experiments

Comparison of EEGLAB and MNE implementations. Our numerical experiments were conducted with the Infomax implementation of MNE-Python [11]. We checked that this implementation matches the reference Infomax implementation in EEGLAB [8] by reproducing the Infomax results published in [9] based on 13 anonymized EEG datasets (publicly available at http://sccn.ucsd. edu/wiki/BSSComparison [9,15]). This comparison is presented in Table 1. It shows the average across the 13 EEG datasets used in [9]. In this table, "MIR" stands for Mutual Information Reduction [9], whereas "ND 90%" denotes the percentage of ICA components with dipolarity larger than $\tau = 90$. In order to fit a single ECD to a topography, we used the same four-sphere model used in [9] for forward computation. The radius of each sphere was equal to 71, 72, 79 and 85 mm, and their corresponding conductivities relative to the cerebrospinal fluid were equal to 0.33, 1.0, 0.0042 and 0.33, respectively.

In the course of this comparison, we found that EEGLAB does not constrain the fitted dipoles to be located *inside* the brain, whereas MNE-Python does. For this reason, EEGLAB tends to report an artificially high number of dipolar components as compared to MNE-Python (second row of Table 1). However, we checked that EEGLAB and MNE-Python agree in the number of high dipolar components for dipoles located inside the brain (third row of Table 1). Even better, we checked that they agree on the *locations* of those dipoles.

Table 1. Comparison of EEGLAB and MNE-Python Infomax implementations.

Metric	EEGLAB	MNE-Python
MIR	43.092901	43.092938
ND 90%	43.445287	31.744312
EEGLAB loc. in ND 90%	22.751896	22.751896

Evaluation of the stochastic gradient approach. We proceed to evaluate the performance of the SG method used by Infomax. We use subject kb77 from the same study [9,15]. The EEG dataset is composed of 306600 samples of 71 channels sampled at 250 Hz. The algorithm is evaluated by monitoring the step size, as well as the Frobenius norm of the relative gradient after each pass on the full data. We consider the following scenarios, differing by the initial step size α_0 and tolerance tol:

- **Infomax a:** EEGLAB defaults: $\alpha_0 = 6.5 \times 10^{-4}/\log(N) \approx 1.5 \times 10^{-4}$ and tol $= 10^{-7}$.

– **Infomax b:** $\alpha_0 = 10^{-3}$ as in [9], and EEGLAB default value for tol $= 10^{-7}$.
– **Infomax c:** same α_0 as in scenario "Infomax b" and tol $= 10^{-14}$.

The left panel of Fig. 2 displays the evolution of the relative gradient norm $\|\widetilde{L}'(\mathbf{W})\|_F$ across passes on the full data. As we can see in this figure, in none of the scenarios the relative gradient ever go to zero. Scenarios **a** and **b**, which use the default tolerance, either reach a plateau or stop early. Yet, reducing the tolerance in scenario **c**, reveals that it is not sufficient to push convergence compared to **b**: the trajectory plateaus just after the stopping point for **b**. However, the two plateaus of cases **a** and **c** are of different natures, as revealed by the right panel of Fig. 2, which displays the step size trajectories. For **a**, the step size remains constant after pass ∼83, hence we observe the known plateau of SG

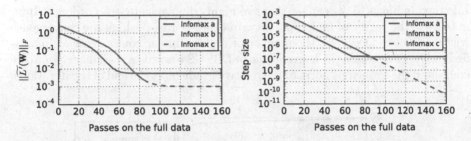

Fig. 2. Left: Frobenius norm of the relative gradient vs. iterations. Right: Step size α vs. iterations.

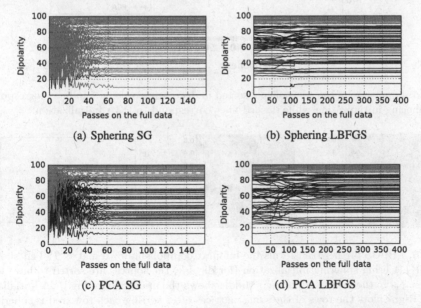

Fig. 3. Evolution for scenario **c** of dipolarity during Infomax (left) followed by LBFGS (right). Initialization with Sphering (top row) and PCA (bottow row). A line is colored red if it exceeds the value of 90% during the iterations. (Color figure online)

methods with fixed step size, whereas for **b** and **c** the annealing policy drives the step size to zero exponentially fast, therefore preventing the algorithm to make any further progress.

Figure 3 shows the evolution of dipolarity of the components across passes on the full data. One can see in Fig. 3(a) that when starting with sphering most of the lines are red which means that they exceed at some point the value of 90. This is due to initialization as explained previously. When initializing Infomax with PCA, see Fig. 3(c), much less components reach this high dipolarity threshold. In both cases, we observe that the dipolarity stops evolving after approximately 70 passes on the dataset, which is consistent with the plateaus of Fig. 2. The two plots on the right column show the same dipolarity metrics, but this time using the quasi-Newton method known as LBFGS (Limited-memory Broyden-Fletcher-Goldfarb-Shanno), taking as initialization the unmixing matrices estimated by Infomax. In Fig. 3(b), we can see that two highly dipolar sources according to Infomax leave the region of high dipolarity. In other words, pushing the convergence with LBFGS reduces here the number of highly dipolar components as quantified in [9]. To evaluate the convergence of LBFGS towards a stationary point, we computed the Frobenius norm of the relative gradient at the end of the iterations.

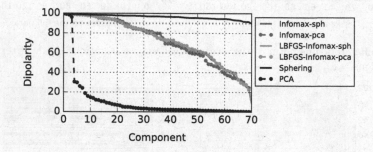

Fig. 4. Dipolarity of the components sorted in decreasing order. Plain lines correspond to sphering initialization while dashed lines correspond to PCA initialization.

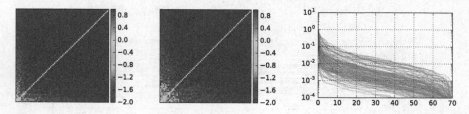

Fig. 5. LBFGS further transforms the Infomax components by a matrix T. Left shows $\log_{10}(|T|)$ using sphering initialization (for display, the sources are sorted to have the largest T_{ij} in the lower left corner). Middle shows the update following PCA initialization. Right plots the rows of the same matrices after sorting each row (red is sphering and black is PCA). (Color figure online)

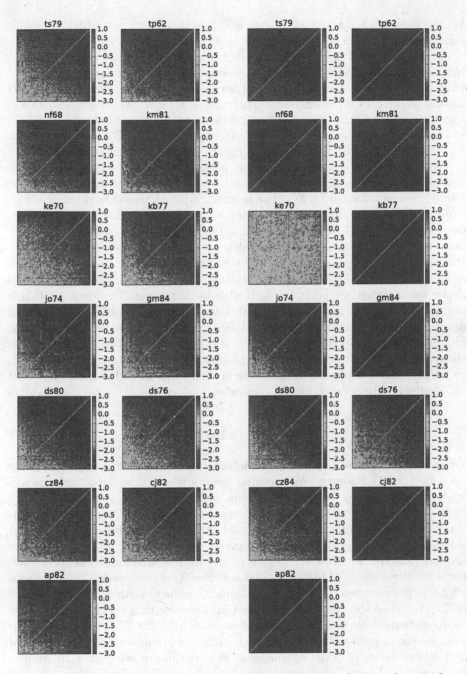

Fig. 6. Changing the initialization changes the local minimum otherwise the transform T linking the sources obtained with the two initializing whiteners (PCA and sphering) would be the identity. Each plot shows $\log_{10}|T|$. Left: Infomax, right: LBFGS. By completing the convergence process, we find that in 8 cases out of 13, the resulting sources do not depend on initialization.

While this norm was about 10^{-4} after SG, it is about 10^{-7} after LBFGS, which confirms that LBFGS does push significantly the convergence.

The dipolarity of the components after sphering or PCA, as well as following Infomax and LBFGS in the same four cases as Fig. 3 is presented in Fig. 4. This plot is an extra evidence that simple sphering already yields almost only highly dipolar sources. PCA, on the contrary, contains far less dipolar sources. This is also in line with Fig. 1. This plot also reveals that Infomax followed by LBFGS reaches almost identical dipolarities. This suggests that LBFGS manages to wash out the effect of initialization by converging to the same local minimum.

To gain further evidence, we ran a number of checks. Figure 5 quantifies how close the unmixing matrix estimated after LBFGS is to the inverse of the mixing matrix obtained by SG. The multiplication of these matrices should be close to the identity (up to permutation). Figure 5 reports that it is far from it, demonstrating that LBFGS deviates non-trivially from the output of Infomax. One can also see that the change operated by LBFGS is larger in the PCA case. In other words, Infomax following PCA brings the estimate further away from a stationary point than the sphering initialization. This conclusion only holds if the same stationary point is reached in both settings. Evidence for this is presented in Fig. 6, where one can see that for this subject (kb77) the estimated unmixing matrix obtained in the PCA condition is close to the inverse of the mixing matrix obtained following sphering (up to a permutation). To assess if there is convergence to the same local minimum when SG is followed by LBFGS, we ran the same computation on all the subjects. Figure 6 shows that for 8 out of 13 subjects the result perfectly replicates.

5   Conclusion

We explored the annealing policy, the initial step size and the stopping criterion used by the SG Infomax. We reported results where this algorithmic choices lead Infomax to stop before reaching an accurate stationary point, despite a high number of iterations and long computation times. We explained theoretically why the sphering initialization used by Infomax produces highly dipolar sources. By further pushing the convergence using a quasi-Newton method, we showed that the initialization influences the output of Infomax, hence overestimating the number of highly dipolar sources. This observation could explain why practitioners tend to avoid dimensionality reduction when Infomax is used on EEG. Indeed sphering cannot be used in this case. This paper should be seen as an instantaneous picture on current usage of ICA for EEG data. Given the massive use of such techniques, we hope that it will motivate the development and dissemination of better optimization schemes in this scientific community.

Acknowledgments. This work was funded by the Paris-Saclay Center for Data Science with partial support from the European Research Council (ERC-YStG-676943).

References

1. Amari, S.-I.: Natural gradient works efficiently in learning. Neural Comput. **10**(2), 251–276 (1998)
2. Bell, A.J., Sejnowski, T.J.: An information-maximization approach to blind separation and blind deconvolution. Neural Comput. **7**(6), 1129–1159 (1995)
3. Bertsekas, D.: Nonlinear Programming. Athena Scientific, Cambridge (1999)
4. Bottou, L.: Large-scale machine learning with stochastic gradient descent. In: Lechevallier, Y., Saporta, G. (eds.) Proceedings of COMPSTAT'2010, pp. 177–186. Springer, Heidelberg (2010)
5. Cardoso, J.F.: Infomax and maximum likelihood for blind source separation. IEEE Signal Process. Lett. **4**, 112–114 (1997)
6. Cardoso, J.F.: Blind signal separation: statistical principles. Proc. IEEE **86**(10), 2009–2025 (1998)
7. Cardoso, J.F., Laheld, B.H.: Equivariant adaptive source separation. IEEE Trans. Sig. Proc. **44**(12), 3017–3030 (1996)
8. Delorme, A., Makeig, S.: EEGLAB: an open source toolbox for analysis of single-trial EEG dynamics including independent component analysis. J. Neurosci. Methods **134**(1), 9–21 (2004)
9. Delorme, A., Palmer, J., Onton, J., Oostenveld, R., Makeig, S., et al.: Independent EEG sources are dipolar. PloS One **7**(2), e30135 (2012)
10. Eldar, Y.C., Oppenheim, A.V.: MMSE whitening and subspace whitening. IEEE Trans. Inf. Theor. **49**(7), 1846–1851 (2003)
11. Gramfort, A., et al.: MNE software for MEG and EEG data. Neuroimage **86**, 446–460 (2014)
12. Jutten, C., Herault, J.: Blind separation of sources, Part I: an adaptive algorithm based on neuromimetic architecture. Signal Process. **24**(1), 1–10 (1991)
13. Makeig, S., Bell, A.J., Jung, T.P., Sejnowski, T.J.: Independent component analysis of electroencephalographic data. In: NIPS, pp. 145–151 (1996)
14. Nunez, S.: Electric Fields of the Brain: The Neurophysics of EEG. Oxford University Press, New York (2006)
15. Onton, J., Delorme, A., Makeig, S.: Frontal midline EEG dynamics during working memory. Neuroimage **27**(2), 341–356 (2005)
16. Pham, D.T., Garat, P.: Blind separation of mixture of independent sources through a quasi-maximum likelihood approach. IEEE Trans. Signal Process. **45**(7), 1712–1725 (1997)
17. Schaul, T., Zhang, S., LeCun, Y.: No more pesky learning rates. In: Proceedings of ICML Conference (2013)
18. Scherg, M., Berg, P.: Use of prior knowledge in brain electromagnetic source analysis. Brain Topogr. **4**(2), 143–150 (1991)

Approximate Joint Diagonalization According to the Natural Riemannian Distance

Florent Bouchard[1]([✉]), Jérôme Malick[2], and Marco Congedo[1]

[1] GIPSA-lab, CNRS, Univ. Grenoble Alpes, Grenoble Institute of Technology,
Grenoble, France
`florent.bouchard@gipsa-lab.fr`
[2] LJK, CNRS, Univ. Grenoble Alpes, Grenoble, France

Abstract. In this paper, we propose for the first time an approximate joint diagonalization (AJD) method based on the natural Riemannian distance of Hermitian positive definite matrices. We turn the AJD problem into an optimization problem with a Riemannian criterion and we developp a framework to optimize it. The originality of this criterion arises from the diagonal form it targets. We compare the performance of our Riemannian criterion to the classical ones based on the Frobenius norm and the log-det divergence, on both simulated data and real electroencephalographic (EEG) signals. Simulated data show that the Riemannian criterion is more accurate and allows faster convergence in terms of iterations. It also performs well on real data, suggesting that this new approach may be useful in other practical applications.

Keywords: Approximate joint diagonalization · Riemannian geometry · Hermitian positive definite matrices · Riemannian optimization

1 Introduction

The approximate joint diagonalization (AJD) of a matrix set is instrumental to solve the blind source separation (BSS) problem. We refer to [1] for a complete review on theory and applications. The AJD of a set $\{C_k\}_{1 \leq k \leq K}$ of K Hermitian positive definite (HPD) matrices of size $n \times n$ consists in finding a full rank matrix B of size $n \times n$ such that the set $\{BC_k B^H\}_{1 \leq k \leq K}$ is composed of matrices as diagonal as possible according to some criterion, where superscript $\cdot^H$ denotes the conjugate transpose.

Criteria of interest are diagonality measures that turn the AJD problem into an optimization problem. The three most popular cost functions are the following ones. A widely used cost function based on the Frobenius distance is given by

$$f_{\mathrm{F}}(B) = \sum_{k=1}^{K} \|BC_k B^H - \mathrm{ddiag}(BC_k B^H)\|_F^2 \,, \tag{1}$$

where $\|\cdot\|_F$ denotes the Frobenius norm and the $\mathrm{ddiag}(\cdot)$ operator returns the matrix with the diagonal elements of its argument. Some variations of (1) have

© Springer International Publishing AG 2017
P. Tichavský et al. (Eds.): LVA/ICA 2017, LNCS 10169, pp. 290–299, 2017.
DOI: 10.1007/978-3-319-53547-0_28

been proposed in [2,3] in order to have better invariance properties of the cost function. In [4], (1) is minimized through an indirect strategy. The idea is that, given B, a matrix A is sought in order to solve the optimization subproblem with cost function

$$\widetilde{f}_{\mathrm{F}}(A) = \sum_{k=1}^{K} \|BC_k B^H - A\,\mathrm{ddiag}(BC_k B^H)A^H\|_F^2 . \tag{2}$$

The joint diagonalizer B is then updated as $B \leftarrow A^{-1}B$. Another popular cost function based on the log-det divergence and introduced in [5,6] is given by

$$f_{\mathrm{LD}}(B) = \sum_{k=1}^{K} \log \frac{\det(\mathrm{ddiag}(BC_k B^H))}{\det(BC_k B^H)}, \tag{3}$$

where $\det(\cdot)$ denotes the determinant of its argument. Recently, a generalization of (3) based on the log-det α-divergence has been proposed in [3] showing promising results.

In this paper, we propose for the first time an AJD method based on the natural Riemannian distance on the cone of HPD matrices [7,8], which has recently attracted much interest in the signal processing and machine learning communities. This new approach exploits the geometrical properties of HPD matrices. Our solution is original compared to those obtained with previous criteria since the diagonal form it targets is profoundly different as we will show.

This paper is divided into four sections including this introduction. In Sect. 2 the Riemannian cost function (Sect. 2.3) is defined along with a framework to optimize it (Sects. 2.3 and 2.4). This new criterion stems from the Riemannian diagonality measure [3] (Sect. 2.2) on the cone of HPD matrices [7,8] (Sect. 2.1). In Sect. 3 we compare the performance of our Riemannian criterion to three state of the art ones on both simulated data and on a real electroencephalographic (EEG) recording. On simulated data (Sect. 3.1) our Riemannian criterion prooves more accurate and allows faster convergence in terms of iterations. It also performs well on the real EEG recording (Sect. 3.2). Finally, in Sect. 4, some conclusions are drawn.

2   Method: Riemannian Distance and Optimization

2.1   Cone of Hermitian Positive Definite Matrices

Let $\mathcal{M}_n(\mathbb{C})$ be the set of $n \times n$ complex matrices and $\mathcal{H}_n = \{C \in \mathcal{M}_n(\mathbb{C}) : C^H = C\}$ be the set of Hermitian matrices. The cone of HPD matrices $\mathcal{H}_n^{++}$ is defined as the set $\{C \in \mathcal{H}_n : C \succ 0\}$. We give a succint introduction of $\mathcal{H}_n^{++}$, the reader is refered to [7] for a complete presentation.

$\mathcal{H}_n^{++}$ is an open subspace of $\mathcal{H}_n$, thus the tangent space $T_C \mathcal{H}_n^{++}$ at $C \in \mathcal{H}_n^{++}$ can be identified as $\mathcal{H}_n$. To turn $\mathcal{H}_n^{++}$ into a Riemannian manifold, we need to endow every tangent space with a Riemannian metric, that is a smooth inner product. At $C \in \mathcal{H}_n^{++}$, the *natural Riemannian metric* on $T_C \mathcal{H}_n^{++}$ is defined for all ξ and η in $\mathcal{H}_n$ as [7]

$$\langle \xi, \eta \rangle_C = \mathrm{tr}(C^{-1} \xi C^{-1} \eta), \tag{4}$$

where $\mathrm{tr}(\cdot)$ denotes the trace operator. Note that this metric is positive definite and yields a norm on $T_C \mathcal{H}_n^{++}$ defined as $\|\xi\|_C = \langle \xi, \xi \rangle_C^{1/2}$. It leads to a *natural Riemannian distance* on $\mathcal{H}_n^{++}$ defined for C_1 and C_2 as [7,8]

$$\mathrm{d}_{\mathcal{H}_n^{++}}(C_1, C_2) = \|\log(C_1^{-1/2} C_2 C_1^{-1/2})\|_F, \tag{5}$$

where $\log(\cdot)$ denotes the matrix logarithm. The function $C_1 \mapsto \mathrm{d}_{\mathcal{H}_n^{++}}(C_1, C_2)$ is convex. Having defined the Riemannian distance $\mathrm{d}_{\mathcal{H}_n^{++}}$ on $\mathcal{H}_n^{++}$, we can define a proper Riemannian diagonality measure.

2.2   Riemannian Diagonality Measure

The subset $\mathcal{D}_n^{++}$ of diagonal positive definite matrices is a closed submanifold of $\mathcal{H}_n^{++}$ with respect to the Riemannian metric (4). The closest diagonal matrix Λ in $\mathcal{D}_n^{++}$ to a matrix C in $\mathcal{H}_n^{++}$ according to a distance or a divergence d is [3]

$$\operatorname*{argmin}_{\Lambda \in \mathcal{D}_n^{++}} d(C, \Lambda). \tag{6}$$

The diagonality measure of C relative to d is therefore the distance or divergence to its closest diagonal matrix Λ. For the distance based on the Frobenius norm (corresponding to functionnal (1)) and for the log-det divergence (corresponding to (3)), the closest diagonal matrix to a matrix C is simply its diagonal part $\mathrm{ddiag}(C)$ [3]. Using the Riemannian distance (5), the closest diagonal matrix Λ is the unique solution to equation [3]

$$\mathrm{ddiag}(\log(C^{-1}\Lambda)) = 0. \tag{7}$$

To our knowledge, there is no closed form solution to this equation, however Λ can be numerically estimated by solving the optimization problem (6) with $d = \mathrm{d}_{\mathcal{H}_n^{++}}$. This can be done as it follows: the directional derivative of $g : \Lambda \mapsto \mathrm{d}_{\mathcal{H}_n^{++}}(C, \Lambda)$ in the direction $\xi \in \mathcal{D}_n$ (set of diagonal matrices) is given by

$$\mathrm{D}\, g(\Lambda)[\xi] = 2\,\mathrm{tr}(\Lambda^{-1}\Lambda\,\mathrm{ddiag}(\log(C^{-1}\Lambda))\Lambda^{-1}\xi). \tag{8}$$

This is a consequence of proposition 2.1 in [8], basic calculations, and of the fact that $\mathcal{D}_n^{++}$ is a closed submanifold of $\mathcal{H}_n^{++}$. From (8), one can obtain the Riemannian gradient of g at Λ with the identification $\langle \mathrm{grad}\, g(\Lambda), \xi \rangle_\Lambda = \mathrm{D}\, g(\Lambda)[\xi]$ [9]. This yields

$$\mathrm{grad}\, g(\Lambda) = 2\Lambda\,\mathrm{ddiag}(\log(C^{-1}\Lambda)). \tag{9}$$

Starting from an initial guess Λ_0 (for example $\mathrm{ddiag}(C)$), one can obtain a sequence of iterates $\{\Lambda_i\}$. Given iterate Λ_i, the Riemannian gradient (9) of g at Λ_i leads to a descent direction ξ_i in $\mathcal{D}_n$ using for example a steepest-gradient or a conjugate gradient scheme (stepsize included in ξ_i if any). Finally, the exponential map of $\mathcal{D}_n^{++}$ yields the iterate Λ_{i+1} as

$$\Lambda_{i+1} = \Lambda_i \exp(\Lambda_i^{-1}\xi_i). \tag{10}$$

2.3   AJD Based on the Riemannian Distance of $\mathcal{H}_n^{++}$

Similarly to (1), the cost function based on the natural Riemannian distance (5) is defined by

$$f_{\mathrm{R}}(B) = \sum_{k=1}^{K} \mathrm{d}_{\mathcal{H}_n^{++}}(BC_kB^H, \Lambda_k)$$
$$= \sum_{k=1}^{K} \|\log((BC_kB^H)^{-1/2}\Lambda_k(BC_kB^H)^{-1/2})\|_F^2 \,, \quad (11)$$

where Λ_k is the closest diagonal matrix to BC_kB^H as per Sect. 2.2. We can minimize (11) by taking an approach similar to the one introduced in [4]. Given the sets $\{BC_kB^H\}$ and $\{\Lambda_k\}$, the idea is to find a matrix A such that the set $\{A\Lambda_kA^H\}$ gets closer (according to (5)) to the set $\{BC_kB^H\}$. This way, matrices $A^{-1}BC_kB^HA^{-H}$ are closer to diagonal form. Further, note that when the best possible matrix A is the identity matrix I_n, $\{BC_kB^H\}$ contains matrices as diagonal as possible according to (5). To find A, we define the optimization subproblem with cost function $\widetilde{f}_{\mathrm{R}}$ as

$$\widetilde{f}_{\mathrm{R}}(A) = \sum_{k=1}^{K} \|\log((BC_kB^H)^{-1/2}A\Lambda_kA^H(BC_kB^H)^{-1/2})\|_F^2 \,. \quad (12)$$

This function has a simpler expression and we can minimize it with a Riemannian gradient based method that we describe in Sect. 2.4. We then update the matrix B according to

$$B \leftarrow A^{-1}B. \quad (13)$$

This leads to Algorithm 1.

Algorithm 1. RD-AJD (Riemannian Distance AJD)

  Input: set of K matrices $\{C_k\}$ in $\mathcal{H}_n^{++}$, initial guess B_0 for B, maximum
        number of iteration $i_{\max}$.
  Output: estimated joint diagonalizer B.
1 Initialization: set $i = 0$ and compute matrices $\{B_0C_kB_0^H\}$.
2 **while** *not convergence* **and** $i < i_{\max}$ **do**
3 | Find diagonal matrices $\{\Lambda_{k,i}\}$ as described in Sect. 2.2.
4 | Set $A_0 = I_n$ and find A_i by performing one step in a descent direction of
    | (12) (see Sect. 2.4).
5 | $B_{i+1} \leftarrow A_i^{-1}B_i$.
6 | $i = i + 1$.

2.4   Riemannian Optimization for the Subproblem

It remains to define a method to optimize the cost function $\widetilde{f}_{\mathrm{R}}$ defined in (12). In order to do so, we perform a Riemannian optimization on the *special polar manifold* [10] defined as

$$\mathcal{SP}_n = \{(U, P) \in \mathcal{H}_n^{++} \times \mathcal{O}_n : \det(P) = 1\} \,. \quad (14)$$

This manifold is a consequence of the following observation: every full rank matrix A admits the *unique polar decomposition* $A = UP$, where $U \in \mathcal{O}_n$ (group of orthogonal matrices) and $P \in \mathcal{H}_n^{++}$, thus the manifold of full rank matrices is equivalent to the product manifold $\mathcal{H}_n^{++} \times \mathcal{O}_n$. To avoid degenerated solutions, we impose $\det(AA^H) = 1$, which is equivalent to $\det(P) = 1$.

In the following, notations $\mathcal{A}$ is used to denote the point (U, P) in $\mathcal{SP}_n$ and A to denote its corresponding full rank matrix UP. Given the initialization $\mathcal{A}_0 = (I_n, I_n)$ (corresponding to $A_0 = I_n$), one can obtain a descent direction ξ_i from the Riemannian gradient of (12) by a steepest-descent or conjugate gradient algorithm (stepsize included in ξ_i if any). The iterate $\mathcal{A}_i$ (corresponding to the matrix A_i) is then obtained through the retraction R of $\mathcal{SP}_n$ as

$$\mathcal{A}_i = R_{\mathcal{A}_0}(\xi_i). \tag{15}$$

From proposition 2.1 in [8] and basic calculations, the directional derivative of $\widetilde{f}_R$ at $\mathcal{A}$ in the direction $\mathcal{Z}$ is

$$\mathrm{D}\,\widetilde{f}_R(\mathcal{A})[\mathcal{Z}] = 4 \sum_{k=1}^{K} \mathrm{tr}(\log((BC_k B^H)^{-1} A\Lambda_k A^H)U Z_U^H) \\ + \mathrm{tr}(P^{-1}\log(A^H(BC_k B^H)^{-1} A\Lambda_k)Z_P^H). \tag{16}$$

Thus, the Euclidean gradient of $\widetilde{f}_R$ on the ambient space of $\mathcal{SP}_n$ is

$$\mathrm{grad}_{\mathcal{E}}\,\widetilde{f}_R(\mathcal{A}) = 4 \sum_{k=1}^{K} \big(\log((BC_k B^H)^{-1} A\Lambda_k A^H)U, \\ P^{-1}\log(A^H(BC_k B^H)^{-1} A\Lambda_k)\big) \tag{17}$$

and the Riemannian gradient of (12) is

$$\mathrm{grad}\,\widetilde{f}_R(U, P) = (\Pi_U(\mathrm{grad}_{\mathcal{E}}\,\widetilde{f}_R(U)),\ \Pi_P(P\,\mathrm{Herm}(\mathrm{grad}_{\mathcal{E}}\,\widetilde{f}_R(P))P)), \tag{18}$$

where $\mathrm{grad}_{\mathcal{E}}\,\widetilde{f}_R(U)$ and $\mathrm{grad}_{\mathcal{E}}\,\widetilde{f}_R(P)$ denote the first and second component of $\mathrm{grad}_{\mathcal{E}}\,\widetilde{f}_R(\mathcal{A})$, respectively. $\Pi_{\mathcal{A}} = (\Pi_U, \Pi_P)$ is the projection map from the ambient space onto the tangent space $T_{\mathcal{A}}\mathcal{SP}_n$ at $\mathcal{A}$. It is given, for $\mathcal{Z} = (Z_U, Z_P) \in \mathbb{R}^{n \times n} \times \mathbb{R}^{n \times n}$, by

$$\Pi_{\mathcal{A}}(\mathcal{Z}) = (Z_U - U\,\mathrm{Herm}(U^H Z_U),\ \mathrm{Herm}(Z_P) - \frac{1}{n}\mathrm{tr}(P^{-1}\,\mathrm{Herm}(Z_P))P), \tag{19}$$

where $\mathrm{Herm}(\cdot)$ returns the Hermitian part of its argument. Finally, the retraction $R_{\mathcal{A}}$ mapping a tangent vector back onto the manifold is, for $\xi_{\mathcal{A}} = (\xi_U, \xi_P)$,

$$R_{\mathcal{A}}(\xi_{\mathcal{A}}) = \left(\mathrm{uf}(U + \xi_U),\ P^{1/2}\exp(P^{-1/2}\xi_P P^{-1/2})P^{1/2}\right) \tag{20}$$

where $\mathrm{uf}(.)$ extracts the orthogonal factor of its argument by Lödwin's orthogonalization [11].

3   Numerical Experiments

We now have all the ingredients to perform RD-AJD. We estimate its performance on simulated data and we compare our approach with those using cost functions (1), (2) and (3). For RD-AJD, we compute the closest diagonal matrices using a Riemannian conjugate gradient on $\mathcal{D}_n^{++}$. Classical algorithms minimizing those cost functions (1), (2) and (3) generally use specific optimization schemes and constraints [4,6]. In order to compare the performance of the criteria, we perform all optimization on $\mathcal{SP}_n$ with the same Riemannian conjugate gradient method. This way, differences in performance are not due to the optimization scheme but solely to the criterion employed. This Riemannian optimization scheme has been shown to perform well in general, see for example [10].

We refer to the resulting algorithms by acronyms FD-AJD (Frobenius distance) for (1), mFD-AJD (modified Frobenius distance) for (2) and LD-AJD (log-det) for (3). We initialize all algorithms with the inverse of the square root of the arithmetic mean of the target matrices. For all of them, the stopping criterion for iterate i is defined as $\|B_{i-1}B_i^{-1} - I_n\|_F^2/n$ and is set to 10^{-6}. Note that when comparing the performance of different algorithms, it is very important to use the same stopping criterion. The Riemannian conjugate gradient method is performed using manopt toolbox [12].

3.1   Simulated Data

We simulate sets of K real valued $n \times n$ matrices $\{C_k\}$ according to the model [13]

$$C_k = A\Lambda_k A^T + \frac{1}{\sigma}E_k\Lambda_k^N E_k^T + \alpha I_n, \tag{21}$$

where matrices A and E_k are random matrices with i.i.d. elements drawn from the normal distribution, σ is a free parameter defining the expected signal to noise ratio, $\alpha = 10^{-3}$ is a free parameter representing uncorrelated noise and matrices Λ_k and Λ_k^N are diagonal matrices with i.i.d. elements respectively corresponding to signal matrices and noise. The pth element $\lambda_{p,k}$ is drawn from a chi-squared distribution with expectation $n/p^{1.5}$.

To estimate how the methods behave, we use two criteria. The first one is the Moreau-Amari index $I_{\text{M-A}}$ [14], which is a measure of accuracy, *i.e.*, of how close to the true one is the estimated solution. It is defined as

$$I_{\text{M-A}}(M) = \frac{1}{2n(n-1)} \sum_{p=1}^{n} \left[\frac{\sum\limits_{q=1}^{n} |M_{pq}|}{\max\limits_{1 \leq q \leq n} |M_{pq}|} + \frac{\sum\limits_{q=1}^{n} |M_{qp}|}{\max\limits_{1 \leq q \leq n} |M_{qp}|} \right] - \frac{1}{(n-1)}, \tag{22}$$

where $M = BA$, with B the estimated joint diagonalizer and A the true mixing matrix of the signal part in (21). The second criterion I_C concerns the convergence speed of the algorithms. It measures the distance between the iterate B_i

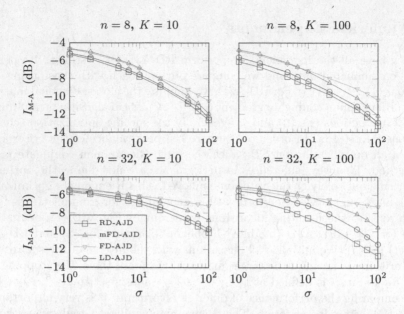

Fig. 1. Mean of the performance of the considered algorithms as a function of the noise parameter σ over 50 trials for different values of n (input matrix dimension) and K (number of matrices). RD-AJD outperforms the other approaches in every cases. The difference with the other methods increases with n and/or K. See text for details.

and the final estimated joint diagonalizer B. It is defined as [13]

$$I_{\text{C}}(B_i) = \frac{\|B_i - B\|_F^2}{\|B\|_F^2}. \tag{23}$$

First, we analyze the quality of the results as a function of the noise parameter σ. One can see in Fig. 1 that RD-AJD outperforms the other methods in all cases. As expected, the performance of the algorithms increases with K and decreases with n. The difference between RD-AJD and the others increases with K and n. This shows that this criterion is more robust with respect to these parameters. This property may be important in practical applications where the size and number of matrices are large.

Concerning the convergence of the algorithms (Fig. 2), RD-AJD generally reaches its final solution faster than the other algorithms in terms of iterations. When it does not ($n = 32$, FD-AJD), it can be explained by the fact that faster methods converge to a less satisfying solution, which is closer to the initial guess B_0. However, in terms computational time, RD-AJD, as performed here, is slower than the other methods since it needs to find the closest diagonal matrices at each iteration. This could be corrected by taking approximations of matrices Λ_k. We will further investigate this in future work.

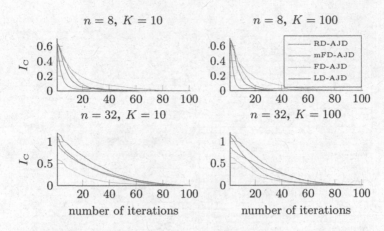

Fig. 2. Mean of the convergence of the considered algorithms as a function of the number of iterations for $\sigma = 50$ over 20 trials. RD-AJD generally converges faster in terms of iterations. When it does not, faster methods converge to a less satisfying solution closer to the initial guess B_0. See text for details.

3.2 Electroencephalographic (EEG) Data

We tested our AJD optimization using the four criteria on an EEG recording acquired on an epileptic patient with 19 electrodes placed according to the international 10–20 system. The sampling rate was 128 Hz and the band-pass was 1–32 Hz. The data comprised 30 s. The BSS of the data was performed using the procedure detailed in [15]. In summary, after a whitening step retaining at least 99.9% of the total variance of the data (reduction to dimension 17 with this data), AJD was performed on the set of Fourier cospectra estimated by 75% overlapping sliding windows of 1 s (Welch method) for frequencies 1 to 32 Hz with 1 Hz resolution.

Figure 3 shows the last 5 s of the original data and the corresponding 17 sources estimated using AJDC [15] performed by the classical AJD algorithm [4], mFD-AJD, mFD-AJD with trace-normalized cospectra, LD-AJD and RD-AJD respectively. Note that FD-AJD gives similar results as compared to mFD-AJD. We are interested here in the three peak-slow wave complexes, often seen in epileptic patients and visible in the original data at frontal locations (Fig. 3, electrode labels starting with F). For all criteria with the exception of mFD-AJD and LD-AJD, the three peak-slow wave complexes are well separated in an unique source (s3 in Fig. 3). This shows that RD-AJD does not need any ad-hoc normalization of the input matrices in order to give satisfying results, whereone to cre sources found by the different methods because the obtained joint diagonalizer are not equivalent. These results demonstrate the accuracy of our optimization procedure and the feasability of using the natural Riemannian distance criterion.

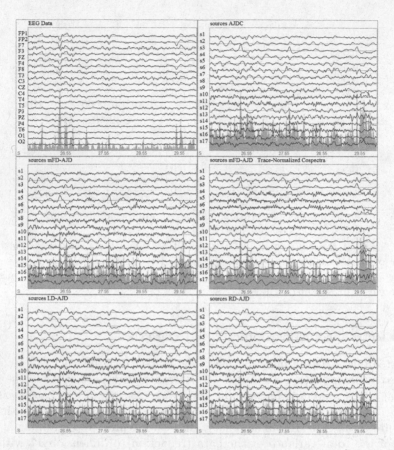

Fig. 3. 5 last seconds of the original EEG data and the 17 sources obtained with classical algorithm AJDC [15], mFD-AJD, mFD-AJD with trace-normalized cospectra, LD-AJD and RD-AJD. See text for details.

4   Conclusions

In this paper, we have provided for the first time an optimization framework for the AJD cost function based on the natural Riemannian distance on the cone of HPD matrices. This problem could not find a solution so far, despite it has been recognized as important [3]. The results obtained on simulated data are promising since our method outperforms the others in terms of accuracy in all cases investigated. Results obtained on real EEG data show that the Riemannian criterion allows to retrieve sources of interest without having to scale the target matrices. Actually, it is invariant by any diagonal scaling which is theoritically advantageous as pointed out in [2,6].

Here, we limited ourselves to a first order optimization method for simplicity. We will investigate second order methods in future works. We will also study if the differences in the performances arise from the properties of the Riemannian distance, the choice of the closest diagonal matrices, or both. Indeed, we can replace the diagonal matrices in (11) by those in (2) and reciprocally. Note that this substitution cannot be operated using (3) due to the properties of the determinant.

Acknowledgment. This work has been partially supported by the LabEx PERSYVAL-Lab (ANR-11-LABX-0025-01) funded by the French program "Investissement d'avenir" and the European Research Council, project CHESS 2012-ERC-AdG-320684.

References

1. Comon, P., Jutten, C.: Handbook of Blind Source Separation: Independent Component Analysis and Applications, 1st edn. Academic Press, New York (2010)
2. Afsari, B.: Sensitivity analysis for the problem of matrix joint diagonalization. SIAM J. Matrix Anal. Appl. **30**(3), 1148–1171 (2008)
3. Alyani, K., Congedo, M., Moakher, M.: Diagonality measures of Hermitian positive-definite matrices with application to the approximate joint diagonalization problem. Linear Algebra its Appl. to be published
4. Tichavský, P., Yeredor, A.: Fast approximate joint diagonalization incorporating weight matrices. IEEE Trans. Signal Process. **57**(3), 878–891 (2009)
5. Flury, B.N., Gautschi, W.: An algorithm for simultaneous orthogonal transformation of several positive definite symmetric matrices to nearly diagonal form. SIAM J. Sci. Stat. Comput. **7**(1), 169–184 (1986)
6. Pham, D.-T.: Joint approximate diagonalization of positive definite Hermitian matrices. SIAM J. Matrix Anal. Appl. **22**(4), 1136–1152 (2000)
7. Bhatia, R.: Positive Definite Matrices. Princeton University Press, Princeton (2009)
8. Moakher, M.: A differential geometric approach to the geometric mean of symmetric positive-definite matrices. SIAM J. Matrix Anal. Appl. **26**(3), 735–747 (2005)
9. Absil, P.-A., Mahony, R., Sepulchre, R.: Optimization Algorithms on Matrix Manifolds. Princeton University Press, Princeton (2008)
10. Bouchard, F., Korczowski, L., Malick, J., Congedo, M.: Approximate joint diagonalization within the Riemannian geometry framework. In: 24th European Signal Processing Conference (EUSIPCO-2016), pp. 210–214 (2016)
11. Carlson, B.C., Keller, J.M.: Orthogonalization procedures and the localization of wannier functions. Phys. Rev. **105**, 102–103 (1957)
12. Boumal, N., Mishra, B., Absil, P.-A., Sepulchre, R.: Manopt, a Matlab toolbox for optimization on manifolds. J. Mach. Learn. Res. **15**, 1455–1459 (2014)
13. Congedo, M., Phlypo, R., Barachant, A.: A fixed-point algorithm for estimating power means of positive definite matrices. In: 24th European Signal Processing Conference (EUSIPCO-2016), pp. 2106–2110 (2016)
14. Moreau, E., Macchi, O.: New self-adaptative algorithms for source separation based on contrast functions. In: IEEE Signal Processing Workshop on Higher-Order Statistics, pp. 215–219. IEEE (1993)
15. Congedo, M., Gouy-Pailler, C., Jutten, C.: On the blind source separation of human electroencephalogram by approximate joint diagonalization of second order statistics. Clin. Neurophysiol. **119**(12), 2677–2686 (2008)

Gaussian Processes for Source Separation in Overdetermined Bilinear Mixtures

Denis G. Fantinato[1]([⊠]), Leonardo T. Duarte[2], Bertrand Rivet[3],
Bahram Ehsandoust[3], Romis Attux[1], and Christian Jutten[3]

[1] School of Electrical and Computer Engineering, University of Campinas,
Campinas, SP, Brazil
{denisgf,attux}@dca.fee.unicamp.br
[2] School of Applied Sciences, University of Campinas, Limeira, SP, Brazil
leonardo.duarte@fca.unicamp.br
[3] GIPSA-Lab, Grenoble INP, CNRS, Grenoble, France
{bertrand.rivet,bahram.ehsandoust,
christian.jutten}@gipsa-lab.grenoble-inp.fr

Abstract. In this work, we consider the nonlinear Blind Source Separation (BSS) problem in the context of overdetermined Bilinear Mixtures, in which a linear structure can be employed for performing separation. Based on the Gaussian Process (GP) framework, two approaches are proposed: the predictive distribution and the maximization of the marginal likelihood. In both cases, separation can be achieved by assuming that the sources are Gaussian and temporally correlated. The results with synthetic data are favorable to the proposal.

Keywords: Blind Source Separation · Bilinear mixtures · Gaussian Process

1 Introduction

In the context of Blind Source Separation (BSS), it is desired to retrieve the original sources signals that were mixed together from a number of observations of these mixtures [1]. Classically, this problem is viewed from the perspective that the mixing process is linear and that the sources are mutually statistically independent. In this context, a number of methods based on Independent Component Analysis (ICA) were successfully employed in many practical applications [1]. However, in certain real world separation problems, e.g., chemical sensor arrays [2], the mixtures are evidently nonlinear. In such cases, the extension of the ICA methods towards a general nonlinear system is not straightforward [3]. In light of this, the studies on nonlinear BSS have been focused on classes of constrained mixing models in which the framework employed in the linear case can be effectively extended. A representative example of these constrained models is the Bilinear mixture equations [4], with actual applications like the show-through effect removal in scanned images [5] and the project of gas sensor arrays [6].

© Springer International Publishing AG 2017
P. Tichavský et al. (Eds.): LVA/ICA 2017, LNCS 10169, pp. 300–309, 2017.
DOI: 10.1007/978-3-319-53547-0_29

The bilinear model is of special interest for us due to three interesting features: (*i*) it can be formulated as a linear system with respect to the mixing coefficients; (*ii*) it is an initial step toward polynomial mixtures; and (*iii*) in the overdetermined case (when the number of mixtures is greater than the number of sources), a linear structure may be enough for separation, under certain conditions on the number of sources and mixtures. In the literature, these features were exploited along with certain statistical properties about the sources, in which we can cite, for instance, circularity [7], finite alphabet [8], sparsity [9] and limited band [10]. In this work, however, we propose a different approach in which the sources are described as independent Gaussian Processes (GP) [11]. Although the GP framework is solely based on second-order statistics (SOS), this idea turns to be promising due to the linear properties found in the overdetermined bilinear model. Moreover, its formulation provides an attractive theoretical approach to perform source separation. The GP method encompassed here is twofold: we consider the GP predictive distribution (a semi-blind approach) and the maximization of the marginal likelihood, respectively.

2   Problem Statement

In the BSS problem, it is considered that a set of N sources are instantaneously mixed, giving rise to M observations according to the following relation:

$$\mathbf{x}(n) = \mathbf{f}\left(\mathbf{s}(n)\right) \tag{1}$$

where $\mathbf{s}(n) = [s_1(n) \cdots s_N(n)]^T$ is the source vector with N elements, $\mathbf{x}(n) = [x_1(n) \cdots x_M(n)]^T$ is the observation vector (of length M) at time instant n and $\mathbf{f}(\cdot)$ is a set of M functions, potentially nonlinear. By assuming the knowledge of only the mixtures $\mathbf{x}(n)$ and certain *a priori* information (e.g., statistical independence among sources), it is desired to recover the original sources $\mathbf{s}(n)$ by means of a separation process, up to scale and permutation factors [1].

Interestingly, for certain types of nonlinear functions $\mathbf{f}(\cdot)$, when the number of mixtures M is larger than that of sources N – which is referred to as the *overdetermined* case –, additional information can be used to systematically simplify the separation process. This attractive feature arises in the context of bilinear mixtures, which will be described in the following.

2.1   Bilinear Mixtures - The Overdetermined Case

The bilinear mixtures belong to a special class of nonlinear mixtures and can be represented by following expression [1,5]:

$$x_i(n) = \sum_{j=1}^{N} a_{ij} s_j(n) + \sum_{k \neq l} b_{ikl} s_k(n) s_l(n), \tag{2}$$

which can be viewed as a linear combination of the sources plus the cross product terms $s_k(n)s_l(n)$, for $k \neq l$. Interestingly, for this type of mixtures, if a given number of additional mixtures is available, it is possible to perform the suppression of the nonlinear terms, reducing the nonlinear problem to a linear one.

For the sake of simplicity, we assume henceforth the overdetermined case of $N = 2$ sources and $M = 3$ mixtures, which is a clarifying and still representative instance for practical scenarios [6]. In this case, by using a vector notation, the bilinear mixtures can be written as

$$\mathbf{x}(n) = \begin{bmatrix} x_1(n) \\ x_2(n) \\ x_3(n) \end{bmatrix} = \begin{bmatrix} a_{11} & a_{12} & b_1 \\ a_{21} & a_{22} & b_2 \\ a_{31} & a_{32} & b_3 \end{bmatrix} \begin{bmatrix} s_1(n) \\ s_2(n) \\ s_1(n)s_2(n) \end{bmatrix}, \tag{3}$$

note that we have used a simplified index notation for b, since there is only a single cross product term – i.e., $s_1(n)s_2(n)$. From a certain perspective, Eq. (3) can be viewed as a linear mixing problem with an additional statistically dependent source $s_1(n)s_2(n) = s_3(n)$ [9]. Regarding the separation task in the overdetermined case, we consider two approaches: (i) the one-stage and (ii) and the two-stage.

In the one-stage approach, a simple linear separating system is considered:

$$\begin{bmatrix} y_1(n) \\ y_2(n) \end{bmatrix} = \begin{bmatrix} w_{11} & w_{12} & w_{13} \\ w_{21} & w_{22} & w_{23} \end{bmatrix} \begin{bmatrix} x_1(n) \\ x_2(n) \\ x_3(n) \end{bmatrix} = \tilde{\mathbf{W}}\mathbf{x}(n), \tag{4}$$

where $\tilde{\mathbf{W}}$ is a non-square matrix of dimension 2×3 ($N \times M$). In this case, $\tilde{\mathbf{W}}$ is adjusted in a single stage.

For the two-stage approach, as shown in [9], a linear combination of the mixtures of the type

$$\begin{bmatrix} z_1(n) \\ z_2(n) \end{bmatrix} = \begin{bmatrix} x_1(n) \\ x_2(n) \end{bmatrix} - \begin{bmatrix} \alpha_1 \\ \alpha_2 \end{bmatrix} x_3(n), \tag{5}$$

is able to suppress the quadratic cross terms in Eq. (2) – i.e., the nonlinear part of the mixture is removed – if the values of $\boldsymbol{\alpha}$ were properly adjusted. Theoretically, it can be shown that the optimum values for $\boldsymbol{\alpha}$ are $\alpha_i = b_i/b_3$ [9] and, when this ideal case is achieved, $\mathbf{z}(n)$ is simply a linear mixture of the sources. We call this the first stage. For the second stage, since we face a linear BSS problem, we can write a 2-by-2 separating system as $\mathbf{y}(n) = \mathbf{W}\mathbf{z}(n)$, whose solution is straightforward to be reached – for example, via standard ICA methods [1]. Although only presented for the case of two sources and three mixtures, this idea can be generalized to N sources, but it will require at least $M = N(N+1)/2$ mixtures to cancel all nonlinear elements [10].

Interestingly, in both cases, a linear structure is able to perform separation. From the literature, for the one-stage approach, the non-square shape of the separating matrix avoids the direct application of certain classical ICA methods. However, it is still possible to use, for example, gradient-based methods for optimization [12]. On the other hand, the efforts in the two-stage approach are mainly aimed at solving the first stage (since the second stage is a well studied problem), usually being assumed certain statistical properties about the sources [7–10].

These approaches are able to encompass a wide range of real world scenarios, but, in order to enlarge its scope, we propose, in this work, a different approach: sources that are described as Gaussian Processes. This idea will lead to two different methods in GP, depending on the approach, one- or two-stage.

3   Gaussian Processes in the Bilinear Mixtures

The main motivation for this work lies in Gaussian processes (GP), which can be defined as a collection of random variables (RV), any finite number of which have a joint Gaussian distribution [11]. GP is able to provide convenient methods, since it can be totally described by second-order statistics (SOS). In the context of bilinear mixtures, a GP can be constructed based on priors about the sources. More specifically, we assume that the sources are stationary, Gaussian distributed and mutually independent. Mathematically, this assumptions can be written, in the case of $N = 2$ sources, as

$$\begin{bmatrix} s_1(n) \\ s_2(n) \end{bmatrix} \sim \mathcal{N} \left(\mu, \begin{bmatrix} \sigma_{s_1}^2 & 0 \\ 0 & \sigma_{s_2}^2 \end{bmatrix} \right) \tag{6}$$

where $\mu = [\mu_1 \ \mu_2]^T$ is the column vector with the sources mean values and $\sigma_{s_i}^2$ the variance of the i-th source $s_i(n)$. For simplicity, we consider in this work that $\mu_i = 0$, for all i. In addition, sources are also assumed to be temporally colored (with different autocorrelation functions).

Based on the model description in Sect. 2 and on the prior given by Eq. (6), it is possible to assert some information about the mixtures $\mathbf{x}(n)$. From Eq. (3), we know that $s_3(n) = s_1(n)s_2(n)$ is the product of two Gaussian distributed RV, which results that $s_3(n)$ can be described by means of a double-sided chi-squared distribution; hence, the probability density function of the mixtures, $p(\mathbf{x})$, is not necessarily Gaussian – it will be approximately Gaussian for small values of b in Eq. (3) or for a large number of sources, due to the central limit theorem [1], however, these cases will not be considered here and $p(\mathbf{x})$ will be assumed to be non-Gaussian.

In our approach, it turns to be interesting to express the estimated sources $\mathbf{y}(n)$ as a GP; in other words, it is desired that a conditional distribution of $\mathbf{y}(n)$ be Gaussian. We consider the two conditional posterior probabilities, $p(\mathbf{y}|\mathbf{x})$ and $p(\mathbf{y}|\mathbf{z})$, which are related to the one- and two-stage approaches, respectively. As we intend to show, they will lead to two representative approaches in the GP formulation, one for each conditional probability: the predictive distribution and the marginal likelihood, as described in the following.

3.1   The Predictive Distribution

In the context of the 'one-stage approach, it is possible to assert a few comments on $p(\mathbf{y}|\mathbf{x})$. Using the Bayes' rule, $p(\mathbf{y}|\mathbf{x}) = p(\mathbf{x}|\mathbf{y})p(\mathbf{y})/p(\mathbf{x})$ and, since $p(\mathbf{x}|\mathbf{y})$ is not necessarily Gaussian and $p(\mathbf{x})$ is definitively not Gaussian, then $p(\mathbf{y}|\mathbf{x})$ is not Gaussian as well. However, from a theoretical standpoint, by conditioning the posterior for a given source, e.g., $\mathbf{s}_i(n)$, we have now $p(\mathbf{y}|\mathbf{x}, s_i) = p(\mathbf{x}|\mathbf{y}, s_i)p(\mathbf{y}|s_i)/p(\mathbf{x}|s_i)$, which is Gaussian distributed: this result becomes more evident by verifying that $p(s_3|s_1)$ or $p(s_3|s_2)$ is Gaussian.

From a GP prediction standpoint, the distribution $p(\mathbf{y}|\mathbf{x}, s_i)$ can be estimated from mixtures $\mathbf{x}(n)$ and a few known source values. This idea can be viewed as an interpolation problem, given certain reference values and the SOS defined by the GP [11]. Once obtained the predictive distribution $p(\mathbf{y}|\mathbf{x}, s_i)$, the estimated sources $\mathbf{y}(n)$ can be obtained using Markov Chain Monte Carlo (MCMC) methods [12].

From the perspective of the bilinear mixture problem, this approach can contribute with an interesting theoretical understanding, but imposes some practical difficulties from the standpoint of blind separation, since the knowledge of certain reference values is difficult to obtain. In light of this, this method is classified as a semi-blind approach. Although this might seem restrictive, in certain cases it can be shown to be feasible, such as in the context of chemical sensor arrays, where the solution concentration measures are preceded by certain calibration points, which can be interpreted here as reference values [6]. Furthermore, the reference values can admit certain degree of error [11], which prompts the employment of this idea along with other methods that are able to provide 'coarse' estimates of sources.

In the GP formulation, the temporal information is crucial. Hence, we compose vectors of the type $\mathbf{y}_i(\mathbf{l}) = [y_i(l_1) \ \ldots \ y_i(l_L)]^T$, for $i = 1, \ldots, N$, which is the i-th output at L time instants, specified by $\mathbf{l} = \{l_1, \ldots, l_L\}$. Suppose now that the sources are known at P time instants, namely $\mathbf{j} = \{j_1, \ldots, j_P\}$, i.e., the vectors $\mathbf{s}_i(\mathbf{j}) = [s_i(j_1) \ \ldots \ s_i(j_P)]^T$, for $i = 1, \ldots, N$, are given (and they will be referred to as 'target' values). Also, the mixtures $\mathbf{x}(l_1), \ldots, \mathbf{x}(l_L), \mathbf{x}(j_1), \ldots, \mathbf{x}(j_P)$ are known (at the time instants $\mathbf{l}$ and $\mathbf{j}$). Then, the goal is to predict the probability density function of $\mathbf{y}_i(\mathbf{l})$, the outputs at time instants $\mathbf{l}$.

For mathematical simplicity, we define the following entities for the case of $N = 2$ sources: $\mathbf{y}(\mathbf{l}) = [\mathbf{y}_1^T(\mathbf{l}) \ \mathbf{y}_2^T(\mathbf{l})]^T$ is the column vector with all outputs at time instants $\mathbf{l}$; similarly, $\mathbf{s}(\mathbf{j}) = [\mathbf{s}_1^T(\mathbf{j}) \ \mathbf{s}_2^T(\mathbf{j})]^T$ is the column vector with the targets; $\mathbf{X}(\mathbf{l}) = [\mathbf{x}(l_1) \ \ldots, \mathbf{x}(l_L)]$ and $\mathbf{X}(\mathbf{j}) = [\mathbf{x}(j_1) \ \ldots, \mathbf{x}(j_P)]$ are the mixtures matrices for time instants $\mathbf{l}$ and $\mathbf{j}$, respectively. Hence, the conditional probability of $\mathbf{y}(\mathbf{l})$ can be denoted as $p(\mathbf{y}(\mathbf{l})|\mathbf{X}(\mathbf{l}), \mathbf{X}(\mathbf{j}), \mathbf{s}(\mathbf{j}))$.

Based on the priors and on the knowledge that $p(\mathbf{y}(\mathbf{l})|\mathbf{X}(\mathbf{l}), \mathbf{X}(\mathbf{j}), \mathbf{s}(\mathbf{j}))$ is Gaussian, we can write the following GP in the case of $N = 2$ sources:

$$\mathbf{y}(\mathbf{l})|\mathbf{X}(\mathbf{j}), \mathbf{s}(\mathbf{j}) \sim \mathcal{GP}(\mathbf{0}, \mathbf{K}(\mathbf{X}(\mathbf{l}), \mathbf{X}(\mathbf{l}))), \qquad (7)$$

where $\mathbf{K}(\mathbf{X}(\mathbf{l}), \mathbf{X}(\mathbf{l}))$ is the covariance matrix of size $2L \times 2L$ in function of $\mathbf{X}(\mathbf{l})$. Equation (7) means that the distribution for $\mathbf{y}(\mathbf{l})$ is jointly Gaussian with zero mean and covariance matrix $\mathbf{K}(\mathbf{X}(l),\mathbf{X}(l))$. However, $\mathbf{K}(\mathbf{X}(l),\mathbf{X}(l))$ must be defined so that it is able to encompass the temporal structure of the sources and the mutual independence information. Hence, in this work, we propose the use of a block-diagonal covariance matrix of the type:

$$\mathbf{K}(\mathbf{X}(l),\mathbf{X}(l)) = \begin{bmatrix} \mathbf{K}_{y_1}(\mathbf{X}(l),\mathbf{X}(l)) & 0 \\ 0 & \mathbf{K}_{y_2}(\mathbf{X}(l),\mathbf{X}(l)) \end{bmatrix} \quad (8)$$

where $\mathbf{K}_{y_i}(\mathbf{X}(l),\mathbf{X}(l))$ is a covariance submatrix of dimension $L \times L$, whose element of the l-th row and l'-th column is given by the squared-exponential (SE) function [11]:

$$\mathbf{K}_{y_i}(\mathbf{X}(l),\mathbf{X}(l')) = \gamma_i \exp\left(\frac{-1}{2}(\mathbf{x}(l) - \mathbf{x}(l'))^T \mathbf{\Sigma}_i^{-1}(\mathbf{x}(l) - \mathbf{x}(l'))\right), \quad (9)$$

with $\mathbf{\Sigma}_i = \sigma_i^2 \mathbf{I}_M$, being σ_i^2 the estimated variance of source i, γ_i a scale factor, and $\mathbf{I}_M$ the identity matrix of order M (here $M = 3$).

Hence, from the GP classical results [11], the predictive distribution can be obtained by

$$p\left(\mathbf{y}(l)|\mathbf{X}(l),\mathbf{X}(j),\mathbf{s}(j)\right) \sim \mathcal{N}\left(\mathbf{K}(\mathbf{X}(j),\mathbf{X}(l))\mathbf{\Gamma}^{-1}\mathbf{s}(j),\right.$$
$$\left.\mathbf{K}(\mathbf{X}(j),\mathbf{X}(j)) - \mathbf{K}(\mathbf{X}(j),\mathbf{X}(l))\mathbf{\Gamma}^{-1}\mathbf{K}(\mathbf{X}(l),\mathbf{X}(j))\right). \quad (10)$$

where $\mathbf{\Gamma} = \mathbf{K}(\mathbf{X}(l),\mathbf{X}(l)) + \mathbf{\Phi}$, with $\mathbf{\Phi} = [\epsilon_1^2 \mathbf{I}_L \ \mathbf{0}; \mathbf{0} \ \epsilon_2^2 \mathbf{I}_L]$, a diagonal matrix which is able to consider the degree of error (or uncertainty) in the target samples by means of noise variances ϵ_i^2 associated to each source. In that sense, the targets $\mathbf{s}(j)$ can admit certain level of error, and the accuracy of the predictive distribution will depend on the estimation of the error variances ϵ_i^2. Equation (10) shows that $p\left(\mathbf{y}(l)|\mathbf{X}(l),\mathbf{X}(j),\mathbf{s}(j)\right)$ is Gaussian distributed with mean $\mathbf{K}(\mathbf{X}(j),\mathbf{X}(l))\mathbf{\Gamma}^{-1}\mathbf{s}(j)$ and covariance matrix $\mathbf{K}(\mathbf{X}(j),\mathbf{X}(j)) - \mathbf{K}(\mathbf{X}(j),\mathbf{X}(l))\mathbf{\Gamma}^{-1}\mathbf{K}(\mathbf{X}(l),\mathbf{X}(j))$.

The complexity of the method exponentially increases with the number of considered time instants (for both target, j, and predicted samples, l). In view of this, for implementation purposes, a Cholesky decomposition can be used to simplify the inversion of $\mathbf{\Gamma}$ and to generate the predicted samples of $\mathbf{y}(l)$ [11]. The variables $\tilde{\boldsymbol{\theta}} = \{\gamma_1,\gamma_2,\sigma_1,\sigma_2,\epsilon_1,\epsilon_2\}$ are called hyperparameters of the GP and can be adjusted by maximizing the marginal likelihood given the targets [11]. Once the predicted source samples are obtained, the separation matrix $\tilde{\mathbf{W}}$ can be directly estimated via supervised approaches [1], if necessary.

3.2   Maximization of the Marginal Likelihood

For the two-stage case, using Bayes' rule, $p(\mathbf{y}|\mathbf{z}) = p(\mathbf{z}|\mathbf{y})p(\mathbf{y})/p(\mathbf{z})$, $p(\mathbf{z})$ – and, consequently, $p(\mathbf{y})$ – can be Gaussian, but not necessarily (note that the output $\mathbf{z}(n)$ will be associated with a Gaussian distribution if the values of $\boldsymbol{\alpha}$ are properly adjusted). Nonetheless, when $p(\mathbf{z})$ is Gaussian, $p(\mathbf{y}|\mathbf{z})$ will also be. In this case, the marginal likelihood $p(\mathbf{y}|\mathbf{z})$ can be maximized with respect to the hyperparameters of the system, forcing it towards the Gaussian distribution.

Using the previously defined notation, for a given set of observations $\mathbf{X}(1)$ and for given values of $\boldsymbol{\alpha}$ and $\mathbf{W}$, it is possible to obtain $\mathbf{Z}(1)$ and, in the sequence, $\mathbf{y}(1)$ – see Eq. (5). Thus, we wish that $p(\mathbf{y}(1)|\mathbf{Z}(1))$ be described according to a GP, i.e.,

$$\mathbf{y}(1) \sim \mathcal{GP}\left(0, \mathbf{K}(\mathbf{Z}(1), \mathbf{Z}(1))\right), \tag{11}$$

being $\mathbf{K}(\mathbf{Z}(1), \mathbf{Z}(1))$ a block-diagonal covariance matrix, as defined in Eq. (8) with inputs $\mathbf{Z}(1)$ instead of $\mathbf{X}(1)$.

By denoting $\boldsymbol{\theta} = \{\gamma_1, \gamma_2, \sigma_1, \sigma_2, \boldsymbol{\alpha}, \mathbf{W}\}$ the vector of all hyperparameters, it is possible to write the log likelihood

$$\mathcal{L} = \frac{-1}{2}\mathbf{y}^T(1)\mathbf{K}^{-1}(\mathbf{Z}(1), \mathbf{Z}(1))\mathbf{y}(1) - \frac{1}{2}\log|\mathbf{K}(\mathbf{Z}(1), \mathbf{Z}(1))| - \frac{(L+1)}{2}\log 2\pi, \tag{12}$$

being $\mathbf{K}(\mathbf{Z}(1), \mathbf{Z}(1))$ a function of $\mathbf{Z}(1)$ and $\boldsymbol{\theta}$.

Hence, by maximizing $\mathcal{L}$ with respect to the hyperparameters $\boldsymbol{\theta}$, we hope to obtain a conditional probability $p(\mathbf{y}(1)|\mathbf{Z}(1))$ that is Gaussian and, in this case, the optimal parameters α and $\mathbf{W}$ are the solution of the separation problem. Note that, in this case, the two-stages are solved simultaneously. As in the predictive case, the complexity grows exponentially with the number of samples and large data sets should be avoided.

The effectiveness of this approach comes from the fact that different time delays are being compared in the GP by construction. In fact, since we are considering only SOS, the temporal structure is essential for separation. This idea is encompassed by the block diagonal covariance function $\mathbf{K}(\mathbf{Z}(1), \mathbf{Z}(1))$ which allows temporal correlation between samples of the same output, but applies decorrelation (in different time delays) for different outputs. This is also valid for the predictive case. In the sequence, we compare the performance of the two proposed GP approaches.

4   Simulation Results

In order to test the two proposed methods, we consider a simulation scenario for the 2 source and 3 mixture case. The sources are assumed to be two colored Gaussian distributed, obtained from $i.i.d.$ Gaussian sampled, that are temporally colored by the finite impulse response (FIR) filters with impulse response $h_1(z) = 1+0.6z^{-1}-0.3z^{-2}$ and $h_2(z) = 1-0.8z^{-1}$, each source separately. We considered the following mixing matrix,

$$\mathbf{A} = \begin{bmatrix} a_{11} & a_{12} & b_1 \\ a_{21} & a_{22} & b_2 \\ a_{31} & a_{32} & b_3 \end{bmatrix} = \begin{bmatrix} -0.8049 & 0.0938 & -0.0292 \\ -0.4430 & 0.9150 & 0.6006 \\ -1.4434 & -0.4997 & 0.5180 \end{bmatrix}, \tag{13}$$

whose mixtures were obtained using Eq. (3).

The predictive distribution is the first approach to be tested. From a set of 100 samples of the mixtures $\mathbf{x}(n)$, we consider the cases in which $P = 10$ and $P = 5$ targets $\hat{\mathbf{s}}(j)$ are available (with and without noise). For this purpose,

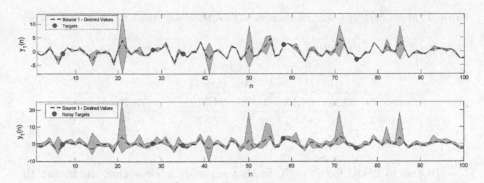

Fig. 1. Predictive - Noise (top) and Noiseless (bottom) Targets. The shaded area denotes the prediction uncertainty.

within the time window of 100 samples, P time instants are randomly picked, for which the sources are assumed to be known. Hence, the objective is to correctly predict the system output $\mathbf{y}(\mathbf{l})$, given $\mathbf{s}(\mathbf{j})$, $\mathbf{X}(\mathbf{j})$ and $\mathbf{X}(\mathbf{l})$. Additionally, a perturbation can be considered with additive white Gaussian noise (to simulate targets with certain degree of error) with variance $\sigma^2 = 1e-1$, resulting a SNR level of 14.9 dB. 1000 independent experiments were considered, from which each realization encompassed a new set of mixtures and targets. We start by adjusting the hyperparameters $\tilde{\boldsymbol{\theta}}$, which were empirically chosen to be $\gamma_1 = 10$, $\gamma_2 = 30$, $\sigma_1 = 4.5$, $\sigma_2 = 5.48$, $\epsilon_1 = \epsilon_2 = 0$. For illustrative purposes, we display in Fig. 1 one realization of the predicted distribution for $P = 5$ and for noiseless (top) and noisy (bottom) targets. The dashed line represents the desired output and the red circle the known targets. The shaded area comprises the region where a realization of $p(y|\mathbf{x}, \mathbf{s})$ could fall (more precisely, the region denotes the predicted mean value ± 3 times the predicted standard deviation for each time instant). Hence, a large shaded area means the prediction is less accurate. It is possible to note for the noiseless targets that, with exception of certain small regions, the predicted region is small and falls really close to the desired output. Although we do not show here, for $P = 10$ noiseless targets, the prediction is very accurate. Notwithstanding, when the provided targets present certain level of error, the accuracy is reduced for all time instants; as shown in Fig. 1 (bottom), the shaded area is increased with respect to the noiseless case. In the noisy case, we have chosen $\epsilon_1^2 = \epsilon_2^2 = \sigma^2 = 1e-1$.

In order to evaluate the quality of the predicted samples, we measured the *signal-to-interference ratio* (SIR), defined as SIR $= 10 \log \left(E[y_i(n)^2] / E[(s_i(n) - y_i(n))^2] \right)$. Since we are interested in the mean performance, it was considered only the predicted mean for computing the SIR (and averaged over all realizations). The results are displayed in Table 1. As previously discussed, in the noiseless case ($\sigma^2 = 0$), we can see high values of SIR for both values of P (above 50 dB), which means that the sources are recovered with small error. However, the reduction of the number of targets causes the performance to slightly decrease. In the case of noisy targets, the SIR is reduced even more, to approximately 40 dB for

Table 1. Mean SIR [dB] for the maximum likelihood and the predictive approaches

	Predictive				Maximum likelihood
	$\sigma^2 = 0$		$\sigma^2 = 1e{-}1$		
	$P = 10$	$P = 5$	$P = 10$	$P = 5$	
Source 1	73.41	64.18	38.08	24.78	13.70
Source 2	74.30	65.35	40.59	27.81	16.26

$P = 10$ and to $25\,\mathrm{dB}$ for $P = 5$. In that sense, it is clear that the higher the number of targets and its quality, the higher is the SIR of the prediction.

For the second approach, we consider the maximization of the log likelihood, in which we wish to adjust the hyperparameters $\boldsymbol{\theta}$ to its optimal values. The adaptation can be performed according to several optimization methods [12]. However, in this work, we adopt a metaheuristic for optimization, which, although not able to guarantee that a global solution will be found, is more robust against local convergence. Particularly, we adopt the *Differential Evolution* (DE) [13] metaheuristic, an efficient technique that exploit the search space information that is available in the current population, instead of using conventional random operators (for more details, please refer to [13]). The chosen DE parameters are $N_P = 300$ (population size), $F = 0.7$, $CR = 0.7$ and 100 iterations.

In this case, we consider a set of 45 samples of $\mathbf{x}(n)$ and performed the adaptation of $\boldsymbol{\theta}$ via the DE method. This procedure was repeated 10 times, for different realizations of $\mathbf{x}(n)$. Even with a reduced number of samples, each optimization leads to good results, with mean value of $\alpha_1 = 0.0456$ and $\alpha_2 = 1.1498$, which are close to the ideal values $\alpha_1 = 0.0564$ and $\alpha_2 = 1.1595$, respectively. The mean SIR values can be seen in Table 1. Although this SIR value is not quite impressive, the performance is good if the reduced number of samples is taken into account.

5   Conclusions

In this work, we proposed two GP formulations to solve the overdetermined bilinear mixing problem concerning the one- and two-stage approaches. Our propositions are both based on the prior that the sources are mutually independent Gaussian distributed, however, they differ in the application of the GP method. We have shown that, for the GP formulation consistency, it was required in the former case the conditioning of the output distribution to the knowledge of certain reference samples, named targets, what prompted us to adopt a predictive based GP approach. In the second case, the two-stage approach allowed the adjustment of the hyperparameters via the maximization of the marginal likelihood. As shown in the simulation results, the predictive approach tends to provide better results, depending on the quality and number of the targets, however, the maximization of the marginal likelihood is also able to perform the separation of the sources. Although these GP methods present certain algorithmic complexity, the computational burden can be reduced by using a relatively small number of samples.

For future works, we consider a deeper analysis of the requirements for separation – e.g., a spectral density analysis of the temporal structure – and the extension of these theoretical analysis to noisy scenarios and to other classes of mixing systems, like the Linear Quadratic.

Acknowledgements. This work was partly supported by FAPESP (2013/14185-2, 2015/23424-6), CNPq and ERC project 2012-ERC-AdG-320684 CHESS.

References

1. Comon, P., Jutten, C.: Handbook of Blind Source Separation: Independent Component Analysis and Applications. Academic Press, Oxford (2010)
2. Duarte, L.T., Jutten, C., Moussaoui, S.: A Bayesian nonlinear source separation method for smart ion-selective electrode arrays. IEEE Sens. J. **9**(12), 1763–1771 (2009)
3. Hosseini, S., Jutten, C.: On the separability of nonlinear mixtures of temporally correlated sources. IEEE Sig. Process. Lett. **10**(2), 43–46 (2003)
4. Hosseini, S., Deville, Y.: Blind separation of linear-quadratic mixtures of real sources using a recurrent structure. In: Mira, J., Álvarez, J.R. (eds.) IWANN 2003. LNCS, vol. 2687, pp. 241–248. Springer, Heidelberg (2003). doi:10.1007/3-540-44869-1_31
5. Merrikh-Bayat, F., Babaie-Zadeh, M., Jutten, C.: Linear-quadratic blind source separating structure for removing show-through in scanned documents. Int. J. Doc. Anal. Recogn. (IJDAR) **14**(4), 319–333 (2011)
6. Duarte, L.T., Jutten, C.: Design of smart ion-selective electrode arrays based on source separation through nonlinear independent component analysis. Oil Gas Sci. Technol. **69**(2), 293–306 (2014)
7. Abed-Meraim, K., Belouchiani, A., Hua, Y.: Blind identification of a linear-quadratic mixture of independent components based on joint diagonalization procedure. IEEE ICASSP **5**, 2718–2721 (1996)
8. Castella, M.: Inversion of polynomial systems and separation of nonlinear mixtures of finite-alphabet sources. IEEE Trans. Sig. Proc. **56**(8), 3905–3917 (2008)
9. Duarte, L.T., Ando, R.A., Attux, R., Deville, Y., Jutten, C.: Separation of sparse signals in overdetermined linear-quadratic mixtures. In: Theis, F., Cichocki, A., Yeredor, A., Zibulevsky, M. (eds.) LVA/ICA 2012. LNCS, vol. 7191, pp. 239–246. Springer, Heidelberg (2012). doi:10.1007/978-3-642-28551-6_30
10. Ando, R.A., Duarte, L.T., Attux, R.R.F.: Blind source separation for overdetermined linear quadratic mixtures of bandlimited signals. In: IEEE International Telecommunications Symposium (ITS) (2014)
11. Rasmussen, C.: Gaussian Processes for Machine Learning. MIT Press, Cambridge (2006)
12. Deville, Y., Duarte, L.T.: An overview of blind source separation methods for linear-quadratic and post-nonlinear mixtures. In: Vincent, E., Yeredor, A., Koldovský, Z., Tichavský, P. (eds.) LVA/ICA 2015. LNCS, vol. 9237, pp. 155–167. Springer, Heidelberg (2015). doi:10.1007/978-3-319-22482-4_18
13. Price, K., Storn, R., Lampinen, J.: Differential Evolution: A Practical Approach to Global Optimization. Springer, Heidelberg (2005)

Model-Independent Method of Nonlinear Blind Source Separation

David N. Levin$^{(\boxtimes)}$

Department of Radiology, University of Chicago,
1310 N. Ritchie Ct., Unit 26 AD, Chicago, IL 60610, USA
d-levin@uchicago.edu
http://radiology.uchicago.edu/directory/david-n-levin

Abstract. Consider a time series of signal measurements $x(t)$, where x has two components. This paper shows how to process the local distributions of measurement velocities in order to construct a two-component mapping, $u(x)$. If the measurements are linear or nonlinear combinations of statistically independent variables, $u(x)$ must be an unmixing function. In other words, the measurement data are separable if and only if $u_1[x(t)]$ and $u_2[x(t)]$ are statistically independent of one another. The method is analytic, constructive, and model-independent. It is illustrated by blindly recovering the separate utterances of two speakers from nonlinear combinations of their waveforms.

Keywords: Blind source separation · Nonlinear signal processing · Invariants · Sensor · Analytic · Model-independent

1 Introduction

Consider an evolving physical system that is being observed by making time-dependent measurements ($x_k(t)$ for $k = 1, 2$), which are invertibly related to the system's state variables. The objective of blind source separation (BSS) is to determine if the measurements are mixtures of the state variables of statistically independent subsystems. Specifically, we want to know if there is an invertible, possibly nonlinear, two-component "unmixing" function, f, that transforms the measurement time series into a time series of separable states:

$$s(t) = f[x(t)]. \tag{1}$$

Here, $s(t)$ denotes a set of components, $s_k(t)$ for $k = 1, 2$, which comprise the state variables of "statistically independent" subsystems. In other words, we want to know if the data can be transformed from the measurement coordinate system, x, into another coordinate system, s, in which the data's components are statistically independent.

There is a variety of methods for solving this blind source separation (BSS) problem for the special case in which f is linear. However, some observed signals

© Springer International Publishing AG 2017
P. Tichavský et al. (Eds.): LVA/ICA 2017, LNCS 10169, pp. 310–319, 2017.
DOI: 10.1007/978-3-319-53547-0_30

(e.g., from biological or economic systems) may be nonlinear functions of the underlying subsystem states. Computational methods of separating such nonlinear mixtures are limited [1], even though humans often seem to separate the data with little effort.

This paper utilizes a criterion for "statistical independence" that differs from the conventional one. Specifically, let $\rho_S(s, \dot{s})$ be the probability density function (PDF) in $(s, \dot{s})$-space, where $\dot{s} = ds/dt$. In this paper, the data are defined to be separable if and only if there is an unmixing function that transforms the measurements so that $\rho_S(s, \dot{s})$ is the product of the density functions of individual components

$$\rho_S(s, \dot{s}) = \prod_{a=1,2} \rho_{Sa}(s_a, \dot{s}_a). \tag{2}$$

where s_a is the state variable of subsystem a. This criterion for separability is consistent with our intuition that the statistical distribution of one subsystem's state and velocity should not depend on the particular state and velocity of any other subsystem.

This criterion for statistical independence should be compared to the conventional criterion, which is formulated in s-space (i.e., state space) instead of $(s, \dot{s})$-space (the space of states and state velocities):

$$\rho_S(s) = \prod_{a=1,2} \rho_{Sa}(s_a). \tag{3}$$

In *every* formulation of BSS, multiple solutions can be created from a given solution by applying "component-wise" transformations, which transform the components of s among themselves. These solutions only differ in their choice of the coordinate systems used to describe each subsystem. However, the criterion in (3) is so weak that it suffers from a much worse non-uniqueness problem: namely, new solutions can almost always be created by mixing the independent state variables of other solutions [1].

There are at least two reasons why (2) is a preferable way of defining "statistical independence":

1. If a physical system is comprised of two independent subsystems, we normally expect that there is a unique way of identifying the subsystems. As mentioned above, (3) is too weak to meet this expectation. On the other hand, (2) is a much stronger constraint than (3). Specifically, (3) can be recovered by integrating both sides of (2) with respect to velocity. This shows that the solutions of (2) are a subset of the solutions of (3). Therefore, it is certainly possible that (2) reformulates the BSS problem so that it has a unique solution (up to component-wise transformations), although this is not proved in this paper.
2. For all systems that obey the laws of classical mechanics and are in thermal equilibrium, the PDF in $(s, \dot{s})$-space is proportional to the Maxwell-Boltzmann distribution. If the system consists of non-interacting subsystems,

its energy is the sum of the subsystem energies, and, therefore, this distribution factorizes exactly as in (2). Thus, for classical physical systems, non-interacting subsystems are statistically independent in the sense of (2).

There are several other ways in which the proposed method of nonlinear BSS differs from methods in the literature:

1. Although there is some earlier work in which BSS is performed with the aid of velocity information [5,7], these papers utilize the *global* distribution of measurement velocities (i.e., the distribution of velocities at all points in state space). In contrast, the method proposed here exploits additional information that is present in the *local* distributions of measurement velocities (i.e., the velocity distributions in each region of state space).
2. Many investigators have attempted to simplify the BSS problem by assuming prior knowledge of the nature of the mixing function; i.e., they have modelled the mixing function. For example, the mixing function has been assumed to have a specific parametric form that describes post-nonlinear mixtures [6], linear-quadratic mixtures [4], and other combinations. In contrast, the present paper proposes a model-independent method that can be used in the presence of any invertible diffeomorphic mixing function.
3. In most other approaches, nonlinear BSS is reduced to the optimization problem of finding the unmixing function that maximizes the independence of the source signals corresponding to the observed mixtures. This usually requires the use of iterative algorithms with attendant issues of convergence and computational cost (e.g. [1]). In contrast, the method proposed in this paper is analytic and constructive.

2   Method

This section describes a five-step procedure for determining if the data are separable and, if so, for constructing an unmixing function.

1. Use the local distribution of measurement velocities to construct two vectors at each point x: $V_{(i)}(x)$ for $i = 1, 2$.
The first step is to construct second-order and fourth-order local correlations of the data's velocity

$$C_{kl}(x) = \langle (\dot{x}_k - \bar{\dot{x}}_k)(\dot{x}_l - \bar{\dot{x}}_l) \rangle_x \tag{4}$$

$$C_{klmn}(x) = \langle (\dot{x}_k - \bar{\dot{x}}_k)(\dot{x}_l - \bar{\dot{x}}_l)(\dot{x}_m - \bar{\dot{x}}_m)(\dot{x}_n - \bar{\dot{x}}_n) \rangle_x \tag{5}$$

where $\bar{\dot{x}} = \langle \dot{x} \rangle_x$, where the bracket denotes the time average over the trajectory's segments in a small neighborhood of x, and where all indices are integers equal to 1 or 2. Because $\dot{x}$ is a contravariant vector, $C_{kl}(x)$ and $C_{klmn}(x)$ are local contravariant tensors of second and fourth rank, respectively. The definition of the PDF implies that $C_{kl}(x)$ and $C_{klmn}(x)$ are two of its moments; e.g.,

$$C_{kl...}(x) = \frac{\int \rho(x, \dot{x})(\dot{x}_k - \bar{\dot{x}}_k)(\dot{x}_l - \bar{\dot{x}}_l)\ldots d\dot{x}}{\int \rho(x, \dot{x}) d\dot{x}}, \tag{6}$$

where $\rho(x, \dot{x})$ is the PDF in the x coordinate system, where "..." denotes possible additional indices on the left side and corresponding factors of $\dot{x} - \bar{\dot{x}}$ on the right side, and where all indices are integers equal to 1 or 2. Although (6) is useful in a formal sense, in practical applications all required correlation functions can be approximated directly from local time averages of the data (e.g., (4) and (5)), without explicitly computing the data's PDF. Also, note that velocity "correlations" with a single subscript vanish identically

$$C_k(x) = 0. \tag{7}$$

Next, let $M(x)$ be any local 2×2 matrix, and use it to define M-transformed velocity correlations, I_{kl} and I_{klmn}

$$I_{kl}(x) = \sum_{1 \leq k', l' \leq 2} M_{kk'}(x) M_{ll'}(x) C_{k'l'}(x), \tag{8}$$

$$I_{klmn}(x) = \sum_{1 \leq k', l', m', n' \leq 2} M_{kk'}(x) M_{ll'}(x) M_{mm'}(x) M_{nn'}(x) C_{k'l'm'n'}(x) \tag{9}$$

Because the $C_{kl}(x)$ are the elements of a matrix that is generically positive definite, it is possible to find a particular form of $M(x)$ that satisfies

$$I_{kl}(x) = \delta_{kl} \tag{10}$$

$$\sum_{1 \leq m \leq 2} I_{klmm}(x) = D_{kl}(x), \tag{11}$$

where δ_{kl} is the Kronecker delta and $D(x)$ is a diagonal 2×2 matrix. Such an $M(x)$ can be constructed from the product of three matrices: (1) a rotation that diagonalizes $C_{kl}(x)$, (2) a diagonal rescaling matrix that transforms this diagonalized correlation into the identity matrix, (3) another rotation that diagonalizes

$$\sum_{1 \leq m \leq 2} \tilde{C}_{klmm}(x),$$

where $\tilde{C}_{klmn}(x)$ is the fourth-order velocity correlation $(C_{klmn}(x))$ after it has been transformed by the first rotation and the rescaling matrix. As long as D is not degenerate, $M(x)$ is unique, up to arbitrary *local* permutations and/or reflections. In almost all applications of interest, the velocity correlations will be continuous functions of x. Therefore, in any neighborhood of state space, there will always be a continuous solution for $M(x)$, and this solution is unique, up to arbitrary *global* permutations and/or reflections.

In any other coordinate system x', the most general solution for M' is given by

$$M'_{kl}(x') = \sum_{1 \leq m, n \leq 2} P_{km} M_{mn}(x) \frac{\partial x_n}{\partial x'_l}, \tag{12}$$

where M is a matrix that satisfies (10) and (11) in the x coordinate system and where P is a product of permutation and reflection matrices. This can be proven by substituting this equation into the definition of $I'_{kl}(x')$ and $I'_{klmn}(x')$ and by noting that these quantities satisfy (10) and (11) in the x' coordinate system because (8) and (9) satisfy them in the x coordinate system. By construction, M is not singular, and, therefore, it has a non-singular inverse.

Notice that (12) shows that the rows of M transform as local covariant vectors, up to global permutations and/or reflections. Likewise, the same equation implies that the columns of M^{-1} transform as local contravariant vectors (denoted as $V_{(i)}(x)$ for $i = 1, 2$), up to global permutations and/or reflections. As shown in the following, these particular vectors contain significant information about the separability of the data. In fact, they can be used to construct a mapping that must be an unmixing function, if one exists.

2. *Use the $V_{(i)}(x)$ to construct a mapping, $u(x) = (u_1(x), u_2(x))$.*
Working in the x coordinate system, we begin by picking any point x_0. We then find a curve $X(\sigma)$ that passes through x_0 and is always tangential to the local vector $V_{(1)}(x)$. Here, σ denotes a variable that parameterizes the curve and increases monotonically as the curve is traversed in one direction. In mathematical terms, $X(\sigma)$ can be chosen to be the solution of the first-order differential equation

$$\frac{dX}{d\sigma} = V_{(1)}(X) \tag{13}$$

that satisfies the boundary condition, $X(0) = x_0$. Then, for each value of σ, we construct a curve, $Y(\tau)$, which passes through the point $X(\sigma)$ and is always tangential to the local vector $V_{(2)}(x)$. Here, τ parameterizes this curve, increasing monotonically as it is traversed in one direction. Mathematically, $Y(\tau)$ is the solution of

$$\frac{dY}{d\tau} = V_{(2)}(Y) \tag{14}$$

that satisfies the boundary condition, $Y(0) = X(\sigma)$. Finally, the function $u_1(x)$ is defined so that it is constant along each of the Y curves. Specifically, $u_1(x) = \sigma$ whenever x is on the Y curve passing through $X(\sigma)$. A function $u_2(x)$ can be defined by following an analogous procedure in which the roles of $V_{(1)}(x)$ and $V_{(2)}(x)$ are switched.

3. *Use $u(x)$ to transform the measured time series $x(t)$ into the u coordinate system.*
This is done by straight-forward substitution: i.e., $u[x(t)] = (u_1[x(t)], u_2[x(t)])$.

4. *Compute the PDF of the transformed time series, $u[x(t)]$, and determine if it factorizes as*

$$\rho_U(u, \dot{u}) = \prod_{a=1,2} \rho_{Ua}(u_a, \dot{u}_a). \tag{15}$$

Here, u denotes $u[x(t)]$, and $\dot{u}$ is its time derivative. Alternatively, we can compute a large set of correlations of the transformed time series and then determine if the higher-order correlations are products of lower-order correlations. If these correlations factorize in this manner, it suggests factorizability of the PDF itself.

5. *Use the result of step 4 to determine if the data are separable, and, if they are, to determine an unmixing function. Specifically, if the PDF is found to be factorizable in step 4, it is obvious that the data are separable and $u(x)$ is an unmixing function. On the other hand, if the PDF is not factorizable, the data are inseparable in any coordinate system.*

This last statement is a consequence of the following fact, which is proved in the next two paragraphs: namely, if the data are separable, the constructed mapping, $u(x)$, must be an unmixing function.

Before proving this, we show that the matrix M and the $V_{(i)}(x)$ have simple forms in the separable coordinate system, s. In particular, we prove that the following diagonal matrix is the M matrix in the s coordinate system

$$M_S(s) = \begin{pmatrix} C_{S11}^{-0.5}(s_1) & 0 \\ 0 & C_{S22}^{-0.5}(s_2) \end{pmatrix}, \tag{16}$$

where $C_{Skl}(s)$ for $k, l = 1, 2$ are the second-order velocity correlations in the s coordinate system. This can be proved by demonstrating that M_S satisfies (10) and (11) in the s coordinate system. To do this, first note that (2), (6), and (7) imply that the second-order velocity correlations are diagonal in the s coordinate system. It follows that (10) is satisfied by M_S in the s coordinate system. Furthermore, it is not difficult to show that (11) is also satisfied by M_S in the s coordinate system. To see this, substitute (16) into the sum in the left side of (11) for $k \neq l$. Because of the diagonality of M_S, each term in this summation is proportional to a fourth-order velocity correlation that has just one index equal to 1 (or 2) and the other three indices all equal to 2 (or 1). Each of these terms must vanish because of (2), (6), and (7). This completes the proof that M_S satisfies both (10) and (11) in the s coordinate system, and, therefore, it is the M matrix in the s coordinate system, as asserted above.

Because M_S is diagonal, the local vectors in the s coordinate system, denoted $V_{S(1)}(s)$ and $V_{S(2)}(s)$, are oriented along the unit vectors, $(1, 0)$ and $(0, 1)$, respectively. Therefore, in the s coordinate system, $X(\sigma)$ is a horizontal straight line passing through the point $s[x_0]$, Similarly, each Y curve is a vertical straight line passing through $s[X(\sigma)]$ for some value of σ. This implies that s_1 is constant along each Y curve, being equal to the value of s_1 at its intersection with the X curve. But, recall that $u_1(x)$ is also constant along each Y curve, being equal to the value of σ at its intersection with the X curve. Therefore, because σ is defined to vary monotonically along the X curve and because the values of s_1 also vary monotonically along that curve, these paired values must be monotonically related to one another; i.e., $\sigma = h_1[s_1]$ where h_1 is a monotonic function. It follows that $u_1(x)$ and $s_1(x)$ must also be monotonically related at each point; i.e., $u_1(x) = h_1[s_1(x)]$. In a similar manner, it can be shown that $u_2(x)$ and $s_2(x)$ are also related by some monotonic function. This means that $u_1(x)$ and $u_2(x)$ are the same as $s_1(x)$ and $s_2(x)$, except for possible component-wise transformations. Because component-wise transformations do not affect separability, it immediately follows that $u(x)$ is an unmixing function, as asserted above.

3   Experiments

In this section, the new BSS technique is illustrated by using it to disentangle nonlinear mixtures of the audio waveforms of two male speakers. Each unmixed waveform was 30 s long and consisted of an excerpt from one of two audio book recordings. The waveform of each speaker, $s_k(t)$ for $k = 1$ or 2, was sampled 16,000 times per second with two bytes of depth. The thick gray lines in Fig. 1 show the two speakers' waveforms during a short (30 ms) interval. These waveforms were then mixed by the nonlinear functions

$$\mu_1(s) = 0.763s_1 + (958 - 0.0225s_2)^{1.5}$$
$$\mu_2(s) = 0.153s_2 + (3.75 * 10^7 - 763s_1 - 229s_2)^{0.5}, \tag{17}$$

where $-2^{15} \leq s_1, s_2 \leq 2^{15}$. This is one of a variety of nonlinear transformations that were tried with similar results. The measurements, $x_k(t)$, were taken to be the variance-normalized, principal components of the sampled waveform mixtures, $\mu_k[s(t)]$. Figure 2a shows how this nonlinear mixing mapped an evenly-spaced Cartesian grid in the s coordinate system onto a warped grid in the x coordinate system. Figure 2b shows a random subset of the measurements $x(t)$, and Fig. 3 shows the time course of $x(t)$ during the same short time interval depicted in Fig. 1. When either waveform mixture, $x_1(t)$ or $x_2(t)$, was played as an audio file, it sounded like a confusing superposition of two voices, which were quite difficult to understand.

The proposed BSS technique was then applied to these measurements as follows:

1. The entire set of 500,000 measurements, consisting of x and $\dot{x}$ at each sampled time, was sorted into a 16×16 array of bins. Then, the $\dot{x}$ distribution in

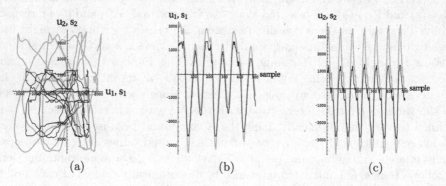

Fig. 1. (a) The thick gray line depicts the trajectory of 30 ms of the two speakers' unmixed speech in the s coordinate system, in which each component is equal to one speaker's speech amplitude. The thin black line depicts the waveforms (u) of the two speakers during the same time interval, recovered by blindly processing their nonlinearly mixed speech. Panels (b) and (c) show the time courses of s_1 and u_1 and of s_2 and u_2, respectively.

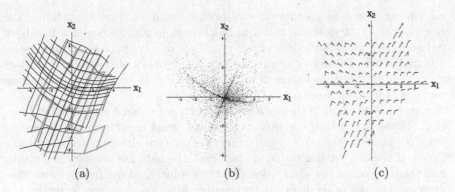

Fig. 2. (a) The thick gray curves comprise a regular Cartesian grid of lines in the s coordinate system, after they were nonlinearly mapped into the x coordinate system by the mixing in (17). The thin black lines depict lines of constant u_1 or of constant u_2, where u denotes a possibly separable coordinate system derived from the measurements. (b) A random subset of the measurements along the trajectory of the mixed waveforms, $x(t)$. (c) The thick gray and thin black lines show the local vectors, $V_{(1)}$ and $V_{(2)}$, respectively, after they have been uniformly scaled for the purpose of display.

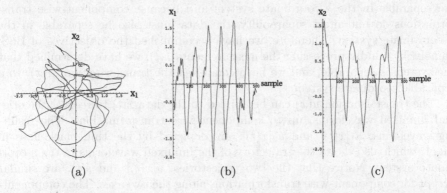

Fig. 3. (a) The trajectory of measurements, $x(t)$, during the 30 ms time interval depicted in Fig. 1. Panels (b) and (c) show the time courses of x_1 and x_2, respectively.

each bin was used to compute local velocity correlations (see (4) and (5)), and these were used to derive M and $V_{(i)}$ for each bin. Figure 2c shows these local vectors at each point.

2. These vectors were used to construct the mapping, $u(x)$. As described in Method, the first step was to choose some point x_0 and then use the vectors $V_{(1)}(x)$ to construct the curvilinear line, $X(\sigma)$. Then, for each point σ on this curve, the local vectors $V_{(2)}(x)$ were used to construct a family of curvilinear lines, $Y(\tau)$. Along each of these Y curves $u_1(x)$ was defined to be constant and equal to the value of σ at the curve's point of intersection with $X(\sigma)$. The mapping, $u_2(x)$, was defined by an analogous procedure. In this way,

each point x was assigned values of both u_1 and u_2, thereby defining the mapping, $u(x)$. A group of the thin black lines in Fig. 2a depict a family of curves having constant values of u_1, which are evenly-spaced and increase as one moves from curve to curve in the family. Figure 2a also shows a family of curves having constant values of u_2, which are evenly-spaced and increase as one moves from curve to curve in the family.

3. As proved in Method, if the data are separable, $u(x)$ must an unmixing function. Therefore, the separability of the data could be determined by seeing if $u[x(t)]$ has a factorizable density function (or factorizable correlation functions). If the density function does factorize, the data are patently separable, and the components of $u[x(t)]$ describe the evolution of the independent subsystems. On the other hand, if the density function does not factorize, the data must be inseparable.

In this illustrative example, the separability of the u coordinate system was verified by a more direct method. Specifically, Fig. 2a shows that the isoclines for increasing values of u_1 (or u_2) nearly coincide with the isoclines for increasing values of s_1 (or s_2). This demonstrates that the u and s coordinate systems differ by component-wise transformations of the form: $(u_1(x), u_2(x)) = (h_1[s_1(x)], h_2[s_2(x)])$ where h_1 and h_2 are monotonic functions. Because the data are separable in the s coordinate system and because component-wise transformations do not affect separability, the data must also be separable in the u coordinate system. Therefore, we have accomplished the objectives of BSS: namely, by blindly processing the measurements $x(t)$, we have determined that the system is separable, and we have computed the transformation, $u(x)$, to a separable coordinate system.

The transformation $u(x)$ can be applied to the data $x(t)$ to recover the original unmixed waveforms, up to component-wise transformations. The resulting waveforms, $u_1[x(t)]$ and $u_2[x(t)]$, are depicted by the thin black lines in Fig. 1, which also shows the trajectory of the unmixed waveforms in the s coordinate system. Notice that the two trajectories, $u[x(t)]$ and $s(t)$, are similar except for component-wise transformations along the two axes. The component-wise transformation is especially noticeable as a stretching of $s_2(t)$ with respect to $u_2[x(t)]$ along the positive s_2 axis. When each of the recovered waveforms, $u_1[x(t)]$ and $u_2[x(t)]$, was played as an audio file, it sounded like a completely intelligible recording of one of the speakers. In each case, the other speaker was not heard, except for a faint "buzzing" sound in the background. Therefore, the component-wise transformations, which related the recovered waveforms to the original unmixed waveforms, did not noticeably reduce intelligibility.

4   Conclusion

This paper describes how to determine if time-dependent signal measurements, $x(t)$, are comprised of linear or nonlinear mixtures of the state variables of statistically independent subsystems. Specifically, the measurement time series is used to derive a mapping, $u(x)$, which must be a transformation to a separable coordinate system, if one exists. Therefore, separability can be determined by testing the statistical independence of the data, after it has been transformed by this mapping. Thus, nonlinear blind source separation has been accomplished.

Some comments on this result:

1. The original problem of looking for an unmixing function $f(x)$ among an *infinite set of functions* was reduced to the simpler problem of constructing a *single mapping*, $u(x)$ and then determining if it transforms the data into separable form.
2. For didactic purposes, the underlying system was assumed to have just two state variables. References [2,3] show how to generalize the method to systems and subsystems having any number of degrees of freedom.
3. The BSS method described in this paper is model-independent in the sense that it can be used to separate data that were mixed by any invertible diffeomorphic mixing function. In contrast, most other approaches to nonlinear BSS are model dependent because they assume that the mixing function has a specific parametric form [4,6].
4. Notice that the proposed method is analytic and constructive, in contrast to the recursive and iterative techniques that are commonly used in the literature [1].

References

1. Comon, P., Jutten, C. (eds.): Handbook of Blind Source Separation, Independent Component Analysis and Applications. Academic Press, Oxford (2010)
2. Levin, D.N.: Nonlinear blind source separation using sensor-independent signal representations. https://arxiv.org/abs/1601.03410
3. Levin, D.N.: Nonlinear blind source separation using sensor-independent signal representations. In: Proceedings ITISE 2016: International Work-Conference on Time Series Analysis, June 27–29, Granada, Spain, pp. 84–95 (2016)
4. Merrikh-Bayat, F., Babaie-Zadeh, M., Jutten, C.: Linear-quadratic blind source separating structure for removing show-through in scanned documents. Int. J. Doc. Anal. Recogn. **14**, 319–333 (2011)
5. Ehsandoust, B., Babaie-Zadeh, M., Jutten, C.: Blind source separation in nonlinear mixture for colored sources using signal derivatives. In: Vincent, E., Yeredor, A., Koldovský, Z., Tichavský, P. (eds.) LVA/ICA 2015. LNCS, vol. 9237, pp. 193–200. Springer, Heidelberg (2015). doi:10.1007/978-3-319-22482-4_22
6. Taleb, A., Jutten, C.: Source separation in post-nonlinear mixtures. IEEE Trans. Sig. Process. **47**, 2807–2820 (1999)
7. Lagrange, S., Jaulin, L., Vigneron, V., Jutten, C.: Analytical solution of the blind source separation problem using derivatives. In: Puntonet, C.G., Prieto, A. (eds.) ICA 2004. LNCS, vol. 3195, pp. 81–88. Springer, Heidelberg (2004). doi:10.1007/978-3-540-30110-3_11

Physics and Bio Signal Processing

The 2016 Signal Separation Evaluation Campaign

Antoine Liutkus[1]([✉]), Fabian-Robert Stöter[2], Zafar Rafii[3], Daichi Kitamura[4],
Bertrand Rivet[5], Nobutaka Ito[6], Nobutaka Ono[7], and Julie Fontecave[8]

[1] Inria, Speech Processing Team, Villers-lès-Nancy, France
antoine.liutkus@inria.fr
[2] International Audio Laboratories Erlangen, Erlangen, Germany
[3] Gracenote, Applied Research, Emeryville, USA
[4] SOKENDAI (The Graduate University for Advanced Studies), Kanagawa, Japan
[5] GIPSA-lab, CNRS, Univ. Grenoble Alpes, Grenoble INP, Grenoble, France
[6] NTT Communication Science Laboratories, NTT Corporation, Tokyo, Japan
[7] National Institute of Informatics, Tokyo, Japan
[8] UJF-Grenoble 1/CNRS/TIMC-IMAG UMR 5525, Grenoble, France

Abstract. In this paper, we report the results of the 2016 community-based Signal Separation Evaluation Campaign (SiSEC 2016). This edition comprises four tasks. Three focus on the separation of speech and music audio recordings, while one concerns biomedical signals. We summarize these tasks and the performance of the submitted systems, as well as provide a small discussion concerning future trends of SiSEC.

1 Introduction

Evaluating source separation algorithms is a challenging topic on its own, as well as finding appropriate datasets on which to train and evaluate various separation systems. In this respect, the Signal Separation Evaluation Campaign (SiSEC) has played an important role. SiSEC was held about every year-and-half since 2008, in conjunction with the LVA/ICA conference. Its purpose is two-fold.

The primary objective of SiSEC is to regularly report the progress of the source separation community, in order to serve as a reference for a comparison of as many methods as possible on the topic of source separation. This involves adapting both the evaluations and the metrics to current trends in the field.

The second important objective of SiSEC is then to provide data the community can use for the design and evaluation of new methods, even outside the scope of the campaign itself. These efforts lead to a significant, although moderate, impact of SiSEC in the community as depicted on Fig. 1.

For the objective evaluation of source separation, two options are now widely accepted and used for SiSEC'2016. First, the BSS Eval toolbox [3] features the signal to distortion ratio (SDR), the source image to spatial distortion ratio (ISR), the signal to interference ratio (SIR), and signal to artifacts ratio (SAR) metrics. All are given in dB and are better with better separation. Second, the PEASS

© Springer International Publishing AG 2017
P. Tichavský et al. (Eds.): LVA/ICA 2017, LNCS 10169, pp. 323–332, 2017.
DOI: 10.1007/978-3-319-53547-0_31

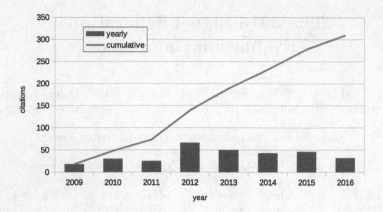

Fig. 1. The number of papers referring to SiSEC. (source: Google Scholar)

toolbox [4] was used in some tasks for providing four perceptually-motivated criteria: the overall perceptual score (OPS), the target-related perceptual score (TPS), the interference-related perceptual score (IPS), and the artifact-related perceptual score (APS).

This sixth SiSEC features the same UND and BGN tasks as proposed last year and summarized in Sects. 2 and 3, respectively. The BIO task presented in Sect. 4 is new. Finally, the MUS task presented in Sect. 5 features new data and accompanying software.

2   UND: Underdetermined-Speech and Music Mixtures

The datasets for the UND task are the same as those described in detail in [1]. The results presented here include those found in previous editions, as well as a new contribution [14], that utilizes both generalized cross correlation (GCC, [21]) and nonnegative matrix factorization (NMF, [22]). GCC was used previously for sound source localization in reverberant environments [23]. NMF is a well-known mathematical framework for many applications, especially in the source separation task. For the acoustic signals, NMF can extract some spectral patterns (bases) and their activations (time-varying gains), and the source separation is achieved by clustering the bases into each source. Wood et al. combined GCC with NMF to localize individual bases over time, such that they may be attributed to individual sources. Computations of Wood's algorithm were between 6 and 7 min per mixture on a dual 2.8 GHz Intel Xeon E5462 quad-core processor with 16 GB of RAM.

From the comparison of the results on Table 1, Wood's algorithm could not outperform the best ever performance on this dataset. Other results for microphone spacings of 5 cm and 1 m with reverberation times of 130 ms and 250 ms may be found on the SiSEC 2016 website[1].

---

[1] http://sisec.inria.fr.

Table 1. Results for the UND task for convolutive mixtures averaged over sources: live-recorded data with 1 m microphone spacing and 250 ms reverberation time in dataset "test"

System	2mic/3src (female)				2mic/4src (female)				2mic/3src (male)				2mic/4src (male)			
	SDR OPS	ISR TPS	SIR IPS	SAR APS	SDR OPS	ISR TPS	SIR IPS	SAR APS	SDR OPS	ISR TPS	SIR IPS	SAR APS	SDR OPS	ISR TPS	SIR IPS	SAR APS
Wood [14]	3.2	6.7	4.7	6.8	2.2	5.0	2.8	4.8	3.1	6.5	4.3	6.6	2.5	5.2	3.1	4.8
(SiSEC 2016)	10.6	8.6	9.0	23.3	27.4	43.7	35.3	47.1	9.7	8.8	9.9	24.2	29.6	47.9	41.7	44.5
Nguyen	**6.1**	9.9	9.3	9.6	4.0	7.5	**7.1**	7.1	5.9	10.1	9.8	8.2	2.5	5.8	4.1	5.4
(SiSEC 2015)	37.1	63.0	48.2	59.0	34.7	60.3	47.6	49.9	40.0	65.8	53.1	53.7	31.8	50.8	43.1	48.0
Cho [15]	5.5	9.5	8.1	9.4	**4.3**	**7.8**	6.8	**7.5**	5.5	9.5	8.2	**9.1**	3.2	6.6	4.7	6.2
(SiSEC 2013)	35.6	62.9	43.4	59.0	33.3	59.0	38.3	52.3	36.0	61.5	44.8	58.7	35.1	57.0	42.8	50.8
Adiloglu [16]	3.0	7.0	5.5	8.1	0.7	4.3	0.9	4.8	3.4	7.1	5.8	8.4	1.5	5.0	2.1	5.2
(SiSEC 2013)	28.4	53.7	35.2	60.8	29.2	46.4	29.4	53.3	26.4	51.4	31.8	63.0	32.7	52.2	36.1	56.1
Hirasawa [17]	2.2	4.2	4.3	4.0	1.2	3.2	0.9	2.6	1.7	3.8	2.8	3.6	0.9	3.0	0.4	1.9
(SiSEC 2011)	22.6	32.6	46.8	38.1	19.5	23.6	41.6	32.8	24.6	36.1	44.0	41.2	20.2	26.3	41.6	34.5
Iso [18]	6.1	9.8	8.7	**10.9**	–	–	–	–	5.5	9.4	8.5	**9.1**	–	–	–	–
(SiSEC 2011)	30.4	59.6	45.1	64.8	–	–	–	–	30.9	54.5	35.0	59.8	–	–	–	–
Cho [19]	3.2	7.4	4.4	8.1	0.0	3.1	-0.7	5.8	4.2	8.8	6.7	8.0	0.9	4.2	1.2	5.2
(SiSEC 2011)	22.0	27.8	20.8	43.6	21.7	24.7	20.0	40.5	37.4	63.3	46.4	55.5	25.2	32.4	25.0	46.4
Nesta (1) [20]	4.3	6.5	7.9	8.4	2.8	5.2	5.3	**6.2**	4.9	7.5	9.1	7.5	3.5	5.9	6.6	5.1
(SiSEC 2011)	38.1	63.1	52.0	56.3	35.5	54.7	49.5	45.8	41.2	63.5	55.0	52.5	35.7	56.3	53.6	42.2
Nesta (2) [20]	6.0	**10.2**	**10.4**	10.2	3.4	6.9	6.3	7.2	**6.2**	10.3	**10.4**	8.6	**4.7**	**8.3**	**8.3**	**6.3**
(SiSEC 2011)	37.3	60.8	50.5	60.2	33.6	49.5	45.0	50.1	39.8	60.1	52.1	55.2	35.7	54.5	51.1	49.6
Ozerov [10]	3.6	8.2	7.4	7.4	1.5	5.1	2.5	4.7	6.0	**10.4**	9.9	8.8	2.2	5.9	3.8	5.4
(SiSEC 2011)	36.0	63.5	48.1	56.2	30.6	47.5	38.1	49.5	39.6	61.3	51.7	58.2	37.4	55.9	50.3	51.7

3 BGN: Two-Channel Mixtures of Speech and Real-World Background Noise

Just like for the UND task, we proposed the same dataset for the task 'two-channel mixtures of speech and real-world background noise (BGN)' as in SiSEC 2013 [1].

Three algorithms were submitted to the BGN task this year, as shown in Table 2. Duong's method [24] is based on NMF with pre-trained speech and noise spectral dictionaries. Liu's method performs Time Difference of Arrival (TDOA) clustering based on GCC-PHAT. Wood's method [14] first applies NMF to the magnitude spectrograms of the mixture signals with channels concatenated in time. Each dictionary atom is then attributed to either the speech or the noise according to its spatial origin.

Considering the results in Table 2, we can see that all methods present some advantages. Whereas Duong's method [24] clearly shows a significant superiority on BSS Eval metrics, this is much less clear when analyzing the PEASS perceptual scores. Wood's method [14] indeed gives the best OPS and IPS scores, suggesting a better overall and interference-related perceptual quality of estimates. Now analyzing APS scores, Liu's method consistently gives results with few annoying artifacts. From all these facts and contradictions, we see the limitations

Table 2. Results for the BGN task

systems	criteria	dev			test					
		Ca1	Sq1	Su1	Ca1	Ca2	Sq1	Sq2	Su1	Su2
(a) Single-channel source estimation										
Duong [24]	SDR	5.6	9.3	4.1	3.7	4.3	10.1	11.6	5.3	4.2
	SIR	14.9	15.4	12.1	13.2	15.0	17.9	18.2	19.3	9.3
	SAR	6.3	10.7	5.3	4.8	4.9	11.1	12.7	5.5	6.6
Liu	SDR	1.9	−3.0	−10.6	1.6	2.7	−4.4	1.9	−12.6	−1.2
	SIR	4.0	−2.9	−9.7	4.5	7.7	−4.3	2.4	−12.2	0.1
	SAR	7.5	16.4	6.9	6.5	5.5	18.8	16.9	10.3	8.0
(b) Multichannel source image estimation (target source)										
systems	criteria	dev			test					
		Ca1	Sq1	Su1	Ca1	Ca2	Sq1	Sq2	Su1	Su2
Duong [24]	SDR	9.4	6.9	4.7	9.6	11.0	9.3	10.2	9.8	7.0
	ISR	23.1	18.0	17.5	23.4	22.6	15.1	18.7	18.5	19.7
	SIR	10.5	9.8	5.4	10.7	12.3	15.6	13.7	12.1	7.4
	SAR	16.9	10.3	11.7	17.6	18.3	11.6	13.5	14.2	19.0
	OPS	14.3	24.1	11.3	10.1	11.5	25.3	16.4	26.0	11.8
	TPS	71.8	65.9	72.4	56.2	58.3	49.2	51.9	73.1	45.3
	IPS	11.3	18.2	5.1	17.3	17.3	49.9	47.0	18.0	29.8
	APS	78.0	66.8	75.1	82.6	81.9	56.1	78.8	57.8	76.0
Liu	SDR	−1.0	−8.5	−12.8	−1.9	0.1	−11.0	−5.6	−16.7	−5.6
	ISR	4.1	1.9	3.8	2.1	2.4	0.6	0.3	2.1	1.4
	SIR	4.9	−2.9	−8.0	5.7	9.1	−4.4	2.2	−11.9	1.1
	SAR	19.7	15.1	7.6	19.3	20.7	17.6	15.9	11.0	13.9
	OPS	9.5	14.2	21.1	10.6	8.9	14.2	17.2	31.3	12.6
	TPS	42.3	38.8	49.5	45.0	43.2	48.3	56.1	62.5	51.0
	IPS	16.8	18.9	15.7	37.0	23.2	47.6	62.5	35.1	50.3
	APS	77.1	70.2	60.1	78.6	79.3	76.0	78.6	50.3	80.1
Wood [14]	SDR	3.0	1.9	0.2	2.9	3.1	−0.7	2.5	−2.6	2.7
	ISR	3.7	7.5	2.5	3.7	3.7	12.7	16.0	3.0	5.5
	SIR	9.4	2.4	−2.6	9.0	12.4	−0.5	3.3	−6.4	3.8
	SAR	5.0	4.0	1.3	5.3	5.2	6.3	8.3	0.3	4.5
	OPS	33.7	38.6	25.9	36.6	35.4	45.1	57.7	26.0	44.1
	TPS	40.5	57.6	24.4	45.4	42.8	60.2	64.6	20.6	57.2
	IPS	60.7	60.5	47.6	66.1	64.5	69.2	74.6	55.4	67.6
	APS	39.0	43.3	31.7	41.0	39.5	47.9	61.4	28.0	48.9

of objective metrics and it seems clear that a real perceptual evaluation would be needed to draw further conclusions.

4   BIO: Separation of Biomedical Signals

Phonocardiography (PCG) is the recording of the sounds generated by the heart. It allows to evaluate some vital functions of the heart. However, the raw recordings of the PCG are not always directly exploitable because of ambient interference (e.g., speech, cough, gastric noise, etc.). Consequently, it is necessary to denoise the raw PCG before their interpretation. An example of clean PCG is plotted on Fig. 2.

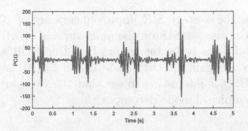

Fig. 2. Phonocardiography signals

The aim of this challenge is to extract the heart activity from raw PCG recordings with a single microphone maintained by a belt on the skin, in front of the heart. 16 sessions have been recorded from 3 healthy participants in different conditions. The quality of the separation process has been evaluated by the BSS Eval toolbox. The SDR, SIR and SAR indexes were computed on sliding windows of 1 s with an overlap of 0.5 s. The performance was only retained for the indexes related to the heart sounds.

Two participants have submitted their results on this specific task:

- The first participant (Part. 1) proposed a method based on the alignment of Empirical Mode Decomposition (EMD) and Lempel-Ziv complexity measure to extract the denoised signal.
- The second participant (Part. 2) proposed a method based on the decomposition of the signal using an ensemble empirical mode decomposition (EEMD) and the selection of some IMFs to filter the signal. Finally, the estimated signal is post-processed to reject additional peaks based on the characteristics of PCG signals.

The results achieved by the submitted methods are plotted on Fig. 3 that shows the distribution of SDR, SIR and SAR for the two participants as well as the noisy data. The red line is the median, the edges of the box are the 25th and 75th percentiles, the whiskers extend to the most extreme values and outliers are plotted by a red cross. In term of SIR, i.e., rejection of noise, Part. 2 is slightly

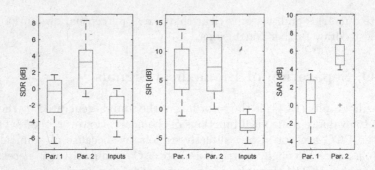

Fig. 3. BIO tasks, results

better than Part. 1: the average SIR improvements are of 10.4 dB and 9.6 dB, respectively, while the average SIR on the noisy data is −3 dB. On the contrary, the Part. 2's method leads to better results based on SDR and SAR than the Part. 1's one: an average gain in SDR of 5.7 dB and 1.4 dB, and an average SAR of 5.5 dB and 0.5 dB. It is interesting to see that the two participants proposed methods based on empirical mode decomposition.

5   MUS: Professionally-Produced Music Recordings

The MUS task attempts at evaluating the performance of music separation methods. In SiSEC 2015 [2], a new dataset was introduced for this task, comprising 100 full-track songs of different musical styles and genres, divided into development and test subsets. This year, this dataset was further heavily remastered so that for each track, it now features a set of four semi-professionally engineered stereo source images (bass, drums, vocals, and other), summing up to realistic mixtures. This corpus was called the Demixing Secret Database (DSD100), as a reference to the'Mixing Secrets' Free Multitrack Download Library it was build from[2]. The duration of the songs ranges from 2 min and 22 s to 7 min and 20 s, with an average duration of 4 min and 10 s.

Additionally, an accompanying software toolbox was developed in Matlab and Python that permits the straightforward processing of the DSD100 dataset. This software is open source and was publicly broadcasted so as to allow the participants to run the evaluation themselves[3].

Similarly to the previous SiSEC editions, MUS was the task attracting the most participants, with 24 systems evaluated. Due to page constraints, we may not detail each method, but encourage the interested reader to refer to SiSEC'2016 website and to the references given therein.

Among the systems evaluated, 10 are blind methods: CHA [5], DUR [6], KAM [8], OZE [10], RAF [11–13], HUA [7], JEO [28]. Then, 14 are supervised methods exploiting variants of deep neural networks: GRA [27], KON [29],

---

[2] www.cambridge-mt.com/ms-mtk.htm.
[3] More info at github.com/faroit/dsdtools.

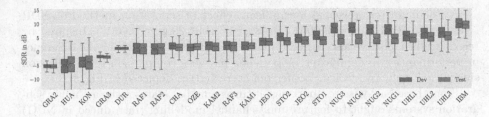

Fig. 4. Results for the SDR of vocals on MUS task for Dev and Test.

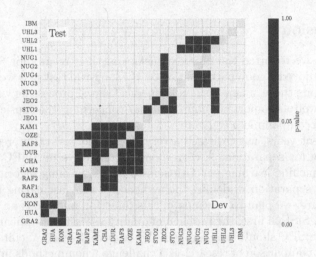

Fig. 5. P-values of Pair-wise difference of Wilcoxon signed-rank test of SDR vocals over method. (upper triangle: Test, lower triangle: Dev). Values $p > 0.05$ indicate no significant differences between the two group results.

UHL [26], NUG [9], and the methods proposed by F.-R. Stöter (STO), consisting of variants of [25, 26] with various representations. Finally, the evaluation also features the scores of Ideal Binary Mask (IBM), computed for left and right channels independently.

Due to space constraints again, Fig. 4 shows the box plots for the SDR of the vocals only, over the whole DSD100 dataset and excluding those few 30 s excerpts for which the IBM method was badly behaved (yielding nan values for its SDR). More results may be found online. For the first time in SiSEC, 30 s excerpts of all separated results may also be found in the webpage dedicated to the results[4]. The striking fact is that most proposed supervised systems considerably outperform blind methods, a trend that is also noticeable on other SIR, SAR metrics. Also, systems like [26] which use additional augmentation data, seem to generalise better, resulting in a smaller gap between Dev and Test.

---

[4] sisec17.audiolabs-erlangen.de.

A Friedman test revealed a significant effect of separation method on SDR (Dev: $\chi^2 = 1083.23, p < 0.0001$, Test: $\chi^2 = 1004.29, p < 0.0001$). Inspired by recent studies [30], we also tested for each pair of method whether the difference in performance was significant. A post-hoc pairwise comparison test (Wilcoxon signed-rank test, two-tailed, Bonferroni corrected) is depicted in Fig. 5.

From these pair-wise comparisons, it turns out that state-of-the art music separation systems ought to feature multichannel modelling (introduced in NUG) and data augmentation (UHL). As depicted by the best scores obtained by UHL3, performing a fusion of different systems is also a promising idea.

6   Conclusion

In this paper, we reported the different tasks and their results for SiSEC'2016. This edition enjoyed a good participation on the long-run tasks, as well as several novelties. Among those, a new task on biomedical signal processing was proposed this year, as well as important improvements concerning the music separation dataset and accompaniment software.

In the recent years, we witnessed a very strong increase of interest in supervised methods for separation. A corresponding objective of SiSEC is to make it easier for machine learning practitioners to adapt learning algorithms to the task of source separation, widening the audience of this fascinating topic.

In the future, we plan to continue in this direction and focus on two important moves for SiSEC: first, the problem of quality assessment appears as largely unsolved and SiSEC should play a role in this respect. Second, facilitating reproducibility and comparison of research is a challenge when methods involve large-scale machine learning systems. SiSEC will shortly host and broadcast separation results of various techniques along datasets to promote easy comparison with state of the art.

References

1. Ono, N., Koldovsky, Z., Miyabe, S., Ito, N.: The 2013 signal separation evaluation campaign. In: Proceedings of MLSP, pp. 1–6, September 2013
2. Ono, N., Rafii, Z., Kitamura, D., Ito, N., Liutkus, A.: The 2015 signal separation evaluation campaign. In: Vincent, E., Yeredor, A., Koldovský, Z., Tichavský, P. (eds.) LVA/ICA 2015. LNCS, vol. 9237, pp. 387–395. Springer, Heidelberg (2015). doi:10.1007/978-3-319-22482-4_45
3. Vincent, E., Griboval, R., Févotte, C.: Performance measurement in blind audio source separation. IEEE Trans. ASLP **14**(4), 1462–1469 (2006)
4. Emiya, V., Vincent, E., Harlander, N., Hohmann, V.: Subjective and objective quality assessment of audio source separation. IEEE Trans. ASLP **19**(7), 2046–2057 (2011)
5. Chan, T-S., Yeh, T-C., Fan, Z-C., Chen, H-W., Su, L., Yang, Y-H., Jang, R.: Vocal activity informed singing voice separation with the iKala dataset. In: Proceedings of ICASSP, pp. 718–722, April 2015
6. Durrieu, J.-L., David, B., Richard, G.: A musically motivated mid-level representation for pitch estimation and musical audio source separation. IEEE J. Sel. Top. Sig. Process. **5**(6), 1180–1191 (2011)

7. Huang, P., Chen, S., Smaragdis, P., Hasegawa-Johnson, M.: Singing-voice separation from monaural recordings using robust principal component analysis. In: Proceedings of ICASSP, pp. 57–60, March 2012
8. Liutkus, A., FitzGerald, D., Rafii, Z., Daudet, L.: Scalable audio separation with light kernel additive modelling. In: Proceedings of ICASSP, pp. 76–80, April 2015
9. Nugraha, A., Liutkus, A., Vincent, E.: Multichannel music separation with deep neural networks. In: Proceedings of EUSIPCO (2016)
10. Ozerov, A., Vincent, E., Bimbot, F.: A general flexible framework for the handling of prior information in audio source separation. IEEE Trans. ASLP **20**(4), 1118–1133 (2012)
11. Rafii, Z., Pardo, B.: REpeating pattern extraction technique (REPET): à simple method for music/voice separation. IEEE Trans. ASLP **21**(1), 71–82 (2013)
12. Liutkus, A., Rafii, Z., Badeau, R., Pardo, B., Richard, G.: Adaptive filtering for music/voice separation exploiting the repeating musical structure. In: Proceedings of ICASSP, pp. 53–56, March 2012
13. Rafii, Z., Pardo, B.: Music/voice separation using the similarity matrix. In: Proceedings of ISMIR, pp. 583–588, October 2012
14. Wood, S., Rouat, J.: Blind speech separation with GCC-NMF. In: Proceedings of Interspeech (2016)
15. Cho, J., Yoo, C.D.: Underdetermined convolutive BSS: Bayes risk minimization based on a mixture of super-Gaussian posterior approximation. IEEE Trans. Audio Speech Lang. Process. **23**(5), 828–839 (2011)
16. Adiloglu, K., Vincent, E.: "Variational Bayesian inference for source separation and robust feature extraction," Technical report, INRIA (2012). https://hal.inria.fr/hal-00726146
17. Hirasawa, Y., Yasuraoka, N., Takahashi, T., Ogata, T., Okuno, H.G.: A GMM sound source model for blind speech separation in under-determined conditions. In: Theis, F., Cichocki, A., Yeredor, A., Zibulevsky, M. (eds.) LVA/ICA 2012. LNCS, vol. 7191, pp. 446–453. Springer, Heidelberg (2012). doi:10.1007/978-3-642-28551-6_55
18. Iso, K., Araki, S., Makino, S., Nakatani, T., Sawada, H., Yamada, T., Nakamura, A.: Blind source separation of mixed speech in a high reverberation environment. In: Proceedings of Hands-free Speech Communication and Microphone Arrays, pp. 36–39 (2011)
19. Cho, J., Choi, J., Yoo, C.D.: Underdetermined convolutive blind source separation using a novel mixing matrix estimation and MMSE-based source estimation. In: Proceedings of IEEE MLSP (2011)
20. Nesta, F., Omologo, M.: Convolutive underdetermined source separation through weighted interleaved ICA and spatio-temporal source correlation. In: Theis, F., Cichocki, A., Yeredor, A., Zibulevsky, M. (eds.) LVA/ICA 2012. LNCS, vol. 7191, pp. 222–230. Springer, Heidelberg (2012). doi:10.1007/978-3-642-28551-6_28
21. Knapp, C.H., Carter, G.C.: The generalized correlation method for estimation of time delay. IEEE Trans. Acousti. Speech Sig. Process. **24**(4), 320–327 (1976)
22. Lee, D.D., Seung, H.S.: Learning the parts of objects by non-negative matrix factorization. Nature **401**, 788–791 (1999)
23. Blandin, C., Ozerov, A., Vincent, E.: Multi-source TDOA estimation in reverberant audio using angular spectra and clustering. Sig. Process. **92**(8), 1950–1960 (2012)
24. Duong, H.-T.T., Nguyen, Q.-C., Nguyen, C.-P., Tran, T.-H., Duong, N.Q.K.: Speech enhancement based on nonnegative matrix factorization with mixed group sparsity constraint. In: Proceedings of ACM International Symposium on Information and Communication Technology, pp. 247–251 (2015)

25. Stöter, F.-R., Liutkus, A., Badeau, R., Edler, B., Magron, P.: Common fate model for unison source separation. In: Proceedings of ICASSP (2016)
26. Uhlich, S., Porcu, M., Giron, F., Enenkl, M., Kemp, T., Takahashi, N., Mitsufuji, Y.: Improving Music Source Separation Based On Deep Neural Networks Through Data Augmentation and Network Blending (2017). Submitted to ICASSP
27. Grais, E., Roma, G., Simpson, A.J., Plumbley, M.: Single-channel audio source separation using deep neural network ensembles. In: Proceedings of AES 140, May 2016
28. Jeong, I.-Y., Lee, K.: Singing voice separation using RPCA with weighted l1-norm. In: Proceedings of LVA/ICA (2017)
29. Huang, P., Kim, M., Hasegawa-Johnson, M., Smaragdis, P.: Joint optimization of masks and deep recurrent neural networks for monaural source separation. IEEE/ACM Trans. Audio Speech Lang. Process. 23(12), 2136–2147 (2015)
30. Simpson, A., Roma, G., Grais, E., Mason, R., Hummersone, C., Plumbley, M., Liutkus, A.: Evaluation of audio source separation models using hypothesis-driven non-parametric statistical methods. In: Proceedings of EUSIPCO (2016)

Multimodality for Rainfall Measurement

Hagit Messer[(✉)]

School of Electrical Engineering, Tel Aviv University, Tel Aviv, Israel
messer@eng.tau.ac.il

Abstract. The need for accurate monitoring of rainfall, essential for many fields such as: hydrology, transportation and agriculture, calls for optimal use of all available resources. However, as the existing monitoring equipment is diverse, and different tools provide measurements of different nature, fusing these measurements is a challenging task. At one extreme, rain gauges provide local, direct measurements of the accumulated rainfall, and at the other end, satellite observations provide remote images of clouds, from which rainfall is estimated. In between, weather radar measures reflectivity which is non-linearly related to rainfall. In light of the new opportunities introduced by the use of physical measurements from cellular communication networks for rainfall monitoring, I first review the approaches for fusion of different rainfall direct and indirect measurements, distinguishing it from data assimilation, widely used in meteorology. I will then suggest a unified approach to the problem, combining parametric and non-parametric tools, and will present preliminary results.

Keywords: Rainfall monitoring · Multimodality · Data fusion

1 Introduction

The importance of accurate rain monitoring arises from many applications. Whether it is required for precisely measuring past precipitation quantities or for generating future predictions, monitoring of rain has been of interest to the human kind since early history. The ground level rain rate at a given time can be modelled as a 2-D stochastic process $r(x, y, t)$. Existing measurements equipment, such as rain-gauges, weather stations, or recently proposed microwave-links [21], sample $r(x, y, t)$ spatially in specific points (or along lines).The rain field $r(x, y, t)$ [mm/hr] can be measured directly or indirectly. Moreover, it can either be sampled instantaneously, or the accumulated rain over a given periodcan be measured:

$$R_t(x, y; T) = \int_{t-T}^{t} r(x, y, z)dz \tag{1}$$

First evidences of intentional rain gauge usage date back to the fourth century B.C. [23]. Yet still, contemporary rain gauges (tipping bucket gauges and electronic ones) are being improved. Development of designated microwave radars dates back to the late 40's of the 20th century [11] and development of cheaper and more precise radars has been a work-in-progress ever since. Satellite based measurements have entered the

© Springer International Publishing AG 2017
P. Tichavský et al. (Eds.): LVA/ICA 2017, LNCS 10169, pp. 333–343, 2017.
DOI: 10.1007/978-3-319-53547-0_32

environmental monitoring turf in the 1960's. Since then, the challenge of gaining precise measurements from these means has been great [1]. Recently much interest has grown around the subject of using existing Commercial Wireless Communication Networks for rain monitoring [8, 12, 17, 21, 22, 28, 29]. Such rain monitoring systems benefit mainly but not solely from the lack of need to deploy any dedicated sensors in the field. Making a use of the existing commercial (e.g., cellular) wireless networks is the equivalent of deploying a very high density of designated sensors but without anyextra cost. Such an amount of sensors, used for environmental monitoring is unprecedented and can provide high temporal and spatial resolution sensing, better area coverage, as well as a diversity of measurements in given points. Moreover, applying advanced signal processing algorithms, which exploit the diversity in the data, while overcoming many of the disadvantages of the previous monitoring methods now seems realizable. Figure 1 presents the leading technologies for rainfall monitoring.

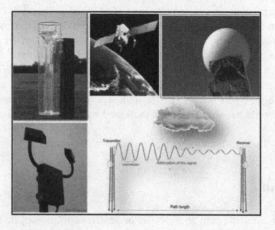

Fig. 1. Rainfall monitoring equipment. Top (left to right): gauge, satellite, radar. Bottom: an electronic gauge - disdrometer (left), microwave link (right).

Comparison between the four rain measurement technologies shows that each has its own advantages and disadvantages, which can be summarized as follows:

1. Rain gauges: direct, near ground measurement of accumulated rain (1) at a given point (x_0, y_0).
2. Microwave links: indirect measurement of the near-ground rain: The instantaneous rain-induced attenuation of the signal along a microwave link of length L is approximated by:

$$A_L(t) = \iint_L ar^b(x, y; t) dx dy \qquad (2)$$

where A_L (in dB) is the logarithmic attenuation per kilometer and a, b are coefficients, depending on the frequency and polarization of the electromagnetic ssignal and on the drop size distribution of the rain.

3. Weather radar and satellites are remote sensing machines. As such, both produce estimates of the rain field over a given area. The radar 2-D maps $\hat{r}_R(x,y,t)$ are limited to a radius of about 100 km around the radar and are based on reflections from clouds at height up to several kilometers above the ground. The satellite maps $\hat{r}_S(x,y,t)$ cover much larger area and are based on mostly optical remote sensing from outer space.

As the phenomenon of interest is the near-ground rain, it is desired to reconstruct $r(x,y,t)$ with the accuracy of a rain gauge, the coverage of a satellite, and at tempo-spatial resolution that will ensure that rainfall is observable everywhere and anytime. Obviously, none of the existing technologies can do it alone, thought the emerging approach of using existing measurements from the widely spread cellular networks shows a great potential [21, 22] to dramatically improve the current situation. However fusion of measurements from all available equipment is definitely the most promising solution for global, accurate reconstruction of $r(x,y,t)$.

In this paper we focus on existing and future approaches for fusing real time measurements from all available equipment to generate an estimate of the instantaneous rain field $r(x,y,t)$ (or the accumulated rain $R_t(x,y;T)$ of (1)) at any point (x,y). We define this problem as fusion of multimodal measurements. In Sect. 2 we explain how this problem differs from other, existing approaches for integration of measurements from different meteorological equipment, e.g., calibration and data assimilation. In Sect. 3 we review parametric and non-parametric fusion methods and propose a unified approach. Section 4 concludes this paper.

2 How Multimodal Measurements Are Currently Used: Calibration and Data Assimilation

Existing approaches for integrating meteorological data from different sources can be classified into two groups: calibration, and data assimilation. While both approaches have been employed for decades, the emerging technology of rain rate monitoring by (commercial) microwave links introduces new avenues to calibration and to data assimilation. This paper is focused on the use of these measurements together with the traditional ones.

As weather radar is the leading technology for estimating rainfall distribution over wide areas, much effort is put in improving the accuracy of this method [7]. However, as it inherently is a remote sensing technology, measuring rain fields much above the ground level, calibration is required to relate $\hat{r}_R(x,y,t)$ to the ground level rain field $r(x,y,t)$. The calibration and the validation or verification of the calibration of weather radar systems is a permanent subject of research and development. This calibration is done by using measurements from other monitoring equipment, mainly rain gauges or satellites [3]. The resulting rain rate map, visually presented for laymen, is a calibrated one. Note, however, that the calibration process integrates measurements from other sources indirectly and periodically. At each time t, the rain map $\hat{r}_R(x,y,t)$ is constructed from radar measurements only, where the mapping algorithm parameters were adjusted with the aid of historical measurement from other monitoring equipment (see Fig. 2a).

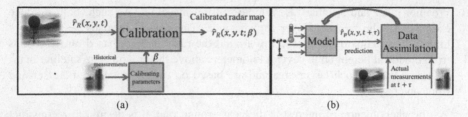

(a)   (b)

Fig. 2. Integrating measurements for Radar calibration (left) and for data assimilation (right)

Data assimilation, on the other hand, is the use of measurements to improve prediction models. As defined by the European Centre for Medium-Range Weather Forecasts: "Data assimilation is typically a sequential time-stepping procedure, in which a previous model forecast is compared with newly received observations, the model state is then updated to reflect the observations, a new forecast is initiated, and so on."[1] Forecasting weather models are based on long term statistics and on measurements of the basic physical building blocks, e.g., temperature, pressure, etc. The predictions are then compared with actual measurements from various equipment to improve the model, which, in turn, will improve future prediction. The data assimilation process is described in Fig. 2b.

Microwave links' data has also been used for radar calibration: to correct for attenuation and other sources of error in ground-based radar rainfall estimates [5, 6] or to adjust radar rainfall images [10, 26]. It has been previously assimilated with data from other sources, mainly radar and gauges, to obtain improved rainfall products [9, 18, 20, 26, 31], and satellites. Recently, geostationary satellite products were suggested for wet-dry classification in link-based rainfall retrievals [26].

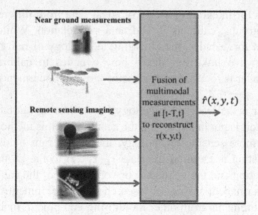

Fig. 3. Fusion of multimodal measurements for rain field reconstruction

---

[1] http://www.ecmwf.int/en/research/data-assimilation.

However, calibration and data assimilation are different from the fusion of multi-modal measurements, defined in Sect. 1 as fusing real time measurements from all available equipment to generate an estimate of the monitored rain field. While for calibration and data assimilation measurements are used to adjust/improve the model which creates the tempo-spatial rainfall presentation, fusion of multimodal measurements is their direct use to improve the rain measurements based on each of them separately [4, 14, 26]. That is, each type of equipment can estimate rain field $r(x, y, t)$ at one point, a line, or in a restricted area. The fusion of the different measurements is used to improve coverage, accuracy or tempo-spatial resolution of the estimated rain field. Figure 3 presents this process.

3   Parametric and Non-parametric Data Fusion Information

Most published works on merging of multimodal measurements for rainfall monitoring do not distinguish between calibration, data assimilation and fusion for improved monitoring. In general, most existing techniques used ground level station for calibrating remote sensing images, or for data assimilation. When the use of microwave links as a new way for ground level measurements was introduced, it attracted attention for new possibilities of data fusion [19]. To the best of our knowledge, no published work has considered real time fusing all 4 types of measurements: local weather stations, microwave links, radar and satellites.

We suggest the following parametric setting of the fusion problem. First, the parameter(s) of interest should be defined. It can be the accumulated rainfall over some observation time T at a given set of K points of interest [13, 14], so the parameter vector is $\theta' = [\theta_1, .., \theta_K]$, where, using (1), for a given time instant t, we have that for $j = 1, 2, \ldots, K$:

$$\theta_j = R_t(x_n, y_m; T) = \int_{t-T}^{t} r(x_n, y_m, z)dz \tag{3}$$

Where x_n, y_m are the coordinates of one of the K points of interest in a given area, which can be on a grid, in which case they represent a pixel in an image. Note that such an image, if normalized by T, will be an estimate of an instantaneous rain field map at instant t, $r(x, y; \theta) = \frac{1}{T}R_t(x_n, y_m; T)$.

Alternatively, one can represent a snapshot of a rain cell, describing the spatial variation of an isolated rainfall, by a parametric model [24], for example, by a 2-D Gaussian shape of unknown parameter vector, whose shape is characterized by 6 parameters: $P, \mu_x, \mu_y, \sigma_x, \sigma_y, \rho$, where P is the rain rate level at the pick of the Gaussian, placed at point (μ_x, μ_y), and σ_x, σ_y, ρ determine its width and the shape, according to:

$$r(x, y; \theta) = P e^{-\frac{1}{2(1-\rho^2)}\left[\frac{(x-\mu_x)^2}{\sigma_x^2} + \frac{(y-\mu_y)^2}{\sigma_y^2} - \frac{2\rho(x-\mu_x)(y-\mu_y)}{\sigma_x \sigma_y}\right]} \tag{4}$$

In this case the instantaneous rain field in a given area, if consists of a single rain-cell, is represented by the vector of 6 parameters $\theta' = [P, \mu_x, \mu_y, \sigma_x, \sigma_y, \rho]$. Having more than one cell increases the number of unknown parameters by products of 6.

Under this parametric presentation, each of the $i - th$ possibly multimodal rain monitoring equipment provides measurements of the form:

$$z_i = Q_i(s_i(r(x, y; \theta)) + n_i) \tag{5}$$

where s_i is the sensing function of the sensor, depending on the specific equipment. For a microwave link, for example, it is a non-linear function of the (unknown) rain rate $(x, y; \theta)$, given by (2). n_i represents the additive noise in the measurement, and Q_i is the quantization operation, which is common in rain measurements [21]. s_i is assumed known and so is the statistics of the noise, and the parametric specific model that relates θ to the measurements. Under these assumptions, assuming a white, normally distributed noise and independent measurements, the likelihood function can be formalized as [16]:

$$L(\mathbf{z}; \theta) = \log\left(\prod_1^N P(z_i; \theta)\right) = \sum_{i=1}^N \log(P(z_i; \theta)) \sum_{i=1}^N \left(\frac{1}{\sqrt{2\pi\sigma_i^2}} \int_{z_i - \frac{\Delta_i}{2}}^{z_i + \frac{\Delta_i}{2}} e^{-\frac{(y-\theta)^2}{2\sigma_i^2}} dy\right)$$

$$= -\frac{1}{2}\sum_{i=1}^N \log(2\pi\sigma_i^2) + \sum_{i=1}^N \log\left(\int_{z_i - \frac{\Delta_i}{2}}^{z_i + \frac{\Delta_i}{2}} e^{-\frac{(y-\theta)^2}{2\sigma_i^2}} dy\right) \tag{6}$$

The resulting maximum likelihood estimator is optimal, but much too complicated. Alternatively, in [13], a sub-optimal, simple parametric estimator has been proposed which is based on linear combination of maximum likelihood estimates of single-modal measurements. In particular, it has been successfully applied for fusing measurements from microwave links and rain gauges [14], previously done with an ad-hoc algorithm [15, 27, 28], and has demonstrated how linearly fused multimodal rain-estimate can perform better than each of them separately. Figure 4 depicts the root mean square error (RMSE) of the estimate of the rainfall in an arbitrary point in a real world scenario (Fig. 4, left), as a function of σ, the variance of the additive noise in the links (assumed to be the same for all links). The variance of the noise in the gauges is assumed small and unchanged. The RMSE has been calculated over 1000 runs with different noise samples, drawn from the same statistics. Under this scenario, it is obvious that the RMSE of the maximum likelihood procedure, applied on measurements from gauges only (green, dash line in Fig. 4), is about the same over all values of σ, while when using only measurements from the microwave links, the RMSE increases as σ increases (blue, dash line in Fig. 4). However, multimodal fusion by linear combination of the two estimates can give RMSE better than each of the individual, single-modal measurements. The solid lines in Fig. 4 show the RMSE of different choice of the superposition parameter, following the theoretical results in [14].

Note that the parametric approach, when applied on a sensors network of sufficient number of gauges and links that are spatially distributed in a given area, can create a 2-D mapping of rain-fields of excellent accuracy and tempo-spatial resolution [21, 27, 28].

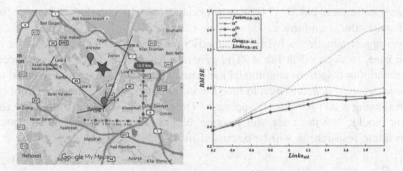

Fig. 4. Simulations of real life scenario (left) with 2 gauges (blue drops) and 7 links (black lines). The RMSE of the rainfall estimate (right) inat a point of interest (the star) using MLE for gauges only (dash, green) or links only (dash, blue) or both (solid, blue), and linear multimodal fusion, as a function of the noise-level (from [13]). (Color figure online)

However, as the near-ground sensors have limited coverage, their merging with remote-sensing measurements is necessary.

The parametric approach can theoretically be generalized to accommodate also remote sensing from radar or satellite. However, as the remote-sensor is actually a single one, which creates a set of tempo-spatial measurements (images or maps), it is inherently different from the case where measurements from distributed sensors are available. In [20] it has been proposed to integrate radar and links' rain maps using a non-parametric approach, where images are merged by pixels, following a certain algorithm. The principles of the algorithm in [20], applied on a given area of interest, are as follows:

1. Create a near-ground rain map based on all available measurements from gauges and microwave links using a parametric approach.
2. Create a calibrated rain map from all available radar systems.
3. Apply a non-parametric algorithm for integrating the maps 1 and 2 such that:
   3.1 If a certain sub-area is covered by only 1 or 2, use it for the resulting map.
   3.2 If a certain sub-area is covered by both 1 and 2, fix the map resolution to the better one, and merge the rain estimate at each pixel by a superposition of the estimates:

$$\widehat{r}(x_i, y_j) = \alpha \widehat{r}_1(x_i, y_j) + (1 - \alpha)\widehat{r}_2(x_i, y_j); 0 \leq \alpha \leq 1 \qquad (7)$$

In [20] excellent results with real data have presented, where α has been set as a constant (per each pixel), according to the distance of a pixel from the radar and the number and density of ground sensors.

The case of [20] is sub-optimal since instead of using a general fusion function $\widehat{r}(x_i, y_j) = f(\widehat{r}_1(x_i, y_j), \widehat{r}_2(x_i, y_j))$, *linear* combination of the single modal estimates is imposed. In general, if α is constant independent on the unknown rainfall, it is not guaranteed that the estimate of the superimposed estimates over-performs the individual estimates, but in [14] it is suggested how to choose α to ensure better performance almost uniformly.

We suggest adopting a hybrid parametric/non-parametric approach for the more general case, as depicted in Fig. 5. The general idea is to produce the most accurate near-ground rain map from multimodal measurements (gauges, microwave links) using a *parametric approach*, and then to fuse it with remote sensing rain maps (from radar, satellites) using a *non-parametric* approach. The proposed solution is built on existing building blocks, but there are still missing blocks which pose new challenges and require future research, as will be discussed in the next section.

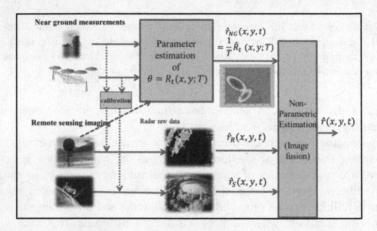

Fig. 5. The proposed approach for suboptimal fusion of multimodal measurements for rain field reconstruction

4 Discussion and Conclusions

While multimodality has been used in meteorology for decades, the recent development of a novel source of near-ground measurements, based on the received signal level in cellular backhaul microwave networks, introduces new challenges. This emerging technology has proven the ability to construct accurate real time 2-D rainfall maps over large areas, as standalone technology or with the integration of measurements from rain gauges [15, 29]. The challenge is, however, to integrate it with all technologies developed especially for precipitation monitoring such as weather radars[2] and satellites[3] In Fig. 5 we summarize our proposed approach for multimodality for precipitation monitoring. It consists of multilayer merging of 3 maps: (i) an accurate but local map based on parametric fusion of near-ground measurements from microwave links and local gauges/weather stations; (ii) a medium-height map created by weather radar; and (iii) a global map constructed by satellites. Note the inverse relation

---

[2] http://www.wmo.int/pages/prog/www/WRO/index_en.html.

[3] http://www.wmo.int/pages/prog/www/WRO/index_en.html.

between coverage and accuracy. Each of the maps employs state of the art technologies and algorithms, including the use of historic measurements for calibration and model verification.

The open challenges in fully implementing the proposed approach include: (a) improving each of the maps separately with all available resources. For example, the use of links' measurements for data assimilation and radar/satellite calibration has not yet commercially deployed. On the other hand, radar measurements can be used for improving near ground maps. These options are depicted by dashed lines in Fig. 5 – the blue line suggests on involving real time radar measurements in the parametric fusion of ground level measurements while the red dashed lines indicate on calibration, as explained in Sect. 2; (b) optimal setting of the fusion between layers. In general, it would have the structure of $\widehat{r}(x_i, y_j) = f\big(\widehat{r}_1(x_i, y_j), \widehat{r}_2(x_i, y_j)\big)$, for each pixel. However, a simplified, linear fusion scheme can be imposed. In this case, it is required to find the superposition parameters (see Eq. (7)) and to generalize it from 2 to 3 maps, to include satellites measurements. Note that imposing linear combination of the maps makes the proposed approach to be suboptimal. It is of great interest to find out the performance loss due to this simplicity constrain.

Moreover, data fusion can also contribute to the robustness of meteorological observations. As each of the available equipment may produce false measurements, multimodal observations can help in detecting outliers. Initial, local results for using microwave links to detect faulty gauges have been presented in [5]. For a complete fusion of multimodal measurements for real time $2D$ rainfall monitoring, this direction has to be further explored, so prior to the processing proposed at Fig. 5, multimodal observations will be used to indicate on the readability of the raw measurements.

Acknowledgments. I deeply thank all past and present members of our research team in Tel Aviv University, co-led by Prof. PinhasAlpet for their fruitful cooperation and discussions. We thank our friends in the Israeli cellular providers: Cellcom, Pelephone, and PHI who providing us datafor more than a decade. Special thanks to InbarFijalkow and to Elad Heiman for fruitful joint work and the preliminary results.

References

1. Barrett, E.C., Beaumont, M.J.: Satellite rainfall monitoring: an overview. Remote Sens. Rev. **11**(1-4), 23–48 (1994)
2. Barthès, L., Mallet, C.: On the opportunistic use of geostationary satellite signals to estimate rain rate in the purpose of radar calibration. In: 7th European Conference on Radar in Meteorology and Hydrology, ERAD 2012 (2012)
3. Berndt, C., Rabiei, E., Haberlandt, U.: Geostatistical merging of rain gauge and radar data for high temporal resolutions and various station density scenarios. J. Hydrol. **508**, 88–101 (2014)
4. Bianchi, B., et al.: A variational approach to retrieve rain rate by combining information from rain gauges, radars, and microwave links. J. Hydrometeorol. **14**(6), 1897–1909 (2013)

5. Bianchi, B., Rieckermann, J., Berne, A.: Detection of faulty rain gauges using telecommunication microwave links. In: 12nd International Conference on Urban Drainage, Porto Alegre/Brazil (2011)
6. Bianchi, B., Rieckermann, J., Berne, A.: Quality control of rain gauge measurements using telecommunication microwave links. J. Hydrol. **492**, 15–23 (2013)
7. Brandes, E.A.: Optimizing rainfall estimates with the aid of radar. J. Appl. Meteorol. **14**(7), 1339–1345 (1975)
8. Chwala, C., et al.: Precipitation observation using microwave backhaul links in the alpine and pre-alpine region of Southern Germany. Hydrol. Earth Syst. Sci. **16**(8), 2647–2661 (2012)
9. Cole, S.J., Moore, R.J.: Hydrological modelling using raingauge-and radar-based estimators of areal rainfall. J. Hydrol. **358**(3), 159–181 (2008)
10. Cummings, R.J., et al.: Using microwave links to adjust the radar rainfall field. Adv. Water Resour. **32**(7), 1003–1010 (2009)
11. Delrieu, G., et al.: Weather radar and hydrology. Adv. Water Resour. **32**(7), 969–974 (2009)
12. Fencl, M., et al.: Commercial microwave links instead of rain gauges: fiction or reality? Water Sci. Technol. **71**(1), 31–37 (2015)
13. Fijalkow, I., Heiman, E., Messer, H.: Rain estimation from heterogeneous sensor networks (2016, submitted)
14. Fijalkow, I., Heiman, E., Messer, H.: Parameter estimation from heterogeneous/multimodal data sets. IEEE Signal Process. Lett. **23**(3), 390–393 (2016)
15. Goldshtein, O., Messer, H., Zinevich, A.: Rain rate estimation using measurements from commercial telecommunications links. IEEE Trans. Signal Process. **57**(4), 1616–1625 (2009)
16. Heiman, E., Messer, H.: Parameter estimation from multiple sensors with mixed resolution of quantization. In: 2014 IEEE 28th Convention of IEEE Electrical and Electronics Engineers in Israel (IEEEI) (2014)
17. Gosset, M., et al.: Improving rainfall measurement in gauge poor regions thanks to mobile telecommunication networks. Bull. Am. Meteorol. Soc. **97**(3), 49–51 (2016)
18. Grum, M., et al.: Combined use of point rain gauges, radar, microwave link and level measurements in urban hydrologicalmodelling. Atmos. Res. **77**(1), 313–321 (2005)
19. Lahat, D., Adali, T., Jutten, C.: Multimodal data fusion: an overview of methods, challenges, and prospects. Proc. IEEE **103**(9), 1449–1477 (2015)
20. Liberman, Y., et al.: New algorithm for integration between wireless microwave sensor network and radar for improved rainfall measurement and mapping. Atmos. Meas. Tech. **7**(10), 3549–3563 (2014)
21. Messer, H., Sendik, O.: A new approach to precipitation monitoring: a critical survey of existing technologies and challenges. IEEE Signal Process. Mag. **32**(3), 110–122 (2015)
22. Messer, H., Zinevich, A., Alpert, P.: Environmental monitoring by wireless communication networks. Science **312**(5774), 713 (2006)
23. Michaelides, S.C. (ed.): Precipitation: Advances in Measurement, Estimation and Prediction. Springer, Heidelberg (2008)
24. Morin, E., David, C., Robert, A.M., Xiaogang, G., Hoshin, V.G., Soroosh, S.: Spatial patterns in thunderstorm rainfall events and their coupling with watershed hydrological response. Adv. Water Resour. **29**(6), 843–860 (2006)
25. Rios Gaona, M.F., et al.: Measurement and interpolation uncertainties in rainfall maps from cellular communication networks. Hydrol. Earth Syst. Sci. **19**(8), 3571–3584 (2015)
26. Sideris, I.V., et al.: Real-time radar–rain-gauge merging using spatio-temporal co-kriging with external drift in the alpine terrain of Switzerland. Q. J. R. Meteorol. Soc. **140**(680), 1097–1111 (2014)

27. Overeem, A., Leijnse, H., Uijlenhoet, R.: Quantitative precipitation estimation using commercial microwave links. IAHS-AISH Publ., 129–134 (2012)
28. Overeem, A., Leijnse, H., Uijlenhoet, R.: Country-wide rainfall maps from cellular communication networks. Proceedings of the National Academy of Sciences **110**(8), 2741–2745 (2013)
29. Overeem, A., Leijnse, H., Uijlenhoet, R.: Retrieval algorithm for rainfall mapping from microwave links in a cellular communication network. Atmos. Meas. Tech. **9**(5), 2425–2444 (2016)
30. Upton, G.J.G., et al.: Microwave links: the future for urban rainfall measurement? Atmos. Res. **77**(1), 300–312 (2005)
31. Vogl, S., et al.: Copula-based assimilation of radar and gauge information to derive bias-corrected precipitation fields. Hydrol. Earth Syst. Sci. **16**(7), 2311–2328 (2012)

Particle Flow SMC-PHD Filter for Audio-Visual Multi-speaker Tracking

Yang Liu[1]($\boxtimes$), Wenwu Wang[1], Jonathon Chambers[2], Volkan Kilic[3],
and Adrian Hilton[1]

[1] Department of Electrical and Electronic Engineering,
University of Surrey, Guildford GU2 7XH, UK
{yangliu,w.wang,a.hilton}@surrey.ac.uk
[2] School of Electrical and Electronic Engineering, Newcastle University,
Newcastle upon Tyne NE1 7RU, UK
Jonathon.Chambers@newcastle.ac.uk
[3] Department of Electrical and Electronics Engineering,
Izmir Katip Celebi University, 35620 Cigli-Izmir, Turkey
volkan.kilic@ikc.edu.tr

Abstract. Sequential Monte Carlo probability hypothesis density (SMC-PHD) filtering has been recently exploited for audio-visual (AV) based tracking of multiple speakers, where audio data are used to inform the particle distribution and propagation in the visual SMC-PHD filter. However, the performance of the AV-SMC-PHD filter can be affected by the mismatch between the proposal and the posterior distribution. In this paper, we present a new method to improve the particle distribution where audio information (i.e. DOA angles derived from microphone array measurements) is used to detect new born particles and visual information (i.e. histograms) is used to modify the particles with particle flow (PF). Using particle flow has the benefit of migrating particles smoothly from the prior to the posterior distribution. We compare the proposed algorithm with the baseline AV-SMC-PHD algorithm using experiments on the AV16.3 dataset with multi-speaker sequences.

Keywords: Audio-visual tracking · PHD filter · SMC implementation · Multi-speaker tracking

1 Introduction

Multi-speaker tracking for indoor environments has received much interest in the fields of computer vision and signal processing [25]. An increasing amount of attention has been paid to the use of audio-visual modalities [2,15], which provide

Y. Liu—This work was supported by the EPSRC Programme Grant S3A: Future Spatial Audio for an Immersive Listener Experience at Home (EP/L000539/1), the BBC as part of the BBC Audio Research Partnership, the China Scholarship Council (CSC), and the EPSRC grant EP/K014307/1 and the MOD University Defence Research Collaboration in Signal Processing.

© Springer International Publishing AG 2017
P. Tichavský et al. (Eds.): LVA/ICA 2017, LNCS 10169, pp. 344–353, 2017.
DOI: 10.1007/978-3-319-53547-0_33

complementary information in addressing several challenges such as occlusion, limited view of cameras, illumination change, and room reverberations.

Several approaches using multi-modal information have been proposed. One such method is based on audio-visual diarization [19], which is only effective when the speakers continuously face the cameras. Kılıç et al. [21] addresses this problem in the framework of audio-visual speaker tracking using a particle filter (PF) and a probability hypothesis density (PHD) filter based on sequential Monte Carlo (SMC) approximation [20]. Different from the Bayesian approaches (Kalman or PF filters) [3,4,27], prior knowledge such as the number of targets is not required in the PHD filter. As for other SMC-PHD filters, the AV-SMC-PHD filter in [20] uses particles to represent the posterior density. However, after some updates, the prior distribution may not overlap with the target distribution [17].

Recently, the particle flow (PF) filter has been proposed for solving the nonlinear and non-Gaussian problem [5,8,9,12]. In this method, particle flow is created by a log-homotopy of the conditional density migrating from the prior to the posterior. Several approaches have been proposed to create the particle flow which can be categorized into five classes: incompressible flow [6], zero diffusion exact flow [7], Coulomb's law particle flow [13], zero-curvature particle flow [9] and non zero diffusion flow [12]. The zero-curvature particle flow has been used widely [18,24,30], as it is straightforward to implement.

Particle flow has been used to improve the accuracy of the particle filter [24], and is denoted as the particle flow particle filter (PFPF). Different from conventional particle filters, the PFPF uses a small number of particles to achieve the similar accuracy as that for particle filters with a higher effective sample size (ESS) [22]. However, for multi-target tracking, a dependent filter needs to be applied to each target, which introduces the model-data association problem [14]. In addition, prior knowledge of the number of targets is needed. In [30], a Gaussian particle flow implementation of the PHD filter (GPF-PHD) is proposed yielding good accuracy in a nonlinear tracking problem. However, in this method, the particles are generated for each target, and the computational cost could be high for a large number of targets and clutter. For non-linear and non-Gaussian problems, the auxiliary particle PHD filter proposed in [1] has better performance than the GPF-PHD filter in terms of Optimal Subpattern Assignment (OSPA) [14], since it efficiently distributes the particles by maximizing the accuracy of the cardinality estimate.

In this paper, we extend the AV-SMC-PHD filter presented in [20] by incorporating particle flow within the particle evolution in order to improve its tracking performance. The major contribution of this paper is a novel particle flow SMC-PHD filtering method for multi-speaker tracking, where the audio data are used to compute the prior distribution and the visual data are applied to compute the particle flow. The posterior distribution is calculated by the color histograms of the visual image and adjusted by the position of the direction of arrival (DOA) lines drawn from the targets. Using audio information, the computational cost for generating the particle flow can be reduced, as only the relevant particles surrounding the DOA line will be chosen; while the influence of the particles, that are likely from the clutter and distant from the DOA line, is mitigated. The proposed method is shown to outperform the baseline AV-SMC-PHD based on evaluations on the AV16.3 dataset.

The reminder of this paper is organized as follows: the next section introduces the AV-SMC-PHD filter and particle flow. Section 3 describes our proposed audio-visual particle flow SMC-PHD (AV-PF-SMC-PHD) filtering algorithm. In Sect. 4, experiments on the AV16.3 dataset are presented to show the performance of the proposed AV-PF-SMC-PHD algorithm as compared with the baseline AV-SMC-PHD algorithm.

2   AV-SMC-PHD Filter and Particle Flow

In this section, the baseline AV-SMC-PHD filter and the particle flow filter are introduced. For the discrete-time and non-linear filtering problems, we assume that the target dynamics and observations are described as a Markov state-space signal model:

$$\widetilde{m}_k = f_{\widetilde{m}}\left(\widetilde{m}_{k-1}, \tau_k\right), \tag{1}$$

where $\widetilde{m}_k$ is the target state vector at time-step k and $\sim$ is used to distinguish the target state from the particle state used later. In this paper, the state vector $m_k = [x_k, y_k, \dot{x}_k, \dot{y}_k]^T$ consists of the target positions (x_k, y_k) and the target velocities $(\dot{x}_k, \dot{y}_k)$, and the observation is a noisy version of the position. The parameter vector τ_k denotes the system excitation and observation noise terms and $f_{\widetilde{m}}$ is the transition density.

2.1   AV-SMC-PHD Filter

In the visual SMC-PHD filter [26], the surviving, spawned and born particles are used to model the existing and new speakers. For detecting the new targets, new particles need to be added randomly, leading to increase in the number of particles and hence to increase in computational load. To address this problem, the AV-SMC-PHD filter is proposed in [20].

Audio information is applied for re-locating existing particles around the DOA lines, since the DOA information shows the approximate direction of the sound emanating from the speakers. The movement distances of the particles $\hat{d}_k$ are calculated as [20]:

$$\hat{d}_k = \frac{d_k}{\|d_k\|_1} \odot d_k \tag{2}$$

where d_k is the perpendicular Euclidean distances between the particles and the DOA line, $\|.\|_1$ is the l_1 norm, and $\odot$ is the element-wise product; $\hat{d}_k$ is applied to relocate the surviving and spawned particles $m_{s,k}$ around the DOA line [20]:

$$m_{s,k} = m_{s,k} \oplus h_k\hat{d}_k \tag{3}$$

where $\boldsymbol{h}_k = [cos(\theta_k), sin(\theta_k), 0, 0]$ and θ_k is the angle from the DOA line. $\oplus$ is the element-wise addition. As such, the particles are modified along the perpendicular movement to the DOA line, and N_Γ born particles are sampled from the new born importance function,

$$\boldsymbol{m}_{k|k-1}^i \sim p_k(\cdot|\boldsymbol{Z}_k). \tag{4}$$

where $\boldsymbol{m}_{k|k-1}^i$ is the i-th predicted particle state at time-step k.

Apart from that, audio information in the AV-SMC-PHD filter can be used to detect the new speakers effectively and the particles are born in particular directions. This reduces the number of particles and hence computational complexity. The DOA lines are determined by the relative delay between pairs of microphone signals [29]. When detecting new targets, the filter compares the number of DOA lines, N_D, with the number of estimated speakers at time $k-1$, N_{k-1}. If $N_D = N_{k-1}$, the number of the speakers remains unchanged. If $N_D < N_{k-1}$, the speakers may walk out of the camera view, or be occluded by other speakers, or the DOA line may not be detected. In this paper, if $N_D < N_{k-1}$ and $N_D \neq 0$, we assume that the number of the speakers reduces to N_D. If $N_D = 0$, we assume that the microphones do not detect the speakers successfully and the number of speakers remains the same as N_{k-1}. If $N_D > N_{k-1}$, a new speaker (or some new speakers) may appear in the scene and hence new born particles should be created. Since born particles are only generated when the detection of a new speaker occurs via audio, the computational complexity is reduced.

The pseudo code of AV-SMC-PHD filter is given in Algorithm 1 where $\{\boldsymbol{m}_k^i, \omega_k^i\}_{i=1}^{N_k}$ is the set of the particle state vectors and weights at time-step k; $\{\widetilde{\boldsymbol{m}}_k^j, \widetilde{\omega}_k^j\}_{j=1}^{\widetilde{N}_k}$ is the target set and $\widetilde{N}_k$ is the number of targets at the time-step k; N_Γ is the number of born particles, which is given as the initial value; $\boldsymbol{Z}_k$ contains observations at time-step k. The weights of the particles are predicted and updated by

$$\omega_{k|k-1}^i = \begin{cases} \dfrac{\phi_{k|k-1}(\boldsymbol{m}_{k|k-1}^i, \boldsymbol{m}_{k-1}^i)\omega_{k-1}^i}{q_k\left(\boldsymbol{m}_{k|k-1}^i|\boldsymbol{m}_{k-1}^i, \boldsymbol{Z}_k\right)} & , i = 1, ..., N \\[3mm] \dfrac{\gamma_k(\boldsymbol{m}_{k|k-1}^i)}{N_\Gamma p_k(\boldsymbol{m}_{k|k-1}^i|\boldsymbol{Z}_k)} & , i = N+1, ..., N+N_\Gamma \end{cases} \tag{5}$$

$$\omega_k^i = \left[1 - p_{D,k}(\boldsymbol{m}_k^i) + \sum_{z \in Z_k} \frac{p_{D,k}(\boldsymbol{m}_k^i)g_k(z|\boldsymbol{m}_k^i)}{\kappa_k(z) + C_k(z)}\right] \omega_{k|k-1}^i \tag{6}$$

where

$$C_k(z) = \sum_{i=1}^{N+N_\Gamma} p_{D,k}(\boldsymbol{m}_k^i)g_k(z|\boldsymbol{m}_k^i)\omega_{k|k-1}^i \tag{7}$$

in which $\omega_{k|k-1}^i$ is the i-th predicted particle weight at time-step k. $\phi_{k|k-1}(.|.)$ is the analogue of the state transition probability with the previous state. $q_k(.|.)$ is the proposal distribution. $\gamma_k(.)$ is the probability of the born particle. $\boldsymbol{Z}_k$ is the observation set at time-step k. $\kappa_k(\boldsymbol{z})$ denotes the clutter intensity of the observation $\boldsymbol{z}$ at time step k. $p_{D,k}(.|.)$ is the probability of detection at time step k. $g_k(.|.)$ is the likelihood of individual targets.

Algorithm 1. AV-SMC-PHD Filter

Input: $\{m_{k-1}^i, \omega_{k-1}^i\}_{i=1}^{N_{k-1}}$, N_Γ, Z_k and DOA line.

Output: $\{\widetilde{m}_k^j, \widetilde{\omega}_k^j\}_{j=1}^{\widetilde{N}_k}$, and $\{m_k^i, \omega_k^i\}_{i=1}^{N_k}$.

  Run:

  Predict existing targets.

  if DOA exists **then**

    Calculate distances d_k.

    Calculate movement distances $\hat{d}_k$ by Eq. (2).

    Concentrate $m_{s,k}$ around the DOA line by (3).

    if new speaker **then**

      Born N_Γ particles uniformly around the DOA line by (4).

    end if

  end if

  Predict the weights of the particles $\omega_{k|k-1}^i$ by Eq. (5).

  (Optional) Update the states and the weights of the particles by the particle flow.

  Update the weights of the particles $\omega_{k|k}^i$ by Eq. (6) and calculate $\widetilde{N}_k = \sum_{i=1}^{N_k} \omega_{k|k}^i$.

  Get $\{\widetilde{m}_k^j, \widetilde{\omega}_k^j\}_{j=1}^{\widetilde{N}_k}$ by the k-means method and get $\{m_k^i, \omega_k^i\}_{i=1}^{N_k}$ by re-sampling.

2.2 Particle Flow

There are several particle flow algorithms. Here we use the zero diffusion exact flow [24], since it is straightforward to implement. Daum and Huang define the flow of the logarithm of the conditional probability density function with respect to step size λ [11]:

$$\log(\psi_k(m, \lambda)) = \log(h_k(m)) + \lambda \log(g_k(m)) \tag{8}$$

where λ takes values from $[0, \triangle\lambda, 2\triangle\lambda, \cdots, N_\lambda\triangle\lambda]$, where $N_\lambda\triangle\lambda = 1$. $g_k(.)$ is the likelihood function. At the start of the flow ($\lambda = 0$), $\psi_k(m_k, \lambda)$ represents the prior density, $h_k(.)$. At the end of the flow ($\lambda = 1$), $\psi_k(m_k, \lambda)$ is translated into the normalized posterior density. This flow simulates the motion of the physical particles as Brownian movement [5] from the prior to the posterior density.

When the prior and the likelihood are unnormalized Gaussian probability densities, the exact solution for the particle flow is given as [10]:

$$\frac{dm}{d\lambda} = A(\lambda) m + b(\lambda) \tag{9}$$

where

$$A(\lambda) = -\frac{1}{2} P H^T (\lambda H P H^T + R)^{-1} H, \tag{10}$$

$$b(\lambda) = (I + 2\lambda A) \left[(I + \lambda A) P H^T R^{-1} z + A\bar{m} \right] \tag{11}$$

in which $\bar{m}$ is the mean of the particle and R is the covariance matrix of the observation noise. For nonlinear problems, the observations need to be linearized for each particle (analogous to an extended Kalman filter). P is the covariance matrix of the particles. H is computed as the Jacobian matrix.

3 Proposed AV-PF-SMC-PHD Filter

In the AV-SMC-PHD filter as already summarised in Sect. 2.1, the particles need to be drawn from a proposal distribution. However, it may not be well matched to the posterior density because of the particle degeneracy issue [7]. To mitigate this problem, we add an adjustment step between the prediction step and update step, where the particle flow Eqs. (9)–(13) are applied to adjust the states and weights of the particles by smoothly migrating them from the prior to the posterior density.

In our proposed filter, audio information is used to calculate the number of particle flows. As in other multi-speaker particle filters, the particles need to be labeled [24] or cluttered [15] before updated. However, the prior information about the number of targets and the association between the targets and particles of the visual SMC-PHD filter is unknown and time-varying in multi-target tracking. In our method, such information could be provided by the DOA lines. We assume that the number of particle flows is the same as the number of DOA lines N_D. Then the particles are classified to N_D sets based on the Euclidean distance between the particles and the DOA lines. These particles are denoted as $\{m_{k|k-1}^i, \omega_{k|k-1}^i\}_{i \in \Lambda(z)}$, where $\Lambda(z)$ is a subset of $E = [1, \cdots, N + N_\Gamma]$. In practice, some particles are created due to clutter and noise. To account for the noise effect, we assume that each flow will only be influenced by the particles in the neighborhood of the DOA lines within a certain distance d.

The mean $\bar{m}(z)$ and covariance $P(z)$ are calculated based on different particle flows. The states of particles are adjusted by Eq. (9) and the weights of the particles also need to be adjusted as

$$\omega_{k|k-1}^i := \frac{q_k(m_{k|k-1}^i | m_{k-1}^i, z)}{q_k(\acute{m}_{k|k-1}^i | m_{k-1}^i, z)} \omega_{k|k-1}^i \tag{12}$$

where $\acute{m}_{k|k-1}^i$ is the updated value of $m_{k|k-1}^i$ by particle flow.

The pseudo-code of the adjustment step of the PF-SMC-PHD filter is presented in Algorithm 2. The observation of particle flow z is calculated by the

---

Algorithm 2. Adjustment Step of the AV-PF-SMC-PHD Filter

Input: $\{m_{k|k-1}^i, \omega_{k|k-1}^i\}_{i=1}^{N_k}$, Z_k, v and the DOA line
Output: $\{m_{k|k-1}^i, \omega_{k|k-1}^i\}_{i=1}^{N_k}$.
  Run:
  for each DOA line **do**
    Calculate z by the reference histogram v and input image Z_k
    for $\lambda \in [0, \triangle\lambda, 2\triangle\lambda, \cdots, N_\lambda \triangle\lambda]$ **do**
      Calculate H via Eq. (13) and A and b by Eq. (10) and Eq. (11), respectively.
      Evaluate flow $\frac{dm_{k|k-1}^i}{d\lambda}$ by Eq. (9) and $m_{k|k-1}^i = m_{k|k-1}^i + \triangle\lambda \frac{dm_{k|k-1}^i}{d\lambda}$.
    end for
    Update the particle weights by Eq. (12).
  end for

---

color histogram matching [16]. The reference histogram v is updated with the estimate from the previous time step $k-1$. Note that m in Eq. (9) should be represented with $m^i_{k|k-1}$, and

$$H = \begin{bmatrix} cos\,(\theta) & -sin\,(\theta) \\ sin\,(\theta) & cos\,(\theta) \end{bmatrix} \tag{13}$$

where $\theta = arctan(\frac{m(2)}{m(1)})$, and $m(1)$ and $m(2)$ are the first and second element of m, respectively.

4   Experimental Results

In this section, the proposed algorithm is compared with the visual SMC-PHD algorithm, the baseline AV-SMC-PHD algorithm in [20] using the AV16.3 dataset [23]. In the visual SMC-PHD filter, the born particles are created randomly in the tracking area but the number of particles is the same as other filters. AV16.3 consists of sequences where speakers are walking and speaking at the same time. Those actions are recorded by three calibrated video cameras at 25 Hz and two circular eight-element microphone arrays at 16 kHz. The audio and video streams are synchronized before running the algorithms. The size of each image frame is 288×360 pixels. All the algorithms are tested with all the three different camera angles of four sequences: Sequences 24, 25, 30 and 45, which correspond to the cases of two and three speakers and are the most challenging sequences in term of movements of the speakers and the number of occlusions.

As in [20], the OSPA metric [28] is employed for measuring the tracking performance. The OSPA is able to evaluate the performance on target number estimation as well as the position estimation, which is suitable for multi-target tracking. A low OSPA implies a better performance. All experiments are run on a computer with Intel i7-3770 CPU with a clock frequency of 3.40 GHz and 8G RAM.

The parameters for the SMC-PHD filter are set as: $p_D = 0.98$, $p_S = 0.99$ and $\sigma_c = 0.1$. The uniform density u is $(360 \times 280)^{-1}$ and the number of particles per speaker is 50. The parameters for particle flow are set empirically as: $\triangle\lambda = 0.01$, $P = [5, 5, 1, 1]$ and $d = 30$. The OSPA metric order parameter a is 2.

Due to page limit, we only show part of the results obtained. First, the OSPA results for Sequence 24 camera #1, as an example, is shown in Fig. 1. The green dotted line is the OSPA for the visual SMC-PHD fitler, the blue dotted line is the OSPA for the AV-SMC-PHD filter, and the red solid line for the AV-PF-SMC-PHD filter. From frame 400 to frame 600 and from frame 800 to frame 1000, there is no occlusion. At most of time such as from frame 350 to frame 700, OSPA for the AV-PF-SMC-PHD filter is the lowest among the three filters. Compared with audio information, the targets can be more quickly tracked with a lower OSPA as compared with the visual filter, especially when a new target appears, such as from frame 0 to frame 350. However, from frame 750 to frame 800, the error of the AV-PF-SMC-PHD filter is larger than that of the AV-SMC-PHD filter, since it is the end of the occlusion, and the particles are modified

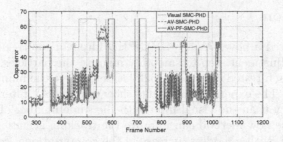

Fig. 1. Performance comparison of the visual SMC-PHD filter, the AV-SMC-PHD filter and the proposed AV-PF-SMC-PHD filters in terms of the OSPA error.

Table 1. Experimental results for the visual SMC-PHD filter, the AV-SMC-PHD filter and the AV-PF-SMC-PHD filter in terms of the OSPA error.

		Visual SMC-PHD	AV-SMC-PHD	AV-PF-SMC-PHD
seq24	cam1	30.46	17.71	16.57
	cam2	35.91	19.83	17.04
	cam3	32.61	18.94	16.71
seq25	cam1	34.96	19.13	16.85
	cam2	31.86	18.47	16.72
	cam3	37.15	21.61	18.58
seq30	cam1	39.35	25.22	20.57
	cam2	35.24	19.37	16.92
	cam3	40.21	25.31	20.57
seq45	cam1	43.17	29.46	27.55
	cam2	43.20	29.47	27.55
	cam3	39.52	28.43	26.07
Average		36.97	22.75	20.14

to the wrong direction by the visual information of the occluded speakers in the previous frame.

Other sequences are also used in the tests and the results of different methods are given in Table 1. The average errors for the visual SMC-PHD filter, the AV-SMC-PHD filter and the AV-PF-SMC-PHD filter are 36.97, 22.75 and 20.14 respectively. With the same number of particles, the visual filter gives a higher OSPA than the audio-visual filters, which means audio information can improve the tracking accuracy of visual SMC-PHD filters. Apart from that, with the particle flow, 12.47% reduction in tracking error has been achieved. However, for the Sequence 24 camera 1, the computational cost has increased from 112 to 703 s. The computational cost for the AV-PF-SMC-PHD filter will also increase with the number of particles and targets. For example, the execution time for the Sequence 45 (1034 s) is larger than that for Sequence 24 (153 s).

5   Conclusion

We have presented a novel AV-PF-SMC-PHD filter for multi-speaker tracking, by adding an adjustment step to smoothly migrate the particles. The proposed algorithm has been tested on the AV16.3 dataset, where the number of speakers varies over time. The experimental results show that the AV filters offer a higher tracking accuracy than the visual filter with the same number of particles. The proposed particle flow method can improve the tracking accuracy over the AV-SMC-PHD filter, with a modest increase in the computational cost.

References

1. Baser, E., Efe, M.: A novel auxiliary particle PHD filter. In: 15th International Conference on Information Fusion, pp. 165–172. IEEE (2012)
2. Bernardin, K., Gehrig, T., Stiefelhagen, R.: Multi-level particle filter fusion of features and cues for audio-visual person tracking. In: Stiefelhagen, R., Bowers, R., Fiscus, J. (eds.) CLEAR/RT -2007. LNCS, vol. 4625, pp. 70–81. Springer, Heidelberg (2008). doi:10.1007/978-3-540-68585-2_5
3. Cevher, V., Sankaranarayanan, A.C., McClellan, J.H., Chellappa, R.: Target tracking using a joint acoustic video system. IEEE Trans. Multimedia **9**(4), 715–727 (2007)
4. Cui, P., Sun, L.F., Wang, F., Yang, S.Q.: Contextual mixture tracking. IEEE Trans. Multimedia **11**(2), 333–341 (2009)
5. Daum, F., Huang, J.: Particle flow for nonlinear filters with log-homotopy. In: Proceedings of SPIE, pp. 696918-1–696918-12 (2008)
6. Daum, F., Huang, J.: Nonlinear filters with log-homotopy. In: Drummond, O.E., Teichgraeber, R.D. (eds.) Optical Engineering + Applications. pp. 669918–669918-15. International Society for Optics and Photonics (2007)
7. Daum, F., Huang, J.: Nonlinear filters with particle flow. In: SPIE Optical Engineering + Applications, pp. 74450R-1–74450R-9. International Society for Optics and Photonics (2009)
8. Daum, F., Huang, J.: Particle flow for nonlinear filters. In: 2011 IEEE International Conference on Acoustics, Speech and Signal Processing, pp. 5920–5923 (2011)
9. Daum, F., Huang, J.: Renormalization group flow and other ideas inspired by physics for nonlinear filters, Bayesian decisions, and transport. In: SPIE Defense + Security, p. 90910I. International Society for Optics and Photonics (2014)
10. Daum, F., Huang, J.: Renormalization group flow in k-space for nonlinear filters, Bayesian decisions and transport. In: 18th International Conference on Information Fusion, pp. 1617–1624. IEEE (2015)
11. Daum, F., Huang, J., Noushin, A.: Exact particle flow for nonlinear filters. In: SPIE Defense, Security, and Sensing, p. 769704 (2010)
12. Daum, F., Huang, J., Noushin, A.: Small curvature particle flow for nonlinear filters. Signal Processing, Sensor Fusion, and Target Recognition XIX **7697**(1), 769704 (2010)
13. Daum, F., Huang, J., Noushin, A.: Coulomb's law particle flow for nonlinear filters. In: Drummond, O.E. (ed.) SPIE Optical Engineering + Applications, pp. 1–15. International Society for Optics and Photonics (2011)
14. Fortmann, T., Bar-Shalom, Y., Scheffe, M.: Sonar tracking of multiple targets using joint probabilistic data association. IEEE J. Oceanic Eng. **8**(3), 173–184 (1983)

15. Gatica-Perez, D., Lathoud, G., Odobez, J.M., McCowan, I.: Audiovisual probabilistic tracking of multiple speakers in meetings. IEEE Trans. Audio Speech Lang. Process. **15**(2), 601–616 (2007)
16. Hafner, J., Sawhney, H.S., Equitz, W., Flickner, M., Niblack, W.: Efficient color histogram indexing for quadratic form distance functions. IEEE Trans. Pattern Anal. Mach. Intell. **17**(7), 729–736 (1995)
17. Khan, M.A., Ulmke, M.: Non-linear and non-Gaussian state estimation using log-homotopy based particle flow filters. In: 2014 Workshop on Sensor Data Fusion: Trends, Solutions, Applications, SDF (2014)
18. Khan, M.A., Ulmke, M.: Non-linear and non-Gaussian state estimation using log-homotopy based particle flow filters. In: Sensor Data Fusion: Trends, Solutions, Applications (SDF), pp. 1–6. IEEE (2014)
19. Kidron, E., Schechner, Y.Y., Elad, M.: Cross-modal localization via sparsity. IEEE Trans. Sig. Process. **55**(4), 1390–1404 (2007)
20. Kilic, V., Barnard, M., Wang, W., Hilton, A., Kittler, J.: Mean-shift and sparse sampling based SMC-PHD filtering for audio informed visual speaker tracking. IEEE Trans. Multimedia **18**(12), 2417–2431 (2016)
21. Kılıç, V., Barnard, M., Wang, W., Kittler, J.: Audio assisted robust visual tracking with adaptive particle filtering. IEEE Trans. Multimedia **17**(2), 186–200 (2015)
22. Kong, A., Liu, J.S., Wong, W.H.: Sequential imputations and Bayesian missing data problems. J. Am. Stat. Assoc. **89**(425), 278–288 (1994)
23. Lathoud, G., Odobez, J.-M., Gatica-Perez, D.: AV16.3: an audio-visual corpus for speaker localization and tracking. In: Bengio, S., Bourlard, H. (eds.) MLMI 2004. LNCS, vol. 3361, pp. 182–195. Springer, Heidelberg (2005). doi:10.1007/978-3-540-30568-2_16
24. Li, Y., Zhao, L., Coates, M.: Particle flow for particle filtering. In: 2016 IEEE International Conference on Acoustics, Speech and Signal Processing, pp. 3979–3983. IEEE (2016)
25. Liu, Q., Rui, Y., Gupta, A., Cadiz, J.J.: Automating camera management for lecture room environments. In: Proceedings of the SIGCHI Conference on Human Factors in Computing Systems, pp. 442–449. ACM (2001)
26. Maggio, E., Piccardo, E., Regazzoni, C., Cavallaro, A.: Particle PHD filtering for multi-target visual tracking. In: 2007 IEEE International Conference on Acoustics, Speech and Signal Processing, vol. 1, pp. I–1101. IEEE (2007)
27. Polat, E., Ozden, M.: A nonparametric adaptive tracking algorithm based on multiple feature distributions. IEEE Trans. Multimedia **8**(6), 1156–1163 (2006)
28. Ristic, B., Vo, B.N., Clark, D., Vo, B.T.: A metric for performance evaluation of multi-target tracking algorithms. IEEE Trans. Sig. Process. **59**(7), 3452–3457 (2011)
29. Talantzis, F., Constantinides, A.G., Polymenakos, L.C.: Estimation of direction of arrival using information theory. IEEE Sig. Process. Lett. **12**(8), 561–564 (2005)
30. Zhao, L., Wang, J., Li, Y., Coates, M.J.: Gaussian particle flow implementation of PHD filter. In: SPIE Defense + Security, p. 98420D (2016)

Latent Variable Analysis in Observation Sciences

Estimation of the Intrinsic Dimensionality in Hyperspectral Imagery via the Hubness Phenomenon

Rob Heylen[1]([✉]), Mario Parente[2], and Paul Scheunders[1]

[1] Visionlab, University of Antwerp, Antwerp, Belgium
{rob.heylen,paul.scheunders}@uantwerpen.be
[2] Electrical and Computer Engineering, University of Massachusetts, Amherst, USA
mparente@ecs.umass.edu

Abstract. As hyperspectral images are high-dimensional data sets containing a lot of redundancy, a first important step in many applications such as spectral unmixing or dimensionality reduction is estimation of the intrinsic dimensionality of the data set. We present a new method for estimation of the intrinsic dimensionality in hyperspectral images based upon the hubness phenomenon, which is the observation that indegree distributions in a K-nearest neighbor graph will become skewed as the intrinsic dimensionality of the data set rises. The proposed technique is based upon comparing the indegree distributions of artificially generated data sets with the one from the target data set, and identifying the best match with some histogram metric. We show that this method obtains superior results compared to many alternatives, and does not suffer from the effects of interband and spectral correlations.

1 Introduction

A pixel in a hyperspectral image samples the full spectrum of the area in its field of view, and typically contains hundreds of spectral bands spanning the visible and infrared regions. Such images are typically obtained from a large distance (aerial or satellite), and the area captured in the field of view of each pixel becomes large, of the order of hundreds of meters squared. Therefore, the spectrum captured in one pixel contains information about all the objects in the field of view of that pixel, and one has to take this spectral mixing into account. A rich literature exists on "spectral unmixing" techniques, which attempt to reverse this spectral mixing operation. See [1,2] for recent overviews.

While several exceptions exist, most unmixing techniques generally assume that a single spectrum, or "endmember", exists for every pure material in the scene, and each observed spectrum can then be considered to be composed of these endmember spectra. In the simplest model, one assumes that each spectrum can be described as a convex linear combination of endmember spectra, leading to the linear mixing model (LMM) [1]. Several more advanced models use higher-order polynomial representations to account for multiple reflections

© Springer International Publishing AG 2017
P. Tichavský et al. (Eds.): LVA/ICA 2017, LNCS 10169, pp. 357–366, 2017.
DOI: 10.1007/978-3-319-53547-0_34

between objects, or use physical modeling via radiative transfer theory or computer simulations to obtain even more accurate mixing models, but with the disadvantage of complex inversion strategies [2].

An important first step in any of these modeling and inversion attempts is to estimate the number of endmembers that one is going to employ. In most spectral mixing models, this number of endmembers (NOE) can be easily related to the intrinsic dimensionality (ID) of the data set. If one assumes that the data set lies on a smooth manifold in spectral space, the ID can be defined as the dimensionality of the Euclidean space to which this manifold is locally homeomorph. Several authors also describe the ID as the minimum number of degrees of freedom one requires to fully describe or model the data set. In the LMM, the NOE p is related to the ID d as $d = p - 1$, as the data points are constrained to a $(p-1)$-dimensional simplex due to the convexity constraints, and in the absence of noise. In nonlinear models, the ID can be larger, as one might have additional hyperparameters. These results indicate that ID estimation can provide a good approximation for the NOE in many cases. Remark that proper estimation of the ID is also important in several other applications, such as dimensionality reduction, classification and clustering.

Many techniques exist in the hyperspectral unmixing literature for estimating the NOE, and these can be roughly divided into two classes: The first class attempts to estimate the dimensionality of the linear subspace that contains the data, while the second class employs manifold techniques. Examples of the former are the Harsanyi-Farrand-Chang virtual dimensionality (VD) technique [4] and hyperspectral signal identification by minimum error (HySime) [5]. An example of the latter is the Grassberger-Procaccia algorithm [6].

Any property that depends strongly on the ID of the data set and is invariant with respect to other parameters might be an appropriate estimator for the ID and the associated NOE in a hyperspectral data set. We present a new method for estimating the ID of a hyperspectral image, based on the hubness phenomenon [7]. The hubness phenomenon is an effect which occurs when the ID of a data set increases: If one creates a K-nearest neighbor (KNN) graph in a random data set, one observes that some of the data points will become highly connected to others when the ID grows, and start to act like hubs in the corresponding graph. The indegree distribution (IDD) of the graph, i.e. the distribution of the number of incoming connections from other data points, will become skewed to higher numbers. This hubness effect is present in numerous real-world high-dimensional data sets, and depends strongly on the ID of the data [7].

The proposed technique is based on generating IDDs for many simulated hyperspectral data sets with known parameters, and comparing the observed IDD with these artificially generated distributions. We show that this approach functions very well on real data sets, and is superior to many alternatives on artificially generated hyperspectral data. The technique possesses only one parameter K used for the KNN graph, but shows no strong dependence on this parameter. Furthermore, we also require a denoising procedure, as the ID depends strongly on the noise in the data set. Remark that techniques which exploit the hubness phenomenon can be considered to belong to the class of manifold techniques, as they employ only a local property of the data, i.e. the KNN graph. This implies that the technique can also be employed for nonlinear mixing situations, where the data lies on a manifold instead of a linear subset. Furthermore, several alternative techniques show a severe underestimation of the NOE in real and simulated data sets due to the inherent correlations which are present between spectra on the one hand, and between adjacent spectral bands on the other hand. The proposed method does not suffer from these effects.

2   The Hubness Phenomenon

Let $X = (x_1, \ldots, x_N)$ be a data matrix containing N data vectors $x_i \in \mathbb{R}^d$ columnwise, with d the embedding dimension of the data set. Let $d_{ij} = \|x_i - x_j\|$ be the Euclidean distance between x_i and x_j. The first or nearest neighbor $\mathrm{nn}_1(x_i)$ of x_i is the point x_j of smallest Euclidean distance. This concept can be extended to the K nearest neighbor $\mathrm{nn}_K(x_i)$, given by the point with the K'th smallest distance to x_i. A KNN graph can be constructed by considering the set of directed edges $(x_i, \mathrm{nn}_j(x_i))$, with $i = 1, \ldots, N$ and $j = 1, \ldots, K$, where the corresponding weights of each edge are given by its Euclidean length. The indegree $N_K(x)$ of a point x is defined as the number of times that x occurs as a neighbor of other points in the KNN graph of the data set X [7]. The distribution of N_K for all data points is the IDD of the data set.

Points with a high indegree relative to the average indegree act like hubs in the graph. The hubness phenomenon describes the observation that the relative number of hubs increases as the ID of the data set increases. In practice, this means that the IDD of the graph becomes skewed, and that more points with very high degrees show up as the ID increases. The cause for the occurrence of hubness is the concentration of measure phenomenon (see [7] for a detailed analysis): The average distance to the data centroid increases linearly with the intrinsic data dimensionality d, but its variance as $\sqrt{d}$. Therefore, the fraction of points which lie significantly closer to the centroid will decrease with increasing d. It is hypothesized that these points will become hubs, as they will lie closer than average to all other points, and hence will have a larger than average probability of being identified as a near neighbor.

As the intended application is estimation of the ID and the associated NOE in hyperspectral imagery, one can use simulated hyperspectral images to illustrate the hubness phenomenon. We employ the popular LMM to generate the artificial data sets, applied to spectra obtained in a laboratory setting. First, obtain p endmembers $E = (e_1, \ldots, e_p)$ randomly from a spectral library, e.g. the USGS spectral library of minerals, and interpolate them at the same d wavelengths, leading to a data set with an ID equal to $\min(p-1, d)$. Next, create N abundance vectors $A = (a_1, \ldots, a_N)$ randomly in a unit simplex by using the symmetric Dirichlet distribution (SDD):

$$P(a|\alpha) = \begin{cases} \frac{\Gamma(\alpha p)}{\Gamma(\alpha)^p} \prod_{i=1}^{p} a_i^{\alpha-1} & \text{if } a \in S_p \\ 0 & \text{otherwise} \end{cases} \tag{1}$$

$\alpha = 0.1$ $\qquad$ $\alpha = 0.5$ $\qquad$ $\alpha = 1$ $\qquad$ $\alpha = 2$ $\qquad$ $\alpha = 10$

Fig. 1. Abundance vectors generated by the SDD for different values of α

with S_p the unit simplex in p dimensions. The SDD will always return abundance vectors that obey convexity: They are non-negative and sum to one. Furthermore, it depends on a single hyperparameter α which can be used to skew the distribution: For $\alpha = 1$ one obtains a uniform distribution over the unit simplex. For $\alpha > 1$ the abundance vectors will be more probable close to the center of the simplex, while for $\alpha < 1$ the abundance vectors will become sparse, and will lie closer to the borders of the simplex. See Fig. 1 for an example.

We use the LMM to generate the data set:

$$X = EA + \boldsymbol{\eta} \qquad (2)$$

where $\boldsymbol{\eta}$ is a noise matrix of size (d, N) containing independent Gaussian noise with a given signal-to-noise ratio (SNR). The IDD is found by determining the KNN graph for a given value of K, counting how many times each point occurs as a neighbor of any other point, and aggregating these degrees in a normalized histogram. These IDDs are shown in Fig. 2 for a noiseless situation with uniform abundance values ($\alpha = 1$ in the SDD), $N = 10^4$, $d = 50$ and $K = 10$, and for several values of p. It can be observed that the shape of these IDDs will indeed shift to heavy-tailed distributions as p, and hence the ID, increases.

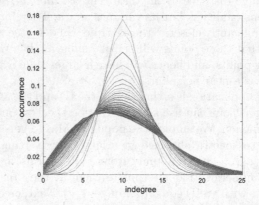

Fig. 2. The IDD for the artificial USGS data sets for $p \in [2, 50]$ (top to bottom), with $K = 10$ and $N = 10000$, each graph averaged over 100 random instances.

3   Properties of Hubness in Hyperspectral Data Sets

In order to better understand the properties of the hubness phenomenon in hyperspectral data sets, we present several results and observations acquired on artificial and simulated hyperspectral data sets. Due to lack of space, we cannot extensively discuss and illustrate these dependencies, but we refer to [8,9] for more information on some of these items.

The hubness phenomenon shows strong dependencies on several aspects of the data set:

- The hubness phenomenon depends strongly on the **noise** in the data set, as noise is typically considered to be independent Gaussian noise, and hence acts as a full-dimensional additional component on top of the signal component of lower ID. This will obscure the actual ID, and stresses the need for a proper denoising procedure before estimation of ID or the NOE.
- Also the **coherence in the spectral library** which was employed to create the artificial data sets has a significant impact: If one uses a library where a high average pairwise correlation (APC) exists between spectra, then a subset of randomly chosen spectra has a large probability of pairs lying close together in spectral space. Such a pair might not introduce enough variability in the data set to be detected as an additional degree of freedom, and hence the hubness phenomenon will not be as explicit as the NOE rises when a highly coherent library is employed. An example of a highly coherent library is the USGS vegetation library, with an APC of 0.78, where the mineral and artificial materials libraries have an APC of 0.17 and 0.18 respectively.
- The value of K which was employed to generate the data set also impacts the hubness phenomenon. For low ID, the IDD will have a modus equal to K, and this modus will decrease for increasing ID. As long as the same value of K is used for all comparisons, this dependence is not an issue.

Several other aspects show only a very small or no dependence:

- The **size of the data set** has only a very small effect on the IDDs.
- Also the **embedding dimension** does not influence the IDDs when the noise is small or non-existent. This can be easily explained due to the independence of local interpoint distances on the embedding dimension when all points are restricted to a lower-dimensional manifold.
- Several data sets show an intrinsic **sparseness**, and also in hyperspectral data sets, it is often more realistic to assume that only a small subset of endmembers contributes to the spectrum observed in a given pixel. This leads to an ID which is locally lower than the ID of the full data set. This effect can be observed for instance by modifying the α parameter used to generate the abundance distributions. This effect stays very small.
- Also the **library size** has no impact on the IDDs, as random subsets of different sizes yield identical IDDs.

– Finally, a surprising result is that also the **metric** which was employed to create the KNN graphs has no influence at all on the IDDs. We employed the L_1, L_2, L_∞, correlation distance, spectral angle distance and spectral information divergence to create the KNN graphs in the same data sets, and obtained virtually identical IDDs in every case.

These results indicate that the IDDs and the associated hubness phenomenon are indicative of the ID of the data set as long as the noise and library coherence are kept under control. Any other parameters used in the data creation will not affect the IDDs.

4   The Algorithm

We want to exploit the strong dependence of the IDD on the ID of the data set to find an estimate of the NOE in a hyperspectral data set. The underlying idea is to generate a library of representative IDDs by artificially generating hyperspectral data sets with a known NOE, and then comparing the IDD of the target data set with the IDDs in this library by using some histogram metric. This way, the artificial data set which corresponds with the target data set can be identified, and the corresponding NOE can be derived.

The results in previous Sect. 3 show that a denoising procedure has to be applied to the target data set in order to avoid overestimation of the NOE, as the presence of noise will typically increase the ID. Because no denoising technique can be expected to function perfectly, the obtained estimates must be considered as maximal values. This leads to the following algorithm:

1. Build an IDD library by generating a large number of random hyperspectral data sets with a known NOE, and determining the average IDD.
2. Denoise the target data set.
3. Determine the IDD of the target data set.
4. Compare this target IDD with those in the IDD library by using a histogram similarity metric to identify the data set with the largest correspondence.

To denoise the target data set, we employed the same method as in the HySime algorithm [5], where each spectral band is recreated from the others by linear regression. Remark that the results might depend significantly on the denoising method used, and that better denoising techniques might exist.

Many histogram or probability distribution metrics can be used to compare two histograms P and Q over the nonnegative integers. We chose to employ the total variation (TV) distance, as it is a simple and intuitive way to compare histograms.

5   Results

5.1   Artificial Data Sets

We have assessed the performance of the proposed algorithm on artificial data sets, and compared the obtained results with those obtained with several

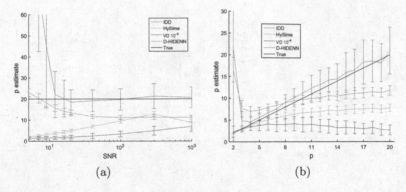

Fig. 3. The estimates for the NOE as (a) a function of the SNR and (b) as a function of p, for data sets mixed with the LMM with $N = 10^4$ pixels and (a) $p = 20$ and (b) SNR = 50.

alternatives from the literature. These are the HySime algorithm [5], the VD algorithm [4], and the denoised hyperspectral intrinsic dimensionality estimation with nearest-neighbor distance ratios (D-HIDENN) method, a technique we developed earlier based on the statistics of nearest-neighbor distance ratios [3]. Remark that the proposed method, HySime and the D-HIDENN method all use the same noise estimation technique. The HySime method has no parameters, while the VD algorithm depends on a false-alarm parameter (chosen to be 10^{-4}), and the D-HIDENN method has two integer parameters, chosen to be 1 and 3. We used $K = 10$ for the KNN graphs in the IDD method.

The data sets were generated using randomly selected spectra from the USGS database, which were mixed according to the LMM (2) with a uniformly and randomly chosen abundance matrix respecting the ASC and ANC (Eq. (1) with $\alpha = 1$). Remark that such data sets were also used for generating the artificial IDDs, with the difference that now noise is present.

For every set of parameters, we averaged over 100 random data sets. The first experiment, displayed in Fig. 3(a), assesses the performance as a function of the noise in the data set. These figures illustrate that for low SNR values, all methods show significant deviations from the true value, and also the variability in the results can be very high. For higher SNR however, the IDD method performs the best on average.

The dependence on the true value of p is illustrated in Fig. 3(b), where the estimates are plotted as a function of p. Note that on average, the IDD method obtains the correct result for all values of p, although with a large variance which can be explained by the creation of these data sets: When a high number of endmembers are randomly drawn from a spectral library, certain subsets of these can be highly correlated. Depending on the number of correlated endmembers, different IDs can be expected in the resulting data sets. The alternative methods show the typical underestimating behavior for increasing p discussed in previous sections. These results indicate that the IDD method yields better results

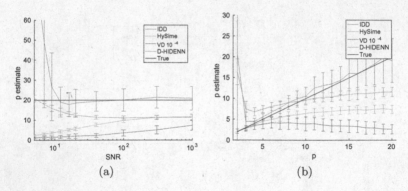

(a)   (b)

Fig. 4. The estimates for the NOE as (a) a function of the SNR and (b) as a function of p, for data sets mixed with the PPNM model with $N = 10^4$ pixels and (a) $p = 20$ and (b) SNR = 50.

over the alternatives on these types of artificial data sets. Remark that the bad performance for the lowest values of p can be attributed to the denoising procedure which does not effectively remove all noise in this case, and that correct results are obtained here in a noiseless scenario. More examples of the p and SNR dependence can be found in [8, 9].

Furthermore, one can assess the performance of these NOE estimation techniques on nonlinearly mixed data as well. An easy way to create a nonlinear data set is to employ the PPNM model, a popular nonlinear mixing model which belongs to the family of bilinear mixing models. This model creates a new data set from an existing, linearly mixed, data set via a polynomial transformation. We employ a second-order transformation of the data set as $x = y + y^2$, with y the result from the LMM. This yields a simple, nonlinearly mixed data set, but with the same ID as the original linearly mixed data set as one does not add degrees of freedom by this transformation. When the data sets employed in Fig. 3 are transformed with the PPNM and the experiments are run again, one obtains Fig. 4. This figure illustrates that the proposed technique obtains the correct answer on average also for nonlinearly mixed data sets as long as the noise is relatively small, while the alternative techniques still show large deviations.

5.2   Real Data Sets

We have also executed the proposed IDD method on several well-known hyperspectral data sets. The data sets under consideration are the Airborne Visible/Infrared Imaging Spectrometer (AVIRIS) Cuprite and Indian Pines data sets. These consist of 220-band images with an IFOV of $20 \times 20\,\mathrm{m}^2$, but we removed the water absorption bands ([1:4], [103:113], [148:166]) before use.

We have plotted the IDDs along with the best match from the library in Fig. 5(a) for the Cuprite data set, and in Fig. 5(b) for the Indian Pines data set. These figures illustrate that the IDD from the data sets indeed closely resembles

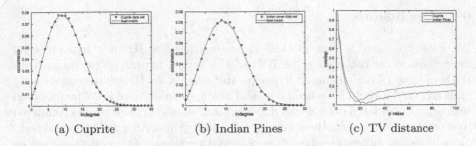

(a) Cuprite   (b) Indian Pines   (c) TV distance

Fig. 5. The IDD and the best match for the (a) Cuprite and (b) Indian Pines data sets. (c) The TV distance as a function of p.

Table 1. The estimates of the NOE obtained by the different techniques, on the AVIRIS Cuprite and Indian Pines data sets.

	Cuprite	Indian Pines
HFC VD 10^{-1}	41	56
HFC VD 10^{-2}	30	34
HFC VD 10^{-3}	22	26
HFC VD 10^{-4}	18	23
HySime	19	14
HIDENN	18	18
D-HIDENN	12	10
IDD method	25	18

one of the artificially calculated IDDs in the library. The TV distance is given in Fig. 5(c) for both data sets. The minimal value of this distance identifies the NOE returned by the algorithm.

The estimates obtained by the different algorithms are displayed in Table 1. The VD technique was used with several values for the false-alarm parameter. The HIDENN and D-HIDENN algorithms used the values $(1, 3)$ for their (k, k') parameter pair. The other algorithms do not depend on a parameter.

This table shows that a wide variety of estimates can be returned. While it is not known what the correct NOE would be in these data sets, classification results on these data sets give a hint to the NOE: 16 classes are employed in the Indian Pines data set, along with a background class. The well-known Tetra-corder classification map of the AVIRIS Cuprite data set employs 25 classes, along with an unknown class. While the number of classes cannot be considered exactly the same as the NOE for several reasons, such as mixed classes, missing classification results and spectral variability, they do indicate that the proposed method obtains similar numbers.

6   Conclusions

We have presented a new method for estimating the ID in a hyperspectral image, based on comparing the IDD of the target hyperspectral image with IDDs obtained from simulated hyperspectral images. As IDDs depends strongly on noise, all data sets have been denoised before estimation of ID. The proposed technique is validated on artificial data sets, and shows superior results compared to several popular alternatives, such as VD and HySime. Also, the proposed technique does not suffer from the often observed underestimation effect caused by spectral correlations between endmembers and between adjacent bands. On real data sets, realistic values seem to be obtained.

Future work involves the inclusion of more advanced denoising procedures, and the inclusion of prior knowledge on the target data set into the procedure (e.g. IDD libraries which depend on the type of data set, or inclusion of expected sparseness into the estimation procedure).

References

1. Bioucas-Dias, J.M., Plaza, A., Dobigeon, N., Parente, M., Du, Q., Gader, P., Chanussot, J.: Hyperspectral unmixing overview: geometrical, statistical, and sparse regression-based approaches. IEEE J. Sel. Top. Appl. Earth Obs. Remote Sens. 5(2), 354–379 (2012)
2. Heylen, R., Parente, M., Gader, P.: A review of nonlinear hyperspectral unmixing methods. IEEE J. Sel. Top. Appl. Earth Obs. Remote Sens. 7(6), 1844–1868 (2014)
3. Heylen, R., Scheunders, P.: Hyperspectral intrinsic dimensionality estimation with nearest-neighbor distance ratios. IEEE J. Sel. Top. Appl. Earth Obs. Remote Sens. 6(2), 570–579 (2013)
4. Chang, C.-I., Du, Q.: Estimation of the number of spectrally distinct signal sources in hyperspectral imagery. IEEE Trans. Geosci. Remote Sens. 42(3), 608–619 (2004)
5. Bioucas-Dias, J.M., Nascimento, J.P.M.: Hyperspectral subspace identification. IEEE Trans. Geosci. Remote Sens. 46(8), 2435–2445 (2008)
6. Du, Q.: Virtual dimensionality estimation for hyperspectral imagery with a fractal-based method. In: IEEE Workshop on Hyperspectral Image and Signal Processing: Evolution in Remote Sensing, pp. 1–4 (2010)
7. Radovanovic, M., Nanopoulos, A., Ivanovic, M.: Hubs in space: popular nearest neighbors in high-dimensional data. J. Mach. Learn. Res. 11, 2487–2531 (2010)
8. Heylen, R., Parente, M., Scheunders, P.: Estimation of the number of endmembers via the hubness phenomenon. In: IEEE Workshop on Hyperspectral Image and Signal Processing: Evolution in Remote Sensing, pp. 1–4 (2016)
9. Heylen, R., Parente, M., Scheunders, P.: Estimation of the number of endmembers in a hyperspectral image via the hubness phenomenon. IEEE Trans. Geosci. Remote Sens. PP(99), 1–10 (2017)

A Blind Identification and Source Separation Method Based on Subspace Intersections for Hyperspectral Astrophysical Data

Axel Boulais[✉], Yannick Deville, and Olivier Berné

Institut de Recherche en Astrophysique et Planétologie (IRAP),
Toulouse University, UPS-OMP, CNRS, Toulouse, France
{axel.boulais,yannick.deville,olivier.berne}@irap.omp.eu

Abstract. This paper presents a geometric method for solving the Blind Source Separation problem. The method is based on a weak sparsity assumption: for each source, there should exist at least one pair of zones that share only this source. The process consists first in finding the pairs of zones sharing a unique source with an original geometric approach. Each pair of zones, having a mono-dimensional intersection, yields an estimate of a column of the mixing matrix up to a scale factor. All intersections are identified by Singular Value Decomposition. The intersections corresponding to the same column of the mixing matrix are then grouped by a clustering algorithm so as to derive a single estimate of each column. The sources are finally reconstructed from the observed vectors and mixing parameters with a least square algorithm. Various tests on synthetic and real hyperspectral astrophysical data illustrate the efficiency of this approach.

1 Introduction

Blind Source Separation (BSS) methods consist in estimating a set of unknown source signals from a set of observed signals which are mixtures of these sources. The type of mixture is partially unknown. The class of the mixing operator is predefined and depends on the data model (linear or nonlinear) but the parameter values of this operator are unknown and are also to be estimated. BSS is a generic signal processing problem. The term "signal" is to be understood in a broad sense since it refers to one-dimensional sources (audio, communications, spectroscopy...), to two-dimensional sources (images) but also to more complex data. BSS may be found in many fields such as acoustics, communications, biomedical engineering, image classification, remote sensing, astrophysics.

BSS methods for linear instantaneous mixtures appeared in the 1980s and since then, three main classes of methods emerged. The first one is Independent Component Analysis (ICA) [7,9,14]. It is based on a probabilistic formalism and requires the source signals to be mutually statistically independent. A second class of methods called Non-negative Matrix Factorization (NMF) [6,9,15] appeared in 1999. It requires the source signals and mixing coefficient values

© Springer International Publishing AG 2017
P. Tichavský et al. (Eds.): LVA/ICA 2017, LNCS 10169, pp. 367–380, 2017.
DOI: 10.1007/978-3-319-53547-0_35

to be non-negative. A third class of methods emerged in the 2000s, namely Sparse Component Analysis (SCA) [7–9,12], *e.g.* using clustering [7]. It requires the source signals to be sparse in the considered representation domain (time, space, frequency, time-frequency, wavelet...).

In this study, we focus on linear mixtures of possibly correlated source signals and unconstrained sign. The non-negativity of the mixing coefficients and source signals is not required but allows one to remove the sign indeterminacy of the results. For this data model, ICA and NMF methods are not suitable. In the case of non-negative data, NMF methods remain appropriate but their decomposition is not unique due to the existence of spurious solutions with the only constraint of non-negativity and standard NMF cost functions yield local minima. To achieve the unmixing of such data, we focus on sparsity requirements. Among the SCA methods, a subclass of methods relies on a geometric interpretation of the BSS problem. These methods achieve the separation by identifying the convex hull containing the mixed data [4]. Although these geometric methods allow different sparsity levels, they require the non-negativity of data and therefore are not suitable in our context. Another SCA method, the TiFROM method [1,8] introduced in 2001 by Deville *et al.*, was originally developed for a linear mixture of mono-dimensional signals. Since then, extensions have been proposed for other types of mixtures and signals [8], especially for possibly correlated images, with the SpaceCorr method [16]. The different versions of this method are all based on the same sparsity assumption. More specifically, they require the presence of "little zones" in mixed data where only one source is non-zero. These zones, called single-source zones, consist of several adjacent sample indices taken in the considered analysis domain. However, the "presence of single-source zone" assumption is not always consistent with the data. Indeed, a large number of source signals or a low number of samples in the data can jeopardise the presence of single-source zones for each source. Thus, it is necessary to relax the sparsity condition in order to separate such data. This approach has already been tried in the context of multispectral image unmixing in Earth observation by Benachir *et al.* with the BiSCorr method [2]. For their needs, authors suppose that there exists at least one two-source analysis zone for each possible pair of sources. They then study the properties of the geometric intersections of lines associated with each pair of two-source zones. Each intersection for pairs with a unique shared source yields the scaled mixing coefficients related to this source.

Other SCA methods available in the literature are to be mentioned. These geometric methods, grouped under the name k-SCA (see for example [13,17,18]), are based on the assumption that a non-negligible fraction of the samples of the observations contains only a few (say k) active sources. This "deficit" in the mixture allows one to identify the mixing coefficients by estimating the hyperplanes spanned by the k-sparse data. Although the method that we propose can be classified in the k-SCA methods, the assumptions that we formulate are different from those required by the previously cited works.

We hereafter detail the sparsity notions introduced in the above-mentioned TiFROM-like and BiSCorr methods. Depending on the methods, the term "sparse" does not designate the same information. We can identify three distinct concepts:

- In the conventional sense, a signal is sparse if a large number of samples are zero or negligible.
- Within the framework of TiFROM-like and BiSCorr methods, the sparsity describes a joint information. The considered source signals are jointly sparse in a zone of the analysis domain if for all observed signals, most source signals are simultaneously inactive in this zone (*i.e.* they are not involved in the observed mixtures). One source is active per sparse zone for TiFROM-like methods and two sources are active per zone for the BiSCorr method.
- A third concept can be introduced from BiSCorr and will be extended in this article. In addition to the previous definitions, sparsity means that a pair of zones have a unique source in common, the others being specific to each zone of the pair.

To summarise, the first concept relates to the number of zero samples of a single signal, the second to the number of source signals simultaneously inactive in a single zone and the last to the number of sources shared by a pair of analysis zones. The method that we propose is based on this last notion.

Unlike the BiSCorr method, our approach does not require a fixed number of sources present in each zone of a pair of zones exploited for sparsity. The only constraint for such a pair is that the zones share only one source. However we can define a necessary but not sufficient condition on the number of sources of each zone of the pair. We assume that the first zone of the pair contains $\ell_1 \in [2, L-1]$ sources (cases $\ell_1 = 1$ and $\ell_1 = L$ are excluded because BSS methods already exist for data which include single-source zones and we therefore only consider the more difficult case when data do not contain such zones). The number of sources ℓ_2 of the second zone then will be in the range $\ell_2 \in [2, L - \ell_1 + 1]$.

From these observations, we introduce some definitions and an assumption:

Definition 1. A pair of zones sharing a unique source is called a "single-source-intersection pair".

Definition 2. A source is said to be "accessible" in the representation domain if there exist at least one single-source-intersection pair having this source in the intersection.

Assumption 1. Each source is accessible in the representation domain.

This paper presents a geometric method, called SIBIS (Subspace-Intersection Blind Identification and Separation) for solving the linear instantaneous BSS problem. SIBIS is based on the search of pairs of subspaces whose intersections contain a unique source signal. The mixing coefficients are identified from these intersections and then the sources are reconstructed. SIBIS requires the linear independence of the sources, for each source there should exist at least one pair of subspaces that share only this source. Section 2 reviews the data model and the geometric framework of the BSS problem in a simple case. In Sect. 3, we describe the various stages of the proposed SIBIS method. Section 4 presents results on synthetic data and real hyperspectral images. Finally, Sect. 5 presents the conclusions and future works.

2   Problem Statement

2.1   Data Model

This section deals with the general context by considering mono-dimensional signals, *i.e.* depending on a single scalar variable (denoted n) that can similarly refer to time, frequency, wavelength. We assume that we have N samples of M observations resulting from linear instantaneous mixtures of L source signals. Using preliminary explicit notations, this reads:

$$obs(m,n) = \sum_{\ell=1}^{L} coef(m,\ell) \times sig(\ell,n) \tag{1}$$

where $obs(m,n)$ is the nth sample of the mth observation, $sig(\ell,n)$ is the nth sample of the ℓth source signal and $coef(m,\ell)$ defines the scale of the contribution of source ℓ in observation m. This can be written in matrix form:

$$Obs = Coef \times Sig \tag{2}$$

where Obs is the observed matrix, $Coef$ the mixing matrix and Sig is the source matrix in this data model, using BSS terminology. The dimensions of these matrices are respectively, $M \times N$, $M \times L$ and $L \times N$. Thus, each row of Obs corresponds to an observation $obs(m,.)$ and each column corresponds to a given sample index n for all observations. Equivalently, the model can be written in its transposed version:

$$Obs^T = Sig^T \times Coef^T \tag{3}$$

where Obs^T is the transpose of the original observed matrix, Sig^T is the mixing matrix and $Coef^T$ is the source matrix of this alternative data model. Using BSS notations, the linear mixing model is:

$$X = AS \tag{4}$$

where X is the observed matrix, A is the matrix of mixing parameters and S is the source matrix.

For the requirements of our method, we suppose that Assumption 1 is satisfied. So there exist analysis zones Z wherein a significant number of source signals are simultaneously inactive. The sparsity property is carried by the source matrix S. Therefore the nature of data (i.e. the properties of matrices Sig and $Coef$) dictates the choice of model (2) or (3), more specifically, which matrix Sig or $Coef$ represents the source matrix S (and respectively matrix of mixing parameters A). Depending on the data, the zones Z are derived from the matrix Sig (model (2)) or from the matrix $Coef^T$ (model (3)).

In the first case, the zones $Z \subset [1, N]$ consist of some adjacent indices of samples of all observations: $X(Z) = \{X(., n) \mid n \in Z\}$. In other words, there exist some adjacent columns of Obs for which a significant number of sources are simultaneously zeros. Sig has zero entries in these corresponding columns: $S(Z) = \{S(., n) \mid n \in Z\}$.

In the second case, the zones $Z \subset [1, M]$ consist of some adjacent indices of samples of all observations: $X(Z) = \{X(., m) \mid m \in Z\}$. In other words, there exist some adjacent columns of Obs^T for which a significant number of sources are simultaneously zeros. $Coef^T$ has zero entries in these corresponding columns: $S(Z) = \{S(., m) \mid m \in Z\}$.

In our astrophysical application (see Sect. 4.2), only model (3) is consistent. To simplify the presentation of the theoretical part of our method (Sects. 2.2 and 3) , we refer only to the model (3) using standard notations (4): $X = Obs^T$, $A = Sig^T$ and $S = Coef^T$. We denote respectively as x_m, a_ℓ and s_m the columns of the matrices X, A and S. We call "observed vectors" the columns of X. Note that to use the model (2), it is sufficient to swap the indices m and n due to the transposition of the model. This case is presented by an application of our method to speech signals (see Sect. 4.1).

2.2 Geometric Framework

We first focus on the simple case of mixtures of $L = 3$ sources and we suppose that Assumption 1 is satisfied. Let us consider each observed vector x_m as an element of an $\mathbb{R}^N$ vector space. We denote Z_1 and Z_2 two zones of the analysis domain, $Z_1 = \{1, 2\}$ and $Z_2 = \{3, 4\}$ which are each restricted to two samples in this toy example. Z_1 and Z_2 respectively correspond to observed vectors $X(Z_1) = [x_1 \ x_2]$ and $X(Z_2) = [x_3 \ x_4]$. The observed vectors in Z_1 are mixtures of the source of indices $\ell = \{1, 2\}$ and those in Z_2 are mixtures of the source of indices $\ell = \{2, 3\}$. So the pair of zones Z_1 and Z_2 only shares source 2. This example is illustrated in Fig. 1. We introduce the different subspaces associated with each zone Z_1 and Z_2. The linear combinations of the observed vectors of the zones Z_1 and Z_2 respectively generate two subspaces, denoted $\mathcal{Z}_1$ and $\mathcal{Z}_2$, each supposedly of dimension 2. In order to ensure the dimension of subspaces $\mathcal{Z}_1$ and $\mathcal{Z}_2$, it is necessary to introduce the following assumption:

Assumption 2. For each analysis zone Z, the number of linearly independent observed vectors x_m is equal to the number of active sources in Z.

We have:

$$\begin{aligned} \mathcal{Z}_1 &= \{z_1 \mid z_1 = \alpha_1 x_1 + \alpha_2 x_2, \quad \alpha_1, \alpha_2 \in \mathbb{R}\}, \\ \mathcal{Z}_2 &= \{z_2 \mid z_2 = \alpha_3 x_3 + \alpha_4 x_4, \quad \alpha_3, \alpha_4 \in \mathbb{R}\}. \end{aligned} \tag{5}$$

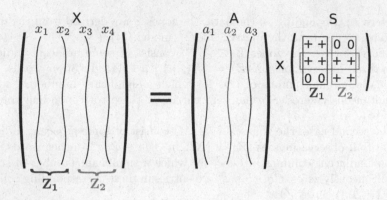

Fig. 1. Schematic representation of a simple example of mixtures of three sources. The zones Z_1 and Z_2 form a single-source-intersection pair. The source shared by the two zones is outlined in red.

We can reformulate the definition of these two subspaces by changing the associated basis. Using Eq. (4), Assumptions 1 and 2, we have:

$$Z_1 = span(a_1, a_2),$$
$$Z_2 = span(a_2, a_3). \tag{6}$$

The matrix A being assumed full column rank, the subspace of $\mathbb{R}^N$ spanned by its columns is of dimension 3. According to the Grassmann relation, we have:

$$dim(Z_1) + dim(Z_2) = dim(Z_1 + Z_2) + dim(Z_1 \cap Z_2) \tag{7}$$

where the sum $Z_1 + Z_2$ coincides with the subspace spanned by A. Therefore, the intersection $Z_1 \cap Z_2$ is of dimension 1. It consists of the elements included in both Z_1 and Z_2. From Eq. (6), a basis of the intersection is obvious. The elements of the intersection are defined by:

$$Z_1 \cap Z_2 = \{v \mid v = \beta a_2, \quad \beta \in \mathbb{R}\}. \tag{8}$$

The geometry of this example is illustrated in Fig. 2. We note that the intersection $Z_1 \cap Z_2$ provides a column of the mixing matrix A up to a scale factor β. This scale indeterminacy, well known in BSS, can be compensated for if a normalisation of the columns of A is possible. The sign indeterminacy can be compensated for if the mixing parameters are known to be non-negative (see Sect. 4).

From this property of the intersection, we deduce a method to estimate the columns of the mixing matrix A only from the data X. The next section presents the different stages of the resulting SIBIS method for a number of sources $L \geqslant 3$.

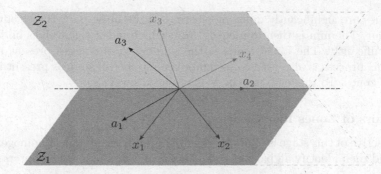

Fig. 2. Geometry of the simple example. The subspaces $\mathcal{Z}_1$ and $\mathcal{Z}_2$ associated with the zones Z_1 and Z_2 intersect at a subspace of dimension 1 containing a_2.

3   BSS Method Based on Subspace Intersection

The SIBIS method aims to decompose the mixed observed signals into a set of L source signals and their relative mixing parameters. It operates in different stages:

– Estimation of the number L of sources,
– Detection of pairs of zones satisfying the single-source-intersection constraint,
– Estimation of the mixing matrix A,
– Reconstruction of the source matrix S.

Each stage is detailed in the following subsections.

3.1   Source Number Estimation

The estimation of the number L of sources present in the observed data is a conventional stage in BSS. For the SIBIS method, it is applied many times, since this process is also used to estimate the number of sources present in the different zones as well as in their unions. Note that the number of sources of a zone Z is equal to the dimension of the subspace $\mathcal{Z}$ associated with this zone if Assumption 2 is satisfied. Due to lack of space we give a brief description of the used process, for more details see [5].

We use a method based on the eigen-decomposition of the covariance matrix Σ_X of the data X. In the data model (4), the mixing matrix A has full column rank by assumption. The observed vectors are derived from a linear combination of the columns of A, so we assume that the matrix X has the same column rank L. Thus, estimating the number of sources present in the data amounts to estimating the rank of X. For data corrupted by an independent and identically distributed noise, the theoretical covariance matrix has $N - L$ eigenvalues equal to the noise variance σ_N^2, corresponding to the noise part and L eigenvalues, greater than σ_N^2, corresponding to the signal part. The curve of the eigenvalues in decreasing order is therefore constituted of two parts. In the signal part, the L

eigenvalues are significantly different whereas in the noise part, the eigenvalues are similar. The aim is then to identify from which index eigenvalues no longer vary significantly. The found index is the number L of sources present in the data. This process is also used to estimate the number of sources present in the different zones. To this end, matrix X is replaced by $X(Z)$.

3.2   Pairs of Zones Identification

The objective of this stage is to first segment the analysis domain in homogeneous zones and then identify all pairs of zones satisfying the single-source-intersection constraint.

Segmentation. The segmentation is based on the Split and Merge algorithm used in image processing. Due to lack of space we give a brief description of the used algorithm, for more details see [11]. The segmentation algorithm proceeds in two distinct steps. At first, the analysis domain is divided into small zones Z (typically 5×5 pixels for an image and 100 samples for mono-dimensional signals). The analysis domain is explored using adjacent or overlapping zones. For each zone, we estimate the dimension of the associated subspace according to the method of Sect. 3.1. Note that this step differs from the original Split and Merge algorithm, the analysis domain is fully divided instead of proceeding by dichotomy. The second step is to group the neighbouring zones Z if their union is homogeneous. To this end, each zone is compared with all its neighbours. If two neighbouring zones have the same active sources, they are merged. Then, we search the new neighbours of this updated zone and the process is repeated. The algorithm stops when none of the remaining zones can be merged. The merging criterion is based on the numbers of sources of the zones Z_1 and Z_2 (i.e. the dimensions of $\mathcal{Z}_1$ and $\mathcal{Z}_2$): if $dim(\mathcal{Z}_1) = dim(\mathcal{Z}_2) = dim(\mathcal{Z}_1 + \mathcal{Z}_2)$, then the union of the zones Z_1 and Z_2 is homogeneous and they are merged. Finally, we obtain the analysis domain segmented into K homogeneous zones from the point of view of the sources present in each zone.

Single-Source-Intersections. The search of the pairs of zones sharing a unique source is once again based on the dimension of the subspace associated with each zone. We aim at finding all pairs of subspaces whose intersection is mono-dimensional. For each of all $\frac{K!}{2!(K-2)!}$ possible pairs, we estimate the dimension of the intersection, from the Grassmann relation Eq. (7). We identify all pairs of zones satisfying:

$$dim(\mathcal{Z}_p \cap \mathcal{Z}_q) = dim(\mathcal{Z}_p) + dim(\mathcal{Z}_q) - dim(\mathcal{Z}_p + \mathcal{Z}_q) = 1 \qquad (9)$$

with $1 \leqslant p < q \leqslant K$.

3.3   Mixing Matrix Estimation

The identification of the columns of the mixing matrix A is performed by estimating all mono-dimensional intersections. Let Z_p and Z_q be a pair of zones with mono-dimensional intersection and $\mathcal{Z}_p$ and $\mathcal{Z}_q$ the two associated subspaces. $\mathcal{Z}_p$ and $\mathcal{Z}_q$ share the source of index $k \in [1, L]$. Before estimating the intersection, it is necessary to obtain a basis of $\mathcal{Z}_p$ and a basis of $\mathcal{Z}_q$. Denote d_p and d_q the dimensions of these subspaces. Any bases are suitable, we simply chose the orthogonal bases provided by Singular Value Decomposition (SVD) of the matrices $X(\mathcal{Z}_p)$ and $X(\mathcal{Z}_q)$. Note that bases of the four fundamental subspaces associated with a matrix can be identified by the SVD [10]. We have:

$$X(\mathcal{Z}_p) = U_p \Sigma_p V_p^T,$$
$$X(\mathcal{Z}_q) = U_q \Sigma_q V_q^T. \tag{10}$$

An orthogonal basis of $\mathcal{Z}_p$ is given by the d_p firsts columns of U_p. The procedure is the same for U_q. We thus obtain two bases whose column vectors p_i and q_j form the matrices P and Q and which are respectively associated with $\mathcal{Z}_p$ and $\mathcal{Z}_q$ according to:

$$\mathcal{Z}_p = span(p_i), \quad \forall i \in [1, d_p]$$
$$\mathcal{Z}_q = span(q_j), \quad \forall j \in [1, d_q]. \tag{11}$$

Let v be an element of the intersection $\mathcal{Z}_p \cap \mathcal{Z}_q$, then $v \in \mathcal{Z}_p$ and $v \in \mathcal{Z}_q$. So v can be expressed in two ways:

$$v = \alpha_1 p_1 + \alpha_2 p_2 + \cdots + \alpha_{d_p} p_{d_p} \tag{12}$$
$$v = \beta_1 q_1 + \beta_2 q_2 + \cdots + \beta_{d_q} q_{d_q} \tag{13}$$

with $\alpha_i, \beta_j \in \mathbb{R}$. We then deduce that:

$$\alpha_1 p_1 + \alpha_2 p_2 + \cdots + \alpha_{d_p} p_{d_p} - (\beta_1 q_1 + \beta_2 q_2 + \cdots + \beta_{d_q} q_{d_q}) = 0. \tag{14}$$

Let the matrix $M = [p_1 \ldots p_{d_p} \quad - q_1 \cdots - q_{d_q}]$ and the vector $\theta = [\alpha_1 \ldots \alpha_{d_p} \quad \beta_1 \ldots \beta_{d_q}]^T$. Equation (14) can be written in the following matrix form:

$$M\theta = 0. \tag{15}$$

We recognize the definition of $null(M)$, the nullspace associated with M. Therefore $(\mathcal{Z}_p \cap \mathcal{Z}_q) \subset null(M)$. Moreover, the rank-nullity theorem [10] for a matrix M ($n \times d$) with $d = d_p + d_q$ gives the relation:

$$rk(M) + dim(null(M)) = d \tag{16}$$

where $rk(M)$ is the column rank of M. The matrix M being composed of the union of two bases P and Q, we have $rk(M) = dim(\mathcal{Z}_p + \mathcal{Z}_q)$. The number of columns d corresponds to the sum of dimensions of the two subspaces $\mathcal{Z}_p$ and $\mathcal{Z}_q$, i.e. $d = dim(\mathcal{Z}_p) + dim(\mathcal{Z}_q)$. According to Eqs. (9) and (16), we have:

$$dim(null(M)) = 1. \tag{17}$$

So we can conclude that:

$$null(M) = \mathcal{Z}_p \cap \mathcal{Z}_q. \tag{18}$$

The intersection $\mathcal{Z}_p \cap \mathcal{Z}_q$ is then estimated from $null(M)$. As mentioned above, the SVD provides bases of the fundamental subspaces associated with matrix $M = U_m \Sigma_m V_m^T$. In our case, $null(M)$ being mono-dimensional, a basis of the nullspace is given by the last column of V_m (denoted v_d). So $\theta = v_d$, this allows us to get v, according to Eq. (12) or (13). v is then used as the basis of the intersection $\mathcal{Z}_p \cap \mathcal{Z}_q$:

$$\mathcal{Z}_p \cap \mathcal{Z}_q = \{v \mid v = \beta a_k, \quad \beta \in \mathbb{R}\}, \tag{19}$$

where $k \in [1, L]$ is the index of the source shared by Z_p and Z_q. This intersection provides an estimate of the column of A (up to a scale factor) associated with the source shared by Z_p and Z_q. This intersection identification is repeated for each pair of zones obtained in the previous steps. Thus, we obtain a set of potential columns of A. Some columns are estimated several times because they are possibly present in several intersections. To compensate for this, we apply clustering (K-means algorithm [19]) to these columns in order to regroup in L clusters the estimates corresponding to the same column of the mixing matrix. The mean of each cluster is retained to form a column of the matrix $\hat{A}$ (the estimate of A). Note that the columns of $\hat{A}$ are obtained in an arbitrary order and up to a scale factor.

3.4   Source Matrix Reconstruction

The final stage of our BSS method is to estimate the source matrix S. This is done by using the classical Least Square algorithm. The source matrix S is estimated column by column by minimizing the cost function:

$$J(\hat{s}_m) = \frac{1}{2}\|x_m - \hat{A}\hat{s}_m\|_2^2 \quad \forall m \in [1, M] \tag{20}$$

where $\hat{s}_m$ is the estimate of the mth column of S. At this stage, we can remove the indeterminacy of the sign of the A columns if the mixing parameters are known to be non-negative.

4   Experimental Results

To evaluate the performance of our method, we perform two different experiments. In a first experiment, we test our method on a synthetic mixture of speech signals. Although an instantaneous linear mixture is not ideal for this type of application, this experiment allows us to evaluate the performance of SIBIS for mono-dimensional signals in the presence of different noise levels. In a second experiment, we use our method on astrophysical hyperspectral images and compare our results with two other methods from the literature. The performance achieved in the first experiment is measured by the following normalized root mean square error (NRMSE):

$$NRMSE = \frac{\|real - estimate\|}{\|real\|}. \tag{21}$$

We use the same measure to compare signals and contribution coefficients. However before performing these measurements, the scale factor indeterminacies specific to the BSS problem must be taken into account. To this end, each estimated or actual signal (and contribution coefficient) is standardized (i.e. is divided by its standard deviation). To compensate for the sign and permutation indeterminacies, we measure the NRMSE of each possible permutation of order and sign. The minimum error provides the correct permutation of the signals.

4.1 Synthetic Data

For this first experiment, we generate a few mixtures of 3 speech signals from the free database VoxForge under the GPL license. To simulate the required sparsity assumption, each of these 3 signals has a time interval in which these samples are set to zero. To this end, the time axis is divided into 3 intervals. In the first interval, only signals 1 and 2 are active. In the second, the signals 2 and 3 are active and in the third the signals 1 and 3 are active. Note that we maintain a transition zone between two intervals where all signals remain non-zero. The contribution coefficients are randomly generated, from a uniform distribution on $[0.5, 1.5]$. We generate 10 different mixtures. This number is constraining for this application but it is necessary to correctly estimate the number of active sources in each zone Z. Finally we add to each mixture a white Gaussian noise to get a Signal to Noise Ratio (SNR) of 30, 40 or 50 dB.

This simulation is modelled by model (2). By construction, the speech signals carry the sparsity property. Therefore, they constitute the sources S and the associated contribution coefficients constitute the mixing parameters A. The results of the different decompositions are given in Table 1.

For the ideal case of noiseless data, the NRMSE is close to the numerical error of Matlab ($2.22e-16$). This performance illustrates the efficiency of our

Table 1. Performance (NRMSE) of SIBIS applied to synthetic mixtures of speech signals.

SNR	Source 1	Source 2	Source 3	Mean
Signals				
Noiseless	2.04e−15	1.63e−14	1.78e−15	6.70e−15
50 dB	3.50e−3	1.29e−2	2.41e−2	1.35e−2
40 dB	8.20e−3	1.80e−2	4.79e−2	2.47e−2
30 dB	3.70e−2	7.48e−2	2.27e−1	1.13e−1
Contribution coefficients				
Noiseless	6.09e−16	5.44e−16	4.85e−16	5.46e−16
50 dB	2.39e−4	1.86e−4	1.90e−3	7.85e−4
40 dB	5.69e−4	1.50e−3	8.78e−4	9.73e−4
30 dB	5.61e−4	9.80e−3	9.50e−3	6.60e−3

method to solve an ideal case. When noise levels are reasonable (40 and 50 dB), SIBIS provides satisfactory results with an NRMSE lower than 5% for the source signals and lower than 0.2% for the mixing parameters. From 30 dB, performance starts to deteriorate. For the first two sources, the unmixing remains acceptable (respectively 3.7% and 7.5% NRMSE for the source signals). However for the third source, the NRMSE becomes substantial (22.7%), the extraction of this source therefore has failed.

Our first conclusions imply that the error is essentially due to the source number estimation method (method known to be sensitive to noise), which yields an incorrect segmentation of the analysis domain. In addition, a small number of observations makes the estimation more difficult. With a larger number of observations (not realistic in this case), the SNR before the method fails to achieve the separation is smaller.

However, if the data are correctly segmented, the construction of a basis of each zone and of each intersection with SVD has an "averaging effect" on the observations in each zone. Thus, noise reduction performed by SVD impacts on results.

4.2   Real Data

In this second experiment, we applied our approach to hyperspectral data in astrophysics. The observed region is NGC 7023-NW, a well studied reflection nebula mapped by the Spitzer space telescope. The hyperspectral image consists of a cube with two spatial dimensions and a spectral one. A hyperspectral cube can be modelled in two different ways: a spectral model (2) where we consider the cube as a set of spectra and a spatial model (3) where we consider the cube as a set of images. For these data, the sparsity property is carried by the spatial domain, Assumption 1 is satisfied by spatial zones of few adjacent pixels. Therefore we use the spatial model (3). The cube is restructured into a matrix X where each column contains an observed pixel spectrum, each column of the matrix of mixing parameters A contains an elementary spectrum and each row of the source matrix S contains a vectorized abundance map. Moreover, these data are derived from a mixture of non-negative possibly correlated elementary spectra and non-negative without sum-to-one constraint abundance maps.

Two other methods have already been used to achieve the decomposition of these data, NMF in a first study [3] and a geometric method (MASS [5]) requiring for each source the presence of a single-source observed pixel (pure pixel) in observations but able to achieve the decomposition without sum-to-one constraint. For these data, the presence of pure pixels is a realistic assumption. However, due to the limited number of pixels (29 × 39), the image does not contain usable single-source zones for the SpaceCorr method [16]. However the spatial distribution of sources suggests that each source is spatially accessible (Assumption 1 is realistic). To cross-validate the results of our method, we compare the spectra extracted by NMF from [3] and MASS from [5] with those obtained by SIBIS (Fig. 3). Note that the extracted spectra are normalized, so that they integrate to one and spectra are presented in an arbitrary order. For the

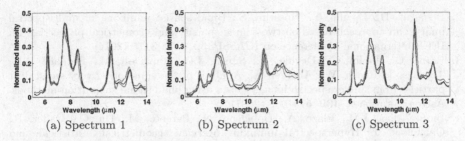

(a) Spectrum 1 (b) Spectrum 2 (c) Spectrum 3

Fig. 3. Extracted spectra in NGC 7023-NW. The blue curves are the spectra extracted with NMF, the black curves those extracted with MASS and the red curves those extracted with SIBIS

considered data, the SIBIS method achieves a decomposition of the data cube with success. The extracted spectra are equivalent to those extracted by NMF or MASS.

5 Conclusion and Future Work

In this paper, we propose a new geometric method using a sparsity assumption for separating data. It is based on finding the subspace intersections which allow one to estimate the parameters of the mixture. The full procedure operates in 4 stages. The method identifies the number of sources present in the considered data, detects the pairs of zones satisfying the single-source-intersection constraint, estimates the mixing matrix and then deduces the source matrix. Our goal was to propose an extension of the methods of the literature using little sparse zones to solve the BSS problem (*i.e.* relax the assumption of presence of single-source (TiFROM-like method) or two-source zones (BiSCorr method)).

Several tests on synthetic data with reasonable SNR show the efficiency of SIBIS. We then experimented our method on real hyperspectral astrophysical images. The obtained results are encouraging since they are comparable with those given by other methods from the literature for data with no ground truth. This motivates us to continue this work by making estimation of the number of active sources more robust to noise. Additional tests will also be conducted to study the case of non-negative data for which the noise introduced negative values.

We will also investigate several possible extensions and variants of the SIBIS method, such as combining SIBIS with methods intended for single-source zones and using a recursive algorithm to relax the single-source-intersection constraint.

References

1. Abrard, F., Deville, Y.: A time-frequency blind signal separation method applicable to underdetermined mixtures of dependent sources. Sig. Process. **85**(7), 1389–1403 (2005)

2. Benachir, D., Deville, Y., Hosseini, S.: Blind spatial unmixing of multispectral images: an approach based on two-source sparsity and geometrical properties. In: IEEE International Conference on ICASSP, pp. 3171–3175 (2014)
3. Berné, O., Joblin, C., Deville, Y., Smith, J.D., Rapacioli, M., Bernard, J.P., Thomas, J., Reach, W., Abergel, A.: Analysis of the emission of very small dust particles from spitzer spectro-imagery data using blind signal separation methods. Astron. Astrophys. **469**, 575–586 (2007)
4. Bioucas-dias, J.M., Plaza, A., Dobigeon, N., Parente, M., Du, Q., Gader, P., Chanussot, J.: Hyperspectral unmixing overview: geometrical, statistical, and sparse regression-based approaches. IEEE J. Sel. Topics Appl. Earth Observ. Remote Sens. **5**(2), 354–379 (2012)
5. Boulais, A., Deville, Y., Berné, O.: A geometrical blind separation method for unconstrained-sum locally dominant sources. In: IEEE International Workshop ECMSM (2015)
6. Cichocki, A., Zdunek, R., Phan, A., Amari, S.I.: Nonnegative matrix and tensor factorizations: applications to exploratory multi-way data analysis and blind source separation. Wiley, New Year (2009)
7. Comon, P., Jutten, C. (eds.): Handbook of Blind Source Separation: Independent Component Analysis and Applications. Elsevier, Oxford (2010)
8. Deville, Y.: Chapter 6, Sparse component analysis: a general framework for linear and nonlinear blind source separation and mixture identification. In: Naik, G.R., Wang, W. (eds.) Blind Source Separation: Advances in Theory, Algorithms and Applications, pp. 151–196. Springer, Heidelberg (2014)
9. Deville, Y.: Blind source separation and blind mixture identification methods. In: Wiley Encyclopedia of Electrical and Electronics Engineering. Wiley, New York (2016).
10. Golub, G.H., Van Loan, C.H.: Matrix Computations, 3rd edn. Johns Hopkins University Press, Baltimore (1996)
11. Gonzalez, R.C., Woods, R.E.: Digital Image Processing, Chapter 10: Image Segmentation. Prentice-Hall Inc., New Jersey (2006)
12. Gribonval, R., Lesage, S.: A survey of sparse component analysis for blind source separation: principles, perspectives, and new challenges. In: ESANN 2006 Proceedings, pp. 323–330 (2006)
13. He, Z., Cichocki, A., Li, Y., Xie, S., Sanei, S.: K-hyperline clustering learning for sparse component analysis. Sig. Process. **89**, 1011–1022 (2009)
14. Hyvärinen, A., Karhunen, J., Oja, E.: Independent Component Analysis. Wiley-Interscience, New Jersey (2001)
15. Lee, D.D., Seung, H.S.: Learning the parts of objects by non-negative matrix factorization. Nature **401**, 788–791 (1999)
16. Meganem, I., Deville, Y., Puigt, M.: Blind separation methods based on correlation for sparse possibly-correlated images. In: IEEE International Conference on ICASSP, pp. 1334–1337 (2010)
17. Naeini, F., Mohimani, H., Babaie-Zadeh, M., Jutten, C.: Estimating the mixing matrix in sparse component analysis (SCA) based on partial k-dimensional subspace clustering. Neurocomputing (Elsevier) **71**, 2330–2343 (2008)
18. Theis, F., Georgiev, P., Cichocki, A.: Robust sparse component analysis based on a generalized hough transform. EURASIP J. Appl. Sig. Process. **2007**(1), 86 (2007)
19. Theodoridis, S., Koutroumbas, K.: Pattern Recognition. Academic Press, London (2009)

Estimating the Number of Endmembers to Use in Spectral Unmixing of Hyperspectral Data with Collaborative Sparsity

Lucas Drumetz[1]([⊠]), Guillaume Tochon[2], Jocelyn Chanussot[1], and Christian Jutten[1]

[1] Univ. Grenoble Alpes, CNRS, GIPSA-lab, 38000 Grenoble, France
{lucas.drumetz,jocelyn.chanussot,christian.jutten}@gipsa-lab.fr
[2] EPITA Research and Development Laboratory (LRDE),
94276 Le Kremlin-Bicêtre, France
guillaume.tochon@lrde.epita.fr

Abstract. Spectral Unmixing (SU) in hyperspectral remote sensing aims at recovering the signatures of the pure materials in the scene (endmembers) and their abundances in each pixel of the image. The usual SU chain does not take spectral variability (SV) into account, and relies on the estimation of the Intrinsic Dimensionality (ID) of the data, related to the number of endmembers to use. However, the ID can be significantly overestimated in difficult scenarios, and sometimes does not correspond to the desired scale and application dependent number of endmembers. Spurious endmembers are then frequently included in the model. We propose an algorithm for SU incorporating SV, using collaborative sparsity to discard the least explicative endmembers in the whole image. We compute an algorithmic regularization path for this problem to select the optimal set of endmembers using a statistical criterion. Results on simulated and real data show the interest of the approach.

Keywords: Hyperspectral images · Remote sensing · Collaborative sparsity · Alternating Direction Method of Multipliers · Regularization path · Bayesian Information Criterion

1 Introduction

The fine spectral resolution of hyperspectral remote sensing images allows to precisely identify and characterize the materials of the observed scene. However, this spectral resolution comes at the price of having a coarser spatial resolution than classical color or even multispectral images. Therefore, there are often several materials of interest present at the same time in the Field of View (FOV) of the sensor during the acquisition of a pixel, and the resulting observed

This work has been partially supported by the European Research Council under the European Community's Seventh Framework Program FP7/2007-2013, under Grant Agreement no. 320684 (CHESS project).

P. Tichavský et al. (Eds.): LVA/ICA 2017, LNCS 10169, pp. 381–391, 2017.
DOI: 10.1007/978-3-319-53547-0_36

spectrum is a mixture of the contributions of these materials. Spectral Unmixing (SU) is a (blind) source separation problem whose goal is to recover the spectral signatures of the pure materials in the scene (called *endmembers*) and to estimate their proportions in each pixel (called *fractional abundances*) [3]. To do that, a Linear Mixing Model (LMM) is often considered, assuming as a first approximation that the contributions of each endmember in each pixel sum up in a linear way, with the abundances as weights. In order to interpret the abundances as fractions, they are usually constrained to be positive and to sum to one in each pixel. The classical linear SU unmixing chain is usually divided into three steps: (i) Estimating the number of endmembers to use, which is a hard scale and application dependent (not to mention somewhat subjective) task, using intrinsic dimensionality (ID) estimation algorithms [2,5], (ii) Extracting the endmembers' signatures, generally with geometric approaches, such as the Vertex Component Analysis (VCA) [14] (usually assuming there are pure pixels in the image), and (iii) Estimating the abundances by constrained least squares, using for instance the Fully Constrained Least Squares Umixing (FCLSU) algorithm of [10]. The main two limitations of this strategy have been identified as nonlinearities [11] and spectral variability (see [6] and references therein). Nonlinear mixtures can occur when each ray of light received by the sensor has interacted with more than one material (e.g. in tree canopies, urban scenarios or in particulate media, such as sand). Dealing with spectral variability amounts to consider that each material possesses a certain intra-class variability, which is not what the usual approach does since it implicitly considers that every material is perfectly represented by a single spectral signature. If these two limitations of the usual SU chain are well identified and currently receiving a lot of attention, much less emphasis is put on the estimation of the number of endmembers to use. This number is often considered to be the same as the ID concept. If these two quantities indeed coincide when the LMM holds on simulated data, there is no such guarantee in nonlinear scenarios or when spectral variability is significant. In addition, the ID can be affected by outliers, which are usually not wanted in SU results. The ID has been shown to be subject to overestimation for several algorithms in difficult scenarios (small spatial dimensions, high spectral dimension, significant noise level) [7]. The errors committed at this step are then propagated to the whole unmixing chain, since spurious endmembers are extracted and incorporated to the model. In this paper, we propose an algorithm to perform linear SU of hyperspectral data, incorporating spectral variability, while automatically identifying the wrongly extracted endmembers and removing them from the pool of endmembers during the SU process, in order to keep the most relevant only. To this end, starting from the likely overestimated ID, we define an optimization problem using collaborative sparsity [13], so that irrelevant endembers, usually associated with sparse and meaningless abundance maps, do not contribute in any pixel of the image. In order to select the appropriate number of endmembers to retain, we compute an algorithmic regularization path [12] for the optimization problem, providing a sequence of smaller and smaller candidate endmember matrices. The sequence goes from the whole initial pool of endmembers to a fully sparse model, each time removing the least explicative endmember in the current matrix. The only step remaining is to select the most appropriate element of this sequence for the problem at hand. We use the Bayesian Information Criterion (BIC) to select the optimal model, favoring models reconstructing the data well with a limited number of parameters.

The remainder of this paper is organized as follows: Sect. 2 presents the proposed approach in detail, then Sect. 3 shows the results of experiments conducted on simulated and real data, and finally Sect. 4 gathers some concluding remarks.

2 Proposed Approach

2.1 Extended Linear Mixing Model

Once the ID of the dataset has been estimated, one usually resorts to an endmember extraction algorithm, such as the VCA to obtain the spectra of the pure materials. The next step is the estimation of the abundances. For a hyperspectral image $\mathbf{X} \in \mathbb{R}^{L \times N}$, where L is the number of spectral bands, and N is the number of pixels, an endmember matrix $\mathbf{S}_0 \in \mathbb{R}^{L \times d}$ has been extracted (where d is the estimated ID). We denote the abundance matrix by $\mathbf{A} \in \mathbb{R}^{d \times N}$. In the usual linear SU setting, the abundances are estimated by nonnegative linear least squares, with the additional abundance sum to one constraint (ASC). However, here, following [17], we change the mixing model to incorporate SV into the unmixing at a negligible cost. We consider the following mixing model:

$$\mathbf{x}_k = \psi_k \sum_{p=1}^{d} a_{pk} \mathbf{s}_{0p} + \mathbf{e}_k, \tag{1}$$

where $\mathbf{x}_k$ is the kth column of $\mathbf{X}$, (i.e. the spectrum of pixel k), and $\mathbf{s}_{0p}$ is the pth column of $\mathbf{S}_0$ (i.e. the reference endmember for material p). $\mathbf{e}_k$ is an additive noise, usually assumed to be Gaussian. a_{pk} is the abundance coefficient of endmember p in pixel k. Finally, ψ_k is a scaling factor, which models SV effects in each pixel, e.g. locally changing illumination conditions in the image, due to the topography of the imaged scene and to the photometric properties of the materials. This model is a simplified version of the Extended Linear Mixing Model (ELMM) [8,17]. This version considers distinct scaling factors for each material. With this model, $\mathbf{S}_0$ is then a reference endmember matrix, and we can define local endmember matrices in each pixel by computing $\mathbf{S}_k \triangleq \psi_k \mathbf{S}_0$. If this model holds, [17] shows that the quantity estimated by nonnegative least squares in each pixel and for each material actually incorporates SV information, via the product $\phi_{kp} \triangleq \psi_k a_{kp}$. The model reduces to the LMM when all scaling factors are equal to 1. CLSU (for Constrained Least Squares Unmixing) solves, for each pixel:

$$\arg\min_{\phi_k \geq 0} \frac{1}{2} \|\mathbf{x}_k - \mathbf{S}_0 \phi_k\|_2^2, \tag{2}$$

where $\phi_k \in \mathbb{R}^d$ collects all the ϕ_{kp} for a given pixel k. If we sum all the entries of ϕ_k, we obtain:

$$\sum_{p=1}^{d} \phi_{pk} = \sum_{p=1}^{d} \psi_k a_{kp} = \psi_k \sum_{p=1}^{d} a_{kp} = \psi_k, \tag{3}$$

by reintroducing the ASC on the *actual* abundances, and not their product with the scaling factor. Then we can easily obtain $\mathbf{a}_k = \frac{\phi_k}{\psi_k}$.

2.2   Collaborative Sparsity for Hyperspectral Unmixing

Our goal is to eliminate the wrongly estimated endmembers from the SU process. To do that, we use collaborative sparsity [13]. This concept, also known as Multiple Measurement Vector, or joint sparsity in the signal processing community, extends regular sparsity to a collection of signals which are encouraged to share the same support. For our application, we would like the abundances of the least explicative endmembers (and hence their product with the scaling factors) to be zero on the whole image support. This can be done by considering the following optimization problem:

$$\arg \min_{\boldsymbol{\Phi} \geq 0} ||\boldsymbol{\Phi}||_{\text{row},0}$$
$$\text{s.t } ||\mathbf{X} - \mathbf{S}_0 \boldsymbol{\Phi}||_F^2 < \delta \tag{4}$$

where $|| \cdot ||_{\text{row},0}$ is the row-wise $\mathcal{L}_0$ norm (computing the number of nonzero rows of its matrix argument) of the whole matrix $\boldsymbol{\Phi} \in \mathbb{R}^{d \times N}$, $|| \cdot ||_F$ is the Frobenius norm, and δ is a desired data fit value. This problem allows us to discard entire rows of the feature matrix $\boldsymbol{\Phi}$, but is nonconvex, combinatorial and NP-hard. In order to obtain a more friendly formulation, we consider the following convex relaxation:

$$\arg \min_{\boldsymbol{\Phi}} \frac{1}{2}||\mathbf{X} - \mathbf{S}_0 \boldsymbol{\Phi}||_F^2 + \lambda ||\boldsymbol{\Phi}||_{2,1} + \mathcal{I}_{\mathbb{R}_+^{d \times N}}(\boldsymbol{\Phi}), \tag{5}$$

where $\mathcal{I}_{\mathbb{R}_+^{d \times N}}$ is the indicator function of the positive orthant of $\mathbb{R}^{d \times N}$, and λ is a regularization parameter. The quantity $|| \cdot ||_{2,1}$ is the mixed $\mathcal{L}_{2,1}$ norm, defined for any matrix $\boldsymbol{\Phi} \in \mathbb{R}^{d \times N}$ as:

$$||\boldsymbol{\Phi}||_{2,1} = \sum_{i=1}^{d} \left(\sum_{j=1}^{N} |\phi_{ij}|^2 \right)^{\frac{1}{2}} = \sum_{i=1}^{d} ||\phi_i||_2, \tag{6}$$

where ϕ_i is the ith row of $\boldsymbol{\Phi}$.

The $\mathcal{L}_{2,1}$ norm encourages row-wise sparsity in the feature matrix, because it is the $\mathcal{L}_1$ norm of a vector made of the $\mathcal{L}_2$ norms of the rows of this matrix. Consequently, many of these $\mathcal{L}_2$ norms will be zero or close to zero, which will produce the desired effect of nulling the coefficients of irrelevant endmembers in all pixels. This problem can be readily be solved using proximal algorithms, for instance the Alternating Direction Method of Multipliers (ADMM) [4]. To use it, we introduce split variables to decouple the different terms in the optimization. We then rewrite problem (5) in an equivalent formulation using linear constraints, which are suitable for the ADMM:

$$\arg \min_{\boldsymbol{\Phi}} \frac{1}{2}||\mathbf{X} - \mathbf{S}_0\boldsymbol{\Phi}||_F^2 + \lambda||\mathbf{U}||_{2,1} + \mathcal{I}_{\mathbb{R}_+^{d \times N}}(\mathbf{V})$$

$$\text{s.t. } \mathbf{U} = \boldsymbol{\Phi}, \ \mathbf{V} = \boldsymbol{\Phi}. \tag{7}$$

The ADMM uses an Augmented Lagrangian (AL) approach to split the hard nondifferentiable problem of Eq. (5) into several easier subproblems w.r.t. the two blocks of variables $\mathbf{U}$ and $\mathbf{V}$, and a so-called *dual update* of the introduced Lagrange mutlipliers, called $\mathbf{C}$ and $\mathbf{D}$ below (they can be initialized to zero). All these subproblems enjoy closed form solutions, which can be iterated until convergence. Collaborative sparsity then seems like a good candidate to discard the unwanted spurious endmembers. However, there are two problems with this approach. The first is that since the linear constraints of the ADMM are only satisfied asymptotically, we have no guarantee that all the entries of the supposedly discarded rows of the feature matrix $\boldsymbol{\Phi}$ will be exactly zero (and this actually happens in practice). Then an arbitrary thresholding step is required to eliminate endmembers with a small contribution [1,13]. The second is that in order to obtain the appropriate sparsity level, the regularization parameter λ needs to be optimized through a grid search, which is computationally very costly, and requires a criterion to select the best run of the algorithm. The next section provides solutions for both issues.

2.3 Computing a Regularization Path

In order to tackle both the regularization parameter issue and the inexact sparsity of the collaborative sparse regression at once, we would like to obtain the regularization path of the solution, as a function of λ. Regularization paths can sometimes be computed cheaply, for instance on the LASSO (for Least Absolute Shrinkage Selection Operator) problem [9]. However, for more complex problems, such as ours, there is no way, to our knowledge, to obtain this regularization path easily. A convenient workaround for this is to compute a so-called ADMM algorithmic regularization path, introduced in [12]. This approach is able to use the ADMM to quickly approximate the sequence of active supports of the variable of interest, when the regularization parameter increases, for certain sparsity regularized least squares problems. Even though there are as of today no theoretical guarantees on the efficiency of this algorithm, it was experimentally shown to be able to efficiently approximate the true sequence of active sets on several problems [12], including the LASSO. Here, we extend this algorithm to collaborative sparsity.

Since exactly solving the optimization problem for a large number of regularization parameters would be too time consuming, we are more interested in finding the active set of endmembers when the weight of the sparsity term increases w.r.t. this of the data fit term. The idea is to find a sequence of end-member matrices, whose number of endmembers is decreasing from d to zero (when the model is fully sparse). Each new matrix contains the same endmembers as the previous one, except for the one which is going to be discarded next, when the weight of the sparsity term gets more important. We modify the ADMM in order to quickly obtain the support of the regularization path. An iteration of the ADMM is carried out for a very small value of the regularization parameter (guaranteeing a fully dense solution). Then, the variables obtained after this iteration are used as a warm start for another iteration with a new slightly higher regularization parameter. By repeating this for several iterations with higher and higher regularization parameters, the split variable $\mathbf{U}$, which undergoes a block soft thresholding (the proximal operator of the $\mathcal{L}_{2,1}$ norm [4]) becomes increasingly sparse. Since we are using warm starts, and because regularization parameters vary slowly, even if the ADMM is not fully converged at each iteration, the support of the active set is encoded in $\mathbf{U}$, often in one iteration only, long before this active set is propagated to $\boldsymbol{\Phi}$ (this will be the case only at convergence, when the constraints of problem (7) are satisfied). With these modifications, we obtain Algorithm 1. ρ is the barrier parameter of the ADMM, which we fix to 1 throughout the paper, so that it does not interfere with the tuning of the regularization parameter. $\mathbf{soft}_\tau$ denotes the block soft thresholding operator with scale parameter in index. If $\mathbf{u} \in \mathbb{R}^N$, $\mathbf{soft}_\tau(\mathbf{u}) = (1 - \frac{\tau}{||\mathbf{u}||_2})_+\mathbf{u}$, where $\cdot_+ = \max(\cdot, 0)$ (and we have $\mathbf{soft}_\tau(\mathbf{0}) = 0$). This operator is applied row-wise to a matrix. Here, we are using a geometric progression for λ (we keep the same notation for the regularization parameter, although it does not match the parameter of Eq. (7), because we do not completely solve the optimization problem), whose common ratio is t. This value should be small to approximate the active sets of the regularization path well enough. The regularization space can be explored very quickly since the algorithm provides around d endmember subsets of the full endmember set extracted by VCA, that need to be tested after this process.

2.4   Selecting the Best Model

Using the active sets $\mathbf{U}^i$, we can recover a sequence of sparser and sparser candidate endmember matrices (whose i^{th} element is denoted as $\mathbf{S}_0^i$). The last step is to select the optimal endmember matrix in the sense of some criterion. We use the BIC [16], which helps choosing from a set of candidate models, by favoring those with an important likelihood, and penalizing those with a high number of parameters. This criterion assumes that the noise is Gaussian, spectrally and spatially white, a strong but still widely used assumption. A candidate model M_i is made of one of the $\mathbf{S}_0^i$ and the corresponding estimated feature matrix with CLSU. For our problem, the BIC writes [15]:

$$BIC_i = \ln(L)P_i + L \ln\left(\frac{||\mathbf{X} - \mathbf{S}_0^i\hat{\boldsymbol{\Phi}}^i||_F^2}{L}\right), \tag{8}$$

Data: X, $\mathbf{S}_0$
Result: The sequence of $\mathbf{U}^i$, $i = 0, ..., i_{max}$
Initialize $\boldsymbol{\Phi}^0$ and choose λ^0 and $t > 0$;
while $||\mathbf{U}^i||_{row,0} \neq 0$ **do**
   $\lambda^i \leftarrow t\lambda^{i-1}$;
   $\mathbf{U}^i \leftarrow \text{soft}_{\lambda^i/\rho}(\boldsymbol{\Phi}^{i-1} - \mathbf{C}^{i-1})$;
   $\boldsymbol{\Phi}^i \leftarrow (\mathbf{S}_0^\top \mathbf{S}_0 + 2\rho \mathbf{I}_d)^{-1}(\mathbf{S}_0^\top \mathbf{X} + \rho(\mathbf{U}^i + \mathbf{V}^{i-1} + \mathbf{C}^{i-1} + \mathbf{D}^{i-1}))$;
   $\mathbf{V}^i \leftarrow (\boldsymbol{\Phi}^i - \mathbf{D}^{i-1})_+$;
   $\mathbf{C}^i \leftarrow \mathbf{C}^{i-1} + \mathbf{U}^i - \boldsymbol{\Phi}^i$;
   $\mathbf{D}^i \leftarrow \mathbf{D}^{i-1} + \mathbf{V}^i - \boldsymbol{\Phi}^i$;
   $i \leftarrow i + 1$
end

Algorithm 1: ADMM algorithmic regularization path for problem (7).

where P_i is the number of endmembers in $\mathbf{S}_0^i \in \mathbb{R}^{L \times P_i}$. $\hat{\boldsymbol{\Phi}}^i$ is the abundance matrix estimated by CLSU using the data and the endmember matrix $\mathbf{S}_0^i$. The best model is simply the one minimizing the BIC value, providing $P \leq d$ endmembers and abundance maps. The endmembers which do not contribute much to the data fit have been discarded.

3 Results

In this section, we show the results of the proposed approach on a simulated and a real dataset. We compare the obtained results to those of the classical SU chain, using the Hyperspectral Subspace Identification by Minimum Error (HySIME) algorithm [2] for ID estimation, VCA for endmember extraction, and nonnegative least squares and the normalization detailed in Sect. 2.1 for the abundance and scaling factor estimation, so that the results follow the ELMM. We call this approach HySIME + S-CLSU (for Scaled CLSU).

3.1 Results on Simulated Data

For the simulated data, we voluntarily put ourselves in a case where ID estimation algorithms are prone to overestimation, namely an image with small spatial dimensions, a high spectral dimension, and non negligible noise values [7]. 6 endmembers were randomly selected from the United States Geological Survey (USGS) specral library, containing in-situ acquired spectra of various minerals. We built synthetic abundance maps for 6 materials using Gaussian Random Fields. We also computed scaling factor maps for each material using mixtures of Gaussians. These two quantities are shown in Fig. 1. Since the actual contribution of a material to a pixel is the product between the abundance and the scaling factor, the effect of either quantity will only be noticeable when the other is sufficient. For example, significant SV for a material with a small abundance

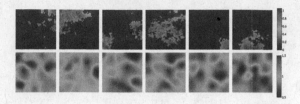

Fig. 1. True abundance (top row) and scaling factor (bottom row) maps for the synthetic data.

will be very hard to recover. We mixed the data using the full ELMM of [8,17] and finally added white Gaussian noise, such that the SNR was 25 dB. This resulted in a $40 \times 40 \times 300$ hyperspectral image. The HySIME algorithm estimated the ID of this dataset to be 16, whereas only 6 endmembers were used in the data generation (as a comparison, the Random Matrix Theory (RMT) based algorithm of [5] returned an ID value of 25). We show the abundance estimation results for our approach and the classical one in Fig. 2. We can see that the HySIME + S-CLSU approach is able to recover correctly 4 of the 6 abundance maps. However, the last two abundances are split in 11 different maps, which correspond to unnecessary (because very close between them) signatures extracted by the VCA. The proposed approach (with the parameters empirically tuned to $\lambda^0 = 10^{-4}$ and $t = 1.01$) only retained 6 endmembers in that case, all of which are associated with abundance maps which are very close to the real ones. We do not show the scaling factor maps here for lack of space, but they are very similar in both cases, and each pixel value accounts for the scaling factor of the predominant material. In a scenario where the full ELMM would be used, the scaling factors would probably be much easier to interpret for our approach, because we would be able to distinguish the contributions of each material, and better explain what happens in heavily mixed pixels. Figure 3 shows the BIC values we get for the obtained sequence of candidate endmember matrices. While the data fit term decreases continually every time we add an endmember, the decrease is marginal after 6 endmembers, while the number of parameters is more and more penalized. The BIC then reaches a minimum for 6 endmembers. This shows that the unnecessary endmembers only fitted the noise, while being difficult to interpret. For each material, we also computed a Root Mean Squared Error (RMSE) on the abundances for material p, with $aRMSE = \frac{1}{N} \sum_{k=1}^{N} ||\mathbf{a}_{\text{true},pk} - \hat{\mathbf{a}}_{pk}||_2$, with $\hat{\mathbf{a}}_{pk}$ the closest abundance map to the true one in the results of one of the two algorithms (see Table 1). We see that except for material 2 (the materials are numbered from left to right in Fig. 1), where the proposed approach recovers a slightly noisy abundance map, the abundances are better recovered by the proposed method.

Table 1. aRMSE values between the true abundance maps and the closest one of the two competing approaches. The best value is in red.

Material	1	2	3	4	5	6
Proposed method	1.0×10^{-3}	1.1×10^{-3}	5.9×10^{-4}	9.2×10^{-4}	4.2×10^{-4}	5.5×10^{-4}
HySIME + S-CLSU	4.6×10^{-3}	8.9×10^{-4}	7.3×10^{-4}	1.4×10^{-3}	5.2×10^{-4}	7.5×10^{-4}

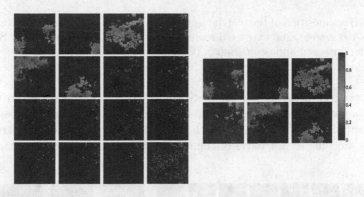

Fig. 2. Abundance maps extracted by HySIME + S-CLSU for the synthetic data (left), and by the proposed approach (right).

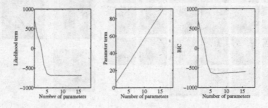

Fig. 3. Data fit term of the BIC (left), parameter term (middle) and BIC value (right), for the simulated data.

3.2   Results on Real Data

To confirm the soundness of our approach, we apply it to a 100×100 subset of the Washington DC mall dataset, acquired over the National Gallery of Art (shown in Fig. 4, with the endmembers extracted by VCA displayed as red crosses) by the HYDICE sensor, comprising 191 spectral bands in the visible and near infrared, with a spatial resolution of 2.8 m. HySIME estimated the ID to be 38 (the RMT algorithm returned 65). We show the estimated abundance maps in Fig. 5. We see that the abundance maps on the left are very hard to interpret, because many are very sparse and related to outlier pixels, while the proposed approach allowed to retain only the most important. Only one visually relevant endmember retained by HySIME+S-CLSU is not present in the proposed approach, corresponding to marble (stairs and dome of the museum). The scaling factor maps, shown in Fig. 4, are relatively similar in both cases, except in the grass part, where some

Fig. 4. RGB composition of the real dataset used (left), with the endmembers extracted by VCA in red crosses, and extracted scaling factor maps for Hysime + S-CLSU (middle) and the proposed approach (right). (Color figure online)

geometrical structures appear for HySIME+S-CLSU, which could correspond to artifacts captured as one of the endmembers. Otherwise, the maps explain well the variability of the scene, with low values for the shadowed trees and structures, and higher for parts of the roofs exposed to the sun, for example.

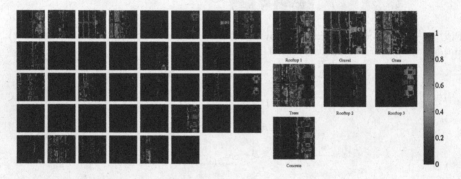

Fig. 5. Abundance maps extracted by HySIME + S-CLSU for the real data (left), and by the proposed approach (right).

4   Conclusion

We have presented a technique to overcome the likely overestimation of the number of endmembers to use in spectral unmixing of hyperspectral data, accounting for spectral variability. It is based on computing an approximate regularization path for a collaborative sparse regression problem, which allows to select the most relevant endmember signatures, and to discard the spurious ones. We confirmed the interest of the proposed approach on a synthetic dataset and a real one. Future work will include the full ELMM (one scaling factor per material) to the framework of the proposed approach.

References

1. Ammanouil, R., Ferrari, A., Richard, C., Mary, D.: Blind and fully constrained unmixing of hyperspectral images. IEEE Trans. Image Process. **23**(12), 5510–5518 (2014)
2. Bioucas-Dias, J., Nascimento, J.: Hyperspectral subspace identification. IEEE Trans. Geosci. Remote Sens. **46**(8), 2435–2445 (2008)
3. Bioucas-Dias, J., Plaza, A., Dobigeon, N., Parente, M., Du, Q., Gader, P., Chanussot, J.: Hyperspectral unmixing overview: Geometrical, statistical, and sparse regression-based approaches. IEEE J. Sel. Top. Appl. Earth Observations Remote Sens. **5**(2), 354–379 (2012)
4. Boyd, S., Parikh, N., Chu, E., Peleato, B., Eckstein, J.: Distributed optimization and statistical learning via the alternating direction method of multipliers. Found. Trends Mach. Learn. **3**(1), 1–122 (2011)
5. Cawse-Nicholson, K., Damelin, S., Robin, A., Sears, M.: Determining the intrinsic dimension of a hyperspectral image using random matrix theory. IEEE Trans. Image Process. **22**(4), 1301–1310 (2013)
6. Drumetz, L., Chanussot, J., Jutten, C.: Endmember variability in spectral unmixing: recent advances. In: Proceedings of IEEE Workshop on Hyperspectral Image and Signal Processing: Evolution in Remote Sensing (WHISPERS), pp. 1–4 (2016)
7. Drumetz, L., Veganzones, M.A., Gómez, R.M., Tochon, G., Dalla Mura, M., Licciardi, G.A., Jutten, C., Chanussot, J.: Hyperspectral local intrinsic dimensionality. IEEE Trans. Geosci. Remote Sens. **54**(7), 4063–4078 (2016)
8. Drumetz, L., Veganzones, M.A., Henrot, S., Phlypo, R., Chanussot, J.: Jutten, C: Blind hyperspectral unmixing using an extended linear mixing model to address spectral variability. IEEE Trans. Image Process. **25**(8), 3890–3905 (2016)
9. Efron, B., Hastie, T., Johnstone, I., Tibshirani, R.: Least angle regression. Ann. Stat. **32**(2), 407–499 (2004)
10. Heinz, D., Chang, C.I.: Fully constrained least squares linear spectral mixture analysis method for material quantification in hyperspectral imagery. IEEE Trans. Geosci. Remote Sens. **39**(3), 529–545 (2001)
11. Heylen, R., Parente, M., Gader, P.: A review of nonlinear hyperspectral unmixing methods. IEEE J. Sel. Top. Appl. Earth Observations Remote Sens. **7**(6), 1844–1868 (2014)
12. Hu, Y., Chi, E., Allen, G.I.: ADMM algorithmic regularization paths for sparse statistical machine learning (2015). arXiv preprint arXiv:150406637
13. Iordache, M.D., Bioucas-Dias, J.M., Plaza, A.: Collaborative sparse regression for hyperspectral unmixing. IEEE Trans. Geosci. Remote Sens. **52**(1), 341–354 (2014)
14. Nascimento, J., Bioucas Dias, J.: Vertex component analysis: a fast algorithm to unmix hyperspectral data. IEEE Trans. Geosci. Remote Sens. **43**(4), 898–910 (2005)
15. Priestley, M.B.: Spectral Analysis and Time Series. Academic Press, London (1981)
16. Schwarz, G.: Estimating the dimension of a model. Ann. Stat. **6**(2), 461–464 (1978)
17. Veganzones, M.A., Drumetz, L., Marrero, R., Tochon, G., Dalla Mura, M., Plaza, A., Bioucas-Dias, J., Chanussot, J.: A new extended linear mixing model to address spectral variability. In: Proceedings of the IEEE Workshop on Hyperspectral Image and Signal Processing: Evolution in Remote Sensing (WHISPERS) (2014)

Sharpening Hyperspectral Images Using Plug-and-Play Priors

Afonso Teodoro[1,2]([envelope]), José Bioucas-Dias[1,2], and Mário Figueiredo[1,2]

[1] Instituto de Telecomunicações, Lisbon, Portugal
[2] Instituto Superior Técnico, Universidade de Lisboa, Lisbon, Portugal
afonso.teodoro@tecnico.ulisboa.pt

Abstract. This paper addresses the problem of fusing hyperspectral (HS) images of low spatial resolution and multispectral (MS) images of high spatial resolution into images of high spatial and spectral resolution. By assuming that the target image lives in a low dimensional subspace, the problem is formulated with respect to the latent representation coefficients. Our major contributions are: (i) using patch-based spatial priors, learned from the MS image, for the latent images of coefficients; (ii) exploiting the so-called plug-and-play approach, wherein a state-of-the-art denoiser is plugged into the iterations of a variable splitting algorithm.

Keywords: Data fusion · Hyperspectral imaging · Multispectral imaging · Latent variables · Plug-and-play · Gaussian mixtures

1 Introduction

In remote sensing imaging, there is a trade off between spectral and spatial resolution, fundamentally due to the limited amount of sunlight energy reaching the sensors. Accordingly, it is common to distinguish between hyperspectral (HS) and multispectral (MS) images: the former usually have hundreds of bands, each with a very narrow range of the electromagnetic spectrum, *i.e.*, high spectral resolution; the latter usually have fewer than 20 bands, with larger bandwidths than in the HS case. The larger bandwidths in MR imaging imply that, for the same SNR, the spatial dimensions of the pixels may be smaller, *i.e.*, the spatial resolution may be higher than in HR sensors. The data fusion problem aims at combining HS and MS sources to produce high resolution HS images.

There are several different approaches to address this data fusion problem (for a comprehensive review, see [6]). The current state-of-the-art methods (*e.g.*, [7, 12]) follow Bayesian or variational approaches, and use the *alternating direction method of multipliers* (ADMM [3]) to solve the resulting optimization problem.

In this paper, we exploit the so-called *plug-and-play* (PnP) approach [11], wherein an off-the-shelf (usually, state-of-the-art) denoiser is *plugged* into the iterations of ADMM or other variable splitting scheme. Most, if not all, state-of-the-art denoisers are patch-based and often can't even be formalized as solutions

© Springer International Publishing AG 2017
P. Tichavský et al. (Eds.): LVA/ICA 2017, LNCS 10169, pp. 392–402, 2017.
DOI: 10.1007/978-3-319-53547-0_37

to an optimization problem (let alone a convex one). The PnP approach chooses to ignore this formal issue, and has recently been shown to yield excellent empirical performance. This approach raises some issues regarding the convergence of the resulting ADMM algorithm, but these won't be addressed here.

This paper is organized as follows. Section 2 formalizes the problem, detailing the models for the HS and MS images. Sections 3 and 4 motivate and present the proposed approach, and experimental results are reported in Sect. 5. Section 6 concludes the paper with some final remarks and directions for future work.

2   Problem Formulation

This paper addresses the problem of fusing HS and MS images. We assume that the HS bands are a blurred and downsampled version of the corresponding bands of the underlying image to be inferred (the *target image*), whereas each MS image is a spectrally degraded version of the target image. Furthermore, both HS and MS images are contaminated with noise, assumed to be additive white Gaussian noise across all bands and pixels. For notation compactness, we organize the HS and MS images into 2D matrices, where each row corresponds to a spectral band, with lexicographically ordered pixels.

Under the above assumptions, the observation model is

$$\mathbf{Y}_h = \mathbf{ZBM} + \mathbf{N}_h, \tag{1}$$
$$\mathbf{Y}_m = \mathbf{RZ} + \mathbf{N}_m, \tag{2}$$

where: $\mathbf{Y}_h \in \mathbb{R}^{L_h \times n_h}$ is the observed HS image; $\mathbf{Z} \in \mathbb{R}^{L_h \times n_m}$ denotes the target image to be estimated; $\mathbf{B} \in \mathbb{R}^{n_m \times n_m}$ is a spatial convolution operator; $\mathbf{M} \in \mathbb{R}^{n_m \times n_h}$ is a uniform subsampling operator; $\mathbf{Y}_m \in \mathbb{R}^{L_m \times n_m}$ is the observed MS image; $\mathbf{R} \in \mathbb{R}^{L_m \times L_n}$ has in its rows the spectral responses of the sensor; $\mathbf{N}_h$ and $\mathbf{N}_m$ are Gaussian noise matrices, with known variances σ_h and σ_m, respectively.

In this paper, we make several simplifying assumptions: (i) that the noise is Gaussian (for other types of noise, see [6,12]); (ii) that the HS sensor has always the same point spread function, which means that each HS band undergoes the same blur; (iii) a fully non-blind scenario (*i.e*, matrices $\mathbf{B}$ and $\mathbf{R}$ are known). Although we could estimate $\mathbf{B}$ from the data, and the spectral responses of the multispectral sensor (matrix $\mathbf{R}$) [7], we don't consider those possibilities in this paper. Finally, we assume periodic boundary conditions, in order to use the *fast Fourier transform* (FFTs) to compute convolutions.

One of the difficulties in fusing HS and MS images is their dimensionality. As mentioned above, the goal is to estimate a high resolution image, with L_h bands and n_m pixels per band, where $L_h > L_m$ and $n_m > n_h$. To lessen this difficulty, a typical approach is to learn a subspace to represent the HS images. Since there is a large correlation between the HS bands, the dimension of this subspace is usually of much lower than L_h [2]. In this paper, we apply *singular value decomposition* (SVD) to the original data, keeping only the p largest singular values, with $p < L_h$. Thus, instead of estimating $\mathbf{Z}$ directly, we estimate $\mathbf{X} \in \mathbb{R}^{p \times n_m}$ (termed *latent images*), and then recover the target image by computing

$$\mathbf{Z} = \mathbf{EX}, \tag{3}$$

where the columns of $\mathbf{E} \in \mathbb{R}^{L_h \times p}$ are the singular vectors corresponding to the p largest singular values of matrix $\mathbf{Y}_h$, thus spanning the same subspace as the target matrix $\mathbf{Z}$. The observation model then becomes

$$\mathbf{Y}_h = \mathbf{EXBM} + \mathbf{N}_h, \tag{4}$$
$$\mathbf{Y}_m = \mathbf{REX} + \mathbf{N}_m. \tag{5}$$

The problem of estimating the image of coefficients $\mathbf{X}$ is, most often, ill-posed. Consequently, solving it satisfactorily demands a regularization term (or prior) to promote specific characteristics on $\mathbf{X}$. The most common approach is to seek a *maximum a posteriori* (MAP) estimate:

$$\widehat{\mathbf{X}} \in \underset{\mathbf{X}}{\operatorname{argmax}} \; p(\mathbf{X} \mid \mathbf{Y}_h, \mathbf{Y}_m) \tag{6}$$

$$= \underset{\mathbf{X}}{\operatorname{argmax}} \; p(\mathbf{Y}_h \mid \mathbf{X})p(\mathbf{Y}_m \mid \mathbf{X})p(\mathbf{X}) \tag{7}$$

$$= \underset{\mathbf{X}}{\operatorname{argmin}} \; \frac{1}{2}\|\mathbf{EXBM} - \mathbf{Y}_h\|_F^2 + \frac{\lambda_m}{2}\|\mathbf{REX} - \mathbf{Y}_m\|_F^2 + \lambda_\phi \phi(\mathbf{X}), \tag{8}$$

where we have used the hypothesis that the noise matrices $\mathbf{N}_h$ and $\mathbf{N}_m$ are independent (conditioned on $\mathbf{X}$) and Gaussian, and $\|\cdot\|_F$ denotes the Frobenius norm. Parameters λ_m and λ_ϕ control the relative weight of each term.

3   Optimization

To address the optimization problem (8), we use ADMM. The first step is a so-called *variable splitting* procedure, which yields a constrained optimization problem equivalent to (8):

$$\widehat{\mathbf{X}}, \widehat{\mathbf{V}}_1, \widehat{\mathbf{V}}_2, \widehat{\mathbf{V}}_3 \in \underset{\mathbf{X}, \mathbf{V}_1, \mathbf{V}_2, \mathbf{V}_3}{\operatorname{argmin}} \; \frac{1}{2}\|\mathbf{EV}_1\mathbf{M} - \mathbf{Y}_h\|_F^2 + \frac{\lambda_m}{2}\|\mathbf{REV}_2 - \mathbf{Y}_m\|_F^2 + \lambda_\phi \phi(\mathbf{V}_3)$$

$$\text{subject to} \quad \mathbf{V}_1 = \mathbf{XB}, \quad \mathbf{V}_2 = \mathbf{X}, \quad \mathbf{V}_3 = \mathbf{X}. \tag{9}$$

The corresponding *augmented Lagrangian* is

$$\mathcal{L}(\mathbf{X}, \mathbf{V}, \mathbf{D}) = \frac{1}{2}\|\mathbf{EV}_1\mathbf{M} - \mathbf{Y}_h\|_F^2 + \frac{\lambda_m}{2}\|\mathbf{REV}_2 - \mathbf{Y}_m\|_F^2 + \lambda_\phi \phi(\mathbf{V}_3)$$

$$+ \frac{\mu}{2}\left(\|\mathbf{XB} - \mathbf{V}_1 - \mathbf{D}_1\|_F^2 + \|\mathbf{X} - \mathbf{V}_2 - \mathbf{D}_2\|_F^2 + \|\mathbf{X} - \mathbf{V}_3 - \mathbf{D}_3\|_F^2 \right),$$

where $\mathbf{D} = (\mathbf{D}_1, \mathbf{D}_2, \mathbf{D}_3)$ are the scaled dual variables (Lagrange multipliers, see [3]), and $\mathbf{V} = (\mathbf{V}_1, \mathbf{V}_2, \mathbf{V}_3)$. For simplicity, we take the same penalty parameter μ for all the constraints. ADMM works by minimizing $\mathcal{L}$ sequentially w.r.t. to each of the primal variables $\mathbf{X}, \mathbf{V}_1, \mathbf{V}_2$, and $\mathbf{V}_3$ (while keeping the others fixed), and then updating the dual variables $\mathbf{D}$ [3]. The resulting algorithm is as follows:

$$\mathbf{X}^{k+1} = \underset{\mathbf{X}}{\operatorname{argmin}} \ \|\mathbf{XB} - \mathbf{V}_1 - \mathbf{D}_1\|_F^2 + \|\mathbf{X} - \mathbf{V}_2 - \mathbf{D}_2\|_F^2 + \|\mathbf{X} - \mathbf{V}_3 - \mathbf{D}_3\|_F^2, \quad (10)$$

$$\mathbf{V}_1^{k+1} = \underset{\mathbf{V}_1}{\operatorname{argmin}} \ \frac{1}{2}\|\mathbf{EV}_1\mathbf{M} - \mathbf{Y}_h\|_F^2 + \frac{\mu}{2}\|\mathbf{X}^{k+1}\mathbf{B} - \mathbf{V}_1 - \mathbf{D}_1^k\|_F^2, \quad (11)$$

$$\mathbf{V}_2^{k+1} = \underset{\mathbf{V}_2}{\operatorname{argmin}} \ \frac{\lambda_m}{2}\|\mathbf{REV}_2 - \mathbf{Y}_m\|_F^2 + \frac{\mu}{2}\|\mathbf{X}^{k+1} - \mathbf{V}_2 - \mathbf{D}_2^k\|_F^2, \quad (12)$$

$$\mathbf{V}_3^{k+1} = \underset{\mathbf{V}_3}{\operatorname{argmin}} \ \lambda_\phi \, \phi(\mathbf{V}_3) + \frac{\mu}{2}\|\mathbf{X}^{k+1} - \mathbf{V}_3 - \mathbf{D}_3^k\|_F^2, \quad (13)$$

$$\mathbf{D}_1^{k+1} = \mathbf{D}_1^k + \mathbf{V}_1^{k+1} - \mathbf{X}^{k+1}\mathbf{B}, \quad (14)$$

$$\mathbf{D}_2^{k+1} = \mathbf{D}_2^k + \mathbf{V}_2^{k+1} - \mathbf{X}^{k+1}, \quad (15)$$

$$\mathbf{D}_3^{k+1} = \mathbf{D}_3^k + \mathbf{V}_3^{k+1} - \mathbf{X}^{k+1}. \quad (16)$$

Subproblems (10)–(12) are quadratic, thus have closed form solutions [7]

$$\mathbf{X}^{k+1} = \left[\left(\mathbf{V}_1^k + \mathbf{D}_1^k\right)\mathbf{B}^T + \left(\mathbf{V}_2^k + \mathbf{D}_2^k\right) + \left(\mathbf{V}_3^k + \mathbf{D}_3^k\right)\right]\left[\mathbf{BB}^T + 2\mathbf{I}\right]^{-1}, \quad (17)$$

$$\mathbf{V}_1^{k+1} = \left[\mathbf{EE}^T + \mu\mathbf{I}\right]^{-1}\left[\mathbf{E}^T\mathbf{Y}_h + \mu\left(\mathbf{X}^{k+1}\mathbf{B} - \mathbf{D}_1^k\right)\right] \odot \mathbf{M}$$
$$+ \left(\mathbf{X}^{k+1}\mathbf{B} - \mathbf{D}_1^k\right) \odot (1 - \mathbf{M}), \quad (18)$$

$$\mathbf{V}_2^{k+1} = \left[\lambda_m\mathbf{E}^T\mathbf{R}^T\mathbf{RE} + \mu\mathbf{I}\right]^{-1}\left[\lambda_m\mathbf{E}^T\mathbf{R}^T\mathbf{Y}_m + \mu\left(\mathbf{X}^{k+1} - \mathbf{D}_2^k\right)\right], \quad (19)$$

with $\odot$ denoting componentwise multiplication.

Subproblem (13) corresponds to the *Moreau proximity operator* [1] of $\lambda_\phi \, \phi$, computed at $\mathbf{X}^{k+1} - \mathbf{D}_3^k$. This can be seen as a denoising operation, with function ϕ acting as the regularizer and $\mathbf{X}^{k+1} - \mathbf{D}_3^k$ as the noisy observed data, where the noise is zero-mean Gaussian with variance λ_ϕ/μ. In the *HySure* method [7], which achieves state-of-the-art results, ϕ is a vector-TV regularizer. In the plug-and-play (PnP) approach, the idea is to use (plug) a state-of-the-art denoiser in the place of this proximity operator of ϕ.

4   Plugging a Gaussian Mixture Model Denoiser

Clean image patches are well modelled by a *Gaussian mixture model* (GMM), learned either from a collection of noiseless patches, or from the noisy input image [8,13]. When the input image is blurred, it may be impossible to obtain a good GMM from its patches, and we need to resort to an external dataset. Unlike other state-of-the-art denoisers, such as BM3D [4], which rely only on the input image, GMM-based denoisers are naturally well suited for this scenario.

Learning a GMM from an external dataset allows obtaining class-specific denoisers [9,10], *i.e.*, we can learn denoisers targeted to particular classes, such as text, faces, or some type of medical images. In principle, these denoisers capture the characteristics of the class better than a general purpose one, hence yielding better performance. In [9,10], we showed that a PnP approach with a GMM-based denoiser achieves state-of-the-art performance on images from specific classes, in several imaging inverse problems.

In the proposed approach, we instantiate the PnP scheme by taking the high spatial resolution MS images as the *"external"* dataset, and use them to learn a GMM-based denoiser. This is a very particular case of class-specific model, where we use an image from the same scene that we are trying to reconstruct. Implicit in this proposal is the assumption that the patches of the latent images are well modelled by the GMM learned from the high spatial resolution images. In the next section, we provide evidence in favour of this assumption.

The GMM-based denoiser that we employ in the PnP scheme is described in detail in [8]. To summarize, we start by extracting all the patches from the MS images, and using the well-known expectation-maximization algorithm to determine the parameters of the GMM. Then, under the GMM prior, we are able to compute the optimal minimum mean squared error estimate (MMSE) of the latent image patches. Letting $\mathbf{x}_i$ and $\mathbf{y}_i$ denote a patch from the clean and noisy latent images, respectively, the resulting MMSE estimate is

$$\hat{\mathbf{x}}_i = \sum_{m=1}^{K} \beta_m(\mathbf{y}_i)\, \mathbf{v}_m(\mathbf{y}_i), \qquad (20)$$

where

$$\mathbf{v}_m(\mathbf{y}_i) = \left(\sigma^2\, \mathbf{C}_m + \mathbf{I}\right)^{-1}\left(\sigma^2\, \mathbf{C}_m^{-1}\mu_m + \mathbf{y}_i\right), \qquad (21)$$

and

$$\beta_m(\mathbf{y}_i) = \frac{\alpha_m\, \mathcal{N}(\mathbf{y}_i; \mu_m, \mathbf{C}_m + \sigma^2\, \mathbf{I})}{\sum_{j=1}^{K} \alpha_j\, \mathcal{N}(\mathbf{y}_i; \mu_j, \mathbf{C}_m + \sigma^2\, \mathbf{I})}, \qquad (22)$$

with $\mathcal{N}(\cdot; \mu, \mathbf{C})$ denoting a Gaussian probability density function of mean μ and covariance $\mathbf{C}$, and σ^2 the variance of the noise. This result has a simple interpretation: $\beta_m(\mathbf{y}_i)$ represents the posterior probability of the i-th patch belonging to the m-th component of the mixture and $\mathbf{v}_m(\mathbf{y}_i)$ is the MMSE estimate of the i-th patch if we knew that it had been generated from the m-th component. Notice that both β_m and $\mathbf{v}_m$ depend on the noisy patches. After computing the MMSE estimate of the patches, they are returned to their location and combined by weighted average.

5   Experimental Results

5.1   Denoising

We begin by showing that a GMM denoiser performs well, not only on the MS images, but also on the HS bands and the latent images, suggesting that it will also perform well in HS sharpening. In these, and all following experiments, the GMM models have 20 components.

(a)   (b)   (c)   (d)

Fig. 1. Denoising: (a) panchromatic image; (b) noisy panchromatic image ($\sigma = 25$); (c) BM3D (PSNR = 28.91 dB); (d) GMM (PSNR = 29.12 dB).

(a)   (b)   (c)   (d)

Fig. 2. Denoising: (a) hyperspectral image; (b) noisy hyperspectral image ($\sigma = 25$); (c) BM3D (PSNR = 33.60 dB); (d) GMM (PSNR = 33.30 dB).

The first experiment (Fig. 1) considers a (cropped) high resolution panchromatic image from the ROSIS Pavia University dataset (messtec.dlr.de/en/technology/dlr-remote-sensing); the GMM is learned from the noisy image itself and performs slightly better than BM3D, in terms of peak SNR.

In the second experiment (Fig. 2), we learn a GMM from a clean high resolution panchromatic image, and use it to denoise one of the low resolution HS bands. In this case, BM3D, which uses only the input image, performs slightly better than the GMM denoiser.

The third experiment (Fig. 3) uses the first latent image (first row of $\mathbf{X}$), with the same GMM as in the second experiment. Again, BM3D performs slightly better, but the GMM still yields very good results.

The results of these experiments show that a denoiser based on an GMM learned from a panchromatic high resolution image, when used to denoise low resolution HS bands or HS latent images (rows of $\mathbf{X}$), performs nearly on par with a state-of-the-art denoiser that uses only the noisy image. This shows that the learned GMM is a good model for these images, which is an important conclusion concerning the use of these denoisers in the PnP approach. Since BM3D relies only on the input image, it may lead to poor results, unless the initialization is very good. On the other hand, the GMM provides a good model for the latent images, thus making the algorithm more robust to initialization.

(a)   (b)   (c)   (d)

Fig. 3. Denoising: (a) latent image ($\mathbf{X}$); (b) noisy latent image ($\sigma = 25$); (c) BM3D (PSNR = 30.74 dB); (d) GMM (PSNR = 30.44 dB).

A way to sidestep the initialization issue with BM3D would be to determine the patch grouping from the panchromatic image and reuse it in the denoising step. Conceptually, this is very similar to what was done in CBM3D [5].

5.2   Fusion

We compare the proposed method with HySure [7], which is representative of the state-of-the-art, and with the PnP method using a BM3D denoiser rather than a GMM-based one. To compare the performance of these algorithms, we use three metrics (see [7]): *erreur relative globale adimensionnelle de synthèse* (ERGAS), *spectral angle mapper* (SAM), and *signal-to-reconstruction error* (SRE),

$$SRE = -10 \log \left(\|\hat{\mathbf{Z}} - \mathbf{Z}\|_F^2 / \|\mathbf{Z}\|_F^2 \right) \text{ (dB)}. \tag{23}$$

Parameters $\mu, \lambda_m, \lambda_{VTV}, \lambda_\phi$ are hand-tuned in order to achieve good results.

Table 1. HS and MS (R, G, B, N-IR) fusion on cropped ROSIS Pavia Univ. dataset.

	HySure iters = 1						HySure iters = 100					
SNR ($\mathbf{Y}_m$)	50 dB			30 dB			50 dB			30 dB		
SNR ($\mathbf{Y}_h$)	50 dB			20 dB			50 dB			20 dB		
Metric	ERGAS	SAM	SRE	ERGAS	SAM	SRE	ERGAS	SAM	SRE	ERGAS	SAM	SRE
BM3D	2.14	4.38	22.30	2.30	4.57	22.43	1.32	1.93	26.77	1.71	2.72	25.63
GMM	1.39	1.94	26.34	1.71	2.75	25.45	1.37	1.95	26.46	1.71	2.76	25.43

Table 2. HS and MS fusion on ROSIS Pavia University dataset.

	Exp. 1 (PAN)			Exp. 2 (PAN)			Exp. 3 (R, G, B, N-IR)			Exp. 4 (R, G, B, N-IR)		
SNR ($\mathbf{Y}_m$)	50 dB			30 dB			50 dB			30 dB		
SNR ($\mathbf{Y}_h$)	50 dB			20 dB			50 dB			20 dB		
Metric	ERGAS	SAM	SRE	ERGAS	SAM	SRE	ERGAS	SAM	SRE	ERGAS	SAM	SRE
HySure	**3.86**	**5.22**	**18.99**	3.99	**5.34**	18.63	1.80	2.95	25.79	2.02	3.34	24.86
Proposed	3.88	5.27	**18.99**	4.02	5.47	**18.65**	**1.61**	**2.55**	**27.48**	1.85	2.97	26.07
ADMM-BM3D	3.92	5.29	18.80	**3.98**	5.43	**18.65**	1.62	**2.55**	27.44	**1.83**	**2.96**	**26.15**

Table 3. HS and MS fusion on RTerrain dataset.

	Exp. 1 (PAN)			Exp. 2 (PAN)			Exp. 3 (R, G, B, N-IR)			Exp. 4 (R, G, B, N-IR)		
SNR ($\mathbf{Y}_m$)	50 dB			30 dB			50 dB			30 dB		
SNR ($\mathbf{Y}_h$)	50 dB			20 dB			50 dB			20 dB		
Metric	ERGAS	SAM	SRE	ERGAS	SAM	SRE	ERGAS	SAM	SRE	ERGAS	SAM	SRE
HySure	2.62	5.34	21.46	2.77	5.35	20.86	1.08	2.68	28.71	1.53	3.42	26.07
Proposed	2.58	**5.15**	**21.69**	2.75	**5.33**	**21.12**	**0.91**	**2.20**	**30.86**	**1.29**	**2.85**	**27.85**
ADMM-BM3D	**2.57**	5.17	21.65	2.76	5.36	21.08	0.93	2.22	30.80	1.31	2.91	27.72

Table 1 illustrates the impact of initialization on the final results. We ran the PnP method both with the GMM and the BM3D denoisers, for 200 iterations. Using the same parameters, we ran 100 iterations of HySure, followed by 100 iterations of PnP with the GMM and BM3D densoisers. As expected, whereas PnP with BM3D varies significantly with the initial estimate, PnP with a GMM prior is more robust. In the remaining experiments, we start by running HySure to obtain a good initialization, then switched to the BM3D or GMM denoisers.

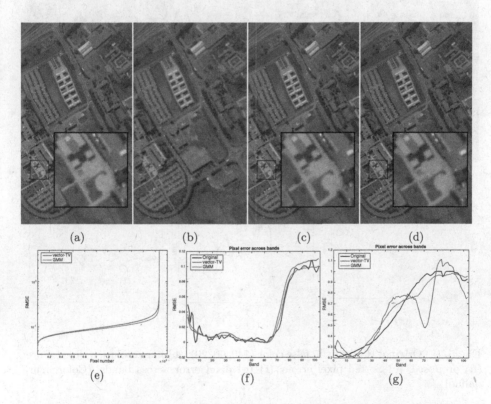

(a)   (b)   (c)   (d)

(e)   (f)   (g)

Fig. 4. (a) Original HS bands in false color $(60, 27, 17)$; (b) low-resolution; (c) HySure; (d) proposed; (e) sorted pixel errors; (f)–(g) pixel error across bands. (Color figure online)

Tables 2 and 3 summarize the results on the full ROSIS Pavia University and RTerrain[1] datasets, respectively. The first two experiments concern PAN-sharpening, a particular case of HS and MS data fusion, and the last two concern MS-sharpening. The PAN and the MS images were simulated from the original HS data using the IKONOS spectral responses[2]. In the former, the three algorithms provide very similar results, yet, in the latter, the PnP approach brings consistent improvements. Figures 4 and 5 show the results in false colour, but the differences are not visually noticeable. However, the plots show that using the GMM prior yields a larger number of pixels with error bellow a given ϵ. This is particularly visible in Fig. 4e where the red (GMM) curve is always below the blue one (vector-TV). In the remaining plots we show the RMSE of each band for given pixels. For most of them, the performance of both methods is similar, as shown in Figs. 4f and 5f, but in some the GMM regularizer works best, Figs. 4g and 5g.

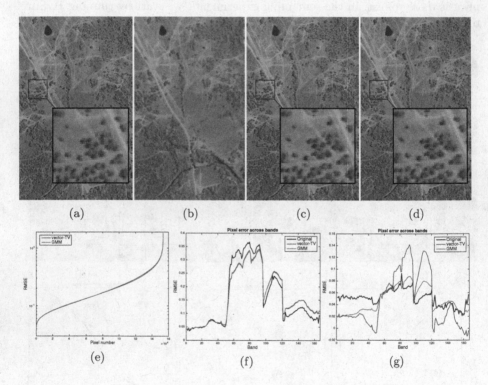

Fig. 5. (a) Original HS bands in false color $(140, 60, 20)$; (b) low-resolution; (c) HySure; (d) proposed; (e) Sorted pixel errors; (f)–(g) pixel error across bands. (Color figure online)

---

[1] Available at http://www.agc.army.mil/Missions/Hypercube.aspx.
[2] Details at http://www.digitalglobe.com/sites/default/files/DG_IKONOS_DS.pdf.

6   Conclusions and Future Work

This paper proposed using the recent PnP approach for fusion of HS and panchromatic or MS images. Moreover, we have proposed using a GMM as a prior for the reconstruction of the HS images, or of the latent images which are a low dimensional representation, with respect to a learned basis, of the original HSI. The GMM was learned from the panchromatic image; we provided empirical evidence that such prior is a good model for the target HS bands. We showed that the proposed approach achieves state-of-the-art results in some of the experiments.

Acknowledgments. This work was partially supported by *Fundação para a Ciência e Tecnologia* (FCT), grants UID/EEA/5008/2013, ERANETMED/0001/2014 and BD/102715/2014.

References

1. Bauschke, H., Combettes, P.: Convex Analysis and Monotone Operator Theory in Hilbert Spaces. Springer, New York (2011)
2. Bioucas-Dias, J., Plaza, A., Dobigeon, N., Parente, M., Du, Q., Gader, P., Chanussot, J.: Hyperspectral unmixing overview: geometrical, statistical, and sparse regression-based approaches. IEEE J. Sel. Topics Appl. Earth Obs. Remote Sens. **5**, 354–379 (2012)
3. Boyd, S., Parikh, N., Chu, E., Peleato, B., Eckstein, J.: Distributed optimization and statistical learning via the alternating direction method of multipliers. Found. Trends Mach. Learn. **3**, 1–122 (2011)
4. Dabov, K., Foi, A., Katkovnik, V., Egiazarian, K.: Image denoising by sparse 3D transform-domain collaborative filtering. IEEE Trans. Image Proc. **16**, 2080–2095 (2007)
5. Dabov, K., Foi, A., Katkovnik, V., Egiazarian, K.: Color image denoising via sparse 3D collaborative filtering with grouping constraint in luminance-chrominance space. In: IEEE ICIP (2007)
6. Loncan, L., Almeida, L., Bioucas-Dias, J., Briottet, X., Chanussot, J., Dobigeon, N., Fabre, S., Liao, W., Licciardi, G., Simões, M., Tourneret, J.-Y., Veganzones, M., Vivone, G., Wei, Q., Yokoya, N.: Hyperspectral pansharpening: a review. IEEE Geosci. Remote Sens. Mag. **3**, 27–46 (2015)
7. Simões, M., Bioucas-Dias, J., Almeida, L., Chanussot, J.: A convex formulation for hyperspectral image superresolution via subspace-based regularization. IEEE Trans. Geosci. Remote Sens. **55**, 3373–3388 (2015)
8. Teodoro, A., Almeida, M., Figueiredo, M.: Single-frame image denoising and inpainting using Gaussian mixtures. In: ICPRAM, pp. 283–288 (2015)
9. Teodoro, A., Bioucas-Dias, J., Figueiredo, M.: Image restoration and reconstruction using variable splitting and class-adapted image priors. In: IEEE-ICIP (2016)
10. Teodoro, A., Bioucas-Dias, J., Figueiredo, M.: Image restoration with locally selected class-adapted models. In: IEEE-MLSP (2016)
11. Venkatakrishnan, S., Bouman, C., Chu, E., Wohlberg, B.: Plug-and-play priors for model based reconstruction. In: IEEE GlobalSIP, pp. 945–948 (2013)

12. Wei, Q., Bioucas-Dias, J., Dobigeon, N., Tourneret, J.-Y.: Hyperspectral and multispectral image fusion based on a sparse representation. IEEE Trans. Geosci. Remote Sens. **53**, 3658–3668 (2015)
13. Zoran, D., Weiss, Y.: From learning models of natural image patches to whole image restoration. In: IEEE-CVPR, pp. 479–486 (2011)

On Extracting the Cosmic Microwave Background from Multi-channel Measurements

Jean-François Cardoso$^{(\boxtimes)}$

Institut d'Astrophysique de Paris (UMR 7095) C.N.R.S., Paris, France
cardoso@iap.fr

Abstract. Extracting a sky map of the Cosmic Microwave Background (CMB) from multi-channel measurements can be seen as a component separation problem in a special context: only one component is of interest (the CMB) and its column in the mixing matrix and its probability distribution are known with high accuracy. The purpose of this paper is *not* to present a new algorithm but rather to discuss, on a purely theoretical basis, the impact of the statistical modeling of the components in a simple case. To do so, we analyze a model of noise-free CMB observations contaminated by coherent components. We show that the maximum likelihood estimate of the CMB in this model does not depend of the model of the contamination.

1 Introduction: Planck Data and the CMB

Observing the sky in the sub-millimeter range offers a wonderful opportunity to cosmologists: capturing the 'Cosmic Microwave Background' (CMB), the light released by the Universe only 380,000 years after the Big Bang and which has traveled almost unperturbed since then, providing us with a snapshot of the Cosmos in its in infancy. Figure 1 shows (bottom left) a full sky rendering of the early Universe: the CMB map released by the Planck ESA collaboration [1]. Such a clean map however is not directly observed: it is a combination of maps of the sky observed in several frequency channels (nine frequencies from 30 GHz to 857 GHz for the Planck satellite), all of them contaminated by *foreground* emissions due to various astrophysical processes. The top row of Fig. 1 shows (in false colors) what the microwave sky looks like in two of the Planck channels.

Getting rid of foreground emissions to access the underlying cosmologic radiation and doing so with high precision is a challenging signal processing task which has received a lot of attention. See [6] for early efforts within the Planck collaboration and [3] for the methods used in the first Planck release.

Key factors making CMB extraction possible are: (**i**) there is no occlusion of the CMB by the foregrounds, (**ii**) the instrument is well calibrated, *i.e.* its response to the CMB is well determined and (**iii**) the CMB is statistically independent of the foregrounds. Those three properties are sufficient for a simple method (the ILC, described next) to extract the CMB but it seems that an optimal processing would require a complete statistical model of the data.

© Springer International Publishing AG 2017
P. Tichavský et al. (Eds.): LVA/ICA 2017, LNCS 10169, pp. 403–413, 2017.
DOI: 10.1007/978-3-319-53547-0_38

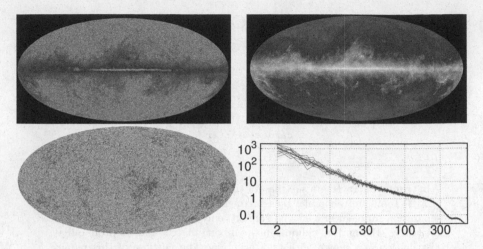

Fig. 1. Top: the sky seen by Planck in the 217 GHz and 857 GHz frequency channels. Bottom left: Planck's combination of all its 9 channels, revealing the underlying map of the Cosmic background, apparently free of contamination by foregrounds. Bottom right: angular spectra $C(\ell)$ in $(\mu K)^2$ versus angular frequency (see text).

It is a purpose of this paper to discuss the importance (or lack thereof) of statistical modeling. For doing so, we simplify as much as possible the model of the instrument (those simplifications are discussed at Sect. 4), as follows.

Denote $d_\nu(\boldsymbol{\eta})$ the sky brightness in direction $\boldsymbol{\eta}$ at frequency ν and

$$\mathbf{d}(\boldsymbol{\eta}) = [d_{\nu_1}(\boldsymbol{\eta}), \ldots, d_{\nu_n}(\boldsymbol{\eta})]^\dagger \tag{1}$$

the $n \times 1$ real vector of those measurements in n frequency channels (for Planck, $n = 9$ and the sky maps are sampled over $N \approx 5 \cdot 10^7$ equal-area pixels). Assuming for simplicity that all channels share a common axi-symmetric point spread function (psf) over the sky, one can write

$$\mathbf{d}(\boldsymbol{\eta}) = \mathbf{a}\, s(\boldsymbol{\eta}) + \mathbf{c}(\boldsymbol{\eta}) \tag{2}$$

where the scalar field $s(\boldsymbol{\eta})$ is the 'true' CMB map convolved with the common psf, and vector $\mathbf{c}(\boldsymbol{\eta})$ represents all contaminations (foregrounds and noise). Assumption **(i)** above means that we can just add $\mathbf{a}\, s(\boldsymbol{\eta})$ to $\mathbf{c}(\boldsymbol{\eta})$; assumption **(ii)** means that the $n \times 1$ vector $\mathbf{a}$ is known and assumption **(iii)** is the statistical independence of the two terms in Eq. (2).

In that context, one looks for the $n \times 1$ vector of weights $\mathbf{w}$, linearly combining the input channels into an estimate $\hat{s}(\boldsymbol{\eta}) = \mathbf{w}^\dagger \mathbf{d}(\boldsymbol{\eta})$, preserving the CMB: $\mathbf{w}^\dagger \mathbf{a} = 1$ and minimizing the residual contamination. Hence, the program is:

$$\hat{s}(\boldsymbol{\eta}) = \mathbf{w}^\dagger \mathbf{d}(\boldsymbol{\eta}) \quad \text{minimize } \mathbf{w}^\dagger \mathbf{c}(\boldsymbol{\eta}) \text{ under the constraint } \mathbf{w}^\dagger \mathbf{a} = 1 \tag{3}$$

ILC: A Simple Method for Cosmic Background Cleanup. The problem
(3) is not well posed without defining a norm for the residual contamination
$\mathbf{w}^\dagger \mathbf{c}(\boldsymbol{\eta})$ and figuring out a way to measure it. For any spherical function $f(\boldsymbol{\eta})$,
denote

$$\langle f(\boldsymbol{\eta}) \rangle_P \overset{\text{def}}{=} \frac{1}{N} \sum_{p=1}^{N} f(\boldsymbol{\eta}_p) \tag{4}$$

its 'pixel-average'. Omitting explicit dependence on $\boldsymbol{\eta}$, one notices that

$$\langle (\mathbf{w}^\dagger \mathbf{d})^2 \rangle_P = (\mathbf{w}^\dagger \mathbf{a})^2 \langle s^2 \rangle_P + 2\,\mathbf{w}^\dagger \langle cs \rangle_P\, \mathbf{w}^\dagger \mathbf{a} + \langle (\mathbf{w}^\dagger \mathbf{c})^2 \rangle_P \tag{5}$$

where the first term is constant under the unit-gain constraint $\mathbf{w}^\dagger \mathbf{a} = 1$ of problem (3) and where the last term is the average power of the residual foreground
contamination. Assume for a moment that $\langle cs \rangle_P = 0$. Then, Eq. (5) would mean
that minimizing $\langle (\mathbf{w}^\dagger \mathbf{d})^2 \rangle_P$ under the constraint $\mathbf{w}^\dagger \mathbf{a} = 1$ is equivalent to minimizing foreground contamination, as measured by $\langle (\mathbf{w}^\dagger \mathbf{c})^2 \rangle_P$. That minimization
problem has a simple explicit solution:

$$\widehat{\mathbf{w}}_P \overset{\text{def}}{=} \arg\min_{\mathbf{w}^\dagger \mathbf{a}=1} \langle (\mathbf{w}^\dagger \mathbf{d})^2 \rangle_P = \frac{\widehat{\mathbf{C}}_P^{-1} \mathbf{a}}{\mathbf{a}^\dagger \widehat{\mathbf{C}}_P^{-1} \mathbf{a}} \quad \text{where} \quad \widehat{\mathbf{C}}_P \overset{\text{def}}{=} \langle \mathbf{d}(\boldsymbol{\eta}) \mathbf{d}(\boldsymbol{\eta})^\dagger \rangle_P \tag{6}$$

known as the ILC (Internal Linear Combination) solution in CMB literature
but also is a classic array processing method known as the 'minimum variance
distortionless response' (MVDR) beamformer, for instance.

That is a strikingly simple method which only requires a good determination
of the instrumental response $\mathbf{a}$ to the signal of interest. However it is derived by
assuming that $\langle cs \rangle_P = 0$ which is true on *ensemble average* because the CMB
signal s is zero-mean and independent from the contaminants $\mathbf{c}$ but is not true as
a pixel-average over the pixels of one CMB sky. Its unavoidable non-zero value is
called 'chance correlation' between CMB and contaminants. It can be mitigated
by considering the likelihood of a simple ICA model. That, however, requires
some tools for dealing with spherical signals, which we now briefly introduce.

Harmonic Decomposition of Spherical Fields. A spherical function $X(\boldsymbol{\eta})$
can be decomposed over the doubly-indexed set $\{Y_{\ell m}(\boldsymbol{\eta}) | \ell \geq 0, |m| \leq \ell\}$ of
spherical harmonics which makes a complete orthonormal basis on the sphere:

$$X(\boldsymbol{\eta}) = \sum_{\ell \geq 0} \sum_{m=-\ell}^{m=\ell} x_{\ell m} Y_{\ell m}(\boldsymbol{\eta}) \quad \longleftrightarrow \quad x_{\ell m} = \int_{S^2} X(\boldsymbol{\eta})\, Y_{\ell m}(\boldsymbol{\eta}) \tag{7}$$

where, in practice, the integral in (7) is approximated by a sum over equal-area
pixels (and where we use *real*-valued spherical harmonics). The discrete index ℓ
is called the 'angular frequency' and is just that: an $Y_{\ell m}$ function with low ℓ has
large (angular) scale variations over the sky while all small (angular) scale details
are carried by high ℓ spherical harmonics. The spherical harmonic coefficients of
an *isotropic* random field are uncorrelated:

$$\mathrm{E}\left(x_{\ell m} x_{\ell' m'} \right) = C(\ell)\, \delta_{ll'}\, \delta_{mm'} \tag{8}$$

and their variance depends only on angular frequency: $\text{Var}(x_{\ell m}) = C(\ell)$ and the collection of variances $\{C(\ell); \ell \geq 0\}$ is called the *angular spectrum*. The natural estimate of the angular spectrum based on a single realization of random field is its *empirical angular spectrum*:

$$\widehat{C}(\ell) \overset{\text{def}}{=} \tfrac{1}{2\ell+1} \sum_{m=-\ell}^{m=+\ell} x_{\ell m}^2 \tag{9}$$

Figure 1 shows the empirical angular spectra of 10 simulated CMB maps (wiggly lines) drawn independently with a typical angular spectrum $C(\ell)$ (smooth black line, taken from the 2015 Planck results). Two important features are to be noted: **(a)** the angular spectrum $C(\ell)$ decreases sharply with ℓ: the power in CMB maps is dominated by the large angular scales and **(b)** since the sum (9) has $2\ell+1$ independent terms, the sample variance of $\widehat{C}(\ell)$ decreases as $1/(2\ell+1)$, hence the larger relative wiggliness seen at lower ℓ in the figure and the more general fact that plain pixel averages involving CMB maps do not have as many degrees of freedom as pixels entering them: the signal is dominated by large scales with relatively few degrees of freedom. See Sect. 4 for a more quantitative discussion.

2   Cleaning Coherent Contamination, ICA Style

In search of a better processing, we investigate the maximum likelihood solution based on an ICA model. This approach allows us to involve the statistical distributions of the components and to discuss their impact.

Coherent Contamination. We discuss cleaning CMB from coherent contamination, that is, we consider an ICA-like linear mixing model

$$\mathbf{d}(\boldsymbol{\eta}) = \mathbf{A}\mathbf{s}(\boldsymbol{\eta}) = [\mathbf{a}\ \mathbf{H}] \begin{bmatrix} s(\boldsymbol{\eta}) \\ \mathbf{f}(\boldsymbol{\eta}) \end{bmatrix} \quad \text{or} \quad \mathbf{D} = \mathbf{A}\mathbf{S} = [\mathbf{a}\ \mathbf{H}] \begin{bmatrix} S \\ \mathbf{F} \end{bmatrix} \tag{10}$$

where matrix $\mathbf{A}$ is $n \times n$ and has vector $\mathbf{a}$ in its first column, where $\mathbf{H}$ is an (unknown) set of $n - 1$ columns and where $\mathbf{f}(\boldsymbol{\eta})$ is a vector of $n - 1$ 'foregrounds' (the second equation concatenates all N pixels to form matrices with N columns).

The contamination in this model is fully coherent in the sense that it can be completely nulled out. Indeed, if $\bar{\mathbf{w}}$ is a vector orthogonal to all columns of $\mathbf{H}$, *i.e.* $\bar{\mathbf{w}}^\dagger \mathbf{H} = 0$, then all foreground contamination vanishes in $\bar{\mathbf{w}}^\dagger \mathbf{D}$. If this vector is normalized into $\mathbf{w} = \bar{\mathbf{w}}/(\bar{\mathbf{w}}^\dagger \mathbf{a})$, then $\mathbf{w}^\dagger \mathbf{D} = S$ *i.e.*, the CMB is perfectly recovered[1]. In other words, perfect CMB recovery only requires the determination of the 'foreground subspace', that is, the linear *subspace* spanned by the columns of $\mathbf{H}$ and not matrix $\mathbf{H}$ itself. We take advantage of this circumstance to introduce a pre-processing step which highlights the impact of knowing the column $\mathbf{a}$ associated with the signal of interest.

---

[1] Of course, such a $\mathbf{w}^\dagger$ would then be nothing else than the first row of $\mathbf{A}^{-1}$.

Preprocessing or Re-parameterization. In all that follows, we pick up a fixed, *arbitrary* $n \times (n-1)$ matrix $\mathbf{T}$ such that $[\mathbf{a}\,\mathbf{T}]$ is invertible and we define a 'pre-processing matrix' $\mathbf{P}$ and pre-processed data $\mathbf{PD}$ as

$$\mathbf{P} \stackrel{\text{def}}{=} [\mathbf{a}\,\mathbf{T}]^{-1}, \qquad \begin{bmatrix} Y \\ \mathbf{G} \end{bmatrix} = \begin{bmatrix} y_1 \cdots y_N \\ \mathbf{g}_1 \cdots \mathbf{g}_N \end{bmatrix} \stackrel{\text{def}}{=} \mathbf{PD} \tag{11}$$

with Y and $\mathbf{G}$ of sizes $1 \times N$ and $(n-1) \times N$ respectively[2].

A simple algebraic manipulation shows that $\mathbf{A}^{-1}$ can be written as

$$\mathbf{A}^{-1} = [\mathbf{a}\,\mathbf{H}]^{-1} = \begin{bmatrix} 1 & -\mathbf{v}^\dagger \\ \mathbf{0}_{(n-1)\times 1} & \mathbf{K}^{-1} \end{bmatrix} \mathbf{P}. \tag{12}$$

Factorization (12) turns the constraint "matrix $\mathbf{A}$ is unknown but for its first column" (which leaves $n(n-1)$ unknowns in $\mathbf{H}$) into the constraint "matrix $\mathbf{A}^{-1}$ has the form (12) with $(n-1)$ unknowns in vector $\mathbf{v}$ and $(n-1)^2$ unknowns in matrix $\mathbf{K}$" which, of course, amounts to the same total of $n(n-1)$. However, only the first row of $\mathbf{A}^{-1}$ matters since we are only interested in recovering the CMB component, hence only $\mathbf{v}$ matters while the matrix parameter $\mathbf{K}$ appears as a pure nuisance parameter. Should it still be estimated? The answer is 'no' because some algebra turns the model (10) for $\mathbf{D}$ turned into the model

$$\begin{bmatrix} Y \\ \mathbf{G} \end{bmatrix} = \begin{bmatrix} S + \mathbf{v}^\dagger \mathbf{KF} \\ \mathbf{KF} \end{bmatrix} \tag{13}$$

for the pre-processed data $\mathbf{PD}$. Hence, the block $\mathbf{G} = \mathbf{KF}$ contains some mixture of the foregrounds $\mathbf{F}$ while the first row Y contains the signal of interest S contaminated by a linear combination (with weights $\mathbf{v}$) of the same mixture $\mathbf{KF}$.

In other words, we can consider only the problem $Y = S + \mathbf{v}^\dagger \mathbf{G}$ where $\mathbf{G}$ is available *deterministically* after pre-processing, leaving the $n-1$ free parameters in $\mathbf{v}$ to be determined. This, of course, is related to the previously discussed fact that only the column space of $\mathbf{H}$ is needed for CMB recovery. Whatever happens *within* that subspace is irrelevant to CMB recovery.

Generic Likelihood. Let us build a likelihood for data $\mathbf{D}$ as a function of the mixing matrix $\mathbf{A}$, assuming only the statistical independence between the signal of interest S and the contaminants $\mathbf{F}$ expressed by

$$p_{\mathbf{S}}(\mathbf{S}) = P_S(S) \cdot P_{\mathbf{F}}(\mathbf{F}). \tag{14}$$

The N samples of $\mathbf{D}$ being related to $\mathbf{S}$ by a linear transform $\mathbf{A}$, one has

$$p_{\mathbf{D}}(\mathbf{D}|\mathbf{A}) = p_{\mathbf{S}}(\mathbf{A}^{-1}\mathbf{D})/|\det(\mathbf{A})|^N \tag{15}$$

---

[2] The pre-processing $\mathbf{P}$ is only introduced here as a 'mathematical device' but a similar idea is actually implemented in the SEVEM algorithm for CMB extraction [3].

Equation (12) yields $\det \mathbf{A} = \det \mathbf{K} / \det \mathbf{P}$ and $\mathbf{A}^{-1}\mathbf{D} = \begin{bmatrix} Y - \mathbf{v}^\dagger \mathbf{G} \\ \mathbf{K}^{-1}\mathbf{G} \end{bmatrix}$ so that:

$$p_\mathbf{D}(\mathbf{D}|\mathbf{A}) = p_S(Y - \mathbf{v}^\dagger \mathbf{G}) \, p_\mathbf{F}(\mathbf{K}^{-1}\mathbf{G}) \det(\mathbf{K})^{-N} \, \det(\mathbf{P})^N. \tag{16}$$

As expected from Eq. (13), the likelihood for $\mathbf{A}$ is now factored with one factor depending only on $\mathbf{v}$ and on the observable quantity $\mathbf{G}$. Thus the maximum likelihood solution for the signal of interest is

$$\widehat{S}_{\mathrm{ML}} = Y - \hat{\mathbf{v}}^\dagger \mathbf{G} \qquad \hat{\mathbf{v}} = \arg\max_{\mathbf{v}} \, p_S(Y - \mathbf{v}^\dagger \mathbf{G}). \tag{17}$$

Again, it means that, for removing noise-free coherent contaminations from a signal S with known mixing $\mathbf{a}$, the maximum likelihood solution depends *only* on the model $p_S(\cdot)$ for the signal S and *not* on the contamination model $p_\mathbf{F}(\cdot)$.

3   Maximum Likelihood Solutions

The maximum likelihood (ML) solution $\hat{\mathbf{v}} = \arg\max_{\mathbf{v}} p_S(Y - \mathbf{v}^\dagger \mathbf{G})$ can be found or characterized in some simple models of interest which we now examine.

If the Signal of Interest is Uncorrelated Gaussian. Let us start with the simplest case when S is modeled as a set of zero-mean uncorrelated Gaussian pixels of variance σ^2. Then

$$\log P_S(S) = -1/2 \sum_{p=1}^N s_p^2 / \sigma^2 + \mathrm{cst} \tag{18}$$

so that the solution of (17) is obtained at the minimum of

$$\langle (y - \mathbf{v}^\dagger \mathbf{g})^2 \rangle_P = \begin{bmatrix} 1 \\ -\mathbf{v} \end{bmatrix}^\dagger \widehat{\mathbf{C}}_{PP} \begin{bmatrix} 1 \\ -\mathbf{v} \end{bmatrix} \text{ where } \widehat{\mathbf{C}}_{PP} \overset{\text{def}}{=} \langle \begin{bmatrix} y \\ \mathbf{g} \end{bmatrix} \begin{bmatrix} y \\ \mathbf{g} \end{bmatrix}^\dagger \rangle_P. \tag{19}$$

The solution for that basic regression problem is:

$$\hat{\mathbf{v}}_P = \widehat{C}_{\mathbf{g}\mathbf{g}}^{-1} \widehat{C}_{\mathbf{g}y} \text{ , with the matrix partitioning } \widehat{\mathbf{C}}_{PP} = \begin{bmatrix} \widehat{C}_{yy} & \widehat{C}_{y\mathbf{g}} \\ \widehat{C}_{\mathbf{g}y} & \widehat{C}_{\mathbf{g}\mathbf{g}} \end{bmatrix}. \tag{20}$$

It is interesting to express that solution in terms of the covariance $\widehat{\mathbf{C}}_P$ of the signals before pre-processing. Chaining the pre-processor $\mathbf{P}$ and the regression $[1 - \hat{\mathbf{v}}_P^\dagger]$ yields (simple calculations omitted):

$$[1 - \hat{\mathbf{v}}_P^\dagger]\mathbf{P} = \frac{\mathbf{a}^\dagger \widehat{\mathbf{C}}_P^{-1}}{\mathbf{a}^\dagger \widehat{\mathbf{C}}_P^{-1}\mathbf{a}} = \hat{\mathbf{w}}_P^\dagger. \tag{21}$$

Hence, it appears (unsurprisingly) that the ILC filter $\hat{\mathbf{w}}_P^\dagger$ of Eq. (6) is the ML solution for the model of uncorrelated Gaussian pixels.

If the Signal of Interest is Non Gaussian i.id. If the samples are modeled as independent with the same marginal density $p(s)$, then $\log P_S(S) = \sum_p \log p(s_p)$ and the likelihood is maximum at a point where its gradient vanishes. This is:

$$\langle \psi(y - \widehat{\mathbf{v}}^\dagger \mathbf{g}) \, \mathbf{g} \rangle_P = 0 \quad \text{with} \quad \psi(s) \overset{\text{def}}{=} -p'(s)/p(s) \tag{22}$$

which gives $n - 1$ conditions of decorrelation between the estimated component $\hat{s} = y - \widehat{\mathbf{v}}^\dagger \mathbf{g}$ passed through the score function ψ and all the other $n - 1$ components collected in vector $\mathbf{g}$. The solution of Eq. (22) improves on $\widehat{\mathbf{w}}_P{}^\dagger$ if the signal of interest is non Gaussian because otherwise, ψ is a linear function and the solution of (22) is easily seen to be identical to $\widehat{\mathbf{w}}_P{}^\dagger$.

If the Signal of Interest is a Gaussian Stationary Field. This is our main focus: the CMB is, to an excellent approximation, the realization of a zero-mean isotropic Gaussian random field. Then, not only are the spherical harmonic coefficients $s_{\ell m}$ of a CMB map S uncorrelated (as in Eq. (8)), they are also normally distributed: $s_{\ell m} \sim \mathcal{N}(0, C(\ell))$. Since the map and its coefficients are linearly related, the log pdf for a map S with coefficients $s_{\ell m}$ is

$$\log P_S(S) = -\tfrac{1}{2} \sum_\ell \sum_m s_{\ell m}^2 / C(\ell) + \text{cst} \tag{23}$$

which is like Eq. (18) with the constant variance σ^2 replaced by the frequency dependent $C(\ell)$ factor. Hence, a simple yet accurate likelihood is readily available in harmonic space. So, we move there, where the data model (10) now reads

$$\mathbf{d}_{\ell m} = \mathbf{A} \mathbf{s}_{\ell m} = [\mathbf{a} \ \mathbf{H}] \begin{bmatrix} s_{\ell m} \\ \mathbf{f}_{\ell m} \end{bmatrix}, \tag{24}$$

involving $\sum_{\ell \leq \ell_{\max}} (2\ell + 1)$ harmonic coefficients instead of N pixels.

The previous analysis, done in pixel space and leading to (20) and (21), can be repeated 'as is' in harmonic space, with σ^2 replaced by the variance of the $s_{\ell m}$ coefficients, namely the angular power spectrum C_ℓ. This straightforward adaptation results in an ML estimate $\widehat{\mathbf{w}}_H$ obtained for the CMB model (23) which takes again the form of an ILC filter as in (6), except that the covariance matrix $\widehat{\mathbf{C}}_H$ is computed in the harmonic domain with an inverse-variance weighting:

$$\widehat{\mathbf{w}}_H = \frac{\widehat{\mathbf{C}}_H{}^{-1} \mathbf{a}}{\mathbf{a}^\dagger \widehat{\mathbf{C}}_H{}^{-1} \mathbf{a}} \quad \text{with} \quad \widehat{\mathbf{C}}_H = \sum_{\ell \leq \ell_{\max}} \sum_{m=-\ell}^{m=+\ell} \mathbf{d}_{\ell m} \mathbf{d}_{\ell m}{}^\dagger / C_\ell. \tag{25}$$

This solution is used for the coarsest scale of the wavelet-based NILC method [4].

4   Discussion

Pixel Space Versus Harmonic Space and Chance Correlation. Let us compare the two versions of the ILC filter obtained in pixel space $\widehat{\mathbf{w}}_P$ and in harmonic space $\widehat{\mathbf{w}}_H$ by comparing the empirical covariance matrices $\widehat{\mathbf{C}}_P$ and $\widehat{\mathbf{C}}_H$. We have

$$\widehat{\mathbf{C}}_P \stackrel{\text{def}}{=} \frac{1}{N} \sum_{p=1}^{N} \mathbf{d}(\boldsymbol{\eta}_p) \mathbf{d}(\boldsymbol{\eta}_p)^\dagger \approx \frac{1}{4\pi} \int_{S^2} \mathbf{d}(\boldsymbol{\eta}) \mathbf{d}(\boldsymbol{\eta})^\dagger = \sum_{\ell} \sum_{m=-\ell}^{m=+\ell} \mathbf{d}_{\ell m} \mathbf{d}_{\ell m}{}^\dagger \quad (26)$$

because the sum over equal area pixels is very close to an integral on the sphere and the last identity is just Parseval on the sphere. Hence, the pixel-average $\widehat{\mathbf{C}}_P$ is a plain average of $\mathbf{d}_{\ell m} \mathbf{d}_{\ell m}{}^\dagger$ terms while $\widehat{\mathbf{C}}_H$ is a weighted average (25) of them, with weights $1/C(\ell)$. For a flat spectrum $C(\ell) = \text{cst}$, we would have $\widehat{\mathbf{w}}_P = \widehat{\mathbf{w}}_H$ but a flat spectrum corresponds to uncorrelated pixels, so we find (again) that the pixel-based ILC is the ML solution in that case. Otherwise, for a correlated field, the two solutions are different. They are *very* different for a field like the CMB with a rapidly decreasing $C(\ell)$. At the end of Sect. 1, we already pointed out the problem of computing correlations in pixel-space for random fields dominated by their large scales, like the CMB. Now, the pixel-average in $\widehat{\mathbf{C}}_P$ can also be expressed as a sum in harmonic space (26) of uncorrelated contributions. The sum includes many terms but only a handful of them dominate because of the shape of the CMB spectrum. The maximum likelihood principle, in its asymptotic wisdom, tells us very clearly how to mitigate this problem: down-weight the summands. Indeed, because of the $1/C(\ell)$ factor in the sum (25), any CMB coefficient $s_{\ell m}$ is affected by a factor $1/\sqrt{C(\ell)}$ so that all $s_{\ell m}/\sqrt{C(\ell)}$ have the same unit variance at all angular frequencies.

The Angular Spectrum as a Statistical Weight. The harmonic version of the ILC is derived assuming that the angular spectrum of the CMB $C(\ell)$ is known in advance. This seems in contradiction with the fact that Planck and other instruments are built to *measure* that spectrum in the first place! Recall however that **(i)** the large scales (low ℓ) of the CMB have already been measured by previous experiments and, more importantly, **(ii)** the $C(\ell)$ only enters as a *weighting* factor in (25) with the implicit role of equalizing the variance of $s_{\ell m}$ versus the frequency ℓ. Even an approximate equalization goes a long way towards counter-balancing the fast-decreasing CMB angular spectrum and would do a plausible job of reducing chance correlation.

Sparsity and CMB Extraction. Blind separation methods based on *sparsity* have been advocated for CMB extraction [2] on the basis that CMB and/or foregrounds are sparse in some transform domain. For one thing, we have seen in Sect. 3 that—at least in the simplified model analyzed in this paper—the statistical distribution of the foregrounds does *not* enter the part of the likelihood which controls CMB extraction: one only needs to take care of the distribution of the CMB. How to do it properly—according to the maximum likelihood principle—is straightforward, as explained above: do it in the harmonic domain. The spherical harmonic coefficients $\{s_{\ell m}\}$ as a whole do have a sparse distribution because $C(\ell)$ varies a lot with ℓ but the variance of any $s_{\ell m}$ is known in advance (it is $C(\ell)$) and its distribution, knowing that, is Gaussian, so that one can use a proper likelihood for it without resorting to sparsity arguments.

Invariance and Chance Correlation. Assume that the model actually holds, so that, after pre-processing, Y does contain the signal of interest plus some linear combination with 'true' weights $\mathbf{v}_\star$ of $n-1$ observed foregrounds $\mathbf{G}$:

$$Y = S + \mathbf{v}_\star{}^\dagger \mathbf{G} \tag{27}$$

The estimated signal is $\widehat{S} = Y - \widehat{\mathbf{v}}^\dagger \mathbf{G}$ with $\widehat{\mathbf{v}}^\dagger = \widehat{C}_{y\mathbf{g}}\widehat{C}_{\mathbf{gg}}{}^{-1}$ where $\widehat{C}_{y\mathbf{g}}$ and $\widehat{C}_{\mathbf{gg}}$ could be sub-blocks of either $\widehat{\mathbf{C}}_{PP}$ or its weighted harmonic version discussed above. In both cases, $\widehat{C}_{y\mathbf{g}} = \widehat{C}_{s\mathbf{g}} + \mathbf{v}_\star{}^\dagger \widehat{C}_{\mathbf{gg}}$ which, combined with (27), yields

$$\widehat{S} - S = -\widehat{C}_{s\mathbf{g}}\widehat{C}_{\mathbf{gg}}{}^{-1}\mathbf{G}. \tag{28}$$

It shows that the error strictly vanishes with the chance correlation $\widehat{C}_{s\mathbf{g}}$ and also that it is strictly invariant with respect to any invertible mixing or rescaling of the observed mixed foregrounds, since that expression does not change if $\mathbf{G}$ is changed into $\mathbf{K}\mathbf{G}$ for any invertible matrix $\mathbf{K}$.

Summary. In summary, this paper addressed a special ICA problem: how to extract one source, the signal of interest (SoI), from a mixture when its column $\mathbf{a}$ in the mixing matrix is known. The ICA problem simplifies so much in that case that it can be solved with a straightforward 2nd-order technique (ILC) which, however, is adversely affected by long range correlations resulting from a fast decaying power spectrum. Looking for a statistically efficient method, we turned to the maximum likelihood approach. We showed that only the pdf of the SoI needs to be used to determine its ML estimate.

The standard (pixel-based) ILC is shown to be the ML estimate when the pdf of the SoI is modeled as i.i.d. Gaussian. For Gaussian stationary processes (on the sphere, the line, the plane), the Fourier coefficients are Gaussian and independently distributed but their variance depends on the power spectrum. When the latter is know (or can be approximated, even roughly), the ML solution retains the ILC structure but the pixel-based covariance matrix used in standard ILC has to be replaced by an inverse-spectrum weighted version which gives the same statistical weight to all the spherical harmonic coefficients $s_{\ell m}$ of the CMB.

5   Conclusion: What's the Point?

This work is motivated by a real problem —extracting the Cosmic Microwave Background from the 9 frequency channels of the Planck mission— but dealing with it in all its complexity is out of the scope of this paper, of course. Instead, this paper provided an analysis based on a simplistic model. Even though this model captures some realistic features of the actual problem (the pdf of the

CMB, the good determination of its signature by vector $\mathbf{a}$, the high coherence of the foregrounds), some other important features were left out. So, what's the point? In this concluding section, we would like to explain why the findings of this paper are relevant to an actual CMB processing, the SMICA method [3,5].

Because the SNR varies very much with frequencies ν and ℓ, it is in fact necessary to let the combination weights depend on ℓ one way or another. One possibility is to compute ℓ-dependent weights $\mathbf{w}_\ell$ by formula (25) with $\widehat{\mathbf{C}}_H$ replaced by (a smoothed version of) $\widehat{\mathbf{C}}_\ell \overset{\text{def}}{=} \sum_m \mathbf{d}_{\ell m}\mathbf{d}_{\ell m}{}^\dagger/(2\ell+1)$. At high ℓ, many $\mathbf{d}_{\ell m}$ are available so that such an 'harmonic ILC' [7] would suffer negligibly from chance correlation and would perform well. At low-ℓ, as stressed above, fewer coefficients are available and this solution would not work. The SMICA method forms ILC weights $\mathbf{w}_\ell = \mathbf{C}_\ell^{-1}\mathbf{a}/\mathbf{a}^\dagger\mathbf{C}_\ell^{-1}\mathbf{a}$ with a spectral covariance matrix parameterized as $\mathbf{C}_\ell = \begin{bmatrix} \mathbf{a} & \mathbf{H} \end{bmatrix} \begin{bmatrix} C_\ell & 0 \\ 0 & \mathbf{\Sigma}_\ell \end{bmatrix} \begin{bmatrix} \mathbf{a} & \mathbf{H} \end{bmatrix}^\dagger + \mathbf{N}_\ell$ where $\mathbf{N}_\ell$ is a (free) diagonal matrix accounting for the noise and where the (free) positive matrix $\mathbf{\Sigma}_\ell$ is the spectral covariance matrix of the foregrounds. The parameters $\mathbf{H}, C_\ell, \mathbf{\Sigma}_\ell$ and $\mathrm{diag}(\mathbf{N}_\ell)$ are determined by fitting this model to $\widehat{\mathbf{C}}_\ell$ jointly over a wide range of ℓ. In that way, the foreground subspace $(\mathrm{Span}(\mathbf{H}))$ is determined from many $\mathbf{d}_{\ell m}$ fighting against chance correlation (and noise is properly taken into account). The fit is by maximum likelihood and the likelihood is made tractable by modeling *all components* as Gaussian isotropic fields. This is an excellent statistical description of the CMB but what about the foregrounds? They are clearly neither Gaussian nor isotropic so the question is: how much accuracy is lost by retaining only second-order spectral information as SMICA implicitly does? Basically, there is no general answer to that question because we lack a statistical model for the foreground distribution in the first place. However, in the limit of vanishing noise (which is almost the case at low ℓ for Planck), since SMICA boils down to a noise-free ML solution for an isotropic Gaussian likelihood, the choice of a statistical model for the foregrounds does *not* affect the CMB solution, as we have shown in Sect. 2. Since we do not expect discontinuity in the impact of the foreground model on the CMB estimate when noise increases, it seems plausible that the impact the foreground model, which is irrelevant without noise, should remain weak when the noise increases. In other words, SMICA may be using a clearly wrong foreground model without paying a large penalty.

References

1. The Planck mission of ESA. http://www.cosmos.esa.int/web/planck/home
2. Abrial, P., Moudden, Y., Starck, J.-L., Fadili, J., Delabrouille, J., Nguyen, M.K.: CMB data analysis and sparsity. Stat. Methodol. **5**(4), 289–298 (2008)
3. The Planck collaboration: Planck 2013 results. XII. Diffuse component separation. Astronomy and Astrophysics, 571, November 2014
4. Delabrouille, J., et al.: A full sky, low foreground, high resolution CMB map from WMAP. A&A **493**(3), 835–857 (2009). arXiv:0807.0773

5. Cardoso, J.-F., et al.: Component separation with flexible models. application to the separation of astrophysical emissions. IEEE J. Sel. Top. Signal Process. **2**, 735–746 (2008). arXiv:0803.1814
6. Leach, S.M., et al.: Component separation methods for the Planck mission. Astron. Astrophys. **491**, 597–615 (2008). arXiv:0805.0269
7. Tegmark, M., de Oliveira-Costa, A., Hamilton, A.J.: High resolution foreground cleaned CMB map from WMAP. Phys. Rev. D, 68(12):123523-+, December 2003

ICA Theory and Applications

Kernel-Based NPLS for Continuous Trajectory Decoding from ECoG Data for BCI Applications

Sarah Engel[1,2], Tetiana Aksenova[1,2], and Andrey Eliseyev[1,2(✉)]

[1] Université Grenoble Alpes, 38000 Grenoble, France
sarah.engel@student.uni-tuebingen.de
[2] CEA, LETI, CLINATEC, MINATEC Campus, 38000 Grenoble, France
tetiana.aksenova@cea.fr, eliseyev.andrey@gmail.com

Abstract. In this paper, nonlinearity is introduced to linear neural activity decoders to improve continuous hand trajectory prediction for Brain-Computer Interface systems. For decoding the high-dimensional data-tensor, a kernel regression was coupled with multilinear PLS (NPLS). Two ways to introduce nonlinearity were studied: a generalized linear model with kernel link function and kernel regression in the NPLS latent variables space (inside or outside the NPLS iterations). The efficiency of these approaches was tested on the publically available database of the simultaneous recordings of three-dimensional hand trajectories and epidural electrocorticogram (ECoG) signals of a Japanese macaque. Compared to linear methods, nonlinearity did not significantly improve the prediction accuracy but did significantly improve the smoothness of the prediction.

Keywords: Brain-computer interface (BCI) · Latent variables · Multilinear partial least squares (NPLS) · Tensors · Kernel regression

1 Introduction

A Brain-Computer Interface (BCI) is a way to transfer brain neural activity into external devices. The main goal of a motor-related BCI is to improve the quality of life of people with severe motor disabilities [1]. For instance, BCI systems aim to provide paralyzed people with mental control of such external devices as robotic arms or exoskeletons [2].

Different brain activity recording methods are currently proposed for use in BCIs. The proposed non-invasive methods include electroencephalography (EEG), magnetencephalography (MEG), and functional magnet resonance imaging (fMEG). The invasive methods include electrocorticography (ECoG) and microelectrode array (ME) recordings. In the invasive approaches, the electrodes are implanted under the cranial bone; non-invasive methods are safe, but invasive ones provide better signal quality. ECoG unites minimal invasiveness with high spatial and temporal signal resolution [3]. Moreover, chronic ECoG recording implants have been recently reported [4]. ECoG data allow continuous recording, whereas non-invasive-based BCIs

© Springer International Publishing AG 2017
P. Tichavský et al. (Eds.): LVA/ICA 2017, LNCS 10169, pp. 417–426, 2017.
DOI: 10.1007/978-3-319-53547-0_39

use mainly binary or multiclass commands [5]. In view of the prospect of motor-related BCI applications aimed at recovering, to some extent, mobility for people with severe motor dysfunctions, this paper considers the problem of decoding the continuous 3D-trajectories of the hand from the ECoG neural activity recordings.

Multi-way analysis [6] was recently reported to be an efficient tool for the identification of a decoder [7] of the hand's trajectory from ECoG data tensor. Tensor (multi-way array) data representation allows simultaneous signal processing in several domains (e.g., temporal, spatial, and frequency). The disadvantage of multiway data analysis is that it generally increases the dimensionality of the task, and dimensionality considerably exceeds the number of observations in the training data set [8]. The Multi-Way Partial Least Squares (NPLS) [6] approach was successfully applied in the past to overcome these problems [9, 10].

Linear models are mainly applied in BCI systems due to their simplicity and robustness; however, linear models are limited to reflecting the complex dependencies in the data. To improve the performance and accuracy of BCI systems, nonlinear techniques, such as kernel functions, neural networks, and genetic algorithms, were applied [7, 8, 11]. At the same time, high-dimensionality features extracted from ECoG data restrict the application of nonlinear methods.

In the current study, we propose a combination of the tensor-based linear (NPLS) and nonlinear (kernel regression) methods to unite the advantages of both approaches. Kernel regression was chosen for nonlinear mapping since it allows the modelling of the nonlinearity of an unknown structure [12]. The algorithm for kernelization of NPLS was reported in [7]; however, direct application of kernels in high-dimensional problems brings unstable solutions [13]. In the current paper, different ways of combining the linear NPLS projection and kernel-based nonlinear mapping are considered.

Accuracy and smoothness are among the crucial requirements for predicting trajectories in BCI studies; inaccurate or unsmooth prediction of the movements could considerably disturb the subject. The current study aims to investigate and compare decoding methods according to their prediction accuracy and smoothness.

2   Methods

2.1   Generic N-way PLS

The high dimensionality of the BCI data complicates the direct application of generic linear regression methods. Partial Least Squares (PLS) regression [14] is a linear approach that is particularly suited for high dimensions of observation. Generic PLS identifies a linear relationship between input $\mathbf{x} \in \mathbb{R}^n$ and output $\mathbf{y} \in \mathbb{R}^m$ variables. The model is built by an iterative projection of the matrixes of N observations $\mathbf{X} \in \mathbb{R}^{N \times n}$ and $\mathbf{Y} \in \mathbb{R}^{N \times m}$ into relatively low-dimensional spaces (spaces of latent variables) in the way that their maximum variation is explained simultaneously.

NPLS is a generalization of the ordinary PLS to the case of tensor [15] input/output data. Similar to PLS, NPLS iteratively projects the tensors of observations $\underline{\mathbf{X}} \in \mathbb{R}^{N \times I_1 \times \ldots \times I_n}$ and $\underline{\mathbf{Y}} \in \mathbb{R}^{N \times J_1 \times \ldots \times J_m}$ into the space of latent variables. A linear relation between the latent variables $\mathbf{t}_f \in \mathbb{R}^N$ and $\mathbf{u}_f \in \mathbb{R}^N$ is built.

$$\underline{\mathbf{X}} = \sum_{f=1}^{F} \mathbf{t}_f \mathbf{w}_1^f \ldots \mathbf{w}_n^f + \underline{\mathbf{E}}, \quad \underline{\mathbf{Y}} = \sum_{f=1}^{F} \mathbf{u}_f \mathbf{q}_1^f \ldots \mathbf{q}_m^f + \underline{\mathbf{F}}, \tag{1}$$
$$\mathbf{u}_f = \mathbf{T}_f \mathbf{b}_f.$$

Here $\mathbf{T}_f = [\mathbf{t}_1, \ldots, \mathbf{t}_f] \in \mathbb{R}^{N \times f}$ and $\mathbf{U}_f = [\mathbf{u}_1, \ldots, \mathbf{u}_f] \in \mathbb{R}^{N \times f}$ are the matrixes of the latent variables after f iterations, $\mathbf{w}_i^f \in \mathbb{R}^{I_i}$ and $\mathbf{q}_j^f \in \mathbb{R}^{J_j}$ are projection vectors, $\mathbf{b}_f$ is the vector of linear coefficients, $\underline{\mathbf{E}}$ and $\underline{\mathbf{F}}$ are residual tensors, and F is the number of factors. The procedure of deflation is applied to the current tensors $\underline{\mathbf{X}}_f$ and $\underline{\mathbf{Y}}_f$ on each iteration f.

Both PLS [16, 17] and NPLS-based methods were reported as efficient in BCIs [9, 10].

2.2 NPLS with Kernel-Based Nonlinear Mapping

Despite a number of advantages, linear models cannot fully reflect the complex nonlinear relations in the analyzed data. The nonlinearity can be introduced to the BCI model in different manners, such as kernel functions [7], neural networks [18], genetic algorithms [19], and generalized additive models [11].

Nonparametric kernel estimators are widely used in many areas to model the nonlinearity of unknown structures [12]. For the kernel regression, the estimation of the conditional mean curve $g(\mathbf{x}) = E(\mathbf{y}|\mathbf{x})$ can be identified as $\hat{g}(\mathbf{x}) = \int \hat{f}(\mathbf{x}, \mathbf{y}) \mathbf{y} d\mathbf{y} / \hat{f}(\mathbf{x})$, where $\hat{f}(\mathbf{x}, \mathbf{y})$ and $\hat{f}(\mathbf{x})$ are estimations of the joint and marginal density functions, respectively. For a given data set $\{\mathbf{x}_i, \mathbf{y}_i\}_{i=1}^{N}$, the Nadaraya-Watson [12] kernel estimators could be used instead of the density functions:

$$\hat{g}_h(\mathbf{x}) = \frac{\sum_{i=1}^{N} \mathbf{y}_i K\left(\frac{\mathbf{x} - \mathbf{x}_i}{h}\right)}{\sum_{i=1}^{N} K\left(\frac{\mathbf{x} - \mathbf{x}_i}{h}\right)},$$

where $K(\cdot)$ is a kernel function and h is a smoothing parameter of the kernel estimator. Following [8], in this paper, Gaussian function is applied as the kernel function.

Application of the nonlinear methods is limited by the dimensionality of the data. In this paper, NPLS is used as a basic method for dimension reduction. Different ways of introducing nonlinearity to the PLS-based algorithms are considered in [20]; in this paper, the nonlinearity is introduced to the NPLS algorithm on each iteration, called the Inner Kernel NPLS (IK-NPLS) algorithm, or after all iterations, called the Outer Kernel NPLS (OK-NPLS) algorithm. In addition, similar to Generalized Linear PLS (GL-PLS) [8], nonlinear mapping can be applied to the result of NPLS (GL-NPLS).

IK-NPLS. For introducing the kernel regression to the NPLS algorithm, the linear inner relation (1) is substituted with

$$\mathbf{u}_f = g_f^{\mathrm{IK}}(\mathbf{T}_f).$$

Here $f = 1, \ldots, F$ is the current NPLS iteration and the nonlinear function $g_f^{IK} : \mathbb{R}^f \to \mathbb{R}$ is identified using kernel regression and applied row-wise to the matrix $\mathbf{T}_f$: $u_f^i = g_f^{IK}\left(t_1^i, \ldots, t_f^i\right)$, $i = 1, \ldots, N$. Then, the standard NPLS procedure of deflation is applied to the current tensors $\underline{\mathbf{X}}_f$ and $\underline{\mathbf{Y}}_f$, and the next iteration is carried out until the number of factors F is reached.

In the IK-NPLS, the nonlinear functions are applied in each iteration and the latent variables $\mathbf{T}_F$ and $\hat{\underline{\mathbf{Y}}}$ are nonlinear functions of $\underline{\mathbf{X}}$.

OK-NPLS. Contrary to IK-NPLS, in OK-NPLS, the nonlinearity is applied only when the whole set of latent variables is identified. Namely, after F iterations, the standard NPLS procedure gives a set of latent variables $\mathbf{T}_F$. Then, the output tensor $\hat{\underline{\mathbf{Y}}}$ is defined as row-wise application of a nonlinear function $g^{OK} : \mathbb{R}^F \to \mathbb{R}^{J_1 \times \ldots \times J_m}$ to the latent variables matrix $\mathbf{T}_F$:

$$\hat{\underline{\mathbf{Y}}} = g^{OK}(\mathbf{T}_F).$$

Latent variables are linear functions of the inputs; nonlinear mapping is applied once to the whole set of the latent variables represented by $\mathbf{T}_F$.

GL-NPLS. GL-NPLS [8] applies a nonlinear link function $g^{GL}(\cdot)$ to the prediction obtained by the NPLS methods. Contrary to OK-NPLS, GL-NPLS nonlinearly maps a set of k previous NPLS predictions, instead of the matrix $\mathbf{T}_F$ of the current latent variables. Thus, the nonlinear function $g^{GL} : \mathbb{R}^{k \times J_1 \times \ldots \times J_m} \to \mathbb{R}^{J_1 \times \ldots \times J_m}$ is defined as:

$$\hat{\underline{\mathbf{y}}}_i = g^{GL}\left(\hat{\underline{\mathbf{y}}}_i^{NPLS}, \hat{\underline{\mathbf{y}}}_{i-1}^{NPLS}, \ldots, \hat{\underline{\mathbf{y}}}_{i-k+1}^{NPLS}\right), \quad \hat{\underline{\mathbf{y}}}_i^{NPLS} = NPLS(\underline{\mathbf{x}}_i).$$

As shown in [8], the Generalized Linear model united with PLS-family methods could significantly improve the smoothness and robustness of the predictions, by means of taking into account not only the current epoch, but also considering the dynamics of the predictions in time.

2.3 Criteria

Different criteria were used in the BCI experiments [21] to evaluate algorithm performance. In this work, a set of criteria proposed in [8] were applied. They allow the assessment of the distance between the original and predicted trajectories in ℓ_1 and ℓ_2 metrics as well as the smoothness of the prediction.

The Pearson Correlation is the most common evaluation criterion $r = \text{corr}(\mathbf{y}, \hat{\mathbf{y}})$, where $\mathbf{y} \in \mathbb{R}^N$ and $\hat{\mathbf{y}} \in \mathbb{R}^N$ are observed and predicted vectors. The Root Mean Square Error (RMSE) criterion is defined by the ℓ_2-norm: $\text{RMSE} = \|\mathbf{y} - \hat{\mathbf{y}}\|_2 / \|\mathbf{y} - \bar{\mathbf{y}}\|_2$, where $\bar{\mathbf{y}}$ is a mean value. Since RMSE is known to be sensitive to outliers, Mean Absolute Error (MAE), based on the ℓ_1-norm, is also considered: $\text{MAE} = \|\mathbf{y} - \hat{\mathbf{y}}\|_1 / \|\mathbf{y} - \bar{\mathbf{y}}\|_1$.

The above criteria do not reflect the smoothness of the predicted trajectory. At the same time, the smoothness of prediction is of great importance for the real-life BCI application. Mean Absolute Differential Error (MADE) is proposed to characterize smoothness [8]. Based on the ℓ_1-norm, it is less sensitive to outliers: $\text{MADE} = \left\| \mathbf{y}' - \hat{\mathbf{y}}' \right\|_1 / \left\| \mathbf{y}' - \overline{\mathbf{y}'} \right\|_1$, where $\mathbf{y}'$ and $\hat{\mathbf{y}}'$ are the first derivatives of the observed and predicted output variables, respectively.

3 Data Description

For comparison and evaluation of the investigated approaches, the problem of reconstruction of the 3D trajectories of the right hand of a non-human primate from its ECoG signals was considered. Ten recordings from a publically available dataset (http://neurotycho.org/data), corresponding to the Japanese macaque, denoted as 'B', were used.

In the experiments, the monkey was trained to receive with its right hand the pieces of food, demonstrated by the experimenter at random locations. The distance from the pieces to the monkey was about 20 cm; the time intervals between demonstrations were random. The ECoG activity of the monkey was recorded by 64 electrodes (Blackrock Microsystems, Salt Lake City, UT, USA) with a sampling rate of 1 kHz per channel. The electrodes were implanted in the epidural space of the left hemisphere of the monkey. The position of the monkey's right hand was recorded simultaneously with the ECoG signal by an optical motion capture system (Vicon Motion System, Oxford, UK) with a sampling rate of 120 Hz. More information about the setup of the experiments can be found in [16, 22].

The 10 data files each contain 15 min of ECoG- and 3D hand coordinate recordings. Each recording was split into two parts: the first 10 min were used for training and the last 5 min were used for the test.

The input feature tensor $\underline{\mathbf{X}}$ was formed from the ECoG-epochs, mapped to temporal-spatial-frequency space by means of the complex Morlet continuous wavelet transform. The epochs were taken continuously with steps equal to 100 ms. Each epoch contained 1 s of the signal. In accordance with [8], the frequencies from 10 to 150 Hz with 10 Hz steps were chosen for analysis. After formation of the input tensor, an artifact filtration method was applied to the data to eliminate the chewing artifacts from the brain activity. After the filtration, the feature tensor was decimated, 100 times along the temporal modality, by averaging the data in the 100-ms sliding windows.

The output tensor (matrix) $\mathbf{Y}$ was formed by the corresponding 3D coordinates of the hand.

Figure 1 represents the experimental scheme (A) as well as the procedure of the data tensors formation [23].

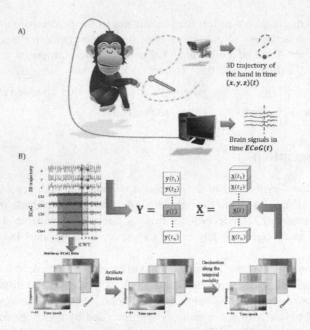

Fig. 1. Experimental scheme and data formation [23]

4   Results

The performance of four methods, namely, NPLS, IK-NPLS, OK-NPLS, and GL-NPLS, were compared. The optimal number of factors was estimated for each recording by the 10-fold cross-validation procedure. The smoothness parameter of kernel regression $h = 25$ and GL-NPLS dimensionality parameter $d = 10$ were taken as described in [8]. The averaged performance criteria for the test sets are represented in Fig. 2.

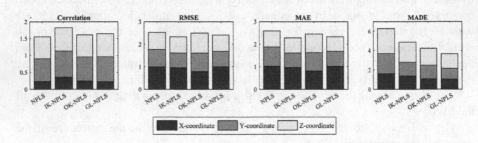

Fig. 2. The prediction quality of the methods, estimated for three coordinates (x, y, z) and averaged over 10 recordings

Figure 3 gives an example of the observed Z-coordinate as well as its prediction by the NPLS, OK-NPLS, IK-NPLS, and GL-NPLS algorithms.

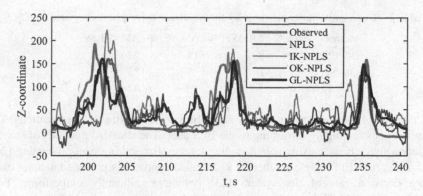

Fig. 3. An example of the predicted Z-coordinate of the trajectory

Table 1. The p-values of the difference for the linear NPLS and nonlinear methods (ANOVA test, significance level $\alpha = 0.05$)

Method	p-value			
	r	RMSE	MAE	MADE
IK-NPLS	0.15	0.13	0.03	0.00
OK-NPLS	0.77	0.84	0.39	0.00
GL-NPLS	0.65	0.35	0.08	0.00

Table 1 represents the p-values of the difference between the evaluation criteria for the NPLS method and nonlinear approaches (ANOVA test, significance level $\alpha = 0.05$).

5 Discussion and Conclusion

The goal of this paper is the study of nonlinear decoders of ECoG brain activity for the reconstruction of continuous hand trajectories for BCI applications. Until now, the majority of proposed BCI decoders were linear. Application of the nonlinear methods to the BCI tasks is restricted due to high dimensionality of the feature space as well as computational time restriction. In the present study, we introduced nonlinearity in a limited manner, combining it with a linear dimension reduction approach.

The NPLS method was chosen as the basic method for comparison due to its efficiency for high-dimensional tasks, as demonstrated in BCI studies [16, 17]. The nonlinearity was introduced to the NPLS methods in different ways, inside (IK-NPLS) and outside (OK-NPLS, GL-NPLS) of NPLS iterations. The nonlinearity was based on Nadaraya-Watson kernel regression with Gaussian kernel, since it allows the modelling of nonlinear dependencies of unknown structures.

Several criteria were applied to evaluate the prediction quality of the methods. Three of them (r, RMSE, and MAE) reflect the prediction accuracy; the last one (MADE) assesses the smoothness as well. The smoothness was studied because this

Table 2. The improvements provided by the nonlinear approaches, compared to NPLS

Method	Δr, %	ΔRMSE, %	ΔMAE, %	ΔMADE, %
IK-NPLS	17	8	12	22
OK-NPLS	4	1	5	33
GL-NPLS	6	5	10	42

characteristic of the predicted trajectory is important for real-life BCI applications. The abrupt motions are highly disturbing to the BCI user. The standard methods of smooth prediction suffer from considerable decoding latency (smoothing post-processing [24], Kalman Filter (KF) [25], etc.), whereas PLS-based approaches provide latencies close to the decision rate of the system [23], which significantly outperforms both post-processing and KF-based approaches. Low system latency is one of the important requirements for real-time BCIs.

The study has demonstrated that introduced nonlinearity gives slight but not significant improvement of prediction accuracy, compared to the linear NPLS method, as shown in Table 2 (ANOVA test, significance level $\alpha = 0.05$). At the same time, the increase of smoothness was significant (up to 42%) for all nonlinear approaches (ANOVA test, significance level $\alpha = 0.05$).

Further studies should: increase the testing database with additional recordings and subjects, add additional variability to the motion trajectories, and validate the results in real-life BCI experiments. The decoders will be tested in human subjects in the frame of the CLINATEC® BCI project, aiming at providing users with neural control over a 4-limb exoskeleton EMY [26] from ECoG data recorded with the WIMAGINE® implant [4].

Acknowledgments. This work was supported in part by grants from the French National Research Agency (ANR-Carnot Institute), Fondation Motrice, Fondation Nanosciences, Fondation de l'Avenir, and Fondation Philanthropique Edmond J. Safra. The authors are grateful to all members of the CEA-LETI-CLINATEC, and especially to Prof. A.-L. Benabid.

References

1. Donoghue, J.P.: Bridging the brain to the world: a perspective on neural interface systems. Neuron **60**, 511–521 (2008)
2. Daly, J.J., Wolpaw, J.R.: Brain-computer interfaces in neurological rehabilitation. Lancet Neurol. **7**, 1032–1043 (2008)
3. Wang, W., Collinger, J.L., Degenhart, A.D., Tyler-Kabara, E.C., Schwartz, A.B., Moran, D. W., Weber, D.J., Wodlinger, B., Vinjamuri, R.K., Ashmore, R.C., Kelly, J.W., Boninger, M. L.: An electrocorticographic brain interface in an individual with tetraplegia. PLoS ONE **8**, e55344 (2013)
4. Mestais, C., Charvet, G., Sauter-Starace, F., Foerster, M., Ratel, D., Benabid, A.L.: WIMAGINE®: wireless 64-channel ECoG recording implant for long term clinical applications. IEEE TNSRE **23**, (2015)

5. Müller-Putz, G.R., Kaiser, V., Solis-Escalante, T., Pfurtscheller, G.: Fast set-up asynchronous brain-switch based on detection of foot motor imagery in 1-channel EEG. Med. Biol. Eng. Comput. **48**, 229–233 (2010)
6. Bro, R.: Multiway calidration. Multilinear PLS. J. Chemom. **10**, 47–61 (1996)
7. Zhao, Q., Zhou, G., Adali, T., Zhang, L., Cichocki, A.: Kernelization of tensor-based models for multiway data analysis: Processing of multidimensional structured data. IEEE Signal Process. Mag. **30**, 137–148 (2013)
8. Eliseyev, A., Aksenova, T.: Stable and artifact-resistant decoding of 3D hand trajectories from ECoG signals using the generalized additive model. J. Neural Eng. **11**, 066005 (2014)
9. Eliseyev, A., Aksenova, T.: Recursive N-way partial least squares for brain-computer interface. PLoS ONE **8**, e69962 (2013)
10. Zhao, Q., Caiafa, C.F., Mandic, D.P., Chao, Z.C., Nagasaka, Y., Fujii, N., Zhang, L., Cichocki, A.: Higher order partial least squares (HOPLS): a generalized multilinear regression method. IEEE Trans. Pattern Anal. Mach. Intell. **35**, 1660–1673 (2013)
11. Gao, Y., Black, M.J., Bienenstock, E., Wu, W., Donoghue, J.P.: A quantitative comparison of linear and non-linear models of motor cortical activity for the encoding and decoding of arm motions. In: The Proceedings of First International IEEE EMBS Conference on Neural Engineering 2003, pp. 189–192. IEEE (2003)
12. Demir, S., Toktamis, Ö.: On the adaptive Nadaraya-Watson kernel regression estimators. Hacet. J. Math. Stat. **39**, 429–437 (2010)
13. Lee, J.A., Verleysen, M.: Nonlinear Dimensionality Reduction. Springer Publishing Company, Incorporated, Heidelberg (2007)
14. Geladi, P., Kowalski, B.R.: Partial least-squares regression: a tutorial. Anal. Chim. Acta **185**, 1–17 (1986)
15. Kolda, T.G., Bader, B.W.: Tensor decompositions and applications. SIAM Rev. **51**, 455–500 (2009)
16. Shimoda, K., Nagasaka, Y., Chao, Z.C., Fujii, N.: Decoding continuous three-dimensional hand trajectories from epidural electrocorticographic signals in Japanese macaques. J. Neural Eng. **9**, 036015 (2012)
17. Chao, Z.C., Nagasaka, Y., Fujii, N.: Long-term asynchronous decoding of arm motion using electrocorticographic signals in monkeys. Front. Neuroengineering **3**, 3 (2010)
18. Hazrati, M.K., Erfanian, A.: An online EEG-based brain–computer interface for controlling hand grasp using an adaptive probabilistic neural network. Med. Eng. Phys. **32**, 730–739 (2010)
19. Fatourechi, M., Bashashati, A., Ward, R.K., Birch, G.E.: A hybrid genetic algorithm approach for improving the performance of the LF-ASD brain computer interface. In: Proceedings of IEEE International Conference on Acoustics, Speech, and Signal Processing, (ICASSP 2005), vol. 5, pp. v/345–v/348. IEEE (2005)
20. Mörtsell, M., Gulliksson, M.: An overview of some non-linear techniques in chemometrics. FSCN, Mitthögskolan (2001)
21. Schlogl, A., Kronegg, J., Huggins, J.E., Mason, S.G.: 19 evaluation criteria for BCI Research. Toward brain-computer interfacing (2007)
22. Nagasaka, Y., Shimoda, K., Fujii, N.: Multidimensional recording (MDR) and data sharing: an ecological open research and educational platform for neuroscience. PLoS ONE **6**, e22561 (2011)
23. Eliseyev, A., Aksenova, T.: Penalized multi-way partial least squares for smooth trajectory decoding from electrocorticographic (ECoG) recording. PLoS ONE **11**, e0154878 (2016)
24. Koyama, S., Chase, S.M., Whitford, A.S., Velliste, M., Schwartz, A.B., Kass, R.E.: Comparison of brain–computer interface decoding algorithms in open-loop and closed-loop control. J. Comput. Neurosci. **29**, 73–87 (2010)

25. Gelb, A.: Applied optimal estimation. MIT press (1974)
26. B. Morinière, Verney, A., Abroug, N., Garrec, P., Perrot, Y.: EMY: a dual arm exoskeleton dedicated to the evaluation of Brain Machine Interface in clinical trials. In: 2015 IEEE/RSJ International Conference on Intelligent Robots and Systems (IROS), pp. 5333–5338 (2015)

On the Optimal Non-linearities for Gaussian Mixtures in FastICA

Joni Virta[1]([⊠]) and Klaus Nordhausen[1,2]

[1] Department of Mathematics and Statistics, University of Turku, Turku, Finland
{joni.virta,klaus.nordhausen}@utu.fi
[2] School of Health Sciences, University of Tampere, Tampere, Finland

Abstract. In independent component analysis we assume that the observed vector is a linear transformation of a latent vector of independent components, our objective being the estimation of the latter. Deflation-based FastICA estimates the components one-by-one by repeatedly maximizing the expected value of some function measuring non-Gaussianity, the derivative of which is called the non-linearity. Under some weak assumptions, the asymptotically optimal non-linearity for extracting sources with a specific density is given by the location score function of the density. In this paper we look into the consequences of this result from the viewpoint of estimating Gaussian location and scale mixtures. As one of our results we justify the common use of hyperbolic tangent, *tanh*, as a non-linearity in blind clustering by showing that it is optimal for estimating certain Gaussian mixtures. Finally, simulations are used to show that the asymptotic optimality results hold in various settings also for finite samples.

Keywords: Asymptotic optimality · Hyperbolic tangent · Independent component analysis

1 Introduction

In independent component analysis (ICA) one assumes that the observed k-vectors $\mathbf{x}_i$, $i = 1, \ldots, n$, are independent realizations of a random vector $\mathbf{x}$ which is a linear transformation of an unobserved vector $\mathbf{z}$ of independent source signals. This corresponds to the model

$$\mathbf{x} = \boldsymbol{\mu} + \boldsymbol{\Omega}\mathbf{z}, \tag{1}$$

where $\boldsymbol{\mu} \in \mathbb{R}^k$, the mixing matrix $\boldsymbol{\Omega} \in \mathbb{R}^{k \times k}$ is non-singular and the latent vector $\mathbf{z}$ has mutually independent components satisfying the following two assumptions: (i) The components of $\mathbf{z}$ are standardized in the sense that $\mathrm{E}(\mathbf{z}) = \mathbf{0}$ and $Cov(\mathbf{z}) = \mathbf{I}$, (ii) at most one of the components of $\mathbf{z}$ is Gaussian.

Assumption (i) fixes both the location $\boldsymbol{\mu}$ and the scales of the columns of $\boldsymbol{\Omega}$ in (1) and (ii) ensures that there are no orthogonally invariant column blocks in the

© Springer International Publishing AG 2017
P. Tichavský et al. (Eds.): LVA/ICA 2017, LNCS 10169, pp. 427–437, 2017.
DOI: 10.1007/978-3-319-53547-0_40

matrix $\boldsymbol{\Omega}$ [7]. After these assumptions the signs and the order of the components of $\boldsymbol{z}$ are still not fixed but this is usually satisfactory in applications.

In ICA one wants to find an estimate for the inverse of the unmixing matrix, $\boldsymbol{\Omega}^{-1} =: \boldsymbol{W} = (\boldsymbol{w}_1, \ldots, \boldsymbol{w}_k)^T$, after which, e.g. the first estimated independent component is obtained as $\boldsymbol{w}_1^T \mathbf{x}$. In FastICA [8] this is done by first standardizing the observed vector, $\boldsymbol{x} \mapsto \boldsymbol{x}_{st} := Cov(\boldsymbol{x})^{-1/2}(\boldsymbol{x} - \mathrm{E}(\boldsymbol{x}))$, which leaves $\boldsymbol{x}_{st}$ a rotation away from the vector $\boldsymbol{z}$ [1]. Then, for estimating this rotation one chooses a non-linearity function $g : \mathbb{R} \mapsto \mathbb{R}$ for which we denote $\mathbf{g}(\mathbf{x}) := (g(x_1), g(x_2), \ldots, g(x_k))^T$. The estimation is formalized in the following definition which, albeit a bit unorthodox way of defining FastICA, nicely captures all variants of it.

Definition 1. L_p-FastICA finds an orthogonal matrix $\boldsymbol{U}$ satisfying

$$\boldsymbol{U} = \underset{\boldsymbol{U}\boldsymbol{U}^T = \boldsymbol{I}}{argmax} \| E(\boldsymbol{g}(\boldsymbol{U}\boldsymbol{x}_{st})) \|_p,$$

where $\|\boldsymbol{x}\|_p = \left(\sum_{i=1}^k |x_i|^p \right)^{\frac{1}{p}}$ is the L_p-norm, $p \geq 1$.

Remark 1. L_1-FastICA is equivalent to the symmetric FastICA [8] and L_2-FastICA is equivalent to the squared symmetric FastICA [11].

Remark 2. Also deflation-based FastICA [8] has a similar formulation using vector norms. Namely, it can be seen as a repeated application of L_∞-FastICA, where $\|\mathbf{x}\|_\infty = \max_i(|x_i|)$. In the first step we search for a single component that maximizes $|\mathrm{E}(g(x))|$ and repeat the process $(k-1)$ times in the orthogonal complement of the already found directions.

The estimating equations of deflation-based FastICA, see e.g. [11,14], show that the non-linearity g is in deflation-based FastICA invariant to its linear part (hence its name) and also to scaling and sign-change of its argument. However, the same does not hold for either symmetric or squared symmetric FastICA.

Lemma 1. *Deflation-based FastICA is invariant under transformations* $g(x) \mapsto ag(sx) + bx + c$, *where* $a, b, c \in \mathbb{R}, a \neq 0, s \in \{-1, 1\}$, *of the used non-linearity* g.

Remark 3. The result of Lemma 1 holds also if one uses the alternative, modified Newton-Raphson algorithm, see [7,11].

If two non-linearities, g_1 and g_2, are equal up to the invariance specified in Lemma 1 we denote it as $g_1 \equiv g_2$. In addition to this invariance, deflation-based FastICA has also another interesting feature; given a component with a regular enough density function, in a certain sense optimal non-linearity for extracting it can be stated. This is formalized in the following lemma, the proof of which can be found in [5,10,11].

Lemma 2. *Let the random variable* z_1 *in* (1) *have a twice continuously differentiable density function* $f : \mathbb{R} \mapsto [0, \infty)$. *Then, assuming that* z_1 *is in deflation-based FastICA extracted first, the non-linearity* $g(x) = -f'(x)/f(x)$ *minimizes the sum of asymptotic variances of the elements of* $\hat{\boldsymbol{w}}_1$.

In the following all uses of the word *optimal* are in the sense of Lemma 2. For other criteria for choosing the non-linearity in deflation-based FastICA see e.g. [2].

The result of Lemma 2 holds conditional on the component of interest being the first to be extracted. This is trivially satisfied in the case the components of z are identically distributed and in a more general case its extraction first can be forced by choosing the starting value of the algorithm appropriately as done for example in reloaded FastICA [13] and adaptive FastICA [10].

In standard FastICA mainly four non-linearity functions are used in practice. They are usually denoted *skew*, *pow3*, *tanh* and *gauss* and correspond to the functions $g(x) = x^2$, $g(x) = x^3$, $g(x) = \tanh(x)$ and $g(x) = x\exp(-x^2/2)$, respectively [6]. The first two are based on the classical use of higher-order cumulants in projection pursuit [4] and the last two provide robust approximations for the negentropy [7], the most popular of the four being *tanh*.

While "robustness" issues are irrelevant when choosing the non-linearity as FastICA will never be robust due to the whitening based on the covariance matrix [14], it is still of interest to ask why some non-linearities seem to work better than others in various situations. Reversing the thinking of Lemma 2 we can then ask, given a non-linearity g, is it possibly optimal for any density f? Solving of the trivial first-order differential equation in combination with Lemma 1 yields the following result.

Lemma 3. *A differentiable and integrable function $g : \mathbb{R} \to \mathbb{R}$ is the optimal non-linearity for independent components with densities $f : \mathbb{R} \to \mathbb{R}_+$ satisfying $f(x) \propto exp(a \int_0^{sx} g(y)dy + bx^2 + cx)$, $\int_{-\infty}^{\infty} xf(x)dx = 0$ and $\int_{-\infty}^{\infty} x^2 f(x)dx = 1$ for some $a, b, c \in \mathbb{R}, a \neq 0, s \in \{-1, 1\}$.*

An analogous result for deflation-based FastICA, symmetric FastICA and EFICA [9] was given already in [16]. However, our version enjoys an extra degree of freedom in its parameters as restricting to deflation-based FastICA only allows, based on Lemma 1, the inclusion of the linear term cx in Lemma 3. The last two conditions in Lemma 3 reflect our assumption (i) that the independent components are standardized. Using Lemma 3 we see that *pow3* is optimal for sources with power exponential density, $f(x) = 2^{5/4}\sqrt{\pi}\Gamma(1/4)^{-2}exp(-2\pi^2\Gamma(1/4)^{-4}x^4)$, where $\Gamma(\cdot)$ is the Gamma function. The non-linearities *skew* and *gauss* are optimal for densities satisfying respectively $f(x) \propto exp(ax^3 + bx^2 + cx)$ and $f(x) \propto exp(a\,exp(-x^2/2) + bx^2 + cx)$ and, to the authors' knowledge, no common probability distributions defined on the whole $\mathbb{R}$ have such densities. However, an interesting remark can still be made. Namely, define sub-Gaussian (super-Gaussian) densities as those $f(x) = exp(-h(x))$ for which $h'(x)/x$ is increasing (decreasing) in $(0, \infty)$ [15]. Then Lemma 2 says that if a non-linearity $g(x)$ is optimal for some density, then that density is sub-Gaussian (super-Gaussian) if $g(x)/x$ is increasing (decreasing) in $(0, \infty)$, verifying the heuristics of using *pow3* for extracting sub-Gaussian sources and *gauss* for extracting super-Gaussian sources [6]. Note however that the definitions of sub- and super-Gaussian densities in [6] are based on kurtosis values and not on density functions.

2 Optimal Non-linearities for Gaussian Mixtures

It is well-known that Gaussian mixture distributions are suitable for approximating other distributions, see e.g. [3] who show that elliptical distributions can be seen as scale mixtures of Gaussian distributions. Motivated by this we will in the following consider two special cases of Gaussian mixture distributions.

2.1 Gaussian Location Mixtures

Consider the following two-parameter mixture distribution family, $\mathcal{L}(\pi, \lambda)$.

$$\pi \mathcal{N}(\frac{\lambda_1}{\sqrt{4 + \lambda_1 \lambda_2}}, \frac{4}{4 + \lambda_1 \lambda_2}) + (1 - \pi)\mathcal{N}(\frac{-\lambda_2}{\sqrt{4 + \lambda_1 \lambda_2}}, \frac{4}{4 + \lambda_1 \lambda_2}), \qquad (2)$$

where the mixing proportion $\pi \in (0, 1)$, the location parameter $\lambda \in (0, \infty)$ and for brevity we denote $\lambda_1 := \lambda/\pi$ and $\lambda_2 := \lambda/(1 - \pi)$. It is easily checked that the random variable $z_1 \sim \mathcal{L}(\pi, \lambda)$ satisfies $\mathrm{E}(z_1) = 0$ and $\mathrm{Var}(z_1) = 1$ for any permissible choices of the parameters and the family $\mathcal{L}(\pi, \lambda)$ then contains every standardized two-group Gaussian location mixture distribution where the two groups have the same variance. Applying then Lemma 2 to this family yields

Theorem 1. *Let $z_1 \sim \mathcal{L}(\pi, \lambda)$ for some $\pi \in (0, 1)$, $\lambda \in (0, \infty)$. Then the optimal non-linearity for extracting z_1 satisfies*

$$g(x) \equiv \left(\pi + \left(e^{t(x)} - 1 \right)^{-1} \right)^{-1},$$

where $t(x) = (\lambda_1 + \lambda_2)(2x\sqrt{4 + \lambda_1 \lambda_2} - \lambda_1 + \lambda_2)/8$.

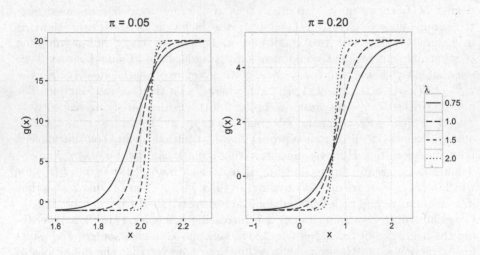

Fig. 1. The optimal non-linearities $g(x)$ for extracting $\mathcal{L}(\pi, \lambda)$-distributed components for various values of π and λ.

The resulting optimal non-linearity in Theorem 1 is quite complex and not of any standard functional form. Its graph for some select choices of parameters is depicted in Fig. 1, all cases exhibiting a sigmoid-like shape. However, considering the symmetric case, $\pi = 1/2$, simplifies the formulae greatly.

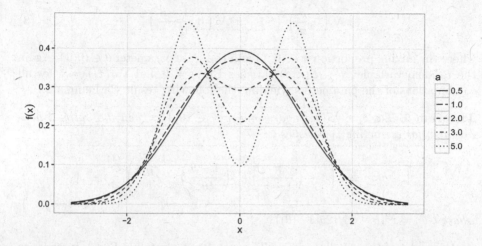

Fig. 2. The plot shows, for different values of $a \in (0, \infty)$, the densities of the symmetric Gaussian location mixtures for which the non-linearity $g(x) = tanh(ax)$ is optimal.

Corollary 1. *Let* $z_1 \sim \mathcal{L}(1/2, \lambda)$ *for some* $\lambda \in (0, \infty)$. *Then the optimal non-linearity for extracting* z_1 *satisfies*

$$g(x) \equiv tanh(\lambda \sqrt{1 + \lambda^2} x).$$

Corollary 1 says that the widely-used hyperbolic tangent is actually optimal for estimating symmetric two-group Gaussian location mixtures, justifying its use in FastICA when we have expectations to find symmetric bimodal components. A similar optimality result for *tanh* was given already in [16] but the resulting family of distributions was not studied further and Corollary 1 now goes to show that the family is for deflation-based FastICA actually $\mathcal{L}(1/2, \lambda)$. As a non-linearity *tanh* is usually given in the form $g(x) = tanh(ax)$ where $a \in (0, \infty)$ is a tuning parameter and we have, using Corollary 1, plotted in Fig. 2 the densities of the distributions for which $g(x) = tanh(ax)$ is the optimal non-linearity for various values of a. The plot implies that the more separated the groups one wants to find, the higher the value of a should be. See Sect. 3 for simulations of this heuristic. Curiously, the standard case $a = 1$ is optimal for components $z_1 \sim \mathcal{L}(1/2, \sqrt{\phi})$, where $\phi := (\sqrt{5} - 1)/2$ is the golden ratio.

2.2 Gaussian Scale Mixtures

We next consider Gaussian scale mixtures via a two-parameter mixture distribution family, $\mathcal{S}(\pi, \theta)$, that contains every standardized two-group Gaussian scale mixture distribution where the two groups have the same expected value:

$$\pi \mathcal{N}\left(0, \frac{\theta}{\pi}\right) + (1 - \pi)\mathcal{N}\left(0, \frac{1 - \theta}{1 - \pi}\right), \tag{3}$$

where the mixing proportion $\pi \in (0, 1)$ and the scale parameter $\theta \in (0, 1)$. Again the random variable $z_1 \sim \mathcal{S}(\pi, \theta)$ satisfies $\mathrm{E}(z_1) = 0$ and $\mathrm{Var}(z_1) = 1$ for all combinations of the parameters, yielding the following result via Lemma 2.

Theorem 2. *Let $z_1 \sim \mathcal{S}(\pi, \theta)$ for some $\pi, \theta \in (0, 1)$. Then the optimal non-linearity for extracting z_1 satisfies*

$$g(x) \equiv x\left(1 + \left(\frac{\pi}{1 - \pi}\right)^{3/2}\left(\frac{1 - \theta}{\theta}\right)^{1/2} e^{t(x)}\right)^{-1},$$

where $t(x) = x^2(\theta - \pi)/(2\theta(1 - \theta))$.

Examples of the non-linearity in Theorem 2 are plotted in Fig. 3. In order to obtain a simpler formula with only one tuning parameter notice that choosing $\theta = 1 - \pi$ corresponds for extreme values of π to a heavy-tail model and in this special case the result of Theorem 2 simplifies as follows.

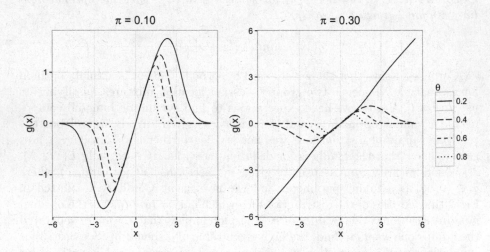

Fig. 3. The optimal non-linearities $g(x)$ for extracting $\mathcal{S}(\pi, \theta)$-distributed components for various values of π and θ.

Corollary 2. *Let $z_1 \sim \mathcal{S}(\pi, 1 - \pi)$ for some $\pi \in (0, 1)$. Then the optimal non-linearity for extracting z_1 satisfies*

$$g(x) \equiv x \left(1 + \left(\frac{\pi}{1 - \pi} \right)^2 e^{t(x)} \right)^{-1},$$

where $t(x) = x^2(1 - 2\pi)/(2\pi(1 - \pi))$.

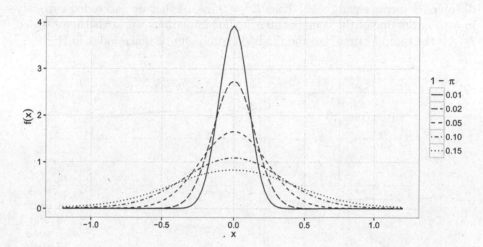

Fig. 4. The plot shows, for different values of $1 - \pi \in (0, 1)$, the densities of the Gaussian scale mixtures for which the non-linearity *tail* is optimal.

In Fig. 4 we have plotted for various values of $1 - \pi$ the densities of those Gaussian scale mixtures for which the non-linearity of Corollary 2 (referred to hereafter as *tail*) is optimal. As distributions $\mathcal{S}(\pi, 1 - \pi)$ with extreme values of π are basically symmetric, heavy-tailed distributions a reasonable guess is that the non-linearity *tail* is useful for extracting also other heavy-tailed symmetric components. This will be investigated in the next section.

3   Simulations

3.1   The Choice of the Tuning Parameter in *tanh(ax)*

The simulations are divided into two parts: the investigation of the tuning parameter a in $tanh(ax)$ and the testing of the non-linearity *tail* of Corollary 2.

For the first we used two different three-variate settings where all components of $z \in \mathbb{R}^3$ were either $\mathcal{L}(0.5, 2)$- or $\mathcal{L}(0.4, 2)$-distributed and we used deflation-based FastICA to estimate one of the components. We considered the non-linearities, *pow3*, *gauss*, *tanh(x)*, *tanh(3x)* and *tanh(5x)*, of which the last one should work the best in the first setting and the second setting investigates how the non-linearities handle small deviations from the distribution they are optimal for. *skew* is not included as it carries no information in symmetric settings. The sample size is taken to be $n = 1000, 2000, 4000, 8000, 16000, 32000$ and the number of repetitions is 10000.

As all three i.i.d. components of z are equally likely to be estimated first, we measured the success of the extraction by the criterion $D^2(\hat{w}_1) = \min\{\|PJ\hat{w}_1 - e_1\|^2\}$, where $\hat{w}_1$ is the estimated first direction, $e_1 = (1, 0, 0)^T$ and the minimum is taken over all 3×3 permutation matrices P and 3×3 diagonal matrices J with diagonal elements equal to ± 1. Thus $D^2 = 0$ means that we succeeded perfectly in estimating one of the components. In the simulations we furthermore scaled D^2 by the sample size n, see the modified minimum distance index in [17].

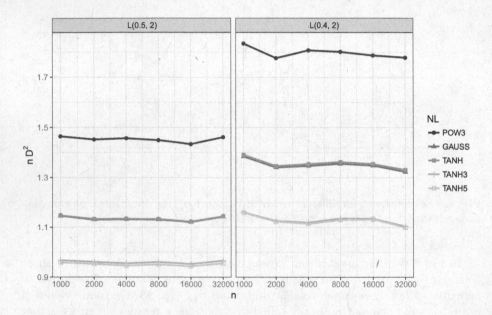

Fig. 5. The results of the first simulation.

The resulted mean criterion values are given in Fig. 5 and show that $tanh(3x)$ and $tanh(5x)$ performed the best in the symmetric setting, *gauss* and $tanh(x)$ not being that far behind. More interestingly, the same conclusions can be drawn also in the asymmetric case. Only the overall level of the extraction is a bit worse.

3.2 Estimating Scale Mixtures and Heavy-Tailed Components

To evaluate the performance of *tail* we considered two three-variate settings where the components of z were all either $\mathcal{S}(0.10, 0.90)$- or t_5-distributed (and standardized), where t_5 denotes a t-distribution with 5 degrees of freedom.

The sample sizes, the number of repetitions and the criterion function were the same as in the previous simulation and we used six non-linearities, *pow3*, *gauss*, $tanh(x)$, *tail* with $\pi = 0.1$, *tail* with $\pi = 0.3$ and *rat3* with $b = 4$, see below. The third-to-last non-linearity should be superior in the first setting and with the second setting we experiment whether *tail* works for other heavy-tailed distributions also. The non-linearity *rat3*, $g(x) = x/(1 + b|x|)^2$, was proposed in [16] to estimate heavy-tailed sources and there $b = 4$ was suggested as a balanced choice for the tuning parameter.

The results in Fig. 6 again show that in the first setting the asymptotically optimal non-linearity, *tail* with $\pi = 0.1$, gives the best separation also for finite samples. In the "experimental" setting with the t_5-distribution $tanh(x)$ proved most useful but also *gauss* and *tail* with $\pi = 0.3$ were quite successful.

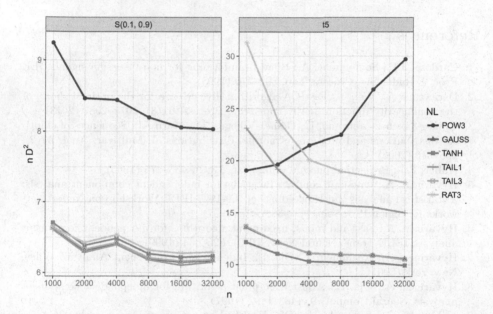

Fig. 6. The results of the second simulation.

4   Discussion

In FastICA the choice of the non-linearity, e.g. the popular *tanh*, is usually motivated with heuristic claims and asymptotic arguments showing that a particular non-linearity is optimal for some class of distributions. However, one is usually not interested in a non-linearity that works well in only a few cases but instead in a multitude of situations – and as also our simulations show, *tanh* performs in general quite well, also with distributions for which it is not optimal. And although there exists cases where *tanh* does not work at all [11,18], this drawback should not be given too much weight; only a few non-linearities are so far shown to work for any combination of sources, assuming that at most one of them has an objective function value of zero, see e.g. [12,17]. Such un-estimable distributions can actually be crafted for any non-linearity [16].

The use of different non-linearities for different components in FastICA has also been considered, see e.g. EFICA [9] and adaptive deflation-based FastICA [10]. While EFICA tries to estimate the optimal non-linearities from the data, adaptive deflation-based FastICA chooses them out of a set of candidates. It seems thus reasonable to include in this set non-linearities which are known to have optimality properties, such as the ones given in our Corollaries 1 and 2.

Acknowledgements. We would like to thank the anonymous referees for their stimulating comments which enhanced the paper and provided us with existing results previously unknown to us. This work was supported by the Academy of Finland Grant 268703.

References

1. Cardoso, J.F., Souloumiac, A.: Blind beamforming for non-Gaussian signals. IEE Proc. F Radar Sig. Process. **140**, 362–370 (1993)
2. Dermoune, A., Wei, T.: FastICA algorithm: five criteria for the optimal choice of the nonlinearity function. IEEE Trans. Sig. Process. **61**(8), 2078–2087 (2013)
3. Gómez-Sánchez-Manzano, E., Gómez-Villegas, M., Marín, J.: Sequences of elliptical distributions and mixtures of normal distributions. J. Multivar. Anal. **97**(2), 295–310 (2006)
4. Huber, P.J.: Projection pursuit. Ann. Stat. **13**(2), 435–475 (1985)
5. Hyvärinen, A.: One-unit contrast functions for independent component analysis: a statistical analysis. In: Proceedings of the 1997 IEEE Workshop on Neural Networks for Signal Processing, pp. 388–397 (1997)
6. Hyvärinen, A.: Fast and robust fixed-point algorithms for independent component analysis. IEEE Trans. Neural Netw. **10**(3), 626–634 (1999)
7. Hyvärinen, A., Karhunen, J., Oja, E.: Independent Component Analysis. Wiley, New York (2001)
8. Hyvärinen, A., Oja, E.: A fast fixed-point algorithm for independent component analysis. Neural Comput. **9**, 1483–1492 (1997)
9. Koldovskỳ, Z., Tichavskỳ, P., Oja, E.: Efficient variant of algorithm FastICA for independent component analysis attaining the Cramer-Rao lower bound. IEEE Trans. Neural Netw. **17**(5), 1265–1277 (2006)
10. Miettinen, J., Nordhausen, K., Oja, H., Taskinen, S.: Deflation-based FastICA with adaptive choices of nonlinearities. IEEE Trans. Sig. Process. **62**(21), 5716–5724 (2014)
11. Miettinen, J., Nordhausen, K., Oja, H., Taskinen, S., Virta, J.: The squared symmetric FastICA estimator. Sig. Process. **131**, 402–411 (2017)
12. Miettinen, J., Taskinen, S., Nordhausen, K., Oja, H.: Fourth moments and independent component analysis. Stat. Sci. **30**, 372–390 (2015)
13. Nordhausen, K., Ilmonen, P., Mandal, A., Oja, H., Ollila, E.: Deflation-based FastICA reloaded. In: Proceedings of 19th European Signal Processing Conference, pp. 1854–1858 (2011)
14. Ollila, E.: The deflation-based FastICA estimator: statistical analysis revisited. IEEE Trans. Sig. Process. **58**(3), 1527–1541 (2010)

15. Palmer, J., Kreutz-Delgado, K., Rao, B.D., Wipf, D.P.: Variational EM algorithms for non-gaussian latent variable models. In: Advances in Neural Information Processing Systems, pp. 1059–1066 (2005)
16. Tichavský, P., Koldovský, Z., Oja, E.: Speed and accuracy enhancement of linear ICA techniques using rational nonlinear functions. In: Davies, M.E., James, C.J., Abdallah, S.A., Plumbley, M.D. (eds.) ICA 2007. LNCS, vol. 4666, pp. 285–292. Springer, Heidelberg (2007). doi:10.1007/978-3-540-74494-8_36
17. Virta, J., Nordhausen, K., Oja, H.: Projection pursuit for non-Gaussian independent components. arXiv (2016). https://arxiv.org/abs/1612.05445
18. Wei, T.: On the spurious solutions of the FastICA algorithm. In: IEEE Workshop on Statistical Signal Processing, pp. 161–164 (2014)

Fast Disentanglement-Based Blind Quantum Source Separation and Process Tomography: A Closed-Form Solution Using a Feedback Classical Adapting Structure

Yannick Deville[1(✉)] and Alain Deville[2]

[1] Université de Toulouse, UPS-CNRS-OMP, IRAP (Institut de Recherche en Astrophysique et Planétologie), 14 Avenue Edouard Belin, 31400 Toulouse, France
yannick.deville@irap.omp.eu
[2] Aix-Marseille Université, CNRS, IM2NP UMR 7334, 13397 Marseille, France
alain.deville@univ-amu.fr

Abstract. We here extend Blind (i.e. unsupervised) Quantum Source Separation and Process Tomography methods. Considering disentanglement-based approaches, we introduce associated optimization algorithms which are much faster than the previous ones, since they reduce the number of source quantum state preparations required for adaptation by a factor of 10^3 typically. This is achieved by unveiling the parametric forms of the optimized cost functions, which allows us to derive a closed-form solution for their optimum.

Keywords: Blind quantum source separation · Unsupervised unmixing · Blind quantum process tomography · Blind quantum system identification and inversion · Disentanglement-based separation principle · Fast algorithms · Quantum bit (qubit) · Qubit uncoupling · Entanglement

1 Prior Work and Problem Statement

Within the information processing (IP) domain, various fields developed very rapidly during the last decades. One of these fields is Blind Source Separation (BSS), which led to various classes of methods, including Independent Component Analysis (ICA) [1]. Until recently, all BSS investigations were performed in a "classical", i.e. non-quantum, framework. Another growing field within the overall IP domain is Quantum Information Processing (QIP) [8]. QIP is closely related to Quantum Physics (QP). It uses abstract representations of systems whose behavior is requested to obey the laws of QP. This already made it possible to develop new and powerful IP methods, which manipulate the states of so-called quantum bits, or qubits.

In 2007, we bridged the gap between classical (B)SS and QIP/QP in [2], by introducing a new field, Quantum Source Separation (QSS), and especially its blind version (BQSS). The QSS problem consists in restoring (the information

P. Tichavský et al. (Eds.): LVA/ICA 2017, LNCS 10169, pp. 438–448, 2017.
DOI: 10.1007/978-3-319-53547-0_41

contained in) individual source quantum states, eventually using only the mixtures (in SS terms [3]) of these states which result from their undesired coupling. The blind (or unsupervised) version of this problem corresponds to the case when the parameter values of the mixing operator are initially unknown and are first estimated by using only mixtures of source quantum states, i.e. without knowing these source states (see also [3] for (B)QSS applications).

A complete BQSS investigation consists of the definition of the same items as in classical BSS, namely: (1) considered mixing model, (2) proposed separating system structure, (3) proposed separation principle (which is the counterpart of e.g. forcing output independence in classical ICA) preferably with an analysis of the resulting so-called "indeterminacies", (4) proposed separation criterion (see e.g. output mutual information minimization in classical ICA), (5) proposed separation algorithm (e.g. gradient-based minimization of cost function).

Using this approach, we initially developed a first class of BQSS methods, which are based on a separation principle that has some relationships with classical ICA (see especially [2,3,5]). More recently, in [4,6], we started to develop another class of BQSS methods, using a new separation principle which is based on the disentanglement of output quantum states of the separating system, that are fed back for adapting that system. This class of methods yields attractive features as compared with the previous one (see details in [4,6]). Our investigations reported in [4,6] cover all five items of the above-defined procedure for developing BQSS methods. However, they required major efforts for the first three of these items, so that we then only resctricted ourselves to a simple approach for the cost function and associated optimization algorithm.

The above specific, iterative, algorithm yields high complexity, especially in terms of the number of source quantum states to be prepared for adapting the separating system, as shown further in this paper. Therefore, after summarizing the concepts from [4,6] which are needed here (see Sects. 2 and 3), a first contribution in this paper consists in introducing a new BQSS algorithm which yields much lower complexity than the above one (see Sect. 4).

Besides, classical BSS is mainly based on the blind inversion of the mixing model. BSS methods therefore typically also perform a blind identification of the mixing model. In [7], we started to develop similar considerations for the quantum counterpart of the above blind system identification problem, that we called "Blind Quantum Process Tomography" (BQPT), since non-blind *quantum* system identification is referred to as "Quantum Process Tomography" by the QIP community (see e.g. [8] p. 389). The second main contribution of the present paper therefore consists of an analysis of the capabilities of the proposed BQSS method from a BQPT point of view (see Sect. 5). Conclusions are eventually drawn from this overall investigation in Sect. 6.

2   Mixing Model

As stated above, computations in the field of QIP use qubits instead of classical bits [8]. In [4], we first detailed the required concepts for a single qubit and then

presented the type of coupling between two qubits that we consider and that
defines the "mixing model", in (B)SS terms, of our investigation. We hereafter
summarize the major aspects of that discussion, which are required in the current
paper.

A qubit with index i considered at a given time t_0 has a quantum state. If
this state is pure, it belongs to a two-dimensional space $\mathcal{E}_i$ and may be expressed
as

$$|\psi_i(t_0)\rangle = \alpha_i| + \rangle + \beta_i| - \rangle \tag{1}$$

in the basis of $\mathcal{E}_i$ defined by the two orthonormal vectors that we hereafter denote
$|+\rangle$ and $|-\rangle$, whereas α_i and β_i are two complex-valued coefficients constrained
to be such that the state $|\psi_i(t_0)\rangle$ is normalized.

In the BQSS configuration studied in this paper, we first consider a system
composed of two qubits, called "qubit 1" and "qubit 2" hereafter, at a given
time t_0. This system has a quantum state. If this state is pure, it belongs to
the four-dimensional space $\mathcal{E}$ defined as the tensor product (denoted $\otimes$) of the
spaces $\mathcal{E}_1$ and $\mathcal{E}_2$ respectively associated with qubits 1 and 2, i.e. $\mathcal{E} = \mathcal{E}_1 \otimes \mathcal{E}_2$.
We hereafter denote $\mathcal{B}_+$ the basis of $\mathcal{E}$ composed of the four orthonormal vectors
$|++\rangle, |+-\rangle, |-+\rangle, |--\rangle$, where e.g. $|+-\rangle$ is an abbreviation for $|+\rangle \otimes |-\rangle$,
with $|+\rangle$ corresponding to qubit 1 and $|-\rangle$ corresponding to qubit 2. Any pure
state of this two-qubit system may then be expressed as

$$|\psi(t_0)\rangle = c_1(t_0)| + +\rangle + c_2(t_0)| + -\rangle + c_3(t_0)| - +\rangle + c_4(t_0)| - -\rangle \tag{2}$$

and has unit norm. It may also be represented by the corresponding vector of
complex-valued components in basis $\mathcal{B}_+$, which reads

$$C_+(t_0) = [c_1(t_0), c_2(t_0), c_3(t_0), c_4(t_0)]^T \tag{3}$$

where T stands for transpose. In particular, we study the case when the two
qubits are independently initialized, with states defined by (1) respectively with
$i = 1$ and $i = 2$. We then have

$$|\psi(t_0)\rangle = |\psi_1(t_0)\rangle \otimes |\psi_2(t_0)\rangle \tag{4}$$

$$= \alpha_1\alpha_2| + +\rangle + \alpha_1\beta_2| + -\rangle + \beta_1\alpha_2| - +\rangle + \beta_1\beta_2| - -\rangle. \tag{5}$$

Besides, we consider the case when the two qubits, which correspond to two
spins $1/2$, have undesired coupling after they have been initialized according to
(4). The considered coupling is based on the Heisenberg model with a cylindrical-
symmetry axis presently collinear to the applied magnetic field. This common
axis is chosen as the "quantization axis", called Oz. This coupling may be represented as

$$C_+(t) = MC_+(t_0) \tag{6}$$

where $C_+(t)$ is the counterpart of (3) at time t and defines the coupled
(or "mixed", in BSS terms) state $|\psi(t)\rangle$ of the two-qubit system at that time,

whereas the unitary matrix M represents the evolution of the system's quantum state from t_0 to t in basis $\mathcal{B}_+$. Our previous calculations show that, for the considered type of coupling

$$M = QDQ^{-1} = QDQ \tag{7}$$

with

$$Q = Q^{-1} = \begin{bmatrix} 1 & 0 & 0 & 0 \\ 0 & \frac{1}{\sqrt{2}} & \frac{1}{\sqrt{2}} & 0 \\ 0 & \frac{1}{\sqrt{2}} & -\frac{1}{\sqrt{2}} & 0 \\ 0 & 0 & 0 & 1 \end{bmatrix} \tag{8}$$

and

$$D = \begin{bmatrix} e^{-i\omega_{1,1}(t-t_0)} & 0 & 0 & 0 \\ 0 & e^{-i\omega_{1,0}(t-t_0)} & 0 & 0 \\ 0 & 0 & e^{-i\omega_{0,0}(t-t_0)} & 0 \\ 0 & 0 & 0 & e^{-i\omega_{1,-1}(t-t_0)} \end{bmatrix} \tag{9}$$

where i is the imaginary unit. The four real (angular) frequencies $\omega_{1,1}$ to $\omega_{1,-1}$ in (9) depend on the physical setup and their values are unknown in practice.

3   Separating System, Separation Principle and Criterion

3.1   Inverting Block of Separating System

The inverting block of the considered separating system is the part of this system which is to be used eventually (i.e. after this block has been adapted) to derive the output quantum state $|\Phi\rangle$ of this system from its input quantum state, which is the above-defined coupled state $|\psi(t)\rangle$. That block here uses quantum processing means only. The output quantum state of that block and therefore of our overall separating system is denoted as

$$|\Phi\rangle = c_1|++\rangle + c_2|+-\rangle + c_3|-+\rangle + c_4|--\rangle. \tag{10}$$

It may also be represented by the corresponding vector of components of $|\Phi\rangle$ in output basis $\mathcal{B}_+$, denoted as

$$C = [c_1, c_2, c_3, c_4]^T. \tag{11}$$

We then have

$$C = UC_+(t) \tag{12}$$

where U defines the unitary quantum-processing operator applied by our separating system to its input $C_+(t)$. As justified below, we choose this operator U to belong to the class defined by

$$U = Q\tilde{D}Q \tag{13}$$

$$\text{with} \quad \tilde{D} = \begin{bmatrix} e^{i\gamma_1} & 0 & 0 & 0 \\ 0 & e^{i\gamma_2} & 0 & 0 \\ 0 & 0 & e^{i\gamma_3} & 0 \\ 0 & 0 & 0 & e^{i\gamma_4} \end{bmatrix} \tag{14}$$

where γ_1 to γ_4 are free real-valued parameters.

3.2 Adapting Block, Separation Principle and Criterion

The above type of inverting block was selected because it can perfectly restore the quantum source state $|\psi(t_0)\rangle$ for adequate values of its free parameters γ_1 to γ_4: setting them so that $\tilde{D} = D^{-1}$ yields $U = M^{-1}$, which results in $C = C_+(t_0)$ and $|\Phi\rangle = |\psi(t_0)\rangle$. However, the condition $\tilde{D} = D^{-1}$ cannot be used as a *practical* procedure for directly assigning $\tilde{D}$, because D is unknown. Instead, a procedure for adapting the parameters γ_1 to γ_4 of $\tilde{D}$ by using only one or several values of the available mixed state $|\psi(t)\rangle$ is therefore required, which corresponds to a *blind* (quantum) source separation problem.

Briefly, the BQSS method that we developed to this end in [4,6] uses the output disentanglement separation principle that we introduced in those papers and that is based on the concept of quantum state entanglement. From this principle, we then derived a two-step adaptation procedure, where each step consists of the global minimization of a cost function expressed with respect to classical-form quantities, namely probabilities of discrete outcomes of spin component measurements performed at the output of the inverting block. The first cost function involves $N_z \geq 2$ (arbitrary but non-redundant) source states, indexed by n, corresponding to source samples in classical BSS. It is defined as

$$F_z = \sum_{n=1}^{N_z} |f_z(n)|^p \tag{15}$$

$$\text{with} \quad f_z(n) = P_{1z}(n)P_{4z}(n) - P_{2z}(n)P_{3z}(n) \tag{16}$$

and e.g. $p = 1$ or 2. $P_{1z}(n)$ to $P_{4z}(n)$ are the above-mentioned probabilities, corresponding to the case when the considered spin components are measured along the above-defined axis Oz. They read

$$P_{1z}(n) = |c_1(n)|^2, P_{2z}(n) = |c_2(n)|^2, P_{3z}(n) = |c_3(n)|^2, P_{4z}(n) = |c_4(n)|^2 \tag{17}$$

where $c_1(n)$ to $c_4(n)$ are the coefficients of (10) for the n-th source state, which is defined by (5) with corresponding parameter values $\alpha_1(n)$ to $\beta_2(n)$.

The global minimization, i.e. cancellation, of F_z is equivalent to enforcing

$$f_z(n) = 0 \tag{18}$$

for n ranging from 1 to N_z. The investigation reported in [4] shows that enforcing (18) for a *single* value of n yields two types of solutions for the γ_j parameters. One of them is the desired solution of the investigated problem and does not depend on the considered source state. On the contrary, the other one is a spurious solution and depends on that source state (that spurious solution was avoided in [4] by enforcing (18) for *several* source states, indexed by n).

The second step of the proposed procedure uses one of the equivalent directions normal to the Oz axis, e.g. the Ox axis, now minimizing a cost function F_x similar to F_z, but using N_x source states and measurements of output spin components along Ox, with associated probabilities $P_{1x}(n)$ to $P_{4x}(n)$:

$$F_x = \sum_{n=1}^{N_x} |f_x(n)|^p \tag{19}$$

with

$$f_x(n) = P_{1x}(n)P_{4x}(n) - P_{2x}(n)P_{3x}(n). \tag{20}$$

4 Separation Algorithms

The original approach of [4,6] uses the following algorithms for minimizing the above-defined cost functions. Due to the parameters upon which these functions depend, the first step of the proposed procedure consists in performing a sweep on one of the parameters γ_2 and γ_3, while the other one, as well as γ_1 and γ_4, are constant. This procedure computes the corresponding (estimated) values of F_z and it eventually keeps the value of the tuned parameter γ_2 or γ_3 which minimizes F_z. It then freezes γ_2 and γ_3. Similarly, the second stage of the proposed procedure then performs a sweep on γ_1 or γ_4, in order to minimize F_x. To tune each parameter γ_j of the quantum circuit which implements the sub-block $\tilde{D}$ of the separating system, what is controlled in practice is not γ_j itself but the value of a corresponding physical quantity, hereafter denoted V_j, which may e.g. be a voltage.

When optimizing F_z with the above algorithm, for each source state with index n and each set of values of all γ_j, the estimation of each set of probabilities $P_{1z}(n)$ to $P_{4z}(n)$ is based on the Repeated Write/Read (RWR) procedure defined e.g. in [2–4]. Briefly, this requires one to repeatedly, that is typically 10^4 times [3], prepare (i.e. initialize) that source quantum state and perform measurements at the output of the inverting block. When using the above sweep-based procedure, these 10^4 preparations must moreover be repeated for each step of the sweep on

γ_2, that is typically 10^3 time. This yields a total number of 10^7 source quantum state preparations, moreover multiplied by the number N_z of states used in output measurements along the Oz axis. The same principle then applies to measurements along the Ox axis used when optimizing F_x.

Since the preparation of each source quantum state is a complicated procedure, the above overall optimization yields a quite high complexity, that we here aim at avoiding. To this end, we introduce a new approach, which is based on the following principle. Instead of estimating the probabilities $P_{1z}(n)$ to $P_{4z}(n)$ and $P_{1x}(n)$ to $P_{4x}(n)$ for many values of each tuned γ_j parameter, we aim at making such estimations only for a few sets of values of these γ_j and then using these estimates to infer the values of all γ_j which correspond to the global minimum of the cost functions F_z and F_x. This approach here becomes possible by taking advantage of the parametric forms of F_z and F_x, which were not unveiled in our previous papers. This new approach assumes that each transform g_j, between a control quantity V_j and the associated parameter γ_j, yielding

$$\gamma_j = g_j(V_j) \qquad \text{with } j \in \{1, \ldots, 4\}, \tag{21}$$

is known and invertible. The first step of the proposed procedure consists in using different values of V_2 (or V_3), and therefore of γ_2 (or γ_3), while the other three γ_j parameters are kept to arbitrary constant values, and in considering all corresponding values $f_z(n)$ involved in F_z, so as to find values of V_2 and V_3 which make F_z equal to zero or, equivalently, which ensure (18) for n ranging from 1 to N_z. In addition, we here take into account the parametric form of $f_z(n)$ as follows. We combine (2)–(5), (6)–(9), (11)–(14), (16) and (17), here with an argument "(n)" since they are applied to the n-th source state of the considered sequence. Lengthy calculations, skipped here, show that the expression of $f_z(n)$ thus obtained may then be transformed into

$$f_z(n) = w_1(n)\cos(2(\gamma_3 - \gamma_2)) + w_2(n)\sin(2(\gamma_3 - \gamma_2)) + w_3(n) \tag{22}$$

where $w_1(n)$, $w_2(n)$ and $w_3(n)$ depend on the source state coefficients $\alpha_1(n)$ to $\beta_2(n)$ and on the mixing parameters. Equation (22) thus yields a parametric expression of $f_z(n)$, with respect to γ_2 and γ_3 (more precisely, with respect to their difference), and with unknown parameter values $w_1(n)$, $w_2(n)$ and $w_3(n)$.

The above parameter values may be derived as follows, considering a single, arbitrary, source state with index n at this stage. When applying known values of V_2 and V_3, and therefore using known values of γ_2 and γ_3, one may estimate the corresponding values of $P_{1z}(n)$ to $P_{4z}(n)$ and derive the corresponding value of $f_z(n)$ from (16). Equation (22) then yields an equation where the only unknowns are $w_1(n)$, $w_2(n)$ and $w_3(n)$. Repeating this procedure for three values of (V_2, V_3) yields three (supposedly linearly independent) linear equations with respect to $w_1(n)$, $w_2(n)$ and $w_3(n)$, that one then just has to solve.

Once the values of $w_1(n)$, $w_2(n)$ and $w_3(n)$ have been obtained, the solutions of (18) in terms of $(\gamma_3 - \gamma_2)$ may easily be derived by using (22). This yields the following two types of solutions:

$$\gamma_3 - \gamma_2 = \frac{1}{2} \arccos \left(\frac{-w_3(n)}{\sqrt{(w_1(n))^2 + (w_2(n))^2}} \right)$$

$$+ \frac{1}{2} \mathrm{sgn}(w_2(n)) \times \arccos \left(\frac{w_1(n)}{\sqrt{(w_1(n))^2 + (w_2(n))^2}} \right) + k_1 \pi \quad (23)$$

$$\gamma_3 - \gamma_2 = -\frac{1}{2} \arccos \left(\frac{-w_3(n)}{\sqrt{(w_1(n))^2 + (w_2(n))^2}} \right)$$

$$+ \frac{1}{2} \mathrm{sgn}(w_2(n)) \times \arccos \left(\frac{w_1(n)}{\sqrt{(w_1(n))^2 + (w_2(n))^2}} \right) + k_2 \pi \quad (24)$$

where k_1 and k_2 are arbitrary integers. These two types of solutions for $(\gamma_3 - \gamma_2)$ are defined up to a multiple of π. We hereafter consider a single value for each of these two types, by setting $k_1 = 0$ and $k_2 = 0$.

Thanks to [4], it was already known that (18) would here yield two solutions, namely a desired one and a spurious one, as explained in Sect. 3.2. Among these solutions (23) and (24), we eventually have to determine which one is the desired solution and which one is the spurious solution. To this end, we take advantage of the property mentioned in Sect. 3.2: the desired solution has the same value (up to estimation errors for $P_{1z}(n)$ to $P_{4z}(n)$) for all used source states, i.e. for all values of n ranging from 1 to N_z; on the contrary, the value of the spurious solution varies with n. The following method may therefore be used to determine the desired solution for $(\gamma_3 - \gamma_2)$. One first determines its two values (23) and (24), with $k_1 = 0$ and $k_2 = 0$, successively for n ranging from 1 to N_z. One then considers all 2^{N_z} possible combinations obtained by selecting one of the above two solutions for each value of n. For each such combination, one derives the variance of all N_z selected values. The combination which yields the lowest variance is considered to be the one for which the selected solution is the desired one for each n (this variance would be zero if all values of $P_{1z}(n)$ to $P_{4z}(n)$ were obtained without estimation errors). Still considering this optimum combination, a single estimate of $(\gamma_3 - \gamma_2)$ is eventually obtained as the mean of all N_z selected solutions which compose this combination.

Corresponding values of V_2 and V_3 are then derived as follows: (i) one fixes one of the parameters γ_2 and γ_3, and hence the corresponding value of V_2 or V_3 by using the inverse of mapping (21) and (ii) one selects the other parameter among γ_2 and γ_3 so that $(\gamma_3 - \gamma_2)$ takes the above estimated value, and one derives the remaining parameter V_2 or V_3 by again using the inverse of mapping (21). One then freezes V_2 and V_3, and thus γ_2 and γ_3.

Similarly, the second step of the overall proposed BQSS algorithm then consists in using different values of V_1 and/or V_4, and therefore of γ_1 and/or γ_4, and in considering all corresponding values $f_x(n)$ involved in F_x, so as to find values of V_1 and V_4 which make F_x equal to zero or, equivalently, which ensure

$$f_x(n) = 0 \qquad \forall\, n \in \{1, \ldots, N_x\}. \tag{25}$$

We here also take into account the parametric form of $f_x(n)$. To this end, we first use (20) and the expressions of $P_{1x}(n)$ to $P_{4x}(n)$ that may be derived from quantum calculations skipped here. This yields

$$f_x(n) = \frac{1}{4}\Re[(c_1^2 + c_4^2 - c_2^2 - c_3^2)(c_1 c_4 - c_2 c_3)^*] \tag{26}$$

where $\Re[.]$ stands for real part. The coefficients in the right-hand term of (26) should here contain an argument "(n)" since they refer to the n-th source state of the considered sequence, but we omit this argument for the sake of readability. These coefficients of (26) are here defined by the expressions obtained for the solution of $F_z = 0$, since the first step of this overall BQSS algorithm was previously used to adapt γ_2 and γ_3 (we here neglect the influence of estimation errors in that first step of the procedure). These coefficient expressions were derived in [4] (see its Eqs. (33)–(38)). Using them, additional tedious calculations here show that, when also setting the constraint $\gamma_1 - \gamma_4 = \gamma_c$ where γ_c is an arbitrary[1] constant, the function $f_x(n)$ of (26) has the parametric form

$$f_x(n) = w_1'(n)\cos(2\gamma_1) + w_2'(n)\sin(2\gamma_1) + w_3'(n) \tag{27}$$

where $w_1'(n)$, $w_2'(n)$ and $w_3'(n)$ have unknown values, which depend on γ_c and on the source state, mixing parameters and value of γ_2 previously fixed in the first step of the procedure. The approach defined above for $f_z(n)$ may then also be applied to $f_x(n)$, so as to determine the values of γ_1 and γ_4, and then of V_1 and V_4, which ensure (25).

As compared with our former, sweep-based, approach summarized at the beginning of this section, the overall improved approach introduced in this paper has the advantage of requiring far fewer qubit preparations, as will now be shown. In the first step of this improved algorithm, corresponding to measurements along the Oz axis, for each source state with index n it requires one to estimate 3 sets of probabilities $P_{1z}(n)$ to $P_{4z}(n)$ in order to derive $w_1(n)$ to $w_3(n)$, instead of typically 10^3 sets of probabilities when performing 10^3 steps in our previous, sweep-based, approach. The resulting number of qubit preparations is therefore typically 3×10^4 here, instead of 10^7 in our previous approach. Similar considerations then apply to the second step of this algorithm, corresponding to measurements along the Ox axis. The only price to pay for this major reduction of the number of qubit preparations is the somewhat more complex principle of the improved algorithm proposed here, the need to know the transforms (21) and the associated computations to be performed (on a classical computer) to derive the adequate values of V_1 to V_4.

---

[1] Considering the approach of this paper alone, γ_c is freely chosen and the simplest choice is $\gamma_c = 0$, i.e. $\gamma_1 = \gamma_4$.

5   Blind Quantum Process Tomography

As explained in Sect. 1, we here analyze the capabilities of the method proposed in Sect. 4 in terms of BQPT, i.e. its ability to blindly identify the mixing model (7)–(9). More precisely, in [7] we explained that the unknown model parameters $\omega_{1,1}$ to $\omega_{1,-1}$ in (9) read

$$\omega_{1,1} = \frac{1}{\hbar}\left[GB - \frac{J_z}{2}\right], \quad \omega_{1,0} = \frac{1}{\hbar}\left[-J_{xy} + \frac{J_z}{2}\right], \tag{28}$$

$$\omega_{0,0} = \frac{1}{\hbar}\left[J_{xy} + \frac{J_z}{2}\right], \quad \omega_{1,-1} = \frac{1}{\hbar}\left[-GB - \frac{J_z}{2}\right] \tag{29}$$

where all quantities are physical parameters defined in [7] and only J_{xy} and J_z have unknown values, which should be blindly estimated.

The investigation reported in [6] was restricted to BQSS, i.e. it did not address the field of BQPT, which was introduced later in [7]. Moreover, the BQSS method of [6] is quite different from the approach introduced here in Sect. 4 , in terms of separation algorithms. However, both methods adapt γ_1 to γ_4 so as to reach the global minimum of the same cost functions (15) and (19). Besides, for the considered coupling model, this minimization may be shown to be equivalent to the disentanglement of the output states of the separating system, for the considered source states. The investigation reported in [7] then entails that, when applying the new method of Sect. 4, γ_1 to γ_4 are tuned to final values which are such that

$$J_{xy} = \frac{\hbar}{2(t - t_0)}(\gamma_3 - \gamma_2 - m\pi) \tag{30}$$

$$J_z = \frac{\hbar}{2(t - t_0)}(\gamma_2 + \gamma_3 - \gamma_1 - \gamma_4 + 2k\pi - m\pi) \tag{31}$$

where k and m are integers. This yields estimates of J_{xy} and J_z (up to the indeterminacies due to $2k\pi$ and $m\pi$).

6   Conclusion

In this paper, we extended the fields of Blind Quantum Source Separation (BQSS) and Process Tomography (BQPT). Whereas our recent papers focused on a separating system structure and associated disentanglement-based separation principle, we here put the emphasis on resulting algorithms. We thus developed much "faster" (in terms of the required number of source quantum state preparations) algorithms than the basic ones that we initially proposed. Since the practical implementation of the considered type of coupled qubits is still quite difficult today, the next steps of this investigation will especially consist in developing a complete classical software emulation of Heisenberg-coupled spins 1/2, to which we will then apply a classical implementation of the methods proposed in this paper, in order to evaluate the performance of these methods.

References

1. Comon, P., Jutten, C. (eds.): Handbook of Blind Source Separation: Independent Component Analysis and Applications. Academic Press, Oxford (2010)
2. Deville, Y., Deville, A.: Blind separation of quantum states: estimating two qubits from an isotropic heisenberg spin coupling model. In: Davies, M.E., James, C.J., Abdallah, S.A., Plumbley, M.D. (eds.) ICA 2007. LNCS, vol. 4666, pp. 706–713. Springer, Heidelberg (2007). doi:10.1007/978-3-540-74494-8_88. Erratum: replace two terms $E\{r_i\}E\{q_i\}$ in (33) of [2] by $E\{r_iq_i\}$, since q_i depends on r_i.
3. Deville, Y., Deville, A.: Classical-processing and quantum-processing signal separation methods for qubit uncoupling. Quantum Inf. Proc. **11**, 1311–1347 (2012)
4. Deville, Y., Deville, A.: A quantum-feedforward and classical-feedback separating structure adapted with monodirectional measurements; blind qubit uncoupling capability and links with ICA. In: Proceedings of MLSP 2013, Track on LVA, Southampton, United Kingdom, 22–25 September 2013
5. Deville, Y., Deville, A.: Quantum-source independent component analysis and related statistical blind qubit uncoupling methods, Chap. 1. In: Naik, G.R., Wang, W. (eds.) Blind Source Separation: Advances in Theory, Algorithms and Applications. Springer, Heidelberg (2014)
6. Deville, Y., Deville, A.: Blind qubit state disentanglement with quantum processing: principle, criterion and algorithm using measurements along two directions. In: Proceedings of ICASSP 2014, Florence, Italy, pp. 6262–6266, 4–9 May 2014
7. Deville, Y., Deville, A.: From blind quantum source separation to blind quantum process tomography. In: Vincent, E., Yeredor, A., Koldovský, Z., Tichavský, P. (eds.) LVA/ICA 2015. LNCS, vol. 9237, pp. 184–192. Springer, Heidelberg (2015). doi:10.1007/978-3-319-22482-4_21
8. Nielsen, M.A., Chuang, I.L.: Quantum Computation and Quantum Information. Cambridge University Press, Cambridge (2000)

Blind Separation of Cyclostationary Sources with Common Cyclic Frequencies

Amine Brahmi[1,2(✉)], Hicham Ghennioui[2], Christophe Corbier[1], M'hammed Lahbabi[2], and François Guillet[1]

[1] Université de Lyon, UJM-Saint-Etienne, LASPI, IUT de Roanne, 42334 Roanne, France
{amine.brahmi,christophe.corbier,guillet}@univ-st-etienne.fr
[2] Université Sidi Mohamed Ben Abdellah, FSTF, LSSC, B.P. 2202, Route d'Immouzzer, Fès, Maroc
{hicham.ghennioui,mhammed.lahbabi}@usmba.ac.ma

Abstract. We propose a new method for blind source separation of cyclostationary sources, whose cyclic frequencies are unknown and may share one or more common cyclic frequencies. The suggested method exploits the second-order cyclostationarity statistics of observation signals to build a set of matrices which has a particular algebraic structure. We also introduce an automatic point selection procedure for the determination of these matrices to be joint diagonalized in order to identify the mixing matrix and recover the source signals as a result. The non-unitary joint diagonalization is ensured by Broyden-Fletcher-Goldfarb-Shanno (BFGS) method which is the most commonly used update strategy for implementing a quasi-newton technique. Numerical simulations are provided to demonstrate the usefulness of the proposed method in the context of digital communications and to compare it with another method based upon an unitary joint diagonalization algorithm.

Keywords: Blind source separation · Joint diagonalization · BFGS method · Second-order cyclostationarity statistics · Automatic point selection procedure · Common cyclic frequencies

1 Introduction

The Blind Source Separation (BSS) is a major problem of signal processing which has been addressed in the last three decades (see [4] for a review) where the objective is to recover the unobserved input signals called sources from their observed unknown mixtures coming to multiple sensors without any pre-knowledge of mixing process. In literature, first researchers have published on this problem and most of their approaches are based on stationarity using the second-order statistics (SOS) or the high-order statistics (HOS) (JADE [3], SOBI [2]). They have proved to establish some limitations in real-world applications where the source signals are very often cyclostationary such as digital communications,

© Springer International Publishing AG 2017
P. Tichavský et al. (Eds.): LVA/ICA 2017, LNCS 10169, pp. 449–458, 2017.
DOI: 10.1007/978-3-319-53547-0_42

mechanics, biomedical engineering and so on. Cyclostationary signals is are frequent random ones with statistical parameters that alter in periodic manner in time. Numerous examples of phenomena producing cyclostationary signals are given in [7]. Recent methods have been proposed to blindly achieve the separation for cyclostationary sources, Abed Meraim et al. [1] suggested to optimize a contrast function built from the cyclic correlation of recovered sources at different time lags, and they presented iterative update equations following from the natural gradient technique. This method is efficient in the case when each source signal has only one cyclic frequency and the number of the source signals which share a common cyclic frequency is known. Ghennioui et al. [9] have proposed a new approach combining a non-unitary joint diagonalization algorithm to a general automatic matrices selection procedure for the case of unknown and different cyclic frequencies. Ghaderi et al. present in [8] a method for blind source extraction of cyclostationary sources, whose cyclic frequencies are known and share some common ones. Ferreol and Chevalier have shown in [6] that the current second or higher order BSS methods perform badly with the stationarity assumption of source signals. Jallon and Chevreuil [13] have given a simple condition on the statistics of the cyclostationary sources which ensures that the maximization of a contrast function performs BSS. Dinh-Tuan Pham [5] have proposed a new approach based of joint diagonalization of a set of cyclic spectral density of observation matrices. The two last approaches are addressed in the simplest mixture model (noise-free data). Despite of the fact theses algorithms are successful under assumed conditions, they have diverse limitations, since in front of real situations, the cyclic frequencies in most of cases are unknown, and may be shared by source signals. The main purpose of this work is to provide a new solution to the blind separation of instantaneous mixtures of cyclostationary source signals which may share one or more unknown common cyclic frequencies. By exploiting the particular structure of cyclic correlation matrices of source signals, we establish that the problem of interest can be rephrased as the problem of joint diagonalization of matrices accordingly chosen using a new detection method. The joint diagonalization algorithm is ensured bythe BFGS Method [15]. Finally, we provide computer simulations in order to point up the proposed approach efficiency in digital telecommunications context.

2   Problem Formulation

The BSS problem can be modelled as a simple linear instantaneous mixture of n emitted source signals that are received by m sensors. The input/output relationship of mixing system is given by:

$$\mathbf{x}(t) = \mathbf{A}\mathbf{s}(t) + \mathbf{b}(t), \tag{1}$$

where $\mathbf{x}(t) \in \mathbb{C}^m$ is the observations vector, $\mathbf{s}(t) \in \mathbb{C}^n$ is the vector of unknown source signals, $\mathbf{b}(t) \in \mathbb{C}^m$ is the additive noise vector and $\mathbf{A} \in \mathbb{C}^{m \times n}$ is the unknown mixing matrix. The purpose of BSS is to find an estimate $\tilde{\mathbf{A}}$ of $\mathbf{A}$ and recover $\mathbf{s}(t)$ from $\mathbf{x}(t)$ only, as: $\tilde{\mathbf{s}}(t) = \tilde{\mathbf{A}}^{\#}\mathbf{x}(t) = \mathbf{P}\mathbf{\Delta}\mathrm{diag}(e^{-j\Phi})\mathbf{s}(t),$

where $j^2 = -1$, $(.)^\#$ denotes pseudo-inverse matrix, $\mathbf{P}$ is a permutation matrix (corresponding to an arbitrary order of restitution of the sources), $\boldsymbol{\Delta}$ is a diagonal matrix (corresponding to arbitrary scaling for the recovered sources), $\boldsymbol{\Phi} = [\phi_1, \ldots, \phi_n]^T$, $\forall \phi_i \in \mathbb{R}$ represents the phase vector (corresponding to phase shift ambiguity in complex domain of the sources) and $\text{diag}(\mathbf{a})$ is square diagonal matrix containing the elements of the vector $\mathbf{a}$. Thus, one should look for a separating matrix $\tilde{\mathbf{A}}^\#$ such that $\tilde{\mathbf{A}}^\# \mathbf{A} = \mathbf{P}\boldsymbol{\Delta}\text{diag}(e^{-j\boldsymbol{\Phi}})$. The following assumptions are held in through this paper:

A1. The mixing matrix $\mathbf{A}$ has full column rank $(m \geq n)$,
A2. The sources are zero-mean, cyclostationary, mutually independent and may share one or more unknown common cyclic frequencies,
A3. The components of noise signals vector $\mathbf{b}(t)$ are stationary, zero-mean random signals and mutually independent from the source signals.

3 Proposed Method

3.1 Preliminaries

Let us define the cyclic correlation matrix of a given signal $\mathbf{x}$

$$\mathbf{R}_{\mathbf{x}}^\xi(\tau) = \lim_{T \to \infty} \frac{1}{T} \int_{\frac{-T}{2}}^{\frac{T}{2}} \mathbf{E}\{\mathbf{x}(t + \frac{\tau}{2})\mathbf{x}^*(t - \frac{\tau}{2})\}e^{-j2\pi\xi t}dt, \tag{2}$$

where $\mathbf{E}\{.\}$ stands for the mathematical expectation. From (1), it can be easily shown that the cyclic correlation matrix of observation signals $\mathbf{x}(t)$ has the following decomposition:

$$\mathbf{R}_{\mathbf{x}}^\xi(\tau) = \mathbf{A}\mathbf{R}_{\mathbf{s}}^\xi(\tau)\mathbf{A}^H + \mathbf{R}_{\mathbf{b}}^\xi(\tau), \tag{3}$$

where $(.)^H$ is the conjugate transpose operator, $\mathbf{R}_{\mathbf{s}}^\xi(\tau)$ (resp. $\mathbf{R}_{\mathbf{b}}^\xi(\tau)$) is the cyclic correlation matrix of source signals (resp. noise signals). The expression of $\mathbf{R}_{\mathbf{b}}^\xi(\tau)$ can be further developed. Under the noise assumption, we have: $\mathbf{R}_{\mathbf{b}}^\xi(\tau) = \mathbf{R}_{\mathbf{b}}(\tau)\delta(\xi)$, where $\delta(\xi) = \lim_{T \to \infty} \frac{1}{T}\int_{\frac{-T}{2}}^{\frac{T}{2}} e^{-j2\pi\xi t}dt$, this implies that: $\delta(\xi) = \begin{cases} 1 & \text{if } \xi = 0 \\ 0 & \text{else} \end{cases}$, therefore, if $\xi \neq 0$ then, $\mathbf{R}_{\mathbf{x}}^\xi(\tau) = \mathbf{A}\mathbf{R}_{\mathbf{s}}^\xi(\tau)\mathbf{A}^H$. Practically, the matrix $\mathbf{R}_{\mathbf{s}}^\xi(\tau)$ has one of the following structures:

S1. If the source signals have pairwise distinct cyclic frequencies then $\mathbf{R}_{\mathbf{s}}^\xi(\tau)$ is a diagonal matrix with only one non-null element corresponding to the i-th source at ξ_i (for this kind of structure, using $\mathbf{R}_{\mathbf{x}}^\xi(\tau)$ a detection procedure has been proposed in [11]),
S2. If the source signals share one or more cyclic frequencies then $\mathbf{R}_{\mathbf{s}}^\xi(\tau)$ at these shared frequencies is a diagonal matrix with n non-null elements which is our case of interest.

Thus, we propose to diagonalize simultaneously the set of cyclic correlation matrices of observation signals $\mathbf{R}_\mathbf{x}^\xi(\tau)$ at different time lags and cyclic frequencies ($\xi \neq 0$). The question now being asked is: how to compose the matrices set $\mathcal{M}_{jd}$ to be joint diagonalized?

3.2   Construction of the Matrices Set

It has to be noted that the pre-knowledge of the cyclic frequencies of source signals $\mathbf{s}(t)$ is optional even if such a thing facilitates the resolution of BSS issue. As matter of fact, we could take advantage of the particular algebraic structure of the matrix $\mathbf{R}_\mathbf{s}^\xi(\tau)$ to perform BSS. When the cyclic frequencies are unknown, we calculate $\mathbf{R}_\mathbf{x}^\xi(\tau)$ for a enough number of frequency bins in order to ensure sweeping almost all the cyclic frequencies of source signals, then, we use a new detection procedure to select the matrices $\mathbf{R}_\mathbf{x}^\xi(\tau)$ which correspond to the structure (S2). In other words, the procedure has to detect matrices $\mathbf{R}_{\mathbf{x}_b}^\xi(\tau) = \mathbf{W}\mathbf{R}_\mathbf{x}^\xi(\tau)\mathbf{W}^H$ ($\mathbf{W}$ is $n \times m$ whitening matrix such that $\mathbf{W}\mathbf{A}\mathbf{A}^H\mathbf{W}^H = \mathbf{I}_n$) which have the same number n of eigenvalues as $\mathbf{R}_\mathbf{s}^\xi(\tau)$ since all source signals are present at a given common cyclic frequency. One possible way to blindly access to diagonal terms of $\mathbf{R}_\mathbf{s}^\xi(\tau)$ consists in computing the eigenvalues of $\mathbf{R}_{\mathbf{x}_b}^\xi(\tau)$. In fact, using the whitening matrix definition in [3], we have:

$$\mathrm{eig}(\mathbf{R}_{\mathbf{x}_b}^\xi(\tau)) = \mathrm{eig}(\mathbf{W}\mathbf{A}\mathbf{R}_\mathbf{s}^\xi(\tau)\mathbf{A}^H\mathbf{W}^H) \tag{4}$$

$$= \mathrm{eig}(\mathbf{R}_\mathbf{s}^\xi(\tau)) = [\theta_1, \ldots, \theta_n]^T \tag{5}$$

where the vector $\mathrm{eig}(\mathbf{M})$ contains the eigenvalues of the square matrix $\mathbf{M}$, $\mathbf{W}\mathbf{A}$ is an unitary matrix and θ_i, for $i = 1, \ldots, n$ are the eigenvalues of $\mathbf{R}_\mathbf{s}^\xi(\tau)$. Therefore, we propose the following criteria:

$$\mathbf{C}_1 = \frac{\sum\limits_{i=1}^{n}(\theta_i)^2}{\|\mathbf{R}_{\mathbf{x}_b}^\xi(\tau)\|_F^2} \geq 1 - \varepsilon, \tag{6}$$

where $\|.\|_F^2$ denotes the Frobenius norm and ε is positive constant. We note that, we also exploit the invariance property of the Frobenius norm under an unitary transformation : $\|\mathbf{R}_{\mathbf{x}_b}^\xi(\tau)\|_F^2 = \|\mathbf{R}_\mathbf{s}^\xi(\tau)\|_F^2$. Ideally $\mathbf{C}_1$ equals 1. However, in practice, the matrices $\mathbf{R}_\mathbf{s}^\xi(\tau)$ can never be strictly diagonal. Thus, matrices $\mathbf{R}_{\mathbf{x}_b}^\xi(\tau)$ should be selected as $\mathbf{C}_1 \geq 1 - \varepsilon$ with ε is close to zero. Furthermore, in order to avoid choosing matrices $\mathbf{R}_{\mathbf{x}_b}^\xi(\tau)$ with low values in the main diagonal, we add to $\mathbf{C}_1$:

$$\mathbf{C}_2 = \det(\mathbf{R}_{\mathbf{x}_b}^\xi(\tau)) = \det(\mathbf{R}_\mathbf{s}^\xi(\tau)) \geq \eta \tag{7}$$

where $\det(.)$ denotes the matrix determinant, and η is a small positive constant. Finally, if a given $\mathbf{R}_{\mathbf{x}_b}^\xi(\tau)$ satisfies $\mathbf{C}_1$ and $\mathbf{C}_2$ then it is retained. Once, the matrices set is built, it is directly joint diagonalized. However, when the cyclic frequencies are a priori known, the operation is much more easier, it is reduced to computing $\mathbf{R}_\mathbf{x}^\xi(\tau)$ at different time lags for each cyclic frequency $\{\xi_i/i = 1,\ldots,n\}$ then to diagonalize simultaneously the set built $\mathcal{M}_{jd} = \{\mathbf{R}_\mathbf{x}^{\xi_i}(\tau_1),\ldots,\mathbf{R}_\mathbf{x}^{\xi_i}(\tau_{max})\}$. The joint diagonalization is ensured by the BFGS method which will be detailed in the following subsection.

3.3 Non-unitary Joint Diagonalization Algorithm

In order to solve the non unitary joint diagonalization problem, we consider the following cost function [10]:

$$\mathcal{F}_{jd}(\mathbf{B}) = \sum_{i=1}^{K} \|\text{Offdiag}\{\mathbf{B}\mathbf{R}_\mathbf{x}^{\xi_i}(\tau_i)\mathbf{B}^H\}\|_F^2, \tag{8}$$

where Offdiag$\{.\}$ denotes the zero-diagonal operator and $\mathbf{R}_\mathbf{x}^{\xi_i}(\tau_i)$ is the i-th matrix which belongs to the set $\mathcal{M}_{jd}$ to be joint diagonalized. We propose an approach based on a BFGS method with an exact computation of the optimal step-size. It estimates the joint diagonalizer matrix $\mathbf{B} \in \mathbb{C}^{n \times m}$ by minimizing problem given in Eq.(8). The BFGS method requires the gradient and the Hessian of the cost function to be computed at each iteration. By successive measurements of the gradient and the Hessian, it builds a quadratic model of the objective function which is sufficiently good that super-linear convergence is achieved.

Algorithm Principle. From initial guesses $\mathbf{B}_{(0)}$, $\mathbf{He}_{(0)}$ and a given number of iterations n_{iter}, the following steps are iterated as $\mathbf{B}_{(k)}$ converges to the solution.

S1. Obtain a search direction $\mathbf{d}_{(k-1)}$ by solving:

$$\mathbf{d}_{(k-1)} = -\mathbf{He}_{(k-1)}^{-1}\nabla_a\mathcal{F}_{jd}(\mathbf{B}_{(k-1)});$$

S2. Perform a line search to find the optimal step-size α (positive a small enough number) in the previous direction;

S3. Update $\mathbf{B}_{(k)} = \mathbf{B}_{(k-1)} + \alpha\mathbf{d}_{(k-1)}$;

S4. Set

$$\mathbf{s}_{(k-1)} = \alpha\mathbf{d}_{(k-1)};$$
$$\mathbf{y}_{(k-1)} = \nabla_a\mathcal{F}_{jd}(\mathbf{B}_{(k)}) - \nabla_a\mathcal{F}_{jd}(\mathbf{B}_{(k-1)});$$

S5. Using rank-one updates specified by gradient evaluations, the Hessian matrix is approximated as follows (see [15]):

$$\mathbf{He}_{(k)} = \mathbf{He}_{(k-1)} + \frac{\mathbf{y}_{(k-1)}\mathbf{y}_{(k-1)}^T}{\mathbf{y}_{(k-1)}^T\mathbf{s}_{(k-1)}} - \frac{\mathbf{He}_{(k-1)}\mathbf{s}_{(k-1)}\mathbf{s}_{(k-1)}^T\mathbf{He}_{(k-1)}}{\mathbf{s}_{(k-1)}^T\mathbf{He}_{(k-1)}\mathbf{s}_{(k-1)}}, \tag{9}$$

$[.]^{-1}$ and $[.]^T$ denote the inverse and transpose of a matrix respectively, $\nabla_a\mathcal{F}_{jd}(\mathbf{B})$ is the complex absolute gradient matrix of the cost function given in Eq.(8) which is defined, as (see [12]): $\nabla_a\mathcal{F}_{jd}(\mathbf{B}) = 2\frac{\partial\mathcal{F}_{jd}(\mathbf{B})}{\partial\mathbf{B}^*}$, where $\mathbf{B}^*$ stands for the complex conjugate of the complex matrix $\mathbf{B}$ and $\frac{\partial}{\partial\mathbf{B}^*}$ is the partial derivative operator. $\nabla_a\mathcal{F}_{jd}(\mathbf{B})$ was calculated earlier in [10]. Practically, $\mathbf{B}_{(0)}$ has to be a full-rank matrix chosen different from the zero matrix as it is a trivial solution of (8) and $\mathbf{He}_{(0)}$ can be initialized with the identity matrix.

Enhanced Line Search. It is taken for granted that finding a good step-size α in search direction is critical issue for decreasing the total number of iterations to reach convergence. The enhanced line search consists of minimization $\mathcal{F}_{jd}(\mathbf{B}_{(k)})$ w.r.t. α. For simplicity, we prefer to hide the dependency upon the iteration k. It is a matter of standard algebraic manipulations to show that $\mathcal{F}_{jd}(\mathbf{B}_{(k)})$ is a 4th degree polynomial in α whose expression is:

$$\mathcal{F}_{jd}(\mathbf{B}) = a\alpha^4 + b\alpha^3 + c\alpha^2 + d\alpha + e. \tag{10}$$

where its coefficients are given by:

$$a = \sum_{i=1}^{K} \mathrm{tr}\left[\mathbf{M}_3 \mathsf{Offdiag}\{\mathbf{M}_3^H\}\right],$$

$$b = -\sum_{i=1}^{K} \mathrm{tr}\left[\mathbf{M}_0 \mathsf{Offdiag}\{\mathbf{M}_3^H\}\right] - \mathrm{tr}\left[\mathbf{M}_3\mathbf{M}_2\right],$$

$$c = \sum_{i=1}^{K} \mathrm{tr}\left[\mathbf{M}_1^H \mathsf{Offdiag}\{\mathbf{M}_3^H\} + \mathbf{M}_3 \mathsf{Offdiag}\{\mathbf{M}_1\}\right] + \mathrm{tr}\left[\mathbf{M}_0\mathbf{M}_2\right], \tag{11}$$

$$d = -\sum_{i=1}^{K} \mathrm{tr}\left[\mathbf{M}_1^H \mathbf{M}_2\right] - \mathrm{tr}\left[\mathbf{M}_0 \mathsf{Offdiag}\{\mathbf{M}_1\}\right],$$

$$e = \sum_{i=1}^{K} \mathrm{tr}\left[\mathbf{M}_1^H \mathsf{Offdiag}\{\mathbf{M}_1\}\right],$$

and,

$$\mathbf{M}_0 = \mathbf{R}_\mathbf{x}^{\xi_i}(\tau_i)^H \mathbf{\Gamma}^H \mathbf{B} + \mathbf{R}_\mathbf{x}^{\xi_i}(\tau_i)^H \mathbf{B}^H \mathbf{\Gamma}, \quad \mathbf{M}_1 = \mathbf{B}\mathbf{R}_\mathbf{x}^{\xi_i}(\tau_i)\mathbf{B}^H,$$

$$\mathbf{M}_2 = \mathsf{Offdiag}\{\mathbf{\Gamma}\mathbf{R}_\mathbf{x}^{\xi_i}(\tau_i)\mathbf{B}^H\} + \mathsf{Offdiag}\{\mathbf{B}\mathbf{R}_\mathbf{x}^{\xi_i}(\tau_i)\mathbf{\Gamma}^H\}, \quad \mathbf{M}_3 = \mathbf{\Gamma}\mathbf{R}_\mathbf{x}^{\xi_i}(\tau_i)^H \mathbf{\Gamma}^H,$$

with $\mathrm{tr}\,[.]$ denotes the trace operator and $\mathbf{\Gamma} = \mathbf{He}^{-1}\nabla_a\mathcal{F}_{jd}(\mathbf{B})$. The optimal α can be found by polynomial rooting of the derivative third-order polynomial, namely by solving w.r.t. α,

$$\frac{\partial\mathcal{F}_{jd}(\mathbf{B})}{\partial\alpha} = 0 \Leftrightarrow 4a\alpha^3 + 3b\alpha^2 + 2c\alpha + d = 0.$$

To which there is three roots, the minimum can be figured out by substituting each root back into the polynomial given in (10) and selecting the solution that provides the littlest value. The Cardano's formula for cubics could also be used to found algebraically the roots. Regarding to the algorithmic complexity, for the gradient matrix whose expression is given in [10], the computational cost approximatively amounts to $4Knm(m+n) + 2Kn^2$ operations (i.e. the cost $\simeq o(4Knm^2)$) if $m \gg n$ or $\simeq o(8Kn^3 + 2Kn^2)$ in the square case $m = n$), the computational cost of the optimal step-size is ruled by the the computation of the 5 coefficients of the 4th degree polynomial, it approximatively amounts to $24Kmn(m+n) + 9Kn^2(1+n)$ operations (i.e. the cost $\simeq o(24Km^2n)$) if $m \gg n$ or $\simeq o(57Kn^3 + 9Kn^2)$ in the square case). Therefore, the computational cost of BFGS algorithm is $24Kmn(m+n) + 9Kn^2(1+n)$ operations (i.e. the cost $\simeq o(24Km^2n)$) if $m \gg n$ or $\simeq o(57Kn^3 + 9Kn^2)$ in the square case). Finally, notice that the global complexity of the BFGS algorithm has to be multiplied by the overall iterations number N_i required to reach the convergence. In practical applications, the computational time necessary to compose the set of the K matrices should be counted too.

3.4 Summary of the Proposed Method

The principle of the proposed is summarized in the following table:

Algorithm 1. Proposed method

Set building :
-Detect the useful matrices $\mathbf{R}_X^\xi(\tau)$ using the selection procedure;
-Compute the set matrices $\mathcal{M}_{jd}$ for each cyclic frequency ξ_i:
$\mathcal{M}_{jd} = \{\mathbf{R}_\mathbf{x}^{\xi_i}(\tau_1), \dots, \mathbf{R}_\mathbf{x}^{\xi_i}(\tau_{max})\}$;
Initialization :
-Given an initial guess $\mathbf{B}_{(0)}$ and an approximate Hessian $\mathbf{He}_{(0)}$;
-Given the number of iterations n_{iter} and a small positive threshold ϵ;
for $k \leftarrow 1, n_{iter}$ **do**
  Compute $\nabla_a \mathcal{F}_{jd}(\mathbf{B})$;
  Perform a line search to find the optimal step-size α;
  Set $\mathbf{B}_{(k)} \leftarrow \mathbf{B}_{(k-1)} - \alpha \mathbf{He}_{(k-1)}^{-1} \nabla_a \mathcal{F}_{jd}(\mathbf{B}_{(k-1)})$;
  Compute $\mathbf{He}_{(k)}$ whose expression is given by Eq. (9);
  if $\mid \mathbf{B}_{(k)} - \mathbf{B}_{(k-1)} \mid \leq \epsilon$ **then**
    break;
  end if
end for

---

4 Numerical Simulations

We present computer simulations to point up the proposed method efficiency in the BSS context. We consider two amplitude-modulated source signals defined as:

$$\mathbf{s}_i(t) = \sum_{k \in \mathbb{Z}} \mathbf{a}_i(k) \mathbf{g}(t - kT_i) \cos(2\pi f_i t + \phi_i),$$

where $\mathbf{a}_{i=1,2}(k)$ are independent and identically distributed zero-mean random binary sequences, $T_{i=1,2}$ represent the period symbol, they are respectively equal to $T_1 = 10$, $T_2 = 8$, $f_{i=1,2} = 0.2$ are the normalized carrier frequencies, $\phi_{i=1,2}$ are the carrier phase, they respectively equal $\phi_1 = \frac{\pi}{6}$, $\phi_2 = \frac{\pi}{8}$ and $\mathbf{g}(n)$ is a triangular waveform such that :

$$\mathbf{g}(n) = \begin{cases} \frac{2}{T}n & \text{if } 0 \leq n \leq \frac{T}{2} \\ -\frac{2}{T}n + 2 & \text{if } \frac{T}{2} + 1 \leq n \leq T - 1 \\ 0 & \text{else.} \end{cases}$$

The source signals share multiple cyclic frequencies $f_1 = f_2 = 0.2$, $5f_1 = \dfrac{10}{T_1} = \dfrac{8}{T_2} = 1$ and their multiples. The mixing matrices were randomly generated, we use $\mathbf{A}_1$ in the first two simulations and $\mathbf{A}_2$ in the last one:

$$\mathbf{A}_1 = \begin{pmatrix} 0.5599 & 0.4070 \\ 0.2899 & 0.8431 \\ 0.9415 & 0.3397 \end{pmatrix}, \quad \mathbf{A}_2 = \begin{pmatrix} 0.8400 & 0.9971 \\ 0.7035 & 0.1965 \end{pmatrix}. \tag{12}$$

The signal to noise ratio (SNR) is computed as $\mathsf{SNR} = -10\log_{10}(\sigma_b^2)$.

We use the detection procedure described in (6) and (7) with $\varepsilon = 10^{-3}$ and $\eta = 10^{-1}$ for all simulations to compose the set of matrices to be joint diagonalized. In order to assess the quality of the estimation, one can measure the Moreau-Amari index presented in [14] defined as:

$$\mathsf{I}_{\mathrm{perf}} = \frac{1}{n(n-1)} \left[\sum_{i=1}^{n} \left(\sum_{j=1}^{n} \frac{\|g_{i,j}\|_F^2}{\max_\ell \|g_{i,\ell}\|_F^2} - 1 \right) + \sum_{j=1}^{n} \left(\sum_{i=1}^{n} \frac{\|g_{i,j}\|_F^2}{\max_\ell \|g_{\ell,j}\|_F^2} - 1 \right) \right], \tag{13}$$

where $g_{i,j}$ is the $(i,j)^{th}$ element of $\mathbf{G} = \mathbf{BA}$. The closer to zero in linear scale ($-\infty$ in logarithmic scale) is, the higher separation accuracy is. Regarding to the

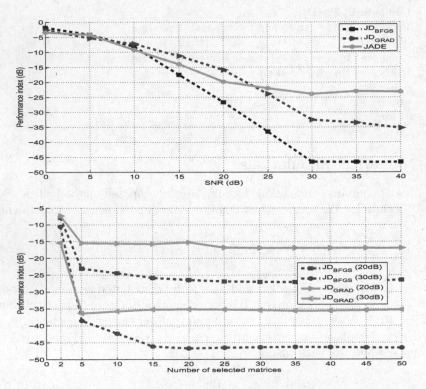

Fig. 1. Performance index versus SNR (top) and versus number of selected matrices (bottom) with $\mathbf{A}_1$.

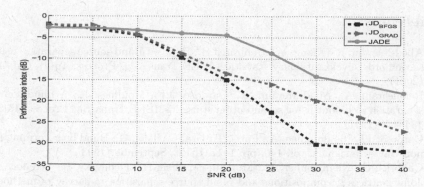

Fig. 2. Performance index versus SNR with $\mathbf{A}_2$.

charts, $I_{\text{perf}}(\cdot)$ is given in decibel and estimated by averaging 100 independent trials. The proposed method JD_{BFGS} is compared to two types of methods. The first one is based on unitary joint diagonalization algorithms: JADE, while the second one is based on the non-unitary joint diagonalization using the optimal gradient descent JD_{GRAD} [9] preceded by the automatic selection procedure of matrices presented previously. Figures. 1 and 2 show that the proposed algorithm takes the lead ahead its competitors in all two cases of the mixing matrix ($m > n$ or $m = n$). In addition, one can notice in the Fig. 1 that when the number of selected matrices to be joint diagonalized gets bigger, the performances are better, the counterpart being that the computational cost increases too. It has to be noted that The performances clearly increase when the signal-to-noise ratio gets higher.

5   Conclusion

To conclude, we have proposed a new approach dedicated to blindly separate instantaneous mixtures of cyclostationary sources. It operates into three steps: first, we compute the cyclic autocorrelation matrices of observation signals, then, a new detection procedure is used to select specific matrices and finally a non unitary joint diagonalization algorithm based on BFGS method is employed to estimate the mixing matrix and recover the sources. One of the main advantages of such an approach is that it applies for source signals which may share one or more common cyclic frequencies. Extensions for further research would be to undertake the blind separation of convolutive mixtures of cyclostationary sources.

Acknowledgment. The work was funded by the Erasmus Mundus Programme of the European Union. We appreciatively acknowledge their financial support.

References

1. Abed-Meraim, K., Xiang, Y., Manton, J.H., Hua, Y.: Blind source separation using second-order cyclostationary statistics. IEEE Trans. Sig. Process. **49**(4), 694–701 (2001)
2. Belouchrani, A., Abed-Meraim, K., Cardoso, J.F., Moulines, E.: A blind source separation technique using second-order statistics. IEEE Trans. Sig. Process. **45**(2), 434–444 (1997)
3. Cardoso, J.F., Souloumiac, A., Paris, T., Ura, C., Tdsi, G.: Blind beamforming for non Gaussian signals, vol. 140, pp. 1–17. IEEE, September 1993
4. Chabriel, G., Kleinsteuber, M., Moreau, E., Shen, H., Tichavsky, P., Yeredor, A.: Joint matrices decompositions and blind source separation: a survey of methods, identification, and applications. IEEE Sig. Process. Mag. **31**(3), 34–43 (2014)
5. Pham, D.T.: Blind separation of cyclostationary sources using joint block approximate diagonalization. In: Davies, M.E., James, C.J., Abdallah, S.A., Plumbley, M.D. (eds.) ICA 2007. LNCS, vol. 4666, pp. 244–251. Springer, Heidelberg (2007). doi:10.1007/978-3-540-74494-8_31
6. Ferreol, A.: On the behavior of current second and higher order blind source separation methods for cyclostationary sources. IEEE Trans. Sig. Process. **48**(6), 1712–1725 (2000)
7. Gardner, W.A., Napolitano, A., Paura, L.: Cyclostationarity: half a century of research. Sign. Process. **86**(4), 639–697 (2006)
8. Ghaderi, F., Makkiabadi, B., McWhirter, J.G., Sanei, S.: Blind source extraction of cyclostationary sources with common cyclic frequencies. In: Proceedings of the IEEE International Conference on Acoustics, Speech and Signal Processing, ICASSP, pp. 4146–4149. IEEE (2010)
9. Ghennioui, H., Thirion-Moreau, N., Moreau, E., Aboutajdine, D., Adib, A.: A novel approach based on non-unitary joint block-diagonalization for the blind MIMO equalization of cyclo-stationary signals. In: European Signal Processing Conference (2008)
10. Ghennioui, H., Thirion-Moreau, N., Moreau, E., Aboutajdine, D.: Gradient-based joint block diagonalization algorithms: application to blind separation of FIR convolutive mixtures. Sig. Process. **90**(6), 1836–1849 (2010)
11. Giulieri, L., Ghennioui, H., Thirion-Moreau, N., Moreau, E.: Nonorthogonal joint diagonalization of spatial quadratic time-frequency matrices for source separation. IEEE Sig. Process. Lett. **12**(5), 415–418 (2005)
12. Hjørungnes, A., Gesbert, D.: Hessians of scalar functions of complex-valued matrices: FLA systematic computational approach. In: Proceedings of the 2007 9th International Symposium on Signal Processing and Its Appllications, ISSPA 2007 (2007)
13. Jallon, P., Chevreuil, A.: Separation of instantaneous mixtures of cyclo-stationary sources. Sig. Process. **87**(11), 2718–2732 (2007)
14. Moreau, E., Macchi, O.: New self-adaptative algorithms for source separation based on contrast functions. In: IEEE Signal Processing Workshop on Higher-Order Statistics, pp. 215–219. IEEE (1993)
15. Walker, H.F.: Quasi-Newton methods. SIAM J. Optim. **141**(10), 135–163 (1978)

Adaptation of a Gaussian Mixture Regressor to a New Input Distribution: Extending the C-GMR Framework

Laurent Girin[1,2(✉)], Thomas Hueber[1], and Xavier Alameda-Pineda[2,3]

[1] CNRS/Grenoble Alpes University/GIPSA-lab, Grenoble, France
{laurent.girin,thomas.hueber}@gipsa-lab.grenoble-inp.fr
[2] INRIA Rhône-Alpes, Grenoble, France
[3] University of Trento, Trento, Italy
xavier.alamedapineda@unitn.it

Abstract. This paper addresses the problem of the adaptation of a Gaussian Mixture Regression (MGR) to a new input distribution, using a limited amount of input-only examples. We propose a new model for GMR adaptation, called Joint GMR (J-GMR), that extends the previously published framework of Cascaded GMR (C-GMR). We provide an exact EM training algorithm for the J-GMR. We discuss the merits of the J-GMR with respect to the C-GMR and illustrate its performance with experiments on speech acoustic-to-articulatory inversion.

Keywords: GMM · Gaussian mixture regression · Adaptation · EM algorithm

1 Introduction

The Gaussian Mixture Regression (GMR) is an efficient regression technique derived from the Gaussian Mixture Model (GMM) [1]. The GMR is widely used in different areas of speech processing, e.g. voice conversion [2,3], acoustic-articulatory mapping [4,5], in image processing, e.g. head pose estimation from depth data [6], and in robotics [7].

Let us consider a GMR that has been trained on a large dataset of *input-output* joint observations. The problem addressed in this paper is the *adaptation* of this GMR to a (moderate) change in the distribution of input data using a limited set of new *input-only* samples. In a practical context, this aims at using a well-estimated GMR with input observations that no more faithfully follow the distribution observed during training. For example, in speech processing, this happens when considering a speaker different from the one used for training. To address this problem, we first proposed in [8] to adapt the model parameters related to input observations using two state-of-the-art adaptation techniques for GMM which are maximum a posteriori (MAP) [9] and maximum likelihood linear regression (MLLR) [10]. Then, we proposed in [11] a general framework called Cascaded GMR (C-GMR) and derived two implementations (see Fig. 1).

© Springer International Publishing AG 2017
P. Tichavský et al. (Eds.): LVA/ICA 2017, LNCS 10169, pp. 459–468, 2017.
DOI: 10.1007/978-3-319-53547-0_43

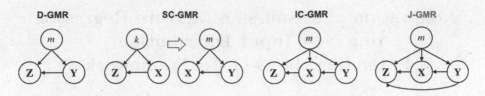

Fig. 1. Graphical representation of the D-, SC-, IC-, and J-GMR.

The first one, referred to as Split-C-GMR (SC-GMR), is a simple chaining of two consecutive GMRs. The second one, referred to as Integrated-C-GMR (IC-GMR) combines the two successive GMRs in a single probabilistic model. The IC-GMR puts into practice the general *missing data* methodology of machine learning [12] to exploit both the small and potentially sparse adaptation dataset and the large and dense training dataset. In [11], this model was shown to provide superior performance to the SC-GMR and to a direct GMR (D-GMR) trained only with new input data and corresponding output data. The D-GMR, SC-GMR and IC-GMR are briefly presented in Sect. 2.

In the present paper, we extend the general framework of C-GMR to a new model, called Joint GMR (J-GMR), presented in Sect. 3. Compared to the IC-GMR, the J-GMR can be considered as more general, in the sense that it aims at modeling the statistical dependencies between all the considered variables. In Sect. 4, we provide the exact associated EM algorithm [13] used to perform the adaptation to new input data. As for the IC-GMR, the J-GMR and associated EM algorithm consider explicitly the incomplete adaptation dataset jointly with the training dataset, using the missing data methodology. In Sect. 5 we illustrate the interest of the new model: The performance of the J-GMR is favorably compared to the D-JMR, SC-GMR and IC-GMR in a speech acoustic-to-articulatory inversion task on simulated data. Section 6 concludes the paper.

2   Cascaded GMR

2.1   Definitions, Notations and Working Hypothesis

Let us consider a GMR between realizations of input $\mathbf{X}$ and output $\mathbf{Y}$ (column) random vectors, of arbitrary finite dimension. Let us define a new input vector $\mathbf{Z}$ to which the GMR is to be adapted. Let us define $\mathbf{V} = [\mathbf{X}^\top, \mathbf{Y}^\top]^\top$ and $\mathbf{O} = [\mathbf{X}^\top, \mathbf{Y}^\top, \mathbf{Z}^\top]^\top$, where $^\top$ denotes the transpose operator. Let $p(\mathbf{X} = \mathbf{x}; \Theta_\mathbf{X})$ denote the probability density function (PDF) of $\mathbf{X}$ (for simplicity, we omit $\mathbf{X}$ and may omit $\Theta_\mathbf{X}$). Let $\mathcal{N}(\mathbf{x}; \mu_\mathbf{X}, \Sigma_\mathbf{XX})$ denote the Gaussian distribution of $\mathbf{X}$ with mean vector $\mu_\mathbf{X}$ and covariance matrix $\Sigma_\mathbf{XX}$. Let $\Sigma_\mathbf{XY}$ denote the cross-covariance matrix between $\mathbf{X}$ and $\mathbf{Y}$, and $\Lambda_\mathbf{XX}$ the precision matrix of $\mathbf{X}$ (similarly for cross-terms). With these notations, the PDF of a GMM on $\mathbf{V}$ writes:

$$p(\mathbf{v}) = \sum_{m=1}^{M} \pi_m \mathcal{N}(\mathbf{v}; \mu_{\mathbf{V},m}, \boldsymbol{\Sigma}_{\mathbf{VV},m}), \tag{1}$$

where M is the number of components, $\pi_m \geq 0$, and $\sum_{m=1}^{M} \pi_m = 1$.

Let us denote $\mathcal{D}_{\mathbf{xy}} = \{\mathbf{x}_n, \mathbf{y}_n\}_{n=1}^{N}$ the large dataset of N i.i.d. vector pairs drawn from the $(\mathbf{X}, \mathbf{Y})$ distribution, that is used for the training of the $\mathbf{X}$-to-$\mathbf{Y}$ GMR. We assume that a limited dataset $\mathcal{D}_{\mathbf{z}}$ of new input vectors $\mathbf{z}$ is available for the adaptation of the GMR. Moreover, we assume that $\mathcal{D}_{\mathbf{z}}$ can be aligned with a subset of the reference input dataset (e.g., in voice conversion, by time-aligning the same sentence pronounced by two speakers). Since reordering of the dataset is arbitrary, we denote $\mathcal{D}_{\mathbf{z}} = \{\mathbf{z}_n\}_{n=1}^{N_0}$, with $N_0 \ll N$.

2.2 D-GMR, SC-GMR and IC-GMR

In this section, we briefly recall three approaches for GMR adaptation considered in [11], which will be used here as a baseline. Their graphical representation is illustrated in Fig. 1. The first one is a direct $\mathbf{Z}$-to-$\mathbf{Y}$ GMR trained using $\mathcal{D}_{\mathbf{zy}} = \{\mathbf{z}_n, \mathbf{y}_n\}_{n=1}^{N_0}$. Inference of $\mathbf{y}$ given an observed value $\mathbf{z}$ is performed by the Minimum Mean Squared Error (MMSE) estimator, which is the posterior mean $\hat{\mathbf{y}} = \mathrm{E}[\mathbf{Y}|\mathbf{z}]$:

$$\hat{\mathbf{y}} = \sum_{m=1}^{M} p(m|\mathbf{z}) \left(\mu_{\mathbf{Y},m} + \boldsymbol{\Sigma}_{\mathbf{YZ},m} \boldsymbol{\Sigma}_{\mathbf{ZZ},m}^{-1} (\mathbf{z} - \mu_{\mathbf{Z},m}) \right), \tag{2}$$

with $p(m|\mathbf{z}) = \frac{\pi_m \mathcal{N}(\mathbf{z}|\mu_{\mathbf{Z},m}, \boldsymbol{\Sigma}_{\mathbf{ZZ},m})}{\sum_{k=1}^{M} \pi_k \mathcal{N}(\mathbf{z}|\mu_{\mathbf{Z},k}, \boldsymbol{\Sigma}_{\mathbf{ZZ},k})}$ and M is the number of mixture compo-nents. This model is referred to as D-GMR. The second and third models are instances of cascaded GMR. The split-cascaded GMR (SC-GMR) consists of chaining two distinct GMRs: a $\mathbf{Z}$-to-$\mathbf{X}$ GMR followed by a $\mathbf{X}$-to-$\mathbf{Y}$ GMR. The inference equation thus consists in chaining $\hat{\mathbf{x}} = \mathrm{E}[\mathbf{X}|\mathbf{z}]$ and $\hat{\mathbf{y}} = \mathrm{E}[\mathbf{Y}|\hat{\mathbf{x}}]$, where both expectations follow (2) with their respective parameters. Note that the two GMRs may have a different number of mixture components. The integrated-cascaded GMR (IC-GMR) combines the $\mathbf{Z}$-to-$\mathbf{X}$ mapping and the $\mathbf{X}$-to-$\mathbf{Y}$ mapping into a single GMR-based mapping process. Importantly, this is made at the component level of the GMR, i.e. *within the mixture*, as opposed to the SC-GMR (see Fig. 1). The corresponding IC mixture model is defined by:

$$p(\mathbf{o}) = \sum_{m=1}^{M} \pi_m p(\mathbf{y}|m) p(\mathbf{x}|\mathbf{y}, m) p(\mathbf{z}|\mathbf{x}, m), \tag{3}$$

where all PDFs are Gaussian. The IC-GMR is given by:

$$\hat{\mathbf{y}} = \sum_{m=1}^{M} p(m|\mathbf{z})[\mu_{\mathbf{Y},m} + \boldsymbol{\Sigma}_{\mathbf{YX},m} \boldsymbol{\Sigma}_{\mathbf{XX},m}^{-1} \boldsymbol{\Sigma}_{\mathbf{XZ},m} \boldsymbol{\Sigma}_{\mathbf{ZZ},m}^{-1} (\mathbf{z} - \mu_{\mathbf{Z},m})]. \tag{4}$$

The above equation is a $\mathbf{Z}$-to-$\mathbf{Y}$ GMR with a specific form of the covariance matrix, i.e. $\boldsymbol{\Sigma}_{\mathbf{YZ},m}$ is not a free parameter:

$$\boldsymbol{\Sigma}_{\mathbf{YZ},m} = \boldsymbol{\Sigma}_{\mathbf{YX},m} \boldsymbol{\Sigma}_{\mathbf{XX},m}^{-1} \boldsymbol{\Sigma}_{\mathbf{XZ},m}. \tag{5}$$

3 Joint GMR

Let us define the *Joint* GMM on $(\mathbf{X}, \mathbf{Y}, \mathbf{Z})$ as:

$$p(\mathbf{o}) = \sum_{m=1}^{M} \pi_m \mathcal{N}(\mathbf{o}; \mu_m, \boldsymbol{\Sigma}_m). \tag{6}$$

The associated J-GMR inference equation is again given by the posterior mean $\hat{\mathbf{y}} = E[\mathbf{Y}|\mathbf{z}]$. Here, we have:

$$p(\mathbf{y}|\mathbf{z}) = \int_{\mathbf{X}} \sum_{m=1}^{M} p(\mathbf{x}, \mathbf{y}, m|\mathbf{z}) d\mathbf{x} = \sum_{m=1}^{M} p(m|\mathbf{z}) p(\mathbf{y}|\mathbf{z}, m). \tag{7}$$

Since the conditional and marginal distributions of a Gaussian are Gaussian as well, (7) is a GMM. Therefore, the J-GMR inference equation turns out to be identical to the usual expression for a direct $\mathbf{Z}$-to-$\mathbf{Y}$ GMR, i.e. (2). This gives the impression of by-passing the information contained in $\mathbf{X}$. However, this is not the case: the complete proposed process for GMR adaptation is not equivalent to a GMR build directly from $(\mathbf{z}, \mathbf{y})$ training data. Indeed, as shown in the next section, the estimation of the J-GMR parameters with the EM algorithm exploits all the available data, i.e. $\mathcal{D}_{\mathbf{xy}}$ and $\mathcal{D}_{\mathbf{z}}$, hence including all $\mathbf{x}$ data.

Remarkably, (5) characterizes the IC-GMR as a particular case of the J-GMR. This is also true at the mixture model level, i.e. (3) is a particular case of (6) with (5). The matrix product $\boldsymbol{\Sigma}_{\mathbf{XZ},m} \boldsymbol{\Sigma}_{\mathbf{ZZ},m}^{-1}$ in (4) enables to go from $\mathbf{z}$ to $\mathbf{x}$, and then $\boldsymbol{\Sigma}_{\mathbf{YX},m} \boldsymbol{\Sigma}_{\mathbf{XX},m}^{-1}$ enables to go from $\mathbf{x}$ to $\mathbf{y}$, so that the IC-GMR goes from $\mathbf{z}$ to $\mathbf{y}$ "passing through $\mathbf{x}$". In contrast, the J-GMR enables to go directly from $\mathbf{z}$ to $\mathbf{y}$, though again, it is not equivalent to the $\mathbf{Z}$-$\mathbf{Y}$ D-GMR since $\mathbf{x}$ data are used at training time, as shown in the next section.

4 EM Algorithm for J-GMR

This section introduces the exact EM algorithm associated to the J-GMR, explicitly handling incomplete adaptation datasets using the general methodology of missing data. The EM iteratively maximizes the expected complete-data log-likelihood, denoted by Q. At iteration $i + 1$, the E-step computes the function $Q(\boldsymbol{\Theta}, \boldsymbol{\Theta}^{(i)})$, where $\boldsymbol{\Theta}^{(i)}$ are the parameters computed at iteration i. The M-step maximizes Q with respect to $\boldsymbol{\Theta}$, obtaining $\boldsymbol{\Theta}^{(i+1)}$. In the following we describe the E-step, the M step, the initialization process, and finally we comment the link between the EM algorithms of the IC-GMR and J-GMR.

4.1 E-step

In order to derive the expected complete-data log-likelihood $Q(\boldsymbol{\Theta}, \boldsymbol{\Theta}^{(i)})$, we follow the general methodology given in [14]–(Sect. 9.4) and [12]. In [11], we have

$$Q(\boldsymbol{\Theta}, \boldsymbol{\Theta}^{(i)}) = \sum_{n=1}^{N_0} \sum_{m=1}^{M} \gamma_{nm}^{(i+1)} \log p(\mathbf{o}_n, m; \boldsymbol{\Theta}_m)$$

$$+ \sum_{n=N_0+1}^{N} \sum_{m=1}^{M} \frac{1}{p(\mathbf{v}_n; \boldsymbol{\Theta}_V^{(i)})} \int p(\mathbf{o}_n, m; \boldsymbol{\Theta}_m^{(i)}) \log p(\mathbf{o}_n, m; \boldsymbol{\Theta}_m) d\mathbf{z}_n. \quad (8)$$

$$Q(\boldsymbol{\Theta}, \boldsymbol{\Theta}^{(i)}) = \sum_{n=1}^{N} \sum_{m=1}^{M} \gamma_{nm}^{(i+1)} \left(\log \pi_m - \frac{\log |\boldsymbol{\Sigma}_m| + (\mathbf{o}'_{nm} - \boldsymbol{\mu}_m)^\top \boldsymbol{\Sigma}_m^{-1} (\mathbf{o}'_{nm} - \boldsymbol{\mu}_m)}{2} \right)$$

$$- \frac{1}{2} \sum_{m=1}^{M} \left(\sum_{n=N_0+1}^{N} \gamma_{nm}^{(i+1)} \right) \mathrm{Tr} \left[\boldsymbol{\Lambda}_{\mathbf{ZZ},m} (\boldsymbol{\Lambda}_{\mathbf{ZZ},m}^{(i)})^{-1} \right]. \quad (9)$$

shown that this leads to the general expression (8), where

$$\gamma_{nm}^{(i+1)} = \frac{p(\mathbf{o}_n, m; \boldsymbol{\Theta}_m^{(i)})}{p(\mathbf{o}_n; \boldsymbol{\Theta}^{(i)})}, \quad n \in [1, N_0], \quad (10)$$

are the so-called *responsibilities* (of component m explaining observation $\mathbf{o}_n$) [14]. Equation (8) is valid for any mixture model on i.i.d. vectors $(\mathbf{Z}, \mathbf{V})$ with partly missing $\mathbf{z}$ data. Here we study the particular case of the J-GMR. For this aim, we denote $\boldsymbol{\mu}_{\mathbf{Z}|\mathbf{v}_n,m}^{(i)}$ the posterior mean of $\mathbf{Z}$ given $\mathbf{v}_n$ for the m-th Gaussian component with parameters $\boldsymbol{\Theta}_m^{(i)}$, i.e.:

$$\boldsymbol{\mu}_{\mathbf{Z}|\mathbf{v}_n,m}^{(i)} = \boldsymbol{\mu}_{\mathbf{Z},m}^{(i)} + \boldsymbol{\Sigma}_{\mathbf{ZV},m}^{(i)} \left(\boldsymbol{\Sigma}_{\mathbf{VV},m}^{(i)} \right)^{-1} (\mathbf{v}_n - \boldsymbol{\mu}_{\mathbf{V},m}^{(i)}). \quad (11)$$

Let us define $\mathbf{o}'_{nm} = [\mathbf{v}_n^\top, \boldsymbol{\mu}_{\mathbf{Z}|\mathbf{v}_n,m}^{(i)\top}]^\top$ if $n \in [N_0+1, N]$, i.e. $\mathbf{o}'_{nm}$ is an "augmented" observation vector in which for $n \in [N_0 + 1, N]$ the missing data vector $\mathbf{z}_n$ is replaced with its estimate $\boldsymbol{\mu}_{\mathbf{Z}|\mathbf{v}_n,m}^{(i)}$. Let us arbitrarily extend $\mathbf{o}'_{nm}$ with $\mathbf{o}'_{nm} = \mathbf{o}_n$ for $n \in [1, N_0]$, and the definition of the responsibilities to the incomplete data vectors $\mathbf{v}_n$:

$$\gamma_{nm}^{(i+1)} = \frac{p(\mathbf{v}_n, m; \boldsymbol{\Theta}_{\mathbf{V},m}^{(i)})}{p(\mathbf{v}_n; \boldsymbol{\Theta}_{\mathbf{V}}^{(i)})}, \quad n \in [N_0 + 1, N]. \quad (12)$$

Then, $Q(\boldsymbol{\Theta}, \boldsymbol{\Theta}^{(i)})$ is given by (9). The proof is provided in [15]. The first double sum in (9) is similar to the one found in the usual EM for GMM (without missing data), except that for $n \in [N_0 + 1, N]$ missing $\mathbf{z}$ data are replaced with their estimate using corresponding $\mathbf{x}$ and $\mathbf{y}$ data and current parameter values, and responsibilities are calculated using available data only. The second term is a correction term that, as seen below, modifies the estimation of the covariance matrices $\boldsymbol{\Sigma}_m$ in the M-step to take into account missing data.

4.2   M-step

Priors: Maximization of $Q(\Theta, \Theta^{(i)})$ with respect to the priors π_m is identical to the classical case of GMM without missing data [14]. For $m \in [1, M]$, we have:

$$\pi_m^{(i+1)} = \frac{1}{N} \sum_{n=1}^{N} \gamma_{nm}^{(i+1)}. \tag{13}$$

Mean vectors: For $m \in [1, M]$, derivating $Q(\Theta, \Theta^{(i)})$ with respect to μ_m and setting the result to zero leads to:

$$\mu_m^{(i+1)} = \frac{\sum_{n=1}^{N} \gamma_{nm}^{(i+1)} \mathbf{o}'_{nm}}{\sum_{n=1}^{N} \gamma_{nm}^{(i+1)}}. \tag{14}$$

This expression is an empirical mean, similar to the classical GMM case, except for the specific definition of observation vectors and responsibilities for $n \in [N_0 + 1, N]$.

Covariance matrices: Let us first express the trace in (9) as a function of Σ_m^{-1} by completing $(\Lambda_{\mathbf{ZZ},m}^{(i)})^{-1}$ with zeros to obtain the matrix $(\Lambda_{\mathbf{ZZ},m}^{0\,(i)})^{-1}$:

$$(\Lambda_{\mathbf{ZZ},m}^{0\,(i)})^{-1} = \begin{bmatrix} \mathbf{0} & \mathbf{0} \\ \mathbf{0} & (\Lambda_{\mathbf{ZZ},m}^{(i)})^{-1} \end{bmatrix}. \tag{15}$$

Thus, $\mathrm{Tr}\left[\Lambda_{\mathbf{ZZ},m}(\Lambda_{\mathbf{ZZ},m}^{(i)})^{-1}\right] = \mathrm{Tr}\left[\Sigma_m^{-1}(\Lambda_{\mathbf{ZZ},m}^{0\,(i)})^{-1}\right]$, and by canceling the derivative of $Q(\Theta, \Theta^{(i)})$ with respect to Σ_m^{-1} we get:

$$\Sigma_m^{(i+1)} = \frac{1}{\sum_{n=1}^{N} \gamma_{nm}^{(i+1)}} \left[\sum_{n=1}^{N} \gamma_{nm}^{(i+1)} (\mathbf{o}'_{nm} - \mu_m)(\mathbf{o}'_{nm} - \mu_m)^{\top} \right.$$
$$\left. + \left(\sum_{n=N_0+1}^{N} \gamma_{nm}^{(i+1)} \right) (\Lambda_{\mathbf{ZZ},m}^{0\,(i)})^{-1} \right]. \tag{16}$$

The first term is the empirical covariance matrix and is similar to the classical GMM without missing data, except again for the specific definition of observation vectors and responsibilities for $n \in [N_0 + 1, N]$. The second term can be seen as an additional correction term that deals with the absence of observed $\mathbf{z}$ data vectors for $n \in [N_0 + 1, N]$. We remark that $\Sigma_m^{(i+1)}$ depends on all the terms of $\Sigma_m^{(i)}$ obtained at previous iteration, since $\Lambda_{\mathbf{ZZ},m}^{(i)} = [\Sigma_m^{(i)\,-1}]_{\mathbf{ZZ}} \neq (\Sigma_{\mathbf{ZZ},m}^{(i)})^{-1}$.

4.3   EM Initialization

Similarly to [11], the initialization of the proposed EM algorithm takes a very peculiar aspect. Indeed, the reference $(\mathbf{X}, \mathbf{Y})$ GMR model is used to initialize the marginal parameters in $(\mathbf{X}, \mathbf{Y})$ of the Joint GMM. The marginal parameters in $\mathbf{Z}$ are initialized using the aligned adaptation data $\mathcal{D}_{\mathbf{z}} = \{\mathbf{z}_n\}_{n=1}^{N_0}$. The cross-term parameters are initialized by constructing the sufficient statistics using $\{\mathbf{z}_n, \mathbf{x}_n, \mathbf{y}_n\}_{n=1}^{N_0}$.

4.4   A Remark on the Link Between the J-GMR EM and the IC-GMR EM

We already noticed that the IC mixture model (3) can be seen as a constrained version of the Joint GMM (6). However, the EM for the IC-GMR presented in [11] (which exploits the linear-Gaussian form of the IC-GMR) is *not* derivable as a particular case of the EM of Sect. 4. More precisely, if one attempts to estimate the IC-GMR parameters with the algorithm we present in this paper, the M-step should be constrained by (5). Naturally, the complexity of the resulting constrained algorithm would be much higher. Consequently, even if the IC-GMR and the J-GMR models are closely related, the two learning algorithms are intrinsically different. This difference arises from the fact that the mixture model underlying the IC-GMR deals with constrained covariance matrices, whereas the Joint GMM uses fully free covariance matrices.

5   Experiments

The performance of the J-GMR was evaluated on a speech acoustic-to-articulatory inversion task which consists in recovering the movements of the tongue, lips, jaw and velum from the speech's acoustics. The goal is to adapt an acoustic-to-articulatory (i.e. $\mathbf{X}$-to-$\mathbf{Y}$) GMR trained on a large dataset $\mathcal{D}_{\mathbf{xy}}$ from a reference speaker given a small amount of audio observations only $\mathcal{D}_{\mathbf{z}}$ from another speaker (referred here to as the *source speaker*). In this study, experiments were conducted on synthetic data obtained using a so-called articulatory synthesizer. This allows us to better understand the behavior of the J-GMR model by controlling finely the structure of the adaptation dataset (as opposed to in-vivo data recorded using motion capture techniques on real human speakers). A synthetic dataset of vowels was thus generated using the Variable Linear Articulatory Model (VLAM) [16]. VLAM consists of a vocal tract model driven by seven control parameters (lips aperture and protrusion; jaw; tongue body, dorsum and apex; velum). For a given articulatory configuration, VLAM deduces the corresponding spectrum using acoustic simulation [17]. Among other articulatory synthesizers, VLAM is of particular interest in our study. Indeed, it integrates a model of the vocal tract growth and enables to generate two different spectra from the same articulatory configuration but different vocal tract length. We used this feature to simulate a parallel acoustic-articulatory dataset for two speakers (reference and source) with different vocal tract length corresponding to speaker age of 25 years and 17 years respectively. We generated 20,000 triplets $(\mathbf{z}, \mathbf{x}, \mathbf{y})$ structured into four clusters simulating the 4 following vowels: /a/, /i/, /u/, /ə/. In our experiments, the spectrum is described by the position and the amplitude of the 4 first formants (i.e. local maxima in the power spectrum), hence 8-dimensional $\mathbf{x}$ and $\mathbf{z}$ observations. Figure 2 displays the 20,000 training acoustic data $\mathbf{x}$ (in the two first formant frequencies plane F1-F2; red points) and a selection of 467 adaptation vectors $\mathbf{z}$ (green points).

The EM algorithms for training the reference $\mathbf{X}$-$\mathbf{Y}$ model (and also the $\mathbf{Z}$-$\mathbf{X}$ model for the SC-GMR) were initialized using a k-means algorithm, repeated 5

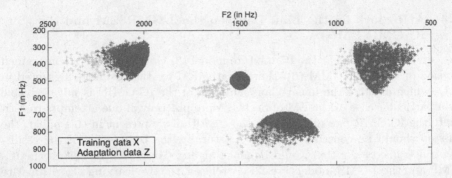

Fig. 2. Synthetic data generated using VLAM, displayed in the F2-F1 acoustic space. (Color figure online)

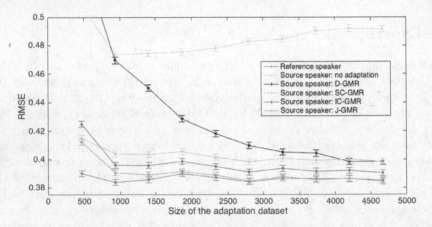

Fig. 3. RMSE of the Z-to-Y mapping as a function of the size of the adaptation data (in number of vectors), for D-GMR, SC-GMR, IC-GMR and J-GMR (lower and upper bounds are given by the X-Y mapping in magenta and the Z-to-Y mapping with no adaption in yellow; error bars represent 95% confidence intervals; RMSE is unitless since articulatory data are arbitrary articulatory control parameters). (Color figure online)

times (only the best initial model was kept for training). For all EMs, the number of iterations was empirically set to 50. All methods were evaluated under a K-fold cross-validation protocol (with $K = 30$). The data was divided in 30 subsets of approximate equal size: 29 subsets for training and 1 subset for test, considering all permutations. In each of the 30 folds, k/30 of the size of the training set was used for adaptation, with $k \in [1, 10]$. For a given value of k, we conducted 10 experiments with a different adaptation dataset. For each experiment, the optimal number of mixture components (within $M = 2, 4, 8, 12, 16, 20$) was determined using cross-validation. The performance was assessed by calculating the average Root Mean Squared Error (RMSE) between the articulatory trajectories $\mathbf{y}$ estimated from the source speaker's acoustics, and the ones generated from the reference speaker. For each RMSE measure, 95% confidence interval (from which statistical significance between two regression techniques can be assessed) was obtained using paired t-test.

The RMSE for the J-GMR, as well as for the D-GMR, SC-GMR and IC-GMR are plotted in Fig. 3, as a function of N_0, the size of the adaptation set. The performance of the J-GMR, SC-GMR and IC-GMR are quite close, and are clearly better than without adaptation and than the D-GMR, especially for low values of N_0. This latter result comes from the fact that the D-GMR exploits only the limited amount of reference speaker's articulatory data that can be associated with the source speaker's audio data. This tends to validate the benefit of exploiting all available $(\mathbf{x}, \mathbf{y})$ observations during the adaptation process, as done in the C-GMR framework. As in [11], the IC-GMR performs better than the SC-GMR, except for the lower N_0. Importantly, we observe a systematic and statistically significant improvement (within the approximate range 1.5%–2.5% of RMSE depending on N_0) of the proposed J-GMR over both the IC-GMR and the SC-GMR (except for the lower N_0 for which the difference between J-GMR and SC-GMR is not significant). This illustrates that the J-GMR is able to better exploit the statistical relations between $\mathbf{z}$, $\mathbf{x}$ and $\mathbf{y}$ data. The link between $\mathbf{x}$ and $\mathbf{y}$ is exploited in the J-GMR, as in the IC-GMR, though the new direct link between $\mathbf{z}$ and $\mathbf{y}$ is also exploited. In short the mapping is not exclusively forced to pass through $\mathbf{x}$, which is shown to be beneficial in the present set of experiments.

6   Conclusion

In this paper, we have extended the general framework of Cascaded-GMR with a new model called J-GMR, for which we provided the exact EM training algorithm explicitly considering missing data. The J-GMR has been shown to perform better than the D-GMR, SC-GMR and IC-GMR in our acoustic-to-articulatory inversion experiments. Altogether, the results show the benefit of considering an intermediate $\mathbf{Z}$-to-$\mathbf{X}$ mapping in the general $\mathbf{Z}$-to-$\mathbf{Y}$ mapping process. This is done explicitly in the SC-GMR and implicitly in both IC-GMR and J-GMR. The relative performance of all these C-GMR models may depend on the latent structure of the data. Hence we believe that all models from this library of GMR adaptation techniques can be of potential interest for other applications. The MATLAB source code of the IC-GMR and J-GMR training and mapping algorithms is available at http://www.gipsa-lab.fr/~thomas.hueber/cgmr/.

As a moderation note, it has to be remembered that, so far, the whole proposed C-GMR framework relies on the assumption that the adaptation data can be aligned with a subset of the training data (see Sect. 2.1), which is a strong hypothesis. In our future work, we will work on relaxing this assumption, e.g. only considering identification of the class of each adaptation data, which will extend the potential applications of the proposed C-GMR models.

Acknowledgements. The authors warmly thank Louis-Jean Boë for his help with the VLAM model.

References

1. McLachlan, G., Peel, D.: Finite Mixture Models. Wiley, New-York (2000)
2. Stylianou, Y., Cappé, O., Moulines, E.: Continuous probabilistic transform for voice conversion. IEEE Trans. Speech Audio Process. **6**(2), 131–142 (1998)
3. Toda, T., Black, A.W., Tokuda, K.: Voice conversion based on maximum-likelihood estimation of spectral parameter trajectory. IEEE Trans. Audio Speech Lang. Process. **15**(8), 2222–2235 (2007)
4. Toda, T., Black, A.W., Tokuda, K.: Statistical mapping between articulatory movements and acoustic spectrum using a Gaussian mixture model. Speech Commun. **50**(3), 215–227 (2008)
5. Zen, H., Nankaku, Y., Tokuda, K.: Continuous stochastic feature mapping based on trajectory HMMs. IEEE Trans. Audio Speech Lang. Process. **19**(2), 417–430 (2011)
6. Tian, Y., Sigal, L., Badino, H., Torre, F., Liu, Y.: Latent gaussian mixture regression for human pose estimation. In: Kimmel, R., Klette, R., Sugimoto, A. (eds.) ACCV 2010. LNCS, vol. 6494, pp. 679–690. Springer, Heidelberg (2011). doi:10.1007/978-3-642-19318-7_53
7. Calinon, S., D'halluin, F., Sauser, E.L., Caldwell, D.G., Billard, A.G.: Learning and reproduction of gestures by imitation: an approach based on hidden Markov model and Gaussian mixture regression. IEEE Rob. Autom. Mag., 17(2): 44–54 (2010)
8. Hueber T, Bailly G, Badin P, Elisei F.: Speaker adaptation of an acoustic-articulatory inversion model using cascaded Gaussian mixture regressions. In: Proceedings of Interspeech, Lyon, France, pp. 2753–2757 (2013)
9. Gauvain, J.-L., Lee, C.-H.: Maximum a posteriori estimation for multivariate Gaussian mixture observations of Markov chains. IEEE Trans. Speech Audio Process. **2**(2), 291–298 (1994)
10. Gales, M.J., Woodland, P.C.: Mean and variance adaptation within the MLLR framework. Comput. Speech Lang. **10**(4), 249–264 (1996)
11. Hueber, T., Girin, L., Alameda-Pineda, X., Bailly, G.: Speaker-adaptive acoustic-articulatory inversion using cascaded Gaussian mixture regression. IEEE/ACM Trans. Audio Speech Lang. Process. **23**(12), 2246–2259 (2015)
12. Ghahramani, Z., Jordan, M.I.: Learning from incomplete data. Technical Report, MIT, Cambridge, MA, USA (1994)
13. McLachlan, G., Thriyambakam, K.: The EM Algorithm and Extensions. Wiley, New York (1997)
14. Bishop, C.M.: Pattern Recognition and Machine Learning. Information Science and Statistics. Springer, New York (2006)
15. Girin, L., Hueber, T., Alameda-Pineda, X.: Appendix to: adaptation of a Gaussian mixture regressor to a new input distribution: extending the C-GMR framework. Technical Report (2016). http://www.gipsa-lab.grenoble-inp.fr/laurent.girin/demo/JGMRappendix.pdf
16. Ménard, L., Schwartz, J.-L., Boë, L.-J., Aubin, J.: Articulatory-acoustic relationships during vocal tract growth for french vowels: analysis of real data and simulations with an articulatory model. J. phonetics **35**(1), 1–19 (2007)
17. Badin, P., Fant, G.: Notes on vocal tract computation, Quarterly Progress and Status Report, pp. 53–108. Department for Speech, Music and Hearing, KTH, Stockholm (1984)

Efficient Optimization of the Adaptive ICA Function with Estimating the Number of Non-Gaussian Sources

Yoshitatsu Matsuda[(✉)] and Kazunori Yamaguchi

Department of General Systems Studies, Graduate School of Arts and Sciences,
The University of Tokyo, 3-8-1, Komaba, Meguro-ku, Tokyo 153-8902, Japan
{matsuda,yamaguch}@graco.c.u-tokyo.ac.jp

Abstract. We propose a new method for efficiently estimating the number of non-Gaussian sources in independent component analysis (ICA). While PCA can find only a few principal components incrementally in the order of significance, ICA has to estimate all the sources after giving the number of them in advance. Then, the appropriate number of sources is determined after the estimation if necessary. Here, we use the adaptive ICA function (AIF), which has been derived by using a simple probabilistic model. It is previously proved that the optimization of AIF with the Gram-Schmidt orthonormalization can find all the sources in descending order of the degree of non-Gaussianity. In this paper, we propose an efficient method for optimizing AIF in the deflation approach by combining fast ICA with the stochastic optimization. In addition, we propose a threshold for determining whether an estimated source is Gaussian or not, which is derived by utilizing the Fisher information of the probabilistic model of AIF. By terminating the optimization when the currently estimated source is Gaussian, the number of sources is estimated efficiently. The experimental results on blind image separation problems verify the usefulness of the proposed method.

1 Introduction

Independent component analysis (ICA) is a widely-used method in many fields such as signal processing [3,4] and feature extraction [6]. ICA estimates unknown sources under the assumptions that the sources are non-Gaussian and they are statistically independent of each other. The linear model of ICA is given as $x = As + n$ where x is the N-dimensional observed signals. A is the $N \times K$ mixing matrix. s and n are the K-dimensional sources and the N-dimensional Gaussian noise, respectively. Here, only x can be observed and the others are unknown. Though ICA is known to be useful, it has still some difficulties. In this paper, we focus on the efficient determination of the number of non-Gaussian sources K. The usual ICA methods estimate the sources after the number of them is given in advance. If necessary, the appropriate number of sources is determined by the estimation results (e.g. the utilization of the information criteria in [7]).

© Springer International Publishing AG 2017
P. Tichavský et al. (Eds.): LVA/ICA 2017, LNCS 10169, pp. 469–478, 2017.
DOI: 10.1007/978-3-319-53547-0_44

On the other hand, PCA can find only the principal components without estimating minor components. Though fast ICA [5] estimates the sources one by one in the deflation approach, the number of sources can not be determined before completing the estimation because it is based on the fixed-point method and the ordering of the estimated sources is not decided [10].

In this paper, we propose a new ICA method estimating the number of sources efficiently by utilizing the adaptive ICA function (AIF). AIF has been proposed in [8], which is derived by a probabilistic model where the Gaussian approximation is applied to the distribution of the estimated sources in the second-order feature space. Consequently, AIF is given as a form of a weighted summation of the 4th-order statistics where the weights depend on the adaptive estimators of kurtoses. One of the most significant property of AIF is that all the sources are estimated in descending order of the degree of non-Gaussianity when AIF is maximized by the Gram-Schmidt process (which is proved in [9]). In this paper, we estimate the number of sources efficiently by utilizing this property. In addition, the Fisher information on the basic probabilistic model of AIF gives an appropriate threshold for determining whether an estimated source is Gaussian or not. The original contribution of this paper consists of the following three parts: (1) Improvement of the optimization of AIF with the Gram-Schmidt orthonormalization by combining fast ICA with the stochastic optimization; (2) Proposal of a threshold for non-Gaussianity by using the Fisher information; (3) Construction of a new ICA method determining the number of non-Gaussian sources efficiently. This paper is organized as follows. In Sect. 2, the derivation and the properties of AIF are described briefly. In Sect. 3, the proposed method is shown. Section 3.1 describes the optimization method for maximizing AIF with the Gram-Schmidt orthonormalization. Section 3.2 derives a threshold for non-Gaussianity. In Sect. 3.3, the complete algorithm of the proposed method is described. Section 4 shows the experimental results on blind image separation problems. Lastly, this paper is concluded in Sect. 5.

2   Objective Function

The adaptive ICA function (AIF) was originally proposed in [8]. The outline of AIF is described below (the details are described in [8,9]). Here, $\boldsymbol{X} = (x_{im})$ denotes observed signals. $\boldsymbol{X}$ is an $N \times M$ matrix (N is the number of signals and M is the sample size). $\boldsymbol{W}$ denotes the $N \times N$ separating matrix and $\boldsymbol{Y} = \boldsymbol{WX}$ denotes the estimated sources. Now, it is assumed that $\boldsymbol{Y} = (y_{im})$ estimates the independent sources accurately. Then, AIF is derived as the likelihood of $\boldsymbol{X}$ by applying the Gaussian approximation to the distribution of the accurately estimated $\boldsymbol{Y}$ in the second-order feature space. Let $\varphi_2(\boldsymbol{X}, m) = (x_{im}x_{jm})$ and $\varphi_2(\boldsymbol{Y}, m) = (y_{im}y_{jm})$ $(i \leq j)$ be the vectors in the second-order polynomial feature space of $\boldsymbol{X}$ and $\boldsymbol{Y}$ for a sample m, respectively. A conditional $N(N+1)/2$-dimensional Gaussian distribution on $\varphi_2(\boldsymbol{X}, m)$ is given as

$$P(\varphi_2(\boldsymbol{X}, m)|\boldsymbol{\alpha}, \boldsymbol{W}) = \prod_{i,j>i} G(y_{im}y_{jm}, 1) \prod_i G(y_{im}^2, \alpha_i) |\boldsymbol{W}|^{N+1} \quad (1)$$

where $\boldsymbol{\alpha} = (\alpha_i)$ is additional unknown parameters. Each α_i is related to the estimator of the kurtosis of the i-th source (as shown later). $G(u, V) = \frac{\exp(-u^2/2V)}{\sqrt{2\pi V}}$ is the Gaussian distribution on u with the mean of 0 and the variance of V. $|\boldsymbol{W}|$ is the determinant of $\boldsymbol{W}$. Then, AIF (denoted as $\Psi(\boldsymbol{\alpha}, \boldsymbol{W})$) is defined as the following log-likelihood function:

$$\Psi(\boldsymbol{\alpha}, \boldsymbol{W}) = \frac{\sum_m \log P(\varphi_2(\boldsymbol{X}, m)|\boldsymbol{\alpha}, \boldsymbol{W})}{M}$$

$$= -\sum_i \log \alpha_i - \sum_{i,j} \left(\frac{1 - \delta_{ij}}{2} + \frac{\delta_{ij}}{\alpha_i} \right) \frac{\sum_m (y_{im} y_{jm} - \delta_{ij})^2}{M} + 2(N+1) \log |\boldsymbol{W}|$$

$$(2)$$

where the constant factor of $M/2$ and some constant additional terms are removed. Though Eq. (2) is the original form of AIF, the following orthonormality constraint is added in [9]:

$$\frac{\sum_m y_{im} y_{jm}}{M} = \delta_{ij} \tag{3}$$

where δ_{ij} is the Kronecker delta for every i and j. Then, AIF of Eq. (2) is simplified as follows:

$$\Psi(\boldsymbol{\alpha}, \boldsymbol{W}) = -\sum_i \log \alpha_i + \sum_i \left(\frac{1}{2} - \frac{1}{\alpha_i} \right) \frac{\sum_m (y_{im}^4 - 1)}{M}. \tag{4}$$

$\Psi(\boldsymbol{\alpha}, \boldsymbol{W})$ can be decomposed into $\sum_i \Psi_i$, each of which is given as

$$\Psi_i(\alpha_i, w_{i1}, \cdots, w_{iN}) = -\log \alpha_i + \left(\frac{1}{2} - \frac{1}{\alpha_i} \right) \frac{\sum_m (y_{im}^4 - 1)}{M} \tag{5}$$

where $y_{im} = \sum_k w_{ik} x_{km}$ and the factor $\frac{1}{2}$ is removed. In the deflation approach, each Ψ_i is maximized one by one under the Gram-Schmidt orthonormalization ($\sum_m y_{im}^2/M = 1$ and $\sum_m y_{im} y_{jm}/M = 0$ for every $j < i$). It is guaranteed that the optimal value of α_i is given as

$$\hat{\alpha}_i = \frac{\sum_m y_{im}^4}{M} - 1 \tag{6}$$

by the Karush-Kuhn-Tucker (KKT) conditions. Note that $\hat{\alpha}_i = \kappa_i + 2$ if y_{im} is the accurate estimation of the i-th source. Here, κ_i is the kurtosis of the i-th source. Therefore, α_i can be regarded as the adaptive estimation of the kurtosis. Note also that $\hat{\alpha}_i = 2$ if the i-th source is Gaussian. The most significant property of AIF to be utilized in this paper is the following theorem:

Theorem 1. *The following three conditions are assumed:*

1. *The linear ICA model $\boldsymbol{x} = \boldsymbol{A}\boldsymbol{s}$ holds, where the mean and the variance of each source $s_i \in \boldsymbol{s}$ are 0 and 1, respectively.*

2. *The sample size M is sufficiently large. In other words, the average over samples is equivalent to the accurate expectation.*
3. *There is no uniform Bernoulli source (in other words, $\kappa_i > -2$ for every i). It is needed for avoiding the divergence of Ψ_i at $\alpha_i = 0$.*

Then, all the sources of $\mathbf{s}$ are extracted in descending order of a value γ_i when each Ψ_i of Eq. (5) is maximized under the Gram-Schmidt orthonormalization, where γ_i is defined as

$$\gamma_i = -\log\left(\kappa_i + 2\right) + \frac{\kappa_i}{2}. \tag{7}$$

The proof is given in [9]. Note that γ_i measures the degree of non-Gaussianity of the i-th source because γ_i has the unique global minimum at $\kappa_i = 0$ (corresponding to the Gaussian distribution) and it is a convex function for $-2 < \kappa_i < \infty$. This theorem guarantees that the maximization of each Ψ_i with the Gram-Schmidt orthonormalization extracts all the sources in descending order of the degree of non-Gaussianity under some reasonable assumptions. We do not discuss the validity of the probabilistic model of AIF in this paper. Nevertheless, note that the above theorem on AIF holds even if the model is not valid.

3   Method

Here, we propose an optimization method maximizing AIF under the Gram-Schmidt orthonormalization. In order to reduce the computational costs for satisfying the orthonormality constraint, $\mathbf{X}$ is assumed to be pre-whitened. Therefore, Gram-Schmidt orthonormalization is applied to the row vectors of $\mathbf{W}$ instead of those of $\mathbf{Y}$.

3.1   Optimization for Each Component

Here, we propose a method maximizing each Ψ_i of Eq. (5) with respect to $(\alpha_i, w_{i1}, \cdots, w_{iN})$. The stochastic optimization is combined with fast ICA for efficient estimation. In addition, a continuous feasible region of α_i is updated gradually in order to avoid the local minima as much as possible.

First, we introduce the following simple optimization using the stochastic optimization and fast ICA on a continuous feasible region of α_i with the lower and the upper bounds ($\alpha_i^{\text{lower}} \leq \alpha_i \leq \alpha_i^{\text{upper}}$):

$$simple\_optimization(\alpha_i^{\text{lower}}, \alpha_i^{\text{upper}})$$

1. Set $(w_{i1}, \cdots, w_{iN})$ randomly under the constraint of $\sum_l w_{il}^2 = 1$.
2. Set α_i to the value nearest to 2 (namely, nearest to the Gaussian distribution) within the feasible region. In other words, α_i is set to

$$\alpha_i = \begin{cases} \alpha_i^{\text{upper}} & \text{if } \alpha_i^{\text{upper}} < 2, \\ \alpha_i^{\text{lower}} & \text{if } \alpha_i^{\text{lower}} > 2, \\ 2 & \text{otherwise.} \end{cases} \tag{8}$$

3. Repeat the following process for a training period T ($= 1000$ in this paper):
   (a) Pick up a sample m randomly.
   (b) Calculate each $y_{im} = \sum_k w_{ik} x_{km}$.
   (c) Update each $w_{ik} := w_{ik} + \rho \frac{\partial \Psi_i}{\partial w_{ik}}$ and each $\alpha_i := \alpha_i + \rho \frac{\partial \Psi_i}{\partial \alpha_i}$ where

$$\frac{\partial \Psi_i}{\partial w_{ik}} = \left(\frac{1}{2} - \frac{1}{\alpha_i} \right) 4 y_{im}^3 x_{km} \tag{9}$$

and

$$\frac{\partial \Psi_i}{\partial \alpha_i} = -\frac{1}{\alpha_i} + \frac{y_{im}^4 - 1}{\alpha_i^2}. \tag{10}$$

Here, ρ is a stepsize (this paper employs a constant $\rho = 0.03$).
   (d) Constrain α_i within $\left[\alpha_i^{\text{lower}}, \alpha_i^{\text{upper}} \right]$ by

$$\alpha_i := \begin{cases} \alpha_i^{\text{upper}} & \text{if } \alpha_i > \alpha_i^{\text{upper}}, \\ \alpha_i^{\text{lower}} & \text{if } \alpha_i < \alpha_i^{\text{lower}}, \\ \alpha_i & \text{otherwise.} \end{cases} \tag{11}$$

   (e) Orthonormalize $(w_{i1}, \cdots, w_{iN})$ in $\boldsymbol{W}$ by the following Gram-Schmidt process: $w_{ik} := w_{ik} - \sum_{j<i} \left(\sum_l w_{il} w_{jl} \right) w_{jk}$ and $w_{ik} := w_{ik} / \sqrt{\sum_l w_{il}^2}$.
4. Optimize $\sum_m y_{im}^4$ with respect to $(w_{i1}, \cdots, w_{iN})$ to the minimum or the maximum at its neighbor by fast ICA, which repeats the following process until the convergence:
   (a) Calculate $y_{im} = \sum_k w_{ik} x_{km}$.
   (b) Update $w_{ik} := \frac{\sum_m x_{km} y_{im}^3}{M} - 3 w_{ik}$ for $k = 1, \cdots, N$.
   (c) Orthonormalize $(w_{i1}, \cdots, w_{iN})$ by the Gram-Schmidt process.
5. Return $(w_{i1}, \cdots, w_{iN})$ estimated by the above optimization, and return the theoretically optimal $\hat{\alpha}_i$ estimated by Eq. (6).

Second, the total optimization method updating the feasible region is described. Initially, there is only one constraint $\alpha_i > 0$, which is desired under the assumption of $\kappa_i > -2$. Therefore, $simple\_optimization(\epsilon, \infty)$ needs to be done at first, where ϵ is a small positive number ($\epsilon = 0.0001$ in this paper). Then, the feasible region is updated by the current $\hat{\alpha}_i$. Let $\hat{\alpha}_i^{\text{cur}}$ be the current $\hat{\alpha}_i$ estimated by the previous optimization. Moreover, we define $F(\hat{\alpha}_i)$ as follows:

$$F(\hat{\alpha}_i) = \Psi_i(\hat{\alpha}_i, w_{i1}, \cdots, w_{iN}) = -\log \hat{\alpha}_i + \frac{\hat{\alpha}_i}{2} - 1. \tag{12}$$

In order to improve $\Psi_i = F(\hat{\alpha}_i)$, the feasible region of $\hat{\alpha}_i$ needs to satisfy $F(\hat{\alpha}_i) \geq F(\hat{\alpha}_i^{\text{cur}})$. Since $F(\hat{\alpha}_i)$ is a convex function, such region is given by the pair of $(0, \hat{\alpha}_i^-]$ and $[\hat{\alpha}_i^+, \infty)$, where $\hat{\alpha}_i^-$ and $\hat{\alpha}_i^+$ are given as

$$\hat{\alpha}_i^- = F_-^{-1}(F(\hat{\alpha}_i^{\text{cur}})), \tag{13}$$

$$\hat{\alpha}_i^+ = F_+^{-1}(F(\hat{\alpha}_i^{\text{cur}})). \tag{14}$$

Here, F_-^{-1} and F_+^{-1} are the inverse functions of F for $\hat{\alpha}_i - 2 \leq 0$ and $\hat{\alpha}_i - 2 \geq 0$, respectively. Either of $\hat{\alpha}_i^-$ and $\hat{\alpha}_i^+$ is equal to $\hat{\alpha}_i^{\mathrm{cur}}$. Utilizing $\hat{\alpha}_i^-$ and $\hat{\alpha}_i^+$, the feasible region can be reduced. Then, the total optimization method updating the feasible region is given as follows:

$total\_optimization(i)$
---

1. Initialize $\hat{\alpha}_i^{\mathrm{cur}}$ to 2 (giving the global minimum of $F(\alpha_i)$).
2. Repeat the process until the update fails L ($= 10$ in this paper) times in a row.
   (a) Calculate $\hat{\alpha}_i^-$ and $\hat{\alpha}_i^+$ by Eqs. (13) and (14).
   (b) Do $simple\_optimization(\epsilon, \hat{\alpha}_i^-)$ and $simple\_optimization(\hat{\alpha}_i^+, \infty)$ independently. Then, select the best solution in the two optimizations.
   (c) Update $\hat{\alpha}_i^{\mathrm{cur}}$ if the best solution in the above simple optimizations is better than $\hat{\alpha}_i^{\mathrm{cur}}$.
3. Return $(w_{i1}, \cdots, w_{iN})$ corresponding to $\hat{\alpha}_i^{\mathrm{cur}}$, and return $\hat{\alpha}_i^{\mathrm{cur}}$.

Though this method searches the solution in both the sub-Gaussian and the super-Gaussian ranges, they are not independent. Both ranges are updated by the common best solution. The optimization parameters (T, ρ, ϵ, and L) were set experimentally.

3.2   Derivation of a Threshold by the Fisher Information

The above method estimates the sources incrementally in descending order of the degree of non-Gaussianity (see Theorem 1). Therefore, if once a Gaussian source is found in the deflation approach, all the rest sources are guaranteed to be Gaussian. In other words, the number of non-Gaussian sources K is determined when the currently estimated source is decided to be Gaussian for the first time. In order to decide whether the i-th source is Gaussian or non-Gaussian, we employ the simple condition $|\alpha_i - 2| < \sigma$, where σ is a positive threshold. In this section, the threshold σ is derived in the following. The variance of α_i is approximated as the Cramer-Rao bound using the Fisher information of α_i. The Fisher information is estimated by the probabilistic model of AIF (Eq. (1) in Sect. 2). It is assumed that each $z_{ij} = y_{im}y_{jm}$ ($j \leq i$) in Eq. (1) is an independent variable. This assumption is the same as in the original probabilistic model. In addition, we assume that W is estimated accurately. Though these assumptions are so bold, the threshold can be estimated easily and it is actually useful as shown in Sect. 4. Thus, the distribution depending on α can be extracted from Eq. (1) as follows:

$$P(z_{11}, z_{22}, \cdots, z_{NN} | \alpha) = \prod_i G(z_{ii}, \alpha_i). \tag{15}$$

Therefore, the Fisher information of α_i is given as

$$\mathcal{I}(\alpha_i) = -\int \frac{\partial^2 \log G(z_{ii}, \alpha_i)}{\partial \alpha_i^2} G(z_{ii}, \alpha_i) \, dz_{ii} = \frac{1}{2\alpha_i^2}, \tag{16}$$

which gives the Cramer-Rao bound on α_i as $\frac{1}{MI(\alpha_i)}$. As the true value of α_i is given as 2 for a Gaussian source, the following threshold is derived:

$$\sigma_{\text{Fisher}} = \tau \sqrt{\frac{1}{MI\left(\alpha_i\right)|_{\alpha_i=2}}} = \tau \sqrt{\frac{8}{M}} \tag{17}$$

where τ is a constant multiplier determining the degree of the statistical significance. $\tau = 20$ is employed in this paper, which means an extreme statistical significance level.

3.3 Complete Algorithm

The complete algorithm is described as follows:

| complete algorithm |
---

1. *Initialization.* Pre-whiten X and let i be 1 (the first component).
2. *Estimation.* Do *total_optimization*(i) for a given i.
3. *Decision and deflation.* If $|\hat{\alpha}_i - 2| < \sigma_{\text{Fisher}}$ (see Eq. (17)), estimate the number of non-Gaussian sources K as $i - 1$ and terminate the algorithm. Otherwise, let i be $i + 1$ (the next component) and return to Step 2 (Estimation) if $i \leq N$.

---

4 Results

Here, the experimental results on blind image separation problems are shown for verifying the proposed method in Sect. 3.3. The original dataset consists of 44 images from the USC-SIPI image database (Volume 3: Miscellaneous). They were transformed into grayscale images of 256×256 pixels. Each pixel corresponds to a sample. In other words, the maximal size of samples was $256 \times 256 = 65,536$. The values in each pixel were normalized over the samples so that their means and their variances are 0 and 1, respectively. Figure 1 shows the histogram of the kurtoses of the 44 images. There are many images with different kurtoses, which consist of 25 super-Gaussian (namely, with positive kurtosis) images and 19 sub-Gaussian (with negative kurtosis) ones. It shows that the employed images are so diversified and they are appropriate for verification. The proposed method using AIF was compared with fast ICA (using the kurtosis or $\log(\cosh)$ as the objective function by the deflation approach) [5] and JADE [2].

First, we show the experimental results where the number of non-Gaussian sources K was set to be equal to the number of signals N (namely, the usual blind image separation). N (and K) was set to 10. Therefore, 10 images were randomly selected from the original 44 images as the sources. The 10 images were sorted in descending order of the degree of non-Gaussianity γ_i. In the same way as in [9] the estimated components were evaluated by the usual Amari's separating error

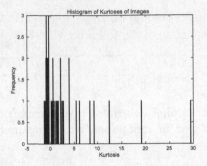

Fig. 1. Histogram of the kurtoses of the 44 original images.

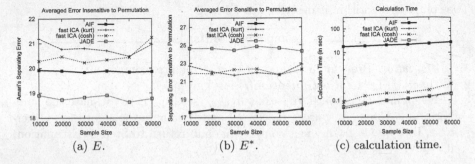

(a) E.   (b) E^*.   (c) calculation time.

Fig. 2. Experimental results in the usual blind image separation (N (the number of signals) $= K$ (the number of non-Gaussian sources) $= 10$): It shows (a) the averaged usual Amari's error E, (b) the averaged error sensitive to permutation E^*, and (c) the median of calculation time (in seconds on a log-scale) for the four ICA methods (the proposed method using AIF (thick solid curves), fast ICA using kurtosis, fast ICA using log (cosh), and JADE) over 100 runs.

E [1] (insensitive to permutation) and the following error sensitive to permutation: $E^* = \sum_{i,j} \left| |b_{ij}| - \delta_{ij} \right|$ where $\boldsymbol{B} = (b_{ij}) = \boldsymbol{WA}$. The square mixing matrix $\boldsymbol{A}$ was given randomly. Experimental runs were carried out over different 100 sets of sources. The averaged errors and the median of the calculation time (for neglecting extremely slow convergences which hardly occurred) along the sample size M (from 10,000 to 60,000) are shown in Fig. 2. Though the proposed method was inferior to JADE for the usual separating error E, it was superior to fast ICA. In addition, the proposed method outperformed all the other methods for the permutation-sensitive error E^*. On the other hand, the calculation time of the proposed method were long in comparison with the other methods. In summary, though the proposed method was much slower, it could extract better components especially in the permutation-sensitive situations.

Second, we show the experimental results on the determination of the number of non-Gaussian sources $K(\leq N)$. Similarly as in the above experiment, the randomly-selected images from the original 44 images were used as the sources.

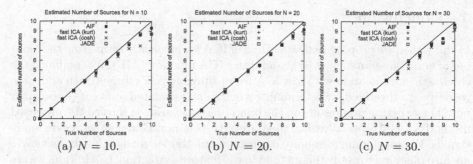

(a) $N = 10$. (b) $N = 20$. (c) $N = 30$.

Fig. 3. Comparison of the true number of sources with the estimated one for $K \leq N$ (averaged over 10 runs): When the estimated number of sources is equal to the true one, the corresponding point is on the diagonal line.

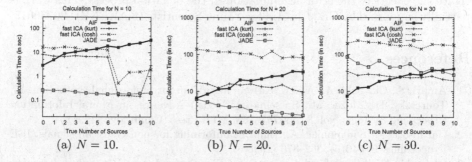

(a) $N = 10$. (b) $N = 20$. (c) $N = 30$.

Fig. 4. Calculation time for estimating the number of non-Gaussian sources (medians over 10 runs in seconds on a log-scale).

M was set to the maximum $65,536$. K was changed from 0 to 10. N was set to 10, 20, and 30. Gaussian noises were added as the sources so that the total number of signals is N. K is estimated by the proposed method using AIF with the threshold σ_{Fisher} of Eq. (17) in Sect. 3.2. 10 runs for different sets of source images were carried out for each K and N. In addition, fast ICA (using the kurtosis or $\log(\cosh)$) and JADE were applied to the same dataset and K was estimated by comparing σ_{Fisher} with the kurtoses of all the extracted components. Fast ICA was terminated if a component did not converge within 5 runs, each of which consists of 1000 updates. Figure 3 compares the true number of sources and the estimated one (averaged over 10 runs). It shows that all the four methods could estimate K almost accurately though the estimated number tended to be slightly smaller. It verifies the usefulness of the proposed threshold σ_{Fisher}. Figure 4 compares the calculation time. Though the proposed method was slower if N is small, it was faster than the other methods including JADE for $K < 5$ and $N = 30$. Considering the complexity of the cumulants-based methods such as JADE, the proposed method is expected to be much faster when N is larger. In summary, the results verify the validity of the proposed threshold and the efficiency of the proposed method for a large number of signals including only a few sources.

5   Conclusion

In this paper, we proposed a new method of ICA for efficiently estimating the number of non-Gaussian sources by the adaptive ICA function (AIF). We combined the stochastic optimization with fast ICA for optimizing AIF efficiently. In addition, we derived a threshold for determining whether an estimated source is Gaussian or not. The experimental results on the blind image separation showed the proposed method is useful especially when there are only a few non-Gaussian sources in many signals. We are planning to apply the proposed method to many other datasets. We are also planning to apply this method to the feature extraction. In addition, we are planning to simplify the current complicated optimization process for improving both the efficiency of calculation and the accuracy of estimation. We are also planning to analyze the properties of the derived threshold σ_{Fisher} further and investigate the effects of the error accumulation in the deflation process. This work is partially supported by Grant-in-Aid for Young Scientists (KAKENHI) 26730013.

References

1. Amari, S., Cichocki, A.: A new learning algorithm for blind signal separation. In: Touretzky, D., Mozer, M., Hasselmo, M. (eds.) Advances in Neural Information Processing Systems, vol. 8, pp. 757–763. MIT Press, Cambridge (1996)
2. Cardoso, J.F., Souloumiac, A.: Blind beamforming for non Gaussian signals. IEE Proceedings-F **140**(6), 362–370 (1993)
3. Cichocki, A., Amari, S.: Adaptive Blind Signal and Image Processing: Learning Algorithms and Applications. Wiley, Chichester (2002)
4. Comon, P., Jutten, C.: Handbook of Blind Source Separation: Independent Component Analysis and Applications. Academic Press, USA (2010)
5. Hyvärinen, A.: Blind source separation by nonstationarity of variance: a cumulant-based approach. IEEE Trans. Neural Netw. **12**(6), 1471–1474 (2001)
6. Hyvärinen, A., Karhunen, J., Oja, E.: Independent Component Analysis. Wiley, Chichester (2001)
7. Li, Y.O., Adali, T., Calhoun, V.D.: Estimating the number of independent components for functional magnetic resonance imaging data. Hum. Brain Mapp. **28**(11), 1251–1266 (2007)
8. Matsuda, Y., Yamaguchi, K.: Adaptive objective function of ICA by gaussian approximation in second-order polynomial feature space. In: Proceedings of IJCNN 2016, pp. 2382–2389. Vancouver, Canada (2016)
9. Matsuda, Y., Yamaguchi, K.: Gram-schmidt orthonormalization to the adaptive ICA function for fixing the permutation ambiguity. In: Hirose, A., Ozawa, S., Doya, K., Ikeda, K., Lee, M., Liu, D. (eds.) ICONIP 2016. LNCS, vol. 9948, pp. 152–159. Springer, Heidelberg (2016). doi:10.1007/978-3-319-46672-9_18
10. Zarzoso, V., Comon, P., Phlypo, R.: A contrast function for independent component analysis without permutation ambiguity. IEEE Trans. Neural Netw. **21**(5), 863–868 (2010)

Feasibility of WiFi Site-Surveying Using Crowdsourced Data

Sylvain Leirens[1], Christophe Villien[1(✉)], and Bruno Flament[2]

[1] Comissariat à l'Énergie Atomique et aux Énergies Alternatives,
17 rue des Martyrs, 38054 Grenoble Cedex, France
{sylvain.leirens,christophe.villien}@cea.fr
[2] InvenSense, 22 Avenue Doyen Louis Weil, Le Doyen, 38000 Grenoble, France
bflament@invensense.com

Abstract. Pedestrian dead reckoning (PDR) trajectories suffer from a significant amount of drift over time, especially when relying on low-cost commercial sensors. For indoor positioning, high level fusion algorithms refine trajectories thanks to kind of map information: e.g. WiFi fingerprinting, and blue prints. Map availability is then of great concern for efficient use of positioning algorithms in practical situations, and could rely on crowdsourced data, i.e. big quantities of data shared by users. In this paper, crowdsourced data include uncertain estimated positions and noisy RSSI (Received Signal Strength Indicator) measurements in order to estimate the spatial distribution of RSSI levels. Using a simple model for a PDR trajectory, we study how a WiFi map can be derived. Simulation results on a corridor use-case illustrate the approach.

Keywords: Crowdsourcing · Site survey · Pedestrian dead reckoning

1 Introduction

Indoor positioning relies mainly on fusing Pedestrian Dead Reckoning (PDR) trajectories with map information: e.g. WiFi fingerprinting, and blue prints with walls and accessibility information (entrances, doors, etc.). Indeed PDR trajectories suffer from a significant amount of drift over time, especially when based on measurements from low-cost and inaccurate commercial sensors. Map availability is of great concern for efficient use of positioning algorithms in practical situations. Site surveying, which is realized manually in practice, is a costly task and should be updated when a change occurs in the wireless local area network configuration, such as the modification of an access point location, or every time the environment is altered in some way, leading to different radio propagation conditions [7].

Consequently map surveying should rely on crowdsourced data more than manual site surveying and/or knowledge of access point locations: outdoor GPS, PDR and RSSI measurements (WiFi, BLE). E.g. a spatial map of RSSI levels can be estimated by fusing big amounts of data including estimated positions

© Springer International Publishing AG 2017
P. Tichavský et al. (Eds.): LVA/ICA 2017, LNCS 10169, pp. 479–488, 2017.
DOI: 10.1007/978-3-319-53547-0_45

and associated RSSI measurements from different users and devices. Positions estimated using PDR are uncertain, as well as RSSI measurements are noisy, leading to a high level of heterogeneity in the data set [10]. Map building using crowdsourced data appears in [6] where trajectory segmentation is used to estimate an indoor hallway structure by considering WiFi and magnetic features and matching similar segments. In [8] a RSSI map maintenance algorithm is proposed based on the assumption that measurements which are close in RSSI space are also close in geometrical space. [11] introduces WiFi-defined landmarks to fuse crowdsourced user trajectories obtained from inertial sensors on users mobile phones.

The issue is to evaluate if a crowdsourced survey is able to approach true site survey as more and more uncertain data is fused. In this paper, the data consists of WiFi RSSI measurements and PDR positions estimates made available by users. Computational resources for surveying are not considered as an issue since high processing power can be easily available in a centralized server-side or cloud-based fashion.

The paper is organized as follows. Section 2 presents the issue of site surveying based on crowdsourced data. Section 3 details the modeling of pedestrian behavior. Section 4 focuses on the surveying algorithm built on a estimating RSSI distributions under PDR uncertainty. Section 5 gives simulation results with a 20-meter corridor case study. Conclusions and future work directions are given in Sect. 6.

2   Problem Statement

WiFi surveying using crowdsourced data is sketched in Fig. 1. Data are collected through users devices (e.g. smartphone) and consist of local measurements (e.g. RSSI levels) and position estimates (e.g. PDR) along trajectories. RSSI measurements are noisy, mainly due to the physical environment where multipath propagation can occur, particular features of each device (sensitivity offset) and device position changes during motion or time. Dead reckoning is by essence a drifting process for is based on accumulating successive displacement estimates through time. Magnetic perturbations make it hard to interpret any absolute heading derived from compass. Bias estimates for a relative heading delivered through accelerometer and gyrometer fusion may lead to heading errors [1]. Moreover the heading information is impacted by the use case, whereby the device held by the user can be in any position on the body and changes through time. Distance traveled is also impacted, as dependent on the user gait.

As a consequence PDR trajectories often look altered when compared to ground truth. Figure 2 illustrates both initial heading offset and step length error issues on a simple rectangular trajectory. A PDR trajectory ends when the user stops walking or when SPC (Smartphone Position Change) is detected, i.e. a significant change in device orientation has occurred. Crowdsourced data imply that large amounts of data are available thanks to numerous users, but in general with a high level of uncertainty.

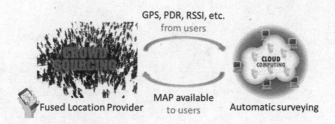

Fig. 1. Site surveying through crowdsourced data.

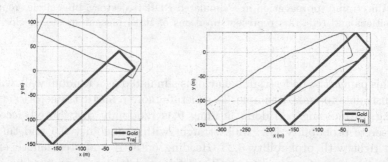

Fig. 2. Estimated trajectories involving heading offset and step length error (red line) compared to ground truth (green line). (Color figure online)

3 Modeling

3.1 Pedestrian Modeling

Cartesian coordinates x and y, at step k $(k = 1, \cdots, N)$, are given by:

$$x_k = x_{k-1} + L(1 + \epsilon_{k-1}) \cos (\theta_{k-1} + \alpha_{k-1}) \tag{1}$$

$$y_k = y_{k-1} + L(1 + \epsilon_{k-1}) \sin (\theta_{k-1} + \alpha_{k-1}) \tag{2}$$

where L is the step length, θ is the heading, ϵ is the step length error and α is the heading offset.

Uncertainty is propagated through linearized PDR equations. The first-order approximate of the covariance of a scalar real-valued function g of random variables u_i $(i = 1, \cdots, n)$:

$$v = g(u_1, u_2, \cdots, u_n) \tag{3}$$

is given by:

$$\sigma_v^2 \approx \sum_{i=1}^{n} \sum_{j=1}^{n} \frac{\partial g}{\partial u_i} \frac{\partial g}{\partial u_j} \mathrm{cov}\,(u_i, u_j) \tag{4}$$

where the partial derivatives are evaluated at the mean value of random variables, and σ stands for standard deviation of subscripted variable.

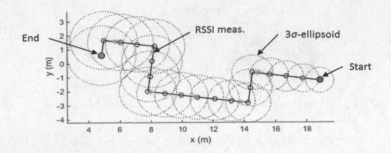

Fig. 3. Uncertainty propagation in a simulated PDR trajectory, blue circles represent step positions and red stars represent positions of RSSI measurements. (Color figure online)

In this paper, the pedestrian behavior is simulated as a random walk without wall constraints to make it simple. To generate more realistic trajectories, taking place e.g. in building corridors, heading θ is randomly distributed according to a discrete probability law: walk straight with probability 0.5 and turn left or turn right with probability 0.25. Heading offset α and step length error ϵ are considered constant and normally distributed (there is no heading drift). Step length L and step frequency are chosen constant for the simulation. These simplifying assumptions make it easy to model a growing uncertainty along a PDR trajectory. A first-order approximate of PDR model accuracy, considering ϵ and α as independent random variables is given by:

$$\sigma^2_{x_k} \approx \sigma^2_{x_{k-1}} + [L\cos(\theta_{k-1})]^2\sigma^2_\epsilon + [L\sin(\theta_{k-1})]^2\sigma^2_\alpha \tag{5}$$

$$\sigma^2_{y_k} \approx \sigma^2_{y_{k-1}} + [L\sin(\theta_{k-1})]^2\sigma^2_\epsilon + [L\cos(\theta_{k-1})]^2\sigma^2_\alpha \tag{6}$$

Figure 3 illustrates uncertainty propagation along a simulated PDR trajectory, where at each step along the trajectory, a 3σ-ellipsoid represents the level of position uncertainty.

3.2   RSSI Map

The 2-dimensional space is partitioned into non-overlapping regions, e.g. rectangular or polytopic cells. Thus any point in the space can be identified to lie in only one of the regions. For each access point, a probability law is associated to every cell in the map, which gives the probability of RSSI level when receiving WiFi signals in a particular region of the space. As a consequence, a RSSI map is a probabilistic look-up table over a discretized space. No analytic modeling assumptions (e.g. based on equations of propagation) are made. The cell dimensions define the spatial resolution of the map. Figure 4 is a heatmap representation of the spatial distribution of RSSI levels obtained in simulation using a simple log-distance path loss model. The map gives the value of RSSI level with maximum probability in each cell.

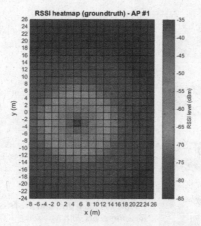

Fig. 4. RSSI heatmap as a graphical representation of RSSI distribution (simulation).

4   Surveying Algorithm

4.1   Admissible Trajectories

Among all available PDR trajectories, the ones with constant orientation of
the device (smartphone) with respect to user body are saved for surveying, e.g.
resorting to heuristics such as SPC detection during motion. A low level of
position uncertainty is considered at the beginning of PDR trajectories, typically
at building entrance with Bluetooth Low Energy (BLE) and proximity sensing.

Figure 5 presents the flow chart of the proposed site surveying approach.
When a new PDR trajectory is available, admissible measurements are selected
according to accuracy in position estimates along trajectory. The level of accuracy in position estimates results from a refinement step based on a previous
survey, such as the map from the previous iteration. Map size is updated if necessary and these new admissible values are used to update RSSI distribution
estimates in each related cell. When a RSSI measurement distribution is accurately estimated, the survey is updated using Bayes' rule to yield the maximum
a posteriori estimate of RSSI levels.

4.2   Admissible Measurements

Whether a measurement is considered to be admissible or not is based on the
computation of the probability $P((x, y) \in \chi_i)$ for a RSSI measurement z at
position (x, y) to belong to a particular cell χ_i:

$$P((x, y) \in \chi_i) = \iint_{(x,y) \in \chi_i} p(x, y) \, dx \, dy \qquad (7)$$

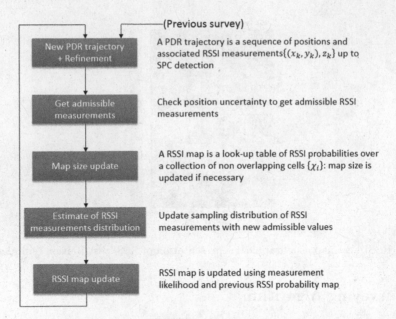

Fig. 5. Flow chart of site surveying.

where $p(x, y)$ is the probability density function of the bivariate normal distribution. Considering rectangular cells, bivariate normal probabilities in the 2-dimensional space are computed using Genz's algorithm [2,3].

A measurement is considered admissible if its probability to belong to the cell is above a specified threshold (e.g. 90%).

4.3   RSSI Map Update

RSSI measurement distributions are computed using a Maximum Likelihood (ML) estimator of unknown mean and variance of a set of measurements. The ML estimate of mean is given by the sample mean, for n measurements z:

$$\mu_{\text{RSSI}}^{\text{ML}} = \frac{1}{n} \sum_{i=1}^{n} z_i \tag{8}$$

The ML estimate of variance is given by the sample variance:

$$(\sigma_{\text{RSSI}}^{\text{ML}})^2 = \frac{1}{n-1} \sum_{i=1}^{n} (z_i - \mu_{\text{RSSI}}^{\text{ML}})^2 \tag{9}$$

If we assume that the sample variance is normally distributed about its mean which is equal to the true variance, it is easy to compute the number of samples (measurements) necessary for a given requirement on estimate accuracy [4]. E.g. to obtain the sample variance within 10% of the true value with probability of 95%, $n \approx 800$ samples are necessary. Alternatively, the achievable estimate accuracy can be computed for a given number of samples.

As soon as a new estimate of RSSI measurement distribution is available in a cell, the RSSI map is updated using Bayesian inference with probability law from previous map as prior PDF. The survey is initialized with uniform PDF in $[-85, -35]$ dBm in every cell of the map. As a consequence, the posterior PDF is a truncated normal PDF.

When the RSSI map has been updated in a particular cell, the RSSI measurement distribution is reset in order to compute a new estimate. The computation of μ_{RSSI}^{ML} and σ_{RSSI}^{ML} can be performed recursively in order to avoid storage of admissible measurements in each cell of the map [5].

4.4 PDR Refinement

The availability of a previous survey makes possible to refine PDR trajectories. In practice, PDR is refined typically using fingerprinting-based algorithms [7,8] or particle filtering [9] by improving position estimates.

Thus we introduce a global correction factor on PDR trajectories as a measure of the performance capability of a given PDR refinement algorithm. From 0 for raw PDR (no correction) up to 1 for ground truth (no uncertainty). This correction level relies mainly on previous survey along with its accuracy (map at previous iteration), number of access points and trajectory size.

5 Simulation Results

5.1 Case Study

The case study is a 20-meter corridor with 2 entrances illustrated in Fig. 6, and the following features:

- Random trajectories of $N = 20$ steps
- Initial position uncertainty: $\sigma_x, \sigma_y = 0.3$ m
- Step length: 0.8 m ($\sigma_\epsilon = 0.2$ m)
- Heading offset: $\sigma_\alpha = 10°$
- Frequency of RSSI measurements in trajectory: 0.74 Hz

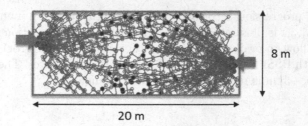

Fig. 6. 20-meter corridor use case with two entrances.

- RSSI measurements in range $[-85, -35]$ dBm (additive white noise ν with σ_ν = 10 dBm)
- Map cell size: 2 m × 2 m

5.2 Surveying Feasibility

The influence of PDR refinement capabilities is crucial, as it bounds the identifiable area for a given set of trajectories. In Fig. 7 are represented performance curves in the plane PDR correction versus quantity of data. E.g., even with 10000 trajectories and PDR correction 0.2, no more than 50% of the corridor can be surveyed.

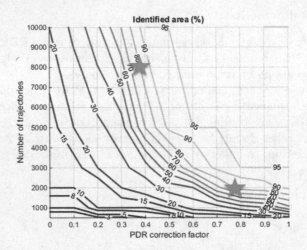

Fig. 7. Surveying performance in the plane number of trajectories versus correction factor. (Color figure online)

In fact, the reachable area increases with quantity of data (PDR trajectories) but remains bounded due to position uncertainty. E.g., a coverage of 80% is achievable with 2000 trajectories along with correction of 0.77, or equivalently with 8000 trajectories along with correction of 0.37 (pointed out with green stars in Fig. 7).

Surveying progression according to the quantity of data (number of PDR trajectories N_{traj}) is illustrated in Fig. 8, for a constant correction factor of 0.6. Error maps allow to evaluate surveying accuracy (absolute error) with respect to ground truth RSSI map when the quantity of data increases. The access point position at $(5, -3)$ m is represented by a blue triangle.

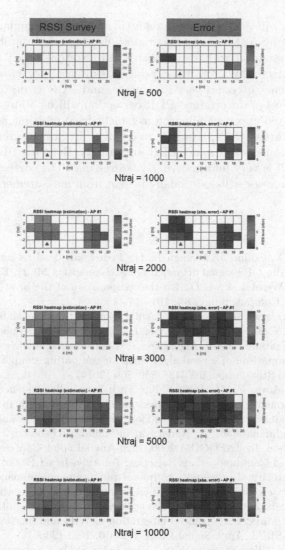

Fig. 8. Surveying for increasing quantities of data (PDR correction of 0.6). (Color figure online)

6   Conclusions

Surveying feasibility of RSSI map has been studied through simulation of PDR trajectories. The space is partitioned into non-overlapping cells. In each cell, the sampling distribution of RSSI measurement is estimated. A large variance of RSSI values is expected in cells where the spatial gradient of RSSI is high valued, typically close to access points. When a previous map is available, PDR trajectories can be corrected to improve accuracy of position estimates.

Future work will deal with more realistic conditions. Simulated trajectories will be replaced by real trajectories from an experimental database, and RSSI levels will be measured with real devices. Initialization based on BLE detection will be replaced by GPS based position estimate: e.g. close to buildings, GPS measurements becomes very inaccurate, and PDR trajectories may start with a higher level of uncertainty. PDR correction will be addressed in a global optimization based approach allowing to benefit from multiple access points.

The key feature of the approach is to estimate RSSI measurements distributions in a partitioned space under PDR uncertainty. Even if a PDR trajectory is uncertain, it still is very informative. By addressing PDR correction in a global approach, future work will additionally benefit from measurement similarity [8].

References

1. Mezentsev, O., Collin, J., Lachapelle, G.: Pedestrian dead reckoning - a solution to navigation in GPS signal degraded areas? Geomatica **59**(2), 175–182 (2005)
2. Drezner, Z., Wesolowsky, G.O.: On the computation of the bivariate normal integral. J. Stat. Comput. Simul. **35**, 101–107 (1989)
3. BVNL. http://www.math.wsu.edu/faculty/genz/software/matlab/bvnl.m
4. Bar-Shalom, Y., Li, X.R., Kirubarajan, T.: Estimation with Applications to Tracking and Navigation: Theory Algorithms and Software. Wiley, New York (2001)
5. Ling, R.: Comparison of several algorithms for computing sample means and variances. J. Am. Stat. Assoc. **69**(348), 859–866 (2014)
6. Singh, J., Mokaya, F., Mori, K., Sidhu, S.: Indoor hallway structure mapping by matching segments from crowd-sourced mobile traces. In: 6th International Conference on Indoor Positioning and Indoor Navigation, Banff, Canada (2015)
7. Bahl, P., Padmanabhan, V.: Radar: an in-building RF-based user location and tracking system. In: INFOCOM 2000, 19th Annual Joint Conference of the IEEE Computer and Communications Societies, Tel Aviv, Israel (2000)
8. Wilk, P., Karciarz, J., Swiatek, J.: Indoor radio map maintenance by automatic annotation of crowdsourced Wi-Fi fingerprints. In: 6th International Conference on Indoor Positioning and Indoor Navigation, Banff, Canada (2015)
9. Evennou, F., Marx, F.: Advanced integration of WiFi and inertial navigation systems. EURASIP J. Appl. Signal Process. **2006**, 1–11 (2006)
10. Hossain, A.M., Jin, Y., Soh, W.-S., Van, H.N.: SSD: A robust RF location fingerprint addressing mobile device's heterogeneity. IEEE Trans. Mobile Comput. **12**(1), 65–77 (2013)
11. Shen, G., Chen, Z., Zhang, P., Moscibroda, T., Zhang, Y.: Walkie-markie: Indoor pathway mapping made easy. In: 10th USENIX Symposium on Networked Systems Design and Implementation (2013)

On Minimum Entropy Deconvolution
of Bi-level Images

K. Nose-Filho[1]([✉]), A.K. Takahata[2], R. Suyama[2], R. Lopes[1],
and J.M.T. Romano[1]

[1] School of Electrical and Computer Engineering,
University of Campinas (UNICAMP), Campinas, Brazil
kenjinose@gmail.com, {rlopes,romano}@decom.fee.unicamp.br
[2] Engineering, Modeling and Applied Social Sciences Center,
Federal University of ABC (UFABC), Santo André, Brazil
{andre.t,ricardo.suyama}@ufabc.edu.br

Abstract. Minimum Entropy Deconvolution (MED) is a sparse blind deconvolution method that searches for a deconvolution filter that leads to the most sparse output, assuming that the desired signal is originally sparse. The present work establishes sufficient conditions for the blind deconvolution of sparse images. Then, based on a measure of sparsity given by the ratio of L_p-norms, we derive a gradient based algorithm for the blind deconvolution of bi-level images, more specifically, for the blind deconvolution of blurred QR Codes. Finally, simulation results are presented considering both synthetic and real data and shows the possibility of achieving really good results by the light of a very simple algorithm.

Keywords: Blind image deconvolution · Bi-level images · QR Codes

1 Introduction

Image deconvolution plays a fundamental role in image processing and has a wide range of applications, which includes seismic images, hyperspectral images, astronomy images, as well as bi-level images. In these applications, the image can be modeled by the convolution of a Point spread Function (PSF) with the *true* image, where the PSF corresponds to the impulse response of an imaging system, and usually results in a blurred image. The goal of image deconvolution is to remove the PSF from the blurred image.

The term blind, in image deconvolution, refers to problems in which there is no explicit knowledge of the *true* image nor of the PSF, but only on a few assumptions about the involved signals. Classically, blind deconvolution methods can be classified into two categories: Linear methods, based on deconvolution filters; and non-linear methods, which aims at estimating both the PSF and the

The authors would like to thanks to CAPES, CNPq and FAPESP (process number 2015/07048-4) for the financial support.

© Springer International Publishing AG 2017
P. Tichavský et al. (Eds.): LVA/ICA 2017, LNCS 10169, pp. 489–498, 2017.
DOI: 10.1007/978-3-319-53547-0_46

true image. A good overview of these methods can be seen in the nice papers of Kundur and Hartzinakos [9,10].

In general, linear methods are simpler (less parameters to be set) and faster (reduced complexity) when compared to non-linear methods. In this sense, we revisit the linear minimum entropy deconvolution (MED) method proposed in [14] and extend it to deal with the blind deconvolution of bi-level images, more specifically, for the identification of QR Codes [22].

Minimum entropy deconvolution methods rely on the fact that the *true* signal has a simple/sparse structure, i.e., it is composed by a few spikes separated by nearly zero terms [24]. By observing that the convolution of such a signal with the PSF would lead to a less sparse signal, MED techniques searches for a deconvolution filter that leads to the most sparse output, which would correspond to the *true* image [3,13,14].

Bi-level images, also known as binary images or two-level images, constitutes a wide range of signals such as texts and bar codes. Each pixel of a bi-level image has only two possible values, usually given by zero or one. The zero pixels, in a conventional grayscale colorbar, are usually related to the black color (absence of color) and the pixels with unitary values are usually related to the white color.

Thus, if the black pixels significantly outnumber the white pixels, these images can be classified as sparse, in which minimum entropy deconvolution methods could be applied [25]. On the other hand, if the white pixels outnumber the black pixels, which is the case of most of real applications [8], the sparsification can be performed by simply flipping the black pixels with the white ones.

The present work is organized as follows: in Sect. 2 we state the image deconvolution problem. Then, in Sect. 3, we establish sufficient conditions for the blind deconvolution of two-dimensional images and, based on a measure of sparsity given by the ratio of L_p-norms [14], we derive a gradient based algorithm for the blind deconvolution of images. In Sect. 4 we analyze the applicability of this algorithm for the blind deconvolution of blurred QR Codes and, in Sect. 5, we present some simulation results with a synthetic and a real image. Finally, in Sect. 6 we state our conclusions.

2 Image Deconvolution

First, let us state the linear image deconvolution problem and set our notation. The discrete two-dimensional convolution model is given by:

$$x(m,n) = h(m,n) * s(m,n) + \nu(m,n),$$
$$= \sum_j \sum_i h(i,j)s(m-i,n-j) + \nu(m,n), \tag{1}$$

where the symbol $*$ stands for the discrete two-dimensional convolution, $\mathbf{S} = (s(m,n))$ is the *true* image, $\mathbf{X} = (x(m,n))$ is the observed image, $\mathbf{H} = (h(m,n))$ is the PSF and $\mathcal{V} = (\nu(m,n))$ is an additive noise term.

Linear deconvolution can be carried out by applying the observed signal $x(m,n)$ into an inverse filter that aims to recover the original image:

$$y(m,n) = w(m,n) * x(m,n), \tag{2}$$

where $w(m,n)$ is the impulse response of the deconvolution filter and $y(m,n)$ is the recovered image. In the noiseless case, perfect deconvolution is said to be achieved when [20]:

$$y(m,n) = cs(m - d_m, n - d_n), \tag{3}$$

where c is a constant scalar, and d_m and d_n are discrete shifts introduced by the deconvolution filter. In this case, the global response of the system is given by:

$$g(m,n) = h(m,n) * w(m,n) = c\delta^2(m - d_m, n - d_n), \tag{4}$$

which is also known as the zero-forcing condition, where $\delta^2(m,n)$ is the two-dimensional delta function.

In the following, we extend the results obtained in [14] for the two-dimensional problem.

3   Minimum Entropy Deconvolution

Minimum entropy deconvolution is a classical sparsity promoting blind deconvolution approach [1,3,24]. It finds applications in ultrasonic inspection [12,17], seismic reflection [2,6,13–16,19,23,24,26], and also on the deconvolution of bi-level images [8,25], where the authors considered the maximization of the fourth order normalized cumulant of the horizontal and vertical differences of the blurred image.

In this work, instead of working with the horizontal and vertical differences, we work on the flipped blurred image, where the black and white pixels are flipped in order to obtain a sparse image. First, let us define the entrywise matrix norm:

$$\|\mathbf{X}\|_r = \left(\sum_j \sum_i |x(i,j)|^r \right)^{1/r}, \tag{5}$$

where $\|\mathbf{X}\|_r$, for $r = 2$, is the Frobenius norm [5]. Then, in order to extend the results obtained in [14] for the two-dimensional problem, we state the following theorem.

Theorem 1. *Let us consider a PSF* $\mathbf{G} = (g(m,n))$ *with at least one non-zero element, and a signal* $\mathbf{S} = (s(m,n))$ *composed of a few spikes of unknown amplitude and position, separated by zero terms, such that the response of the PSF to each of these spikes does not overlap. Also, let us consider that* $y(m,n) = s(m,n) * g(m,n)$ *and that* $\|\mathbf{Y}\|_2 = \|\mathbf{S}\|_2$, *then:*

1. $\|\mathbf{Y}\|_p \geq \|\mathbf{S}\|_p$, *for* $p \in [1,2[$ *and* $\|\mathbf{Y}\|_q \leq \|\mathbf{S}\|_q$, *for* $q \in]2,\infty]$.
2. $\|\mathbf{Y}\|_r = \|\mathbf{S}\|_r$, *for* $r \neq 2$, *if and only if* $g(m,n) = \delta^2(m - d_m, n - d_n)$,

Proof. Since the matrices $\mathbf{G}$ and $\mathbf{S}$ are such that the response of the PSF to each of the spikes in $\mathbf{S}$ does not overlap, then, in a similar way as to [14]:

$$\|\mathbf{Y}\|_r = \|\mathbf{S}\|_r \|\mathbf{G}\|_r. \tag{6}$$

Thus, if $\|\mathbf{Y}\|_2 = \|\mathbf{S}\|_2$, then $\|\mathbf{G}\|_2 = 1$ and, due to the norm inequality:

$$\|\mathbf{G}\|_p \geq 1 \geq \|\mathbf{G}\|_q, \tag{7}$$

where, $p \in [1, 2[$ and $q \in]2, \infty]$. So, as a direct consequence, we have that:

$$\|\mathbf{Y}\|_p \geq \|\mathbf{S}\|_p, \text{ and } \|\mathbf{Y}\|_q \leq \|\mathbf{S}\|_q, \tag{8}$$

where the equality holds if and only if, $g(m, n)$ assumes the form of $\delta^2(m - d_m, n - d_n)$. □

In other words, the convolution of a PSF with a sufficiently sparse image, such that equality (6) holds, reduce, or at most conserve, the sparsity degree of the *true* image. Thus, deconvolution can be performed by a linear filter able to retrieve the most sparse signal.

3.1 Gradient Based Algorithm

Theorem 1 indicates that deconvolution can be performed by solving the following optimization problem[1]:

$$\underset{\mathbf{W}}{\text{minimize}} \ \ \|\mathbf{Y}\|_p = \left(\sum_n \sum_m \left| \sum_j \sum_i w(i,j) x(m - i, n - j) \right|^p \right)^{1/p} \tag{9}$$
$$\text{subject to } \|\mathbf{G}\|_2 = 1.$$

One of the issues of having $\|\mathbf{G}\|_2 = 1$ is that $\mathbf{H}$ is unknown. Assuming that $\|\mathbf{S}\|_2 = 1$, this restriction can be incorporated in the cost function, by normalizing it, obtaining the following unconstrained optimization problem [21]:

$$\underset{\mathbf{W}}{\text{minimize}} \ \ \frac{\|\mathbf{Y}\|_p}{\|\mathbf{Y}\|_2}. \tag{10}$$

In order to solve this problem, we propose the usage of a simple gradient based method. For that, let us consider the auxiliary variables $u = \left(\sum_n \sum_m |y(m, n)|^p \right)^{1/p}$ and $v = \left(\sum_n \sum_m |y(m, n)|^2 \right)^{1/2}$. The derivative of u/v with respect to $w(i, j)$ is given by:

$$\frac{\partial \frac{u}{v}}{\partial w(i,j)} = \frac{1}{v} \frac{\partial u}{\partial w(i,j)} - \frac{u}{v^2} \frac{\partial v}{\partial w(i,j)}, \tag{11}$$

---

[1] For simplicity, we derive the algorithm for $p \in [1, 2[$ only. The difference is that, for $q \in]2, \infty[$, we obtain a maximization problem.

where

$$\frac{\partial u}{\partial w(i,j)} = u^{1/p-1} \sum_{n} \sum_{m} \mathrm{sgn}(y(m,n))|y(m,n)|^{p-1}x(m-i,n-j),$$

$$\frac{\partial v}{\partial w(i,j)} = \frac{1}{v} \sum_{n} \sum_{m} y(m,n)x(m-i,n-j), \tag{12}$$

and sgn($\cdot$) denotes the sign function.

Finally, we choose the following update rule:

$$\mathbf{W}(k+1) = \mathbf{W}(k) - \mu(k)\frac{\nabla J(\mathbf{W}(k))}{\|\nabla J(\mathbf{W}(k))\|_2},$$

$$\mathbf{W}(k+1) = \frac{\mathbf{W}(k+1)}{\|\mathbf{W}(k+1)\|_2},$$

$$\mu(k+1) = \lambda * \mu(k), \tag{13}$$

where k denotes the iteration index, $\mu(k)$ is a variable step size, λ is a forgetting factor [18] and the normalization is performed in order to improve the convergence of the algorithm [4].

It is important to observe that the cost function in (10) may present some local minima and a good initialization must be provided in order to obtain a reliable estimation of the true image. For most of the cases, a good initialization is given by a single spike of unit magnitude located at the center of W.

4   Blind Deconvolution of QR Codes

The blind deconvolution of bi-level images is of great interest in several applications involving text and bar code identification [8,11,22,25]. Bi-level images composed of texts and bar codes have naturally a simple structure appearance, in the sense that the pixels containing one of the two possible values usually outnumber the pixels containing the other one. So, if the black pixels outnumber the white pixels, the image has naturally a sparse appearance. Instead, if white pixels outnumber the black pixels, which is the case of QR Codes, a sparse image can be obtained by simply flipping the black pixels with the white ones.

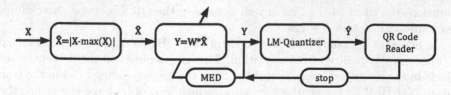

Fig. 1. Block diagram of the processing of a blurred QR Code, for the proposed MED algorithm.

Considering that the image is composed of non-negative pixels, this can be done by simply taking the absolute value of the image subtracted from its maximum value:

$$\bar{\mathbf{X}} = |\mathbf{X} - \|\mathbf{X}\|_{\max}|, \tag{14}$$

where $\|\mathbf{X}\|_{\max} = \max_{ij}|x(i,j)|$.

By observing that after each iteration the effect of the PSF is significantly attenuated, we propose a procedure that consists in flipping back the deconvolved image after each iteration, quantizing it with a Loyd-max quantizer [7] and presenting it to a QR Code reader for smartphones that stops the process when the code is read. This procedure is summarized by the block diagram of Fig. 1.

5   Simulation Results

In this section, we present some simulation results considering two different scenarios. In the first one, the observed image is obtained by the convolution of the QR Code with a 2D Gaussian function, in presence of additive white Gaussian noise. In the second scenario, we consider a photograph of a QR Code captured with a conventional smartphone camera.

5.1   Synthetic Scenario

The first QR Code corresponds to the *url* of the wikipedia home page (http://en.m.wikipedia.org), as presented by Fig. 2a. The QR Code was convolved with a Gaussian PSF with 51 pixels of variance. Then, we performed 20 Monte Carlo (MC) simulations for a SNR equal to 15 dB in order to evaluate the proposed method. In this scenario, it was considered an initial step size $\mu(0) = 0.001$, a forgetting factor $\lambda = 0.9999$ and an 9×9-tap deconvolution filter.

Figure 2b presents the blurred QR Code $\mathbf{X}$ and Fig. 2c presents the flipped QR Code $\bar{\mathbf{X}}$, which is the input of the deconvolution filter. Figure 3a presents the evolution of the cost function for one MC simulation. Figure 4a presents the image of Fig. 2c after the flipping and quantization process, and corresponds to the image presented to the QR Code reader at iteration $k = 0$. Finally, Figs. 4b and c present the deconvolved, flipped, and quantized images after iterations $k = 60$ and $k = 120$, respectively. As we can see, the MED algorithm works in the sense of increasing the sparsity degree of the image by attenuating, after each iteration, the effect of the PSF. In all the cases, the QR Code was successfully read around iteration $k = 120$.

In addition, we computed the Bit Error Rate (BER) of the two-pixel image to measure the quality of the algorithm. This is computed as the percentage of pixels in the estimated image that differ from the corresponding pixel in the true image. The BER for the synthetic scenario, which was computed disregarding the borders of the QR Codes, is shown in Fig. 3b, for several different Monte Carlo trials. For comparison purposes, we also show the BER of the quantized version of the observed, unprocessed image, and the BER obtained by the quantized output

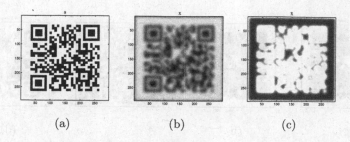

(a)   (b)   (c)

Fig. 2. (a) Original QR Code **S**, (b) blurred QR Code **X**, and (c) flipped blurred QR Code $\bar{\mathbf{X}}$.

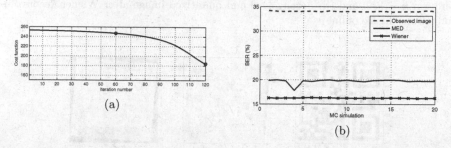

Fig. 3. (a) Evolution of the cost function for one MC simulation of the synthetic s, where the circles indicates the iteration number considered in Fig. 4. (b) The BER of the quantized blurred image, the quantized deconvolved image obtained with the proposed MED algorithm and the quantized deconvolved image obtained with the Wiener filter

of a Wiener filter. Note that this filter requires the knowledge about the blurring PSF and the SNR ratio. Due to that fact and the way the filter is designed, its BER can be seen as a "lower" bound to the achievable BER. On the other hand, the quantization of the observed image yields an "upper" bound on the BER, as this image did not go through any deconvolution process.

From the results in Fig. 3b, it is possible to observe that the proposed MED algorithm achieves a BER close to those obtained with the Wiener filter. To further contrast the performance of the proposed method, in Fig. 4d we show the quantized deconvolved QR Code with the Wiener filter, while the quantized deconvolved QR Code with the proposed MED algorithm is presented in Fig. 4c. The red rectangles in Fig. 4c point to some parts of the image that are more faithfully reproduced by the MED filter output. When the three images (unprocessed Fig. 4a, MED Fig. 4c and Wiener Fig. 4d) are fed to a QR Code reader, only the MED filter output can be decoded.

5.2 Real Scenario

The second QR Code, presented by Fig. 5a, corresponds to the text message: 123456789-B. The blurred QR Code, presented by Fig. 5b, corresponds to a photograph captured with a conventional smartphone camera, where a 86° digital

(a)   (b)   (c)   (d)

Fig. 4. (a) Figure 2c after flipping and quantization, (b) deconvolved, flipped and quantized image after iteration $k = 60$, (c) deconvolved, flipped and quantized image after iteration $k = 120$, and (d) deconvolved and quantized image after Wiener deconvolution. (Color figure online)

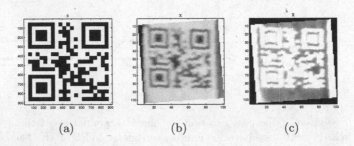

(a)   (b)   (c)

Fig. 5. (a) Original QR Code $\mathbf{S}$, (b) observed QR Code $\mathbf{X}$, (c) flipped QR Code $\bar{\mathbf{X}}$.

Fig. 6. Evolution of the cost function for the real QR Code, where the circles indicates the iteration number considered in Fig. 4.

(a)   (b)   (c)

Fig. 7. (a) Figure 5c after flipping and quantization, (b) deconvolved, flipped and quantized image after iteration $k = 10$, and deconvolved, flipped and quantized image after iteration $k = 25$.

rotation was applied for the correct orientation of the image. Figure 5c presents the flipped QR Code $\bar{\mathbf{X}}$ and Fig. 6 presents the evolution of the cost function for this scenario. In this case, it was considered an initial step size $\mu(0) = 0.001$, a forgetting factor $\lambda = 0.9999$ and an 5×5-tap deconvolution filter.

Figure 7a presents the image of Fig. 5c after the flipping and quantization process, which corresponds to the image presented to the QR Code reader at iteration zero. Finally, Figs. 7b and c present the deconvolved, flipped and quantized image after iterations $k = 10$ and $k = 25$, respectively. In this case, the QR Code successfully was read at iteration $k = 25$.

6   Conclusions

In this work, we addressed the minimum entropy deconvolution problem by establishing sufficient conditions for the blind deconvolution of sparse images. Then, based on a measure of sparsity given by the ratio of L_p-norms, we derived a gradient based algorithm for the blind deconvolution of bi-level images, more specifically, for the blind deconvolution of blurred QR Codes.

Instead of working directly on the observed image, we proposed to work on the flipped image, by simply flipping the black with the white pixels in order to obtain a sparse image. After each iteration, the image was flipped back, quantized and then presented to a QR Code reader that stopped the process when the code was read.

The algorithm was evaluated in two different scenarios, a synthetic one, in which we consider a 2D Gaussian PSF, and a real one, resulted from the employ of a conventional smartphone camera. The presented results have shown the possibility of achieving really good results by the light of a very simple algorithm that has the potential to be implemented in smartphones applications.

References

1. Cabrelli, C.A.: Minimum entropy deconvolution and simplicity: a noniterative algorithm. Geophysics **50**, 394–413 (1984)
2. Claerbout, J.F.: Minimum information deconvolution. Technical report, SEP-15 (1978)
3. Donoho, D.L.: On minimum entropy deconvolution, pp. 656–608. Academic Press (1981)
4. Douglas, S.C., Amari, S.I., Kung, S.Y.: On gradient adaptation with unit-norm constraints. IEEE Trans. Sig. Process. **48**, 1843–1847 (2000)
5. Golub, G.H., Van Loan, C.F.: Matrix Computations, 3rd edn. John Hopkins, Baltimore (1996)
6. Gray, W.: A theory for variable norm deconvolution. Technical report, SEP-15 (1978)
7. Jayant, N.S., Noll, P.: Digital Coding of Waveforms: Principles and Applications to Speech and Video. Prentice-Hall, Englewood Cliffs (1984)
8. Kim, J., Jang, S.: High order statistics based blind deconvolution of bi-level images with unknown intensity values. Opt. Express **18**(12), 12872–12889 (2010)

9. Kundur, D., Hatzinakos, D.: Blind image deconvolution. Sig. Process. Mag. **13**, 43–64 (1996)

10. Kundur, D., Hatzinakos, D.: Blind image deconvolution revisited. Sig. Process. Mag. **13**(6), 61–63 (1996)

11. Li, T., Lii, K.: A joint estimation approach for two-tone image deblurring by blind deconvolution. IEEE Trans. Image Process. **11**, 847–858 (2002)

12. Nandi, A.K., Mampel, D., Roscher, B.: Blind deconvolution of ultrasonic signal in nondestructive testing applications. IEEE Trans. Sig. Process. **45**, 1382–1390 (1997)

13. Nose-Filho, K., Jutten, C., Romano, J.M.T.: Sparse blind deconvolution based on scale invariant smoothed ℓ_0-norm. In: Proceedings of the 2014 EUSIPCO (2014)

14. Nose-Filho, K., Romano, J.M.T.: On ℓ_p-norm sparse blind deconvolution. In: Proceedings of the 2014 MLSP (2014)

15. Nose-Filho, K., Takahata, A., Lopes, R., Romano, J.: A fast algorithm for sparse multichannel blind deconvolution. Geophysics (2016)

16. Ooe, M., Ulrych, T.J.: Minimum entropy deconvolution with an exponential transformation. Geophys. Prospect. **27**, 458–473 (1979)

17. Pflug, L.A., Broadhead, M.K.: Alternate norms for the Cabrelli and Wiggins blind deconvolution algorithms. Technical report, Naval Research Laboratory (1998)

18. Sayed, A.H.: Fundamentals of Adaptive Filtering. Wiley, New York (2003)

19. Scargle, J.D.: Absolute value optimization to estimate phase properties of stochastic time series. IEEE Trans. Inf. Theory **23**(1), 140–143 (1977)

20. Shalvi, O., Weinstein, E.: New criteria for blind deconvolution of nonminimun phase systems (channels). IEEE Trans. Inf. Theory **36**, 312–321 (1990)

21. Shalvi, O., Weinstein, E.: Super-exponential methods for blind deconvolution. IEEE Trans. Inf. Theory **39**(2), 504–519 (1993)

22. Sörös, G., Semmler, S., Humair, L., Hilliges, O.: Fast blur removal for wearable QR code scanners. In: Proceedings of the 2015 ACM International Symposium on Wearable Computers, pp. 117–124. ACM (2015)

23. Takahata, A.K., Nadalin, E.Z., Ferrari, R., Duarte, L.T., Suyama, R., Lopes, R.R., Romano, J.M.T., Tygel, M.: Unsupervised processing of geophysical signals. IEEE Sig. Process. Mag. **27**, 27–35 (2012)

24. Wiggins, R.A.: Minimum entropy deconvolution. Geoexploration **16**(1,2), 21–35 (1978)

25. Wu, H.: Minimum entropy deconvolution for restoration of blurred two-tone images. Electron. Lett. **26**, 1137–1146 (1990)

26. Zhang, Y., Claerbout, J.: A new bidirectional deconvolution method that overcomes the minimum phase assumption. Technical report, SEP-142 (2011)

A Joint Second-Order Statistics and Density Matching-Based Approach for Separation of Post-Nonlinear Mixtures

Denis G. Fantinato[1]([✉]), Leonardo T. Duarte[2], Paolo Zanini[3], Bertrand Rivet[3], Romis Attux[1], and Christian Jutten[3]

[1] School of Electrical and Computer Engineering, University of Campinas, Campinas, SP, Brazil
{denisgf,attux}@dca.fee.unicamp.br
[2] School of Applied Sciences, University of Campinas, Campinas, SP, Brazil
leonardo.duarte@fca.unicamp.br
[3] GIPSA-Lab, Grenoble INP, CNRS, Grenoble, France
{paolo.zanini,bertrand.rivet,christian.jutten}@gipsa-lab.grenoble-inp.fr

Abstract. In the context of Post-Nonlinear (PNL) mixtures, source separation can be performed in a two-stage approach, which encompasses a nonlinear and a linear compensation part. In the former part, it is usually required the knowledge of all the source distributions. In this work, we propose a less restrictive approach, where only one source distribution is needed to be known – here, chosen to be a colored Gaussian. The other sources are only required to present a time structure. The method combines, in a joint-based approach, the use of the second-order statistics (SOS) and the matching of distributions, which shows to be less costly than the classical method of computing the marginal entropy for all sources. The simulation results are favorable to the proposal.

Keywords: Blind source separation · Post-Nonlinear mixtures · Second-order statistics · Density matching

1 Introduction

In the area of signal processing, the problem of retrieving a set of source signals from their mixtures has been intensively studied for three decades. Since this task is performed with only the knowledge of some samples of the mixtures, this problem is named Blind Source Separation (BSS) [1]. The majority of the initial efforts were aimed at the standard linear and instantaneous mixture problem, with the assumption that the sources are mutually independent. These studies resulted in a well-founded and solid theoretical framework known as Independent Component Analysis (ICA) [1]. Although it can count with a vast number of practical applications, there are certain cases in which the linear assumption is insufficient – e.g., smart chemical sensor arrays [2] and hyperspectral imaging [3] – and nonlinear mixing models must be considered. Notwithstanding,

© Springer International Publishing AG 2017
P. Tichavský et al. (Eds.): LVA/ICA 2017, LNCS 10169, pp. 499–508, 2017.
DOI: 10.1007/978-3-319-53547-0_47

from a general nonlinear standpoint, the ICA framework may not provide the sufficient information for performing source separation. Thus, the studies on this topic were focused on a constrained set of nonlinear models in which the ICA methods are still valid [4], like the so-called Post-Nonlinear (PNL) models [5].

The approaches for solving the PNL mixing problem can be roughly divided into the *joint* and the *two-stage* approaches [6]. In the former case, an ICA-based method is usually employed [5]. In the second case, the nonlinear part is solved in a first step – e.g., via a Gaussianization method [7] – and, for the subsequent step, there remains a linear BSS problem, which is a well studied issue [1]. Additionally, if the sources present a temporal structure, a second-order statistics (SOS)-based approach can be employed in the second stage [7]. Notwithstanding, these approaches may suffer some drawbacks: in the joint approach, it is usually necessary to estimate the mutual information, which may be computationally costly and also be susceptible to local minima convergence. In the two-stage approach, the nonlinear compensation methods, for achieving accurate enough results, may require prior assumptions, which can not be available in certain scenarios [6]. In this work, we consider a less restrictive approach by assuming the knowledge of the distribution shape of a single source, e.g., a Gaussian distribution, and that the sources present temporal structure. In this case, we propose a joint approach which allies a SOS-based cost function to a density (Gaussian) matching which can be simply performed via kernel estimators [8]. We also consider a robust metaheuristic known as Differential Evolution (DE) [9] to avoid suboptimal convergence.

2   The Post-Nonlinear Mixtures

In the blind source separation (BSS) problem, the main objective is to retrieve the original sources $\mathbf{s}(n)$ from the observed mixtures $\mathbf{x}(n) = \boldsymbol{\Phi}(\mathbf{s}(n))$, where $\mathbf{x}(n) = [x_1(n) \; \cdots \; x_M(n)]^T$ is the observation vector of length M, $\mathbf{s}(n) = [s_1(n) \; \cdots \; s_N(n)]^T$ is the source vector with N elements and $\boldsymbol{\Phi}(\cdot)$ is the mixing function [1]. Classically, it is assumed that the mixing function can be described as linear and instantaneous system of the type $\mathbf{x}(n) = \mathbf{A}\mathbf{s}(n)$, where $\mathbf{A}$ is a $M \times N$ matrix. However, this model is not sufficient for certain applications. In that sense, the *Post-Nonlinear* (PNL) model rises as an emblematic and significant step in nonlinear BSS [1,5].

The PNL system comprises two stages of mixing: the linear and the nonlinear stages. As illustrated in Fig. 1, the mixtures can be written as $\mathbf{x}(n) = \mathbf{f}(\mathbf{A}\mathbf{s}(n))$, being $\mathbf{f}(\cdot)$ a set of M component-wise functions. The separation system is a mirrored version of the mixing system, being its output given by $\mathbf{y}(n) = \mathbf{W}\mathbf{g}(\mathbf{x}(n))$, where $\mathbf{W}$ is a $N \times M$ matrix and $\mathbf{g}(\cdot)$ is a set of M component-wise functions, ideally the inverse of $\mathbf{f}(\cdot)$ [1].

2.1   Separation Techniques for PNL Mixtures

In the context of PNL mixtures, it is possible to classify the separation techniques into two main classes: the *joint* and the *two-stage* approaches [6].

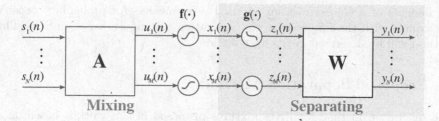

Fig. 1. Mixing and separating systems in the PNL model.

In the former, the main idea is to jointly adjust $\mathbf{g}(\cdot)$ and $\mathbf{W}$ by minimizing a given statistical dependence measure; generally, the use of the ICA framework represents an efficient methodology for performing separation, but issues like local convergence and constrained adaptation of the nonlinearities require special attention – e.g., it is necessary that $\mathbf{f}(\cdot)$ and $\mathbf{g}(\cdot)$ be bijective pairs [6].

On the other hand, for the two-stage approach, the linear and the nonlinear mixing stages are addressed separately, i.e., two different but "simpler" problems need to be solved: $\mathbf{g}(\cdot)$ is adapted so that the nonlinear part of the mixtures are completely suppressed and, then, $\mathbf{W}$ is adjusted to solve the classic linear BSS problem. There are a number of methods for adapting $\mathbf{g}(\cdot)$ – the first stage –, as those based on some *a priori* information [10], but the most common approach is that based on Gaussianization: from the perspective of the central limit theorem, the resultant random variables after the linear mixing stage will tend to be "more" Gaussian. Thus, the most intuitive idea for adapting $\mathbf{g}(\cdot)$ is to make its output $\mathbf{z}(n)$ Gaussian again [7]. This strategy reveals to be more effective when the number of sources N is large – according to the central limit theorem – or when the sources are Gaussian distributed. One can also include among these ideas the notion of the matching of probability distributions, which was one of the first methods in the PNL two stage approaches [11]. In this case, the nonlinearity compensation is accomplished when the distributions associated with $\mathbf{u}(n)$ and with $\mathbf{z}(n)$ are matched – note, however, that the *a priori* knowledge of the distribution of $\mathbf{u}(n)$ is required. This idea will also be relevant for the present work.

The second stage – i.e., the adaptation of the linear term $\mathbf{W}$ – is usually solved with classical ICA methods, which encompass higher-order statistics (HOS) [1,6]. However, when the sources are temporally colored, methods based on second-order statistics (SOS) can be applied, since they are known for its robustness and reliable simplicity. This idea is exploited in [7] by using a Gaussianization method in the first stage followed by a temporal decorrelation separation (TDSEP) method [1] in the second stage. In fact, this approach is interesting because it merges the simplicity of the second-order framework with simple source priors, for solving the complex nonlinear mixtures.

Although each approach presents its own particular advantages, in this work, we propose the use of a joint approach which is able, to a certain extent, to mix the benefits of a Gaussianization method – by means of a probability density matching – with the simplicity of the separation techniques based on SOS. The method will be described in the next section.

3   Proposed Separation Method

The separation method for PNL mixtures proposed in this work is based on a criterion that mixes the use of SOS and the matching of a (Gaussian) probability density. We start with the following assumptions: (*i*) there is at least one Gaussian source; (*ii*) the sources are jointly wide-sense stationary, present a temporal structure with different autocorrelation functions and are mutually independent; (*iii*) $\mathbf{f}(\cdot)$ is a set of invertible nonlinear functions; and (*iv*) the linear mixing matrix have, at least, two nonzero entries per row and per column.

Since we aim at the joint approach, we seek a single separation criterion which should be able to jointly adapt $\mathbf{g}(\cdot)$ and $\mathbf{W}$. However, this criterion will be composed of two parts, whose concepts can be understood separately – as we intend to show – but not its *modus operandi*.

3.1   Second-Order Statistics for Blind Separation

The first part of the criterion is based on the temporal structure of the sources. More precisely, we make use of the classical second-order joint diagonalization methods for linear BSS, which were the starting points for approaches and algorithms like SOBI, AMUSE, TDSEP and modified versions [1].

In this case, the SOS are exploited through time lagged covariance matrices:

$$\mathbf{R}_{\mathbf{y},d_s} = E\left[\mathbf{y}(n)\mathbf{y}^T(n-d_s)\right],\tag{1}$$

being d_s a constant lag. The main idea is to simultaneously (approximately) diagonalize the lagged covariance matrices for different values of d_s previously chosen, which can be summarized in the following cost [1]:

$$J_{SOS}(\boldsymbol{\theta}) = \sum_{d_s \in \mathcal{S}} \text{off}\left(\mathbf{R}_{\mathbf{y},d_s}\right) = \sum_{d_s \in \mathcal{S}} \sum_{i \neq j} \left(E\left[y_i(n)y_j(n-d_s)\right]\right)^2,\tag{2}$$

where off($\cdot$) the sum of the squares of the off-diagonal elements of a given matrix; $\mathcal{S}$ the set of chosen delays and $\boldsymbol{\theta}$ the set of parameters to be adjusted, i.e., $\boldsymbol{\theta} = \{\mathbf{g}(\cdot), \mathbf{W}\}$. An additional normalization term $(E\left[y_i^2(n)\right] - 1)^2$ for $i = \{1, \ldots, N\}$ is considered, since there are no whitening step for the nonlinear case. To solve the problem, $J_{SOS}(\boldsymbol{\theta})$ has to be minimized under a constraint over the linear separating matrix $\mathbf{W}$, in order to avoid convergence to the trivial solution.

Source separation based only on SOS is known to provide sufficient statistical information in the linear mixing case. However, in the nonlinear problem, additional statistics might be necessary.

3.2   Matching of Gaussian Distributions

Since we consider that at least one of the sources is Gaussian, this statistical information can be used in the second part of the criterion. Instead of using Gaussianization methods [6,7], a multidimensional density matching approach can be employed, such as the quadratic divergence between densities via kernel density estimators [8].

Basically, the idea is to force one of the recovered sources, say $y_1(n)$, to be Gaussian with a given temporal correlation (from a covariance matrix). In order to use the temporal information, we consider the following vector, related to the first output $y_1(n)$:

$$\overline{\mathbf{y}}_1(n) = [y_1(n)\, y_1(n-1)\, \ldots\, y_1(n-d_m)]^T,\tag{3}$$

where d_m is the maximum number of delays considered. In this case, the temporal covariance matrix of $\overline{\mathbf{y}}_1(n)$ is $\mathbf{R}_{\overline{\mathbf{y}}_1} = E\left[\overline{\mathbf{y}}_1(n)\overline{\mathbf{y}}_1^T(n)\right]$. Hence, we can formulate a criterion that aims at the match of an estimated multivariate density to a multivariate Gaussian distribution with zero mean and covariance matrix $\mathbf{R}_{\overline{\mathbf{y}}_1}$.

$$
\begin{aligned}
J_{GM}(\boldsymbol{\theta}_1) &= \int_D \left(f_{\overline{Y}_1}(\mathbf{v}) - G_{\mathbf{R}_{\overline{\mathbf{y}}_1}}(\mathbf{v})\right)^2 d\mathbf{v}\\
&= \int_D f_{\overline{Y}_1}^2(\mathbf{v})d\mathbf{v} + \int_D G_{\mathbf{R}_{\overline{\mathbf{y}}_1}}^2(\mathbf{v})d\mathbf{v} - 2\int_D f_{\overline{Y}_1}(\mathbf{v})G_{\mathbf{R}_{\overline{\mathbf{y}}_1}}(\mathbf{v})d\mathbf{v}
\end{aligned}\tag{4}
$$

where $f_{\overline{Y}_1}(\mathbf{v})$ is the multivariate density associated with the vector $\overline{\mathbf{y}}_1(n)$ at point $\mathbf{v}$; $G_{\mathbf{R}_{\overline{\mathbf{y}}_1}}(\mathbf{v})$ is a Gaussian distribution with covariance matrix $\mathbf{R}_{\overline{\mathbf{y}}_1}$, $D \subseteq \mathbb{R}^{d_m+1}$ and $\boldsymbol{\theta}_1 = \{\mathbf{g}(\cdot), \mathbf{w}_1\}$, being $\mathbf{w}_1$ the vector corresponding to the first row of $\mathbf{W}$.

To estimate $f_{\overline{Y}_1}(\mathbf{v})$, we consider a kernel density estimation method [12] using Gaussian kernels, which will lead to further simplifications in our case. Hence, the kernel estimate of $f_{\overline{Y}_1}(\mathbf{v})$ is:

$$\hat{f}_{\overline{Y}_1}(\mathbf{v}) = \frac{1}{L}\sum_{i=1}^{L} G_{\Sigma}\left(\mathbf{v} - \overline{\mathbf{y}}_1(i)\right),\tag{5}$$

where L is the number of vector samples of $\overline{\mathbf{y}}_1(n)$ and

$$G_{\Sigma}\left(\mathbf{v} - \overline{\mathbf{y}}_1(i)\right) = \frac{1}{\sqrt{(2\pi)^{d_m+1}|\Sigma|}}exp\left[\frac{-1}{2}(\mathbf{v} - \overline{\mathbf{y}}_1(i))^T \Sigma^{-1}(\mathbf{v} - \overline{\mathbf{y}}_1(i))\right],\tag{6}$$

is the multivariate symmetric Gaussian kernel with covariance matrix $\Sigma = \sigma^2\mathbf{I}$, where $\mathbf{I}$ is the identity matrix of order d_m+1 and σ^2 the kernel size; $|\Sigma|$ is the determinant of Σ. Replacing the estimate $f_{\overline{Y}_1}(\mathbf{v})$ into Eq. (4), it is possible to write:

$$\hat{J}_{GM}(\boldsymbol{\theta}_1) = \frac{1}{L^2}\sum_{i=1}^{L}\sum_{j=1}^{L}G_{2\Sigma}\left(\overline{\mathbf{y}}_1(i) - \overline{\mathbf{y}}_1(j)\right) + G_{2\mathbf{R}_{\overline{\mathbf{y}}_1}}(0) - \frac{2}{L}\sum_{i=1}^{L}G_{\Sigma+\mathbf{R}_{\overline{\mathbf{y}}_1}}\left(\overline{\mathbf{y}}_1(i)\right)\tag{7}$$

where the following relation was used [8]:

$$\int_D G_\Sigma \left(\mathbf{v} - \overline{\mathbf{y}}_1(i)\right) G_\Sigma \left(\mathbf{v} - \overline{\mathbf{y}}_1(j)\right) d\mathbf{v} = G_{2\Sigma} \left(\overline{\mathbf{y}}_1(i) - \overline{\mathbf{y}}_1(j)\right). \tag{8}$$

The goal is to minimize the cost $\hat{J}_{GM}(\boldsymbol{\theta}_1)$. It is expected that, in the optimization process, $\mathbf{R}_{\overline{\mathbf{y}}_1}$ converges to a scaled version of $\mathbf{R}_{\overline{\mathbf{s}}_k}$, the temporal covariance matrix of a Gaussian source $s_k(n)$, as will be explained ahead.

It is worth mentioning that this method requires the adjustment of the kernel size σ, which, for Gaussian distributions, can be done using the Silverman's rule [13], i.e., $\sigma_o = \sigma_{y_1} \left(4/(L\left(2(d_m+1)+1\right))\right)^{1/(d_m+5)}$, where σ_{y_1} is the standard deviation of $y_1(n)$. The number of delays, d_m, should be a trade-off between the amount of temporal information used and the computational cost.

3.3   The Combined Approach

With both costs $J_{SOS}(\boldsymbol{\theta})$ and $J_{GM}(\boldsymbol{\theta}_1)$ at hand, we are able to analyze some illustrative cases that might be further clarifying. Nonetheless, for the sake of briefness, we appeal to certain intuitive properties within the BSS problem.

We start by considering the sole minimization of $J_{GM}(\boldsymbol{\theta}_1)$ and, for simplicity, we suppose the $N = 2$ sources case with the following possible types of sources: (i) only one of the sources is Gaussian distributed and (ii) both sources are Gaussian and temporally colored (with different autocorrelation functions). In the scenario (i), we know that, at the end of the linear mixing problem (with all linear coefficients non-null), $\mathbf{u}(n)$ will tend to have a joint Gaussian distribution, but not exactly Gaussian due to one of the sources being not Gaussian. After the nonlinearities $\mathbf{f}(\cdot)$, it is expected that $\mathbf{x}(n)$ will be even farther from the Gaussian distribution. By forcing $y_1(n)$ to be Gaussian via minimization of $J_{GM}(\boldsymbol{\theta}_1)$, it is expected that the nonlinear separating functions $\mathbf{g}(\cdot)$ are able to produce a Gaussian-like distribution for $\mathbf{z}(n)$, so that the linear separating structure $\mathbf{W}_1$ will be able to extract a Gaussian source, but not necessarily the desired one. Hence, in the case (i), if considered the additional minimization of the cost $J_{SOS}(\boldsymbol{\theta})$, it might be able to recover the correct Gaussian and, consequently, the other source, since their lagged covariance matrices will be jointly diagonalized. In case (ii), since the linear mixtures of Gaussian distributions remains Gaussian, we have that $\mathbf{u}(n)$ would be jointly Gaussian. The nonlinearity $\mathbf{f}(\cdot)$, again, will drive the distribution of $\mathbf{x}(n)$ away from Gaussianity. By minimizing $J_{GM}(\boldsymbol{\theta}_1)$ in this case, it is expected that nonlinearities be compensated, but the linear part will be unable to separate between the two Gaussian sources. Now, if we also consider the minimization of $J_{SOS}(\boldsymbol{\theta})$, we know from the linear BSS theory that Gaussian distributions can be separated and the estimation of the temporal covariance matrix $\mathbf{R}_{\overline{\mathbf{y}}_1}$ will be more precise. Undoubtedly, it is not possible to determine which of the Gaussian sources will be recovered at $y_1(n)$, but, since the BSS problem admits permutation of the solutions, this is not an issue.

In fact, a bond between both SOS and GM criteria emerges in the temporal information used by both costs, where there is an important synergy: the diagonalization of $\mathbf{R}_{\mathbf{y},d_s}$ aids the convergence of $\mathbf{R}_{\overline{\mathbf{y}}_1}$ to $\mathbf{R}_{\overline{\mathbf{s}}_k}$ – the temporal covariance matrix of a Gaussian source – in addition, the information that the source $y_1(n)$ is Gaussian can also contribute to it; in turn, when $\mathbf{R}_{\overline{\mathbf{y}}_1}$ tends to $\mathbf{R}_{\overline{\mathbf{s}}_k}$, it can aid with the separation of the other sources when diagonalizing $\mathbf{R}_{\mathbf{y},d_s}$.

These illustrative cases reveal how the joint minimization of the SOS and GM costs might aid the separation task. Hence, we propose the following combined cost:

$$J_{SOS+GM}(\boldsymbol{\theta}) = J_{SOS}(\boldsymbol{\theta}) + \alpha J_{GM}(\boldsymbol{\theta}_1), \tag{9}$$

where α is a trade-off parameter between costs. The other parameters that require (pre-)adjustment are the number of samples and of time delays.

In the following, we present some performance analysis of the proposed SOS+GM criterion in simulation scenarios.

4   Simulation Results

In this section, we analyze the performance of the SOS+GM criterion and compare it with two other methods: the minimization of only the SOS cost (joint approach) and the Gaussianization process followed by the minimization of the SOS cost (the two-stage approach proposed in [7]). For the Gaussianization method, the maximization of Shannon's entropy was considered, using (univariate) Gaussian kernel estimators [14].

The analyses were conducted in two scenarios. In the first one, we consider two Gaussian sources that are temporally colored by the finite impulse response (FIR) filters $h_1(z) = 1 + 0.5z^{-1} + 0.2z^{-2}$ and $h_2(z) = 1 - 0.8z^{-1}$, one for each source. For the second scenario, one of the sources is a temporally correlated Gaussian (by the filter $h_1(z)$) and the other is a uniformly distributed signal (from -1 to $+1$) with no temporal structure. In both scenarios, the mixtures were the result of $\mathbf{x}(n) = (\mathbf{As}(n))^3$, being $\mathbf{A} = [0.25\ 0.86; -0.86\ 0.25]$. For the separating structure, we considered, in place of $\mathbf{g}(\cdot)$, parametric functions of the type $z_i(n) = g_{i,1}x_i(n) + g_{i,2}\,\text{sign}(x_i(n))\,\sqrt[3]{|x_i(n)|}$, where the operator $\text{sign}(\cdot)$ returns a $+1$ if $x_i(n) \geq 0$ or a -1 if $x_i(n) < 0$; followed by a 2×2 matrix $\mathbf{W}$.

In all cases, the number of delays and the number of samples considered remained fixed. For the SOS cost, common to all considered methods, we adopted 3 delays with $\mathcal{S} = \{0, 1, 2\}$ and 500,000 samples of $\mathbf{y}(n)$ (the SOS cost demanded a higher accuracy in its estimation, hence the large number of samples). For the Gaussianization method, 500 samples of the vector $\mathbf{z}(n)$ were used to estimate the marginal entropies. For the GM cost (part of the SOS+GM criterion), we considered $d_m = 1$ (delays 0 and 1), 500 samples of the vector $\overline{\mathbf{y}}_1(n)$ and $\alpha = 1$.

To perform the optimization of the weights (nonlinear and linear), we adopted the metaheuristic known as Differential Evolution (DE) [9], which is an efficient technique to avoid local convergence. The DE parameters were chosen to be $N_P = 300$ (population size), $F = 0.7$, $CR = 0.7$ and 100 iterations – for more details, please refer to [9]. For the joint approaches, a single run of the DE adapts all coefficients, while, for the two-stage approach, two DE runs are necessary, one for the nonlinear and other for the linear part. After training, the performance

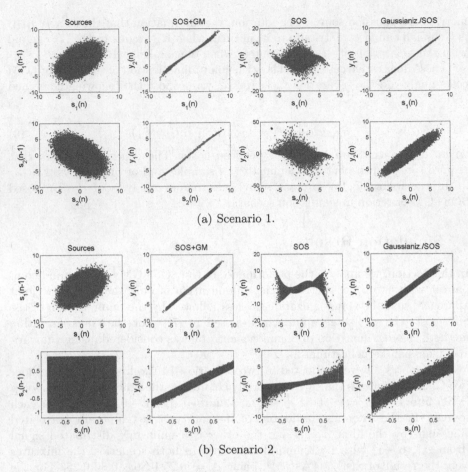

(a) Scenario 1.

(b) Scenario 2.

Fig. 2. Scatter plots of the sources and of the outputs for each method. (Color figure online)

of the best individual in the population was measured in terms of SIR, defined as $\text{SIR} = 10 \log \left(E[y_i(n)^2] / E[(s_i(n) - y_i(n))^2] \right)$, after sign and variance correction.

Figure 2 shows, for both scenarios, the scatter plot $s_i(n) \times s_i(n-1)$ of each source and the outputs of the SOS+GM, the SOS and the Gaussianization/SOS methods through the plots $s(n) \times y(n)$, where a diagonal line means that a perfect separation was achieved (the red dots are the output samples used to estimate the GM cost). The measured SIR values for each case are displayed in Table 1.

In scenario 1, the proposed SOS+GM method was able to recover both sources with high SIR values. The output $y_1(n)$, in this case, recovered the source $s_2(n)$ and preserved its temporal structure with higher precision, being, consequently, associated with a higher SIR level. The second method (sole SOS criterion) has not performed well, being its plots $s_i(n) \times y_i(n)$ in Fig. 2(a) far from a diagonal line and with outputs associated with low values of SIR. For the

Table 1. Performance in terms of SIR [dB]

	Sources	GM+SOS	SOS	Gauss./SOS
Scenario 1	Gaussian $h_1(z)$ - Source 1	61.56	10.37	70.28
	Gaussian $h_2(z)$ - Source 2	75.01	−3.33	19.49
Scenario 2	Gaussian $h_1(z)$ - Source 1	38.50	−0.51	21.36
	Uniform - Source 2	39.66	5.76	34.07

two-stage Gaussianization/SOS method, Gaussian signals were recovered, but just one of them preserved the temporal structure of the source $(s_1(n))$. In fact, this result comes from a drawback of the two-stage approach: in the Gaussianization step, the outputs $\mathbf{z}(n)$ can be very close to Gaussian distributions, but may carry a small nonlinear residue (due to precision issues on estimation, for example); then, in the linear separation step, this residue can not be treated. Indeed, in the simulations, the optimization of the SOS cost in the second-stage was not able to achieve its lowest value, since the nonlinear residue could not be treated by a linear structure. Even though, the SIR value of 19.49 dB obtained in the estimation of $s_2(n)$ can be considered acceptable in nonlinear scenarios (note that, in Fig. 2(a), the deviation of the output $y_2(n)$ from $s_2(n)$ are not severe).

In the second scenario, our SOS+GM method performed as expected and was able to recover the Gaussian source and its temporal structure at output $y_1(n)$, as indicated by the plot $s_1(n) \times y_1(n)$ of Fig. 2(b). Also, the second source was recovered with an SIR value of 39.66 dB, being a reliable estimate. For the SOS method, again, the performance measures were far from the desired, indicating that the sole minimization of SOS cost is not sufficient for nonlinear separation. Finally, the Gaussianization/SOS method could provide reasonable estimates of the sources, as shown in Fig. 2(b). However, due to the presence of the uniform distribution in the mixture, the Gaussianization step was not able to completely compensate the nonlinearities, causing a reduction on the performance. Indeed, in this scenario, since the proposed method does not encompass any assumption on the distribution shape of the sources different from the one that is Gaussian, it can obtain better results.

5   Conclusions

In this work, we have proposed a joint approach for source separation in the PNL model. The method allies the use of the second-order statistics to the density matching approach. By only assuming temporally colored sources and, at least, one Gaussian source, this method is able to perform the separation based on less restrictive requirements than the usual two-stage methods, whose assumptions apply to all sources. Also, it can be computationally simpler than estimating mutual independence in the classical ICA framework. Along with the use of the DE metaheuristic, the simulations indicated that the proposed method is more robust than the Gaussianization method in the case of two Gaussian sources and in the case of one Gaussian and one uniformly distributed source.

Since this work is still in its initial stage, there are plenty of possibilities for future works. We consider, for instance, the analysis of the conditions for the extension to a higher number of sources; the assumption of a known source distribution which is not Gaussian; and, finally, the proposition of a gradient-based algorithm.

Acknowledgements. This work was partly supported by FAPESP (2013/14185-2, 2015/23424-6), CNPq and ERC project 2012-ERC-AdG-320684 CHESS.

References

1. Comon, P., Jutten, C.: Handbook of Blind Source Separation: Independent Component Analysis and Applications. Academic Press, Oxford (2010)
2. Duarte, L.T., Jutten, C., Moussaoui, S.: A Bayesian nonlinear source separation method for smart ion-selective electrode arrays. IEEE Sens. J. **9**(12), 1763–1771 (2009)
3. Meganem, I., Deville, Y., Hosseini, S., Déliot, P., Briottet, X., Duarte, L.T.: Linear-quadratic and polynomial non-negative matrix factorization; application to spectral unmixing. In: 19th IEEE European Signal Processing Conference, pp. 1859–1863 (2011)
4. Hosseini, S., Jutten, C.: On the separability of nonlinear mixtures of temporally correlated sources. IEEE Sig. Process. Lett. **10**(2), 43–46 (2003)
5. Taleb, A., Jutten, C.: Source separation in post-nonlinear mixtures. IEEE Trans. Sig. Process. **47**(10), 2807–2820 (1999)
6. Deville, Y., Duarte, L.T.: An overview of blind source separation methods for linear-quadratic and post-nonlinear mixtures. In: Vincent, E., Yeredor, A., Koldovský, Z., Tichavský, P. (eds.) LVA/ICA 2015. LNCS, vol. 9237, pp. 155–167. Springer, Heidelberg (2015). doi:10.1007/978-3-319-22482-4_18
7. Ziehe, A., Kawanabe, M., Harmeling, S., Müller, K.R.: Blind separation of post-nonlinear mixtures using gaussianizing transformations and temporal decorrelation. J. Mach. Learn. Res. **4**, 1319–1338 (2003)
8. Fantinato, D., Boccato, L., Neves, A., Attux, R.: Multivariate PDF matching via kernel density estimation. In: Proceedings of the Symposium Series on Computational Intelligence (2014)
9. Price, K., Storn, R., Lampinen, J.: Differential Evolution: A Practical Approach to Global Optimization. Springer, Heidelberg (2005)
10. Duarte, L.T., Suyama, R., Attux, R., Rivet, B., Jutten, C., Romano, J.M.T.: Blind compensation of nonlinear distortions: application to source separation of post-nonlinear mixtures. IEEE Trans. Sig. Process. **60**, 5832–5844 (2012)
11. White, S.A.: Restoration of nonlinearly distorted audio by histogram equalization. J. Audio Eng. Soc. **30**(11), 828–832 (1982)
12. Parzen, E.: On the estimation of a probability density function and the mode. Ann. Math. Statist. **33**, 1065–1067 (1962)
13. Silverman, B.: Density Estimation for Statistics and Data Analysis. Chapman and Hall, London (1986)
14. Silva, L., de Sá, J., Alexandre, L.: Neural network classification using Shannon's entropy. In: European Symposium on Artificial Neural Networks, pp. 217–222 (2005)

Optimal Measurement Times for Observing a Brownian Motion over a Finite Period Using a Kalman Filter

Alexandre Aksenov[✉], Pierre-Olivier Amblard, Olivier Michel, and Christian Jutten

GIPSA-lab, 11, rue des Mathmatiques, 38240 Saint-Martin d'Hères, France
alexander1aksenov@gmail.com

Abstract. This article deals with the optimization of the schedule of measures for observing a random process in time using a Kalman filter, when the length of the process is finite and fixed, and a fixed number of measures are available. The measure timetable plays a critical role for the accuracy of this estimator. Two different criteria of optimality of a timetable (not necessarily regular) are considered: the maximal and the mean variance of the estimator. Both experimental and theoretical methods are used for the problem of minimizing the mean variance. The theoretical methods are based on studying the cost function as a rational function. An analytical formula of the optimal instant of measure is obtained in the case of one measure. Its properties are studied. An experimental solution is given for a particular case with $n > 1$ measures.

Keywords: Random walk · Wiener process · Kalman filter

1 Introduction

When a latent phenomenon is observed through different acquisition methods, more information can be acquired than from a single method, but making the most of these measurements is a challenge [5]. This is due to discrepancies in the nature of data, in particular in the sampling. The observer often cannot control the instants of measure and makes regular measures with each of the available sensors. In this case, controlling the delays between measurements with different sensors can lead to a consequent gain in the quality of the estimator [1]. One may also ask: what is the optimal (not necessarily regular) timetable of measurements?

A model, where sensors are active during an interval of time, has been considered [3]. On the other hand, the model with instantaneous measures is considered in this article. In this model one observes a *scalar* continuous latent variable on

This work has been partly supported by the European project ERC-2012-AdG-320684-CHESS.

P. Tichavský et al. (Eds.): LVA/ICA 2017, LNCS 10169, pp. 509–518, 2017.
DOI: 10.1007/978-3-319-53547-0_48

a *finite* interval of time with noisy sensors, each having access to *only one* measurement at one time instant and having its own measurement noise variance. A Kalman filter based approach for estimating the hidden state is taken in this paper, and two ways of evaluating the quality of estimation are considered: the maximal and the mean variance of the estimator over time.

The main theoretical result of this text is the optimal instant of measure (with respect to the mean variance of the estimation error) given by (16) in the case of one measure. The problem whether the variance of the estimator can be maintained below a fixed limit over the whole interval is essentially solved. The main experimental result is the numerical computation of the optimal schedules in a particular case where 2 measures are available.

The paper is organized as follows. The general (multimodal, irregularly scheduled) Kalman estimation model is defined in Sect. 2. The maximal variance of the estimator is used as the cost function in Sect. 3. In the formal treatment of this problem, the maximal variance of the estimator is considered as a bound. It is proved that the intuitive algorithm "measure when the error variance reaches the bound" is globally optimal in this sense. An algorithm of this kind has been proposed for a model of linear filtering in discrete time [2]. The more challenging problem, where the mean variance of the estimator is used as the cost function, is stated in Sect. 4. The only case where the optimal solution can be written in closed form is the case of one measure, and this case is studied in detail in Subsects. 4.1 and 4.2.

2   Model Description

We assume that the estimation of the system state is done by computing the time evolution of a parameter, and that the variance of the estimation grows linearly between measurements. This simple assumption models the fact that decreasing the measure frequency decreases the accuracy on the system state estimation. In this purpose, we consider a real Brownian motion $\theta(t)$ ($t \in [0, T]$), satisfying for $t > s$, $\theta(t) - \theta(s) \stackrel{d}{\sim} \mathcal{N}(0, \sigma^2(t-s))$ i.e., the increments are Gaussian with mean 0 and variance $\sigma^2(t - s)$.

Suppose n sensors can make measurements at moments $t_1, \ldots, t_n$. It is assumed that each sensor k returns a measured value equal to X_k at time t_k and that $0 \leqslant t_1 \leqslant \cdots \leqslant t_n \leqslant T$. No subsequence of the sequence $(t_1, \ldots, t_n)$ is constrained to be regular in any sense. Suppose, the initial state $\theta(0)$ is a Gaussian random variable of mean $\bar{\theta}_0$ and variance v_0. Suppose that $\theta(0)$, the measurement noise and the evolution of the Brownian motion $\theta(t)$ are independent. The Kalman filter framework can apply with the state and measurement equations:

$$\theta(t_k) = \theta(t_{k-1}) + w_k, \ w_k \stackrel{d}{\sim} \mathcal{N}(0, \sigma^2(t_k - t_{k-1})) \tag{1}$$

$$X_k = \theta(t_k) + n_k, \ n_k \stackrel{d}{\sim} \mathcal{N}(0, v_k). \tag{2}$$

By the theory of Kalman filtering (see [4]), the maximum likelihood estimate $\hat{\theta}_{t_k}^{t_k}$ of $\theta(t_k)$ and its variance $\Gamma_{t_k}^{t_k}$ are defined by the following recursive equations:

$$
\begin{cases}
\hat{\theta}_{t_k}^{t_k} = \hat{\theta}_{t_k}^{t_{k-1}} + K(t_k)\left(X_k - \hat{\theta}_{t_{k-1}}^{t_{k-1}}\right) & (3) \\[2mm]
\hat{\theta}_{t_k}^{t_{k-1}} = \hat{\theta}_{t_{k-1}}^{t_{k-1}} & (4) \\[2mm]
\Gamma_{t_k}^{t_k} = \Gamma_{t_k}^{t_{k-1}} - K(t_k)\Gamma_{t_k}^{t_{k-1}} & (5) \\[2mm]
K(t_k) = \Gamma_{t_k}^{t_{k-1}}\left(\Gamma_{t_k}^{t_{k-1}} + v_k\right)^{-1} & (6) \\[2mm]
\Gamma_{t_k}^{t_{k-1}} = \Gamma_{t_{k-1}}^{t_{k-1}} + \sigma^2(t_k - t_{k-1}), & (7)
\end{cases}
$$

where $\hat{\theta}_{t_k}^{t_l}$ ($l \in \{k-1, k\}$) is the maximum likelihood estimate of $\theta(t_k)$ conditionally to the data available at time t_l, and $\Gamma_{t_k}^{t_l}$ is the variance of the estimate $\hat{\theta}_{t_k}^{t_l}$. $K(t_k)$ is the Kalman gain used for the update at time t_k. In order for (7) to make sense for $k = 1$, define $t_0 = 0$ and $\Gamma_{t_0}^{t_0} = v_0$.

Remark that, by (5) and (6), using the fact that all quantities are scalar,

$$
\Gamma_{t_k}^{t_k} = \Gamma_{t_k}^{t_{k-1}} - \frac{\left(\Gamma_{t_k}^{t_{k-1}}\right)^2}{\Gamma_{t_k}^{t_{k-1}} + v_k} = \frac{\left(\Gamma_{t_k}^{t_{k-1}} + v_k\right)\Gamma_{t_k}^{t_{k-1}} - \left(\Gamma_{t_k}^{t_{k-1}}\right)^2}{\Gamma_{t_k}^{t_{k-1}} + v_k} = \frac{v_k\Gamma_{t_k}^{t_{k-1}}}{v_k + \Gamma_{t_k}^{t_{k-1}}},
$$
$$(8)$$

which is equivalent (by (7)) to

$$
\left(\Gamma_{t_k}^{t_k}\right)^{-1} = v_k^{-1} + \left(\Gamma_{t_k}^{t_{k-1}}\right)^{-1} = v_k^{-1} + \left(\Gamma_{t_{k-1}}^{t_{k-1}} + \sigma^2(t_k - t_{k-1})\right)^{-1}. \tag{9}
$$

Therefore, each $\Gamma_{t_k}^{t_k}$ is a rational function of $\sigma^2, t_1, \ldots, t_k, v_0, \ldots, v_k$.

For each $t \in [0, T]$, denote $v(t)$ the variance of $\hat{\theta}(t)$, i.e. the variance when the last measurement was taken plus the uncertainty due to the time without new feedbacks. It equals:

$$
v(t) = \Gamma_{t_k}^{t_k} + \sigma^2(t - t_k) \text{ where } k = \max\{i | t_i \leqslant t\}. \tag{10}
$$

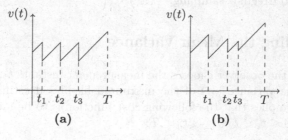

Fig. 1. The function $v(t)$ in particular cases. In **(a)**, $v_0 = \frac{1}{2}, v_1 = v_2 = v_3 = 1, T = 1, \sigma^2 = 1$ and $t_1 = 0.128, t_2 = 0.369, t_3 = 0.611$. In **(b)**, $v_0 = \frac{1}{2}, v_1 = 1, v_2 = 2, v_3 = 3, T = 1, \sigma^2 = 1$ and $t_1 = 0.241, t_2 = 0.494, t_3 = 0.641$. The values of v_1, v_2, v_3 control the differences of the variance before and after the measurement.

$v(t)$ is a piecewise linear function composed of line intervals of slope σ^2. Two examples of functions $v(t)$ are shown Fig. 1. In the first example, v_1, v_2, v_3 are equal, in the second example they are different.

In these and in all the other examples of this article, the value $\sigma^2 = 1$ has been taken. This can be done without loss of generality.

3   Controlling the Maximal Variance

In this section, the observer chooses the measurement instants $t_1, \ldots, t_n$ so that the variance $v(t)$ of the maximum likelihood estimator of $\theta(t)$ does not exceed a fixed bound V:

$$\forall t \in [0, T] \ v(t) \leqslant V. \tag{11}$$

The question is: how large can the length T of the process be so that the constraint (11) can be satisfied? The following lemma answers this question.

Lemma 1. *Suppose that $v_0 \leqslant V$. The constraint (11) can be satisfied if and only if*

$$T \leqslant \frac{V - v_0}{\sigma^2} + \left(\sum_{k=1}^{n} \delta(V, v_k) \right) \ where \ \sigma^2 \delta(V, v) = V - \frac{Vv}{V + v}. \tag{12}$$

One can remark that the time defined by (12) increases when V increases and decreases when any of v_i increases (i.e., if the sensors or the estimate of the initial state are less accurate).

The proof of Lemma 1 (see Appendix) also implies that the intuitive algorithm "measure when $v(t) \geqslant V$ is reached" (which consists in iteratively applying (32) to define the instants of measure) keeps the estimation error variance bounded by V during as long time as possible. Remark that this algorithm is also optimal in this sense when $v_0 > V$. Indeed, in this case the first measure has to be done at the instant 0 in order to achieve $v(0) \leqslant V$.

It is interesting to remark that if $v_1 = \ldots = v_n$, this algorithm defines a regular sampling. Conversely, if measurement accuracies differ, the optimal solution leads to irregular sampling.

4   Controlling the Mean Variance

In this section, the observer chooses the measurement instants $t_1, \ldots, t_n$ so that the mean of the variance $v(t)$ of the maximum likelihood estimator of $\theta(t)$ is minimal. This implies that the following cost function is to be minimized under the constraint $0 \leqslant t_1 \leqslant t_2 \leqslant \ldots \leqslant t_n \leqslant T$:

$$J_{\sigma^2, T, v_0, v_1, \ldots, v_n}(t_1, \ldots, t_n) = \int_0^T v(t) dt$$

$$= \frac{\sigma^2 t_1^2}{2} + v_0 t_1 + \frac{\sigma^2 (t_2 - t_1)^2}{2} + \Gamma_{t_1}^{t_1}(t_2 - t_1) + \cdots + \frac{\sigma^2 (T - t_n)^2}{2} + \Gamma_{t_n}^{t_n}(T - t_n). \tag{13}$$

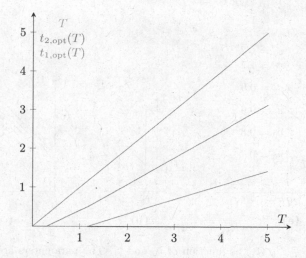

Fig. 2. Experimentally computed values of $t_{1,\text{opt}}(1, T, 1, 1, 1)$ and $t_{2,\text{opt}}(1, T, 1, 1, 1)$ ($n = 2$) for T varying from 0.01 to 5. For example, $t_{1,\text{opt}}(1, 3, 1, 1, 1) = 0.696$ and $t_{2,\text{opt}}(1, 3, 1, 1, 1) = 1.763$ define irregularly spaced measures.

One can remark that the cost function (13) is rational in its $2n + 3$ parameters $\sigma^2, T, v_0, \ldots, v_n, t_1, \ldots, t_n$.

Suppose that this function is minimized in a unique point

$$(t_{1,\text{opt}}(\sigma^2, T, v_0, v_1, \ldots, v_n), \ldots, t_{n,\text{opt}}(\sigma^2, T, v_0, v_1, \ldots, v_n)). \tag{14}$$

These values are the optimal measurement instants. We can wonder where these instants are located, and especially if some of them are equal to zero. The minimizer is indeed unique in the case $n = 1$, which is proved in Subsect. 4.2.

We are also interested in the behavior of the optimal measurement times as functions of T: monotonicity, asymptotic properties, etc. The dependency of the optimal instants on T for $n = 2$ and fixed values of σ^2, v_0, v_1, v_2 is shown Fig. 2. It suggests that the optimal instants vary continuously, are monotonically increasing and are close to piecewise-linear functions of T. It also suggests that some optimal instants are located at zero when T is small enough. When T is large, the optimal instants are approximately equally spaced in time. These properties are proved for $n = 1$ measure in Subsects. 4.1 and 4.2.

4.1 The Optimal Instant in Case of One Measure: Qualitative Results

In this and the next subsections, the above problem is studied for the particular case where $n = 1$ measure can be performed. All questions listed above are solved in terms of explicit formulas in Sect. 4.2.

The cost function (13) takes the form

$$J_{\sigma^2, T, v_0, v_1}(t_1) = \frac{\sigma^2 t_1^2}{2} + v_0 t_1 + \frac{\sigma^2 (T - t_1)^2}{2} + \frac{(\sigma^2 t_1 + v_0) v_1 (T - t_1)}{\sigma^2 t_1 + v_0 + v_1}. \tag{15}$$

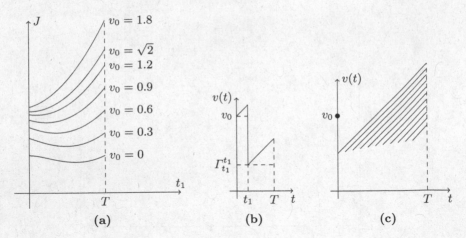

Fig. 3. (a): $J_{\sigma^2,T,v_0,v_1}(t_1)$ as function of v_0 and t_1. The parameters are $v_1 = 1, T = 1, \sigma^2 = 1$. The cost function is minimized at $t_1 = 0$ if and only if $v_0 \geqslant \sqrt{2}$. **(b)**: An example of a function $v(t)$ showing the geometric interpretation of the rectangular and the triangular terms of the expression (15) of the integral cost function. **(c)**: The dependency of the function $v(t)$ in the interval $t \in [t_1, T]$ on the choice of t_1. In this example, $n = 1$, $v_0 = v_1 = 1$, $T = 1.2$, $\sigma^2 = 1$, and t_1 takes values in $[0, 1]$. Each straight line represents one possible function $v(t)$. The slopes of all lines equal σ^2 and their left endpoints have coordinates $(t_1, \Gamma_{t_1}^{t_1})$.

Its behavior is shown Fig. 3(a). Remark that the RHS term in Eq. (15) can be split into two terms: the "rectangular term" $\left(v_0 t_1 + \frac{(\sigma^2 t_1 + v_0)v_1(T-t_1)}{\sigma^2 t_1 + v_0 + v_1}\right)$ and the "triangular term" $\left(\frac{\sigma^2 t_1^2}{2} + \frac{\sigma^2(T-t_1)^2}{2}\right)$, respectively accounting for the contributions of the rectangular and triangular shaped area in the integral of $v(t)$, and shown on Fig. 3(b). Minimizing the cost function $J_{T,v_0,v_1}(t_1)$ constitutes a trade-off between minimizing these two terms.

Different situations are possible as it can be seen on Fig. 3(a). One can define the **regime 1** as the set of situations when $t_1 = 0$ is the optimum. Similarly, define the **regime 2** as the set of situations where the optimal t_1 is in the interior of the interval $[0, T]$. Then, the optimal t_1 is the point where the derivative of the cost function (15) vanishes. Its value is given by (20). Remark that in the regime 2, the optimal t_1 can be larger than $\frac{T}{2}$.

The distinction between these two regimes is justified by the following analytic property: the function, which defines the optimal t_1,

$$t_{1,\text{opt}}(\sigma^2, T, v_0, v_1) = \arg\min_{t_1} J_{\sigma^2,T,v_0,v_1}(t_1)$$

$$= \max\left(0, \frac{-3v_0 - 3v_1 + \sigma^2 T + \sqrt{(\sigma^2 T + v_0 + 5v_1)^2 - (4v_1)^2}}{4\sigma^2}\right), \quad (16)$$

is differentiable everywhere except at the border between the regions which correspond to the two regimes.

4.2   Derivation and Properties of (16)

One can decide whether a local minimum is achieved at $t_1 = 0$ by computing the corresponding partial derivative:

$$\frac{\partial J_{\sigma^2,T,v_0,v_1}(t_1)}{\partial t_1}\bigg|_{t_1=0} = \frac{v_0}{v_0 + v_1}\left(v_0 - \sigma^2 T\left(\frac{v_1}{v_0 + v_1} + 1\right)\right). \quad (17)$$

The partial derivative at a point $t_1 = t'$ has a similar expression:

$$\frac{\partial J_{\sigma^2,T,v_0,v_1}(t_1)}{\partial t_1}\bigg|_{t_1=t'} = \frac{v_0 + \sigma^2 t'}{v_0 + v_1 + \sigma^2 t'}$$

$$\times \left(v_0 + \sigma^2 t' - \sigma^2(T - t')\left(\frac{v_1}{v_0 + v_1 + \sigma^2 t'} + 1\right)\right). \quad (18)$$

Remark that the RHS of (18) is a product of two increasing (with respect to t') factors, the first of which $\left(\frac{v_0+\sigma^2 t'}{v_0+v_1+\sigma^2 t'}\right)$ is positive. Therefore, the locus of nonnegativity of $\frac{\partial J_T(t_1)}{\partial t_1}\big|_{t_1=t'}$ is an interval of the form $[t_{1,\mathrm{opt}}, T]$, where $t_{1,\mathrm{opt}}$ may equal zero or be strictly positive.

Consequently, two different behaviors of the cost function are possible. In the first case (regime 1), it is increasing near $t_1 = 0$ (its derivative at zero (17) is nonnegative). Then, the cost function $J_T(t_1)$ is increasing and convex on the whole interval $[0, T]$, and its global minimum is $t_{1,\mathrm{opt}}(T) = 0$. According to (17), this situation corresponds to

$$T \leqslant T_{\mathrm{crit}} = \frac{v_0}{\sigma^2\left(\frac{v_1}{v_0+v_1} + 1\right)}. \quad (19)$$

Remark that T_{crit} is an increasing function of v_0 and a decreasing function of v_1 and of σ^2.

Intuitively, this behavior is observed when T is small or v_0 is large, which means that the prior information is poor. In this case, it is penalizing not to take a measure immediately in order to get better information. More formally, the rectangular term has an order of magnitude $O(T)$ when T tends to zero, while the triangular term has an order of magnitude $O(T^2)$. Therefore, when T is small enough, choosing $t_1 = 0$ should minimize both the rectangular term and the sum.

In the second case (regime 2), the cost function is decreasing near $t_1 = 0$. This is observed when (19) does not hold, i.e. T is large or v_0 is small. Then, the minimum of the cost function is reached at the only nonzero point $(t_{1,\mathrm{opt}})$, where its derivative (18) equals zero. By equating the derivative (18) to zero, one gets the following expressions for $t_{1,\mathrm{opt}}$ and for T:

$$t_{1,\mathrm{opt}} = \frac{-3v_0 - 3v_1 + \sigma^2 T + \sqrt{(\sigma^2 T + v_0 + 5v_1)^2 - (4v_1)^2}}{4\sigma^2} \quad (20)$$

and

$$\sigma^2 T = 2\sigma^2 t_{1,\text{opt}} + v_0 - v_1 + \frac{2v_1^2}{v_0 + \sigma^2 t_{1,\text{opt}} + 2v_1}. \tag{21}$$

Remark that the function $t_{1,\text{opt}}(T)$ defined by (20) is concave and increasing. When $T \to \infty$, one gets the asymptotic expansion

$$t_{1,\text{opt}}(T) = \frac{\sigma^2 T + v_1 - v_0}{2\sigma^2} + o\left(\frac{1}{T}\right), \tag{22}$$

the function being always smaller than its asymptote:

$$t_{1,\text{opt}}(T) < \frac{\sigma^2 T + v_1 - v_0}{2\sigma^2}. \tag{23}$$

The following intuitive argument can be given for the order of magnitude of the optimal instant: $t_{1,\text{opt}}(T) \sim \frac{T}{2}$ (by (22)). When T is large, the triangular terms become more important than the "rectangular terms". Therefore, the minimum of the sum should be close to the value $\frac{T}{2}$, which minimizes the triangular term.

Using (19) and (20), it is easy to check that

$$t_{1,\text{opt}}(T_{\text{crit}}) = \frac{-3v_0 - 3v_1 + \sigma^2 T_{\text{crit}} + \sqrt{(\sigma^2 T_{\text{crit}} + v_0 + 5v_1)^2 - (4v_1)^2}}{4\sigma^2} = 0, \tag{24}$$

i.e., both formulas of regime 1 and regime 2 coincide if the values of the parameters lie on the boundary. This proves (16).

Remark that the dependence of $t_{1,\text{opt}}$ in σ^2 and T is simplified by the relation

$$t_{1,\text{opt}}(\frac{\sigma^2}{\alpha}, \alpha T, v_0, v_1) = \alpha t_{1,\text{opt}}(\sigma^2, T, v_0, v_1), \tag{25}$$

therefore, the ratio $t_{1,\text{opt}}/T$ depends only on $\sigma^2 T, v_0$ and v_1.

5   Conclusion and Perspectives

A simple model is studied, where the variance about the system parameters (here a single parameter) evolving over a finite period of time grows linearly in the absence of measure. The properties of the optimal measure timetable are studied.

In Sect. 4, the particular case, where the instant of exactly 1 measure is to be chosen, is studied in detail. The system can behave in one of the two different regimes. If the duration of the process is larger than a critical value, the optimal instant is inside the interval and is asymptotically close (when T is large) to $\frac{T}{2}$. If the duration of the process is smaller than the critical value, the sampling at time 0 turns out to be optimal. These results are in a closed form, therefore they are more precise than the results obtained for the model considered in [3].

One goal of the future research is to find the optimal measurement instants when the number of measures is $n > 1$. Partial analytic results, which explain some properties of Fig. 2, are available. These extend the case ($n = 1$).

The main method is the following: in a subproblem where the instants $t_1, \ldots, t_{n-1}$ are fixed, one can find the optimal instant (t_n) using the results of this article, then check whether the given instants $t_1, \ldots, t_{n-1}$ are locally optimal.

In this problem, the order of the measures is fixed. It is also possible to allow it to vary. The main property of this problem is the fact that the cost function is no longer rational, but piecewise-rational.

Another objective of the future research is to consider more complex models than the real Brownian motion considered presently.

Appendix: Proof of Lemma 1

The following intuitively clear result is important for the proof.

Lemma 2. *The variance* $\Gamma_{t_1}^{t_1} = \frac{1}{\frac{1}{v_1} + \frac{1}{\sigma^2 t_1 + v_0}}$ *(see (9) for* $k = 1$*) satisfies*

$$\frac{\partial \Gamma_{t_1}^{t_1}}{\partial t_1} < \sigma^2. \tag{26}$$

The proof is done by a direct computation. An interpretation of this inequality is the following: in a setting with $n = 1$ measure, the variance of the estimator of $\theta(T)$ equals $v(T) = \sigma^2(T - t_1) + \Gamma_{t_1}^{t_1}$ and decreases when the instant t_1 of the measure approaches T. This is represented Fig. 3(c).

We prove Lemma 1 by induction on n (the number of measures). If $n = 0$, the function $v(t)$ has a simple form: $v(t) = v_0 + \sigma^2 t$. Therefore, the constraint (11) is expressed as $v(T) = v_0 + \sigma^2 T \leqslant V$, which is equivalent to (12).

Now we are going to prove the Lemma for $n + 1$ measures supposing that it is valid for n measures. The first instant t_1 must be chosen in the interval $[0, \frac{V - v_0}{\sigma^2}]$ in order for (11) to hold for $t \in [0, t_1[$. The function $v(t)$ in the interval $[t_1, T]$ is also defined by (3)–(10), but with the following parameters:

$$n_{\text{new}} = n, \ T_{\text{new}} = T_{\text{old}} - t_1 \tag{27}$$

$$v_{0,\text{new}} = \Gamma_{t_1,\text{old}}^{t_1} = \frac{(v_{0,\text{old}} + \sigma^2 t_{1,\text{old}}) v_{1,\text{old}}}{v_{0,\text{old}} + \sigma^2 t_{1,\text{old}} + v_{1,\text{old}}}, \tag{28}$$

$$v_{k,\text{new}} = v_{k+1,\text{old}} \text{ and } t_{k,\text{new}} = t_{k+1,\text{old}} - t_1 \text{ for } k \in \{1, \ldots, n\} \tag{29}$$

By applying the induction hypothesis to the new function, we obtain that the constraint (11) can be satisfied if and only if

$$\sigma^2(T - t_1) \leqslant V - \Gamma_{t_1}^{t_1} + \sigma^2 \left(\sum_{k=2}^{n+1} \delta(V, v_k) \right) \tag{30}$$

(all quantities here are the "old" ones, i.e. with respect to the initial problem), which is equivalent to

$$\Gamma_{t_1}^{t_1} - \sigma^2 t_1 \leqslant V - \sigma^2 T + \sigma^2 \left(\sum_{k=2}^{n+1} \delta(V, v_k) \right). \tag{31}$$

The RHS of (31) is independent of t_1. By Lemma 2, the LHS is a strictly decreasing function of t_1. Therefore, (31) can be satisfied if and only if it holds for

$$t_1 = \frac{V - v_0}{\sigma^2}. \tag{32}$$

After replacing t_1 by its value (32) in (31) and applying the definition (12) of $\delta(V, v_1)$, one gets

$$\sigma^2 T \leqslant V - v_0 + \sigma^2 \left(\sum_{k=1}^{n+1} \delta(v_k) \right), \tag{33}$$

which is Lemma 1.

References

1. Bourrier, A., Amblard, P.-O., Michel, O., Jutten, C.: Multimodal Kalman filtering. In: IEEE International Conference on Acoustics, Speech and Signal Processing (ICASSP), Shanghai, China, March 2016
2. Bruni, C., Koch, G., Papa, F.: A measurement policy in stochastic linear filtering problems. Comput. Math. Appl. **61**, 546–566 (2011)
3. Herring, K., Melsa, J.: Optimum measurements for estimation. IEEE Trans. Autom. Control **19**(3), 264–266 (1974)
4. Jazwinski, A.H.: Stochastic Processes and Filtering Theory. Mathematics in Science and Engineering. Academic Press, New York (1970). UKM
5. Lahat, D., Adalı, T., Jutten, C.: Multimodal data fusion: an overview of methods, challenges and prospects. Proc. IEEE **103**(9), 1449–1477 (2015)

On Disjoint Component Analysis

K. Nose-Filho[1]([✉]), L.T. Duarte[2], and J.M.T. Romano[1]

[1] School of Electrical and Computer Engineering,
University of Campinas (UNICAMP), Campinas, Brazil
kenjinose@gmail.com, romano@decom.fee.unicamp.br
[2] School of Applied Sciences, University of Campinas (UNICAMP), Limeira, Brazil
leonardo.duarte@fca.unicamp.br

Abstract. Disjoint Component Analysis (DCA) is a recent blind source separation approach which is based on the assumption that the original sources have disjoint supports. In DCA, the recovery process is carried out by maximizing the disjoint support of the estimated sources. In the present work, we provide sufficient conditions for the separation of both disjoint and *quasi*-disjoint signals. In addition, we propose an effective DCA criterion to evaluate the level of superposition of the recovered sources. The minimization of such criterion is implemented by an algorithm based on Givens rotations. Finally, simulation results are presented in order to assess the performance of the proposed method.

Keywords: Blind source separation · Disjoint component analysis · Givens rotations

1 Introduction

Blind Source Separation (BSS) is a problem of great relevance in signal processing due to its interesting theoretical and practical aspects. Its unsupervised characteristic constitutes a real challenge but, on other hand, opens the way for applications in different areas such as audio, speech, image processing, seismic reflection, communications, astronomy, chemistry and biomedicine [9].

In BSS, the main goal is to recover a set of N original signals (sources) based exclusively on the observation of M mixed versions of these sources and on a few assumptions about the mixing process and the sources. Since the mixing process and the sources are unknown, BSS is an ill-posed problem, and, thus, its solution requires additional hypotheses.

The usual approach for tackling BSS is known as Independent Component Analysis (ICA), which is based on the hypothesis that the sources are mutually statistical independent and at most one source is Gaussian distributed [8,9,15,25]. In ICA, separation is usually performed by the maximization of criteria that are related to statistical independence, such as mutual information, higher-order statistics and nonlinear decorrelation-based criteria.

The authors would like to thanks to CAPES, CNPq and FAPESP (process number 2015/07048-4) for the financial support.

© Springer International Publishing AG 2017
P. Tichavský et al. (Eds.): LVA/ICA 2017, LNCS 10169, pp. 519–528, 2017.
DOI: 10.1007/978-3-319-53547-0_49

A significant attention has also been given to the hypothesis that the sources have a sparse support [3,9,12,13,17,20,22]. In BSS, sparsity has been mainly addressed to solve the under-determined case, i.e., when the number of sensors is smaller than the number of sources $M < N$. Separation of sparse sources is usually performed in two steps: first, the mixing matrix is estimated [2,5,14, 21,24,26–28]. Then, the estimation of the sources is performed. In this second stage, the use of a linear transformation is not possible and, thus, some non-linear techniques are required for the estimation of the sources [5,7,18,19].

More recently, a new method known as Disjoint Component Analysis (DCA) has been introduced in [1]. DCA is based on the hypothesis that the sources have disjoint supports. By the observation that mixing disjoint signals results in signals that are less disjoint with respect to the sources, separation can be achieved by estimating a separating matrix that maximizes the disjointness of the estimated sources.

In the present work, we deal with several questions around DCA. In Sect. 2 we introduce a theorem that provides sufficient conditions over the involved signals in order to ensure the separation of disjoint signals. We also present a DCA criterion that evaluates the level of superposition of the estimated sources. In Sect. 3, we show that, for disjoint sources, the proposed criterion presents its minima at separating solutions. Moreover, since disjointness seems to be a very strong assumption when it comes to real world signals, we show that the non-overlapping condition, which is inherent to DCA, is sufficient but not strictly necessary. We show that, depending on the level of superposition of the sources, the proposed criterion is still a contrast function, i.e., the global minima are still placed at separating solutions. In Sect. 4, we provide an algorithm for DCA, which is based on Givens separating rotations. In Sect. 5, simulations results are presented in order to illustrate the performance of the proposed algorithm and, in Sect. 6, we state our conclusions.

2   Disjoint Component Analysis

In the standard linear and instantaneous mixing model, the j-th mixture is given by a linear combination of N sources, as follows [25]:

$$\mathbf{x}_j = a_{j,1}\mathbf{s}_1 + a_{j,2}\mathbf{s}_2 + ... + a_{j,N}\mathbf{s}_N, \ j \in \{1,...,M\}, \tag{1}$$

where $\mathbf{s}_i = [s_{i,0} \ s_{i,1} \ ... \ s_{i,K-1}]^T$ is the vector containing a set of K samples of the i-th source. By representing the sets of N sources and M mixtures by the matrices $\mathbf{S} = [\mathbf{s}_1 \ \mathbf{s}_2 \ ... \ \mathbf{s}_N]^T$ and $\mathbf{X} = [\mathbf{x}_1 \ \mathbf{x}_2 \ ... \ \mathbf{x}_M]^T$, respectively, (1) can be rewritten in a matrix notation

$$\mathbf{X} = \mathbf{AS}, \tag{2}$$

where $\mathbf{A} = (a_{i,j}) \in \mathbb{R}^{M \times N}$ corresponds to the mixing matrix. In BSS, both the mixing matrix $\mathbf{A}$ and the source matrix $\mathbf{S}$ are unknown, and, thus, separation is performed based only on the knowledge of the observation matrix $\mathbf{X}$.

BSS can be conducted in several ways. In ICA [8], when $M \geq N$, separation is accomplished in two steps: (1) the data is multiplied by a whitening matrix $\mathbf{W}$

$$\tilde{\mathbf{X}} = \mathbf{W}\mathbf{X}, \tag{3}$$

and (2) the pre-whitened data $\tilde{\mathbf{X}}$ is multiplied by an orthogonal separating matrix $\mathbf{B}$, i.e., $\mathbf{B}^T\mathbf{B} = \mathbf{B}\mathbf{B}^T = \mathbf{I}$, in order to retrieve statistically independent signals

$$\mathbf{Y} = \mathbf{B}\tilde{\mathbf{X}}. \tag{4}$$

Ideally, $\mathbf{BW} = \mathbf{A}^{-1}$, but, due to scaling and permutation ambiguities, BSS also admits solutions given by $\mathbf{BW} = \mathbf{DPA}^{-1}$, where $\mathbf{D}$ and $\mathbf{P}$ correspond to a diagonal and a permutation matrix, respectively [9].

On the other hand, disjoint component analysis exploits the hypothesis that the sources have disjoint supports, as defined as follows:

Definition 1. *Let* $\mathbf{S} = [\mathbf{s}_1, \mathbf{s}_2, \ldots, \mathbf{s}_N]^T$ *be the source matrix containing the vectors of sources. The sources* $\mathbf{s}_i$ *are said to be disjoint if at most one of them is non-zero for each sample* k*, i.e.,*

$$\mathbf{s}_i \odot \mathbf{s}_j = \mathbf{0}, \ \forall i \neq j, \tag{5}$$

where the symbol $\odot$ *denotes the Hadamard product between two vectors and* $\mathbf{0}$ *is the null vector.*

By relying on the observation that disjointness is lost after the mixing process, it is possible to show that separation can be carried out by retrieving disjoint signals. Sufficient conditions for that are summarized by the following theorem.

Theorem 1. *If the signals* $\mathbf{s}_i$ *are non-null vectors, real-valued, and disjoint and* $\mathbf{G} = (g_{i,j}) \in \mathbb{R}^{N \times N}$ *is an orthogonal matrix, then, the signals* $\mathbf{y}_i = \sum_{l=1}^{N} g_{i,l}\mathbf{s}_l, i \in \{1, \ldots, N\}$ *are disjoint if, and only if, there is exactly one non-null element in each row and column of* $\mathbf{G}$*.*

Proof. The Hadamard product $\mathbf{y}_i \odot \mathbf{y}_j$, for $i \neq j$, is given by:

$$\sum_{l=1}^{N} g_{i,l}\mathbf{s}_l \odot \sum_{m=1}^{N} g_{j,m}\mathbf{s}_m = \sum_{l=1}^{N}\sum_{m=1}^{N} g_{i,l}g_{j,m}\left(\mathbf{s}_l \odot \mathbf{s}_m\right). \tag{6}$$

Due to the disjointness condition of the sources, (6) can be reduced by:

$$\sum_{l=1}^{N} g_{i,l}g_{j,l}\left(\mathbf{s}_l \odot \mathbf{s}_l\right). \tag{7}$$

Since the signals $\mathbf{s}_i$ are non-null vectors and $\mathbf{G}$ is an orthogonal matrix, Eq. (7) is equal to the null vector if, and only if, there is exactly one non-null element in each row and column of $\mathbf{G}$. □

In BSS, the orthogonal matrix $\mathbf{G}$ can be written in terms of the combined response of the mixing matrix $\mathbf{A}$, the whitening matrix $\mathbf{W}$, and the separating matrix $\mathbf{B}$:

$$\mathbf{G} = \mathbf{BWA}. \tag{8}$$

Since the sources have disjoint supports, separation is achieved by estimating an orthogonal separating matrix $\mathbf{B}$ in order to make the columns of $\mathbf{Y}$ disjoint supports. Note that having exactly one non-zero element in each row and column of $\mathbf{G}$ leads to $\mathbf{G} = \mathbf{DP}$. In other words, if the sources have disjoint supports, the recovery of this property leads to a scaled and permuted version of the original sources.

Theorem 1 suggests that a cost function for BSS could be given by the sum of the ℓ_p-norm of the pairwise Hadamard products (PHP) of the recorded signals $\mathbf{y}_i$, leading to the following optimization problem:

$$\underset{\mathbf{B}}{\text{minimize}} \quad J_{PHP_p}(\mathbf{B}) = \sum_{i=1}^{N} \sum_{j>i}^{N} \|\mathbf{y}_i \odot \mathbf{y}_j\|_p^p, \tag{9}$$

$$\text{subject to} \quad \mathbf{BB}^{\mathbf{T}} = \mathbf{I}$$

In the following, we analyze the convexity of this criterion for the case of disjoint and *quasi*-disjoint signals, i.e., the case in which there might be superposition between the sources but the degree of superposition is limited.

3   Analysis of the Minima of the Criterion

In this section, we present two results related to the criterion (9)—these results are valid for the case of binary mixtures of two sources. First, we show that if Theorem 1 is satisfied, then the criterion given by Eq. (9) presents minima at separating solutions. Second, we show that if the signals are *quasi*-disjoint signals, the global minima are still separating solutions depending on the level of superposition of the sources.

As discussed in Sect. 2, a matrix $\mathbf{G}$ can be used to represent the global mapping between the actual sources and the retrieved ones. In the case of two mixtures and two sources, matrix $\mathbf{G}$ can be further simplified by considering a single parametrization variable. Indeed, let us consider the Givens rotation for the matrices $\mathbf{AW}$, and $\mathbf{B}$:

$$\mathbf{AW} = \begin{bmatrix} \cos\alpha & -\sin\alpha \\ \sin\alpha & \cos\alpha \end{bmatrix}, \quad \mathbf{B} = \begin{bmatrix} \cos\beta & -\sin\beta \\ \sin\beta & \cos\beta \end{bmatrix}. \tag{10}$$

In this case, the combined response $\mathbf{G}$ is given by:

$$\mathbf{G} = \begin{bmatrix} \cos(\alpha+\beta) & -\sin(\alpha+\beta) \\ \sin(\alpha+\beta) & \cos(\alpha+\beta) \end{bmatrix}, \tag{11}$$

in which the separating solutions are given by $\beta = -\alpha + c\pi/2$, $c \in \mathbb{Z}$. Therefore, the criterion introduced in (9) can be rewritten as:

$$
\begin{aligned}
J_{PHP_p}(\beta) &= \|(\cos(\alpha+\beta)\mathbf{s}_1 - \sin(\alpha+\beta)\mathbf{s}_2) \odot (\sin(\alpha+\beta)\mathbf{s}_1 + \cos(\alpha+\beta)\mathbf{s}_2)\|_p^p, \\
&= \sum_k \left| 0.5\sin(2\alpha+2\beta)\left(s_{1,k}^2 - s_{2,k}^2\right) + \cos(2\alpha+2\beta)(s_{1,k}s_{2,k})\right|^p,
\end{aligned}
$$

$$(12)$$

which leads to the following lemma.

Lemma 1. *If the signals $\mathbf{s}_1$ and $\mathbf{s}_2$ are non-null vectors, real-valued and disjoint and $\mathbf{G}$ is given by (11), then $J_{PHP_p}(\beta)$ is a periodic function with minima at the separating solutions $\beta = -\alpha + c\pi/2$, $c \in \mathbb{Z}$.*

Proof. Since the sources are disjoint, it asserts that $\mathbf{s}_1 \odot \mathbf{s}_2 = \mathbf{0}$ and, thus, (12) can be simplified as follows:

$$
J_{PHP_p}(\beta) = 0.5^p \left(\sum_k \left|s_{1,k}^2\right|^p + \left|s_{2,k}^2\right|^p \right) |\sin(2\alpha+2\beta)|^p, \tag{13}
$$

where $\mathbf{s}_1^2 = \mathbf{s}_1 \odot \mathbf{s}_1$. This is a periodic function of frequency $\pi/2$ with minima at the separating solutions $\beta = -\alpha + c\pi/2$, $c \in \mathbb{Z}$. $\square$

In the sequence, we analyze the proposed criterion for the case of two *quasi-disjoint* signals, which are defined as follows:

Definition 2. *Two quasi-disjoint signals s_1 and s_2 can be split in three different supports. A support in which only s_1 is different from zero $S_1 = \{k \in \{0, \dots, K-1\} : s_{1,k} \neq 0, s_{2,k} = 0\}$, a support in which only s_2 is different from zero $S_2 = \{k \in \{0, \dots, K-1\} : s_{1,k} = 0, s_{2,k} \neq 0\}$ and a common support in which both signals are different from zero $S_c = \{k \in \{0, \dots, K-1\} : s_{1,k} \neq 0, s_{2,k} \neq 0\}$.*

Our analysis searches for the ratio between the level of superposition of the common and disjoint supports, in terms of its energy and magnitudes, so that the global minima of (12) correspond to the separating solutions. For that we propose the following lemma:

Lemma 2. *If the signals $\mathbf{s}_1$ and $\mathbf{s}_2$ are quasi-disjoint signals such that $\sum_{k \in S_1} \left|s_{1,k}^2\right|^p + \sum_{k \in S_2} \left|s_{2,k}^2\right|^p > \sum_{k \in S_c} \left|s_{1,k}^2 + s_{2,k}^2\right|^p$ and $\mathbf{G}$ is given by (11), then $J_{PHP_p}(\beta)$ is a periodic function with global minima at the separating solutions $\beta = -\alpha + c\pi/2$, $c \in \mathbb{Z}$.*

Proof. Let us split the criterion in (12) in three different terms:

$$
J_{PHP_p}^{S_1}(\beta) + J_{PHP_p}^{S_2}(\beta) + J_{PHP_p}^{S_c}(\beta), \tag{14}
$$

where

$$J^{S_1}_{PHP_p}(\beta) = 0.5^p \left(\sum_{k \in S_1} \left| s^2_{1,k} \right|^p \right) |\sin(2\alpha + 2\beta)|^p,$$

$$J^{S_2}_{PHP_p}(\beta) = 0.5^p \left(\sum_{k \in S_2} \left| s^2_{2,k} \right|^p \right) |\sin(2\alpha + 2\beta)|^p,$$

$$J^{S_c}_{PHP_p}(\beta) = \sum_{k \in S_c} \left| 0.5 \sin(2\alpha + 2\beta) \left(s^2_{1,k} - s^2_{2,k} \right) + \cos(2\alpha + 2\beta)(s_{1,k}s_{2,k}) \right|^p.$$

$$(15)$$

The term $J^{S_1}_{PHP_q}(\alpha + \beta) + J^{S_2}_{PHP_q}(\alpha + \beta)$ is a periodic function of frequency $\pi/2$ and amplitude

$$0.5^p \left(\sum_{k \in S_1} \left| s^2_{1,k} \right|^p + \sum_{k \in S_2} \left| s^2_{2,k} \right|^p \right),$$

with minima at the separating solutions $\beta = -\alpha + c\pi/2$, $c \in \mathbb{Z}$.

For the common support, if we consider the points given by $(s_{1,k}, s_{2,k})$ in their polar coordinates:

$$s_{1,k} = r_k \cos(\theta_k),$$
$$s_{2,k} = r_k \sin(\theta_k). \qquad (16)$$

it is possible to rewrite the term $J^{S_c}_{PHP_p}(\beta)$ as

$$0.5^p \sum_{k \in S_c} \left| r^2_k \right|^p |\sin(2\alpha + 2\beta + 2\theta_k)|^p, \qquad (17)$$

which corresponds to the sum of periodic functions of frequency $\pi/2$ and amplitudes $0.5^p \left| r^2_k \right|^p$ with minima at $\beta = -\alpha - \theta_k + c\pi/2$, $c \in \mathbb{Z}$, with $c \in \mathbb{Z}$. Thus, if

$$\sum_{k \in S_1} \left| s^2_{1,k} \right|^p + \sum_{k \in S_2} \left| s^2_{2,k} \right|^p > \sum_{k \in S_c} \left| r^2_k \right|^p \qquad (18)$$

where $r_k = \sqrt{s^2_{1,k} + s^2_{2,k}}$, then $J_{PHP_p}(\beta)$ is a periodic function with global minima at the separating solutions $\beta = -\alpha + c\pi/2$, $c \in \mathbb{Z}$. Moreover, possible local minima are given by $\beta = -\alpha - \theta_k + c\pi/2$, depending on the distribution of θ_k.$\square$

In the sequel, an algorithm based on Givens rotations is introduced in order to address the optimization problem expressed in (9).

4   An Algorithm Based on Givens Rotations

Algorithms based on Givens rotations are very popular in BSS [4,10,11,16] and rely on the idea that the separation criterion can be optimized by iteratively

rotating the whitened data in one given direction at each time. The update rule
for such algorithm is given by:

$$\mathbf{B}(l+1) = \mathbf{T}(m, n, \beta)\mathbf{B}(l), \tag{19}$$

where $\mathbf{T}(m, n, \beta) = (t_{i,j})$ is a Givens rotation matrix, that is, for a given $m \neq n$,
we have

$$t_{i,i} = 1, \text{for } i \neq m, n,$$
$$t_{m,m} = t_{n,n} = \cos(\beta),$$
$$t_{m,n} = -t_{n,m} = \sin(\beta), \tag{20}$$

and the remainders are 0. This matrix is orthogonal for all $m \neq n$ and conse-
quently $\mathbf{T}(m, n, \beta)\mathbf{B}(l)$ remains orthogonal if $\mathbf{B}(0)$ is initialized by an orthogonal
matrix, e.g., $\mathbf{B}(0) = \mathbf{I}$.

Since the cost function in (9) is continuous with respect to the parameters
of the separating matrix, there is a value of $\beta \in \mathbb{R}$ such that:

$$J\Big(\mathbf{T}(m, n, \beta)\mathbf{B}(l)\Big) \leq J\Big(\mathbf{B}(l)\Big), \tag{21}$$

which ensures the convergence of the proposed method.

A complete iteration of the algorithm is done when all the $N(N-1)/2$
rotation angles are updated. Since the search is restricted to a bounded interval
$[0 \ \pi]$, we consider the Golden section method in [23] in order to estimate β.
Figure 1 summarizes the algorithm based on Givens rotations.

Algorithm 1:

1. Pre-whitening step
       Compute the EVD of $\mathbf{XX^T}$;
       $\mathbf{XX^T} = \mathbf{E_X D_X E_X^T}$;
       $\mathbf{W} = \mathbf{D_X^{-1/2} E_X^T}$;
       $\mathbf{\tilde{X}} = \mathbf{WX}$;
2. Set $\mathbf{B}(0) \leftarrow \mathbf{I}$;
3. For $l = 0$ to $L - 1$ step 1 do
       For $m = 1$ to $N - 1$ step 1 do
           For $n = m + 1$ to N step 1 do
           Find β that minimizes (9)
               $\mathbf{B}(l+1) \leftarrow \mathbf{T}(m, n, \beta)\mathbf{B}(l)$;
           End;
       End;
       End;

Fig. 1. Algorithm based on Givens rotations.

5   Simulation Results

In our simulations, we perform 20 Monte Carlo simulations for five *quasi*-disjoint sources generated by 3000 samples of a Bernoulli Gaussian random variable in a noisy scenario. We control the level of superposition of the sources by varying the probability of zero occurrence p_0 of the Bernoulli Gaussian random variable. We consider three different values for p_0: $p_0 = 0.8, 0.6,$ and 0.4. The noise is given by an additive white Gaussian noise (AWGN) for different signal-to-noise (SNR) ratios. The mixing matrix is generated by samples of a Gaussian random variable.

For the proposed algorithm, we considered a total of five iterations and $p = 1$ and the results are compared with the JADE algorithm [6]. In order to evaluate the results, we considered the signal-to-interference ratio (SIR), which is given by

$$\text{SIR}_{dB} = \frac{1}{N} \sum_{i=1}^{N} 10 \log \left(\frac{\|\mathbf{s}_i\|^2}{\|\mathbf{s}_i - \hat{\mathbf{s}}_i\|^2} \right), \tag{22}$$

where $\mathbf{s}_i$ is the i-th source and $\hat{\mathbf{s}}_i$ is the corresponding recovered source, after the removal of scale and permutation ambiguities.

In Fig. 2, we present the results obtained by the proposed method and the JADE algorithm for different SNRs and different values of p_0. The proposed

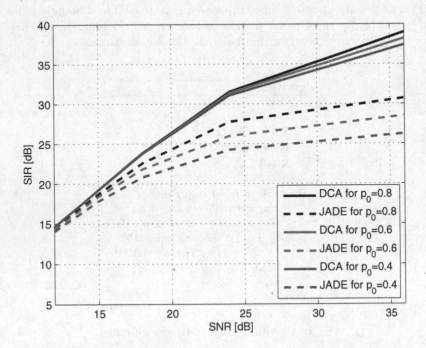

Fig. 2. SIR in dB of the proposed method and the JADE algorithm, for different SNRs and sources with different levels of sparsity.

algorithm outperforms the JADE algorithm and is less sensitive to the level of superposition of the sources than the JADE algorithm. These results indicates the good convergence and the viability of the proposed method for disjoint and *quasi*-disjoint signals even in noisy scenarios with moderate SNRs.

6   Conclusions

In this work, we addressed several questions around BSS based on disjoint component analysis. We presented a theorem stating that, if the sources are originally disjoint, then retrieving disjoint signals ensures the recovery of the actual sources up to permutation and scale ambiguities.

Based on the stated theorem, we developed a separation criterion and demonstrated that this criterion presents minima at separating solutions. Moreover, given the fact that disjointness seems to be a very strong assumption when it comes to real world signals, we also showed that the proposed criterion remains a contrast function for the case of *quasi*-disjoint signals, i.e., depending on the level of the superposition of the sources, the global minima of the studied DCA criterion remains at the separating solutions.

Finally, we proposed a DCA algorithm based on Givens rotations and provided numerical experiments to illustrate the performance of the proposed method. The obtained results are promising as they indicated the viability of the proposed method for both disjoint and *quasi*-disjoint signals even in noisy scenarios.

References

1. Anemüller, J.: Maximization of component disjointness: a criterion for blind source separation. In: Davies, M.E., James, C.J., Abdallah, S.A., Plumbley, M.D. (eds.) ICA 2007. LNCS, vol. 4666, pp. 325–332. Springer, Heidelberg (2007). doi:10.1007/978-3-540-74494-8_41
2. Babaie-Zadeh, M., Jutten, C., Mansour, A.: Sparse ICA via cluster-wise PCA. Neurocomputing **69**, 1458–1466 (2006)
3. Batany, Y.-M., Donno, D., Duarte, L., Chauris, H., Deville, Y., Romano, J.: A necessary and sufficient condition for the blind extraction of the sparsest source in convolutive mixtures. In: Proceedings of the 2016 EUSIPCO (2016)
4. Bienati, N., Spagnolini, U., Zecca, M.: An adaptive blind signal separation based on the joint optimization of Givens rotations. In: Proceedings of the 2001 ICASSP (2001)
5. Bofill, P., Zibulevsky, M.: Underdetermined blind source separation using sparse representations. Signal Process. **81**, 2353–2362 (2001)
6. Cardoso, J.F., Souloumiac, A.: Blind beamforming for non Gaussian signals. IEE Proc. F **140**, 362–370 (1993)
7. Chen, S.S., Donoho, D.L., Saunders, M.A.: Atomic decomposition by basis pursuit. SIAM Rev. **43**, 129–159 (2001)
8. Comon, P.: Independent component analysis, a new concept? Signal Process. **36**(3), 287–314 (1994). http://www.sciencedirect.com/science/article/pii/0165168494900299

9. Comon, P., Jutten, C. (eds.): Handbook of Blind Source Seapartion. Academic Press, Oxford (2010)

10. Congedo, M., Phlypo, R., Pham, D.T.: Approximate joint singular value decomposition of an asymmetric rectangular matrix set. IEEE Trans. Signal Process. **59**, 415–424 (2011)

11. Delfosse, N., Loubaton, P.: Adaptive blind separation of independent sorces: a deflation approach. Signal Process. **45**, 59–83 (1995)

12. Duarte, L.T., Suyama, R., Attux, R., Romano, J.M.T., Jutten, C.: Blind extraction of sparse components based on ℓ_0-norm minimization. In: Proceedings of the 2011 SSP, pp. 617–620 (2011)

13. Gribonval, R., Lesage, S.: A survey of sparse component analysis for blind source separation: principles, perspectives, and new challenges. In: Proceedings of the 2006 ESANN, pp. 323–330. Bruges (2006)

14. Hulle, M.M.V.: Clustering approach to square and non-square blind source separation. In: Proceedings of the 1999 IEEE Workshop on Neural Networks for Signal Processing, pp. 315–323 (1999)

15. Hyvarinen, A., Karhunen, J., Oja, E.: Independent Component Analysis. Wiley, New York (2001)

16. Min, Z., Weijun, L., Guoxu, Z., Zhiheng, Z.: Blind source separation algorithm based on adaptive givens rotations. In: Proceedings of the 2009 ICNC (2009)

17. Mishali, M., Eldar, Y.C.: Sparse source separation from orthogonal mixtures. In: Proceedings of the 2009 ICASSP (2009)

18. Mohimani, H., Babaie-Zadeh, M., Jutten, C.: A fast approach for overcomplete decomposition based on smoothed ℓ^0 norm. IEEE Trans. Signal Process. **57**, 289–301 (2009)

19. Mourad, N., Reilly, J.P.: ℓp minimization for sparse vector reconstruction. In: Proceedings of the 2009 ICASSP (2009)

20. Nadalin, E.Z., Takahata, A.K., Duarte, L.T., Suyama, R., Attux, R.: Blind extraction of the sparsest component. In: Vigneron, V., Zarzoso, V., Moreau, E., Gribonval, R., Vincent, E. (eds.) LVA/ICA 2010. LNCS, vol. 6365, pp. 394–401. Springer, Heidelberg (2010). doi:10.1007/978-3-642-15995-4_49

21. O'Grady, P.D., Pearlmutter, B.A.: Hard-LOST: modified k-means for oriented lines. In: Proceedings of the 2004 ISSC, Belfast (2004)

22. O'grady, P.D., Pearlmutter, B.A., Rickard, S.: Survey of sparse and non-sparse methods in source separation. Int. J. Imaging Syst. Technol. **15**, 18–33 (2005)

23. Press, W.H., Teukolsky, S.A., Vetterling, W.T., Flannery, B.P.: Numerical Recipes: The Art of Scientific Computing. Cambridge University Press, Cambridge (2007)

24. Rickard, S., Yilmaz, O.: On the approximate W-disjoint orthogonality of speech. In: Proceedings of the 2002 ICASSP, vol. 1, pp. I-529–I-532 (2002)

25. Romano, J.M.T., Attux, R., Cavalcante, C.C., Suyama, R.: Unsupervised Signal Processing: Channel Equalization and Source Separation. CRC Press, Boca Raton (2011)

26. Theis, F.J.: A geometric algorithm for overcomplete linear ICA. Neurocomputing **56**, 381–398 (2004)

27. Thiagarajan, J.J., Ramamurthy, K.N., Spanias, A.: Mixing matrix estimation using discriminative clustering for blind source separation. Digit. Signal Process. **23**, 9–18 (2012)

28. Zibulevsky, M., Kisilev, P., Zeevi., Y.Y., Pearlmutter, A. : Blind source separation via multinode sparse representation. In: Proceedings of the 2002 Advances in Neural Information Processing Systems, vol. 12. MIT Press (2002)

Sparsity-Aware Signal Processing

Accelerated Dictionary Learning
for Sparse Signal Representation

Fateme Ghayem[1(✉)], Mostafa Sadeghi[1], Massoud Babaie-Zadeh[1],
and Christian Jutten[2]

[1] Department of Electrical Engineering, Sharif University of Technology, Tehran, Iran
`fateme.ghayem@gmail.com`, `m.saadeghii@gmail.com`, `mbzadeh@yahoo.com`
[2] GIPSA-Lab, Institut Universitaire de France, Grenoble, France
`christian.jutten@gipsa-lab.grenoble-inp.fr`

Abstract. Learning sparsifying dictionaries from a set of training signals has been shown to have much better performance than pre-designed dictionaries in many signal processing tasks, including image enhancement. To this aim, numerous practical dictionary learning (DL) algorithms have been proposed over the last decade. This paper introduces an accelerated DL algorithm based on iterative proximal methods. The new algorithm efficiently utilizes the iterative nature of DL process, and uses accelerated schemes for updating dictionary and coefficient matrix. Our numerical experiments on dictionary recovery show that, compared with some well-known DL algorithms, our proposed one has a better convergence rate. It is also able to successfully recover underlying dictionaries for different sparsity and noise levels.

Keywords: Sparse representation · Compressed sensing · Dictionary learning · Proximal algorithms

1 Introduction

The information contents of natural signals are usually significantly less than their ambient dimensions. This fact has been extensively used in many applications, including compressed sensing [1]. Let $\mathbf{y} \in \mathbb{R}^m$ be a given (natural) signal. Then, its representation over a set of signal building blocks, $\{\mathbf{d}_1, \cdots, \mathbf{d}_n\}$ (called atoms) is written as $\mathbf{y} = \sum_{i=1}^{n} x_i \cdot \mathbf{d}_i = \mathbf{D}\mathbf{x}$, where $\mathbf{D} \in \mathbb{R}^{m \times n}$ is called dictionary, which contains $\mathbf{d}_i$'s as its columns, and $\mathbf{x} \in \mathbb{R}^n$ is the vector of coefficients. If the dictionary is chosen appropriately, then the coefficients vector $\mathbf{x}$ is expected to be very sparse. So, an important question is how to choose the sparsifying dictionary $\mathbf{D}$. Discrete cosine transform (DCT) and wavelets are two well-known predesigned dictionaries which can be used in sparsity-based applications. However, such fixed transforms (dictionaries) are suitable for only particular class of signals. An alternative and more efficient way to choose the dictionary is to

This work has been funded by ERC project 2012-ERC-AdG-320684 CHESS and by the Center for International Scientific Studies and Collaboration (CISSC).

© Springer International Publishing AG 2017
P. Tichavský et al. (Eds.): LVA/ICA 2017, LNCS 10169, pp. 531–541, 2017.
DOI: 10.1007/978-3-319-53547-0_50

learn it from a set of training signals. This process is called *dictionary learning* (DL), which has received a lot of attention over the last decade [2]. Given a set of training signals $\{y_i\}_{i=1}^{L}$ collected as the columns of the matrix $\mathbf{Y} \in \mathbb{R}^{m \times L}$, a dictionary $\mathbf{D}$ and the corresponding coefficient matrix $\mathbf{X}$ are optimized such that the representation error $\|\mathbf{Y} - \mathbf{DX}\|_F^2$ becomes small and $\mathbf{X}$ has sufficiently sparse columns. This problem can be formulated as

$$\min_{\mathbf{D} \in \mathcal{D}, \mathbf{X}} \left\{ \frac{1}{2} \|\mathbf{Y} - \mathbf{DX}\|_F^2 + \lambda \|\mathbf{X}\|_0 \right\}, \tag{1}$$

where $\lambda > 0$ is a regularization parameter, $\|\cdot\|_0$ denotes the so-called ℓ_0 pseudo-norm which counts the number of non-zero entries, and $\mathcal{D}$ is defined as follows:

$$\mathcal{D} \triangleq \left\{ \mathbf{D} \in \mathbb{R}^{m \times n} \mid \forall i : \|d_i\|_2 = 1 \right\}. \tag{2}$$

Many algorithms have been proposed to solve (1) or its variants [3–8]. Most of these algorithms follow an alternating minimization approach, consisting of two main steps: sparse representation (SR) and dictionary update (DU). In the first step, $\mathbf{D}$ is kept fixed and the minimization is done over $\mathbf{X}$. There exist many efficient algorithms to perform this step, see e.g., [9,10] and the references therein. In the DU step, $\mathbf{X}$ is set to its current estimate obtained in the SR step, and $\mathbf{D}$ is updated. We refer to one round of performing these two steps as one *DL iteration*.

When SR and DU steps are solved using iterative algorithms, the iterations of DL can be efficiently utilized to reduce the work needed for updating $\mathbf{D}$ and $\mathbf{X}$. In other words, the final estimates of $\mathbf{D}$ and $\mathbf{X}$ obtained in each DL iteration can be used to initialize the SR and DU steps of the next DL iteration. Besides accelerating the whole DL procedure, this so-called *warm-starting* may also avoid undesired local minima. Earlier works, including method of optimal directions (MOD) [3] and K-singular value decomposition (K-SVD) [4], do not take full advantage of this fact. In fact, both of them use orthogonal matching pursuit (OMP) [11] to perform the SR step, which does not efficiently use current estimate of the coefficient matrix[1]. Moreover, MOD finds the unconstrained least-squares solution of the DU step

$$\mathbf{D} = \mathbf{YX}^T (\mathbf{XX}^T)^{-1} \tag{3}$$

which is followed by a column normalization. So, the previous estimate of $\mathbf{D}$ is not used to find the next one. K-SVD, on the other hand, updates the dictionary atom-by-atom together with the non-zero entries of the coefficient matrix. In this way, the previous estimates of the atoms are used, in someway, to get the new estimates. Instances of the DL algorithms that utilize iterative algorithms to update $\mathbf{D}$ and $\mathbf{X}$ include [7], which uses iterative majorization minimization,

---

[1] It should be mentioned that, a modification to OMP has been proposed in [12], which reuses the coefficients obtained in each DL iteration in order to initialize OMP for the next DL iteration.

and [8], which proposes a multi-block hybrid proximal alternating (MBHPA) algorithm to solve the DL problem in (1).

In this paper, new iterative schemes based on proximal gradient algorithms [13] are proposed to perform SR and DU steps. Unlike the algorithm proposed in [8], which is also based on proximal approach, our algorithm is equipped with accelerated extrapolation and inertial techniques [14,15]. Moreover, ℓ_1 norm is used here, as the sparsity measure in contrast to [8] that uses ℓ_0 pseudo-norm. Another difference between our algorithm and the one proposed in [8] is that, we update the whole dictionary using iterative proximal gradient method, while [8] updates the atoms of the dictionary sequentially. As will be shown in Sect. 3, our proposed algorithm, which we call accelerated dictionary learning (ADL), outperforms K-SVD and the algorithm introduced in [8].

The rest of the paper is organized as follows. In Sect. 2, our new iterative algorithms for performing SR and DU steps are introduced. Then, Sect. 3 presents the simulation results.

2 Proposed Method

2.1 Main Problem

We target the following problem in order to learn overcomplete dictionaries:

$$\min_{\mathbf{D} \in \mathcal{D}, \mathbf{X}} \left\{ \|\mathbf{Y} - \mathbf{D}\mathbf{X}\|_F^2 + \lambda \|\mathbf{X}\|_1 \right\}. \tag{4}$$

In the same way as usual methods, solving (4) consists of SR and DU steps, which are performed alternatively. In the rest of this section, we will separately illustrate how each step is realized to update the coefficient matrix, $\mathbf{X}$, and the dictionary, $\mathbf{D}$. Before proceeding, let us review some notations and terminologies related to proximal algorithms [13].

Definition 1 ([13]). *The projection of a point* $\mathbf{x} \in \mathbb{R}^n$ *onto a non-empty set* $\mathcal{S} \subseteq \mathbb{R}^n$ *is defined as*

$$P_{\mathcal{S}}\{\mathbf{x}\} \triangleq \operatorname*{argmin}_{\mathbf{u} \in \mathcal{S}} \frac{1}{2} \|\mathbf{x} - \mathbf{u}\|_2^2.$$

Definition 2 ([13]). *The proximal mapping of a convex function* $g : \operatorname{dom}_g \longrightarrow \mathbb{R}$ *is defined as*

$$\operatorname{Prox}_g\{\mathbf{x}\} \triangleq \operatorname*{argmin}_{\mathbf{u} \in \operatorname{dom}_g} \left\{ \frac{1}{2} \|\mathbf{u} - \mathbf{x}\|_2^2 + g(\mathbf{u}) \right\}.$$

Let $\delta_{\mathcal{S}}(\mathbf{x})$ denote the indicator function of the set $\mathcal{S}$, i.e.,

$$\delta_{\mathcal{S}}(\mathbf{x}) \triangleq \begin{cases} 0 & \mathbf{x} \in \mathcal{S} \\ \infty & \mathbf{x} \notin \mathcal{S}. \end{cases} \tag{5}$$

The proximal mapping of $\delta_{\mathcal{S}}$ is, then, the projection onto $\mathcal{S}$ [13]

$$\operatorname{Prox}_{\delta_{\mathcal{S}}}\{\mathbf{x}\} = P_{\mathcal{S}}\{\mathbf{x}\}. \tag{6}$$

2.2 Sparse Representation

To perform the SR step, let us define

$$f(\mathbf{D}, \mathbf{X}) \triangleq \frac{1}{2}\|\mathbf{Y} - \mathbf{D}\mathbf{X}\|_F^2. \tag{7}$$

The SR step for the i-th DL iteration is then

$$\mathbf{X}^{(i)} = \underset{\mathbf{X}}{\operatorname{argmin}} \; \left\{ f(\mathbf{D}^{(i-1)}, \mathbf{X}) + \lambda\|\mathbf{X}\|_1 \right\}. \tag{8}$$

To solve the above problem using iterative proximal gradient algorithms [13], $f(\mathbf{D}^{(i-1)}, \mathbf{X})$ is replaced with its quadratic approximation around the previous estimate of $\mathbf{X}^{(i)}$ [13]. It is straightforward to show that the final problem is

$$\begin{aligned} \mathbf{X}_{k+1}^{(i)} &= \underset{\mathbf{X}}{\operatorname{argmin}} \left\{ \frac{1}{2\mu_x}\|\mathbf{X} - \widehat{\mathbf{X}}_k^{(i)}\|_F^2 + \lambda\|\mathbf{X}\|_1 \right\} \\ &= \operatorname{Prox}_{\mu_x \lambda\|.\|_1} \left\{ \widehat{\mathbf{X}}_k^{(i)} \right\}, \end{aligned} \tag{9}$$

where, $\widehat{\mathbf{X}}_k^{(i)} \triangleq \mathbf{X}_k^{(i)} - \mu_x \nabla_X f(\mathbf{D}^{(i-1)}, \mathbf{X}_k^{(i)})$, k stands for the iteration index, and μ_x is a step-size which is set as $\mu_x = 1/\|(\mathbf{D}^{(i-1)})^T \mathbf{D}^{(i-1)}\|$, with $\|.\|$ being the matrix spectral norm [16]. The proximal mapping of the ℓ_1 norm is the so-called soft-thresholding operation [13]. The component-wise soft-thresholding function, denoted by $\operatorname{Soft}_\lambda$, is defined as [17]

$$\operatorname{Soft}_\lambda\{x\} \triangleq \operatorname{sgn}(x) \cdot \max(|x| - \lambda, 0). \tag{10}$$

The iterative proximal gradient algorithm to solve (8) can be compactly written as

$$\mathbf{X}_{k+1}^{(i)} = \operatorname{Soft}_{\mu_x \lambda} \left\{ \mathbf{X}_k^{(i)} - \mu_x \nabla_X f(\mathbf{D}^{(i-1)}, \mathbf{X}_k^{(i)}) \right\}. \tag{11}$$

In order to accelerate the algorithm, we add an extrapolation step [15] to the above algorithm as follows,

$$\begin{cases} \mathbf{Z}_k^{(i)} = \mathbf{X}_k^{(i)} + w_x(\mathbf{X}_k^{(i)} - \mathbf{X}_{k-1}^{(i)}) \\ \mathbf{X}_{k+1}^{(i)} = \operatorname{Soft}_{\mu_x \lambda} \left\{ \mathbf{Z}_k^{(i)} - \mu_x \nabla_X f(\mathbf{D}^{(i-1)}, \mathbf{Z}_k^{(i)}) \right\}, \end{cases} \tag{12}$$

in which, $w_x \geq 0$ is a weighting parameter which controls the convergence rate of the algorithm. The above iterations are repeated until $\|\mathbf{X}_{k+1}^{(i)} - \mathbf{X}_k^{(i)}\|_F \leq \tau_x$, where τ_x is a given tolerance. This accelerated iterative scheme has already been discussed in some previous works, including [14,18] to solve the vector form of the ℓ_1-based sparse representation problem.

2.3   Dictionary Update

The problem to be solved in the DU step is as fallows:

$$D^{(i+1)} = \underset{D \in \mathcal{D}}{\operatorname{argmin}} \ f(D, X^{(i)}), \tag{13}$$

which can be equivalently written as

$$D^{(i+1)} = \underset{D}{\operatorname{argmin}} \ \left\{ f(D, X^{(i)}) + \delta_{\mathcal{D}}(D) \right\}. \tag{14}$$

Following the same approach used to solve (8), the iterative proximal gradient algorithm to solve (14) becomes as

$$D_{k+1}^{(i+1)} = \underset{D}{\operatorname{argmin}} \ \left\{ \frac{1}{2\mu_d} \|D - \widehat{D}_k^{(i+1)}\|_F^2 + \delta_{\mathcal{D}}(D) \right\}$$

$$= \operatorname{Prox}_{\mu_d \delta_{\mathcal{D}}} \left\{ \widehat{D}_k^{(i+1)} \right\}, \tag{15}$$

where, $\widehat{D}_k^{(i+1)} \triangleq D_k^{(i+1)} - \mu_d \nabla_D f(D_k^{(i+1)}, X^{(i)})$, and μ_d is a step-size, which is set as $\mu_d = 1/\|(X^{(i)})^T X^{(i)}\|$. According to (6), the proximal mapping in (15) is the projection onto $\mathcal{D}$. So, the iterative proximal gradient algorithm to solve (14) can be compactly written as

$$D_{k+1}^{(i+1)} = \mathcal{P}_{\mathcal{D}} \left\{ D_k^{(i+1)} - \mu_d \nabla_D f(D_k^{(i+1)}, X^{(i)}) \right\}. \tag{16}$$

To accelerate the above algorithm, we add an inertial term [15] to the above algorithm as follows:

$$\begin{cases} C_k^{(i+1)} = D_k^{(i+1)} - \mu_d \nabla_D f(D_k^{(i+1)}, X^{(i)}) \\ D_{k+1}^{(i+1)} = \mathcal{P}_{\mathcal{D}} \left\{ C_k^{(i+1)} \right\} + w_d (D_k^{(i+1)} - D_{k-1}^{(i+1)}), \end{cases} \tag{17}$$

in which, $w_d \geq 0$ is a weighting parameter, which controls the convergence rate of the algorithm. Similar to the X update step, the above iterations are repeated until $\|D_{k+1}^{(i+1)} - D_k^{(i+1)}\|_F \leq \tau_d$, where τ_d is a given tolerance. Here, one may wonder why an inertial scheme is not used for the SR step, or similarly, why an extrapolated scheme, like the one used in (12), is not used instead of (17). In fact, it was observed in our simulations that for the SR step, the extrapolation, and for the DU step, the inertial technique result in the fastest convergence.

2.4   Non-zero Coefficients Update

Similar to K-SVD, in order to further accelerate the whole DL algorithm, the non-zero entries of X are also updated after the DU step. Let $x_{[l]}$ denote the

Algorithm 1. ADL

Require: $\mathbf{Y}$, $\mathbf{D}^{(1)} = \mathbf{D}_0$, $\mathbf{X}^{(1)} = \mathbf{X}_0$, λ, I, τ_X, τ_D, w_x, w_d.
for $i = 1, 2, \cdots, I$ **do**
   1. Sparse representation:
   $\mathbf{D} = \mathbf{D}^{(i)}$, $k = 1$.
   $\mu_x = 1/\|\mathbf{D}^T\mathbf{D}\|$.
   while $\|\mathbf{X}_{k+1}^{(i)} - \mathbf{X}_k^{(i)}\|_F > \tau_X$ **do**
      $\widehat{\mathbf{X}}_k^{(i)} = \mathbf{X}_k^{(i)} + w_x(\mathbf{X}_k^{(i)} - \mathbf{X}_{k-1}^{(i)})$
      $\mathbf{X}_{k+1}^{(i)} = \mathrm{Soft}_{\lambda\mu_X}\left\{\widehat{\mathbf{X}}_k^{(i)} - \mu_x\nabla_{\mathbf{X}}f(\mathbf{D}, \widehat{\mathbf{X}}_k^{(i)})\right\}$
      $k = k + 1$
   end while
   $\mathbf{X}^{(i)} = \mathbf{X}_{k+1}^{(i)}$.
   2. Dictionary update:
   $\mathbf{X} = \mathbf{X}^{(i)}$, $k = 1$.
   $\mu_d = 1/\|\mathbf{X}^T\mathbf{X}\|$.
   while $\|\mathbf{D}_{k+1}^{(i)} - \mathbf{D}_k^{(i)}\|_F > \tau_d$ **do**
      $\widehat{\mathbf{D}}_k^{(i)} = \mathbf{D}_k^{(i)} - \mu_d\nabla_{\mathbf{D}}f(\mathbf{D}_k^{(i)}, \mathbf{X})$
      $\mathbf{D}_{k+1}^{(i)} = \mathcal{P}_{\mathcal{D}}(\widehat{\mathbf{D}}_k^{(i)}) + w_d(\mathbf{D}_k^{(i)} - \mathbf{D}_{k-1}^{(i)})$
      $k = k + 1$
   end while
   $\mathbf{D}^{(i)} = \mathbf{D}_{k+1}^{(i)}$.
   3. Non-zero coefficients update:
   for $l = 1, 2, \cdots, n$ **do**
      $\Omega_l = \{i \mid \mathbf{x}_{[l]}(i) \neq 0\}$
      $\mathbf{E}_l = \mathbf{Y} - \sum_{i \neq l} \mathbf{d}_i\mathbf{x}_{[i]}$
      $\mathbf{x}_{[l]}(\Omega_l) = \mathbf{d}_l^T\mathbf{E}_l(\Omega_l)$
   end for
end for
Output: $\mathbf{D} = \mathbf{D}^i$, $\mathbf{X} = \mathbf{X}^i$.

l-th row of $\mathbf{X}$, and $\Omega_l \triangleq \{j \mid \mathbf{x}_{[l]}(j) \neq 0\}$ be the indexes of non-zeros in $\mathbf{x}_{[l]}$. Then, $\mathbf{x}_{[l]}(\Omega_l)$ is updated as follows

$$\mathbf{x}_{[l]}(\Omega_l) = \underset{\mathbf{x}_{[l]}^r}{\mathrm{argmin}}\frac{1}{2}\|\mathbf{E}(\Omega_l) - \mathbf{d}_l\mathbf{x}_{[l]}^r\|_F^2 = \mathbf{d}_l^T\mathbf{E}(\Omega_l) \tag{18}$$

in which, $\mathbf{E}(\Omega_l)$ contains those columns of $\mathbf{E} = \mathbf{Y} - \sum_{i \neq l} \mathbf{d}_i\mathbf{x}_{[i]}$ indexed in Ω_l, and $\mathbf{x}_{[l]}^r$ is a row vector of length $|\Omega_l|$. This process is repeated for all the rows of $\mathbf{X}$.

A detailed description of the proposed DL algorithm, which we call accelerated dictionary learning (ADL), is given in Algorithm 1.

3   Simulations

In this section, the performance of our proposed DL algorithm is compared with K-SVD[2] and the algorithm proposed in [8], which is referred to as MBHPA-DL[3]. We consider a dictionary recovery experiment, in which, given a set of training signals generated by sparse linear combinations of the atoms in a known dictionary, the goal is to recover the underlying dictionary. Our simulations were performed in MATLAB R2013a environment on a system with 3.8 GHz Intel cori 7 CPU and 8 GB RAM, under Microsoft Windows 7 operating system. As a rough measure of complexity, we will report the runtimes of the algorithms.

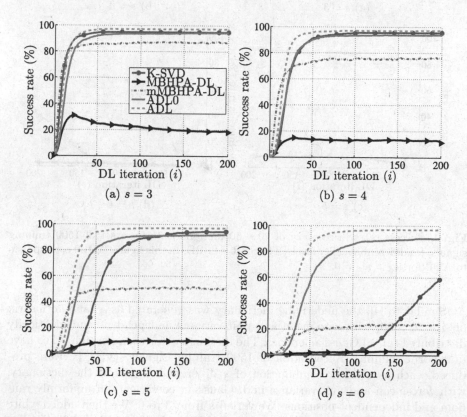

Fig. 1. Success rates in recovery of 20×50 dictionaries from a set of 1500 training signals, for different sparsity levels. The SNR is 100 dB. In this figure, ADL0 corresponds to ADL for $w_d = w_x = 0$.

---

[2] For K-SVD and OMP, we have used K-SVD-Box v10 and OMP-Box v10 available at http://www.cs.technion.ac.il/ronrubin/software.html.

[3] The MATLAB implementation of our proposed algorithm together with those of the other compared algorithms will be made available at https://sites.google.com/site/fatemeghayem/.

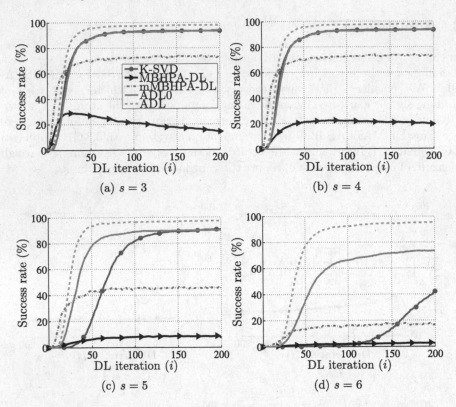

Fig. 2. Success rates in recovery of 20×50 dictionaries from a set of 1500 training signals, for different sparsity levels. The SNR is 20 dB. In this figure, ADL0 corresponds to ADL for $w_d = w_x = 0$.

Similar to [4], the underlying dictionary was generated as a random matrix of size 20×50, with zero-mean and unit-variance independent and identically distributed (i.i.d.) Gaussian entries. The dictionary was then normalized to have unit-norm columns. A collection of 1500 training signals, $\{\mathbf{y}_i\}_{i=1}^{1500}$, were produced, each as a linear combination of s different columns of the dictionary, with zero-mean and unit-variance i.i.d. Gaussian coefficients in uniformly random and independent positions. We varied s from 3 to 6. We then added white Gaussian noise with signal to noise ratio (SNR) levels of 100 dB and 20 dB. The algorithms were then applied onto these noisy training signals, and the resulting recovered dictionaries were compared to the generating dictionary as follows. Assume that $\mathbf{d}_i$ is a known atom and $\bar{\mathbf{d}}_i$ is the atom in the recovered dictionary that best matches $\mathbf{d}_i$ among the others. We say that the recovery is successful if $|\mathbf{d}_i^T \bar{\mathbf{d}}_i|$ is above 0.99 [4]. To evaluate the performance of the algorithms, we calculated the percentage of recovered atoms. We performed 200 iterations between the SR and DU steps for all the algorithms. The initial dictionary was made by randomly choosing different columns of the training signals followed by a normalization.

It should be noted that the original MBHPA-DL algorithm proposed in [8] performs only one iteration to update the coefficient matrix in SR step. This leads to a very slow convergence rate, usually ending up with a dictionary far away from the underlying one, which is confirmed by our simulations. As a solution, we modified MBHPA-DL's SR step by running it until a stopping criterion, similar to the one used in Algorithm 1, is satisfied. We call this version of the algorithm mMBHPA-DL, for modified MBHPA-DL.

The initial coefficient matrix for both ADL, MBHPA-DL, and mMBHPA-DL was set to the zero matrix. It was also observed that, in average, for all values of s, $\lambda = 0.1$ works well. The tolerances for terminating the SR and DU steps were set as $\tau_x = \tau_d = 0.005$. The weight parameters, w_x and w_d, in ADL were set to 0.85. In addition, to see the effect of using extrapolation and inertial accelerating schemes, we executed ADL for $w_x = w_d = 0$. We refer to this version of ADL as ADL0.

The resulting success rates of the algorithms, averaged over 50 trials, versus DL iterations, and for two different noise levels of 100 dB and 20 dB, are shown in Figs. 1 and 2, respectively. The averaged runtimes are also reported in Table 1. Inspection of these results leads to several conclusions as follows:

- As clearly demonstrated in Figs. 1 and 2, mMBHPA-DL considerably outperforms the original MBHPA-DL algorithm.
- The proposed algorithm, for both cases of zero (ADL0) and non-zero weight parameters (ADL), is more successful in recovery of underlying dictionaries than K-SVD and mMBHPA-DL. Moreover, the use of non-zero weight parameters in ADL increases the convergence rate of the algorithm, especially for $s = 5$ and $s = 6$ and high noise level.
- While the performances of K-SVD and mMBHPA-DL deteriorate for $s = 5$ and $s = 6$, with that of mMBHPA-DL being more severe, ADL and ADL0 have promising recovery performances for all values of s. Indeed, ADL recovers the underlying dictionaries almost perfectly for all values of s, and both low and high noise levels.
- In terms of runtimes, K-SVD is the fastest algorithm. This is mainly due to the fact that, it uses the batch OMP algorithm [19] in its SR step, which is optimized for large training matrices. Moreover, ADL runs faster than mMBHPA-DL. Another noticeable remark is that, ADL0 has higher runtime than ADL. In fact, it takes more time for ADL0 to satisfy the stopping criteria of the SR and DU steps, because it does not utilize accelerated schemes.

Table 1. Average runtimes (in second). In this table, ADL0 corresponds to ADL for $w_d = w_x = 0$.

Algorithm	K-SVD	mMBHPA-DL	ADL0	ADL
Runtime (s)	2.69	15.52	18.29	11.80

4   Conclusion

This paper presented an accelerated dictionary learning (DL) algorithm based on iterative proximal algorithms to be used in sparse representation-based applications. Our proposed approach combines first-order proximal algorithms with accelerating inertial and extrapolation schemes to update coefficient matrix and dictionary alternatively, resulting in a simple, yet efficient DL algorithm. It was demonstrated through a dictionary recovery experiment that, compared with the well-known K-SVD [4] and a recently introduced algorithm [8], the proposed algorithm is more successful in recovering underlying dictionaries from a set of, possibly noisy, training signals with different sparsity levels. Future works include applying the proposed algorithm to real-world applications such as image denoising, and establishing its convergence.

References

1. Candès, E.J., Wakin, M.B.: An introduction to compressive sampling. IEEE Sig. Proc. Mag. **25**(2), 21–30 (2008)
2. Rubinstein, R., Bruckstein, A.M., Elad, M.: Dictionaries for sparse representation modeling. Proc. IEEE **98**(6), 1045–1057 (2010)
3. Engan, K., Aase, S.O., Hakon Husoy, J.: Method of optimal directions for frame design. In: Proceedings of IEEE ICASSP (1999)
4. Aharon, M., Elad, M., Bruckstein, A.: K-SVD: an algorithm for designing over-complete dictionaries for sparse representation. IEEE Trans. Sig. Process. **54**(11), 4311–4322 (2006)
5. Sadeghi, M., Babaie-Zadeh, M., Jutten, C.: Learning over-complete dictionaries based on atom-by-atom updating. IEEE Trans. Sig. Proc. **62**(4), 883–891 (2014)
6. Sadeghi, M., Babaie-Zadeh, M., Jutten, C.: Dictionary learning for sparse representation: a novel approach. IEEE Sig. Proc. Lett. **20**(12), 1195–1198 (2013)
7. Yaghoobi, M., Blumensath, T., Davies, M.E.: Dictionary learning for sparse approximations with the majorization method. IEEE Trans. Sig. Process. **57**(6), 2178–2191 (2009)
8. Bao, C., Ji, H., Quan, Y., Shen, Z.: Dictionary learning for sparse coding: algorithms and convergence analysis. IEEE Trans. Pattern Anal. Mach. Intell. **38**(7), 1356–1369 (2015)
9. Mousavi, H.S., Monga, V., Tran, T.D.: Iterative convex refinement for sparse recovery. IEEE Sig. Proc. Lett. **22**(11), 1903–1907 (2015)
10. Sadeghi, M., Babaie-Zadeh, M.: Iterative sparsification-projection: fast and robust sparse signal approximation. IEEE Trans. Sig. Proc. **64**(21), 5536–5548 (2016)
11. Pati, Y.C., Rezaiifar, R., Krishnaprasad, P.S.: Orthogonal matching pursuit: recursive function approximation with applications to wavelet decomposition. In: Proceedings of Asilomar Conference on Signals, Systems, and Computers (1993)
12. Smith, L.N., Elad, M.: Improving dictionary learning: multiple dictionary updates and coefficient reuse. IEEE Sig. Proc. Lett. **20**(1), 79–82 (2013)
13. Parikh, N., Boyd, S.: Proximal algorithms. Found. Trends Optim. **1**(3), 123–231 (2014)
14. Xu, Y., Yin, W.: A globally convergent algorithm for nonconvex optimization based on block coordinate update (2015)

15. Ochs, P., Chen, Y., Brox, T., Pock, T.: iPiano: inertial proximal algorithm for nonconvex optimization. SIAM J. Imag. Sci. **7**(2), 388–1419 (2014)
16. Meyer, C.D.: Matrix Analysis and Applied Linear Algebra. Society for Industrial and Applied Mathematics (SIAM), Philadelphia (2000)
17. Elad, M.: Sparse and Redundant Representations. Springer, New York (2010)
18. Beck, A., Teboulle, M.: A fast iterative shrinkage-thresholding algorithm for linear inverse problems. SIAM J. Imag. Sci. **2**(1), 183–202 (2009)
19. Rubinstein, R., Zibulevsky, M., Elad, M.: Efficient implementation of the K-SVD algorithm using batch orthogonal matching pursuit. Technical report, Technion University (2008)

BSS with Corrupted Data
in Transformed Domains

Cécile Chenot[(✉)] and Jérôme Bobin

CEA, IRFU, Service D'Astrophysique - SEDI, 91191 Gif-sur-Yvette Cedex, France
{cecile.chenot,jerome.bobin}@cea.fr

Abstract. Most techniques of Blind Source Separation (BSS) are highly sensitive to the presence of gross errors while these last are ubiquitous in many real-world applications. This mandates the development of *robust* BSS methods, especially to handle the determined case for which there is currently no strategy able to separate the outliers from the sources contributions. We propose a new method which exploits the difference of structural contents that is naturally exhibited by the sources and the outliers in many applications to accurately separate the two contributions. More precisely, we exploit the sparse representations of the signals in two adapted and different dictionaries to estimate jointly the mixing matrix, the sources and the outliers. Preliminary results show the good accuracy of the proposed algorithm in various settings.

Keywords: Blind source separation · Robust recovery · Outliers · Sparse signal modeling · Morphological diversity

1 Introduction

Multichannel data are nowadays encountered in various domains such as astrophysics [4] or remote sensing [8]. Recovering the underlying signals in these data is generally necessary to analyze them. This extraction of the meaningful information can be done using Blind Source Separation (BSS). The standard instantaneous linear mixture model assumes that BSS aims at recovering the n sources $\{\mathbf{S}_i\}_{i=1..n}$ linearly mixed into $m \geq n$ observations $\{\mathbf{X}_j\}_{j=1..m}$ with $t > n$ samples. This model can be conveniently recast in the following matrix form:

$$\mathbf{X} = \mathbf{A}\mathbf{S} + \mathbf{N}, \tag{1}$$

where $\mathbf{X} \in \mathbf{R}^{m \times t}$ designates the linear observations, $\mathbf{A} \in \mathbf{R}^{m \times n}$ the unknown mixing matrix, $\mathbf{S} \in \mathbf{R}^{n \times t}$ the sources and $\mathbf{N} \in \mathbf{R}^{m \times t}$ a Gaussian noise term accounting for model imperfections.

This model is too simplistic to represent satisfactorily complex real-world applications. Indeed, the data can be corrupted by localized and large errors,

J. Bobin—This work is supported by the European Community through the grants PHySIS (contract no. 640174), DEDALE (contract no. 665044) and LENA (ERC StG no. 678282) within the H2020 Framework Program.

P. Tichavský et al. (Eds.): LVA/ICA 2017, LNCS 10169, pp. 542–552, 2017.
DOI: 10.1007/978-3-319-53547-0_51

designated in the following as *outliers* **O**. These deviations from the linear model (1) encompass unexpected physical events such as the presence of spectral variability in hyperspectral unmixing [8], the presence of point-source emissions in astrophysics [15], and also malfunctions of captors [12], to name only a few. In the following, we will assume that the data can be better expressed by:

$$\mathbf{X} = \mathbf{AS} + \mathbf{O} + \mathbf{N}, \tag{2}$$

where $\mathbf{O} \in \mathbf{R}^{m \times t}$ stands for the outliers.

Robust BSS methods in the literature

Despite the unavoidable presence of outliers in some applications, most of the BSS methods in the literature are highly sensitive to their presence [7] and only few strategies dedicated to this problem have been developed. They can mainly be divided into three classes:

- Within the ICA-framework, the authors of [13] promote the mutual independence of the sources by using the robust β-divergence instead of the standard and sensitive Kullback-Leibler divergence. However, since this method only estimates $\mathbf{A}$, no separation between $\mathbf{O}$ and $\mathbf{S}$ is performed.
- The "two-step methods" reside in: (i) eliminating $\mathbf{O}$ from the data and (ii) performing the separation on the "outliers-free" observations. This strategy has been particularly popularized in hyperspectral imaging [12], [19] for which a precise separation between $\mathbf{O}$ and the low-rank matrix $\mathbf{AS}$ has been shown to be possible with the algorithm PCP [6].
- The component separation techniques aim at recovering simultaneously $\mathbf{A}, \mathbf{S}$ and $\mathbf{O}$. It has essentially been used in the NMF framework [1,8,10,11]. The efficiency of these methods strongly depends on the non-negativity assumption, which is not valid in a large number of applications.

In [7], we proposed a component separation method exploiting the sparse representations of $\mathbf{S}$ and $\mathbf{O}$ in a same dictionary to jointly estimate $\mathbf{A}, \mathbf{S}$ and $\mathbf{O}$. Even though $\mathbf{A}$ is well estimated, this method is unable to accurately separate $\mathbf{O}$ from the sources contributions, especially when the number of sources is close to the number of observations [7]. Indeed, even if $\mathbf{A}$ is perfectly known, separating $\mathbf{O}$ from $\mathbf{S}$ is an ill-posed problem since it amounts recovering the sought-after signals $\left[\mathbf{S}^T \mathbf{O}^T\right]^T$ from the observations $\mathbf{X}$ obtained with the sensing matrix $[\mathbf{A I}]$:

$$\text{Recover } \mathbf{S} \text{ and } \mathbf{O} \text{ given } \mathbf{A} \text{ and } \mathbf{X} \text{ such that } \mathbf{X} = \left(\mathbf{A I}\right)\begin{pmatrix}\mathbf{S}\\\mathbf{O}\end{pmatrix}, \tag{3}$$

where $\mathbf{I}$ denotes the identity matrix of size $m \times m$.
Solving (3) requires additional assumptions on the signals such as:

- The outliers do not lie in the span of $\mathbf{A}$ or $\mathbf{AS}$ is low rank while $\mathbf{O}$ is sparse and broadly distributed [6]. Consequently, if $m \gg n$, the outliers can be separated precisely from the sources contribution but this is not valid if m is close to n.

– This ill-posed problem can also be handled using sparsity-based regularization [2,7]. Nonetheless, the compressibility of $\mathbf{S}$ and $\mathbf{O}$ in a same dictionary is not sufficient to solve (3): it also necessary that every sample of $\left[\mathbf{S}^T\mathbf{O}^T\right]^T$ be sparse. This condition is rarely verified in practice (*e.g.* $\mathbf{O}$ is column-sparse such as in [7,8]). However, if the structural contents of the sources and the outliers are different, it is possible to separate the two signals by representing sparsely each signal in one specific dictionary (morphological diversity principle [14]). The two dictionaries then help discriminating between the two contributions.

In the following, we will assume that the morphologies of the outliers and the sources are different in order to separate the two contributions [14]. This additional assumption is usually valid in imaging problems. For instance, in hyperspectral imaging, stripping lines created by malfunctions of captors (the outliers) have a different geometry than the spatial distributions of the observed components (the sources) [12]. Similarly in astrophysics, point source emissions (outliers) have a different morphology than the components of interest which are more broadly distributed [4,15].

Contributions

We introduce a new robust BSS algorithm, coined rGMCA, enforcing the sparsity of the sources and the one of the outliers in different transformed domains. It exploits the difference of morphology between outliers and sources to separate the two contributions and estimates precisely the mixing matrix, the sources and outliers, without restrictive hypothesis on low-rankness or non-negativity. A review of the morphological diversity principle is provided in Sect. 2. The algorithm rGMCA is detailed in Sect. 3. Last, numerical experiments are presented in Sect. 4, showing the good performances of the rGMCA algorithm.

Notations

The Moore-Penrose pseudo-inverse of the matrix $\mathbf{M}$ is designated by $\mathbf{M}^\dagger$ and its transpose by $\mathbf{M}^T$. The jth column of $\mathbf{M}$ is denoted $\mathbf{M}^j$, the ith row $\mathbf{M}_i$, and the i,jth entry $\mathbf{M}_{i,j}$. The norm $\|\mathbf{M}\|_2$ denotes the Frobenius norm of $\mathbf{M}$, and more generally $\|\mathbf{M}\|_p$ designates the p-norm of the matrix $\mathbf{M}$ seen as a long vector. The soft-thresholding operator is denoted $\mathcal{S}_\lambda(\mathbf{M})$, where

$$\left[\mathcal{S}_\lambda(\mathbf{M})\right]_{i,j} = \begin{cases} \mathbf{M}_{i,j} - \text{sign}(\mathbf{M}_{i,j}) * \lambda_i \text{ if } |\mathbf{M}_{i,j}| > \lambda_i \\ 0 \text{ otherwise} \end{cases}$$

2   Sparsity and Morphological Diversity

We aim at separating the outliers from the sources by assuming that their morphological/structural contents are different. For this purpose, we introduce two appropriate dictionaries: $\mathbf{\Phi_O}$ and $\mathbf{\Phi_S}$. These dictionaries are key to separating the two contributions. They are chosen so that the corresponding expansion coefficients of $\mathbf{O}$ and $\mathbf{S}$ are sparse:

$$\mathbf{O}_j = \alpha_{\mathbf{O}_j}\mathbf{\Phi_O}, \forall j \in \{1..m\} \quad \text{and} \quad \mathbf{S}_i = \alpha_{\mathbf{S}_i}\mathbf{\Phi_S}, \forall i \in \{1..n\},$$

where $\{\alpha_{\mathbf{O}_j}\}_{j=1..m}$ and $\{\alpha_{\mathbf{S}_i}\}_{i=1..n}$ are composed of few significant samples. For instance, wavelets can be used to represent sparsely natural images and curvelets for smooth curves to cite only two [14].

The morphological diversity between the sources and the outliers implies that each component $\{\mathbf{O}_j\}_{j=1..m}$ or $\{\mathbf{S}_i\}_{i=1..n}$ has its sparsest expansion coefficients in $\mathbf{\Phi_O}$ or $\mathbf{\Phi_S}$ respectively:

$$\forall i \in \{1..n\}, \forall j \in \{1..m\}, \left\|\mathbf{O}_j\mathbf{\Phi_O}^T\right\|_0 < \left\|\mathbf{O}_j\mathbf{\Phi_S}^T\right\|_0 \text{ and } \left\|\mathbf{S}_i\mathbf{\Phi_S}^T\right\|_0 < \left\|\mathbf{S}_i\mathbf{\Phi_O}^T\right\|_0.$$

Therefore, it is possible to solve (3) by seeking for the sparsest representations, in the spirit of the MCA (*Morphological Component Analysis*) algorithm. The latest aims at separating k different morphological components of a monochannel signal, given k appropriate dictionaries, by maximizing the sparsity of the expansion coefficients of each morphological component in its corresponding dictionary. The good performances of MCA support the utilization of the sparsity to separate different morphological components [14].

In the next section, we will present how, we exploit the morphological diversity between the sources and the outliers to separate the two contributions. Besides, the sparse representations of $\mathbf{S}$ will be also used to discriminate between the sources. Indeed, sparsity has been shown to be a powerful criterion to unmix the sources [5].

3 The Algorithm rGMCA

The use of sparsity in our strategy is twofold: it allows for an accurate separation between the outliers and the sources by exploiting their morphological diversity and it is also used to discriminate between the sources. In order to exploit simultaneously these two aspects, we propose to estimate jointly $\mathbf{A}$, $\mathbf{S}$ and $\mathbf{O}$ by minimizing the following cost function:

$$\underset{\mathbf{A,S,O}}{\text{minimize}} \frac{1}{2} \|\mathbf{X} - \mathbf{AS} - \mathbf{O}\|_2^2 + \lambda \left\|\mathbf{S}\mathbf{\Phi_S}^T\right\|_1 + \beta \left\|\mathbf{O}\mathbf{\Phi_O}^T\right\|_{p,q}, \tag{4}$$

where the first term designates the data fidelity term, well suited to deal with the remaining Gaussian noise, and the second and third terms enforce respectively the sparsity of $\mathbf{S}$ and $\mathbf{O}$ in their corresponding dictionary. In the following, we will assume that the outliers corrupt entire columns of the data such as in [8]. Consequently, we will promote this structure by using the $\ell_{2,1}$ norm ($p = 2, q = 1$).

Despite the non-convexity of the proposed problem, it can be tackled using Block Coordinate Relaxation [16]. Alternatively estimating $\mathbf{A}$, $\mathbf{O}$ and $\mathbf{S}$ propagates the errors from one variable to the others, and thus performs poorly if not

---

Algorithm 1. rGMCA Algorithm

1: **procedure** RGMCA($\mathbf{X}, n$)
2:     Initialize $\tilde{\mathbf{A}}^{(0)}$ (randomly or with a PCA), $\tilde{\mathbf{S}}^{(0)} = 0$ and $\tilde{\mathbf{O}}^{(0)} = 0$.
3:     **while** $k < K$ **do**
4:       Set $\tilde{\mathbf{S}}^{(0,k)} \leftarrow \tilde{\mathbf{S}}^{(k-1)}$ and $\tilde{\mathbf{A}}^{(0,k)} \leftarrow \tilde{\mathbf{A}}^{(k-1)}$
5:       **while** $i < I$ **do**   ▷ Joint estimation of $\mathbf{A}$ and $\mathbf{S}$
6:         Update $\tilde{\mathbf{S}}^{(i,k)}$ with (6)
7:         Update $\tilde{\mathbf{A}}^{(i,k)}$ with (7)
8:       Set $\tilde{\mathbf{S}}^{(k)} \leftarrow \tilde{\mathbf{S}}^{(i-1,k)}$ and $\tilde{\mathbf{A}}^{(k)} \leftarrow \tilde{\mathbf{A}}^{(i-1,k)}$
9:       Set $\tilde{\mathbf{S}}^{(0,k)} \leftarrow \tilde{\mathbf{S}}^{(k)}$ and $\tilde{\mathbf{O}}^{(0,k)} \leftarrow \tilde{\mathbf{O}}^{(k-1)}$
10:      **while** $j < J$ **do**   ▷ Joint estimation of $\mathbf{S}$ and $\mathbf{O}$
11:        Update $\tilde{\mathbf{S}}^{(j,k)}$ with (5)
12:        Update $\tilde{\mathbf{O}}^{(j,k)}$ with (8)
13:       Set $\tilde{\mathbf{S}}^{(k)} \leftarrow \tilde{\mathbf{S}}^{(i-1,k)}$ and $\tilde{\mathbf{O}}^{(k)} \leftarrow \tilde{\mathbf{O}}^{(i-1,k)}$
     return $\mathbf{S}^{(k-1)}, \tilde{\mathbf{A}}^{(k-1)}, \tilde{\mathbf{O}}^{(k-1)}$.

---

initialized with a good accuracy. We propose instead to fully exploit the structure of the problem by using the scheme presented in Algorithm 1 to minimize (4):

- Estimating $\mathbf{A}$ and $\mathbf{S}$ jointly for fixed $\mathbf{O}$: it exploits the joint sparsity of the sources to retrieve more precisely $\mathbf{A}$ from the denoised observations $\mathbf{X} - \mathbf{O}$.
- Estimating $\mathbf{O}$ and $\mathbf{S}$ for fixed $\mathbf{A}$ such as in (3): it provides a precise separation of the two contributions by using their morphological diversity.

We found that this scheme was the less prone to be trapped into local minima.

3.1 Estimation of A and S

Estimating $\mathbf{A}$ and $\mathbf{S}$ for fixed $\mathbf{O}$ amounts to minimize the following cost function:

$$\underset{\mathbf{S}, \mathbf{A}}{\operatorname{argmin}} \frac{1}{2} \|\mathbf{X} - \mathbf{O} - \mathbf{AS}\|_2^2 + \lambda \|\mathbf{S}\Phi_\mathbf{S}^T\|_1 .$$

This problem is similar to the GMCA algorithm [5], performed on the residual $\mathbf{X} - \mathbf{O}$. This algorithm was first proposed in [5], as well as its fast version that we use to speed-up rGMCA. This fast version seeks directly for the sparse coefficients $\alpha_\mathbf{S}$ and $\mathbf{A}$:

- The estimate of $\alpha_\mathbf{S}$, for fixed $\mathbf{A}$, is obtained by minimizing:

$$\underset{\alpha_\mathbf{S}}{\operatorname{argmin}} \frac{1}{2} \|(\mathbf{X} - \mathbf{O})\Phi_\mathbf{S}^T - \mathbf{A}\alpha_\mathbf{S}\|_2^2 + \lambda \|\alpha_\mathbf{S}\|_1 . \tag{5}$$

This can be solved using ISTA or FISTA [3] or with a projected least square as it is proposed in [5] (generally faster than using a proximal method):

$$\alpha_\mathbf{S} = \mathcal{S}_\lambda(\mathbf{A}^\dagger ((\mathbf{X} - \mathbf{O})\Phi_\mathbf{S}^T)) . \tag{6}$$

– The estimate of $\mathbf{A}$ is given by:

$$\mathbf{A} = \left((\mathbf{X} - \mathbf{O})\mathbf{\Phi}_\mathbf{S}^T\right) \alpha_\mathbf{S}^\dagger. \tag{7}$$

Details on GMCA can be found in [5].

3.2   Estimation of S and O

Estimating $\mathbf{S}$ and $\mathbf{O}$ for fixed $\mathbf{A}$ corresponds to the ill-posed problem presented in (3). In the spirit of the MCA algorithm [14], we estimate alternatively the sparse coefficients $\alpha_\mathbf{S}$ and $\alpha_\mathbf{O}$ by working directly in their associated transformed domains with the following updates:

– The estimation of the $\alpha_\mathbf{S}$ is given by (5), which is solved using FISTA.
– The estimation of the $\alpha_\mathbf{O}$ is given by :

$$\underset{\alpha_\mathbf{O}}{\operatorname{argmin}} \frac{1}{2} \left\| (\mathbf{X} - \mathbf{AS})\, \mathbf{\Phi}_\mathbf{O}^T - \alpha_\mathbf{O} \right\|_2^2 + \beta \left\| \alpha_\mathbf{O} \right\|_{2,1}.$$

Every entry $k \in \{1..t\}$ is obtained with the closed form:

$$\tilde{\alpha}_\mathbf{O}^k = \left((\mathbf{X} - \mathbf{AS})\,\mathbf{\Phi}_\mathbf{O}^T\right)^k \times \max\left(0, 1 - \frac{\beta}{\left\|\left((\mathbf{X} - \mathbf{AS})\,\mathbf{\Phi}_\mathbf{O}^T\right)^k\right\|_2}\right). \tag{8}$$

3.3   Choice of the Parameters

The parameters λ and β are automatically set.

Strategy for λ: It has been shown in [5] that using a decreasing strategy for λ in the GMCA algorithm increases its robustness against local minima. Practically, an increasing number of entries are selected. The final threshold λ_i for each source $\mathbf{S}_i$ is $k\sigma_i$, where $k \in (1,3)$ and σ_i is the standard deviation of the noise contaminating the ith source. If σ_i is not known, it can be estimated with the MAD (median absolute deviation) operator. A large k prevents the incorporation of Gaussian noise in the source estimate.
When estimating jointly $\mathbf{O}$ and $\mathbf{S}$, the values of λ_i are directly set to the final thresholds $k\sigma_i$.

Strategy for β: The value of β is fixed to the value $\sigma \times \sqrt{2} \times \frac{\Gamma\left(\frac{m+1}{2}\right)}{\Gamma\left(\frac{m}{2}\right)}$ which corresponds to an estimation of $\mathbb{E}\left\{\left\|(\mathbf{N}\mathbf{\Phi}_\mathbf{O}^T)^k\right\|_2\right\}$, $(\operatorname{mad}((\mathbf{X} - \mathbf{AS} - \mathbf{O})\mathbf{\Phi}_\mathbf{O}^T)$ corresponds to a good estimate of the standard deviation of $\mathbf{N}\mathbf{\Phi}_\mathbf{O}^T$ if it is not known), and thus limits the impact of the Gaussian noise.

4 Numerical Experiments

In this section, we compare rGMCA with the standard robust BSS methods presented in the introduction: minimization of the β-divergence in Φ_O [13] (tuned implementation from [9]), the combination PCP+GMCA (the outliers are first discarded from the observations with a tuned implementation of PCP in Φ_O [6] and then $\mathbf{A}$ and $\mathbf{S}$ are estimated with GMCA in Φ_S [5]) and also with GMCA to illustrate the benefits of using robust strategies. We investigate the performances of the algorithms with respect to the following criteria:

- The unmixing precision is measured with the global criterion $\Delta_A = \frac{\|\tilde{\mathbf{A}}^\dagger \mathbf{A} - \mathbf{I}\|_1}{n^2}$ [5] and the maximal angle made between the estimated and true columns of $\mathbf{A}$ defined as $\max_{i, i \in \{1,..,n\}} \arccos < \mathbf{A}^i, \tilde{\mathbf{A}}^i >$ in degree.
- The accuracy of the separation between $\mathbf{S}$ and $\mathbf{O}$ is assessed with the minimal SDR (signal distortion ratio [17]) obtained for each estimation of $\mathbf{S}$.

Since the minimization of the β-divergence does not estimate $\mathbf{S}$, we will compute the SDR from the estimated sources $\mathcal{S}_\lambda \left(\tilde{\mathbf{A}}^\dagger \mathbf{X}_{\Phi_S} \right) \Phi_S$, where $\tilde{\mathbf{A}}$ is the mixing matrix estimated with the algorithm.

4.1 Monte Carlo Simulations

In this first part, we investigate the robustness of the algorithms with Monte-Carlo simulation (80 runs) on 1D signals with varying parameters (the amplitude and the percentage of the corrupted data) and the following setting:

- A total of 8 sources, sparse in DCT, are mixed into 20 observations which are corrupted with the Gaussian noise $\mathbf{N}$ (standard deviation of 0.1, SNR around 40 dB), and the outliers which are sparse in the direct domain.
- The columns of $\mathbf{O}$ follow a Bernoulli-Gaussian law with an activation parameter ρ (default value 10%) and a standard deviation σ_O (default value 100).
- The sparse coefficients of $\mathbf{S}$ are also drawn from a Bernoulli-Gaussian law (activation parameter of 5%, standard deviation of 100).

Percentage of corrupted data. As shown in Fig. 1, the minimization of the β-divergence and GMCA are highly sensitive to the increasing percentage of outliers. Not only $\mathbf{S}$ is poorly estimated Fig. 1a, but also the sources are not correctly unmixed Fig. 1c: the unmixing process is challenging without the explicit estimation of $\mathbf{O}$. The combination PCP+GMCA provides the most robust unmixing process Fig. 1c, but cannot separate precisely the outliers from the source contribution without further hypothesis [18], Fig. 1a. The algorithm rGMCA returns the most accurate estimations of $\mathbf{S}$ and $\mathbf{A}$ while ρ is lower than 30%, what should be achieved in practice if the dictionary Φ_O is wisely chosen.

Amplitude of the outliers. The minimization of the β-divergence and GMCA have similar performances: they fail whenever the amplitude of $\mathbf{O}$ and the one

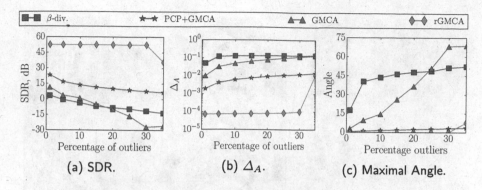

Fig. 1. Influence of the percentage of corrupted entries.

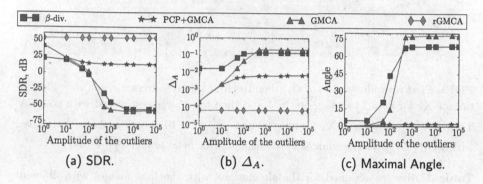

Fig. 2. Influence of the amplitude of the outliers.

of **S** are of the same order of magnitude. The outliers which are detrimental for the unmixing are discarded from **X** with PCP (Fig. 2c: the maximal angle is small), but on the overall, PCP struggles to separate accurately **O** from **AS** (Fig. 2a: the SDR is lower than the one obtained with rGMCA). Besides, whenever the amplitude of **O** is larger than the one of the sources, the performances of PCP+GMCA are constant. Last, rGMCA is not influence by the amplitude of the outliers with this setting since the precision reached for the estimation of **A** and **S** stays constant.

4.2   2D Simulations

In this section, we compare PCP+GMCA and rGMCA which were significantly the most successful in Sect. 4.1 on 2D applications. The fist row of Fig. 3 shows the sources (four 128×128 images, approximately sparse in wavelets [14]) and the outliers corresponding to a high-frequencies texture (approximately sparse in DCT). We observe the influence of the amplitude of **O** and the number of observations m. When varying m, the maximal amplitude of **O** is set to the

Fig. 3. First row: illustration of $\mathbf{O}_1$, then the four initial sources. Second row: illustration of $\mathbf{X}_1$ for $m = 34$ and $\frac{\|\mathbf{O}\|_\infty}{\|\mathbf{AS}\|_\infty} = 1$ and then the sources estimated with rGMCA. Third row: illustration of $\mathbf{X}_1$ for $m = 20$ and $\frac{\|\mathbf{O}\|_\infty}{\|\mathbf{AS}\|_\infty} = 10$ and then sources estimated with PCP+GMCA. The sources were estimated for $m = 34$ and $\frac{\|\mathbf{O}\|_\infty}{\|\mathbf{AS}\|_\infty} = 1$.

Table 1. Results obtained for the simulations with the four images with different numbers of observations m or amplitudes of the outliers.

Method	Errors	m					$\frac{\|\mathbf{O}\|_\infty}{\|\mathbf{AS}\|_\infty}$				
		4	6	10	18	34	0.01	0.1	1	10	100
PCP+GMCA	SDR, dB	−4.3	−3.5	−2.6	−2.4	−1.9	−2.1	−2.1	−1.6	−3.1	−4.7
	Max. angle	1.1	1.5	1.3	0.8	0.7	0.8	0.8	0.8	1.2	19.6
rGMCA	SDR, dB	9.3	12.6	13.8	14.1	14.0	13.6	13.7	14.0	14.9	15.7
	Max. angle	3.6	2.4	1.3	0.8	0.7	0.9	0.9	0.9	0.9	0.9

maximal amplitude of $\mathbf{AS}$, and respectively, when varying the amplitude of $\mathbf{O}$, we set $m = 20$. The metrics are averaged for four experiments.

Amplitude of the outliers. Contrary to the previous 1D-case, PCP+GMCA is shown to be sensitive to the amplitude of $\mathbf{O}$ since it becomes unable to estimate $\mathbf{A}$ for the largest amplitude Table 1. Moreover, even if $\mathbf{A}$ is correctly retrieved, the SDR of the sources estimated with PCP+GMCA is very low: the separation between outliers and sources is not correct (see the second row of Fig. 3). The method rGMCA is more reliable as it returns fair estimates of the mixing matrix and the sources for almost all the experiments. More surprisingly, the SDR obtained with rGMCA increases with the amplitude of $\mathbf{O}$: it becomes easier to distinguish the contribution of $\mathbf{O}$ from the one of $\mathbf{N}$. It is also proportionally less influenced by the bias introduced by the different thresholding processes. This improved estimation of $\mathbf{O}$ leads to accurate estimates of $\mathbf{S}$.

Number of observations. It has been emphasized in [7] and in the introduction that the ratio $\frac{m}{n}$ is crucial for BSS in the presence of outliers. The unmixing process and the estimation of $\mathbf{S}$ should be easier if $m \gg n$ for both algorithms. The results obtained by the two strategies are indeed improved for a larger m Table 1. Besides, even if the estimated $\mathbf{A}$ is slightly more precise for PCP+GMCA, the sources returned by rGMCA are much more accurate (see Table 1 and Fig. 3).

5   Conclusion

The BSS problem in the presence of outliers is challenging since it requires a robust sources unmixing but also a precise separation of the outliers from the sources contributions. This task is not properly handled by the standard robust BSS methods without restrictive hypothesis. We propose a new method coined rGMCA that estimates jointly the sources, the outliers and the mixing matrix. It exploits the difference of morphology between the sources and the outliers to separate precisely the two contributions, including in the challenging determined case. Preliminary experiments show that rGMCA yields a precise estimation of the mixing matrix and also of the sources in various settings. The discrepancy between rGMCA and the standard robust methods is particularly important for the sources estimations in the proposed experiments. This supports the use of the morphological diversity to discriminate efficiently between the outliers and the sources.

References

1. Altmann, Y., McLaughlin, S., Hero, A.: Robust linear spectral unmixing using anomaly detection. IEEE Trans. Comput. Imag. **1**(2), 74–85 (2015)
2. Amini, S., Sadeghi, M., Joneidi, M., Babaie-Zadeh, M., Jutten, C.: Outlier-aware dictionary learning for sparse representation. In: 2014 IEEE International Workshop on Machine Learning for Signal Processing (MLSP), pp. 1–6. IEEE (2014)
3. Beck, A., Teboulle, M.: A fast iterative shrinkage-thresholding algorithm for linear inverse problems. SIAM J. Imag. Sci. **2**(1), 183–202 (2009)
4. Bobin, J., Sureau, F., Starck, J.L., Rassat, A., Paykari, P.: Joint Planck and WMAP CMB map reconstruction. Astron. Astrophys. **563**, A105 (2014)
5. Bobin, J., Starck, J.L., Fadili, J., Moudden, Y.: Sparsity and morphological diversity in blind source separation. IEEE Trans. Image Process. **16**(11), 2662–2674 (2007)
6. Candès, E.J., Li, X., Ma, Y., Wright, J.: Robust principal component analysis? J. ACM (JACM) **58**(3), 11 (2011)
7. Chenot, C., Bobin, J., Rapin, J.: Robust sparse blind source separation. IEEE Sig. Process. Lett. **22**(11), 2172–2176 (2015)

8. Fevotte, C., Dobigeon, N.: Nonlinear hyperspectral unmixing with robust nonnegative matrix factorization. IEEE Trans. Image Process. **24**(12), 4810–4819 (2015)
9. Gadhok, N., Kinsner, W.: An implementation of β-divergence for blind source separation. In: Canadian Conference on Electrical and Computer Engineering, 2006, CCECE 2006, pp. 1446–1449. IEEE (2006)
10. Halimi, A., Bioucas-Dias, J., Dobigeon, N., Buller, G.S., McLaughlin, S.: Fast hyperspectral unmixing in presence of nonlinearity or mismodelling effects. arXiv preprint arXiv:1607.05336 (2016)
11. Li, C., Ma, Y., Mei, X., Liu, C., Ma, J.: Hyperspectral unmixing with robust collaborative sparse regression. Remote Sens. **8**(7), 588 (2016)
12. Li, Q., Li, H., Lu, Z., Lu, Q., Li, W.: Denoising of hyperspectral images employing two-phase matrix decomposition. IEEE J. Sel. Top. Appl. Earth Observations Remote Sens. **7**(9), 3742–3754 (2014)
13. Mihoko, M., Eguchi, S.: Robust blind source separation by beta divergence. Neural comput. **14**(8), 1859–1886 (2002)
14. Starck, J.L., Murtagh, F., Fadili, J.M.: Sparse Image and Signal Processing Wavelets, Curvelets, Morphological Diversity. Cambridge University Press, Cambridge (2010)
15. Sureau, F., Starck, J.L., Bobin, J., Paykari, P., Rassat, A.: Sparse point-source removal for full-sky CMB experiments: application to WMAP 9-year data. Astron. Astrophys. **566**, A100 (2014)
16. Tseng, P.: Convergence of a block coordinate descent method for nondifferentiable minimization. J. Optim. Theor. Appl. **109**(3), 475–494 (2001)
17. Vincent, E., Gribonval, R., Févotte, C.: Performance measurement in blind audio source separation. IEEE Trans. Audio Speech Lang. Process. **14**(4), 1462–1469 (2006)
18. Xu, H., Caramanis, C., Sanghavi, S.: Robust PCA via outlier pursuit. In: Advances in Neural Information Processing Systems, pp. 2496–2504 (2010)
19. Zhang, H., He, W., Zhang, L., Shen, H., Yuan, Q.: Hyperspectral image restoration using low-rank matrix recovery. IEEE Trans. Geosci. Remote Sens. **52**(8), 4729–4743 (2014)

Singing Voice Separation Using RPCA with Weighted l_1-norm

Il-Young Jeong and Kyogu Lee[✉]

Music and Audio Research Group, Seoul National University,
1 Gwanak-ro, Gwanak-gu, Seoul 08826, Korea
kglee@snu.ac.kr

Abstract. In this paper, we present an extension of robust principal component analysis (RPCA) with weighted l_1-norm minimization for singing voice separation. While the conventional RPCA applies a uniform weight between the low-rank and sparse matrices, we use different weighting parameters for each frequency bin in a spectrogram by estimating the variance ratio between the singing voice and accompaniment. In addition, we incorporate the results of vocal activation detection into the formation of the weighting matrix, and use it in the final decomposition framework. From the experimental results using the DSD100 dataset, we found that proposed algorithm yields a meaningful improvement in the separation performance compared to the conventional RPCA.

Keywords: Singing voice separation · Robust principal component analysis · Weighted l_1-norm minimization

1 Introduction

Singing voice separation (SVS), or separating singing voice and accompaniment from a musical mixture is a challenging task. Many of the previous studies have attempted to use the distinctive characteristics of each source: fundamental frequency (f_0) and its harmonic structure of singing voice [11], repeatability [12], spectral/temporal continuity [5,14], and so on.

Huang *et al.*, on the other hand, proposed to use a low-rank/sparse model for singing voice separation [4]. Approaches based on the low-rank/sparse model assume that accompaniment in music is usually repetitive because the number of instruments and notes in the accompaniment is limited. It is therefore presumed that the spectrogram of the accompaniment can be represented as a low-rank matrix. On the other hand, singing voice can be expressed as a sparse matrix because most of energy is concentrated on the f_0 trajectory and its harmonics. Based on these observations, robust principal component analysis (RPCA) [2] that decomposes a matrix into low-rank and sparse parts, was applied to separate singing voice and accompaniment in a mixture [4].

Although RPCA has been successfully applied to SVS, there is still plenty of room for improvement. Numerous studies have tried to extend the basic

© Springer International Publishing AG 2017
P. Tichavský et al. (Eds.): LVA/ICA 2017, LNCS 10169, pp. 553–562, 2017.
DOI: 10.1007/978-3-319-53547-0_52

RPCA-based approach. Sprechmann *et al.* presented a robust nonnegative matrix factorization, where an accompaniment spectrogram is represented by a combination of a few nonnegative spectra [13]. Jeong and Lee tried to extend RPCA by generalizing the nuclear norm and l_1-norm to Schatten-p norm and l_p-norm, respectively, and suggested the appropriate value of p, for SVS in particular [6]. Chan *et al.* imposed additional vocal activation information to RPCA to remove the singing voice in the non-vocal frames [3].

In this paper we focus on the fact that minimization of the nuclear norm and l_1-norm affects not only the low-rankness and sparsity of two decomposed matrices, but also their relative scale. Therefore, if prior information of their relative scale is known, it can be utilized in matrix decomposition by controlling the relative importance between the nuclear and l_1-norm minimization terms. Furthermore, each time-frequency component of the spectrogram might have different prior, so we have to apply different weights to each element.

In our work, we construct a weighting matrix using two distinctive features: (1) frequency-dependent variance ratio between accompaniment and singing voice, and (2) the presence of singing voice, which is obtained by conducting a simple vocal activity detection (VAD) algorithm. In doing so, we go through a two-stage process that VAD is performed on the pre-separated singing voice, followed by the re-separation stage using updated the weighting matrix.

2   Algorithm

2.1   Robust Principal Component Analysis

Ideally, the low-rank and the sparse components can be decomposed from their mixture by solving the following optimization problem:

$$\begin{aligned} \text{minimize} \quad & \text{rank}(L) + \lambda\,\text{nonzero}(S), \\ s.t. \quad & L + S = M, \end{aligned} \tag{1}$$

where $M \in \mathbb{R}^{F \times T}$, $L \in \mathbb{R}^{F \times T}$, and $S \in \mathbb{R}^{F \times T}$ are the mixture, low-rank, and sparse matrix, respectively. rank($\cdot$) and nonzero($\cdot$) denote the rank and the number of nonzero components in a matrix, respectively. λ denotes the relative weight between two terms. Since above objective function is difficult to solve, Candès *et al.* presented its convex relaxation, or RPCA, as follows [2]:

$$\begin{aligned} \text{minimize} \quad & |L|_* + \lambda|S|_1, \\ s.t. \quad & L + S = M, \end{aligned} \tag{2}$$

where $|\cdot|_*$ and $|\cdot|_1$ denote the nuclear norm (sum of singular values) and l_1-norm (sum of the absolute values of matrix elements), respectively. These properly approximate rank($\cdot$) and nonzero($\cdot$) in Eq. (1) and allow to solve it in a convex formulation. As in Eq. (1), λ decides the relative importance between two norms. Candès *et al.* suggested $\lambda = 1/\sqrt{\max(F, T)}$ [2], and Huang *et al.* generalized it as $\lambda = k/\sqrt{\max(F, T)}$ with a parameter k [4].

2.2   RPCA with Weighted l_1-norm

Since λ in Eq. (2) is a global parameter for all the element of M, or $M_{f,t}$, once its value is decided then all $M_{f,t}$ have the same importance for the low-rankness of $L_{f,t}$ and the sparsity of $S_{f,t}$. However, it is not always proper in actual situation, and might be too simple. For example, if we know that $L_{f,t} = 0$ for some (f,t), we may able to choose the value of λ to be $\lambda = 0$ for those element. If $S_{f,t} = 0$, on the contrary, we may set $\lambda \to \infty$. To apply the different weight for each element, we present RPCA with weighted l_1-norm, or weighted RPCA (wRPCA), which replace λ to the weighting matrix Λ as:

$$\begin{aligned} \text{minimize} \quad & |L|_* + |\Lambda \otimes S|_1, \\ s.t. \quad & L + S = M, \end{aligned} \tag{3}$$

where $\otimes$ denotes the element-wise multiplication operator. Note that $|\Lambda \otimes S|_1$ is a weighted l_1-norm of S, which has been presented in a number of previous studies [1,7]. To solve Eq. (3), optimization method for RPCA such as augmented Lagrangian multiplier (ALM) method can be directly used, just by replacing λ to Λ.

3   Singing Voice Separation

3.1   SVS Using RPCA

Huang *et al.* suggested that RPCA can be applied to separate the singing voice and the accompaniment from music signal [4]. In the case of music accompaniment, instruments often reproduce the same sounds in the same music, therefore its magnitude spectrogram can be represented as a low-rank matrix. On the contrary, singing voice has a sparse distribution in the spectrogram domain due to its strong harmonic structure. Therefore, M, L, and S in Eq. (2) can be considered as a spectrogram of the input music, accompaniment, and singing voice, respectively. After the separation is done in the spectrogram domain, the waveform for each source is obtained by directly applying the phase of the original mixture.

3.2   Proposed Method: SVS Using wRPCA

We extended previous RPCA-based SVS framework, by using wRPCA instead of RPCA in particular. We refer several previous studies to design the separation framework [3,8,9].

Nonnegativity Constraint. At first, we added a nonnegativity constraint in Eq. (3) as follows:

$$\begin{aligned} \text{minimize} \quad & |L|_* + |\Lambda \otimes S|_1, \\ s.t. \quad & L + S = M, \quad L \geq 0, \quad S \geq 0. \end{aligned} \tag{4}$$

This constraint prevent that large value of $\Lambda_{f,t}$ makes large negative value for S. The optimization of Eq. (4) is similar as of Eq. (2) or Eq. (3) but L and S are rectified as $x \leftarrow \max(x, 0)$ in every iteration.

Two-Stage Framework Using VAD. There were two opposite studies on SVS and VAD. Chan *et al.* suggested that additional vocal activity information can improve SVS [3]. On the other hand, Lehner and Widmer suggested that SVS can improve the accuracy of VAD algorithm [10]. To apply both of these suggestions, we conducted the two-stage framework as follows. At the first stage, the sources are separated without vocal activity information. Next, vocal activity is detected using the separated singing voice. In the second separation stage, the sources are separated again with detected vocal activity information. We basically used VAD algorithm presented by Lehner *et al.* which uses well-designed mel-frequency cepstral coefficients (MFCC) as features [8]. In addition, we also used the vocal variance features which were also proposed in their other studies [9]. For the classification, we used random forest with 500 trees, and used threshold of 0.55. As a post-processing step, median filtering was applied to the frame-wise classification results with 7 frames filter length (1.4s). Note that above framework is also based on the previous study [8]. Because the temporal resolution of spectrogram and VAD might be different, we aligned them by considering those absolute time indices so that we can obtain the frame-wise VAD results.

Choosing the Value for Λ. We choose the value of Λ as follows. At first, we decompose Λ as

$$\Lambda = k\lambda\Delta, \tag{5}$$

where λ is $1/\sqrt{max(F,T)}$ suggested by Candès *et al.* [2], and k is a global parameter used by Huang *et al.* [4]. In this work, we empirically set it to be $k = 0.6$. Δ is a element-wise weighting matrix which is our main interest.

To select the appropriate value for Δ, we basically focused on the fact that Δ should be smaller when singing voice is relatively stronger than accompaniment, and be larger in the opposite case. If we try to set the frequency-wise weight, therefore it might be reasonable to use the ratio of their variance as

$$\Delta_{f,t} = \frac{b_A(f)}{b_V(f)}, \tag{6}$$

where $b_A(f)$ and $b_V(f)$ are the variances of the accompaniment and singing voice, respectively, in f-th frequency bin. Assuming both singing voice and accompaniment have the Laplacian distribution, they can be estimated by calculating the l_1-norm for each frequency bin in the training data as follows:

$$b_A(f) = \sum_t |A_{f,t}|,$$

$$b_V(f) = \sum_t |V_{f,t}|, \tag{7}$$

where A and V are the training data of the accompaniment and singing voice, respectively, that all the spectrograms of tracks in the training set are concatenated over time. Note that we assume that both accompaniment and singing voice for training are from the same music, those therefore have the same time length.

This variance ratio might be different when only vocal-activated frames are estimated. At least it will be smaller than Eq. (6) in overall, since all the non-vocal frames where singing voice is absent are excluded. In addition, since we know that there is no singing voice in the non-vocal frames, we can set the weight for those frames to infinite so the singing voice can be successfully eliminated. Consequently, we set $\hat{\Delta}$ for the second separation stage as follows:

$$\hat{\Delta}_{f,t} = \begin{cases} \frac{\hat{b}_A(f)}{\hat{b}_V(f)}, & \text{if } p(t) = 1, \\ \infty, & \text{otherwise,} \end{cases} \tag{8}$$

where $p(t)$ is the vocal activity information for the t-th frame: $p(t) = 1$ for the vocal-activated frames and 0 for the non-vocal ones. $\hat{b}_A(f)$ and $\hat{b}_V(f)$ are similar as $b_A(f)$ and $b_V(f)$, respectively, but estimated from the vocal-activated frames only as

$$\hat{b}_A(f) = \sum_t |\hat{A}_{f,t}|,$$
$$\hat{b}_V(f) = \sum_t |\hat{V}_{f,t}|, \tag{9}$$

where $\hat{A}$ and $\hat{V}$ are the excerpts of A and V, respectively, which include the vocal-activated frames ($p(t) = 1$) only.

Handling Multi-channel Signals. Real-world music data are mostly provided in a multi-channel format *e.g.* stereo. Although the spatial information is helpful for better separation results, it is beyond the scope of this work. Therefore, the tracks are mixed down to a single-channel format. We simply took an average of spectrograms over channel and perform RPCA (or wRPCA) to this averaged spectrogram. We were concerned that the data is spatially biased if we take an average of waveform (center enhanced) or perform the algorithms to each channel separately (left/right enhanced). After the separation of $M = L + S$ is done, the separated singing voice and accompaniment of original multi-channel signal is obtained by using the Wiener-like filter (or soft mask) as $L/(L + S)$ for the accompaniment or $S/(L + S)$ for the singing voice for each channel.

4   Experimental Results

We applied our SVS algorithm to the dataset and the evaluation criteria from sixth community-based signal separation evaluation campaign (SiSEC 2016): professionally-produced music recordings (MUS) [16]. This campaign provided

Demixing Secrets Dataset 100 (DSD100), which consist 50 tracks for training ('dev') and other 50 for testing ('test'). All the tracks are sampled at 44.1 kHz and have stereo channels. Because there are 4 sources (vocals, bass, drums, and others) for each track, we considered the sum of bass, drums, and others as accompaniment. We used the dev set only to set Λ and $\hat{\Lambda}$, and even to train the VAD algorithm. In our experiments, VAD scores 0.87 F-score and 84% accuracy from the test set. As the evaluation criteria, it measures signal-to-distortion ratio (SDR), image-to-spatial distortion ratio (ISR), source-to-interference ratio (SIR), and source-to-artifacts ratio (SAR) based on BSS-Eval [15]. To generate the spectrogram of music, we took the magnitude of short-time Fourier transform with Hanning window of 4096 samples and half overlap.

Figure 1 shows the comparison of conventional RPCA, wRPCA, and two-stage wRPCA with VAD, and Table 1 shows the numerical values of the median of SDR. From this result, we can find that the proposed wRPCA improve SDR score

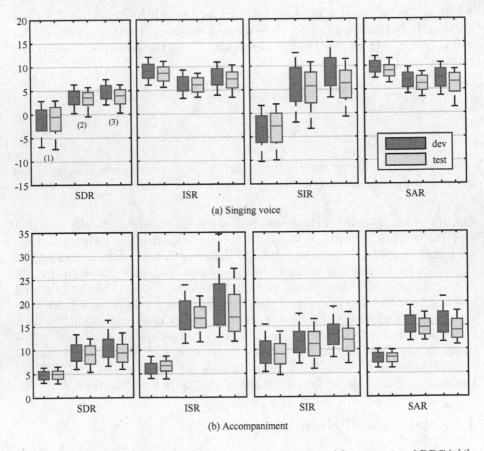

(a) Singing voice

(b) Accompaniment

Fig. 1. Comparison of singing voice separation results using (1) conventional RPCA [4], (2) proposed wRPCA, and (3) wRPCA with VAD.

Table 1. Numerical values of median SDR in Fig. 1.

SDR(dB)	dev			test		
	RPCA	wRPCA	wRPCA w/VAD	RPCA	wRPCA	wRPCA w/VAD
Singing voice	−0.83	3.80	4.74	−0.51	3.54	3.92
Accompaniment	4.78	9.68	10.52	5.00	9.13	9.45

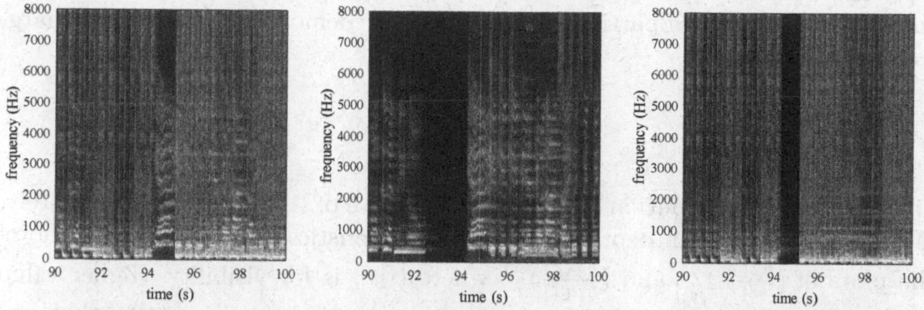

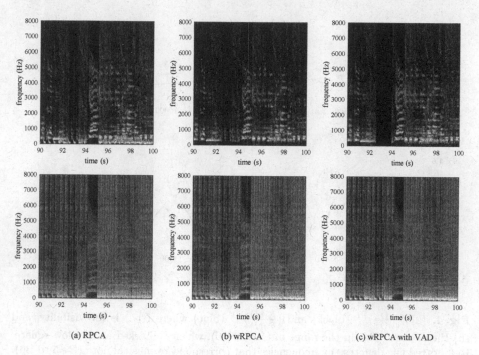

(a) Mixture (b) Singing voice (c) Accompaniment

Fig. 2. Log-spectrograms of example mixture, singing voice, and accompaniment. Audio clips are excerpted from 'AM Contra - Heart Peripheral' in the dev set of DSD100.

(a) RPCA (b) wRPCA (c) wRPCA with VAD

Fig. 3. Log-spectrograms of separated singing voice (top) and accompaniment (bottom). Input mixture is same as in Fig. 2.

for both singing voice and accompaniment, and even VAD does. However, the improvement from VAD is considerably degraded in the test set compared to the dev set. Considering that VAD for dev data makes almost perfect accuracy since it is trained by itself, we can expect that the better VAD algorithm is required to maximize its effectiveness. Example results are shown in Figs. 2 and 3. Compared to the conventional RPCA, it is observed that wRPCA successfully improve the separation quality, especially in the low-frequency region, and even VAD does in the non-vocal frames in particular. Audio files are demonstrated at http://marg. snu.ac.kr/svs_wrpca.

5   Discussion

Since the main contribution of our work is the use of Λ and $\hat{\Lambda}$, more accurately, Δ and $\hat{\Delta}$, we discuss in depth about the characteristics of them. Figure 4 shows the plots of $(\frac{b_A(f)}{b_V(f)})^{-1}$ and $(\frac{\hat{b}_A(f)}{\hat{b}_V(f)})^{-1}$ where $(\cdot)^{-1}$ is for visibility. Higher value means that the singing voice is stronger than the accompaniment in that frequency bin. What follows are several interesting insights we found from these plots.

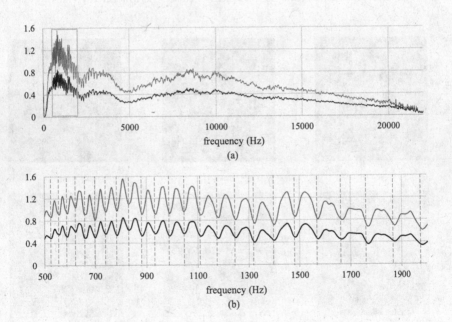

Fig. 4. (a) $(\frac{b_A(f)}{b_V(f)})^{-1}$ (black) and $(\frac{\hat{b}_A(f)}{\hat{b}_V(f)})^{-1}$ (blue) where $(\cdot)^{-1}$ is for visibility, and (b) the enlarged plot in the range of (500, 2000), which is marked as a yellow square. Red dotted line denotes the frequencies that correspond to musical note (C#5 to B6). (Color figure online)

- $(\frac{b_A(f)}{b_V(f)})^{-1}$ and $(\frac{\hat{b}_A(f)}{b_V(f)})^{-1}$ both show similar trends but only the scales are different, and we expect it means that the spectral characteristics of accompaniment are similar between in vocal and non-vocal frames.
- Singing voice is extremely weaker than accompaniment in very low frequency range (lower than 100 Hz). It is reasonable because singing voice is mostly distributed in f_0 and its harmonics, which is rarely occur in those range, while some instruments such as bass and drums can be. Some previous studies for SVS have applied this characteristics by using high-pass filtering [5,14].
- Some peaks can be found from the envelope, that are located around 0.7, 1.5, 3, and 8 kHz. we expect it is related with the formants of singing voice.
- From Fig. 4(b), we found an interesting phenomena that the singing voice is relatively weak in the frequency bins which correspond to the musical notes compared to those neighbor frequency bins. Although it needs more experiments to clarify the reason, we made some possible hypotheses as follows: (1) the mainlobe of singing voice may wider than that of accompaniment, (2) singing voice has stronger vibrato in general, and it may cause the 'blurred peak' in a long window length, or (3) singers frequently fail to sound exact note frequency, and make more errors than the instrumental players.

6 Conclusion

A novel framework for RPCA-based SVS was presented. In particular, we replaced the l_1-norm term to the weighted l_1-norm, and proposed to use the frequency-dependent variance ratio between singing voice and accompaniment to make the weighting matrix. In addition, we apply VAD for SVS by conducting a two-stage separation framework. In future works, we will investigate a method for finding a better weighting matrix Λ. The spatial information that is discarded in the current study also will be tried to be applied in the separation procedure.

Acknowledgments. This research was supported by the MSIP (Ministry of Science, ICT and Future Planning), Korea, under the ITRC (Information Technology Research Center) support program (IITP-2016-H8501-16-1016) supervised by the IITP (Institute for Information & communications Technology Promotion)

References

1. Candès, E.J., Wakin, M.B., Boyd, S.P.: Enhancing sparsity by reweighted l_1 minimization. J. Fourier Anal. Appl. **14**(5–6), 877–905 (2008)
2. Candès, E.J., Li, X., Ma, Y., Wright, J.: Robust principal component analysis? J. ACM **58**(3), 11 (2011)
3. Chan, T.S., Yeh, T.C., Fan, Z.C., Chen, H.W., Su, L., Yang, Y.H., Jang, R.: Vocal activity informed singing voice separation with the iKala dataset. In: IEEE International Conference on Acoustics, Speech and Signal Processing, pp. 718–722 (2015)

4. Huang, P.S., Chen, S.D., Smaragdis, P., Hasegawa-Johnson, M.: Singing-voice separation from monaural recordings using robust principal component analysis. In: IEEE International Conference on Acoustics, Speech and Signal Processing, pp. 57–60 (2012)

5. Jeong, I.Y., Lee, K.: Vocal separation from monaural music using temporal/spectral continuity and sparsity constraints. IEEE Signal Process. Lett. **21**(10), 1197–1200 (2014)

6. Jeong, I.Y., Lee, K.: Vocal separation using extended robust principal component analysis with Schatten p/l_p-norm and scale compression. In: IEEE International Workshop on Machine Learning for Signal Processing, pp. 1–6 (2014)

7. Khajehnejad, M.A., Xu, W., Avestimehr, A.S., Hassibi, B.: Analyzing weighted minimization for sparse recovery with nonuniform sparse models. IEEE Trans. Sig. Process. **59**(5), 1985–2001 (2011)

8. Lehner, B., Sonnleitner, R., Widmer, G.: Towards light-weight, real-time-capable singing voice detection. In: International Society for Music Information Retrieval Conference, pp. 53–58 (2013)

9. Lehner, B., Widmer, G., Sonnleitner, R.: On the reduction of false positives in singing voice detection. In: IEEE International Conference on Acoustics, Speech and Signal Processing, pp. 7480–7484 (2014)

10. Lehner, B., Widmer, G.: Monaural blind source separation in the context of vocal detection. In: International Society for Music Information Retrieval Conference, pp. 309–316 (2015)

11. Li, Y., Wang, D.: Separation of singing voice from music accompaniment for monaural recordings. IEEE Trans. Audio Speech Lang. Process. **15**(4), 1475–1487 (2007)

12. Rafii, Z., Pardo, B.: Repeating pattern extraction technique (REPET): a simple method for music/voice separation. IEEE Trans. Audio Speech Lang. Process. **21**(1), 73–84 (2013)

13. Sprechmann, P., Bronstein, A.M., Sapiro, G.: Real-time online singing voice separation from monaural recordings using robust low-rank modeling. In: International Society for Music Information Retrieval Conference, pp. 67–72 (2012)

14. Tachibana, H., Ono, N., Sagayama, S.: Singing voice enhancement in monaural music signals based on two-stage harmonic/percussive sound separation on multiple resolution spectrograms. IEEE/ACM Trans. Audio Speech Lang. Process. **22**(1), 228–237 (2014)

15. Vincent, E., Gribonval, R., Fvotte, C.: Performance measurement in blind audio source separation. IEEE Trans. Audio Speech Lang. Process. **14**(4), 1462–1469 (2006)

16. Vincent, E., Araki, S., Theis, F.J., Nolte, G., Bofill, P., Sawada, H., Ozerov, A., Gowreesunker, B.V., Lutter, D., Duong, N.Q.K.: The signal separation evaluation campaign (2007–2010): achievements and remaining challenges. Sig. Process. **92**(8), 1928–1936 (2012)

Multimodal Approach to Remove Ocular Artifacts from EEG Signals Using Multiple Measurement Vectors

Victor Maurandi[1(✉)], Bertrand Rivet[1], Ronald Phlypo[1], Anne Guérin–Dugué[1], and Christian Jutten[1,2]

[1] CNRS, UMR 5216, Univ. Grenoble Alpes, GIPSA-Lab, 38000 Grenoble, France
{victor.maurandi,bertrand.rivet,ronald.phlypo,anne.guerin,
christian.jutten}@gipsa-lab.grenoble-inp.fr
[2] Institut Universitaire de France, 75231 Paris, France

Abstract. This paper deals with the extraction of eye-movement artifacts from EEG data using a multimodal approach. The gaze signals, recorded by an eye-tracker, share a similar temporal structure with the artifacts induced in EEG recordings by ocular movements. The proposed approach consists in estimating this specific common structure using Multiple Measurement Vectors which is then used to denoise the EEG data. This method can be used on single trial data and can be extended to multitrial data subject to some additional preprocessing. Finally, the proposed method is applied to gaze and EEG experimental data and is compared with some popular algorithms for eye movement artifact correction from the literature.

Keywords: Ocular artifact extraction · EEG · Gaze · Multiple measurement vectors · Multimodality

1 Introduction

Scalp electroencephalography (EEG) is a popular non-invasive method to monitor cerebral activity. It allows to measure the effect of electrical brain activity on the potential field at the scalp using surface electrodes. However, interpreting the recordings is challenging, in part due to different kinds of noise [1]. Among them, one finds the ocular artifacts that are induced by blinks or eye-movements, see, e.g., Iwasaki *et al.* (2005) [2] for an in-depth study on the topic. The most straightforward method to avoid these artifacts is to restrict subjects to move their eyes during the experimental recordings. However, this excludes experimental protocols where visual scene exploration or reading is a key aspect of the cognitive study.

For the last thirty years, a number of numerical methods to remove ocular artifacts have been considered in the literature. Among those, one finds the

This work has been partly funded by ERC-AdG-2012-320864 CHESS project.

P. Tichavský et al. (Eds.): LVA/ICA 2017, LNCS 10169, pp. 563–573, 2017.
DOI: 10.1007/978-3-319-53547-0_53

regression approach [3,4] and Independent Component Analysis (ICA) [1,5]. The regression approach requires a reference for the ocular artifact. Usually, this method uses the electrooculogram (EOG) that provides a measurement of the electric field associated with the ocular activity which is recorded by electrodes localized in the vicinity of the eyes. In the regression approach, it is assumed that the EOG matches with the ocular artifacts contained in the EEG observations up to scaling factors. The goal is then to identify these factors and substract weighted reference channels from the EEG. Despite its simplicity, this method presents some major drawbacks. First, it needs additional electrodes for the reference channels to be available. Second, since EOG electrodes are also placed on the skin, volume conductivity of the latter results in cross-talk of cerebral activity and ocular artefact, even on the EOG electrodes. This implies a bias in the regression toward closeby electrodes.

Under the ICA model, a linear, instantaneous mixing model is estimated, since this is in line with the linearized Maxwell equations at the frequencies of interest. The latent sources constituting the EEG observations are assumed to be statistically independent. The goal of this method is to estimate the linear mixing operator and the latent sources through maximization of the source independence. Once the sources and the linear operator estimated, one can identify the ocular artifact components among these sources and remove their contribution from the EEG [1]. ICA has shown its efficiency and it is still widely used in the EEG community. Nevertheless, ICA also suffers from some drawbacks. First, it needs a large number of observations (large with respect to the number of electrodes) to accurately estimate the probability density functions or their approximations used in the computation of the independence criterion. In addition, since the sources are not truly independent, removing identified ocular artifact source components may result in the removal of cerebral activity, thus losing information of interest. Finally, since we consider only linear operators, suppression of the contribution of a source results in a decrease of the dimension of the signal subspace.

In this paper, we propose a novel method for the denoising of EEG data contaminated by eye-movement artifacts based on the multimodal nature of the gaze and EEG [6]. We focus on saccades, which are the eye-movements related to the action of moving from one fixation point to another. During a saccade, the EEG observations can be decomposed as a linear superposition of the electrical brain activity and a potential induced by the eye-movement (ocular artifact) [1]. In the meantime, an eye-tracker provides a measurement of the eye-movement (gaze direction relative to a screen). These gaze signals present a main advantage compared to EOG as they contain no brain activity. Since the gaze signals share a similar temporal structure (up to temporal filtering) with the ocular artifacts in the EEG observations [2], we consider the eye-tracker observations as reference signals for saccade denoising of EEG data. Due to this temporal filtering, the regression approach does not seem to be an optimal option. Motivated by the temporal similarity of eye-movement (artifacts) signals recorded from both modalities, we propose to use a Multiple Measurement

Vectors method (MMV) [7] (also called Collaborative Lasso [8] or Multichannel Sparse Recovery [9]). MMV aims at exploiting the structure shared by gaze and EEG recordings, sparsely representing them in a single well-chosen dictionary. Our hypothesis is that only the part of the EEG observations related to saccades will be estimated by the sparse joint decomposition. It can then be substracted from EEG channels to recover a clean brain activity. Although a linear super-position of the temporal signals is considered, this method will not suffer from data dimension reduction as is the case for regression or ICA.

This paper is organized as follows: in Sect. 2, the proposed method and data preprocessings are described. In Sect. 3, numerical processings on gaze and EEG experimental data are presented and comparison with some other methods are provided. Finally, the conclusion and some perspectives are detailed in Sect. 4.

2   Proposed Method

In this section, we present the MMV approach and we describe how to preprocess gaze and EEG data for using the proposed method.

2.1   Multiple Measurement Vectors

The purpose of MMV is to obtain a sparse representation of multiple observed signals in a single, well-chosen dictionary, exploiting redundancy in these signals. The considered model is the following

$$\mathbf{Y} = \boldsymbol{\Phi}\mathbf{X} + \mathbf{R}, \tag{1}$$

where $\mathbf{Y} = [\mathbf{y}_1 \dots \mathbf{y}_N]$ is a data matrix containing N signals $\mathbf{y}_n$ ($n \in \{1, \dots, N\}$) stored in N_s-dimensional column vectors (where N_s is the number of samples). $\boldsymbol{\Phi} \in \mathbb{R}^{N_s \times M}$ is a (finite) dictionary of M atoms chosen to extract the particular structure shared among the $\mathbf{y}_i$, maximaly capturing its redundancy. There is no assumption about orthogonality among the atoms. $\mathbf{X} \in \mathbb{R}^{M \times N}$ is a row sparse code matrix in which the nonzero coefficients model the particular signal shape in the dictionary. $\mathbf{R} \in \mathbb{R}^{N_s \times N}$ is the residual, *i.e.*, all $\mathbf{Y}$ components that do not present the particular shape we are looking for. For the considered application, $\boldsymbol{\Phi}\mathbf{X}$ models the ocular artifacts we want to remove and $\mathbf{R}$ corresponds to the clean brain activity. In this work, the goal is to estimate $\mathbf{X}$ optimizing the following proxy cost function

$$\Psi(\mathbf{X}) = \frac{1}{2} \parallel \mathbf{Y} - \boldsymbol{\Phi}\mathbf{X} \parallel_F^2 + \lambda \parallel \mathbf{X} \parallel_{2,1}, \tag{2}$$

with $\parallel \cdot \parallel_F$ the Frobenius norm and $\parallel \cdot \parallel_{2,1}$ the 2, 1–mixed norm [10] defined as

$$\parallel \mathbf{X} \parallel_{2,1} = \sum\nolimits_{m=1}^{M} \left(\sum\nolimits_{n=1}^{N} \mid \mathbf{X}_{m,n} \mid^2 \right)^{1/2}. \tag{3}$$

This mixed norm is used to keep or discard entire rows of coefficients from the matrix $\mathbf{X}$ in order to represent each signal from $\mathbf{Y}$ with the same atoms. Thus, we extract a common structure shared among all $\mathbf{y}_n$. Finally, λ (2) is a regularization parameter inducing row sparsity on $\mathbf{X}$. In this paper, λ is arbitrarily fixed, however it is important to notice that this parameter can be chosen by using for example cross-validation.

For this estimation issue, we consider a variable splitting and an augmented Lagrangian as follows

$$\overline{\Psi}(\mathbf{X}, \mathbf{Z}, \mathbf{U}) = \frac{1}{2} \parallel \mathbf{Y} - \mathbf{\Phi X} \parallel_F^2 + \lambda \parallel \mathbf{Z} \parallel_{2,1} + \mathbf{U}^t(\mathbf{X} - \mathbf{Z}) + \frac{\rho}{2} \parallel \mathbf{X} - \mathbf{Z} \parallel_F^2, \quad (4)$$

where $\mathbf{Z}$ is the split variable, $\mathbf{U}$ is the matrix of Lagrangian multipliers, and ρ is a regularization constant linked to the converge speed [8]. The optimization problem reads

$$\widehat{\mathbf{X}} = \arg \min_{\mathbf{X}} \min_{\mathbf{Z}} \max_{\mathbf{U}} \overline{\Psi}(\mathbf{X}, \mathbf{Z}, \mathbf{U}) \approx \arg \min_{\mathbf{X}} \Psi(\mathbf{X}), \quad (5)$$

and we use the Alternating Direction Method of Multipliers algorithm (ADMM) [8] as a solver. The ADMM convergence is guaranteed since the functions in (2) respect the assumptions described in, *e.g.*, [11].

The main question remains how to build the data matrix $\mathbf{Y}$ from the available observations and the choice of a dictionary $\mathbf{\Phi}$ adapted to the problem.

2.2   Data Matrix Building

This MMV method can be used to denoise either only one saccade or several saccades (respectively, single trial and multitrial processing). The first preprocessing step is to epoch the recordings in order to keep only the interesting parts of the signals (see Fig. 1). To do so, we localize the saccades on gaze channels (at the dash-dot line in Fig. 1). Then, we extract a predefined time window containing only one saccade (in dash lines in Fig. 1). These constitute the trials. Each trial is made of N_s samples and contains a first fixation, then the saccade and finally a second fixation.

We consider K trials. For each trial, we have P recordings from the eye-tracker (the number of gaze channels) and Q recordings from the EEG sensors (the number of electrodes). For the kth trial, $k \in \{1, \ldots, K\}$, we can build, respectively (resp.), a gaze matrix $\mathbf{G}^{(k)} \in \mathbb{R}^{N_s \times P}$ and an EEG matrix $\mathbf{E}^{(k)} \in \mathbb{R}^{N_s \times Q}$ defined for all $n \in \{1, \ldots, N_s\}$ by

$$\mathbf{G}^{(k)} = [\mathbf{g}_1^{(k)}(n), \ldots, \mathbf{g}_P^{(k)}(n)] \quad \text{and} \quad \mathbf{E}^{(k)} = [\mathbf{e}_1^{(k)}(n), \ldots, \mathbf{e}_Q^{(k)}(n)], \quad (6)$$

where $\mathbf{g}_p^{(k)}(n) \in \mathbb{R}^{N_s \times 1}$, $p \in \{1, \ldots, P\}$ and $\mathbf{e}_q^{(k)}(n) \in \mathbb{R}^{N_s \times 1}$, $q \in \{1, \ldots, Q\}$ are column vectors representing the signals from, resp., the pth gaze channel and the qth EEG channel. An optional preprocessing is the downsampling. Indeed, the number of samples directly impacts the computational time. Thus, if the epoched signals are made of too many samples, then one can downsample them

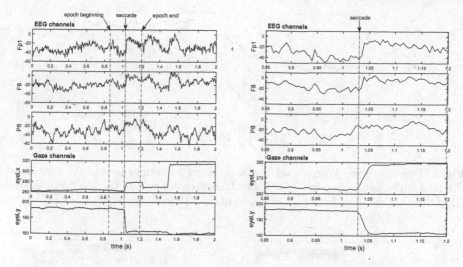

Fig. 1. Selection and epochage of one saccade from gaze and EEG data. On the left side: observed signals (from three EEG channels and two gaze ones) containing at least one interesting saccade to extract. On the right side: the epoched saccade on the same five channels during the previously selected time segment.

respecting the Nyquist-Shannon sampling theorem. As, in the same trial, the gaze signals and the ocular artifacts contained in EEG signals share a similar structure, one may directly use the MMV on data matrix $\mathbf{Y} = [\mathbf{G}^{(k)}, \mathbf{E}^{(k)}]$.

The case of multitrial processing raises a main issue. Indeed, in each trial, data present a common temporal structure linked to the ocular artifacts that is important for the proposed method efficiency. Among trials, one can find similar structures (or shape) but this time with some temporal distortions due to the difference among saccades or among subjects moving their eyes. These distorsions may impact the MMV performance in the considered application. In order to fix this issue, we propose to align the different trials using an extension of the Dynamic Time Warping (DTW) [12] called the Generalized Time Warping (GTW) [13]. DTW is an algorithm for measuring similarity between two time series. This method can be used to compute an optimal match between two temporal signals. As it is shown in Fig. 2, DTW calculates nonlinear functions for each time serie, such that the sum of the distances between their points is smallest and so the correlation between both signals is maximum. The considered distance depends on which algorithm is used. GTW generalizes DTW method for more than two sets of time series. Whatever the saccade orientation, GTW aims at matching the shape of signals from different trials. For that, GTW computes, for each set, a nonlinear bijective function that warps time and allows to minimize the shape difference among the set of time series. Due to their very similar shape (see Fig. 3), gaze signals seems simpler to align. Thus, we directly apply the GTW on all matrices $\mathbf{G}^{(k)}, k \in \{1, \ldots, K\}$. Once the nonlinear functions are computed, they are applied to gaze and EEG matrices of the corresponding trial.

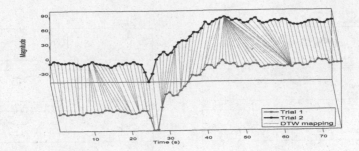

Fig. 2. Dynamic Time Warping method associating to each sample of the trial 1 and the trial 2 the point of, resp., trial 2 and trial 1 that presents the smallest distance (thin lines are mapping between the points of each time serie).

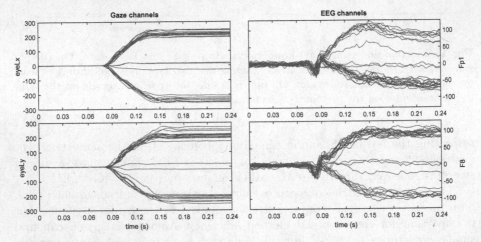

Fig. 3. All trials for some gaze and EEG channels after downsampling and preprocessing by GTW method.

After this shape matching step, we can build two new matrices. They contain the concatenation of each channel of each trial, resp., for the gaze and for the EEG observations. They are defined as follows

$$\mathbf{G} = [\mathsf{GTW}(\mathbf{G}^{(1)}), \ldots, \mathsf{GTW}(\mathbf{G}^{(K)})] \quad \text{and} \quad \mathbf{E} = [\mathsf{GTW}(\mathbf{E}^{(1)}), \ldots, \mathsf{GTW}(\mathbf{E}^{(K)})],$$

$$(7)$$

where $\mathsf{GTW}(\cdot)$ is the generalized time warping operator, $\mathbf{G}$ and $\mathbf{E}$ are, resp., made of KP and KQ channels. These dimensions may be very large and have a major impact on the computational complexity. We propose to reduce the size of these matrices. As we have induced a similar shape among the gaze trials using the GTW, we can expect that the contribution of the ocular artifact components is also similar in EEG observations for all the trials. Hence, we propose to do a Principal Component Analysis (PCA) on $\mathbf{G}$ and an other one on $\mathbf{E}$. For the gaze matrix $\mathbf{G}$, all the channels share a smooth step shape and thus the first principal component should explain almost entirely the signal power. We propose to threshold the principal components keeping the most powerful ones and dropping the others when we have enough to explain 99% of the original signal power. For the EEG, we only want to extract the saccade components. As saccades induce an electrical potential much larger in magnitude than the brain activity, we propose to keep the first principal components explaining 95% of the entire power. Both these empirical thresholds highly reduce the size of $\mathbf{G}$ and $\mathbf{E}$. Finally, we use MMV on the new data matrix $\mathbf{Y} = [\mathrm{PC}_{99\%}(\mathbf{G}), \mathrm{PC}_{95\%}(\mathbf{E})]$, where $\mathrm{PC}_{99\%}(G)$ and $\mathrm{PC}_{95\%}(E)$ are the operators extracting the principal components.

It remains to explain how to select the dictionary $\mathbf{\Phi}$.

2.3 Dictionary Selection

Since we aim at decomposing only the ocular artifact components, we consider a dictionary containing atoms that match with the gaze signals. As these signals look like smooth steps, our choice is to use the following sigmoidal function

$$f_{\alpha,\beta}(t) = \frac{1}{1 + e^{-\alpha(t-\beta)}}, \tag{8}$$

where α and β are, resp., the scale and the translation parameters. In order to take into account the gaze signals overshoots and the side effects due to the signal finite support, we include the derivative of (8) in the dictionary

$$g_{\alpha,\beta}(t) = \frac{\partial f_{\alpha,\beta}(t)}{\partial t} = \alpha f_{\alpha,\beta}(t) f_{-\alpha,\beta}(t). \tag{9}$$

We also add the constant function which acts as an offset and we normalize all the atoms. Finally, all the atoms are seen as column vectors and we concatenate them in the dictionary $\mathbf{\Phi}$ for all considered scales α and translations β.

Hereafter, we summarize the outline of the proposed novel MMV method for gaze and EEG multimodal approach (called MMV-G&E).

---

1. Preprocessing from gaze and EEG observations
   - Epoch $\rightarrow$ build $\mathbf{G}^{(k)}$ and $\mathbf{E}^{(k)}$, $k \in \{1, \dots, K\}$ (6)
   - Downsample (optional)
   - If $K > 1$: perform GTW $\rightarrow$ build $\mathbf{G}$ and $\mathbf{E}$ (7)
   - Build $\mathbf{Y} = [\mathrm{PC}_{99\%}(\mathbf{G}), \mathrm{PC}_{95\%}(\mathbf{E})]$
2. MMV method: optimize the cost function $\Psi(\mathbf{X})$ (4)
   - Choose the dictionary $\mathbf{\Phi}$ with respect to the data
   - Fix the regularization parameters λ and ρ
   - Solve (5) for $\widehat{\mathbf{X}}$, e.g., using ADMM
3. Remove the ocular artifact estimates: $\widehat{\mathbf{R}} = \mathbf{Y} - \mathbf{\Phi}\widehat{\mathbf{X}}$.

---

3 Experiments

In this section, we assess the MMV-G&E performance on gaze and EEG real-data. These come from an experiment in visual exploration where participants had to search a target from a set of distractors [14]. Sixty four active electrodes (BrainProductsGmbH) were mounted on an EEG cap (BrainCap^{TM}) placed on the scalp in compliance with the international 10–20 system. To be compatible with the EEG acquisition, eye-movements were recorded by a remote binocular infrared eye-tracker EyeLink 1000 (SR Research) to track the gaze direction of the left eye while the observer was looking at the stimuli. The EyeLink system was used in the Pupil-Corneal Reflection tracking mode. For both acquisition devices, the sampling frequency was 1000 Hz. Off-line, EEG signals and gaze samples were synchronized using hardware triggers signals sent in parallel to the EEG recorder and the eye-tracker, along the experiment. Let note that the EEG electrode F3 was defective during the experiment and has been removed from the data ($Q = 63$). Concerning the gaze information, we take into account the vertical and the horizontal channels ($P = 2$). We consider $K = 26$ epoched trials. Each signal, downsampled at 333 Hz, is composed of $N_s = 75$ samples and lasts about 225 ms. After the GTW preprocessing, we obtain the data displayed in Fig. 3 for both gaze channels and two EEG channels. In Sect. 3.1, we describe the selected parameters for using the proposed method and we show some qualitative results obtained on real-data. Finally, a validation method is proposed to assess the performance of MMV-G&E and comparisons with standard methods from the literature are provided in Sect. 3.2.

3.1 MMV-G&E Parameters and Qualitative Results

For this experiment, the MMV-G&E regularization parameters have been heuristically fixed: $\lambda = 42$ and $\rho = 1$. Future work will consist in optimizing λ. Concerning the dictionary, the atoms defined, for $t = -10, \ldots, 10$, with a step of $20/(N_s - 1)$, by (8) and (9) are concatenated with the constant function as explained in Sect. 2.3. The scale and translation parameters are empirically chosen: $\alpha \in \{1, \ldots, 10\}$ and $\beta = -10, \ldots, 10$, with a step of $20/(N_s - 1)$. Figure 4 shows the denoising by the proposed method on the considered experimental data. After removing the saccade contribution estimates (in dashed lines) from the observations (in dotted curves), we obtain the denoised signals (in solid lines) which seem to conserve the pre-saccadic behavior that corresponds to pure brain activity. We can observe that, as expected, the saccade contribution estimates depend on the considered electrodes. Thus, MMV-G&E method derives high magnitude saccades for Fp1 and F8 and very low magnitude ones for P8.

3.2 Comparisons and Validation

Here, we compare MMV-G&E to some algorithms from the state of the art:

- the regression method [3,4] with the gaze taken as reference,
- Infomax algorithm (ICA) [1] applied to a matrix in which gaze and EEG channels are concatenated for each trial and then all trials are stacked,

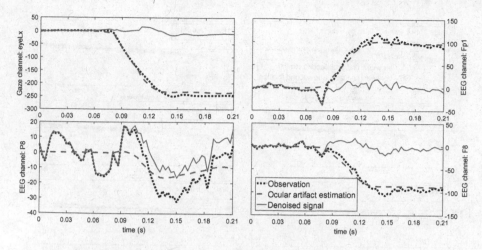

Fig. 4. MMV-G&E denoising effect for one trial and some gaze and EEG channels.

- CCA [15] that finds projections on a common space, maximizing the correlation between gaze and EEG,
- the coupled tensor factorization method RACMTF [6].

In order to assess the efficiency of these methods, we propose the following validation. From each EEG trial $\mathbf{E}^{(k)}$, we extract three windows of 20 samples representing, resp., the pre-saccadic fixation, the saccade and the post-saccadic fixation, stored in three matrices, resp., $\mathbf{E}_{pr}^{(k)}$, $\mathbf{E}_{sa}^{(k)}$ and $\mathbf{E}_{po}^{(k)}$ of size 20×63. Each extracted signal is centered. Then, we stack the trials such that $\mathbf{E}_{pr} = [\mathbf{E}_{pr}^{(1)\ T}, \ldots, \mathbf{E}_{pr}^{(K)\ T}]^T$ where $(\cdot)^T$ is the transpose operator. We do the same for $\mathbf{E}_{sa}$ and $\mathbf{E}_{po}$. Finally, we compute two vectors of generalized eigenvalues (GEV):

$$\mathbf{d_1} = \mathsf{GEV}(\mathsf{Cov}(\mathbf{E}_{pr}), \mathsf{Cov}(\mathbf{E}_{po})) \text{ and } \mathbf{d_2} = \mathsf{GEV}(\mathsf{Cov}(\mathbf{E}_{sa}), \mathsf{Cov}(\mathbf{E}_{po})), \quad (10)$$

where $\mathsf{Cov}(\cdot)$ is the covariance operator. The vectors $\mathbf{d_1}$ and $\mathbf{d_2}$ are displayed in Fig. 5. The brain activity can be assumed to be stationary for long time segments. Here, as the trials are stacked in the matrices $\mathbf{E}_{pr}$, $\mathbf{E}_{sa}$ and $\mathbf{E}_{po}$, we can expect that each $\mathbf{d_1}$ entry should tend to 1. Due to the saccade power, we have $\mathbf{d}_{2,i} \geq \mathbf{d}_{1,i}$ ($i \in \{1, \ldots, 63\}$). This is confirmed before denoising (see Fig. 5). After this processing, we expect to reduce the distance between each pair of generalized eigenvalues (ideally $\mathbf{d}_{2,i} = \mathbf{d}_{1,i}$). For indicative information, a measurement between $\mathbf{d_2}$ and $\mathbf{d_1}$ is provided, in Fig. 5, using the mean square error in logarithmic scale ($\mathrm{MSE}_{\log_{10}}$). In this figure, we can observe that the proposed method obtains slightly better results than regression one and outperforms the three other algorithms on this example.

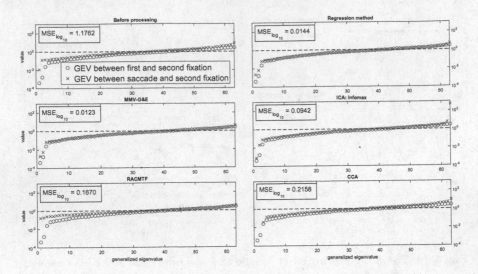

Fig. 5. Validation comparing the GEV between the covariance matrices of pre and post-saccadic fixations and GEV between the covariance matrices of pre-saccadic fixation and denoised saccade. MMV-G&E is confronted with four popular method.

4   Conclusions and Perspectives

In this paper, we propose a multimodal approach to tackle the eye-movement artifact removal in EEG recordings. The gaze signals, used as a reference, share a similar shape with the ocular artifacts. The considered MMV method allows to exploit this property decomposing the data in a row sparse way in a same well-chosen dictionary. Only the structure shared by gaze and EEG recordings is estimated and is used to extract the ocular artifacts from the EEG data. One may notice that the use of MMV-G&E for single trial processing is straightforward, yet, it is more complicated for multitrial processing. Indeed the signals between different trials have to share the sought similar temporal structure. In order to enforce this constraint, we propose to use the GTW method that warps time in order to align the signals. The experiments on gaze and EEG real-data have shown the proposed method efficiency for the ocular artifact extraction. Moreover MMV-G&E compares favorably to classical methods from the literature. Future work will consider other extensions as a clever choice for the MMV thresholding parameter or some additional constraints for the sparse representation linked to the temporal structure of gaze and EEG data. It will also consist in testing MMV-G&E performance on various criteria.

References

1. Jung, T.-P., Makeig, S., Humphries, C., Lee, T.-W., Mckeown, M.-J., Iragui, V., Sejnowski, T.: Removing electroencephalographic artifacts by blind source separation. Psychophysiology **37**(2), 163–178 (2000)

2. Iwasaki, M., Kellinghaus, C., Alexopoulos, A.-V., Burgess, R.-C., Kumar, A.-N., Han, Y.-H., Lüders, H.-O., Leigh, R.-J.: Effects of eyelid closure, blinks, and eye movements on the electroencephalogram. Clin. Neurophysiol. **116**(4), 878–885 (2005)
3. Semlitsch, H.-V., Anderer, P., Schuster, P., Presslich, O.: A solution for reliable and valid reduction of ocular artifacts, applied to the P300 ERP. Psychophysiology **23**(6), 695–703 (1986)
4. Schlögl, A., Keinrath, C., Zimmermann, D., Scherer, R., Leeb, R., Pfurtscheller, G.: A fully automated correction method of EOG artifacts in EEG recordings. Clin. Neurophysiol. **118**(1), 98–104 (2007)
5. Comon, P., Jutten, C.: Handbook of Blind Source Separation: Independent Component Analysis and Applications. Academic Press, Burlington (2000)
6. Rivet, B., Duda, M., Guérin-Dugué, A., Jutten, C., Comon, P.: Multimodal approach to estimate the ocular movements during EEG recordings: a coupled tensor factorization method. In: Proceedings of 37th Annual International Conference of the IEEE Engineering in Medicine and Biology Society (EMBC), pp. 6983–6986 (2015)
7. Cotter, S.-F., Rao, B.-D., Engan, K., Kreutz-Delgado, K.: Sparse solutions to linear inverse problems with multiple measurement vectors. IEEE Trans. Sig. Process. **53**(7), 2477–2488 (2006)
8. Boyd, S., Parikh, N., Chu, E., Peleato, B., Eckstein, J.: Distributed optimization and statistical learning via the alternating direction method of multipliers. In: Foundations and Trends® in Machine Learning, vol. 3, no. 1, pp. 1–122 (2011)
9. Eldar, Y.-C., Rauhut, H.: Average case analysis of multichannel sparse recovery using convex relaxation. IEEE Trans. Inf. Theory **51**(1), 505–519 (2010)
10. Kowalski, M.: Sparse regression using mixed norms. Appl. Comput. Harmonic Anal. **27**(3), 303–324 (2009)
11. Nishihara, R., Lessard, L., Recht, B., Packard, A., Jordan, M.-I.: A General Analysis of the Convergence of ADMM. arXiv preprint (2015)
12. Berndt, D.-J., Clifford, J.: Using dynamic time warping to find patterns in time series. In: Proceedings of 2012 IEEE Conference on KDD Workshop, vol. 10, no. 16, pp. 359–370 (1994)
13. Zhou, F., De la Torre, F.: Generalized time warping for multi-modal alignment of human motion. In: Proceedings of IEEE Conference on Computer Vision and Pattern Recognition (CVPR), pp. 1282–1289 (2012)
14. Kristensen, E., Guerin-Dugué, A., Rivet, B.: Comparison between Adjar and Xdawn algorithms to estimate eye-fixation related potentials distorted by overlapping. In: Proceedings of 7th International IEEE/EMBS Conference on Neural Engineering (NER), pp. 976–979 (2015)
15. Hardoon, D., Szedmak, S., Shawe-Taylor, J.: Canonical correlation analysis: an overview with application to learning methods. Neural Comput. **16**(12), 2639–2664 (2004)

Author Index

Printed in the United States
By Bookmasters